T0181514

Advanced
Engineering
Mathematics
with MATLAB®

Advances in Applied Mathematics

Series Editors:
Daniel Zwillinger, H. T. Banks

Advanced Engineering Mathematics with MATLAB, 4th Edition
Dean G. Duffy

CRC Standard Mathematical Tables and Formulas, 33rd Edition
Edited by Daniel Zwillinger

Advanced Mathematical Modeling with Technology
William P. Fox, Robert E. Burks

Introduction to Quantum Control and Dynamics
Domenico D'Alessandro

Handbook of Radar Signal Analysis
Bassem R. Mahafza, Scott C. Winton, Atef Z. Elsherbeni

Separation of Variables and Exact Solutions to Nonlinear PDEs
Andrei D. Polyanin, Alexei I. Zhurov

Boundary Value Problems on Time Scales, Volume I
Svetlin Georgiev, Khaled Zennir

Boundary Value Problems on Time Scales, Volume II
Svetlin Georgiev, Khaled Zennir

Observability and Mathematics
Fluid Mechanics, Solutions of Navier-Stokes Equations, and Modeling
Boris Khots

Handbook of Differential Equations, 4th Edition
Daniel Zwillinger, Vladimir Dobrushkin

Experimental Statistics and Data Analysis for Mechanical and Aerospace Engineers
James Middleton

Advanced Engineering Mathematics with MATLAB, 5th Edition
Dean G. Duffy

https://www.routledge.com/Advances-in-Applied-Mathematics/book-series/CRCADVAPPMTH?pd=published,forthcoming&pg=1&pp=12&so=pub&view=list

Advances in Applied Mathematics

FIFTH EDITION

Advanced Engineering Mathematics
with MATLAB®

Dean G. Duffy

CRC Press
Taylor & Francis Group
Boca Raton London New York

CRC Press is an imprint of the
Taylor & Francis Group, an **informa** business

A CHAPMAN & HALL BOOK

Fifth edition published 2022
by CRC Press
6000 Broken Sound Parkway NW, Suite 300, Boca Raton, FL 33487-2742

and by CRC Press
2 Park Square, Milton Park, Abingdon, Oxon, OX14 4RN

© 2022 Dean G. Duffy

First edition published by CRC Press 1997
Second edition published by CRC Press 2003
Third edition published by CRC Press 2010
Fourth edition published by CRC Press 2016

CRC Press is an imprint of Taylor & Francis Group, LLC

Library of Congress Cataloging-in-Publication Data

Names: Duffy, Dean G., author.
Title: Advanced engineering mathematics with MATLAB / Dean G. Duffy.
Description: Fifth edition. | Boca Raton : CRC Press, 2022. | Series:
   Advances in applied mathematics | Includes bibliographical references
   and index. | Summary: "In the four previous editions the author
   presented a text firmly grounded in the mathematics that engineers and
   scientists must understand and know how to use. Tapping into decades of
   teaching at the US Navy Academy and the US Military Academy and serving
   for twenty-five years at (NASA) Goddard Space Flight, he combines a
   teaching and practical experience that is rare among authors of advanced
   engineering mathematics books. This edition offers a smaller, easier to
   read, and useful version of this classic textbook. While competing
   textbooks continue to grow, the book presents a slimmer, more concise
   option. Instructors and students alike are rejecting the encyclopedic
   tome with its higher and higher price aimed at undergraduates. To assist
   in the choice of topics included in this new edition, the author
   reviewed the syllabi of various engineering mathematics courses that are
   taught at a wide variety of schools. Due to time constraints an
   instructor can select perhaps 3 to 4 topics from the book, the most
   likely being ordinary differential equations, Laplace transforms,
   Fourier series and separation of variables to solve the wave, heat, or
   Laplace's equation. Laplace transforms are occasionally replaced by
   linear algebra or vector calculus. Sturm-Liouville problem and special
   functions (Legendre and Bessel functions) are included for completeness.
   Topics such as z-transforms and complex variables are now offered in a
   companion book, Advanced Engineering Mathematics: A Second Course by the
   same author. MATLAB is still employed to reinforce the concepts that are
   taught. Of course, this Edition continues to offer a wealth of examples
   and applications from the scientific and engineering literature, a
   highlight of previous editions. Worked solutions are given in the back
   of the book"-- Provided by publisher.
Identifiers: LCCN 2021038560 (print) | LCCN 2021038561 (ebook) | ISBN
   9780367624057 (hardback) | ISBN 9781032164984 (paperback) | ISBN
   9781003109303 (ebook)
Subjects: LCSH: Engineering mathematics--Data processing. | MATLAB.
Classification: LCC TA345 .D84 2022 (print) | LCC TA345 (ebook) | DDC
   620.001/51--dc23
LC record available at https://lccn.loc.gov/2021038560
LC ebook record available at https://lccn.loc.gov/2021038561

ISBN: 9780367624057 (hbk)
ISBN: 9781032164984 (pbk)
ISBN: 9781003109303 (ebk)

DOI: 10.1201/9781003109303

Publisher's note: This book has been prepared from camera-ready copy provided by the authors

Dedicated to the Brigade of Midshipmen

and the Corps of Cadets

Contents

Chapter 1:
First-Order Ordinary
Differential Equations   1

Chapter 2:
Higher-Order Ordinary
Differential Equations   47

Chapter 3:
Linear Algebra   101

Chapter 7:
The Laplace Transform   **295**

Chapter 8:
The Wave Equation   **347**

Chapter 12:
Special Functions   **493**

Acknowledgments

I would like to thank the many midshipmen and cadets who have taken engineering mathematics from me. They have been willing or unwilling guinea pigs in testing out many of the ideas and problems in this book. Special thanks go to Dr. Mike Marcozzi for his many useful suggestions for improving this book. Most of the plots and calculations were done using MATLAB®.

MATLAB is a registered trademark of
The MathWorks Inc.
24 Prime Park Way
Natick, MA 01760-1500
Phone: (508) 647-7000
Email: info@mathworks.com
www.mathworks.com

Author

Dean G. Duffy received his bachelor of science in geophysics from Case Institute of Technology (Cleveland, Ohio) and his doctorate of science in meteorology from the Massachusetts Institute of Technology (Cambridge, Massachusetts). He served in the United States Air Force from September 1975 to December 1979 as a numerical weather prediction officer. After his military service, he began a twenty-five-year (1980 to 2005) association with NASA at the Goddard Space Flight Center (Greenbelt, Maryland) where he focused on numerical weather prediction, oceanic wave modeling, and dynamical meteorology. He also wrote papers in the areas of Laplace transforms, antenna theory, railroad tracks, and heat conduction. In addition to his NASA duties, he taught engineering mathematics, differential equations, and calculus at the United States Naval Academy (Annapolis, Maryland) and the United States Military Academy (West Point, New York). Drawing from his teaching experience, he has written several books on transform methods, engineering mathematics, Green's functions, and mixed-boundary-value problems.

Introduction

The new edition of my *Advanced Engineering Mathematics with MATLAB* is dramatically smaller than previous editions. The encyclopedic engineering mathematics tome is dead, killed by the student's ability to look up a particular topic on the Web. Because of time constraints in engineering curriculum, the mathematics instructor can select perhaps 3 to 4 topics from the book, the most likely being ordinary differential equations, Laplace transforms, Fourier series, and separation of variables to solve one of the fundamental partial differential equations of mathematical physics. Sometimes Laplace transforms are replaced by linear algebra or vector calculus.

To assist me in the choice of topics to include in this new edition, I reviewed the syllabi of various engineering mathematics courses that are taught at a wide variety of schools. There were a few who had courses that taught the Sturm-Liouville problem and special functions (Legendre and Bessel functions). They are included for completeness.

Certain topics, such as z-transforms and complex variables, did not warrant (sadly) inclusion. Because many engineering students still need this material, I have written a companion book, *Advanced Engineering Mathematics: A Second Course*, that still presents this material.

The book begins with first- and higher-order ordinary differential equations, Chapters 1 and 2, respectively. After some introductory remarks, Chapter 1 devotes itself to presenting general methods for solving first-order ordinary differential equations. These methods include separation of variables, employing the properties of homogeneous, linear, and exact differential equations, and finding and using integrating factors.

The reason most students study ordinary differential equations is for their use in elementary physics, chemistry, and engineering courses. Because these differential equations contain constant coefficients, we focus on how to solve them in Chapter 2, along with a detailed analysis of the simple, damped, and forced harmonic oscillator. Furthermore, we include the commonly employed techniques of undetermined coefficients and variation of parameters for finding particular solutions. Finally, the special equation of Euler and Cauchy is included because of its use in solving partial differential equations in spherical coordinates.

The next few chapters are given in arbitrary order because they require no additional knowledge beyond calculus and ordinary differential equations. Chapter 3 presents linear algebra as a method for solving systems of linear equations and includes such topics as matrices, determinants, Cramer's rule, and the solution of systems of ordinary differential equations via the classic eigenvalue problem. Vector calculus is presented in Chapter 4 and focuses on the gradient operator as it applies to line integrals, surface integrals, the divergence theorem, and Stokes' theorem.

We now turn to a trio of chapters built off the concept of Fourier series. Fourier series allow us to reconstruct any arbitrary, well-behaved function defined over the interval $(-L, L)$ in terms of cosines and sines. We then use this concept to solve ordinary differential equations, Section 5.6. Finally we show how these ideas may be applied when the functions is given as data points, Section 5.7.

Fourier series are expanded to functions that are defined over the interval $(-\infty, \infty)$ in Chapter 6. We then use Fourier integrals to find the particular solution for ordinary differential equations in Section 6.6.

Having presented the general aspects of differential equations, we introduce in Chapter 7 the technique of Laplace transforms. Derived from Fourier transforms when the function is defined over the interval $[0, \infty)$, this method allows us to solve initial-value problems where the forcing function turns "on" and "off." Engineers use it extensively in their disciplines.

The reason why Fourier series are so important is their application in solving partial differential equations using separation of variables. We explore this topic using the prototypical wave equation (Section 8.3), heat equation (Section 9.3) and Laplace's equation (Section 10.3). We also investigate numerical solutions for each of these equations in Sections 8.5, 9.5 and 10.5, respectively. Having taught the method of separation of variables using Fourier series, we expand the student's knowledge by generalizing Fourier series to eigenfunction expansions. Sections 11.1–11.3 explain how any piece-wise continuous function can be re-expressed in an eigenfunction expansion using eigenfunctions from the classic Sturm-Liouville problem. Then separation of variables is retaught from this generalized viewpoint.

We conclude the book by focusing on Bessel functions (Section 12.2) and Legendre polynomials (Section 12.1). These eigenfunctions appear in the solution of the wave, heat, and Laplace's equations in cylindrical and spherical coordinates, respectively.

MATLAB is still employed to reinforce the concepts that are taught. Of course, this book still continues my principle of including a wealth of examples from the scientific and engineering literature. Worked solutions are given in the back of the book.

List of Definitions

Function	Definition
$\delta(t-a)$	$= \begin{cases} \infty, & t = a, \\ 0, & t \neq a, \end{cases} \qquad \int_{-\infty}^{\infty} \delta(t-a)\,dt = 1$
$\mathrm{erf}(x)$	$= \dfrac{2}{\sqrt{\pi}} \int_{0}^{x} e^{-y^2}\,dy$
$\Gamma(x)$	gamma function
$H(t-a)$	$= \begin{cases} 1, & t > a, \\ 0, & t < a. \end{cases}$
$\Im(z)$	imaginary part of the complex variable z
$I_n(x)$	modified Bessel function of the first kind and order n
$J_n(x)$	Bessel function of the first kind and order n
$K_n(x)$	modified Bessel function of the second kind and order n
$P_n(x)$	Legendre polynomial of order n
$\Re(z)$	real part of the complex variable z
$\mathrm{sgn}(t-a)$	$= \begin{cases} -1, & t < a, \\ 1, & t > a. \end{cases}$
$Y_n(x)$	Bessel function of the second kind and order n

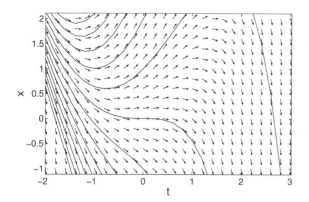

Chapter 1

First-Order Ordinary

Differential Equations

A *differential equation* is any equation that contains the derivatives or differentials of one or more dependent variables with respect to one or more independent variables. Because many of the known physical laws are expressed as differential equations, a sound knowledge of how to solve them is essential. In the next two chapters we present the fundamental methods for solving *ordinary differential equations*—a differential equation that contains only ordinary derivatives of one or more dependent variables. Later, we show in Section 7.7 how Laplace transform methods can be used to solve ordinary differential equations, while systems of linear ordinary differential equations are treated in Section 3.6. Solutions for *partial differential equations*—a differential equation involving partial derivatives of one or more dependent variables of two or more *independent* variables—are given in Chapters 8, 9, and 10.

1.1 CLASSIFICATION OF DIFFERENTIAL EQUATIONS

Differential equations are classified three ways: by *type*, *order*, and *linearity*. There are two *types*: *ordinary* and *partial differential equations*, which have already been defined. Examples of ordinary differential equations include

$$\frac{dy}{dx} - 2y = x, \tag{1.1.1}$$

$$(x - y)\,dx + 4y\,dy = 0, \tag{1.1.2}$$

$$\frac{du}{dx} + \frac{dv}{dx} = 1 + 5x, \tag{1.1.3}$$

and

$$\frac{d^2y}{dx^2} + 2\frac{dy}{dx} + y = \sin(x). \tag{1.1.4}$$

On the other hand, examples of partial differential equations include

$$\frac{\partial u}{\partial x} + \frac{\partial u}{\partial y} = 0, \tag{1.1.5}$$

$$y\frac{\partial u}{\partial x} + x\frac{\partial u}{\partial y} = 2u, \tag{1.1.6}$$

and

$$\frac{\partial^2 u}{\partial t^2} + 2\frac{\partial u}{\partial t} = \frac{\partial^2 u}{\partial x^2}. \tag{1.1.7}$$

In the examples that we have just given, we have explicitly written out the differentiation operation. However, from calculus we know that dy/dx can also be written y'. Similarly, the partial differentiation operator $\partial^4 u/\partial x^2 \partial y^2$ is sometimes written u_{xxyy}. We will also use this notation from time to time.

The *order* of a differential equation is given by the highest-order derivative. For example,

$$\frac{d^3y}{dx^3} + 3\frac{d^2y}{dx^2} + \left(\frac{dy}{dx}\right)^2 - y = \sin(x) \tag{1.1.8}$$

is a third-order ordinary differential equation. Because we can rewrite

$$(x+y)\,dy - x\,dx = 0 \tag{1.1.9}$$

as

$$(x+y)\frac{dy}{dx} = x \tag{1.1.10}$$

by dividing Equation 1.1.9 by dx, we have a first-order ordinary differential equation here. Finally

$$\frac{\partial^4 u}{\partial x^2 \partial y^2} = \frac{\partial^2 u}{\partial t^2} \tag{1.1.11}$$

is an example of a fourth-order partial differential equation. In general, we can write an nth-order, ordinary differential equation as

$$f\left(x, y, \frac{dy}{dx}, \cdots, \frac{d^n y}{dx^n}\right) = 0. \tag{1.1.12}$$

The final classification is according to whether the differential equation is *linear* or *nonlinear*. A differential equation is *linear* if it can be written in the form:

$$a_n(x)\frac{d^n y}{dx^n} + a_{n-1}(x)\frac{d^{n-1} y}{dx^{n-1}} + \cdots + a_1(x)\frac{dy}{dx} + a_0(x)y = f(x). \tag{1.1.13}$$

Note that the linear differential equation, Equation 1.1.13, has two properties: (1) The dependent variable y and *all* of its derivatives are of first degree (the power of each term involving y is 1). (2) Each coefficient depends only on the independent variable x. Examples of linear first-, second-, and third-order ordinary differential equations are

$$(x+1)\,dy - y\,dx = 0, \tag{1.1.14}$$

$$y'' + 3y' + 2y = e^x, \tag{1.1.15}$$

and

$$x\frac{d^3y}{dx^3} - (x^2 + 1)\frac{dy}{dx} + y = \sin(x), \tag{1.1.16}$$

respectively. If the differential equation is not linear, then it is *nonlinear*. Examples of nonlinear first-, second-, and third-order ordinary differential equations are

$$\frac{dy}{dx} + xy + y^2 = x, \tag{1.1.17}$$

$$\frac{d^2y}{dx^2} - \left(\frac{dy}{dx}\right)^5 + 2xy = \sin(x), \tag{1.1.18}$$

and

$$yy''' + 2y = e^x, \tag{1.1.19}$$

respectively.

At this point it is useful to highlight certain properties that all differential equations have in common regardless of their type, order, and whether they are linear or not. First, it is not obvious that just because we can write down a differential equation, a solution exists. The *existence* of a solution to a class of differential equations constitutes an important aspect of the theory of differential equations. Because we are interested in differential equations that arise from applications, their solution should exist. In Section 1.2 we address this question further.

Quite often a differential equation has the solution $y = 0$, a *trivial* solution. For example, if $f(x) = 0$ in Equation 1.1.13, a quick check shows that $y = 0$ is a solution. Trivial solutions are generally of little value.

Another important question is how many solutions does a differential equation have? In physical applications, *uniqueness* is not important because, if we are lucky enough to actually find a solution, then its ties to a physical problem usually suggest uniqueness. Nevertheless, the question of uniqueness is of considerable importance in the theory of differential equations. Uniqueness should not be confused with the fact that many solutions to ordinary differential equations contain arbitrary constants, much as indefinite integrals in integral calculus. A solution to a differential equation that has no arbitrary constants is called a *particular solution*.

- **Example 1.1.1**

Consider the differential equation

$$\frac{dy}{dx} = x + 1, \qquad y(1) = 2. \tag{1.1.20}$$

This condition $y(1) = 2$ is called an *initial condition* and the differential equation plus the initial condition constitute an *initial-value problem*. Straightforward integration gives

$$y(x) = \int (x + 1)\, dx + C = \tfrac{1}{2}x^2 + x + C. \tag{1.1.21}$$

Equation 1.1.21 is the *general solution* to the differential equation, Equation 1.1.20, because it is a solution to the differential equation for *every* choice of C. However, if we now satisfy the initial condition $y(1) = 2$, we obtain a *particular solution*. This is done by substituting the corresponding values of x and y into Equation 1.1.21, or

$$2 = \tfrac{1}{2}(1)^2 + 1 + C = \tfrac{3}{2} + C, \quad \text{or} \quad C = \tfrac{1}{2}. \tag{1.1.22}$$

Therefore, the solution to the initial-value problem Equation 1.1.20 is the particular solution

$$y(x) = (x + 1)^2/2. \tag{1.1.23}$$

$\square$

Finally, it must be admitted that most differential equations encountered in the "real" world cannot be written down either explicitly or implicitly. For example, the simple differential equation $y' = f(x)$ does not have an analytic solution unless you can integrate $f(x)$. This begs the question of why it is useful to learn analytic techniques for solving differential equations that often fail us. The answer lies in the fact that differential equations that we can solve share many of the same properties and characteristics of differential equations which we can only solve numerically. Therefore, by working with and examining the differential equations that we can solve exactly, we develop our intuition and understanding about those that we can only solve numerically.

Problems

Find the order and state whether the following ordinary differential equations are linear or nonlinear:

1. $y'/y = x^2 + x$

2. $y^2 y' = x + 3$

3. $\sin(y') = 5y$

4. $y''' = y$

5. $y'' = 3x^2$

6. $(y^3)' = 1 - 3y$

7. $y''' = y^3$

8. $y'' - 4y' + 5y = \sin(x)$

9. $y'' + xy = \cos(y'')$

10. $(2x + y)\,dx + (x - 3y)\,dy = 0$

11. $(1 + x^2)y' = (1 + y)^2$

12. $yy'' = x(y^2 + 1)$

13. $y' + y + y^2 = x + e^x$

14. $y''' + \cos(x)y' + y = 0$

15. $x^2 y'' + x^{1/2}(y')^3 + y = e^x$

16. $y''' + xy'' + e^y = x^2$

1.2 SEPARATION OF VARIABLES

The simplest method of solving a first-order ordinary differential equation, if it works, is *separation of variables*. It has the advantage of handling both linear and nonlinear problems, especially *autonomous equations*.[1] From integral calculus, we already met this technique when we solved the first-order differential equation

$$\frac{dy}{dx} = f(x). \tag{1.2.1}$$

---

[1] An autonomous equation is a differential equation where the independent variable does not explicitly appear in the equation, such as $y' = f(y)$.

By multiplying both sides of Equation 1.2.1 by dx, we obtain

$$dy = f(x)\,dx. \tag{1.2.2}$$

At this point we note that the left side of Equation 1.2.2 contains only y while the right side is purely a function of x. Hence, we can integrate directly and find that

$$y = \int f(x)\,dx + C. \tag{1.2.3}$$

For this technique to work, we must be able to rewrite the differential equation so that all of the y dependence appears on one side of the equation while the x dependence is on the other. Finally, we must be able to carry out the integration on both sides of the equation.

One of the interesting aspects of our analysis is the appearance of the arbitrary constant C in Equation 1.2.3. To evaluate this constant, we need more information. The most common method is to require that the dependent variable give a particular value for a particular value of x. Because the independent variable x often denotes time, this condition is usually called an *initial condition*, even in cases when the independent variable is not time.

• **Example 1.2.1**

Let us solve the ordinary differential equation

$$\frac{dy}{dx} = \frac{e^y}{xy}. \tag{1.2.4}$$

Because we can separate variables by rewriting Equation 1.2.4 as

$$ye^{-y}\,dy = \frac{dx}{x}, \tag{1.2.5}$$

its solution is simply

$$-ye^{-y} - e^{-y} = \ln|x| + C \tag{1.2.6}$$

by direct integration.   □

• **Example 1.2.2**

Let us solve

$$\frac{dy}{dx} + y = xe^x y, \tag{1.2.7}$$

subject to the initial condition $y(0) = 1$.

Multiplying Equation 1.2.7 by dx, we find that

$$dy + y\,dx = xe^x y\,dx, \tag{1.2.8}$$

or

$$\frac{dy}{y} = (xe^x - 1)\,dx. \tag{1.2.9}$$

A quick check shows that the left side of Equation 1.2.9 contains only the dependent variable y while the right side depends solely on x and we have separated the variables onto one side or the other. Finally, integrating both sides of this equation, we have

$$\ln(y) = xe^x - e^x - x + C. \tag{1.2.10}$$

Since $y(0) = 1$, $C = 1$ and

$$y(x) = \exp[(x-1)e^x + 1 - x]. \tag{1.2.11}$$

In addition to the tried-and-true method of solving ordinary differential equations by hand, scientific computational packages such as MATLAB provide symbolic toolboxes that are designed to do the work for you. In the present case, typing:

```
dsolve('Dy+y=x*exp(x)*y','y(0)=1','x')
```

yields:

```
ans =
1/exp(-1)*exp(-x+x*exp(x)-exp(x))
```

which is equivalent to Equation 1.2.11.

Our success here should not be overly generalized. Sometimes these toolboxes give the answer in a rather obscure form or they fail completely. For example, in the previous example, MATLAB gives the answer:

```
ans =
-lambertw((log(x)+C1)*exp(-1))-1
```

The MATLAB function `lambertw` is Lambert's W function, where `w = lambertw(x)` is the solution to $we^w = x$. Using this definition, we can construct the solution as expressed in Equation 1.2.6.   □

• **Example 1.2.3**

Consider the nonlinear differential equation

$$x^2 y' + y^2 = 0. \tag{1.2.12}$$

Separating variables, we find that

$$-\frac{dy}{y^2} = \frac{dx}{x^2}, \quad \text{or} \quad \frac{1}{y} = -\frac{1}{x} + C, \quad \text{or} \quad y = \frac{x}{Cx - 1}. \tag{1.2.13}$$

Equation 1.2.13 shows the wide variety of solutions possible for an ordinary differential equation. For example, if we require that $y(0) = 0$, then there are infinitely many different solutions satisfying this initial condition because C can take on any value. On the other hand, if we require that $y(0) = 1$, there is no solution because we cannot choose *any* constant C such that $y(0) = 1$. Finally, if we have the initial condition that $y(1) = 2$, then there is only one possible solution corresponding to $C = \frac{3}{2}$.

Consider now the trial solution $y = 0$. Does it satisfy Equation 1.2.12? Yes, it does. On the other hand, there is no choice of C that yields this solution. The solution $y = 0$ is called a *singular solution* to this equation. *Singular solutions* are solutions to a differential equation that cannot be obtained from a solution with arbitrary constants.

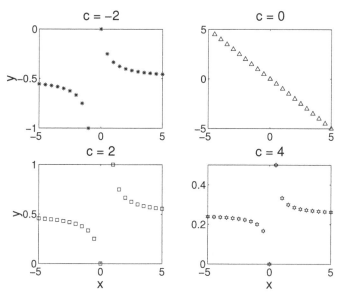

Figure 1.2.1: The solution to Equation 1.2.13 when $C = -2, 0, 2, 4$.

Finally, we illustrate Equation 1.2.13 using MATLAB. This is one of MATLAB's strengths —the ability to convert an abstract equation into a concrete picture. Here the MATLAB script:

```
clear
hold on
x = -5:0.5:5;
for c = -2:2:4
  y = x ./ (c*x-1);
  if (c== -2) subplot(2,2,1), plot(x,y,'*')
    axis tight; title('c = -2'); ylabel('y','Fontsize',20); end
  if (c== 0) subplot(2,2,2), plot(x,y,'^')
    axis tight; title('c = 0'); end
  if (c== 2) subplot(2,2,3), plot(x,y,'s')
    axis tight; title('c = 2'); xlabel('x','Fontsize',20);
    ylabel('y','Fontsize',20); end
  if (c== 4) subplot(2,2,4), plot(x,y,'h')
    axis tight; title('c = 4'); xlabel('x','Fontsize',20); end
end
```

yields Figure 1.2.1, which illustrates Equation 1.2.13 when $C = -2, 0, 2$, and 4. □

The previous example showed that first-order ordinary differential equations may have a unique solution, no solution, or many solutions. From a complete study[2] of these equations, we have the following theorem:

---

[2] The proof of the existence and uniqueness of first-order ordinary differential equations is beyond the scope of this book. See Ince, E. L., 1956: *Ordinary Differential Equations*. Dover Publications, Inc., Chapter 3.

Theorem: Existence and Uniqueness

Suppose some real-valued function $f(x, y)$ is continuous on some rectangle in the xy-plane containing the point (a, b) in its interior. Then the initial-value problem

$$\frac{dy}{dx} = f(x, y), \qquad y(a) = b, \tag{1.2.14}$$

has at least one solution on the same open interval I containing the point $x = a$. Furthermore, if the partial derivative $\partial f / \partial y$ is continuous on that rectangle, then the solution is unique on some (perhaps smaller) open interval I_0 containing the point $x = a$. □

- **Example 1.2.4**

Consider the initial-value problem $y' = 3y^{1/3}/2$ with $y(0) = 1$. Here $f(x, y) = 3y^{1/3}/2$ and $f_y = y^{-2/3}/2$. Because f_y is continuous over a small rectangle containing the point $(0, 1)$, there is a unique solution around $x = 0$, namely $y = (x + 1)^{3/2}$, which satisfies the differential equation and the initial condition. On the other hand, if the initial condition reads $y(0) = 0$, then f_y is *not* continuous on *any* rectangle containing the point $(0, 0)$ and there is no unique solution. For example, two solutions to this initial-value problem, valid on any open interval that includes $x = 0$, are $y_1(x) = x^{3/2}$ and

$$y_2(x) = \begin{cases} (x - 1)^{3/2}, & x \geq 1, \\ 0, & x < 1. \end{cases} \tag{1.2.15}$$

□

- **Example 1.2.5: Hydrostatic equation**

Consider an atmosphere where its density varies only in the vertical direction. The pressure at the surface equals the weight per unit horizontal area of all of the air from sea level to outer space. As you move upward, the amount of air remaining above decreases and so does the pressure. This is why we experience pressure sensations in our ears when ascending or descending in an elevator or airplane. If we rise the small distance dz, there must be a corresponding small decrease in the pressure, dp. This pressure drop must equal the loss of weight in the column per unit area, $-\rho g \, dz$. Therefore, the pressure is governed by the differential equation

$$dp = -\rho g \, dz, \tag{1.2.16}$$

commonly called the *hydrostatic equation*.

To solve Equation 1.2.16, we must express ρ in terms of pressure. For example, in an isothermal atmosphere at constant temperature T_s, the ideal gas law gives $p = \rho R T_s$, where R is the gas constant. Substituting this relationship into our differential equation and separating variables yields

$$\frac{dp}{p} = -\frac{g}{RT_s} \, dz. \tag{1.2.17}$$

Integrating Equation 1.2.17 gives

$$p(z) = p(0) \exp\left(-\frac{gz}{RT_s}\right). \tag{1.2.18}$$

Thus, the pressure (and density) of an isothermal atmosphere decreases exponentially with height. In particular, it decreases by e^{-1} over the distance RT_s/g, the so-called "scale height." □

• **Example 1.2.6: Terminal velocity**

As an object moves through a fluid, its viscosity resists the motion. Let us find the motion of a mass m as it falls toward the earth under the force of gravity when the drag varies as the square of the velocity.

From Newton's second law, the equation of motion is

$$m\frac{dv}{dt} = mg - C_D v^2, \tag{1.2.19}$$

where v denotes the velocity, g is the gravitational acceleration, and C_D is the drag coefficient. We choose the coordinate system so that a downward velocity is positive.

Equation 1.2.19 can be solved using the technique of separation of variables if we change from time t as the independent variable to the distance traveled x from the point of release. This modification yields the differential equation

$$mv\frac{dv}{dx} = mg - C_D v^2, \tag{1.2.20}$$

since $v = dx/dt$. Separating the variables leads to

$$\frac{v\,dv}{1 - kv^2/g} = g\,dx, \tag{1.2.21}$$

or

$$\ln\left(1 - \frac{kv^2}{g}\right) = -2kx, \tag{1.2.22}$$

where $k = C_D/m$ and $v = 0$ for $x = 0$. Taking the inverse of the natural logarithm, we finally obtain

$$v^2(x) = \frac{g}{k}\left(1 - e^{-2kx}\right). \tag{1.2.23}$$

Thus, as the distance that the object falls increases, so does the velocity, and it eventually approaches a constant value $\sqrt{g/k}$, commonly known as the *terminal velocity*.

Because the drag coefficient C_D varies with the superficial area of the object while the mass depends on the volume, k increases as an object becomes smaller, resulting in a smaller terminal velocity. Consequently, although a human being of normal size will acquire a terminal velocity of approximately 120 mph, a mouse, on the other hand, can fall any distance without injury. □

• **Example 1.2.7: Interest rate**

Consider a bank account that has been set up to pay out a constant rate of P dollars per year for the purchase of a car. This account has the special feature that it pays an annual interest rate of r on the current balance. We would like to know the balance in the account at any time t.

Although financial transactions occur at regularly spaced intervals, an excellent approximation can be obtained by treating the amount in the account $x(t)$ as a continuous function of time governed by the equation

$$x(t + \Delta t) \approx x(t) + rx(t)\Delta t - P\Delta t, \tag{1.2.24}$$

where we have assumed that both the payment and interest are paid in time increments of Δt. As the time between payments tends to zero, we obtain the first-order ordinary differential equation

$$\frac{dx}{dt} = rx - P. \tag{1.2.25}$$

If we denote the initial deposit into this account by $x(0)$, then at any subsequent time

$$x(t) = x(0)e^{rt} - P\left(e^{rt} - 1\right)/r. \tag{1.2.26}$$

Although we could compute $x(t)$ as a function of P, r, and $x(0)$, there are only three separate cases that merit our close attention. If $P/r > x(0)$, then the account will eventually equal zero at $rt = \ln\{P/\left[P - rx(0)\right]\}$. On the other hand, if $P/r < x(0)$, the amount of money in the account will grow without bound. Finally, the case $x(0) = P/r$ is the equilibrium case where the amount of money paid out balances the growth of money due to interest so that the account always has the balance of P/r.   ☐

• Example 1.2.8: Steady-state flow of heat

When the inner and outer walls of a body, for example the inner and outer walls of a house, are maintained at *different constant* temperatures, heat will flow from the warmer wall to the colder one. When each surface parallel to a wall has attained a constant temperature, the flow of heat has reached a steady state. In a steady-state flow of heat, each surface parallel to a wall, because its temperature is now constant, is referred to as an isothermal surface. Isothermal surfaces at different distances from an interior wall will have different temperatures. In many cases the temperature of an isothermal surface is only a function of its distance x from the interior wall, and the rate of flow of heat Q in a unit time across such a surface is proportional both to the area A of the surface and to dT/dx, where T is the temperature of the isothermal surface. Hence,

$$Q = -\kappa A \frac{dT}{dx}, \tag{1.2.27}$$

where κ is called the thermal conductivity of the material between the walls.

In place of a flat wall, let us consider a hollow cylinder whose inner and outer surfaces are located at $r = r_1$ and $r = r_2$, respectively. At steady state, Equation 1.2.27 becomes

$$Q_r = -\kappa A \frac{dT}{dr} = -\kappa(2\pi rL)\frac{dT}{dr}, \tag{1.2.28}$$

assuming no heat generation within the cylindrical wall.

We can find the temperature distribution inside the cylinder by solving Equation 1.2.28 along with the appropriate conditions on $T(r)$ at $r = r_1$ and $r = r_2$ (the boundary conditions). To illustrate the wide choice of possible boundary conditions, let us require that the inner surface is maintained at the temperature T_1. We assume that along the outer surface, heat is lost by convection to the environment, which has the temperature T_∞. This heat loss is usually modeled by the equation

$$\kappa \left.\frac{dT}{dr}\right|_{r=r_2} = -h(T - T_\infty), \tag{1.2.29}$$

where $h > 0$ is the convective heat transfer coefficient. Upon integrating Equation 1.2.28,

$$T(r) = -\frac{Q_r}{2\pi\kappa L} \ln(r) + C, \tag{1.2.30}$$

where Q_r is also an unknown. Substituting Equation 1.2.30 into the boundary conditions, we obtain

$$T(r) = T_1 + \frac{Q_r}{2\pi\kappa L} \ln(r_1/r), \tag{1.2.31}$$

with

$$Q_r = \frac{2\pi\kappa L(T_1 - T_\infty)}{\kappa/r_2 + h\ln(r_2/r_1)}. \tag{1.2.32}$$

As r_2 increases, the first term in the denominator of Equation 1.2.32 decreases while the second term increases. Therefore, Q_r has its largest magnitude when the denominator is smallest, assuming a fixed numerator. This occurs at the critical radius $r_{cr} = \kappa/h$, where

$$Q_r^{max} = \frac{2\pi\kappa L(T_1 - T_\infty)}{1 + \ln(r_{cr}/r_1)}. \tag{1.2.33}$$

$\square$

- **Example 1.2.9: Population dynamics**

Consider a population $P(t)$ that can change only by a birth or death but not by immigration or emigration. If $B(t)$ and $D(t)$ denote the number of births or deaths, respectively, as a function of time t, the *birth rate* and *death rate* (in births or deaths per unit time) is

$$b(t) = \lim_{\Delta t \to 0} \frac{B(t + \Delta t) - B(t)}{P(t)\Delta t} = \frac{1}{P}\frac{dB}{dt}, \tag{1.2.34}$$

and

$$d(t) = \lim_{\Delta t \to 0} \frac{D(t + \Delta t) - D(t)}{P(t)\Delta t} = \frac{1}{P}\frac{dD}{dt}. \tag{1.2.35}$$

Now,

$$P'(t) = \lim_{\Delta t \to 0} \frac{P(t + \Delta t) - P(t)}{\Delta t} \tag{1.2.36}$$

$$= \lim_{\Delta t \to 0} \frac{[B(t + \Delta t) - B(t)] - [D(t + \Delta t) - D(t)]}{\Delta t} \tag{1.2.37}$$

$$= B'(t) - D'(t). \tag{1.2.38}$$

Therefore,

$$P'(t) = [b(t) - d(t)]P(t). \tag{1.2.39}$$

When the birth and death rates are constants, namely $\bar{b}$ and $\bar{d}$, respectively, the population evolves according to

$$P(t) = P(0) \exp\left[(\bar{b} - \bar{d})\, t\right].\tag{1.2.40}$$

□

• Example 1.2.10: Logistic equation

The study of population dynamics yields an important class of first-order, nonlinear, ordinary differential equations: the logistic equation. This equation arose in Pierre François Verhulst's (1804–1849) study of animal populations.[3] If $x(t)$ denotes the number of species in the population and k is the (constant) environment capacity (the number of species that can simultaneously live in the geographical region), then the logistic or Verhulst's equation is

$$x' = ax(k - x)/k,\tag{1.2.41}$$

where a is the population growth rate for a small number of species.

To solve Equation 1.2.41, we rewrite it as

$$\frac{dx}{(1 - x/k)x} = \frac{dx}{x} + \frac{x/k}{1 - x/k}\, dx = r\, dt.\tag{1.2.42}$$

Integration yields

$$\ln|x| - \ln|1 - x/k| = rt + \ln(C),\tag{1.2.43}$$

or

$$\frac{x}{1 - x/k} = Ce^{rt}.\tag{1.2.44}$$

If $x(0) = x_0$,

$$x(t) = \frac{kx_0}{x_0 + (k - x_0)e^{-rt}}.\tag{1.2.45}$$

As $t \to \infty$, $x(t) \to k$, the asymptotically stable solution.   □

• Example 1.2.11: Chemical reactions

Chemical reactions are often governed by first-order ordinary differential equations. For example, first-order reactions, which describe reactions of the form A $\xrightarrow{k}$ B, yield the differential equation

$$-\frac{1}{a}\frac{d[A]}{dt} = k[A],\tag{1.2.46}$$

where k is the rate at which the reaction is taking place. Because for every molecule of A that disappears one molecule of B is produced, $a = 1$ and Equation 1.2.46 becomes

$$-\frac{d[A]}{dt} = k[A].\tag{1.2.47}$$

Integration of Equation 1.2.47 leads to

$$-\int \frac{d[A]}{[A]} = k \int\, dt.\tag{1.2.48}$$

---

[3] Verhulst, P. F., 1838: Notice sur la loi que la population suit dans son accroissement. *Correspond. Math. Phys.*, **10**, 113–121.

If we denote the initial value of [A] by $[A]_0$, then integration yields

$$-\ln[A] = kt - \ln[A]_0, \tag{1.2.49}$$

or

$$[A] = [A]_0 e^{-kt}. \tag{1.2.50}$$

The exponential form of the solution suggests that there is a *time constant* τ, which is called the *decay time* of the reaction. This quantity gives the time required for the concentration of decrease by $1/e$ of its initial value $[A]_0$. It is given by $\tau = 1/k$.

Turning to second-order reactions, there are two cases. The first is a reaction between two identical species: $A + A \xrightarrow{k}$ products. The rate expression here is

$$-\frac{1}{2}\frac{d[A]}{dt} = k[A]^2. \tag{1.2.51}$$

The second case is an overall second-order reaction between two unlike species, given by A + B $\xrightarrow{k}$ X. In this case, the reaction is first order in each of the reactants A and B and the rate expression is

$$-\frac{d[A]}{dt} = k[A][B]. \tag{1.2.52}$$

Turning to Equation 1.2.51 first, we have by separation of variables

$$-\int_{[A]_0}^{[A]} \frac{d[A]}{[A]^2} = 2k\int_0^t d\tau, \tag{1.2.53}$$

or

$$\frac{1}{[A]} = \frac{1}{[A]_0} + 2kt. \tag{1.2.54}$$

Therefore, a plot of the inverse of A versus time will yield a straight line with slope equal to $2k$ and intercept $1/[A]_0$.

With regard to Equation 1.2.52, because an increase in X must be at the expense of A and B, it is useful to express the rate equation in terms of the concentration of X, $[X] = [A]_0 - [A] = [B]_0 - [B]$, where $[A]_0$ and $[B]_0$ are the initial concentrations. Then, this equation becomes

$$\frac{d[X]}{dt} = k([A]_0 - [X])([B]_0 - [X]). \tag{1.2.55}$$

Separation of variables leads to

$$\int_{[X]_0}^{[X]} \frac{d\xi}{([A]_0 - \xi)([B]_0 - \xi)} = k\int_0^t d\tau. \tag{1.2.56}$$

To integrate the left side, we rewrite the integral

$$\int \frac{d\xi}{([A]_0 - \xi)([B]_0 - \xi)} = \int \frac{d\xi}{([A]_0 - [B]_0)([B]_0 - \xi)} - \int \frac{d\xi}{([A]_0 - [B]_0)([A]_0 - \xi)}. \tag{1.2.57}$$

Carrying out the integration,

$$\frac{1}{[A]_0 - [B]_0} \ln\left(\frac{[B]_0[A]}{[A]_0[B]}\right) = kt. \tag{1.2.58}$$

Again the reaction rate constant k can be found by plotting the data in the form of the left side of Equation 1.2.58 against t.

Problems

For Problems 1–9, solve the following ordinary differential equations by separation of variables. Then use MATLAB to plot your solution. Try and find the symbolic solution using MATLAB's `dsolve`.

1. $\dfrac{dy}{dx} = xe^y$

   2. $(1 + y^2)\, dx - (1 + x^2)\, dy = 0$

   3. $\ln(x)\dfrac{dx}{dy} = xy$

4. $\dfrac{y^2}{x}\dfrac{dy}{dx} = 1 + x^2$

   5. $\dfrac{dy}{dx} = \dfrac{2x + xy^2}{y + x^2 y}$

   6. $\dfrac{dy}{dx} = (xy)^{1/3}$

7. $\dfrac{dy}{dx} = e^{x+y}$

   8. $\dfrac{dy}{dx} = (x^3 + 5)(y^2 + 1)$

   9. $x \sin(y)\dfrac{dy}{dx} = \sec(y)$

10. Solve the initial-value problem

$$\frac{dy}{dt} = -ay + \frac{b}{y^2}, \qquad y(0) = y_0,$$

where a and b are constants.

11. Setting $u = y - x$, solve the first-order ordinary differential equation

$$\frac{dy}{dx} = \frac{y - x}{x^2} + 1.$$

12. Using the hydrostatic equation, show that the pressure within an atmosphere where the temperature decreases uniformly with height, $T(z) = T_0 - \Gamma z$, varies as

$$p(z) = p_0 \left(\frac{T_0 - \Gamma z}{T_0}\right)^{g/(R\Gamma)},$$

where p_0 is the pressure at $z = 0$.

13. Using the hydrostatic equation, show that the pressure within an atmosphere with the temperature distribution

$$T(z) = \begin{cases} T_0 - \Gamma z, & 0 \le z \le H, \\ T_0 - \Gamma H, & H \le z, \end{cases}$$

is

$$p(z) = p_0 \begin{cases} \left(\dfrac{T_0 - \Gamma z}{T_0}\right)^{g/(R\Gamma)}, & 0 \le z \le H, \\[2ex] \left(\dfrac{T_0 - \Gamma H}{T_0}\right)^{g/(R\Gamma)} \exp\left[-\dfrac{g(z - H)}{R(T_0 - \Gamma H)}\right], & H \le z, \end{cases}$$

where p_0 is the pressure at $z = 0$.

14. The voltage V as a function of time t within an electrical circuit[4] consisting of a capacitor with capacitance C and a diode in series is governed by the first-order ordinary differential equation

$$C\frac{dV}{dt} + \frac{V}{R} + \frac{V^2}{S} = 0,$$

where R and S are positive constants. If the circuit initially has a voltage V_0 at $t = 0$, find the voltage at subsequent times.

15. A glow plug is an electrical element inside a reaction chamber, which either ignites the nearby fuel or warms the air in the chamber so that the ignition will occur more quickly. An accurate prediction of the wire's temperature is important in the design of the chamber.

Assuming that heat convection and conduction are not important,[5] the temperature T of the wire is governed by

$$A\frac{dT}{dt} + B(T^4 - T_a^4) = P,$$

where A equals the specific heat of the wire times its mass, B equals the product of the emissivity of the surrounding fluid times the wire's surface area times the Stefan-Boltzmann constant, T_a is the temperature of the surrounding fluid, and P is the power input. The temperature increases due to electrical resistance and is reduced by radiation to the surrounding fluid.

Show that the temperature is given by

$$\frac{4B\gamma^3 t}{A} = 2\left[\tan^{-1}\left(\frac{T}{\gamma}\right) - \tan^{-1}\left(\frac{T_0}{\gamma}\right)\right] - \ln\left[\frac{(T-\gamma)(T_0+\gamma)}{(T+\gamma)(T_0-\gamma)}\right],$$

where $\gamma^4 = P/B + T_a^4$ and T_0 is the initial temperature of the wire.

16. Let us denote the number of tumor cells by $N(t)$. Then a widely used deterministic tumor growth law[6] is

$$\frac{dN}{dt} = bN\ln(K/N),$$

where K is the largest tumor size and $1/b$ is the length of time required for the specific growth to decrease by $1/e$. If the initial value of $N(t)$ is $N(0)$, find $N(t)$ at any subsequent time t.

17. The drop in laser intensity in the direction of propagation x due to one- and two-photon absorption in photosensitive glass is governed[7] by

$$\frac{dI}{dx} = -\alpha I - \beta I^2,$$

---

[4] See Aiken, C. B., 1938: Theory of the diode voltmeter. *Proc. IRE*, **26**, 859–876.

[5] See Clark, S. K., 1956: Heat-up time of wire glow plugs. *Jet Propulsion*, **26**, 278–279.

[6] See Hanson, F. B., and C. Tier, 1982: A stochastic model of tumor growth. *Math. Biosci.*, **61**, 73–100.

[7] See Weitzman, P. S., and U. Österberg, 1996: Two-photon absorption and photoconductivity in photosensitive glasses. *J. Appl. Phys.*, **79**, 8648–8655.

where I is the laser intensity, α and β are the single-photon and two-photon coefficients, respectively. Show that the laser intensity distribution is

$$I(x) = \frac{\alpha I(0)e^{-\alpha x}}{\alpha + \beta I(0)\left(1 - e^{-\alpha x}\right)},$$

where $I(0)$ is the laser intensity at the entry point of the media, $x = 0$.

18. The third-order reaction A + B + C $\xrightarrow{k}$ X is governed by the kinetics equation

$$\frac{d[\mathrm{X}]}{dt} = k\left([\mathrm{A}]_0 - [\mathrm{X}]\right)\left([\mathrm{B}]_0 - [\mathrm{X}]\right)\left([\mathrm{C}]_0 - [\mathrm{X}]\right),$$

where $[\mathrm{A}]_0$, $[\mathrm{B}]_0$, and $[\mathrm{C}]_0$ denote the initial concentration of A, B, and C, respectively. Find how $[\mathrm{X}]$ varies with time t.

19. The reversible reaction A $\underset{k_2}{\overset{k_1}{\rightleftarrows}}$ B is described by the kinetics equation[8]

$$\frac{d[\mathrm{X}]}{dt} = k_1\left([\mathrm{A}]_0 - [\mathrm{X}]\right) - k_2\left([\mathrm{B}]_0 + [\mathrm{X}]\right),$$

where $[\mathrm{X}]$ denotes the increase in the concentration of B while $[\mathrm{A}]_0$ and $[\mathrm{B}]_0$ are the initial concentrations of A and B, respectively. Find $[\mathrm{X}]$ as a function of time t. Hint: Show that this differential equation can be written

$$\frac{d[\mathrm{X}]}{dt} = (k_1 - k_2)\left(\alpha + [\mathrm{X}]\right), \qquad \alpha = \frac{k_1[\mathrm{A}]_0 - k_2[\mathrm{B}]_0}{k_1 + k_2}.$$

1.3 HOMOGENEOUS EQUATIONS

A *homogeneous ordinary differential equation* is a differential equation of the form

$$M(x,y)\,dx + N(x,y)\,dy = 0, \tag{1.3.1}$$

where both $M(x,y)$ and $N(x,y)$ are homogeneous functions of the same degree n. That means: $M(tx,ty) = t^n M(x,y)$ and $N(tx,ty) = t^n N(x,y)$. For example, the ordinary differential equation

$$(x^2 + y^2)\,dx + (x^2 - xy)\,dy = 0 \tag{1.3.2}$$

is a homogeneous equation because both coefficients are homogeneous functions of degree 2:

$$M(tx,ty) = t^2x^2 + t^2y^2 = t^2(x^2 + y^2) = t^2 M(x,y), \tag{1.3.3}$$

and

$$N(tx,ty) = t^2x^2 - t^2xy = t^2(x^2 - xy) = t^2 N(x,y). \tag{1.3.4}$$

[8] See Küster, F. W., 1895: Ueber den Verlauf einer umkehrbaren Reaktion erster Ordnung in homogenem System. *Z. Physik. Chem.*, **18**, 171–179.

Why is it useful to recognize homogeneous ordinary differential equations? Let us set $y = ux$ so that Equation 1.3.2 becomes

$$(x^2 + u^2 x^2)\,dx + (x^2 - ux^2)(u\,dx + x\,du) = 0. \tag{1.3.5}$$

Then,

$$x^2(1 + u)\,dx + x^3(1 - u)\,du = 0, \tag{1.3.6}$$

$$\frac{1 - u}{1 + u}\,du + \frac{dx}{x} = 0, \tag{1.3.7}$$

or

$$\left(-1 + \frac{2}{1 + u}\right)du + \frac{dx}{x} = 0. \tag{1.3.8}$$

Integrating Equation 1.3.8,

$$-u + 2\ln|1 + u| + \ln|x| = \ln|c|, \tag{1.3.9}$$

$$-\frac{y}{x} + 2\ln\left|1 + \frac{y}{x}\right| + \ln|x| = \ln|c|, \tag{1.3.10}$$

$$\ln\left[\frac{(x + y)^2}{cx}\right] = \frac{y}{x}, \tag{1.3.11}$$

or

$$(x + y)^2 = cxe^{y/x}. \tag{1.3.12}$$

Problems

First show that the following differential equations are homogeneous and then find their solution. Then use MATLAB to plot your solution. Try and find the symbolic solution using MATLAB's dsolve.

1. $(x + y)\dfrac{dy}{dx} = y$
2. $(x + y)\dfrac{dy}{dx} = x - y$
3. $2xy\dfrac{dy}{dx} = -(x^2 + y^2)$

4. $x(x + y)\dfrac{dy}{dx} = y(x - y)$
5. $x\dfrac{dy}{dx} = y + 2\sqrt{xy}$
6. $x\dfrac{dy}{dx} = y - \sqrt{x^2 + y^2}$

7. $\dfrac{dy}{dx} = \sec(y/x) + y/x$
8. $\dfrac{dy}{dx} = e^{y/x} + y/x.$
9. $x^2\dfrac{dy}{dx} = x^2 + y^2$

1.4 EXACT EQUATIONS

Consider the multivariable function $z = f(x, y)$. Then the total derivative is

$$dz = \frac{\partial f}{\partial x}\,dx + \frac{\partial f}{\partial y}\,dy = M(x, y)\,dx + N(x, y)\,dy. \tag{1.4.1}$$

If the solution to a first-order ordinary differential equation can be written as $f(x, y) = c$, then the corresponding differential equation is

$$M(x, y)\,dx + N(x, y)\,dy = 0. \tag{1.4.2}$$

How do we know if we have an *exact equation*, Equation 1.4.2? From the definition of $M(x, y)$ and $N(x, y)$,

$$\frac{\partial M}{\partial y} = \frac{\partial^2 f}{\partial y \partial x} = \frac{\partial^2 f}{\partial x \partial y} = \frac{\partial N}{\partial x}, \tag{1.4.3}$$

if $M(x, y)$ and $N(x, y)$ and their first-order partial derivatives are continuous. Consequently, if we can show that our ordinary differential equation is exact, we can integrate

$$\frac{\partial f}{\partial x} = M(x, y) \qquad \text{and} \qquad \frac{\partial f}{\partial y} = N(x, y) \tag{1.4.4}$$

to find the solution $f(x, y) = c$.

• **Example 1.4.1**

Let us check and see if

$$[y^2 \cos(x) - 3x^2 y - 2x]\, dx + [2y \sin(x) - x^3 + \ln(y)]\, dy = 0 \tag{1.4.5}$$

is exact.

Since $M(x, y) = y^2 \cos(x) - 3x^2 y - 2x$, and $N(x, y) = 2y \sin(x) - x^3 + \ln(y)$, we find that

$$\frac{\partial M}{\partial y} = 2y \cos(x) - 3x^2, \tag{1.4.6}$$

and

$$\frac{\partial N}{\partial x} = 2y \cos(x) - 3x^2. \tag{1.4.7}$$

Because $N_x = M_y$, Equation 1.4.5 is an exact equation.   □

• **Example 1.4.2**

Because Equation 1.4.5 is an exact equation, let us find its solution. Starting with

$$\frac{\partial f}{\partial x} = M(x, y) = y^2 \cos(x) - 3x^2 y - 2x, \tag{1.4.8}$$

direct integration gives

$$f(x, y) = y^2 \sin(x) - x^3 y - x^2 + g(y). \tag{1.4.9}$$

Substituting Equation 1.4.9 into the equation $f_y = N$, we obtain

$$\frac{\partial f}{\partial y} = 2y \sin(x) - x^3 + g'(y) = 2y \sin(x) - x^3 + \ln(y). \tag{1.4.10}$$

Thus, $g'(y) = \ln(y)$, or $g(y) = y \ln(y) - y + C$. Therefore, the solution to the ordinary differential equation, Equation 1.4.5, is

$$y^2 \sin(x) - x^3 y - x^2 + y \ln(y) - y = c. \tag{1.4.11}$$

□

● **Example 1.4.3**

Consider the differential equation

$$(x + y)\, dx + x \ln(x)\, dy = 0 \qquad (1.4.12)$$

on the interval $(0, \infty)$. A quick check shows that Equation 1.4.12 is not exact since

$$\frac{\partial M}{\partial y} = 1, \quad \text{and} \quad \frac{\partial N}{\partial x} = 1 + \ln(x). \qquad (1.4.13)$$

However, if we multiply Equation 1.4.12 by $1/x$ so that it becomes

$$\left(1 + \frac{y}{x}\right) dx + \ln(x)\, dy = 0, \qquad (1.4.14)$$

then this modified differential equation is exact because

$$\frac{\partial M}{\partial y} = \frac{1}{x}, \quad \text{and} \quad \frac{\partial N}{\partial x} = \frac{1}{x}. \qquad (1.4.15)$$

Therefore, the solution to Equation 1.4.12 is

$$x + y \ln(x) = C. \qquad (1.4.16)$$

This mysterious function that converts an inexact differential equation into an exact one is called an *integrating factor*. Unfortunately there is no general rule for finding one unless the equation is linear.

Problems

Show that the following equations are exact. Then solve them, using MATLAB to plot them. Finally, try and find the symbolic solution using MATLAB's `dsolve`.

1. $2xyy' = x^2 - y^2$

2. $(x + y)y' + y = x$

3. $(y^2 - 1)\, dx + [2xy - \sin(y)]\, dy = 0$

4. $[\sin(y) - 2xy + x^2]\, dx$
$+ [x\cos(y) - x^2]\, dy = 0$

5. $-y\, dx/x^2 + (1/x + 1/y)\, dy = 0$

6. $(3x^2 - 6xy)\, dx - (3x^2 + 2y)\, dy = 0$

7. $y\sin(xy)\, dx + x\sin(xy)\, dy = 0$

8. $(2xy^2 + 3x^2)\, dx + 2x^2y\, dy = 0$

9. $(2xy^3 + 5x^4y)\, dx$
$+ (3x^2y^2 + x^5 + 1)\, dy = 0$

10. $(x^3 + y/x)\, dx + [y^2 + \ln(x)]\, dy = 0$

11. $[x + e^{-y} + x\ln(y)]\, dy$
$+ [y\ln(y) + e^x]\, dx = 0$

12. $\cos(4y^2)\, dx - 8xy\sin(4y^2)\, dy = 0$

13. $\sin^2(x + y)\, dx - \cos^2(x + y)\, dy = 0$

14. $\cos(4y^2)\, dx - 8xy\sin(4y^2)\, dy = 0$

15. Show that the integrating factor for $(x-y)y' + \alpha y(1-y) = 0$ is $\mu(y) = y^a/(1-y)^{a+2}$, $a+1 = 1/\alpha$. Then show that the solution is

$$\alpha x \frac{y^{a+1}}{(1-y)^{a+1}} - \int_0^y \frac{\xi^{a+1}}{(1-\xi)^{a+2}}\, d\xi = C.$$

1.5 LINEAR EQUATIONS

In the case of first-order ordinary differential equations, any differential equation of the form

$$a_1(x)\frac{dy}{dx} + a_0(x)y = f(x) \tag{1.5.1}$$

is said to be linear.

Consider now the linear ordinary differential equation

$$x\frac{dy}{dx} - 4y = x^6 e^x \tag{1.5.2}$$

or

$$\frac{dy}{dx} - \frac{4}{x}y = x^5 e^x. \tag{1.5.3}$$

Let us now multiply Equation 1.5.3 by x^{-4}. (How we knew that it should be x^{-4} and not something else will be addressed shortly.) This magical factor is called an *integrating factor* because Equation 1.5.3 can be rewritten

$$\frac{1}{x^4}\frac{dy}{dx} - \frac{4}{x^5}y = xe^x, \tag{1.5.4}$$

or

$$\frac{d}{dx}\left(\frac{y}{x^4}\right) = xe^x. \tag{1.5.5}$$

Thus, our introduction of the integrating factor x^{-4} allows us to use the differentiation product rule in reverse and collapse the right side of Equation 1.5.4 into a single x derivative of a function of x times y. If we had selected the incorrect integrating factor, the right side would not have collapsed into this useful form.

With Equation 1.5.5, we may integrate both sides and find that

$$\frac{y}{x^4} = \int xe^x \, dx + C, \tag{1.5.6}$$

or

$$\frac{y}{x^4} = (x-1)e^x + C, \tag{1.5.7}$$

or

$$y = x^4(x-1)e^x + Cx^4. \tag{1.5.8}$$

From this example, it is clear that finding the integrating factor is crucial to solving first-order, linear, ordinary differential equations. To do this, let us first rewrite Equation 1.5.1 by dividing through by $a_1(x)$ so that it becomes

$$\frac{dy}{dx} + P(x)y = Q(x), \tag{1.5.9}$$

or

$$dy + [P(x)y - Q(x)]\, dx = 0. \tag{1.5.10}$$

If we denote the integrating factor by $\mu(x)$, then

$$\mu(x)dy + \mu(x)[P(x)y - Q(x)]\, dx = 0. \tag{1.5.11}$$

Clearly, we can solve Equation 1.5.11 by direct integration if it is an exact equation. If this is true, then

$$\frac{\partial \mu}{\partial x} = \frac{\partial}{\partial y} \left\{ \mu(x)[P(x)y - Q(x)] \right\}, \tag{1.5.12}$$

or

$$\frac{d\mu}{dx} = \mu(x)P(x), \qquad \text{and} \qquad \frac{d\mu}{\mu} = P(x)\, dx. \tag{1.5.13}$$

Integrating Equation 1.5.13,

$$\mu(x) = \exp\left[\int^x P(\xi)\, d\xi\right]. \tag{1.5.14}$$

Note that we do not need a constant of integration in Equation 1.5.14 because Equation 1.5.11 is unaffected by a constant multiple. It is also interesting that the integrating factor only depends on $P(x)$ and not $Q(x)$.

We can summarize our findings in the following theorem.

Theorem: Linear First-Order Equation

If the functions $P(x)$ and $Q(x)$ are continuous on the open interval I containing the point x_0, then the initial-value problem

$$\frac{dy}{dx} + P(x)y = Q(x), \qquad y(x_0) = y_0,$$

has a unique solution $y(x)$ on I, given by

$$y(x) = \frac{C}{\mu(x)} + \frac{1}{\mu(x)} \int^x Q(\xi)\mu(\xi)\, d\xi$$

with an appropriate value of C, and $\mu(x)$ is defined by Equation 1.5.14. □

The procedure for implementing this theorem is as follows:

- **Step 1:** If necessary, divide the differential equation by the coefficient of dy/dx. This gives an equation of the form Equation 1.5.9 and we can find $P(x)$ by inspection.

- **Step 2:** Find the integrating factor by Equation 1.5.14.

- **Step 3:** Multiply the equation created in Step 1 by the integrating factor.

- **Step 4:** Run the derivative product rule in reverse, collapsing the left side of the differential equation into the form $d[\mu(x)y]/dx$. If you are unable to do this, you have made a mistake.

- **Step 5:** Integrate both sides of the differential equation to find the solution.

The following examples illustrate the technique.

• **Example 1.5.1**

Let us solve the linear, first-order ordinary differential equation

$$xy' - y = 4x \ln(x). \tag{1.5.15}$$

We begin by dividing through by x to convert Equation 1.5.15 into its canonical form. This yields

$$y' - \frac{1}{x}y = 4\ln(x). \tag{1.5.16}$$

From Equation 1.5.16, we see that $P(x) = 1/x$. Consequently, from Equation 1.5.14, we have that

$$\mu(x) = \exp\left[\int^x P(\xi)\,d\xi\right] = \exp\left(-\int^x \frac{d\xi}{\xi}\right) = \frac{1}{x}. \tag{1.5.17}$$

Multiplying Equation 1.5.16 by the integrating factor, we find that

$$\frac{y'}{x} - \frac{y}{x^2} = \frac{4\ln(x)}{x}, \tag{1.5.18}$$

or

$$\frac{d}{dx}\left(\frac{y}{x}\right) = \frac{4\ln(x)}{x}. \tag{1.5.19}$$

Integrating both sides of Equation 1.5.19,

$$\frac{y}{x} = 4\int \frac{\ln(x)}{x}\,dx = 2\ln^2(x) + C. \tag{1.5.20}$$

Multiplying Equation 1.5.20 through by x yields the general solution

$$y = 2x\ln^2(x) + Cx. \tag{1.5.21}$$

Although it is nice to have a closed-form solution, considerable insight can be gained by graphing the solution for a wide variety of initial conditions. To illustrate this, consider the MATLAB script:

```
clear
% use symbolic toolbox to solve Equation 1.5.15
y = dsolve('x*Dy-y=4*x*log(x)','y(1) = c','x');
% take the symbolic version of the solution
%     and convert it into executable code
solution = inline(vectorize(y),'x','c');
close all; axes; hold on
% now plot the solution for a wide variety of initial conditions
x = 0.1:0.1:2;
for c = -2:4
   if (c==-2) plot(x,solution(x,c),'.'); end
   if (c==-1) plot(x,solution(x,c),'o'); end
   if (c== 0) plot(x,solution(x,c),'x'); end
   if (c== 1) plot(x,solution(x,c),'+'); end
   if (c== 2) plot(x,solution(x,c),'*'); end
```

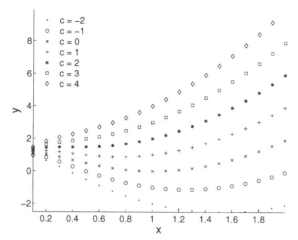

Figure 1.5.1: The solution to Equation 1.5.15 when the initial condition is $y(1) = c$.

```
  if (c== 3) plot(x,solution(x,c),'s'); end
  if (c== 4) plot(x,solution(x,c),'d'); end
end
axis tight
xlabel('x','Fontsize',20); ylabel('y','Fontsize',20)
legend('c = -2','c = -1','c = 0','c = 1',...
       'c = 2','c = 3','c = 4'); legend boxoff
```

This script does two things. First, it uses MATLAB's symbolic toolbox to solve Equation 1.5.15. Alternatively, we could have used Equation 1.5.21 and introduced it as a function. The second portion of this script plots this solution for $y(1) = C$ where $C = -2, -1, 0, 1, 2, 3, 4$. Figure 1.5.1 shows the results. As $x \to 0$, we note how all of the solutions behave like $2x \ln^2(x)$. □

• **Example 1.5.2**

Let us solve the first-order ordinary differential equation

$$\frac{dy}{dx} = \frac{y}{y - x} \tag{1.5.22}$$

subject to the initial condition $y(2) = 6$.

Beginning as before, we rewrite Equation 1.5.22 in the canonical form

$$(y - x)y' - y = 0. \tag{1.5.23}$$

Examining Equation 1.5.23 more closely, we see that it is a nonlinear equation in y. On the other hand, if we treat x as the *dependent* variable and y as the *independent variable*, we can write Equation 1.5.23 as the *linear* equation

$$\frac{dx}{dy} + \frac{x}{y} = 1. \tag{1.5.24}$$

Proceeding as before, we have that $P(y) = 1/y$ and $\mu(y) = y$, so that Equation 1.5.24 can be rewritten

$$\frac{d}{dy}(yx) = y \tag{1.5.25}$$

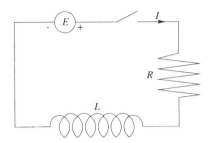

Figure 1.5.2: Schematic diagram for an electric circuit that contains a resistor of resistance R and an inductor of inductance L.

or

$$yx = \tfrac{1}{2}y^2 + C. \tag{1.5.26}$$

Introducing the initial condition, we find that $C = -6$. Solving for y, we obtain

$$y = x \pm \sqrt{x^2 + 12}. \tag{1.5.27}$$

We must take the positive sign in order that $y(2) = 6$ and

$$y = x + \sqrt{x^2 + 12}. \tag{1.5.28}$$

$\square$

• Example 1.5.3: Electric circuits

A rich source of first-order differential equations is the analysis of simple electrical circuits. These electrical circuits are constructed from three fundamental components: the resistor, the inductor, and the capacitor. Each of these devices gives the following voltage drop: In the case of a resistor, the voltage drop equals the product of the resistance R times the current I. For the inductor, the voltage drop is $L\, dI/dt$, where L is called the inductance, while the voltage drop for a capacitor equals Q/C, where Q is the instantaneous charge and C is called the capacitance.

How are these voltage drops applied to mathematically describe an electrical circuit? This question leads to one of the fundamental laws in physics, **Kirchhoff's law**: *The algebraic sum of all the voltage drops around an electric loop or circuit is zero.*

To illustrate Kirchhoff's law, consider the electrical circuit shown in Figure 1.5.2. By Kirchhoff's law, the electromotive force E, provided by a battery, for example, equals the sum of the voltage drops across the resistor RI and $L\, dI/dt$. Thus the (differential) equation that governs this circuit is

$$L\frac{dI}{dt} + RI = E. \tag{1.5.29}$$

Assuming that E, I, and R are constant, we can rewrite Equation 1.5.29 as

$$\frac{d}{dt}\left[e^{Rt/L}I(t)\right] = \frac{E}{L}e^{Rt/L}. \tag{1.5.30}$$

Integrating both sides of Equation 1.5.30,

$$e^{Rt/L}I(t) = \frac{E}{R}e^{Rt/L} + C_1, \tag{1.5.31}$$

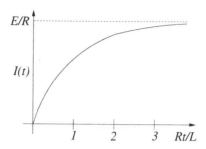

Figure 1.5.3: The temporal evolution of current $I(t)$ inside an electrical circuit shown in Figure 1.5.2 with a constant electromotive force E.

or

$$I(t) = \frac{E}{R} + C_1 e^{-Rt/L}. \qquad (1.5.32)$$

To determine C_1, we apply the initial condition. Because the circuit is initially dead, $I(0) = 0$, and

$$I(t) = \frac{E}{R}\left(1 - e^{-Rt/L}\right). \qquad (1.5.33)$$

Figure 1.5.3 illustrates Equation 1.5.33 as a function of time. Initially the current increases rapidly but the growth slows with time. Note that we could also have solved this problem by separation of variables.

Quite often, the solution is separated into two parts: the *steady-state solution* and the *transient solution*. The steady-state solution is that portion of the solution which remains as $t \to \infty$. It can equal zero. Presently it equals the constant value, E/R. The transient solution is that portion of the solution which vanishes as time increases. Here it equals $-Ee^{-Rt/L}/R$.

Although our analysis is a useful approximation to the real world, a more realistic one would include the nonlinear properties of the resistor.[9] To illustrate this, consider the case of an RL circuit without any electromotive source ($E = 0$) where the initial value for the current is I_0. Equation 1.5.29 now reads

$$L\frac{dI}{dt} + RI(1 - aI) = 0, \qquad I(0) = I_0. \qquad (1.5.34)$$

Separating the variables,

$$\frac{dI}{I(aI - 1)} = \frac{dI}{I - 1/a} - \frac{dI}{I} = \frac{R}{L}dt. \qquad (1.5.35)$$

Upon integrating and applying the initial condition, we have that

$$I = \frac{I_0 e^{-Rt/L}}{1 - aI_0 + aI_0 e^{-Rt/L}}. \qquad (1.5.36)$$

---

[9] For the analysis of

$$L\frac{dI}{dt} + RI + KI^\beta = 0,$$

see Fairweather, A., and J. Ingham, 1941: Subsidence transients in circuits containing a non-linear resistor, with reference to the problem of spark-quenching. *J. IEE, Part 1*, **88**, 330–339.

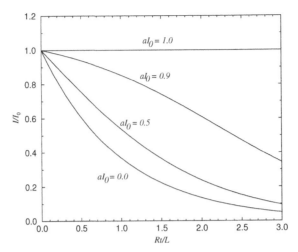

Figure 1.5.4: The variation of current I/I_0 as a function of time Rt/L with different values of aI_0.

Figure 1.5.4 shows $I(t)$ for various values of a. As the nonlinearity reduces resistance, the decay in the current is reduced. If $aI_0 > 1$, Equation 1.5.36 predicts that the current would grow with time. The point here is that nonlinearity can have a dramatic influence on a physical system.

Consider now the electrical circuit shown in Figure 1.5.5, which contains a resistor with resistance R and a capacitor with capacitance C. Here the voltage drop across the resistor is still RI while the voltage drop across the capacitor is Q/C. Therefore, by Kirchhoff's law,

$$RI + \frac{Q}{C} = E. \tag{1.5.37}$$

Equation 1.5.37 is *not* a differential equation. However, because current is the time rate of change in charge $I = dQ/dt$, our differential equation becomes

$$R\frac{dQ}{dt} + \frac{Q}{C} = E, \tag{1.5.38}$$

which is the differential equation for the instantaneous charge.

Let us solve Equation 1.5.38 when the resistance and capacitance are constant but the electromotive force equals $E_0 \cos(\omega t)$. The corresponding differential equation is now

$$R\frac{dQ}{dt} + \frac{Q}{C} = E_0 \cos(\omega t). \tag{1.5.39}$$

The differential equation has the integrating factor $e^{t/(RC)}$ so that it can be rewritten

$$\frac{d}{dt}\left[e^{t/(RC)}Q(t)\right] = \frac{E_0}{R}e^{t/(RC)}\cos(\omega t). \tag{1.5.40}$$

Integrating Equation 1.5.40,

$$e^{t/(RC)}Q(t) = \frac{CE_0}{1 + R^2C^2\omega^2}e^{t/(RC)}\left[\cos(\omega t) + RC\omega \sin(\omega t)\right] + C_1 \tag{1.5.41}$$

or

$$Q(t) = \frac{CE_0}{1 + R^2C^2\omega^2}\left[\cos(\omega t) + RC\omega \sin(\omega t)\right] + C_1 e^{-t/(RC)}. \tag{1.5.42}$$

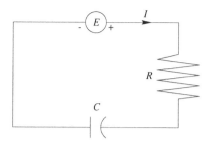

Figure 1.5.5: Schematic diagram for an electric circuit that contains a resistor of resistance R and a capacitor of capacitance C.

If we take the initial condition as $Q(0) = 0$, then the final solution is

$$Q(t) = \frac{CE_0}{1 + R^2 C^2 \omega^2} \left[\cos(\omega t) - e^{-t/(RC)} + RC\omega \sin(\omega t) \right]. \qquad (1.5.43)$$

Figure 1.5.6 illustrates Equation 1.5.43. Note how the circuit eventually supports a purely oscillatory solution (the steady-state solution) as the exponential term decays to zero (the transient solution). Indeed, the purpose of the transient solution is to allow the system to adjust from its initial condition to the final steady state. $\qquad \square$

• **Example 1.5.4: Terminal velocity**

When an object passes through a fluid, the viscosity of the fluid resists the motion by exerting a force on the object proportional to its velocity. Let us find the motion of a mass m that is initially thrown upward with the speed v_0.

If we choose the coordinate system so that it increases in the vertical direction, then the equation of motion is

$$m \frac{dv}{dt} = -kv - mg \qquad (1.5.44)$$

with $v(0) = v_0$ and $k > 0$. Rewriting Equation 1.5.44, we obtain the first-order linear differential equation

$$\frac{dv}{dt} + \frac{k}{m} v = -g. \qquad (1.5.45)$$

Its solution in *nondimensional* form is

$$\frac{kv(t)}{mg} = -1 + \left(1 + \frac{kv_0}{mg} \right) e^{-kt/m}. \qquad (1.5.46)$$

The displacement from its initial position is

$$\frac{k^2 x(t)}{m^2 g} = \frac{k^2 x_0}{m^2 g} - \frac{kt}{m} + \left(1 + \frac{kv_0}{mg} \right) \left(1 - e^{-kt/m} \right). \qquad (1.5.47)$$

As $t \to \infty$, the velocity tends to a constant downward value, $-mg/k$, the so-called "terminal velocity," where the aerodynamic drag balances the gravitational acceleration. This is the steady-state solution.

Why have we written Equation 1.5.46 and Equation 1.5.47 in this nondimensional form? There are two reasons. First, the solution reduces to three fundamental variables, a *nondimensional* displacement $x_* = k^2 x(t)/(m^2 g)$, velocity $v_* = kv(t)/(mg)$, and time

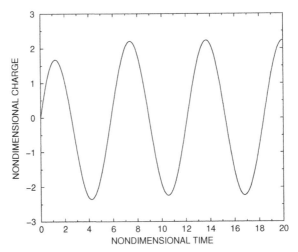

Figure 1.5.6: The temporal evolution of the nondimensional charge $(1 + R^2 C^2 \omega^2) Q(t) /(C E_0)$ in the electric circuit shown in Figure 1.5.4 as a function of nondimensional time ωt when the circuit is driven by the electromotive force $E_0 \cos(\omega t)$ and $RC\omega = 2$.

$t_* = kt/m$, rather than the six original parameters and variables: g, k, m, t, v, and x. Indeed, if we had substituted t_*, v_*, and x_* into Equation 1.5.45, we would have obtained the following simplified initial-value problem:

$$\frac{dv_*}{dt_*} + v_* = -1, \quad \frac{dx_*}{dt_*} = v_*, \quad v_*(0) = \frac{kv_0}{mg}, \quad x_*(0) = \frac{k^2 x_0}{m^2 g} \tag{1.5.48}$$

right from the start. The second advantage of the nondimensional form is the compact manner in which the results can be displayed, as Figure 1.5.7 shows.

From Equation 1.5.46 and Equation 1.5.47, the trajectory of the ball is as follows: If we define the coordinate system so that $x_0 = 0$, then the object will initially rise to the height H given by

$$\frac{k^2 H}{m^2 g} = \frac{kv_0}{mg} - \ln\left(1 + \frac{kv_0}{mg}\right) \tag{1.5.49}$$

at the time

$$\frac{kt_{max}}{m} = \ln\left(1 + \frac{kv_0}{mg}\right), \tag{1.5.50}$$

when $v(t_{max}) = 0$. It will then fall toward the earth. Given sufficient time $kt/m \gg 1$, it would achieve terminal velocity.   □

• **Example 1.5.5: The Bernoulli equation**

Bernoulli's equation,

$$\frac{dy}{dx} + p(x)y = q(x)y^n, \qquad n \neq 0, 1, \tag{1.5.51}$$

is a first-order, nonlinear differential equation. This equation can be transformed into a first-order, linear differential equation by introducing the change of variable $z = y^{1-n}$. Because

$$\frac{dz}{dx} = (1 - n)y^{-n}\frac{dy}{dx}, \tag{1.5.52}$$

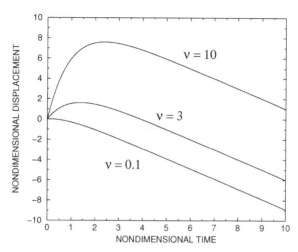

Figure 1.5.7: The nondimensional displacement $k^2x(t)/(m^2g)$ as a function of nondimensional time kt/m of an object of mass m thrown upward at the initial nondimensional speed $\nu = kv_0/(mg)$ in a fluid that retards its motion as $-kv$.

the transformed Bernoulli equation becomes

$$\frac{dz}{dx} + (1-n)p(x)z = (1-n)q(x). \tag{1.5.53}$$

This is now a first-order linear differential equation for z and can be solved using the methods introduced in this section. Once z is known, the solution is found by transforming back from z to y.

To illustrate this procedure, consider the nonlinear ordinary differential equation

$$x^2y\frac{dy}{dx} - xy^2 = 1, \tag{1.5.54}$$

or

$$\frac{dy}{dx} - \frac{y}{x} = \frac{y^{-1}}{x^2}. \tag{1.5.55}$$

Equation 1.5.55 is a Bernoulli equation with $p(x) = -1/x$, $q(x) = 1/x^2$, and $n = -1$. Introducing $z = y^2$, it becomes

$$\frac{dz}{dx} - \frac{2z}{x} = \frac{2}{x^2}. \tag{1.5.56}$$

This first-order linear differential equation has the integrating factor $\mu(x) = 1/x^2$ and

$$\frac{d}{dx}\left(\frac{z}{x^2}\right) = \frac{2}{x^4}. \tag{1.5.57}$$

Integration gives

$$\frac{z}{x^2} = C - \frac{2}{3x^3}. \tag{1.5.58}$$

Therefore, the general solution is

$$y^2 = z = Cx^2 - \frac{2}{3x}. \tag{1.5.59}$$

Problems

Find the solution for the following differential equations. State the interval on which the general solution is valid. Then use MATLAB to examine their behavior for a wide class of initial conditions.

1. $y' + y = e^x$

2. $y' + 2xy = x$

3. $x^2 y' + xy = 1$

4. $(2y + x^2)\, dx = x\, dy$

5. $y' - 3y/x = 2x^2$

6. $y' + 2y = 2\sin(x)$

7. $y' + 2\cos(2x)y = 0$

8. $xy' + y = \ln(x)$

9. $y' + 3y = 4, \quad y(0) = 5$

10. $y' - y = e^x/x, \quad y(e) = 0$

11. $\sin(x)y' + \cos(x)y = 1$

12. $[1 - \cos(x)]y' + 2\sin(x)y = \tan(x)$

13. $y' + [a\tan(x) + b\sec(x)]y = c\sec(x)$

14. $(xy + y - 1)\, dx + x\, dy = 0$

15. $y' + 2ay = \dfrac{x}{2} - \dfrac{\sin(2\omega x)}{4\omega}, \quad y(0) = 0.$

16. $y' + \dfrac{2k}{x^3}y = \ln\left(\dfrac{x+1}{x}\right), \quad k > 0, \ y(1) = 0.$

17. Solve the following initial-value problem:

$$kxy\frac{dy}{dx} = y^2 - x, \qquad y(1) = 0.$$

Hint: Introduce the new dependent variable $p = y^2$.

18. If $x(t)$ denotes the equity capital of a company, then under certain assumptions[10] $x(t)$ is governed by

$$\frac{dx}{dt} = (1 - N)rx + S,$$

where N is the dividend payout ratio, r is the rate of return of equity, and S is the rate of net new stock financing. If the initial value of $x(t)$ is $x(0)$, find $x(t)$.

19. The assimilation[11] of a drug into a body can be modeled by the chemical reaction A $\xrightarrow{k_1}$ B $\xrightarrow{k_2}$ C, which is governed by the chemical kinetics equations

$$\frac{d[A]}{dt} = -k_1[A], \qquad \frac{d[B]}{dt} = k_1[A] - k_2[B], \qquad \frac{d[C]}{dt} = k_2[B],$$

where [A] denotes the concentration of the drug in the gastrointestinal tract or in the site of injection, [B] is the concentration of the drug in the body, and [C] is either the amount of drug eliminated by various metabolic functions or the amount of the drug utilized by

---

[10] See Lebowitz, J. L., C. O. Lee, and P. B. Linhart, 1976: Some effects of inflation on a firm with original cost depreciation. *Bell J. Economics*, **7**, 463–477.

[11] See Calder, G. V., 1974: The time evolution of drugs in the body: An application of the principle of chemical kinetics. *J. Chem. Educ.*, **51**, 19–22.

various action sites in the body. If $[A]_0$ denotes the initial concentration of A, find $[A]$, $[B]$, and $[C]$ as a function of time t.

20. Find the current in an RL circuit when the electromotive source equals $E_0 \cos^2(\omega t)$. Initially the circuit is dead.

Find the general solution for the following Bernoulli equations:

21. $\dfrac{dy}{dx} + \dfrac{y}{x} = -y^2$ 22. $x^2\dfrac{dy}{dx} = xy + y^2$ 23. $\dfrac{dy}{dx} - \dfrac{4y}{x} = x\sqrt{y}$

24. $\dfrac{dy}{dx} + \dfrac{y}{x} = -xy^2$ 25. $2xy\dfrac{dy}{dx} - y^2 + x = 0$ 26. $x\dfrac{dy}{dx} + y = \frac{1}{2}xy^3$

1.6 GRAPHICAL SOLUTIONS

In spite of the many techniques developed for their solution, many ordinary differential equations cannot be solved analytically. In the next two sections, we highlight two alternative methods when analytical methods fail. Graphical methods seek to understand the nature of the solution by examining the differential equations at various points and infer the complete solution from these results. In the last section, we highlight the numerical techniques that are now commonly used to solve ordinary differential equations on the computer.

• *Direction fields*

One of the simplest numerical methods for solving first-order ordinary differential equations follows from the fundamental concept that the derivative gives the *slope* of a straight line that is tangent to a curve at a given point.

Consider the first-order differential equation

$$y' = f(x,y), \tag{1.6.1}$$

which has the initial value $y(x_0) = y_0$. For any (x,y) it is possible to draw a short line segment whose slope equals $f(x,y)$. This graphical representation is known as the *direction field* or *slope field* of Equation 1.6.1. Starting with the initial point (x_0, y_0), we can then construct the *solution curve* by extending the initial line segment in such a manner that the tangent of the solution curve parallels the direction field at each point through which the curve passes.

Before the days of computers, it was common to first draw lines of constant slope (*isoclines*) or $f(x,y) = c$. Because along any isocline all of the line segments had the same slope, considerable computational savings were realized. Today, computer software exists that performs these graphical computations with great speed.

To illustrate this technique, consider the ordinary differential equation

$$\frac{dx}{dt} = x - t^2. \tag{1.6.2}$$

Its exact solution is

$$x(t) = Ce^t + t^2 + 2t + 2, \tag{1.6.3}$$

where C is an arbitrary constant. Using the MATLAB script:

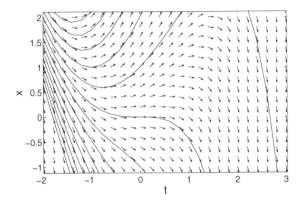

Figure 1.6.1: The direction field for Equation 1.6.2. The solid lines are plots of the solution with various initial conditions.

```
clear
% create grid points in t and x
[t,x] = meshgrid(-2:0.2:3,-1:0.2:2);
% load in the slope
slope = x - t.*t;
% find the length of the vector (1,slope)
length = sqrt(1 + slope .* slope);
% create and plot the vector arrows
quiver(t,x,1./length,slope./length,0.5)
axis equal tight
hold on
% plot the exact solution for various initial conditions
tt = [-2:0.2:3];
for cval = -10:1:10
  x_exact = cval * exp(tt) + tt.*tt + 2*tt + 2;
  plot(tt,x_exact)
  xlabel('t','Fontsize',20)
  ylabel('x','Fontsize',20)
end
```

we show in Figure 1.6.1 the directional field associated with Equation 1.6.2 along with some of the particular solutions. Clearly the vectors are parallel to the various particular solutions. Therefore, without knowing the solution, we could choose an arbitrary initial condition and sketch its behavior at subsequent times. The same holds true for nonlinear equations.

• *Rest points and autonomous equations*

In the case of autonomous differential equations (equations where the independent variable does not explicitly appear in the equation), considerable information can be gleaned from a graphical analysis of the equation.

Consider the nonlinear ordinary differential equation

$$x' = \frac{dx}{dt} = x(x^2 - 1). \tag{1.6.4}$$

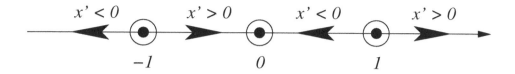

Figure 1.6.2: The phase line diagram for the ordinary differential equation, Equation 1.6.4.

The time derivative x' vanishes at $x = -1, 0, 1$. Consequently, if $x(0) = 0$, $x(t)$ will remain zero forever. Similarly, if $x(0) = 1$ or $x(0) = -1$, then $x(t)$ will equal 1 or -1 for all time. For this reason, values of x for which the derivative x' is zero are called *rest points*, *equilibrium points*, or *critical points* of the differential equation.

The behavior of solutions near rest points is often of considerable interest. For example, what happens to the solution when x is near one of the rest points $x = -1, 0, 1$?

Consider the point $x = 0$. For x slightly greater than zero, $x' < 0$. For x slightly less than 0, $x' > 0$. Therefore, for any initial value of x near $x = 0$, x will tend to zero. In this case, the point $x = 0$ is an asymptotically *stable critical point* because whenever x is perturbed away from the critical point, it tends to return there again.

Turning to the point $x = 1$, for x slightly greater than 1, $x' > 0$; for x slightly less than 1, $x' < 0$. Because any x near $x = 1$, but not equal to 1, will move away from $x = 1$, the point $x = 1$ is called an *unstable critical point*. A similar analysis applies at the point $x = -1$. This procedure of determining the behavior of an ordinary differential equation near its critical points is called a *graphical stability analysis*.

- *Phase line*

A graphical representation of the results of our graphical stability analysis is the *phase line*. On a phase line, the equilibrium points are denoted by circles. See Figure 1.6.2. Also on the phase line we identify the sign of x' for all values of x. From the sign of x', we then indicate whether x is increasing or decreasing by an appropriate arrow. If the arrow points toward the right, x is increasing; toward the left x decreases. Then, by knowing the *sign* of the derivative for all values of x, together with the starting value of x, we can determine what happens as $t \to \infty$. Any solution that is approached asymptotically as $t \to \infty$ is called a *steady-state output*. In our present example, $x = 0$ is a steady-state output.

Problems

In previous sections, you used various techniques to solve first-order ordinary differential equations. Now check your work by using MATLAB to draw the direction field and plot your analytic solution for the following problems taken from previous sections:

1. Section 1.2, Problem 5

2. Section 1.3, Problem 1

3. Section 1.4, Problem 5

4. Section 1.5, Problem 3

For the following autonomous ordinary differential equations, draw the phase line. Then classify each equilibrium solution as either stable or unstable.

5. $x' = \alpha x(1 - x)(x - \frac{1}{2})$

6. $x' = (x^2 - 1)(x^2 - 4)$

7. $x' = -4x - x^3$ 8. $x' = 4x - x^3$

1.7 NUMERICAL METHODS

By now you have seen most of the exact methods for finding solutions to first-order ordinary differential equations. The methods have also given you a view of the general behavior and properties of solutions to differential equations. However, it must be admitted that in many instances exact solutions cannot be found and we must resort to numerical solutions.

In this section we present the two most commonly used methods for solving differential equations: Euler and Runge-Kutta methods. There are many more methods and the interested student is referred to one of countless numerical methods books. A straightforward extension of these techniques can be applied to systems of first-order and higher-order differential equations.

• *Euler and modified Euler methods*

Consider the following first-order differential equation and initial condition:

$$\frac{dy}{dx} = f(x, y), \qquad y(x_0) = y_0. \tag{1.7.1}$$

Euler's method is based on a Taylor series expansion of the solution about x_0 or

$$y(x_0 + h) = y(x_0) + hy'(x_0) + \tfrac{1}{2}y''(\xi)h^2, \qquad x_0 < \xi < x_0 + h, \tag{1.7.2}$$

where h is the step size. Euler's method consists of taking a sufficiently small h so that only the first two terms of this Taylor expansion are significant.

Let us now replace $y'(x_0)$ by $f(x_0, y_0)$. Using subscript notation, we have that

$$y_{i+1} = y_i + hf(x_i, y_i) + O(h^2). \tag{1.7.3}$$

Equation 1.7.3 states that if we know the values of y_i and $f(x_i, y_i)$ at the position x_i, then the solution at x_{i+1} can be obtained with an error[12] $O(h^2)$.

The trouble with Euler's method is its lack of accuracy, often requiring an extremely small time step. How might we improve this method with little additional effort?

One possible method would retain the first three terms of the Taylor expansion rather than the first two. This scheme, known as the *modified Euler method*, is

$$y_{i+1} = y_i + hy'(x_i) + \tfrac{1}{2}h^2 y_i'' + O(h^3). \tag{1.7.4}$$

This is clearly more accurate than Equation 1.7.3.

An obvious question is how do we evaluate y_i'', because we do not have any information on its value. Using the forward derivative approximation, we find that

$$y_i'' = \frac{y_{i+1}' - y_i'}{h}. \tag{1.7.5}$$

---

[12] The symbol O is a mathematical notation indicating relative magnitude of terms, namely that $f(\epsilon) = O(\epsilon^n)$ provided $\lim_{\epsilon \to 0} |f(\epsilon)/\epsilon^n| < \infty$. For example, as $\epsilon \to 0$, $\sin(\epsilon) = O(\epsilon)$, $\sin(\epsilon^2) = O(\epsilon^2)$, and $\cos(\epsilon) = O(1)$.

Substituting Equation 1.7.5 into Equation 1.7.4 and simplifying

$$y_{i+1} = y_i + \frac{h}{2}\left(y_i' + y_{i+1}'\right) + O(h^3). \tag{1.7.6}$$

Using the differential equation,

$$y_{i+1} = y_i + \frac{h}{2}\left[f(x_i, y_i) + f(x_{i+1}, y_{i+1})\right] + O(h^3). \tag{1.7.7}$$

Although $f(x_i, y_i)$ at (x_i, y_i) are easily calculated, how do we compute $f(x_{i+1}, y_{i+1})$ at (x_{i+1}, y_{i+1})? For this we compute a first guess via the Euler method, Equation 1.7.3; Equation 1.7.7 then provides a refinement on the value of y_{i+1}.

In summary then, the simple Euler scheme is

$$y_{i+1} = y_i + k_1 + O(h^2), \quad k_1 = hf(x_i, y_i), \tag{1.7.8}$$

while the modified Euler method is

$$y_{i+1} = y_i + \tfrac{1}{2}(k_1 + k_2) + O(h^3), \; k_1 = hf(x_i, y_i), \; k_2 = hf(x_i + h, y_i + k_1). \tag{1.7.9}$$

• **Example 1.7.1**

Let us illustrate Euler's method by numerically solving

$$x' = x + t, \qquad x(0) = 1. \tag{1.7.10}$$

A quick check shows that Equation 1.7.10 has the exact solution $x_{\text{exact}}(t) = 2e^t - t - 1$. Using the MATLAB script:

```
clear
for i = 1:3
% set up time step increment and number of time steps
  h = 1/10^i; n = 10/h;
% set up initial conditions
  t=zeros(n+1,1); t(1) = 0;
  x_euler=zeros(n+1,1); x_euler(1) = 1;
  x_modified=zeros(n+1,1); x_modified(1) = 1;
  x_exact=zeros(n+1,1); x_exact(1) = 1;
% set up difference arrays for plotting purposes
  diff1 = zeros(n,1); diff2 = zeros(n,1); tplot = zeros(n,1);
% define right side of differential equation, Equation 1.7.10
  f = inline('xx+tt','tt','xx');
  for k = 1:n
    t(k+1) = t(k) + h;
% compute exact solution
    x_exact(k+1) = 2*exp(t(k+1)) - t(k+1) - 1;
% compute solution via Euler's method
    k1 = h * f(t(k),x_euler(k));
    x_euler(k+1) = x_euler(k) + k1;
    tplot(k) = t(k+1);
    diff1(k) = x_euler(k+1) - x_exact(k+1);
```

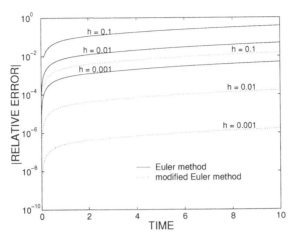

Figure 1.7.1: The relative error $[x(t) - x_{exact}(t)]/x_{exact}(t)$ of the numerical solution of Equation 1.7.10 using Euler's method (the solid line) and modified Euler's method (the dotted line) with different time steps h.

```
    diff1(k) = abs(diff1(k) / x_exact(k+1));
% compute solution via modified Euler method
    k1 = h * f(t(k),x_modified(k));
    k2 = h * f(t(k+1),x_modified(k)+k1);
    x_modified(k+1) = x_modified(k) + 0.5*(k1+k2);
    diff2(k) = x_modified(k+1) - x_exact(k+1);
    diff2(k) = abs(diff2(k) / x_exact(k+1));
  end
% plot relative errors
  semilogy(tplot,diff1,'-',tplot,diff2,':')
  hold on
  xlabel('TIME','Fontsize',20)
  ylabel('|RELATIVE ERROR|','Fontsize',20)
  legend('Euler method','modified Euler method')
  legend boxoff;
  num1 = 0.2*n; num2 = 0.8*n;
  text(3,diff1(num1),['h = ',num2str(h)],'Fontsize',15,...
      'HorizontalAlignment','right',...
      'VerticalAlignment','bottom')
  text(9,diff2(num2),['h = ',num2str(h)],'Fontsize',15,...
      'HorizontalAlignment','right',...
      'VerticalAlignment','bottom')
end
```

Both the Euler and modified Euler methods have been used to numerically integrate Equation 1.7.10 and the absolute value of the relative error is plotted in Figure 1.7.1 as a function of time for various time steps. In general, the error grows with time. The decrease of error with smaller time steps, as predicted in our analysis, is quite apparent. Furthermore, the superiority of the modified Euler method over the original Euler method is clearly seen. □

• *Runge-Kutta method*

As we have just shown, the accuracy of numerical solutions of ordinary differential equations can be improved by adding more terms to the Taylor expansion. The Runge-Kutta method[13] builds upon this idea, just as the modified Euler method did.

Let us assume that the numerical solution can be approximated by

$$y_{i+1} = y_i + ak_1 + bk_2, \tag{1.7.11}$$

where

$$k_1 = hf(x_i, y_i) \quad \text{and} \quad k_2 = hf(x_i + A_1 h, y_i + B_1 k_1). \tag{1.7.12}$$

Here a, b, A_1, and B_1 are four unknowns. Equation 1.7.11 was suggested by the modified Euler method that we just presented. In that case, the truncated Taylor series had an error of $O(h^3)$. We anticipate such an error in the present case.

Because the Taylor series expansion of $f(x + h, y + k)$ about (x, y) is

$$f(x + h, y + k) = f(x, y) + (hf_x + kf_y) + \tfrac{1}{2}\left(h^2 f_{xx} + 2hk f_{xy} + k^2 f_{yy}\right)$$
$$+ \tfrac{1}{6}\left(h^3 f_{xxx} + 3h^2 k f_{xxy} + 3hk^2 f_{xyy} + k^3 f_{yyy}\right) + \cdots, \tag{1.7.13}$$

k_2 can be rewritten

$$k_2 = hf[x_i + A_1 h, y_i + Bhf(x_i, y_i)] \tag{1.7.14}$$
$$= h\left[f(x_i, y_i) + (A_1 h f_x + B_1 h f f_y)\right] \tag{1.7.15}$$
$$= hf + A_1 h^2 f_x + B_1 h^2 f f_y, \tag{1.7.16}$$

where we have retained only terms up to $O(h^2)$ and neglected all higher-order terms. Finally, substituting Equation 1.7.16 into Equation 1.7.11 gives

$$y_{i+1} = y_i + (a + b)hf + (A_1 b f_x + B_1 b f f_y)h^2. \tag{1.7.17}$$

This equation corresponds to the second-order Taylor expansion:

$$y_{i+1} = y_i + hy_i' + \tfrac{1}{2}h^2 y_i''. \tag{1.7.18}$$

Therefore, if we wish to solve the differential equation $y' = f(x, y)$, then

$$y'' = f_x + f_y y' = f_x + f f_y. \tag{1.7.19}$$

Substituting Equation 1.7.19 into Equation 1.7.18, we have that

$$y_{i+1} = y_i + hf + \tfrac{1}{2}h^2(f_x + f f_y). \tag{1.7.20}$$

A direct comparison of Equation 1.7.17 and Equation 1.7.20 yields

$$a + b = 1, \quad A_1 b = \tfrac{1}{2}, \quad \text{and} \quad B_1 b = \tfrac{1}{2}. \tag{1.7.21}$$

---

[13] Runge, C., 1895: Ueber die numerische Auflösung von Differentialgleichungen. *Math. Ann.*, **46**, 167–178; Kutta, W., 1901: Beitrag zur näherungsweisen Integration totaler Differentialgleichungen. *Zeit. Math. Phys.*, **46**, 435–453. For a historical review, see Butcher, J. C., 1996: A history of Runge-Kutta methods. *Appl. Numer. Math.*, **20**, 247–260 and Butcher, J. C., and G. Wanner, 1996: Runge-Kutta methods: Some historical notes. *Appl. Numer. Math.*, **22**, 113–151.

Although Carl David Tolmé Runge (1856–1927) began his studies in Munich, his friendship with Max Planck led him to Berlin and pure mathematics with Kronecker and Weierstrass. It was his professorship at Hanover beginning in 1886 and subsequent work in spectroscopy that led him to his celebrated paper on the numerical integration of ordinary differential equations. Runge's final years were spent in Göttingen as a professor in applied mathematics. (Portrait taken with permission from Reid, C., 1976: *Courant in Göttingen and New York: The Story of an Improbable Mathematician.* Springer-Verlag, 314 pp. ©1976, by Springer-Verlag New York Inc.)

These three equations have four unknowns. If we choose $a = \frac{1}{2}$, we immediately calculate $b = \frac{1}{2}$ and $A_1 = B_1 = 1$. Hence the second-order Runge-Kutta scheme is

$$y_{i+1} = y_i + \tfrac{1}{2}(k_1 + k_2), \tag{1.7.22}$$

where $k_1 = hf(x_i, y_i)$ and $k_2 = hf(x_i + h, y_i + k_1)$. Thus, the second-order Runge-Kutta scheme is identical to the modified Euler method.

Although the derivation of the second-order Runge-Kutta scheme yields the modified Euler scheme, it does provide a framework for computing higher-order and more accurate schemes. A particularly popular one is the fourth-order Runge-Kutta scheme

$$y_{i+1} = y_i + \tfrac{1}{6}(k_1 + 2k_2 + 2k_3 + k_4), \tag{1.7.23}$$

where

$$k_1 = hf(x_i, y_i), \tag{1.7.24}$$

$$k_2 = hf(x_i + \tfrac{1}{2}h, y_i + \tfrac{1}{2}k_1), \tag{1.7.25}$$

$$k_3 = hf(x_i + \tfrac{1}{2}h, y_i + \tfrac{1}{2}k_2), \tag{1.7.26}$$

and

$$k_4 = hf(x_i + h, y_i + k_3). \tag{1.7.27}$$

Martin Wilhelm Kutta (1867–1944) was an academic (primarily in mathematics) who held positions at Munich, Jena, Aachen and Stuttgart between 1894 and 1936. It was during his doctorial thesis (*Beiträge zur näherungsweisen Integration totaler Differentialgleichungen*) at the University of Munich that he developed the Runge-Kutta method for the numerical integration of ordinary differential equations based upon an 1895 publication by Runge.

- **Example 1.7.2**

Let us illustrate the fourth-order Runge-Kutta by redoing the previous example using the MATLAB script:

```
clear
% test out different time steps
for i = 1:4
% set up time step increment and number of time steps
  if i==1 h = 0.50; end; if i==2 h = 0.10; end;
  if i==3 h = 0.05; end; if i==4 h = 0.01; end;
  n = 10/h;
% set up initial conditions
  t=zeros(n+1,1); t(1) = 0;
  x_rk=zeros(n+1,1); x_rk(1) = 1;
  x_exact=zeros(n+1,1); x_exact(1) = 1;
% set up difference arrays for plotting purposes
  diff = zeros(n,1); tplot = zeros(n,1);
% define right side of differential equation
  f = inline('xx+tt','tt','xx');
  for k = 1:n
    x_local = x_rk(k); t_local = t(k);
    k1 = h * f(t_local,x_local);
    k2 = h * f(t_local + h/2,x_local + k1/2);
    k3 = h * f(t_local + h/2,x_local + k2/2);
    k4 = h * f(t_local + h,x_local + k3);
    t(k+1) = t_local + h;
```

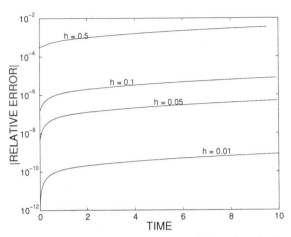

Figure 1.7.2: Same as Figure 1.7.1 except that we have used the fourth-order Runge-Kutta method.

```
    x_rk(k+1) = x_local + (k1+2*k2+2*k3+k4) / 6;
    x_exact(k+1) = 2*exp(t(k+1)) - t(k+1) - 1;
    tplot(k) = t(k);
    diff(k) = x_rk(k+1) - x_exact(k+1);
    diff(k) = abs(diff(k) / x_exact(k+1));
  end
% plot relative errors
  semilogy(tplot,diff,'-')
  hold on
  xlabel('TIME','Fontsize',20)
  ylabel('|RELATIVE ERROR|','Fontsize',20)
  num1 = 2*i; num2 = 0.2*n;
  text(num1,diff(num2),['h = ',num2str(h)],'Fontsize',15,...
      'HorizontalAlignment','right',...
      'VerticalAlignment','bottom')
end
```

The error growth with time is shown in Figure 1.7.2. Although this script could be used for any first-order ordinary differential equation, the people at MATLAB have an alternative called `ode45`, which combines a fourth-order and a fifth-order method that are similar to our fourth-order Runge-Kutta method. Their scheme is more efficient because it varies the step size, choosing a new time step at each step in an attempt to achieve a given desired accuracy.   □

• *Adams-Bashforth method*

All of the methods presented so far (Euler, modified Euler, Runge-Kutta) are single point methods; the solution at $i+1$ depends solely on a single point i. A popular alternative to these schemes are multistep methods that compute y_{i+1} by reusing previously obtained values of y_n where $n < i$.

We begin our derivation of a multistep method by rewriting Equation 1.7.1 as

$$dy = f(x,y)\, dx. \tag{1.7.28}$$

Integrating both sides of Equation 1.7.28, we obtain

$$y(x_{i+1}) - y(x_i) = \int_{x_i}^{x_{i+1}} dy = \int_{x_i}^{x_{i+1}} f(x, y)\, dx. \qquad (1.7.29)$$

The Adams-Bashforth method[14] replaces the integrand in Equation 1.7.29 with an approximation derived from Newton's backward difference formula:

$$f(x, y) \approx f_i + \xi \nabla f_i + \tfrac{1}{2}\xi(\xi + 1)\nabla^2 f_i + \tfrac{1}{6}\xi(\xi + 1)(\xi + 2)\nabla^3 f_i, \qquad (1.7.30)$$

where $\xi = (x - x_i)/h$ or $x = x_i + h\xi$,

$$\nabla f_i = f(x_i, y_i) - f(x_{i-1}, y_{i-1}), \qquad (1.7.31)$$

$$\nabla^2 f_i = f(x_i, y_i) - 2f(x_{i-1}, y_{i-1}) + f(x_{i-2}, y_{i-2}), \qquad (1.7.32)$$

and

$$\nabla^3 f_i = f(x_i, y_i) - 3f(x_{i-1}, y_{i-1}) + 3f(x_{i-2}, y_{i-2}) - f(x_{i-3}, y_{i-3}). \qquad (1.7.33)$$

Substituting Equation 1.7.30 into Equation 1.7.29 and carrying out the integration, we find that

$$y(x_{i+1}) = y(x_i) + \frac{h}{24}\left[55f(x_i, y_i) - 59f(x_{i-1}, y_{i-1}) + 37f(x_{i-2}, y_{i-2}) - 9f(x_{i-3}, y_{i-3})\right]. \qquad (1.7.34)$$

Thus, the Adams-Bashforth method is an explicit finite difference formula that has a global error of $O(h^4)$. Additional computational savings can be realized if the old values of the slope are stored and used later. A disadvantage is that some alternative scheme (usually Runge-Kutta) must provide the first three starting values.

• **Example 1.7.3**

The flight of projectiles provides a classic application of first-order differential equations. If the projectile has a mass m and its motion is opposed by the drag $mgkv^2$, where g is the acceleration due to gravity and k is the quadratic drag coefficients, Newton's law of motion gives

$$\frac{dv}{dt} = -g\sin(\theta) - gkv^2, \qquad (1.7.35)$$

where θ is the slope of the trajectory to the horizon. From kinematics,

$$\frac{dx}{dt} = v\cos(\theta), \quad \frac{dy}{dt} = v\sin(\theta), \quad \frac{d\theta}{dt} = -\frac{g\cos(\theta)}{v}. \qquad (1.7.36)$$

An interesting aspect of this problem is the presence of a *system* of ordinary differential equations.

---

[14] Bashforth, F., and J. C. Adams, 1883: *An Attempt to Test the Theories of Capillary Action by Comparing the Theoretical and Measured Forms of Drops of Fluid. With an Explanation of the Method of Integration Employed in Constructing the Tables Which Give the Theoretical Forms of Such Drops.* Cambridge University Press, 139 pp.

Although we can obtain an exact solution to this problem,[15] let us illustrate the Adams-Bashforth method to compute the solution to Equation 1.7.35 and Equation 1.7.36. We begin by computing the first three time steps using the Runge-Kutta method. Note that we first compute the k_1 for *all* of the dependent variables before we start computing the values of k_2. Similar considerations hold for k_3 and k_4.

```
clear
a = 0; b = 7.85; N = 100; g = 9.81; c = 0.000548;
h = (b-a)/N; t = (a:h:b+h);
% set initial conditions
v(1) = 44.69; theta(1) = pi/3; x(1) = 0; y(1) = 0;
for i = 1:3
  angle = theta(i); vv = v(i);
  k1_vel = -g*sin(angle) - g*c*vv*vv;
  k1_angle = -g*cos(angle) / vv;
  k1_x = vv * cos(angle);
  k1_y = vv * sin(angle);
  angle = theta(i)+h*k1_angle/2; vv = v(i)+h*k1_vel/2;
  k2_vel = -g*sin(angle) - g*c*vv*vv;
  k2_angle = -g*cos(angle) / vv;
  k2_x = vv * cos(angle);
  k2_y = vv * sin(angle);
  angle = theta(i)+h*k2_angle/2; vv = v(i)+h*k2_vel/2;
  k3_vel = -g*sin(angle) - g*c*vv*vv;
  k3_angle = -g*cos(angle) / vv;
  k3_x = vv * cos(angle);
  k3_y = vv * sin(angle);
  angle = theta(i)+h*k3_angle; vv = v(i)+h*k3_vel;
  k4_vel = -g*sin(angle) - g*c*vv*vv;
  k4_angle = -g*cos(angle) / vv;
  k4_x = vv * cos(angle);
  k4_y = vv * sin(angle);
  v(i+1) = v(i) + h*(k1_vel+2*k2_vel+2*k3_vel+k4_vel)/6;
  x(i+1) = x(i) + h*(k1_x+2*k2_x+2*k3_x+k4_x)/6;
  y(i+1) = y(i) + h*(k1_y+2*k2_y+2*k3_y+k4_y)/6;
  theta(i+1) = theta(i) + h*(k1_angle+2*k2_angle ...
          +2*k3_angle+k4_angle)/6;
end
```

Having computed the first three values of each of the dependent variables, we turn to the Adams-Bashforth method to compute the remaining portion of the numerical solution:

```
for i = 4:N
  angle = theta(i); vv = v(i);
  k1_vel = -g*sin(angle) - g*c*vv*vv;
  k1_angle = -g*cos(angle) / vv;
  k1_x = vv * cos(angle);
  k1_y = vv * sin(angle);
```

---

[15] Tan, A., C. H. Frick, and O. Castillo, 1987: The fly ball trajectory: An older approach revisited. *Am. J. Phys.*, **55**, 37–40; Chudinov, P. S., 2001: The motion of a point mass in a medium with a square law of drag. *J. Appl. Math. Mech.*, **65**, 421–426.

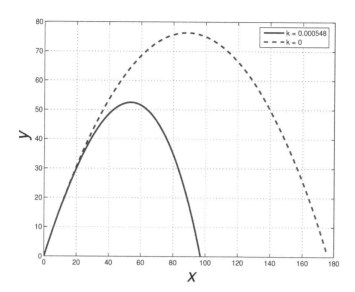

Figure 1.7.3: The trajectory of a projectile with and without air resistance when it is initially fired with a muzzle velocity of 44.69 m/s and an elevation of $\theta = 60°$. All units are in the MKS system.

```
   angle = theta(i-1); vv = v(i-1);
   k2_vel = -g*sin(angle) - g*c*vv*vv;
   k2_angle = -g*cos(angle) / vv;
   k2_x = vv * cos(angle);
   k2_y = vv * sin(angle);
   angle = theta(i-2); vv = v(i-2);
   k3_vel = -g*sin(angle) - g*c*vv*vv;
   k3_angle = -g*cos(angle) / vv;
   k3_x = vv * cos(angle);
   k3_y = vv * sin(angle);
   angle = theta(i-3); vv = v(i-3);
   k4_vel = -g*sin(angle) - g*c*vv*vv;
   k4_angle = -g*cos(angle) / vv;
   k4_x = vv * cos(angle);
   k4_y = vv * sin(angle);
% Use Equation 1.7.35 and Equation 1.7.36 for v, x, y and θ
   v(i+1) = v(i) + h*(55*k1_vel-59*k2_vel+37*k3_vel-9*k4_vel)/24;
   x(i+1) = x(i) + h*(55*k1_x-59*k2_x+37*k3_x-9*k4_x)/24;
   y(i+1) = y(i) + h*(55*k1_y-59*k2_y+37*k3_y-9*k4_y)/24;
   theta(i+1) = theta(i) + h*(55*k1_angle-59*k2_angle ...
                +37*k3_angle-9*k4_angle)/24;
end
```

Figure 1.7.3 illustrates this numerical solution when $k = 0$ and $k = 0.000548$ s^2/m^2 and the shell is fired with the initial velocity $v(0) = 44.69$ m/s and elevation $\theta(0) = \pi/3$ with $x(0) = y(0) = 0$.

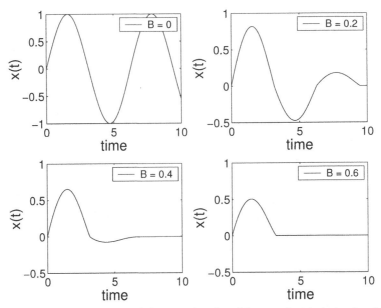

Figure 1.7.4: The numerical solution of the equation describing an electrical circuit with a nonlinear resistor. Here $\beta = 0.2$ and $\Delta t = 0.01$.

Problems

Using Euler's, Runge-Kutta, or the Adams-Bashforth method for various values of $h = 10^{-n}$, find the numerical solution for the following initial-value problems. Check your answer by finding the exact solution.

1. $x' = x - t$, $\quad x(0) = 2$

2. $x' = tx$, $\quad x(0) = 1$

3. $x' = x^2/(t+1)$, $\quad x(0) = 1$

4. $x' = x + e^{-t}$, $\quad x(1) = 0$

5. Consider the integro-differential equation

$$\frac{dx}{dt} + \int_0^t x(\tau)\,d\tau + B\,\mathrm{sgn}(x)|x|^{\beta} = 1, \qquad B, \beta \geq 0,$$

where the signum function is defined by

$$\mathrm{sgn}(t) = \begin{cases} 1, & t > 0, \\ 0, & t = 0, \\ -1, & t < 0. \end{cases}$$

This equation describes the (nondimensional) current,[16] $x(t)$, within an electrical circuit that contains a capacitor, inductor, and nonlinear resistor. Assuming that the circuit is initially dead, x(0) = 0, write a MATLAB script that uses Euler's method to compute $x(t)$. Use a simple Riemann sum to approximate the integral. See Figure 1.7.4. Examine the solution for various values of B and β as well as time step Δt.

---

[16] Monahan, T. F., 1960: Calculation of the current in non-linear surge-current-generation circuits. *Proc. IEE, Part C*, **107**, 288–291.

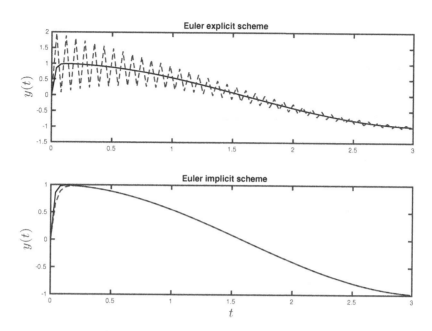

Figure 1.7.5: The numerical solution of a *stiff* differential equation (stated in Problem 6) using Euler's explicit and implicit schemes with $\Delta t = 1/25.5$. The solid line is the exact solution while the dashed line gives the numerical solution.

6. Consider the initial-value problem:

$$\frac{dy(t)}{dt} = -50\left[y(t) - \cos(t)\right], \qquad y(0) = 0.$$

Step 1: Solve this problem exactly and show that

$$y(t) = \frac{50}{2501}\left[\sin(t) + 50\cos(t)\right] - \frac{2500}{2501}\exp(-50t).$$

Step 2: Let us introduce a time step Δt so that $t_n = n\Delta t$ with $n = 0, 1, 2, \ldots$. Then we can find a numerical solution to our initial-value problem using the *explicit Euler method*:

$$\frac{y(t_n + \Delta t) - y(t_n)}{\Delta t} = -50\left[y(t_n) - \cos(t_n)\right],$$

with $y(t_0) = 0$, or

$$y_{n+1} = y_n - 50\Delta t\left[y_n - \cos(t_n)\right], \qquad y_0 = 0,$$

where $y_n = y(t_n)$. Write code and compare this numerical solution with the exact solution for various values of Δt.

Step 3: An alternative method to the explicit scheme is the (implicit) *backward Euler scheme*:

$$\frac{y(t_n + \Delta t) - y(t_n)}{\Delta t} = -50\left[y(t_n + \Delta t) - \cos(t_{n+1})\right],$$

or

$$(1 + 50\Delta t)\, y_{n+1} = y_n + 50\Delta t \cos(t_{n+1}), \qquad y_0 = 0.$$

Redo Step 2 and compare this numerical scheme with the exact solution as a function of time step Δt.

This problem involves a *stiff differential equation*, a differential equation for which certain *numerical* methods used to solve the equation (here the explicit Euler method) are *unstable* unless the time step is taken to be extremely small. Stiff differential equations are characterized by solutions that decay rapidly or oscillate rapidly.

Further Readings

Boyce, W. E., and R. C. DiPrima, 2004: *Elementary Differential Equations and Boundary Value Problems.* Wiley, 800 pp. Classic textbook.

Ince, E. L., 1956: *Ordinary Differential Equations.* Dover, 558 pp. The source book on ordinary differential equations.

Zill, D. G., and M. R. Cullen, 2008: *Differential Equations with Boundary-Value Problems.* Brooks Cole, 640 pp. Nice undergraduate textbook.

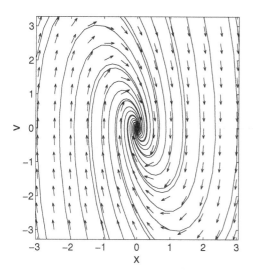

Chapter 2
Higher-Order Ordinary
Differential Equations

Although first-order ordinary differential equations exhibit most of the properties of differential equations, higher-order ordinary differential equations are more ubiquitous in the sciences and engineering. This chapter is devoted to the most commonly employed techniques for their solution.

A *linear* nth-order ordinary differential equation is a differential equation of the form

$$a_n(x)\frac{d^n y}{dx^n} + a_{n-1}(x)\frac{d^{n-1} y}{dx^{n-1}} + \cdots + a_1(x)\frac{dy}{dx} + a_0(x)y = f(x). \qquad (\mathbf{2.0.1})$$

If $f(x) = 0$, then Equation 2.0.1 is said to be *homogeneous*; otherwise, it is *nonhomogeneous*. A linear differential equation is *normal* on an interval I if its coefficients and $f(x)$ are continuous, and the value of $a_n(x)$ is never zero on I.

Solutions to Equation 2.0.1 generally must satisfy not only the differential equations but also certain specified conditions at one or more points. *Initial-value problems* are problems where *all* of the conditions are specified at a single point $x = a$ and have the form: $y(a) = b_0$, $y'(a) = b_1$, $y''(a) = b_2$, ..., $y^{(n-1)}(a) = b_{n-1}$, where b_0, b_1, b_2, ..., b_{n-1} are arbitrary constants. A quick check shows that if Equation 2.0.1 is homogeneous and normal on an interval I and *all* of the initial conditions equal zero at the point $x = a$ that lies in I, then $y(x) \equiv 0$ on I. This follows because $y = 0$ is a solution of Equation 2.0.1 and satisfies the initial conditions.

At this point, a natural question would be whether the solution exists for this initial-value problem and, if so, whether it is unique. From a detailed study of this question,[1] we have the following useful theorem.

Theorem: Existence and Uniqueness

Suppose that the differential equation, Equation 2.0.1, is normal on the open interval I containing the point $x = a$. Then, given n numbers $b_0, b_1, \ldots, b_{n-1}$, the nth-order linear equation, Equation 2.0.1, has a unique solution on the entire interval I that satisfies the n initial conditions $y(a) = b_0, y'(a) = b_1, \ldots, y^{(n-1)}(a) = b_{n-1}$. □

• Example 2.0.1

The solution $y(x) = \frac{4}{3}e^x - \frac{1}{3}e^{-2x}$ to the ordinary differential equation $y''' + 2y'' - y' - 2y = 0$ satisfies the initial conditions $y(0) = 1$, $y'(0) = 2$, and $y''(0) = 0$ at $x = 0$. Our theorem guarantees us that this is the *only* solution with *these* initial values. □

Another class of problems, commonly called (two-point) *boundary-value problems*, occurs when conditions are specified at two *different* points $x = a$ and $x = b$ with $b > a$. An important example, in the case of second-order ordinary differential equations, is the Sturm-Liouville problem where the boundary conditions are $\alpha_1 y(a) + \beta_1 y'(a) = 0$ at $x = a$ and $\alpha_2 y(b) + \beta_2 y'(b) = 0$ at $x = b$. The Sturm-Liouville problem is treated in Chapter 11.

Having introduced some of the terms associated with higher-order ordinary linear differential equations, how do we solve them? One way is to recognize that these equations are really a set of linear, first-order ordinary differential equations. For example, the linear second-order linear differential equation

$$y'' - 3y' + 2y = 3x \tag{2.0.2}$$

can be rewritten as the following system of first-order ordinary differential equations:

$$y' - y = v, \quad \text{and} \quad v' - 2v = 3x \tag{2.0.3}$$

because

$$y'' - y' = v' = 2v + 3x = 2y' - 2y + 3x, \tag{2.0.4}$$

which is the same as Equation 2.0.2. This suggests that this equation can be solved by applying the techniques from the previous chapter. Proceeding along this line, we first find that

$$v(x) = C_1 e^{2x} - \tfrac{3}{2}x - \tfrac{3}{4}. \tag{2.0.5}$$

Therefore,

$$y' - y = C_1 e^{2x} - \tfrac{3}{2}x - \tfrac{3}{4}. \tag{2.0.6}$$

Again, applying the techniques from the previous chapter, we have that

$$y = C_1 e^{2x} + C_2 e^x + \tfrac{3}{2}x + \tfrac{9}{4}. \tag{2.0.7}$$

Note that the solution to this second-order ordinary differential equation contains *two* arbitrary constants.

---

[1] The proof of the existence and uniqueness of solutions to Equation 2.0.1 is beyond the scope of this book. See Ince, E. L., 1956: *Ordinary Differential Equations*. Dover Publications, Inc., Section 3.32.

• Example 2.0.2

In the case of linear, second-order ordinary differential equations, a similar technique, called *reduction in order*, provides a method for solving differential equations if we know one of its solutions.

Consider the second-order ordinary differential equation

$$x^2 y'' - 5xy' + 9y = 0. \tag{2.0.8}$$

A quick check shows that $y_1(x) = x^3 \ln(x)$ is a solution of Equation 2.0.8. Let us now assume that the *general* solution can be written $y(x) = u(x)x^3 \ln(x)$. Then

$$y' = u'(x)x^3 \ln(x) + u(x) \left[3x^2 \ln(x) + x^2 \right], \tag{2.0.9}$$

and

$$y'' = u''(x)x^3 \ln(x) + 2u'(x) \left[3x^2 \ln(x) + x^2 \right] + u(x) \left[6x \ln(x) + 5x \right]. \tag{2.0.10}$$

Substitution of $y(x)$, $y'(x)$, and $y''(x)$ into Equation 2.0.8 yields

$$x^5 \ln(x)u'' + \left[x^4 \ln(x) + 2x^4 \right] u' = 0. \tag{2.0.11}$$

Setting $u' = w$, separation of variables leads to

$$\frac{w'}{w} = -\frac{1}{x} - \frac{2}{x \ln(x)}. \tag{2.0.12}$$

Note how our replacement of $u'(x)$ with $w(x)$ has reduced the second-order ordinary differential equation to a first-order one. Solving Equation 2.0.12, we find that

$$w(x) = u'(x) = -\frac{C_1}{x \ln^2(x)}, \tag{2.0.13}$$

and

$$u(x) = \frac{C_1}{\ln(x)} + C_2. \tag{2.0.14}$$

Because $y(x) = u(x)x^3 \ln(x)$, the complete solution is

$$y(x) = C_1 x^3 + C_2 x^3 \ln(x). \tag{2.0.15}$$

Substitution of Equation 2.0.15 into Equation 2.0.8 confirms that we have the correct solution.

We can verify our answer by using the symbolic toolbox in MATLAB. Typing the command:

```
dsolve('x*x*D2y-5*x*Dy+9*y=0','x')
```

yields:

```
ans =
C1*x^3+C2*x^3*log(x)
```
□

In summary, we can reduce (in principle) any higher-order, linear ordinary differential equations into a system of first-order ordinary differential equations. This system of differential equations can then be solved using techniques from the previous chapter. In Chapter 3 we will pursue this idea further. Right now, however, we will introduce methods that allow us to find the solution in a more direct manner.

• **Example 2.0.3**

An *autonomous differential equation* is one where the independent variable does not appear explicitly. In certain cases we can reduce the order of the differential equation and then solve it.

Consider the autonomous ordinary differential equation

$$y'' = 2y^3. \tag{2.0.16}$$

The trick here is to note that

$$y'' = \frac{dv}{dx} = v\frac{dv}{dy} = 2y^3, \tag{2.0.17}$$

where $v = dy/dx$. Integrating both sides of Equation 2.0.17, we find that

$$v^2 = y^4 + C_1. \tag{2.0.18}$$

Solving for v,

$$\frac{dy}{dx} = v = \sqrt{C_1 + y^4}. \tag{2.0.19}$$

Integrating once more, we have the final result that

$$x + C_2 = \int \frac{dy}{\sqrt{C_1 + y^4}}. \tag{2.0.20}$$

Problems

For the following differential equations, use reduction of order to find a second solution. Can you obtain the general solution using `dsolve` in MATLAB?

1. $xy'' + 2y' = 0, \quad y_1(x) = 1$

2. $y'' + y' - 2y = 0, \quad y_1(x) = e^x$

3. $x^2 y'' + 4xy' - 4y = 0, \quad y_1(x) = x$

4. $xy'' - (x+1)y' + y = 0, \quad y_1(x) = e^x$

5. $(2x - x^2)y'' + 2(x-1)y' - 2y = 0,$
   $\qquad y_1(x) = x - 1$

6. $y'' + \tan(x)y' - 6\cot^2(x)y = 0,$
   $\qquad y_1(x) = \sin^3(x)$

7. $4x^2 y'' + 4xy' + (4x^2 - 1)y = 0,$
   $\qquad y_1(x) = \cos(x)/\sqrt{x}$

8. $y'' + ay' + b(1 + ax - bx^2)y = 0,$
   $\qquad y_1(x) = e^{-bx^2/2}$

Solve the following autonomous ordinary differential equations:

9. $yy'' = y'^2$

10. $y'' = 2yy', \quad y(0) = y'(0) = 1$

11. $yy'' = y' + y'^2$ 12. $2yy'' = 1 + y'^2$

13. $y'' = e^{2y}, \quad y(0) = 0, y'(0) = 1$ 14. $y''' = 3yy', \quad y(0) = y'(0) = 1, y''(0) = \frac{3}{2}$

15. Solve the nonlinear second-order ordinary differential equation

$$\frac{d^2 y}{dx^2} - \frac{1}{x}\frac{dy}{dx} - \frac{1}{2}\left(\frac{dy}{dx}\right)^2 = 0$$

by (1) reducing it to the Bernoulli equation

$$\frac{dv}{dx} - \frac{v}{x} - \frac{v^2}{2} = 0, \qquad v(x) = u'(x),$$

(2) solving for $v(x)$, and finally (3) integrating $u' = v$ to find $u(x)$.

16. Consider the differential equation

$$a_2(x)y'' + a_1(x)y' + a_0(x)y = 0, \qquad a_2(x) \neq 0.$$

Show that this ordinary differential equation can be rewritten

$$u'' + f(x)u = 0, \qquad f(x) = \frac{a_0(x)}{a_2(x)} - \frac{1}{4}\left[\frac{a_1(x)}{a_2(x)}\right]^2 - \frac{1}{2}\frac{d}{dx}\left[\frac{a_1(x)}{a_2(x)}\right],$$

using the substitution

$$y(x) = u(x)\exp\left[-\frac{1}{2}\int^x \frac{a_1(\xi)}{a_2(\xi)}\,d\xi\right].$$

2.1 HOMOGENEOUS LINEAR EQUATIONS WITH CONSTANT COEFFICIENTS

In our drive for more efficient methods to solve higher-order, linear, ordinary differential equations, let us examine the simplest possible case of a homogeneous differential equation with constant coefficients:

$$a_n\frac{d^n y}{dx^n} + a_{n-1}\frac{d^{n-1}y}{dx^{n-1}} + \cdots + a_2 y'' + a_1\frac{dy}{dx} + a_0 y = 0. \qquad (\mathbf{2.1.1})$$

Although we could explore Equation 2.1.1 in its most general form, we will begin by studying the second-order version, namely

$$ay'' + by' + cy = 0, \qquad (\mathbf{2.1.2})$$

since it is the next step up the ladder in complexity from first-order ordinary differential equations.

Motivated by the fact that the solution to the first-order ordinary differential equation $y' + ay = 0$ is $y(x) = C_1 e^{-ax}$, we make the educated guess that the solution to Equation 2.1.2 is $y(x) = Ae^{mx}$. Direct substitution into Equation 2.1.2 yields

$$\left(am^2 + bm + c\right)Ae^{mx} = 0. \qquad (\mathbf{2.1.3})$$

The constant A cannot equal 0 because that would give $y(x) = 0$ and we would have a trivial solution. Furthermore, since $e^{mx} \neq 0$ for arbitrary x, Equation 2.1.3 simplifies to

$$am^2 + bm + c = 0. \tag{2.1.4}$$

Equation 2.1.4 is called the *auxiliary* or *characteristic equation*. At this point we must consider three separate cases.

• *Distinct real roots*

In this case the roots to Equation 2.1.4 are real and unequal. Let us denote these roots by $m = m_1$, and $m = m_2$. Thus, we have the two solutions:

$$y_1(x) = C_1 e^{m_1 x}, \qquad \text{and} \qquad y_2(x) = C_2 e^{m_2 x}. \tag{2.1.5}$$

We will now show that the most general solution to Equation 2.1.2 is

$$y(x) = C_1 e^{m_1 x} + C_2 e^{m_2 x}. \tag{2.1.6}$$

This result follows from the *principle of (linear) superposition*.

Theorem: *Let $y_1, y_2, \ldots, y_k$ be solutions of the homogeneous equation, Equation 2.1.1, on an interval I. Then the linear combination*

$$y(x) = C_1 y_1(x) + C_2 y_2(x) + \cdots + C_k y_k(x), \tag{2.1.7}$$

where C_i, $i = 1, 2, \ldots, k$, are arbitrary constants, is also a solution on the interval I.

Proof: We will prove this theorem for second-order ordinary differential equations; it is easily extended to higher orders. By the superposition principle, $y(x) = C_1 y_1(x) + C_2 y_2(x)$. Upon substitution into Equation 2.1.2, we have that

$$a\left(C_1 y_1'' + C_2 y_2''\right) + b\left(C_1 y_1' + C_2 y_2'\right) + c\left(C_1 y_1 + C_2 y_2\right) = 0. \tag{2.1.8}$$

Recombining the terms, we obtain

$$C_1\left(a y_1'' + b y_1' + c y_1\right) + C_2\left(a y_2'' + b y_2' + c y_2\right) = 0, \tag{2.1.9}$$

or

$$0 C_1 + 0 C_2 = 0. \tag{2.1.10}$$

$\square$

• **Example 2.1.1**

A quick check shows that $y_1(x) = e^x$ and $y_2(x) = e^{-x}$ are two solutions of $y'' - y = 0$. Our theorem tells us that *any* linear combination of these solutions, such as $y(x) = 5e^x - 3e^{-x}$, is also a solution.

How about the converse? Is *every* solution to $y'' - y = 0$ a linear combination of $y_1(x)$ and $y_2(x)$? We will address this question shortly.

$\square$

• **Example 2.1.2**

Let us find the general solution to

$$y'' + 2y' - 15y = 0. \tag{2.1.11}$$

Assuming a solution of the form $y(x) = Ae^{mx}$, we have that

$$(m^2 + 2m - 15)Ae^{mx} = 0. \tag{2.1.12}$$

Because $A \neq 0$ and e^{mx} generally does not equal zero, we obtain the auxiliary or characteristic equation

$$m^2 + 2m - 15 = (m + 5)(m - 3) = 0. \tag{2.1.13}$$

Therefore, the general solution is

$$y(x) = C_1 e^{3x} + C_2 e^{-5x}. \tag{2.1.14}$$

□

• *Repeated real roots*

When $m = m_1 = m_2$, we have only the single exponential solution $y_1(x) = C_1 e^{m_1 x}$. To find the second solution we apply the reduction of order technique shown in Example 2.0.2. Performing the calculation, we find

$$y_2(x) = C_2 e^{m_1 x} \int \frac{e^{-bx/a}}{e^{2m_1 x}} \, dx. \tag{2.1.15}$$

Since $m_1 = -b/(2a)$, the integral simplifies to $\int dx$ and

$$y(x) = C_1 e^{m_1 x} + C_2 x e^{m_1 x}. \tag{2.1.16}$$

• **Example 2.1.3**

Let us find the general solution to

$$y'' + 4y' + 4y = 0. \tag{2.1.17}$$

Here the auxiliary or characteristic equation is

$$m^2 + 4m + 4 = (m + 2)^2 = 0. \tag{2.1.18}$$

Therefore, the general solution is

$$y(x) = (C_1 + C_2 x)e^{-2x}. \tag{2.1.19}$$

□

• *Complex conjugate roots*

When $b^2 - 4ac < 0$, the roots become the complex pair $m_1 = \alpha + i\beta$ and $m_2 = \alpha - \beta i$, where α and β are real and $i^2 = -1$. Therefore, the general solution is

$$y(x) = C_1 e^{(\alpha+i\beta)x} + C_2 e^{(\alpha-\beta i)x}. \tag{2.1.20}$$

Although Equation 2.1.20 is quite correct, most engineers prefer to work with real functions rather than complex exponentials. To this end, we apply Euler's formula to eliminate $e^{i\beta x}$ and $e^{-i\beta x}$ since

$$e^{i\beta x} = \cos(\beta x) + i\sin(\beta x), \tag{2.1.21}$$

and

$$e^{-i\beta x} = \cos(\beta x) - i\sin(\beta x). \tag{2.1.22}$$

Therefore,

$$y(x) = C_1 e^{\alpha x}[\cos(\beta x) + i\sin(\beta x)] + C_2 e^{\alpha x}[\cos(\beta x) - i\sin(\beta x)] \tag{2.1.23}$$
$$= C_3 e^{\alpha x}\cos(\beta x) + C_4 e^{\alpha x}\sin(\beta x), \tag{2.1.24}$$

where $C_3 = C_1 + C_2$, and $C_4 = iC_1 - iC_2$.

• **Example 2.1.4**

Let us find the general solution to

$$y'' + 4y' + 5y = 0. \tag{2.1.25}$$

Here the auxiliary or characteristic equation is

$$m^2 + 4m + 5 = (m+2)^2 + 1 = 0, \tag{2.1.26}$$

or $m = -2 \pm i$. Therefore, the general solution is

$$y(x) = e^{-2x}[C_1\cos(x) + C_2\sin(x)]. \tag{2.1.27}$$

$\square$

So far we have only dealt with second-order differential equations. When we turn to higher-order ordinary differential equations, similar considerations hold. In place of Equation 2.1.4, we now have the nth-degree polynomial equation

$$a_n m^n + a_{n-1} m^{n-1} + \cdots + a_2 m^2 + a_1 m + a_0 = 0 \tag{2.1.28}$$

for its auxiliary equation.

When we treated second-order ordinary differential equations, we were able to classify the roots to the auxiliary equation as distinct real roots, repeated roots, and complex roots. In the case of higher-order differential equations, such classifications are again useful, although all three types may occur with the same equation. For example, the auxiliary equation

$$m^6 - m^5 + 2m^4 - 2m^3 + m^2 - m = 0 \tag{2.1.29}$$

has the distinct roots $m = 0$ and $m = 1$ with the twice repeated, complex roots $m = \pm i$.

Although the possible combinations increase with higher-order differential equations, the solution technique remains the same. For each distinct real root $m = m_1$, we have a corresponding homogeneous solution $e^{m_1 x}$. For each complex pair $m = \alpha \pm \beta i$, we have the corresponding pair of homogeneous solutions $e^{\alpha x} \cos(\beta x)$ and $e^{\alpha x} \sin(\beta x)$. For a repeated root $m = m_1$ of multiplicity k, regardless of whether it is real or complex, we have either $e^{m_1 x}, x e^{m_1 x}, x^2 e^{m_1 x}, \ldots, x^k e^{m_1 x}$ in the case of real m_1 or

$$e^{\alpha x} \cos(\beta x), e^{\alpha x} \sin(\beta x), x e^{\alpha x} \cos(\beta x), x e^{\alpha x} \sin(\beta x),$$

$$x^2 e^{\alpha x} \cos(\beta x), x^2 e^{\alpha x} \sin(\beta x), \ldots, x^k e^{\alpha x} \cos(\beta x), x^k e^{\alpha x} \sin(\beta x)$$

in the case of complex roots $\alpha \pm \beta i$. For example, the general solution for the roots to Equation 2.1.29 is

$$y(x) = C_1 + C_2 e^x + C_3 \cos(x) + C_4 \sin(x) + C_5 x \cos(x) + C_6 x \sin(x). \tag{2.1.30}$$

- **Example 2.1.5**

Let us find the general solution to

$$y''' + y' - 10y = 0. \tag{2.1.31}$$

Here the auxiliary or characteristic equation is

$$m^3 + m - 10 = (m - 2)(m^2 + 2m + 5) = (m - 2)[(m + 1)^2 + 4] = 0, \tag{2.1.32}$$

or $m = -2$ and $m = -1 \pm 2i$. Therefore, the general solution is

$$y(x) = C_1 e^{-2x} + e^{-x}[C_2 \cos(2x) + C_3 \sin(2x)]. \tag{2.1.33}$$

$\square$

Having presented the technique for solving constant coefficient, linear, ordinary differential equations, an obvious question is: How do we know that we have captured all of the solutions? Before we can answer this question, we must introduce the concept of linear dependence.

A set of functions $f_1(x), f_2(x), \ldots, f_n(x)$ is said to be *linearly dependent* on an interval I if there exist constants $C_1, C_2, \ldots, C_n$, not all zero, such that

$$C_1 f_1(x) + C_2 f_2(x) + C_3 f_3(x) + \cdots + C_n f_n(x) = 0 \tag{2.1.34}$$

for each x in the interval; otherwise, the set of functions is said to be *linearly independent*. This concept is easily understood when we have only two functions $f_1(x)$ and $f_2(x)$. If the functions are linearly dependent on an interval, then there exist constants C_1 and C_2 that are not both zero, where

$$C_1 f_1(x) + C_2 f_2(x) = 0 \tag{2.1.35}$$

for every x in the interval. If $C_1 \neq 0$, then

$$f_1(x) = -\frac{C_2}{C_1} f_2(x). \tag{2.1.36}$$

In other words, if two functions are linearly dependent, then one is a constant multiple of the other. Conversely, two functions are linearly independent when neither is a constant multiple of the other on an interval.

• **Example 2.1.6**

Let us show that $f(x) = 2x$, $g(x) = 3x^2$, and $h(x) = 5x - 8x^2$ are linearly dependent on the real line.

To show this, we must choose three constants, C_1, C_2, and C_3, such that

$$C_1 f(x) + C_2 g(x) + C_3 h(x) = 0, \tag{2.1.37}$$

where not all of these constants are nonzero. A quick check shows that

$$15 f(x) - 16 g(x) - 6 h(x) = 0. \tag{2.1.38}$$

Clearly, $f(x)$, $g(x)$, and $h(x)$ are linearly dependent. $\qquad\qquad\square$

• **Example 2.1.7**

This example shows the importance of defining the interval on which a function is linearly dependent or independent. Consider the two functions $f(x) = x$ and $g(x) = |x|$. They are linearly dependent on the interval $(0, \infty)$ since $C_1 x + C_2 |x| = C_1 x + C_2 x = 0$ is satisfied for any nonzero choice of C_1 and C_2 where $C_1 = -C_2$. What happens on the interval $(-\infty, 0)$? They are still linearly dependent but now $C_1 = C_2$. $\qquad\square$

Although we could use the fundamental concept of linear independence to check and see whether a set of functions is linearly independent or not, the following theorem introduces a procedure that is very straightforward.

Theorem: Wronskian Test of Linear Independence

Suppose $f_1(x), f_2(x), \ldots, f_n(x)$ possess at least $n-1$ derivatives. If the determinant[2]

$$\begin{vmatrix} f_1 & f_2 & \cdots & f_n \\ f_1' & f_2' & \cdots & f_n' \\ \vdots & \vdots & & \vdots \\ f_1^{(n-1)} & f_2^{(n-1)} & \cdots & f_n^{(n-1)} \end{vmatrix}$$

is not zero for at least one point in the interval I, then the functions $f_1(x)$, $f_2(x)$, $\ldots$, $f_n(x)$ are linearly independent on the interval. The determinant in this theorem is denoted by $W[f_1(x), f_2(x), \ldots, f_n(x)]$ and is called the *Wronskian* of the functions.

Proof: We prove this theorem by contradiction when $n = 2$. Let us assume that $W[f_1(x_0), f_2(x_0)] \neq 0$ for some fixed x_0 in the interval I and that $f_1(x)$ and $f_2(x)$ are linearly dependent on the interval. Since the functions are linearly dependent, there exists C_1 and C_2, both not zero, for which

$$C_1 f_1(x) + C_2 f_2(x) = 0 \tag{2.1.39}$$

---

[2] If you are unfamiliar with determinants, see Section 3.2.

for every x in I. Differentiating Equation 2.1.39 gives

$$C_1 f_1'(x) + C_2 f_2'(x) = 0. \tag{2.1.40}$$

We may view Equation 2.1.39 and Equation 2.1.40 as a system of equations with C_1 and C_2 as the unknowns. Because the linear dependence of f_1 and f_2 implies that $C_1 \neq 0$ and/or $C_2 \neq 0$ for each x in the interval,

$$W[f_1(x), f_2(x)] = \begin{vmatrix} f_1 & f_2 \\ f_1' & f_2' \end{vmatrix} = 0 \tag{2.1.41}$$

for every x in I. This contradicts the assumption that $W[f_1(x_0), f_2(x_0)] \neq 0$ and f_1 and f_2 are linearly independent. $\qquad\square$

• **Example 2.1.8**

Are the functions $f(x) = x$, $g(x) = xe^x$, and $h(x) = x^2 e^x$ linearly dependent on the real line? To find out, we compute the Wronskian or

$$W[f(x), g(x), h(x)] = \begin{vmatrix} e^x & xe^x & x^2 e^x \\ e^x & (x+1)e^x & (x^2+2x)e^x \\ e^x & (x+2)e^x & (x^2+4x+2)e^x \end{vmatrix} = e^{3x} \begin{vmatrix} 1 & x & x^2 \\ 0 & 1 & 2x \\ 0 & 0 & 2 \end{vmatrix} = 2e^{3x} \neq 0. \tag{2.1.42}$$

Therefore, x, xe^x, and $x^2 e^x$ are linearly *independent*. $\qquad\square$

Having introduced this concept of linear independence, we are now ready to address the question of how many linearly independent solutions a homogeneous linear equation has.

Theorem:

On any interval I over which an n-th order homogeneous linear differential equation is normal, the equation has n linearly independent solutions $y_1(x), y_2(x), \dots, y_n(x)$ and any particular solution of the equation on I can be expressed as a linear combination of these linearly independent solutions.

Proof: Again for convenience and clarity we prove this theorem for the special case of $n = 2$. Let $y_1(x)$ and $y_2(x)$ denote solutions on I of Equation 2.1.2. We know that these solutions exist by the existence theorem and have the following values:

$$y_1(a) = 1, \quad y_2(a) = 0, \quad y_1'(a) = 0, \quad y_2'(a) = 1 \tag{2.1.43}$$

at some point a on I. To establish the linear independence of y_1 and y_2 we note that, if $C_1 y_1(x) + C_2 y_2(x) = 0$ holds identically on I, then $C_1 y_1'(x) + C_2 y_2'(x) = 0$ there too. Because $x = a$ lies in I, we have that

$$C_1 y_1(a) + C_2 y_2(a) = 0, \tag{2.1.44}$$

and

$$C_1 y_1'(a) + C_2 y_2'(a) = 0, \tag{2.1.45}$$

which yields $C_1 = C_2 = 0$ after substituting Equation 2.1.43. Hence, the solutions y_1 and y_2 are linearly independent.

To complete the proof we must now show that any particular solution of Equation 2.1.2 can be expressed as a linear combination of y_1 and y_2. Because y, y_1, and y_2 are all solutions of Equation 2.1.2 on I, so is the function

$$Y(x) = y(x) - y(a)y_1(x) - y'(a)y_2(x), \qquad (2.1.46)$$

where $y(a)$ and $y'(a)$ are the values of the solution y and its derivative at $x = a$. Evaluating Y and Y' at $x = a$, we have that

$$Y(a) = y(a) - y(a)y_1(a) - y'(a)y_2(a) = y(a) - y(a) = 0, \qquad (2.1.47)$$

and

$$Y'(a) = y'(a) - y(a)y_1'(a) - y'(a)y_2'(a) = y'(a) - y'(a) = 0. \qquad (2.1.48)$$

Thus, Y is the trivial solution to Equation 2.1.2. Hence, for every x in I,

$$y(x) - y(a)y_1(x) - y'(a)y_2(x) = 0. \qquad (2.1.49)$$

Solving Equation 2.1.49 for $y(x)$, we see that y is expressible as the linear combination

$$y(x) = y(a)y_1(x) + y'(a)y_2(x) \qquad (2.1.50)$$

of y_1 and y_2, and the proof is complete for $n = 2$.

Problems

Find the general solution to the following differential equations. Check your general solution by using `dsolve` in MATLAB.

1. $y'' + 6y' + 5y = 0$ 2. $y'' - 6y' + 10y = 0$ 3. $y'' - 2y' + y = 0$

4. $y'' - 3y' + 2y = 0$ 5. $y'' - 4y' + 8y = 0$ 6. $y'' + 6y' + 9y = 0$

7. $y'' + 6y' - 40y = 0$ 8. $y'' + 4y' + 5y = 0$ 9. $y'' + 8y' + 25y = 0$

10. $4y'' - 12y' + 9y = 0$ 11. $y'' + 8y' + 16y = 0$ 12. $y''' + 4y'' = 0$

13. $y'''' + 4y'' = 0$ 14. $y'''' + 2y''' + y'' = 0$ 15. $y''' - 8y = 0$

16. $y'''' - y''' - 2y'' = 0$ 17. $y'''' - 3y''' + 3y'' - y' = 0$ 18. $y'''' - 4y''' + 6y'' - 4y' + y = 0$

19. The simplest differential equation with "memory"— its past behavior affects the present —is

$$y' = -\frac{A}{2\tau} \int_{-\infty}^{t} e^{-(t-x)/\tau} y(x) \, dx.$$

Solve this integro-differential equation by differentiating it with respect to t to eliminate the integral.

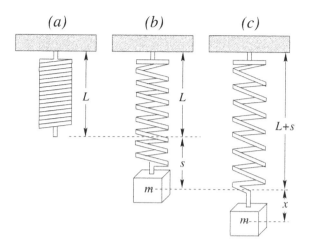

Figure 2.2.1: Various configurations of a mass/spring system. The spring alone has a length L, which increases to $L+s$ when the mass is attached. During simple harmonic motion, the length of the mass/spring system varies as $L + s + x$.

2.2 SIMPLE HARMONIC MOTION

Second-order, linear, ordinary differential equations often arise in mechanical or electrical problems. The purpose of this section is to illustrate how the techniques that we just derived may be applied to these problems.

We begin by considering the mass-spring system illustrated in Figure 2.2.1 where a mass m is attached to a flexible spring suspended from a rigid support. If there were no spring, then the mass would simply fall downward due to the gravitational force mg. Because there is no motion, the gravitational force must be balanced by an upward force due to the presence of the spring. This upward force is usually assumed to obey Hooke's law, which states that the restoring force is opposite to the direction of elongation and proportional to the amount of elongation. Mathematically the equilibrium condition can be expressed $mg = ks$.

Consider now what happens when we disturb this equilibrium. This may occur in one of two ways: We could move the mass either upward or downward and then release it. Another method would be to impart an initial velocity to the mass. In either case, the motion of the mass/spring system would be governed by Newton's second law, which states that the acceleration of the mass equals the imbalance of the forces. If we denote the downward displacement of the mass from its equilibrium position by positive x, then

$$m\frac{d^2x}{dt^2} = -k(s+x) + mg = -kx, \qquad (\mathbf{2.2.1})$$

since $ks = mg$. After dividing Equation 2.2.1 by the mass, we obtain the second-order differential equation

$$\frac{d^2x}{dt^2} + \frac{k}{m}x = 0, \qquad (\mathbf{2.2.2})$$

or

$$\frac{d^2x}{dt^2} + \omega^2 x = 0, \qquad (\mathbf{2.2.3})$$

where $\omega^2 = k/m$ and ω is the *circular frequency*. Equation 2.2.3 describes *simple harmonic motion* or *free undamped motion*. The two initial conditions associated with this differential

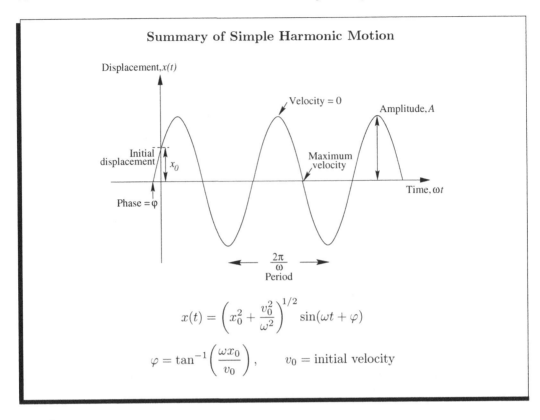

equation are

$$x(0) = \alpha, \qquad x'(0) = \beta. \tag{2.2.4}$$

The first condition gives the initial amount of displacement while the second condition specifies the initial velocity. If $\alpha > 0$ while $\beta < 0$, then the mass starts from a point below the equilibrium position with an initial upward velocity. On the other hand, if $\alpha < 0$ with $\beta = 0$, the mass is at rest when it is released $|\alpha|$ units above the equilibrium position. Similar considerations hold for other values of α and β.

To solve Equation 2.2.3, we note that the solutions of the auxiliary equation $m^2 + \omega^2 = 0$ are the complex numbers $m_1 = \omega i$, and $m_2 = -\omega i$. Therefore, the general solution is

$$x(t) = A\cos(\omega t) + B\sin(\omega t). \tag{2.2.5}$$

The (natural) *period* of free vibrations is $T = 2\pi/\omega$ while the (natural) frequency is $f = 1/T = \omega/(2\pi)$.

- **Example 2.2.1**

Let us solve the initial-value problem

$$\frac{d^2x}{dt^2} + 4x = 0, \qquad x(0) = 10, \qquad x'(0) = 0. \tag{2.2.6}$$

The physical interpretation is that we have pulled the mass on a spring down 10 units *below* the equilibrium position and then release it from rest at $t = 0$. Here, $\omega = 2$ so that

$$x(t) = A\cos(2t) + B\sin(2t) \tag{2.2.7}$$

from Equation 2.2.5.

Because $x(0) = 10$, we find that

$$x(0) = 10 = A \cdot 1 + B \cdot 0 \tag{2.2.8}$$

so that $A = 10$. Next, we note that

$$\frac{dx}{dt} = -20\sin(2t) + 2B\cos(2t). \tag{2.2.9}$$

Therefore, at $t = 0$,

$$x'(0) = 0 = -20 \cdot 0 + 2B \cdot 1 \tag{2.2.10}$$

and $B = 0$. Thus, the equation of motion is $x(t) = 10\cos(2t)$.

What is the physical interpretation of our equation of motion? Once the system is set into motion, it stays in motion with the mass oscillating back and forth 10 units above and below the equilibrium position $x = 0$. The period of oscillation is $2\pi/2 = \pi$ units of time.□

• Example 2.2.2

A weight of 45 N stretches a spring 5 cm. At time $t = 0$, the weight is released from its equilibrium position with an upward velocity of 28 cm s^{-1}. Determine the displacement $x(t)$ that describes the subsequent free motion.

From Hooke's law,

$$F = mg = 45\,\mathrm{N} = k \times 5\,\mathrm{cm} \tag{2.2.11}$$

so that $k = 9$ N cm^{-1}. Therefore, the differential equation is

$$\frac{d^2x}{dt^2} + 196\,\mathrm{s}^{-2}x = 0. \tag{2.2.12}$$

The initial displacement and initial velocity are $x(0) = 0$ cm and $x'(0) = -28$ cm s^{-1}. The negative sign in the initial velocity reflects the fact that the weight has an initial velocity in the negative or upward direction.

Because $\omega^2 = 196$ s^{-2} or $\omega = 14$ s^{-1}, the general solution to the differential equation is

$$x(t) = A\cos(14\,\mathrm{s}^{-1}t) + B\sin(14\,\mathrm{s}^{-1}t). \tag{2.2.13}$$

Substituting for the initial displacement $x(0)$ in Equation 2.2.13, we find that

$$x(0) = 0\,\mathrm{cm} = A \cdot 1 + B \cdot 0, \tag{2.2.14}$$

and $A = 0$ cm. Therefore,

$$x(t) = B\sin(14\,\mathrm{s}^{-1}t) \tag{2.2.15}$$

and

$$x'(t) = 14\,\mathrm{s}^{-1}B\cos(14\,\mathrm{s}^{-1}t). \tag{2.2.16}$$

Substituting for the initial velocity,

$$x'(0) = -28\,\mathrm{cm\,s}^{-1} = 14\,\mathrm{s}^{-1}B, \tag{2.2.17}$$

and $B = -2$ cm. Thus the equation of motion is

$$x(t) = -2\,\mathrm{cm}\,\sin(14\,\mathrm{s}^{-1}t). \tag{2.2.18}$$

□

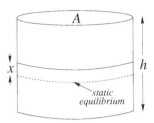

Figure 2.2.2: Schematic of a floating body partially submerged in pure water.

● **Example 2.2.3: Vibration of floating bodies**

Consider a solid cylinder of radius a that is partially submerged in a bath of pure water as shown in Figure 2.2.2. Let us find the motion of this cylinder in the vertical direction assuming that it remains in an upright position.

If the displacement of the cylinder from its static equilibrium position is x, the weight of water displaced equals $Ag\rho_w x$, where ρ_w is the density of the water and g is the gravitational acceleration. This is the restoring force according to the Archimedes principle. The mass of the cylinder is $Ah\rho$, where ρ is the density of cylinder. From Newton's second law, the equation of motion is

$$\rho Ahx'' + Ag\rho_w x = 0, \tag{2.2.19}$$

or

$$x'' + \frac{\rho_w g}{\rho h}x = 0. \tag{2.2.20}$$

From Equation 2.2.20 we see that the cylinder will oscillate about its static equilibrium position $x = 0$ with a frequency of

$$\omega = \left(\frac{\rho_w g}{\rho h}\right)^{1/2}. \tag{2.2.21}$$

$\square$

When both A and B are nonzero, it is often useful to rewrite the homogeneous solution, Equation 2.2.5, as

$$x(t) = C\sin(\omega t + \varphi) \tag{2.2.22}$$

to highlight the amplitude and phase of the oscillation. Upon employing the trigonometric angle-sum formula, Equation 2.2.22 can be rewritten

$$x(t) = C\sin(\omega t)\cos(\varphi) + C\cos(\omega t)\sin(\varphi) = A\cos(\omega t) + B\sin(\omega t). \tag{2.2.23}$$

From Equation 2.2.23, we see that $A = C\sin(\varphi)$ and $B = C\cos(\varphi)$. Therefore,

$$A^2 + B^2 = C^2\sin^2(\varphi) + C^2\cos^2(\varphi) = C^2, \tag{2.2.24}$$

and $C = \sqrt{A^2 + B^2}$. Similarly, $\tan(\varphi) = A/B$. Because the tangent is positive in both the first and third quadrants and negative in both the second and fourth quadrants, there are two possible choices for φ. The proper value of φ satisfies the equations $A = C\sin(\varphi)$ and $B = C\cos(\varphi)$.

If we prefer the amplitude/phase solution

$$x(t) = C\cos(\omega t - \varphi), \tag{2.2.25}$$

we now have

$$x(t) = C\cos(\omega t)\cos(\varphi) + C\sin(\omega t)\sin(\varphi) = A\cos(\omega t) + B\sin(\omega t). \tag{2.2.26}$$

Consequently, $A = C\cos(\varphi)$ and $B = C\sin(\varphi)$. Once again, we obtain $C = \sqrt{A^2 + B^2}$. On the other hand, $\tan(\varphi) = B/A$.

Problems

Solve the following initial-value problems and write their solutions in terms of amplitude and phase:

1. $x'' + 25x = 0,$ $\quad x(0) = 10,$ $\quad x'(0) = -10$

2. $4x'' + 9x = 0,$ $\quad x(0) = 2\pi,$ $\quad x'(0) = 3\pi$

3. $x'' + \pi^2 x = 0,$ $\quad x(0) = 1,$ $\quad x'(0) = \pi\sqrt{3}$

4. A 4-kg mass is suspended from a 100 N/m spring. The mass is set in motion by giving it an initial downward velocity of 5 m/s from its equilibrium position. Find the displacement as a function of time.

5. A spring hangs vertically. A weight of mass M kg stretches it L m. This weight is removed. A body weighing m kg is then attached and allowed to come to rest. It is then pulled down s_0 m and released with a velocity v_0. Find the displacement of the body from its point of rest and its velocity at any time t.

6. A particle of mass m moving in a straight line is *repelled* from the origin by a force F. (a) If the force is proportional to the distance from the origin, find the position of the particle as a function of time. (b) If the initial velocity of the particle is $a\sqrt{k}$, where k is the proportionality constant and a is the distance from the origin, find the position of the particle as a function of time. What happens if $m < 1$ and $m = 1$?

2.3 DAMPED HARMONIC MOTION

Free harmonic motion is unrealistic because there are always frictional forces that act to retard motion. In mechanics, the drag is often modeled as a resistance that is proportional to the instantaneous velocity. Adopting this resistance law, it follows from Newton's second law that the harmonic oscillator is governed by

$$m\frac{d^2x}{dt^2} = -kx - \beta\frac{dx}{dt}, \tag{2.3.1}$$

where β is a positive *damping constant*. The negative sign is necessary since this resistance acts in a direction opposite to the motion.

Dividing Equation 2.3.1 by the mass m, we obtain the differential equation of *free damped motion*,

$$\frac{d^2x}{dt^2} + \frac{\beta}{m}\frac{dx}{dt} + \frac{k}{m}x = 0, \tag{2.3.2}$$

or

$$\frac{d^2x}{dt^2} + 2\lambda\frac{dx}{dt} + \omega^2 x = 0. \tag{2.3.3}$$

We have written 2λ rather than just λ because it simplifies future computations. The auxiliary equation is $m^2 + 2\lambda m + \omega^2 = 0$, which has the roots

$$m_1 = -\lambda + \sqrt{\lambda^2 - \omega^2}, \qquad \text{and} \qquad m_2 = -\lambda - \sqrt{\lambda^2 - \omega^2}. \tag{2.3.4}$$

From Equation 2.3.4 we see that there are three possible cases which depend on the algebraic sign of $\lambda^2 - \omega^2$. Because all of the solutions contain the damping factor $e^{-\lambda t}$, $\lambda > 0$, $x(t)$ vanishes as $t \to \infty$.

- *Case I: $\lambda > \omega$*

Here the system is *overdamped* because the damping coefficient β is large compared to the spring constant k. The corresponding solution is

$$x(t) = Ae^{m_1 t} + Be^{m_2 t}, \tag{2.3.5}$$

or

$$x(t) = e^{-\lambda t}\left(Ae^{t\sqrt{\lambda^2-\omega^2}} + Be^{-t\sqrt{\lambda^2-\omega^2}}\right). \tag{2.3.6}$$

In this case the motion is smooth and nonoscillatory.

- *Case II: $\lambda = \omega$*

The system is *critically damped* because any slight decrease in the damping force would result in oscillatory motion. The general solution is

$$x(t) = Ae^{m_1 t} + Bte^{m_1 t}, \tag{2.3.7}$$

or

$$x(t) = e^{-\lambda t}(A + Bt). \tag{2.3.8}$$

The motion is quite similar to that of an overdamped system.

- *Case III: $\lambda < \omega$*

In this case the system is *underdamped* because the damping coefficient is small compared to the spring constant. The roots m_1 and m_2 are complex:

$$m_1 = -\lambda + i\sqrt{\omega^2 - \lambda^2}, \qquad \text{and} \qquad m_2 = -\lambda - i\sqrt{\omega^2 - \lambda^2}. \tag{2.3.9}$$

The general solution now becomes

$$x(t) = e^{-\lambda t}\left[A\cos\left(t\sqrt{\omega^2 - \lambda^2}\right) + B\sin\left(t\sqrt{\omega^2 - \lambda^2}\right)\right]. \tag{2.3.10}$$

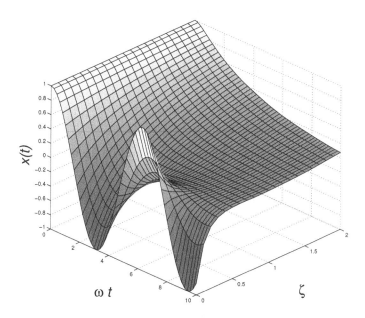

Figure 2.3.1: The displacement $x(t)$ of a damped harmonic oscillator as a function of time and $\zeta = \lambda/\omega$.

Equation 2.3.10 describes oscillatory motion that decays as $e^{-\lambda t}$. Equations 2.3.6, 2.3.8, and 2.3.10 are illustrated in Figure 2.3.1 when the initial conditions are $x(0) = 1$ and $x'(0) = 0$.

Just as we could write the solution for the simple harmonic motion in the amplitude/phase format, we can write any damped solution Equation 2.3.10 in the alternative form

$$x(t) = Ce^{-\lambda t}\sin\left(t\sqrt{\omega^2 - \lambda^2} + \varphi\right), \qquad (\mathbf{2.3.11})$$

where $C = \sqrt{A^2 + B^2}$ and the phase angle φ is given by $\tan(\varphi) = A/B$ such that $A = C\sin(\varphi)$ and $B = C\cos(\varphi)$. The coefficient $Ce^{-\lambda t}$ is sometimes called the *damped coefficient of vibrations*. Because Equation 2.3.11 is *not* a periodic function, the quantity $2\pi/\sqrt{\omega^2 - \lambda^2}$ is called the *quasi period* and $\sqrt{\omega^2 - \lambda^2}$ is the *quasi frequency*. The quasi period is the time interval between two successive maxima of $x(t)$.

● **Example 2.3.1**

A body with mass $m = \frac{1}{2}$ kg is attached to the end of a spring that is stretched 2 m by a force of 100 N. Furthermore, there is also attached a dashpot[3] that provides 6 N of resistance for each m/s of velocity. If the mass is set in motion by further stretching the spring $\frac{1}{2}$ m and giving it an upward velocity of 10 m/s, let us find the subsequent motion.

We begin by first computing the constants. The spring constant is $k = (100 \text{ N})/(2 \text{ m})$ $= 50$ N/m. Therefore, the differential equation is

$$\tfrac{1}{2}x'' + 6x' + 50x = 0 \qquad (\mathbf{2.3.12})$$

with $x(0) = \frac{1}{2}$ m and $x'(0) = -10$ m/s. Here the units of $x(t)$ are meters. The characteristic or auxiliary equation is

$$m^2 + 12m + 100 = (m+6)^2 + 64 = 0, \qquad (\mathbf{2.3.13})$$

---

[3] A mechanical device—usually a piston that slides within a liquid-filled cylinder—used to damp the vibration or control the motion of a mechanism to which it is attached.

**Review of the Solution of the
Underdamped Homogeneous Oscillator Problem**

$mx'' + \beta x' + kx = 0$ subject to $x(0) = x_0$, $x'(0) = v_0$ has the solution

$$x(t) = Ae^{-\lambda t}\sin(\omega_d t + \varphi),$$

where

$\omega = \sqrt{k/m}$ is the undamped natural frequency,

$\lambda = \beta/(2m)$ is the damping factor,

$\omega_d = \sqrt{\omega^2 - \lambda^2}$ is the damped natural frequency,

and the constants A and φ are determined by

$$A = \sqrt{x_0^2 + \left(\frac{v_0 + \lambda x_0}{\omega_d}\right)^2}$$

and

$$\varphi = \tan^{-1}\left(\frac{x_0 \omega_d}{v_0 + \lambda x_0}\right).$$

---

or $m = -6 \pm 8i$. Therefore, we have an underdamped harmonic oscillator and the general solution is

$$x(t) = e^{-6t}\left[A\cos(8t) + B\sin(8t)\right]. \tag{2.3.14}$$

Consequently, each cycle takes $2\pi/8 = 0.79$ second. This is longer than the 0.63 second that would occur if the system were undamped.

From the initial conditions,

$$x(0) = A = \tfrac{1}{2}, \quad\text{and}\quad x'(0) = -10 = -6A + 8B. \tag{2.3.15}$$

Therefore, $A = \tfrac{1}{2}$ and $B = -\tfrac{7}{8}$. Consequently,

$$x(t) = e^{-6t}\left[\tfrac{1}{2}\cos(8t) - \tfrac{7}{8}\sin(8t)\right] = \frac{\sqrt{65}}{8}e^{-6t}\cos(8t + 2.62244). \tag{2.3.16}$$

$\square$

• Example 2.3.2: Design of a wind vane

In its simplest form, a wind vane is a flat plate or airfoil that can rotate about a vertical shaft. See Figure 2.3.2. In static equilibrium it points into the wind. There is usually a counterweight to balance the vane about the vertical shaft.

A vane uses a combination of the lift and drag forces on the vane to align itself with the wind. As the wind shifts direction from θ_0 to the new direction θ_i, the direction θ in which the vane currently points is governed by the equation of motion[4]

$$I\frac{d^2\theta}{dt^2} + \frac{NR}{V}\frac{d\theta}{dt} = N(\theta_i - \theta), \tag{2.3.17}$$

---

[4] For a derivation of Equation 2.3.12 and Equation 2.3.13, see subsection 2 of Section 3 in Barthelt, H. P., and G. H. Ruppersberg, 1957: Die mechanische Windfahne, eine theoretische und experimentelle Untersuchung. *Beitr. Phys. Atmos.*, **29**, 154–185.

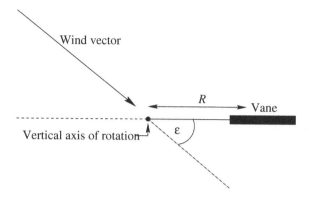

Figure 2.3.2: Schematic of a wind vane. The counterbalance is not shown.

where I is the vane's moment of inertia, N is the aerodynamic torque per unit angle, and R is the distance from the axis of rotation to the effective center of the aerodynamic force on the vane. The aerodynamic torque is given by

$$N = \tfrac{1}{2} C_L \rho A V^2 R, \qquad (2.3.18)$$

where C_L is the lift coefficient, ρ is the air density, A is the vane area, and V is the wind speed.

Dividing Equation 2.3.17 by I, we obtain the second-order ordinary differential equation

$$\frac{d^2(\theta - \theta_i)}{dt^2} + \frac{NR}{IV} \frac{d(\theta - \theta_i)}{dt} + \frac{N}{I}(\theta - \theta_i) = 0. \qquad (2.3.19)$$

The solution to Equation 2.3.19 is

$$\theta - \theta_i = A \exp\left(-\frac{NRt}{2IV}\right) \cos(\omega t + \varphi), \qquad (2.3.20)$$

where

$$\omega^2 = \frac{N}{I} - \frac{N^2 R^2}{4 I^2 V^2}, \qquad (2.3.21)$$

and A and φ are the two arbitrary constants that would be determined by presently unspecified initial conditions. Consequently an ideal wind vane is a damped harmonic oscillator where the wind torque should be large and its moment of inertia should be small.

Problems

For the following values of m, β, and k, find the position $x(t)$ of a damped oscillator for the given initial conditions:

1. $m = \tfrac{1}{2}$, $\quad \beta = 3$, $\quad k = 4$, $\quad x(0) = 2$, $\quad x'(0) = 0$

2. $m = 1$, $\quad \beta = 10$, $\quad k = 125$, $\quad x(0) = 3$, $\quad x'(0) = 25$

3. $m = 4$, $\quad \beta = 20$, $\quad k = 169$, $\quad x(0) = 4$, $\quad x'(0) = 16$

4. For an underdamped oscillator, show that the time required for its amplitude to drop to half of its original value equals $\ln(2)/\lambda$. For a fixed value of λ/ω, what is the minimum number of cycles that has occurred?

5. For what values of c does $x'' + cx' + 4x = 0$ have critically damped solutions?

6. For what values of c are the motions governed by $4x'' + cx' + 9x = 0$ (a) overdamped, (b) underdamped, and (c) critically damped?

7. For an overdamped mass-spring system, prove that the mass can pass through its equilibrium position $x = 0$ at most once.

2.4 METHOD OF UNDETERMINED COEFFICIENTS

Homogeneous ordinary differential equations become nonhomogeneous when the right side of Equation 2.0.1 is nonzero. How does this case differ from the homogeneous one that we have treated so far?

To answer this question, let us begin by introducing a function $y_p(x)$—called a *particular solution*—whose only requirement is that it satisfies the differential equation

$$a_n(x)\frac{d^n y_p}{dx^n} + a_{n-1}(x)\frac{d^{n-1} y_p}{dx^{n-1}} + \cdots + a_1(x)\frac{dy_p}{dx} + a_0(x)y_p = f(x). \qquad (2.4.1)$$

Then, by direct substitution, it can be seen that the general solution to any nonhomogeneous, linear, ordinary differential equation is

$$y(x) = y_H(x) + y_p(x), \qquad (2.4.2)$$

where $y_H(x)$—the *homogeneous* or *complementary solution*—satisfies

$$a_n(x)\frac{d^n y_H}{dx^n} + a_{n-1}(x)\frac{d^{n-1} y_H}{dx^{n-1}} + \cdots + a_1(x)\frac{dy_H}{dx} + a_0(x)y_H = 0. \qquad (2.4.3)$$

Why have we introduced this complementary solution? Because the particular solution already satisfies the ordinary differential equation. The purpose of the complementary solution is to introduce the arbitrary constants that any general solution of an ordinary differential equation must have. Thus, because we already know how to find $y_H(x)$, we must only invent a method for finding the particular solution to have our general solution.

• **Example 2.4.1**

Let us illustrate this technique with the second-order, linear, nonhomogeneous ordinary differential equation

$$y'' - 4y' + 4y = 2e^{2x} + 4x - 12. \qquad (2.4.4)$$

Taking $y(x) = y_H(x) + y_p(x)$, direct substitution yields

$$y_H'' + y_p'' - 4(y_H' + y_p') + 4(y_H + y_p) = 2e^{2x} + 4x - 12. \qquad (2.4.5)$$

If we now require that the particular solution $y_p(x)$ satisfies the differential equation

$$y_p'' - 4y_p' + 4y_p = 2e^{2x} + 4x - 12, \qquad (2.4.6)$$

Equation 2.4.5 simplifies to the homogeneous ordinary differential equation

$$y_H'' - 4y_H' + 4y_H = 0. \tag{2.4.7}$$

A quick check[5] shows that the particular solution to Equation 2.4.6 is $y_p(x) = x^2 e^{2x} + x - 2$. Using techniques from the previous section, the complementary solution is $y_H(x) = C_1 e^{2x} + C_2 x e^{2x}$.

$\square$

In general, finding $y_p(x)$ is a formidable task. In the case of constant coefficients, several techniques have been developed. The most commonly employed technique is called the *method of undetermined coefficients*, which is used with linear, constant coefficient, ordinary differential equations when $f(x)$ is a constant, a polynomial, an exponential function $e^{\alpha x}$, $\sin(\beta x)$, $\cos(\beta x)$, or a finite sum and products of these functions. Thus, this technique applies when the function $f(x)$ equals $e^x \sin(x) - (3x - 2)e^{-2x}$ but not when it equals $\ln(x)$.

Why does this technique work? The reason lies in the set of functions that we have allowed to be included in $f(x)$. They enjoy the remarkable property that derivatives of their sums and products yield sums and products that are also constants, polynomials, exponentials, sines, and cosines. Because a linear combination of derivatives such as $ay_p'' + by_p' + cy_p$ must equal $f(x)$, it seems reasonable to assume that $y_p(x)$ has the same form as $f(x)$. The following examples show that our conjecture is correct.

- **Example 2.4.2**

Let us illustrate the method of undetermined coefficients by finding the particular solution to

$$y'' - 2y' + y = x + \sin(x). \tag{2.4.8}$$

From the form of the right side of Equation 2.4.8, we guess the particular solution

$$y_p(x) = Ax + B + C\sin(x) + D\cos(x). \tag{2.4.9}$$

Therefore,

$$y_p'(x) = A + C\cos(x) - D\sin(x), \tag{2.4.10}$$

and

$$y_p''(x) = -C\sin(x) - D\cos(x). \tag{2.4.11}$$

Substituting into Equation 2.4.8, we find that

$$y_p'' - 2y_p' + y_p = Ax + B - 2A - 2C\cos(x) + 2D\sin(x) = x + \sin(x). \tag{2.4.12}$$

Since Equation 2.4.12 must be true for all x, the constant terms must sum to zero or $B - 2A = 0$. Similarly, all of the terms involving the polynomial x must balance, yielding $A = 1$ and $B = 2A = 2$. Turning to the trigonometric terms, the coefficients of $\sin(x)$ and $\cos(x)$ give $2D = 1$ and $-2C = 0$, respectively. Therefore, the particular solution is

$$y_p(x) = x + 2 + \tfrac{1}{2}\cos(x), \tag{2.4.13}$$

and the general solution is

$$y(x) = y_H(x) + y_p(x) = C_1 e^x + C_2 x e^x + x + 2 + \tfrac{1}{2}\cos(x). \tag{2.4.14}$$

---

[5] We will show how $y_p(x)$ was obtained momentarily.

We can verify our result by using the symbolic toolbox in MATLAB. Typing the command:

```
dsolve('D2y-2*Dy+y=x+sin(x)','x')
```

yields:

```
ans =
x+2+1/2*cos(x)+C1*exp(x)+C2*exp(x)*x
```
       □

• **Example 2.4.3**

Let us find the particular solution to

$$y'' + y' - 2y = xe^x \qquad (2.4.15)$$

by the method of undetermined coefficients.

From the form of the right side of Equation 2.4.15, we guess the particular solution

$$y_p(x) = Axe^x + Be^x. \qquad (2.4.16)$$

Therefore,

$$y_p'(x) = Axe^x + Ae^x + Be^x, \qquad (2.4.17)$$

and

$$y_p''(x) = Axe^x + 2Ae^x + Be^x. \qquad (2.4.18)$$

Substituting into Equation 2.4.15, we find that

$$3Ae^x = xe^x. \qquad (2.4.19)$$

Clearly we cannot choose a constant A such that Equation 2.4.19 is satisfied. What went wrong?

To understand why, let us find the homogeneous or complementary solution to Equation 2.4.15; it is

$$y_H(x) = C_1e^{-2x} + C_2e^x. \qquad (2.4.20)$$

Therefore, one of the assumed particular solutions, Be^x, is also a homogeneous solution and cannot possibly give a nonzero left side when substituted into the differential equation. Consequently, it would appear that the method of undetermined coefficients does not work when one of the terms on the right side is also a homogeneous solution.

Before we give up, let us recall that we had a similar situation in the case of linear homogeneous second-order ordinary differential equations when the roots from the auxiliary equation were equal. There we found one of the homogeneous solutions was e^{m_1x}. We eventually found that the second solution was xe^{m_1x}. Could such a solution work here? Let us try.

We begin by modifying Equation 2.4.16 by multiplying it by x. Thus, our new guess for the particular solution reads

$$y_p(x) = Ax^2e^x + Bxe^x. \qquad (2.4.21)$$

Then,

$$y_p' = Ax^2e^x + 2Axe^x + Bxe^x + Be^x, \qquad (2.4.22)$$

and

$$y_p'' = Ax^2 e^x + 4Axe^x + 2Ae^x + Bxe^x + 2Be^x. \tag{2.4.23}$$

Substituting Equation 2.4.21 into Equation 2.4.15 gives

$$y_p'' + y_p' - 2y_p = 6Axe^x + 2Ae^x + 3Be^x = xe^x. \tag{2.4.24}$$

Grouping together terms that vary as xe^x, we find that $6A = 1$. Similarly, terms that vary as e^x yield $2A + 3B = 0$. Therefore,

$$y_p(x) = \tfrac{1}{6}x^2 e^x - \tfrac{1}{9}xe^x, \tag{2.4.25}$$

so that the general solution is

$$y(x) = y_H(x) + y_p(x) = C_1 e^{-2x} + C_2 e^x + \tfrac{1}{6}x^2 e^x - \tfrac{1}{9}xe^x. \tag{2.4.26}$$

$\square$

In summary, the method of finding particular solutions to higher-order ordinary differential equations by the method of undetermined coefficients is as follows:

- **Step 1:** Find the homogeneous solution to the differential equation.

- **Step 2:** Make an initial guess at the particular solution. The form of $y_p(x)$ is a linear combination of all linearly independent functions that are generated by repeated differentiations of $f(x)$.

- **Step 3:** If any of the terms in $y_p(x)$ given in Step 2 duplicate any of the homogeneous solutions, then that particular term in $y_p(x)$ must be multiplied by x^n, where n is the smallest positive integer that eliminates the duplication.

- **Example 2.4.4**

  Let us apply the method of undetermined coefficients to solve

  $$y'' + y = \sin(x) - e^{3x}\cos(5x). \tag{2.4.27}$$

  We begin by first finding the solution to the homogeneous version of Equation 2.4.27:

  $$y_H'' + y_H = 0. \tag{2.4.28}$$

Its solution is

$$y_H(x) = A\cos(x) + B\sin(x). \tag{2.4.29}$$

To find the particular solution we examine the right side of Equation 2.4.27 or

$$f(x) = \sin(x) - e^{3x}\cos(5x). \tag{2.4.30}$$

Taking a few derivatives of $f(x)$, we find that

$$f'(x) = \cos(x) - 3e^{3x}\cos(5x) + 5e^{3x}\sin(5x), \tag{2.4.31}$$

$$f''(x) = -\sin(x) - 9e^{3x}\cos(5x) + 30e^{3x}\sin(5x) + 25e^{3x}\cos(5x), \tag{2.4.32}$$

and so forth. Therefore, our guess at the particular solution is

$$y_p(x) = Cx\sin(x) + Dx\cos(x) + Ee^{3x}\cos(5x) + Fe^{3x}\sin(5x). \qquad (2.4.33)$$

Why have we chosen $x\sin(x)$ and $x\cos(x)$ rather than $\sin(x)$ and $\cos(x)$? Because $\sin(x)$ and $\cos(x)$ are homogeneous solutions to Equation 2.4.27, we must multiply them by a power of x.

Since

$$y_p''(x) = 2C\cos(x) - Cx\sin(x) - 2D\sin(x) - Dx\cos(x)$$
$$+ (30F - 16E)e^{3x}\cos(5x) - (30E + 16F)e^{3x}\sin(5x), \qquad (2.4.34)$$

$$y_p'' + y_p = 2C\cos(x) - 2D\sin(x)$$
$$+ (30F - 15E)e^{3x}\cos(5x) - (30E + 15F)e^{3x}\sin(5x) \qquad (2.4.35)$$
$$= \sin(x) - e^{3x}\cos(5x). \qquad (2.4.36)$$

Therefore, $2C = 0$, $-2D = 1$, $30F - 15E = -1$, and $30E + 15F = 0$. Solving this system of equations yields $C = 0$, $D = -\frac{1}{2}$, $E = \frac{1}{75}$, and $F = -\frac{2}{75}$. Thus, the general solution is

$$y(x) = A\cos(x) + B\sin(x) - \tfrac{1}{2}x\cos(x) + \tfrac{1}{75}e^{3x}[\cos(5x) - 2\sin(5x)]. \qquad (2.4.37)$$

Problems

Use the method of undetermined coefficients to find the general solution of the following differential equations. Verify your solution by using `dsolve` in MATLAB.

1. $y'' + 4y' + 3y = x + 1$ 2. $y'' - y = e^x - 2e^{-2x}$

3. $y'' + 2y' + 2y = 2x^2 + 2x + 4$ 4. $y'' + y' = x^2 + x$

5. $y'' + 2y' = 2x + 5 - e^{-2x}$ 6. $y'' - 4y' + 4y = (x+1)e^{2x}$

7. $y'' + 4y' + 4y = xe^x$ 8. $y'' - 4y = 4\sinh(2x)$

9. $y'' + 9y = x\cos(3x)$ 10. $y'' + y = \sin(x) + x\cos(x)$

11. Solve
$$y'' + 2ay' = \sin^2(\omega x), \qquad y(0) = y'(0) = 0,$$

by (a) the method of undetermined coefficients and (b) integrating the ordinary differential equation so that it reduces to

$$y' + 2ay = \frac{x}{2} - \frac{\sin(2ax)}{4a},$$

and then using the techniques from the previous chapter to solve this first-order ordinary differential equation.

2.5 FORCED HARMONIC MOTION

Let us now consider the situation when an external force $f(t)$ acts on a vibrating mass on a spring. For example, $f(t)$ could represent a driving force that periodically raises and lowers the support of the spring. The inclusion of $f(t)$ in the formulation of Newton's second law yields the differential equation

$$m\frac{d^2x}{dt^2} = -kx - \beta\frac{dx}{dt} + f(t), \tag{2.5.1}$$

$$\frac{d^2x}{dt^2} + \frac{\beta}{m}\frac{dx}{dt} + \frac{k}{m}x = \frac{f(t)}{m}, \tag{2.5.2}$$

or

$$\frac{d^2x}{dt^2} + 2\lambda\frac{dx}{dt} + \omega^2 x = F(t), \tag{2.5.3}$$

where $F(t) = f(t)/m$, $2\lambda = \beta/m$, and $\omega^2 = k/m$. To solve this nonhomogeneous equation we will use the method of undetermined coefficients.

- **Example 2.5.1**

Let us find the solution to the nonhomogeneous differential equation

$$y'' + 2y' + y = 2\sin(t), \tag{2.5.4}$$

subject to the initial conditions $y(0) = 2$ and $y'(0) = 1$.
  The homogeneous solution is easily found and equals

$$y_H(t) = Ae^{-t} + Bte^{-t}. \tag{2.5.5}$$

From the method of undetermined coefficients, we guess that the particular solution is

$$y_p(t) = C\cos(t) + D\sin(t), \tag{2.5.6}$$

so that

$$y_p'(t) = -C\sin(t) + D\cos(t), \tag{2.5.7}$$

and

$$y_p''(t) = -C\cos(t) - D\sin(t). \tag{2.5.8}$$

Substituting $y_p(t)$, $y_p'(t)$, and $y_p''(t)$ into Equation 2.5.4 and simplifying, we find that

$$-2C\sin(t) + 2D\cos(t) = 2\sin(t) \tag{2.5.9}$$

or $D = 0$ and $C = -1$.
  To find A and B, we now apply the initial conditions on the general solution

$$y(t) = Ae^{-t} + Bte^{-t} - \cos(t). \tag{2.5.10}$$

The initial condition $y(0) = 2$ yields

$$y(0) = A + 0 - 1 = 2, \tag{2.5.11}$$

or $A = 3$. The initial condition $y'(0) = 1$ gives

$$y'(0) = -A + B = 1, \tag{2.5.12}$$

or $B = 4$, since

$$y'(t) = -Ae^{-t} + Be^{-t} - Bte^{-t} + \sin(t). \tag{2.5.13}$$

Therefore, the solution that satisfies the differential equation and initial conditions is

$$y(t) = 3e^{-t} + 4te^{-t} - \cos(t). \tag{2.5.14}$$

$\square$

• **Example 2.5.2**

Let us solve the differential equation for a weakly damped harmonic oscillator when the constant forcing F_0 "turns on" at $t = t_0$. The initial conditions are that $x(0) = x_0$ and $x'(0) = v_0$. Mathematically, the problem is

$$x'' + 2\lambda x' + \omega^2 x = \begin{cases} 0, & 0 < t < t_0, \\ F_0, & t_0 < t, \end{cases} \tag{2.5.15}$$

with $x(0) = x_0$ and $x'(0) = v_0$.

To solve Equation 2.5.15, we first divide the time domain into two regions: $0 < t < t_0$ and $t_0 < t$. For $0 < t < t_0$,

$$x(t) = Ae^{-\lambda t}\cos(\omega_d t) + Be^{-\lambda t}\sin(\omega_d t), \tag{2.5.16}$$

where $\omega_d^2 = \omega^2 - \lambda^2$. Upon applying the initial conditions,

$$x(t) = x_0 e^{-\lambda t}\cos(\omega_d t) + \frac{v_0 + \lambda x_0}{\omega_d} e^{-\lambda t}\sin(\omega_d t), \tag{2.5.17}$$

as before.

For the region $t_0 < t$, we write the general solution as

$$x(t) = Ae^{-\lambda t}\cos(\omega_d t) + Be^{-\lambda t}\sin(\omega_d t) + \frac{F_0}{\omega^2}$$
$$+ Ce^{-\lambda(t-t_0)}\cos[\omega_d(t-t_0)] + De^{-\lambda(t-t_0)}\sin[\omega_d(t-t_0)]. \tag{2.5.18}$$

Why have we written our solution in this particular form rather than the simpler

$$x(t) = Ce^{-\lambda t}\cos(\omega_d t) + De^{-\lambda t}\sin(\omega_d t) + \frac{F_0}{\omega^2}? \tag{2.5.19}$$

Both solutions satisfy the differential equation, as direct substitution verifies. However, the algebra is greatly simplified when Equation 2.5.18 rather than Equation 2.5.19 is used in matching the solution from each region at $t = t_0$. There, both the solution and its first derivative must be continuous or

$$x(t_0^-) = x(t_0^+), \quad \text{and} \quad x'(t_0^-) = x'(t_0^+), \tag{2.5.20}$$

**Review of the Solution of the
Forced Harmonic Oscillator Problem**

The undamped system $mx'' + kx = F_0 \cos(\omega_0 t)$ subject to the initial conditions $x(0) = x_0$ and $x'(0) = v_0$ has the solution

$$x(t) = \frac{v_0}{\omega} \sin(\omega t) + \left(x_0 - \frac{f_0}{\omega^2 - \omega_0^2} \right) \cos(\omega t) + \frac{f_0}{\omega^2 - \omega_0^2} \cos(\omega_0 t),$$

where $f_0 = F_0/m$ and $\omega = \sqrt{k/m}$. The underdamped system $mx'' + \beta x' + kx = F_0 \cos(\omega_0 t)$ has the *steady-state* solution

$$x(t) = \frac{f_0}{\sqrt{(\omega^2 - \omega_0^2)^2 + (2\lambda\omega_0)^2}} \cos\left[\omega_0 t - \tan^{-1}\left(\frac{2\lambda\omega_0}{\omega^2 - \omega_0^2} \right) \right],$$

where $2\lambda = \beta/m$.

---

where t_0^- and t_0^+ are points just below and above t_0, respectively. When Equation 2.5.17 and Equation 2.5.18 are substituted, we find that $C = -F_0/\omega^2$, and $\omega_d D = \lambda C$. Thus, the solution for the region $t_0 < t$ is

$$x(t) = x_0 e^{-\lambda t} \cos(\omega_d t) + \frac{v_0 + \lambda x_0}{\omega_d} e^{-\lambda t} \sin(\omega_d t) + \frac{F_0}{\omega^2} \tag{2.5.21}$$

$$- \frac{F_0}{\omega^2} e^{-\lambda(t-t_0)} \cos[\omega_d(t - t_0)] - \frac{\lambda F_0}{\omega_d \omega^2} e^{-\lambda(t-t_0)} \sin[\omega_d(t - t_0)].$$

In a book that treats Laplace transforms, you will see that they are particularly useful for this type of problem when the forcing function changes abruptly one or more times. ☐

As noted earlier, nonhomogeneous solutions consist of the homogeneous solution plus a particular solution. In the case of a damped harmonic oscillator, another, more physical, way of describing the solution involves its behavior over an extended time. That portion of the solution which eventually becomes negligible as $t \to \infty$ is often referred to as the *transient term*, or *transient solution*. In Equation 2.5.14 the transient solution equals $3e^{-t} + 4te^{-t}$. On the other hand, the portion of the solution that remains as $t \to \infty$ is called the *steady-state solution*. In Equation 2.5.14 the steady-state solution equals $-\cos(t)$.

One of the most interesting forced oscillator problems occurs when $\beta = 0$ and the forcing function equals $F_0 \sin(\omega_0 t)$, where F_0 is a constant. Then the initial-value problem becomes

$$\frac{d^2 x}{dt^2} + \omega^2 x = F_0 \sin(\omega_0 t). \tag{2.5.22}$$

Let us solve this problem when $x(0) = x'(0) = 0$.

The homogeneous solution to Equation 2.5.22 is

$$x_H(t) = A \cos(\omega t) + B \sin(\omega t). \tag{2.5.23}$$

To obtain the particular solution, we assume that

$$x_p(t) = C \cos(\omega_0 t) + D \sin(\omega_0 t). \tag{2.5.24}$$

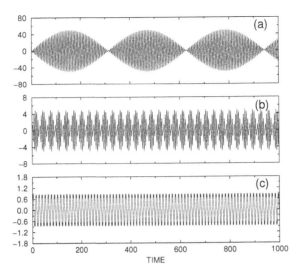

Figure 2.5.1: The solution, Equation 2.5.31, as a function of time when $\omega = 1$ and ω_0 equals (a) 1.02, (b) 1.2, and (c) 2.

This leads to

$$x_p'(t) = -C\omega_0 \sin(\omega_0 t) + D\omega_0 \cos(\omega_0 t), \qquad (2.5.25)$$

$$x_p''(t) = -C\omega_0^2 \cos(\omega_0 t) + D\omega_0^2 \sin(\omega_0 t), \qquad (2.5.26)$$

and

$$x_p'' + \omega^2 x_p = C(\omega^2 - \omega_0^2) \cos(\omega_0 t) + D(\omega^2 - \omega_0^2) \sin(\omega_0 t) = F_0 \sin(\omega_0 t). \qquad (2.5.27)$$

We immediately conclude that $C(\omega^2 - \omega_0^2) = 0$, and $D(\omega^2 - \omega_0^2) = F_0$. Therefore,

$$C = 0, \quad \text{and} \quad D = \frac{F_0}{\omega^2 - \omega_0^2}, \qquad (2.5.28)$$

provided that $\omega \neq \omega_0$. Thus,

$$x_p(t) = \frac{F_0}{\omega^2 - \omega_0^2} \sin(\omega_0 t). \qquad (2.5.29)$$

To finish the problem, we must apply the initial conditions to the general solution

$$x(t) = A \cos(\omega t) + B \sin(\omega t) + \frac{F_0}{\omega^2 - \omega_0^2} \sin(\omega_0 t). \qquad (2.5.30)$$

From $x(0) = 0$, we find that $A = 0$. On the other hand, $x'(0) = 0$ yields $B = -\omega_0 F_0 / [\omega(\omega^2 - \omega_0^2)]$. Thus, the final result is

$$x(t) = \frac{F_0}{\omega(\omega^2 - \omega_0^2)} \left[\omega \sin(\omega_0 t) - \omega_0 \sin(\omega t) \right]. \qquad (2.5.31)$$

Figure 2.5.1 illustrates Equation 2.5.31 as a function of time.

The most arresting feature in Figure 2.5.1 is the evolution of the uniform amplitude of the oscillation shown in frame (c) into the one shown in frame (a) where the amplitude exhibits a sinusoidal variation as $\omega_0 \to \omega$. In acoustics these fluctuations in the amplitude are called *beats*, the loud sounds corresponding to the larger amplitudes.

As our analysis indicates, Equation 2.5.31 does not apply when $\omega = \omega_0$. As we shall shortly see, this is probably the most interesting configuration. We can use Equation 2.5.31 to examine this case by applying L'Hôpital's rule in the limiting case of $\omega_0 \to \omega$. This limiting process is analogous to "tuning in" the frequency of the driving frequency $[\omega_0/(2\pi)]$ to the frequency of free vibrations $[\omega/(2\pi)]$. From experience, we expect that given enough time we should be able to substantially increase the amplitudes of vibrations. Mathematical confirmation of our physical intuition is as follows:

$$x(t) = \lim_{\omega_0 \to \omega} F_0 \frac{\omega \sin(\omega_0 t) - \omega_0 \sin(\omega t)}{\omega(\omega^2 - \omega_0^2)} = F_0 \lim_{\omega_0 \to \omega} \frac{d[\omega \sin(\omega_0 t) - \omega_0 \sin(\omega t)]/d\omega_0}{d[\omega(\omega^2 - \omega_0^2)]/d\omega_0} \quad (2.5.32)$$

$$= F_0 \lim_{\omega_0 \to \omega} \frac{\omega t \cos(\omega_0 t) - \sin(\omega t)}{-2\omega_0 \omega} = F_0 \frac{\omega t \cos(\omega t) - \sin(\omega t)}{-2\omega^2} \quad (2.5.33)$$

$$= \frac{F_0}{2\omega^2} \sin(\omega t) - \frac{F_0 t}{2\omega} \cos(\omega t). \quad (2.5.34)$$

As we suspected, as $t \to \infty$, the displacement grows without bounds. This phenomenon is known as *pure resonance*. We could also have obtained Equation 2.5.34 directly using the method of undetermined coefficients involving the initial value problem

$$\frac{d^2 x}{dt^2} + \omega^2 x = F_0 \sin(\omega t), \qquad x(0) = x'(0) = 0. \quad (2.5.35)$$

Because there is almost always some friction, pure resonance rarely occurs and the more realistic differential equation is

$$\frac{d^2 x}{dt^2} + 2\lambda \frac{dx}{dt} + \omega^2 x = F_0 \sin(\omega_0 t). \quad (2.5.36)$$

Its solution is

$$x(t) = C e^{-\lambda t} \sin\left(t\sqrt{\omega^2 - \omega_0^2} + \varphi\right) + \frac{F_0}{\sqrt{(\omega^2 - \omega_0^2)^2 + 4\lambda^2 \omega_0^2}} \sin(\omega_0 t - \theta), \quad (2.5.37)$$

where

$$\sin(\theta) = \frac{2\lambda \omega_0}{\sqrt{(\omega^2 - \omega_0^2)^2 + 4\lambda^2 \omega_0^2}}, \qquad \cos(\theta) = \frac{\omega^2 - \omega_0^2}{\sqrt{(\omega^2 - \omega_0^2)^2 + 4\lambda^2 \omega_0^2}}, \quad (2.5.38)$$

and C and φ are determined by the initial conditions. To illustrate Equation 2.5.37 we rewrite the amplitude and phase of the particular solution as

$$\frac{F_0}{\sqrt{(\omega^2 - \omega_0^2)^2 + 4\lambda^2 \omega_0^2}} = \frac{F_0}{\omega^2 \sqrt{(1 - r^2)^2 + 4\beta^2 r^2}} \qquad \text{and} \qquad \tan(\theta) = \frac{2\beta r}{1 - r^2}, \quad (2.5.39)$$

where $r = \omega_0/\omega$ and $\beta = \lambda/\omega$. Figures 2.5.2 and 2.5.3 graph Equation 2.5.37 as functions of r for various values of β.

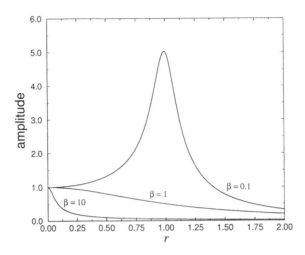

Figure 2.5.2: The amplitude of the particular solution Equation 2.5.37 for a forced, damped simple harmonic oscillator (normalized with F_0/ω^2) as a function of $r = \omega_0/\omega$.

● **Example 2.5.3: Electrical circuits**

In the previous chapter, we saw how the mathematical analysis of electrical circuits yields first-order linear differential equations. In those cases we only had a resistor and capacitor or a resistor and inductor. One of the fundamental problems of electrical circuits is a circuit where a resistor, capacitor, and inductor are connected in series, as shown in Figure 2.5.4.

In this RCL circuit, an instantaneous current flows when the key or switch K is closed. If $Q(t)$ denotes the instantaneous charge on the capacitor, Kirchhoff's law yields the differential equation

$$L\frac{dI}{dt} + RI + \frac{Q}{C} = E(t), \qquad (2.5.40)$$

where $E(t)$, the electromotive force, may depend on time, but where L, R, and C are constant. Because $I = dQ/dt$, Equation 2.5.40 becomes

$$L\frac{d^2Q}{dt^2} + R\frac{dQ}{dt} + \frac{Q}{C} = E(t). \qquad (2.5.41)$$

Consider now the case when resistance is negligibly small. Equation 2.5.41 will become identical to the differential equation for the forced simple harmonic oscillator, Equation 2.5.3, with $\lambda = 0$. Similarly, the general case yields various analogs to the damped harmonic oscillator:

Case 1	Overdamped	$R^2 > 4L/C$
Case 2	Critically damped	$R^2 = 4L/C$
Case 3	Underdamped	$R^2 < 4L/C$

In each of these three cases, $Q(t) \to 0$ as $t \to \infty$. (See Problem 7.) Therefore, an RLC electrical circuit behaves like a damped mass-spring mechanical system, where inductance acts like mass, resistance is the damping coefficient, and $1/C$ is the spring constant.

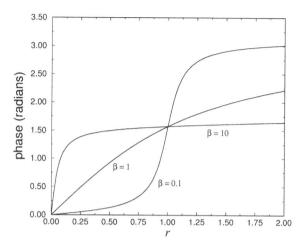

Figure 2.5.3: The phase of the particular solution, Equation 2.5.37, for a forced, damped simple harmonic oscillator as a function of $r = \omega_0/\omega$.

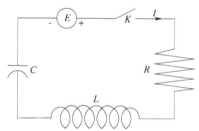

Figure 2.5.4: A simple electrical circuit containing a resistor of constant resistance R, capacitor of constant capacitance C, and inductor of constant inductance L driven by a time-dependent electromotive force $E(t)$.

Problems

1. Find the values of γ so that $x'' + 6x' + 18 = \cos(\gamma t)$ is in resonance.

2. The differential equation $x'' + 2x' + 2x = 10\sin(2t)$ describes a damped, forced oscillator. If the initial conditions are $x(0) = x_0$ and $x'(0) = 0$, find its solution by hand and by using MATLAB. Plot the solution when $x_0 = -10, -9, \ldots, 9, 10$. Give a physical interpretation to what you observe.

3. At time $t = 0$, a mass m is suddenly attached to the end of a hanging spring with a spring constant k. Neglecting friction, find the subsequent motion if the coordinate system is chosen so that $x(0) = 0$.

Step 1: Show that the differential equation is $mx'' + kx = mg$ with the initial conditions $x(0) = x'(0) = 0$.

Step 2: Show that the solution to Step 1 is $x(t) = mg\left[1 - \cos(\omega t)\right]/k$ where $\omega^2 = k/m$.

4. At what driving frequency does a forced, damped, simple harmonic oscillator have its maximum amplitude? At what driving frequency does the *velocity* have its maximum amplitude?

5. Consider the electrical circuit shown in Figure 2.5.4, which now possesses negligible resistance and has an applied voltage $E(t) = E_0[1 - \cos(\omega t)]$. Find the *current* if the circuit is initially dead.

6. Find the general solution to the differential equation governing a forced, damped harmonic equation $mx'' + cx' + kx = F_0 \sin(\omega t)$, where m, c, k, F_0, and ω are constants. Write the particular solution in amplitude/phase format.

7. Prove that the *transient* solution to Equation 2.5.41 tends to zero as $t \to \infty$ if R, C, and L are greater than zero.

2.6 VARIATION OF PARAMETERS

As the previous section has shown, the method of undetermined coefficients can be used when the right side of the differential equation contains constants, polynomials, exponentials, sines, and cosines. On the other hand, when the right side contains terms other than these, variation of parameters provides a method for finding the particular solution.

To understand this technique, let us return to our solution of the first-order ordinary differential equation

$$\frac{dy}{dx} + P(x)y = f(x). \tag{2.6.1}$$

Its solution is

$$y(x) = C_1 e^{-\int P(x)\,dx} + e^{-\int P(x)\,dx} \int e^{\int P(x)\,dx} f(x)\,dx. \tag{2.6.2}$$

The solution, Equation 2.6.2, consists of two parts: The first term is the homogeneous solution and can be written $y_H(x) = C_1 y_1(x)$, where $y_1(x) = e^{-\int P(x)\,dx}$. The second term is the particular solution and equals the product of some function of x, say $u_1(x)$, times $y_1(x)$:

$$y_p(x) = e^{-\int P(x)\,dx} \int e^{\int P(x)\,dx} f(x)\,dx = u_1(x)y_1(x). \tag{2.6.3}$$

This particular solution bears a striking resemblance to the homogeneous solution if we replace $u_1(x)$ with C_1.

Variation of parameters builds upon this observation by using the homogeneous solution $y_1(x)$ to construct a guess for the particular solution $y_p(x) = u_1(x)y_1(x)$. Upon substituting this guessed $y_p(x)$ into Equation 2.6.1, we have that

$$\frac{d}{dx}(u_1 y_1) + P(x)u_1 y_1 = f(x), \tag{2.6.4}$$

$$u_1 \frac{dy_1}{dx} + y_1 \frac{du_1}{dx} + P(x)u_1 y_1 = f(x), \tag{2.6.5}$$

or

$$y_1 \frac{du_1}{dx} = f(x), \tag{2.6.6}$$

since $y_1' + P(x)y_1 = 0$.

Using the technique of separating the variables, we have that

$$du_1 = \frac{f(x)}{y_1(x)}\,dx, \quad \text{and} \quad u_1(x) = \int \frac{f(x)}{y_1(x)}\,dx. \tag{2.6.7}$$

Consequently, the particular solution equals

$$y_p(x) = u_1(x)y_1(x) = y_1(x) \int \frac{f(x)}{y_1(x)}\,dx. \tag{2.6.8}$$

Upon substituting for $y_1(x)$, we obtain Equation 2.6.3.

How do we apply this method to the linear second-order differential equation

$$a_2(x)y'' + a_1 y'(x) + a_0(x)y = g(x), \tag{2.6.9}$$

or

$$y'' + P(x)y' + Q(x)y = f(x), \tag{2.6.10}$$

where $P(x)$, $Q(x)$, and $f(x)$ are continuous on some interval I?

Let $y_1(x)$ and $y_2(x)$ denote the homogeneous solutions of Equation 2.6.10. That is, $y_1(x)$ and $y_2(x)$ satisfy

$$y_1'' + P(x)y_1' + Q(x)y_1 = 0, \tag{2.6.11}$$

and

$$y_2'' + P(x)y_2' + Q(x)y_2 = 0. \tag{2.6.12}$$

Following our previous example, we now seek two functions $u_1(x)$ and $u_2(x)$ such that

$$y_p(x) = u_1(x)y_1(x) + u_2(x)y_2(x) \tag{2.6.13}$$

is a particular solution of Equation 2.6.10. Once again, we replaced our arbitrary constants C_1 and C_2 by the "variable parameters" $u_1(x)$ and $u_2(x)$. Because we have two unknown functions, we require two equations to solve for $u_1(x)$ and $u_2(x)$. One of them follows from substituting $y_p(x) = u_1(x)y_1(x) + u_2(x)y_2(x)$ into Equation 2.6.10. The other equation is

$$y_1(x)u_1'(x) + y_2(x)u_2'(x) = 0. \tag{2.6.14}$$

This equation is an assumption that is made to simplify the first and second derivative, which is clearly seen by computing

$$y_p' = u_1 y_1' + y_1 u_1' + u_2 y_2' + y_2 u_2' = u_1 y_1' + u_2 y_2', \tag{2.6.15}$$

after applying Equation 2.6.14. Continuing to the second derivative,

$$y_p'' = u_1 y_1'' + y_1' u_1' + u_2 y_2'' + y_2' u_2'. \tag{2.6.16}$$

Substituting these results into Equation 2.6.10, we obtain

$$\begin{aligned} y_p'' + P(x)y_p' + Q(x)y_p &= u_1 y_1'' + y_1' u_1' + u_2 y_2'' + y_2' u_2' \\ &\quad + P u_1 y_1' + P u_2 y_2' + Q u_1 y_1 + Q u_2 y_2, \tag{2.6.17} \\ &= u_1 \left[y_1'' + P(x)y_1' + Q(x)y_1 \right] + u_2 \left[y_2'' + P(x)y_2' + Q(x)y_2 \right] \\ &\quad + y_1' u_1' + y_2' u_2' = f(x). \tag{2.6.18} \end{aligned}$$

Hence, $u_1(x)$ and $u_2(x)$ must be functions that also satisfy the condition

$$y_1' u_1' + y_2' u_2' = f(x). \tag{2.6.19}$$

It is important to note that the differential equation must be written so that it conforms to Equation 2.6.10. This may require the division of the differential equation by $a_2(x)$ so that you have the correct $f(x)$.

Equations 2.6.14 and 2.6.19 constitute a linear system of equations for determining the unknown derivatives u_1' and u_2'. By Cramer's rule,[6] the solutions of Equation 2.6.14 and Equation 2.6.19 equal

$$u_1'(x) = \frac{W_1}{W}, \qquad \text{and} \qquad u_2'(x) = \frac{W_2}{W}, \tag{2.6.20}$$

where

$$W = \begin{vmatrix} y_1 & y_2 \\ y_1' & y_2' \end{vmatrix}, \quad W_1 = \begin{vmatrix} 0 & y_2 \\ f(x) & y_2' \end{vmatrix}, \quad \text{and} \quad W_2 = \begin{vmatrix} y_1 & 0 \\ y_1' & f(x) \end{vmatrix}. \tag{2.6.21}$$

The determinant W is the Wronskian of y_1 and y_2. Because y_1 and y_2 are linearly independent on I, the Wronskian will never equal zero for every x in the interval.

These results can be generalized to any nonhomogeneous, nth-order, linear equation of the form

$$y^{(n)} + P_{n-1}(x)y^{(n-1)} + P_1(x)y' + P_0(x) = f(x). \tag{2.6.22}$$

If $y_H(x) = C_1 y_1(x) + C_2 y_2(x) + \cdots + C_n y_n(x)$ is the complementary function for Equation 2.6.22, then a particular solution is

$$y_p(x) = u_1(x)y_1(x) + u_2(x)y_2(x) + \cdots + u_n(x)y_n(x), \tag{2.6.23}$$

where the u_k', $k = 1, 2, \ldots, n$, are determined by the n equations:

$$y_1 u_1' + y_2 u_2' + \cdots + y_n u_n' = 0, \tag{2.6.24}$$
$$y_1' u_1' + y_2' u_2' + \cdots + y_n' u_n' = 0, \tag{2.6.25}$$
$$\vdots$$
$$y_1^{(n-1)} u_1' + y_2^{(n-1)} u_2' + \cdots + y_n^{(n-1)} u_n' = f(x). \tag{2.6.26}$$

The first $n - 1$ equations in this system, like Equation 2.6.14, are assumptions made to simplify the first $n - 1$ derivatives of $y_p(x)$. The last equation of the system results from substituting the n derivative of $y_p(x)$ and the simplified lower derivatives into Equation 2.6.22. Then, by Cramer's rule, we find that

$$u_k' = \frac{W_k}{W}, \qquad k = 1, 2, \ldots, n, \tag{2.6.27}$$

where W is the Wronskian of $y_1, y_2, \ldots, y_n$, and W_k is the determinant obtained by replacing the kth column of the Wronskian by the column vector $[0, 0, 0, \cdots, f(x)]^T$.

• **Example 2.6.1**

Let us apply variation of parameters to find the general solution to

$$y'' + y' - 2y = xe^x. \tag{2.6.28}$$

We begin by first finding the homogeneous solution that satisfies the differential equation

$$y_H'' + y_H' - 2y_H = 0. \tag{2.6.29}$$

---

[6] If you are unfamiliar with Cramer's rule, see Section 3.3.

Applying the techniques from Section 2.1, the homogeneous solution is

$$y_H(x) = Ae^x + Be^{-2x}, \tag{2.6.30}$$

yielding the two independent solutions $y_1(x) = e^x$, and $y_2(x) = e^{-2x}$. Thus, the method of variation of parameters yields the particular solution

$$y_p(x) = e^x u_1(x) + e^{-2x} u_2(x). \tag{2.6.31}$$

From Equation 2.6.14, we have that

$$e^x u_1'(x) + e^{-2x} u_2'(x) = 0, \tag{2.6.32}$$

while

$$e^x u_1'(x) - 2e^{-2x} u_2'(x) = xe^x. \tag{2.6.33}$$

Solving for $u_1'(x)$ and $u_2'(x)$, we find that

$$u_1'(x) = \tfrac{1}{3}x, \qquad \text{or} \qquad u_1(x) = \tfrac{1}{6}x^2, \tag{2.6.34}$$

and

$$u_2'(x) = -\tfrac{1}{3}xe^{3x}, \qquad \text{or} \qquad u_2(x) = \tfrac{1}{27}(1 - 3x)e^{3x}. \tag{2.6.35}$$

Therefore, the general solution is

$$y(x) = Ae^x + Be^{-2x} + e^x u_1(x) + e^{-2x} u_2(x) \tag{2.6.36}$$
$$= Ae^x + Be^{-2x} + \tfrac{1}{6}x^2 e^x + \tfrac{1}{27}(1 - 3x)e^x \tag{2.6.37}$$
$$= Ce^x + Be^{-2x} + \left(\tfrac{1}{6}x^2 - \tfrac{1}{9}x\right)e^x. \tag{2.6.38}$$

$\square$

• **Example 2.6.2**

Let us find the general solution to

$$y'' + 2y' + y = e^{-x}\ln(x) \tag{2.6.39}$$

by variation of parameters on the interval $(0, \infty)$.

We start by finding the homogeneous solution that satisfies the differential equation

$$y_H'' + 2y_H' + y_H = 0. \tag{2.6.40}$$

Applying the techniques from Section 2.1, the homogeneous solution is

$$y_H(x) = Ae^{-x} + Bxe^{-x}, \tag{2.6.41}$$

yielding the two independent solutions $y_1(x) = e^{-x}$ and $y_2(x) = xe^{-x}$. Thus, the particular solution equals

$$y_p(x) = e^{-x}u_1(x) + xe^{-x}u_2(x). \tag{2.6.42}$$

From Equation 2.6.14, we have that

$$e^{-x}u_1'(x) + xe^{-x}u_2'(x) = 0, \tag{2.6.43}$$

while
$$-e^{-x}u_1'(x) + (1-x)e^{-x}u_2'(x) = e^{-x}\ln(x). \qquad (2.6.44)$$

Solving for $u_1'(x)$ and $u_2'(x)$, we find that

$$u_1'(x) = -x\ln(x), \qquad \text{or} \qquad u_1(x) = \tfrac{1}{4}x^2 - \tfrac{1}{2}x^2\ln(x), \qquad (2.6.45)$$

and

$$u_2'(x) = \ln(x), \qquad \text{or} \qquad u_2(x) = x\ln(x) - x. \qquad (2.6.46)$$

Therefore, the general solution is

$$y(x) = Ae^{-x} + Bxe^{-x} + e^{-x}u_1(x) + xe^{-x}u_2(x) \qquad (2.6.47)$$
$$= Ae^{-x} + Bxe^{-x} + \tfrac{1}{2}x^2\ln(x)e^{-x} - \tfrac{3}{4}x^2e^{-x}. \qquad (2.6.48)$$

We can verify our result by using the symbolic toolbox in MATLAB. Typing the command:

```
dsolve('D2y+2*Dy+y=exp(-x)*log(x)','x')
```

yields:

```
ans =
1/2*exp(-x)*x^2*log(x)-3/4*exp(-x)*x^2+C1*exp(-x)+C2*exp(-x)*x
```
□

● **Example 2.6.3**

So far, all of our examples have yielded closed-form solutions. To show that this is not necessarily so, let us solve
$$y'' - 4y = e^{2x}/x \qquad (2.6.49)$$

by variation of parameters.

Again we begin by solving the homogeneous differential equation

$$y_H'' - 4y_H = 0, \qquad (2.6.50)$$

which has the solution

$$y_H(x) = Ae^{2x} + Be^{-2x}. \qquad (2.6.51)$$

Thus, our two independent solutions are $y_1(x) = e^{2x}$ and $y_2(x) = e^{-2x}$. Therefore, the particular solution equals

$$y_p(x) = e^{2x}u_1(x) + e^{-2x}u_2(x). \qquad (2.6.52)$$

From Equation 2.6.14, we have that

$$e^{2x}u_1'(x) + e^{-2x}u_2'(x) = 0, \qquad (2.6.53)$$

while

$$2e^{2x}u_1'(x) - 2e^{-2x}u_2'(x) = e^{2x}/x. \qquad (2.6.54)$$

Solving for $u_1'(x)$ and $u_2'(x)$, we find that

$$u_1'(x) = \frac{1}{4x}, \qquad \text{or} \qquad u_1(x) = \tfrac{1}{4}\ln|x|, \qquad (2.6.55)$$

and

$$u_2'(x) = -\frac{e^{4x}}{4x}, \qquad \text{or} \qquad u_2(x) = -\frac{1}{4}\int_{x_0}^{x}\frac{e^{4t}}{t}\,dt. \qquad (\textbf{2.6.56})$$

Therefore, the general solution is

$$y(x) = Ae^{2x} + Be^{-2x} + e^{2x}u_1(x) + e^{-2x}u_2(x) \qquad (\textbf{2.6.57})$$

$$= Ae^{2x} + Be^{-2x} + \frac{1}{4}\ln|x|e^{2x} - \frac{1}{4}e^{-2x}\int_{x_0}^{x}\frac{e^{4t}}{t}\,dt. \qquad (\textbf{2.6.58})$$

Problems

Use variation of parameters to find the general solution for the following differential equations. Then see if you can obtain your solution by using `dsolve` in MATLAB.

1. $y'' - 4y' + 3y = e^{-x}$ 2. $y'' - y' - 2y = x$ 3. $y'' - 4y = xe^{x}$

4. $y'' + 9y = 2\sec(x)$ 5. $y'' + 4y' + 4y = xe^{-2x}$ 6. $y'' + 2ay' = \sin^2(\omega x)$

7. $y'' - 4y' + 4y = (x+1)e^{2x}$ 8. $y'' - 4y = \sin^2(x)$ 9. $y'' - 2y' + y = e^x/x$

10. $y'' + y = \tan(x)$ 11. $y'' - 4y' + 3y = 2\cos(x+3)$ 12. $y'' - 3y' + 2y = \cos(e^{-x})$

2.7 EULER-CAUCHY EQUATION

The Euler-Cauchy or equidimensional equation is a linear differential equation of the form

$$a_n x^n \frac{d^n y}{dx^n} + a_{n-1}x^{n-1}\frac{d^{n-1}y}{dx^{n-1}} + \cdots + a_1 x\frac{dy}{dx} + a_0 y = f(x), \qquad (\textbf{2.7.1})$$

where a_n, a_{n-1}, ..., a_0 are constants. The important point here is that in each term the power to which x is raised equals the *order* of differentiation.

To illustrate this equation, we will focus on the homogeneous, second-order, ordinary differential equation

$$ax^2\frac{d^2 y}{dx^2} + bx\frac{dy}{dx} + cy = 0. \qquad (\textbf{2.7.2})$$

The solution of higher-order ordinary differential equations follows by analog. If we wish to solve the nonhomogeneous equation

$$ax^2\frac{d^2 y}{dx^2} + bx\frac{dy}{dx} + cy = f(x), \qquad (\textbf{2.7.3})$$

we can do so by applying variation of parameters using the complementary solutions that satisfy Equation 2.7.2.

Our analysis starts by trying a solution of the form $y = x^m$, where m is presently undetermined. The first and second derivatives are

$$\frac{dy}{dx} = mx^{m-1}, \quad \text{and} \quad \frac{d^2 y}{dx^2} = m(m-1)x^{m-2}, \qquad (\textbf{2.7.4})$$

respectively. Consequently, substitution yields the differential equation

$$ax^2 \frac{d^2y}{dx^2} + bx \frac{dy}{dx} + cy = ax^2 \cdot m(m-1)x^{m-2} + bx \cdot mx^{m-1} + cx^m \qquad \textbf{(2.7.5)}$$

$$= am(m-1)x^m + bmx^m + cx^m \qquad \textbf{(2.7.6)}$$

$$= [am(m-1) + bm + c]\, x^m. \qquad \textbf{(2.7.7)}$$

Thus, $y = x^m$ is a solution of the differential equation whenever m is a solution of the *auxiliary equation*

$$am(m-1) + bm + c = 0, \qquad \text{or} \qquad am^2 + (b-a)m + c = 0. \qquad \textbf{(2.7.8)}$$

At this point we must consider three different cases that depend upon the values of a, b, and c.

• *Distinct real roots*

Let m_1 and m_2 denote the real roots of Equation 2.7.8 such that $m_1 \neq m_2$. Then,

$$y_1(x) = x^{m_1} \qquad \text{and} \qquad y_2(x) = x^{m_2} \qquad \textbf{(2.7.9)}$$

are homogeneous solutions to Equation 2.7.2. Therefore, the general solution is

$$y(x) = C_1 x^{m_1} + C_2 x^{m_2}. \qquad \textbf{(2.7.10)}$$

• *Repeated real roots*

If the roots of Equation 2.7.8 are repeated $[m_1 = m_2 = -(b-a)/2]$, then we presently have only one solution, $y = x^{m_1}$. To construct the second solution y_2, we use reduction in order. We begin by first rewriting the Euler-Cauchy equation as

$$\frac{d^2y}{dx^2} + \frac{b}{ax} \frac{dy}{dx} + \frac{c}{ax^2} y = 0. \qquad \textbf{(2.7.11)}$$

Letting $P(x) = b/(ax)$, we have

$$y_2(x) = x^{m_1} \int \frac{e^{-\int [b/(ax)]\, dx}}{(x^{m_1})^2}\, dx = x^{m_1} \int \frac{e^{-(b/a)\ln(x)}}{x^{2m_1}}\, dx \qquad \textbf{(2.7.12)}$$

$$= x^{m_1} \int x^{-b/a} x^{-2m_1}\, dx = x^{m_1} \int x^{-b/a} x^{(b-a)/a}\, dx \qquad \textbf{(2.7.13)}$$

$$= x^{m_1} \int \frac{dx}{x} = x^{m_1} \ln(x). \qquad \textbf{(2.7.14)}$$

The general solution is then

$$y(x) = C_1 x^{m_1} + C_2 x^{m_1} \ln(x). \qquad \textbf{(2.7.15)}$$

For higher-order equations, if m_1 is a root of multiplicity k, then it can be shown that

$$x^{m_1}, x^{m_1} \ln(x), x^{m_1} [\ln(x)]^2, \ldots, x^{m_1} [\ln(x)]^{k-1}$$

are the k linearly independent solutions. Therefore, the general solution of the differential equation equals a linear combination of these k solutions.

• *Conjugate complex roots*

If the roots of Equation 2.7.8 are the complex conjugate pair $m_1 = \alpha + i\beta$, and $m_2 = \alpha - i\beta$, where α and β are real and $\beta > 0$, then a solution is

$$y(x) = C_1 x^{\alpha + i\beta} + C_2 x^{\alpha - i\beta}. \qquad (2.7.16)$$

However, because $x^{i\theta} = [e^{\ln(x)}]^{i\theta} = e^{i\theta \ln(x)}$, we have by Euler's formula

$$x^{i\theta} = \cos[\theta \ln(x)] + i \sin[\theta \ln(x)], \quad \text{and} \quad x^{-i\theta} = \cos[\theta \ln(x)] - i \sin[\theta \ln(x)]. \qquad (2.7.17)$$

Substitution into Equation 2.7.16 leads to

$$y(x) = C_3 x^{\alpha} \cos[\beta \ln(x)] + C_4 x^{\alpha} \sin[\beta \ln(x)], \qquad (2.7.18)$$

where $C_3 = C_1 + C_2$, and $C_4 = iC_1 - iC_2$.

• **Example 2.7.1**

Let us find the general solution to

$$x^2 y'' + 5xy' - 12y = \ln(x) \qquad (2.7.19)$$

by the method of undetermined coefficients and variation of parameters.

In the case of undetermined coefficients, we begin by letting $t = \ln(x)$ and $y(x) = Y(t)$. Substituting these variables into Equation 2.7.19, we find that

$$Y'' + 4Y' - 12Y = t. \qquad (2.7.20)$$

The homogeneous solution to Equation 2.7.20 is

$$Y_H(t) = A'e^{-6t} + B'e^{2t}, \qquad (2.7.21)$$

while the particular solution is

$$Y_p(t) = Ct + D \qquad (2.7.22)$$

from the method of undetermined coefficients. Substituting Equation 2.7.22 into Equation 2.7.20 yields $C = -\frac{1}{12}$ and $D = -\frac{1}{36}$. Therefore,

$$Y(t) = A'e^{-6t} + B'e^{2t} - \tfrac{1}{12}t - \tfrac{1}{36}, \quad \text{or} \quad y(x) = \frac{A}{x^6} + Bx^2 - \tfrac{1}{12}\ln(x) - \tfrac{1}{36}. \qquad (2.7.23)$$

To find the particular solution via variation of parameters, we use the homogeneous solution

$$y_H(x) = \frac{A}{x^6} + Bx^2 \qquad (2.7.24)$$

to obtain $y_1(x) = x^{-6}$ and $y_2(x) = x^2$. Therefore,

$$y_p(x) = x^{-6} u_1(x) + x^2 u_2(x). \qquad (2.7.25)$$

Substitution of Equation 2.7.25 in Equation 2.7.19 yields the system of equations:

$$x^{-6}u_1'(x) + x^2 u_2'(x) = 0, \quad \text{and} \quad -6x^{-7}u_1'(x) + 2xu_2'(x) = \ln(x)/x^2. \qquad (\mathbf{2.7.26})$$

Solving for $u_1'(x)$ and $u_2'(x)$,

$$u_1'(x) = -\frac{x^5 \ln(x)}{8}, \quad \text{and} \quad u_2'(x) = -\frac{\ln(x)}{8x^3}. \qquad (\mathbf{2.7.27})$$

The solutions of these equations are

$$u_1(x) = -\frac{x^6 \ln(x)}{48} + \frac{x^6}{288}, \quad \text{and} \quad u_2(x) = -\frac{\ln(x)}{16x^2} - \frac{1}{32x^2}. \qquad (\mathbf{2.7.28})$$

The general solution then equals

$$y(x) = \frac{A}{x^6} + Bx^2 + x^{-6}u_1(x) + x^2 u_2(x) = \frac{A}{x^6} + Bx^2 - \tfrac{1}{12}\ln(x) - \tfrac{1}{36}. \qquad (\mathbf{2.7.29})$$

We can verify this result by using the symbolic toolbox in MATLAB. Typing the command:

```
dsolve('x^2*D2y+5*x*Dy-12*y=log(x)','x')
```

yields:

```
ans =
-1/12*log(x)-1/36+C1*x^2+C2/x^6
```

Problems

Find the general solution for the following Euler-Cauchy equations valid over the domain $(-\infty, \infty)$. Then check your answer by using `dsolve` in MATLAB.

1. $x^2 y'' + xy' - y = 0$
2. $x^2 y'' + 2xy' - 2y = 0$
3. $x^2 y'' - 2y = 0$

4. $x^2 y'' - xy' + y = 0$
5. $x^2 y'' + 3xy' + y = 0$
6. $x^2 y'' - 3xy' + 4y = 0$

7. $x^2 y'' - y' + 5y = 0$
8. $4x^2 y'' + 8xy' + 5y = 0$
9. $x^2 y'' + xy' + y = 0$

10. $x^2 y'' - 3xy' + 13y = 0$
11. $x^3 y''' - 2x^2 y'' - 2xy' + 8y = 0$
12. $x^2 y'' - 2xy' - 4y = x$

2.8 PHASE DIAGRAMS

In Section 1.6 we showed how solutions to first-order ordinary differential equations could be *qualitatively* solved through the use of the phase line. This concept of qualitatively studying differential equations showed promise as a method for deducing many of the characteristics of the solution to a differential equation without actually solving it. In this section we extend these concepts to second-order ordinary differential equations by introducing the *phase plane*.

Consider the differential equation

$$x'' + \text{sgn}(x) = 0, \qquad (\mathbf{2.8.1})$$

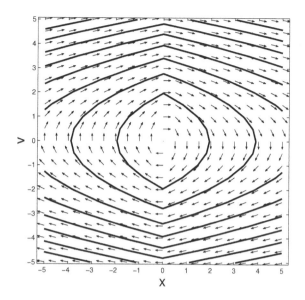

Figure 2.8.1: Phase diagram for the differential equation, Equation 2.8.1.

where the signum function is defined by

$$\operatorname{sgn}(t) = \begin{cases} 1, & t > 0, \\ 0, & t = 0, \\ -1, & t < 0. \end{cases}$$

Equation 2.8.1 describes, for example, the motion of an infinitesimal ball rolling in a "V"-shaped trough in a constant gravitational field.[7]

Our analysis begins by introducing the new dependent variable $v = x'$ so that Equation 2.8.1 can be written

$$v \frac{dv}{dx} + \operatorname{sgn}(x) = 0, \tag{2.8.2}$$

since

$$x'' = \frac{d^2 x}{dt^2} = \frac{dv}{dt} = \frac{dx}{dt}\frac{dv}{dx} = v \frac{dv}{dx}. \tag{2.8.3}$$

Equation 2.8.2 relates v to x and t has disappeared explicitly from the problem. Integrating Equation 2.8.2 with respect to x, we obtain

$$\int v\, dv + \int \operatorname{sgn}(x)\, dx = C, \qquad \text{or} \qquad \tfrac{1}{2}v^2 + |x| = C. \tag{2.8.4}$$

Equation 2.8.4 expresses conservation of energy because the first term on the left side of this equation expresses the kinetic energy while the second term gives the potential energy. The value of C depends upon the initial condition $x(0)$ and $v(0)$. Thus, for a specific initial condition, our equation gives the relationship between x and v for the motion corresponding to the initial condition.

Although there is a closed-form solution for Equation 2.8.1, let us imagine that there is none. What could we learn from Equation 2.8.4?

Equation 2.8.4 can be represented in a diagram, called a *phase plane*, where x and v are its axes. A given pair of (x, v) is called a *state* of the system. A given state determines all subsequent states because it serves as initial conditions for any subsequent motion.

---

[7] See Lipscomb, T., and R. E. Mickens, 1994: Exact solution to the axisymmetric, constant force oscillator equation. *J. Sound Vib.*, **169**, 138–140.

For each different value of C, we will obtain a curve, commonly known as *phase paths,
trajectories*, or *integral curves*, on the phase plane. In Figure 2.8.1, we used the MATLAB
script:

```
clear
% set up grid points in the (x,v) plane
[x,v] = meshgrid(-5:0.5:5,-5:0.5:5);
% compute slopes
dxdt = v; dvdt = -sign(x);
% find magnitude of vector [dxdt,dydt]
L = sqrt(dxdt.*dxdt + dvdt.*dvdt);
% plot scaled vectors
quiver(x,v,dxdt./L,dvdt./L,0.5); axis equal tight
hold
% contour trajectories
contour(x,v,v.*v/2 + abs(x),8)
h = findobj('Type','patch'); set(h,'Linewidth',2);
xlabel('x','Fontsize',20); ylabel('v','Fontsize',20)
```

to graph the phase plane for Equation 2.8.1. Here the phase paths are simply closed, oval-
shaped curves that are symmetric with respect to both the x and v *phase space axes*. Each
phase path corresponds to a particular possible motion of the system. Associated with each
path is a direction, indicated by an arrow, showing how the state of the system changes as
time increases.

An interesting feature on Figure 2.8.1 is the point $(0,0)$. What is happening there? In
our discussion of the phase line, we sought to determine whether there were any *equilibrium*
or *critical points*. Recall that at an equilibrium or critical point, the solution is constant
and was given by $x' = 0$. In the case of second-order differential equations, we again have
the condition $x' = v = 0$. For this reason, equilibrium points are always situated on the
abscissa of the phase diagram.

The condition $x' = 0$ is insufficient for determining critical points. For example, when
a ball is thrown upward, its velocity equals zero at the peak height. However, this is clearly
not a point of equilibrium. Consequently, we must impose the additional constraint that
$x'' = v' = 0$. In the present example, equilibrium points occur where $x' = v = 0$ and
$v' = -\text{sgn}(x) = 0$ or $x = 0$. Therefore, the point $(0,0)$ is the critical point for Equation
2.8.1.

The closed curves immediately surrounding the origin in Figure 2.8.1 show that we
have periodic solutions there because on completing a circuit, the original state returns and
the motion simply repeats itself indefinitely.

Once we have found an equilibrium point, an obvious question is whether it is stable
or not. To determine this, consider what happens if the initial state is displaced slightly
from the origin. It lands on one of the nearby closed curves and the particle oscillates with
small amplitude about the origin. Thus, this critical point is *stable*.

In the following examples, we further illustrate the details that may be gleaned from a
phase diagram.

- **Example 2.8.1**

The equation describing a simple pendulum is

$$ma^2\theta'' + mga\sin(\theta) = 0, \tag{2.8.5}$$

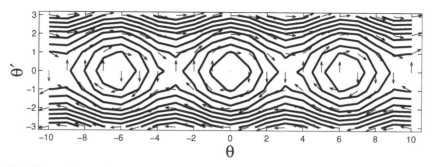

Figure 2.8.2: Phase diagram for a simple pendulum.

where m denotes the mass of the bob, a is the length of the rod or light string, and g is the acceleration due to gravity. Here the conservation of energy equation is

$$\tfrac{1}{2}ma^2\theta'^2 - mga\cos(\theta) = C. \qquad (\mathbf{2.8.6})$$

Figure 2.8.2 is the phase diagram for the simple pendulum. Some of the critical points are located at $\theta = \pm 2n\pi$, $n = 0, 1, 2, \ldots$, and $\theta' = 0$. Near these critical points, we have closed patterns surrounding these critical points, just as we did in the earlier case of an infinitesimal ball rolling in a "V"-shaped trough. Once again, these critical points are *stable* and the region around these equilibrium points corresponds to a pendulum swinging to and fro about the vertical. On the other hand, there is a new type of critical point at $\theta = \pm(2n-1)\pi$, $n = 0, 1, 2, \ldots$ and $\theta' = 0$. Here the trajectories form hyperbolas near these equilibrium points. Thus, for any initial state that is near these critical points, we have solutions that move away from the equilibrium point. This is an example of an *unstable* critical point. Physically these critical points correspond to a pendulum that is balanced on end. Any displacement from the equilibrium results in the bob falling from the inverted position.

Finally, we have a wavy line as $\theta' \to \pm\infty$. This corresponds to whirling motions of the pendulum where θ' has the same sign and θ continuously increases or decreases. $\square$

- **Example 2.8.2: Damped harmonic oscillator**

Consider the ordinary differential equation

$$x'' + 2x' + 5x = 0. \qquad (\mathbf{2.8.7})$$

The exact solution to this differential equation is

$$x(t) = e^{-t}\left[A\cos(2t) + B\sin(2t)\right], \qquad (\mathbf{2.8.8})$$

and

$$x'(t) = 2e^{-t}\left[B\cos(2t) - A\sin(2t)\right] - e^{-t}\left[A\cos(2t) + B\sin(2t)\right]. \qquad (\mathbf{2.8.9})$$

To construct its phase diagram, we again define $v = x'$ and replace Equation 2.8.7 with $v' = -2v - 5x$. The MATLAB script:

```
clear
% set up grid points in the x,x' plane
[x,v] = meshgrid(-3:0.5:3,-3:0.5:3);
% compute slopes
```

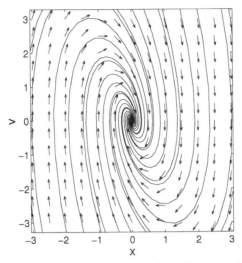

Figure 2.8.3: Phase diagram for the damped harmonic oscillator, Equation 2.8.7.

```
dxdt = v; dvdt = -2*v - 5*x;
% find length of vector
L = sqrt(dxdt.*dxdt + dvdt.*dvdt);
% plot direction field
quiver(x,v,dxdt./L,dvdt./L,0.5); axis equal tight
hold
% compute x(t) and v(t) at various times and a's and b's
for b = -3:2:3; for a = -3:2:3;
  t = [-5:0.1:5];
  xx = exp(-t) .* (a*cos(2*t) + b*sin(2*t));
  vv = 2 * exp(-t) .* (b*cos(2*t) - a*sin(2*t)) - xx;
% plot these values
  plot(xx,vv)
end; end;
xlabel('x','Fontsize',20); ylabel('v','Fontsize',20)
```

was used to construct the phase diagram for Equation 2.8.7 and is shown in Figure 2.8.3. Here the equilibrium point is at $x = v = 0$. This is a new type of critical point. It is called a *stable node* because all slight displacements from this critical point eventually return to this equilibrium point.

Problems

1. Using MATLAB, construct the phase diagram for $x'' - 3x' + 2x = 0$. What happens around the point $x = v = 0$?

2. Consider the nonlinear differential equation $x'' = x^3 - x$. This equation arises in the study of simple pendulums with swings of moderate amplitude.

(a) Show that the conservation law is

$$\tfrac{1}{2}v^2 - \tfrac{1}{4}x^4 + \tfrac{1}{2}x^2 = C.$$

What is special about $C = 0$ and $C = \tfrac{1}{4}$?

(b) Show that there are three critical points: $x = 0$ and $x = \pm 1$ with $v = 0$.

(c) Using MATLAB, graph the phase diagram with axes x and v.

For the following ordinary differential equations, find the equilibrium points and then classify them. Use MATLAB to draw the phase diagrams.

3. $x'' = 2x'$ 4. $x'' + \operatorname{sgn}(x)x = 0$ 5. $x'' = \begin{cases} 1, & |x| > 2, \\ 0, & |x| < 2. \end{cases}$

2.9 NUMERICAL METHODS

When differential equations cannot be integrated in closed form, numerical methods must be employed. In the finite difference method, the discrete variable x_i or t_i replaces the continuous variable x or t and the differential equation is solved progressively in increments h starting from known initial conditions. The solution is approximate, but with a sufficiently small increment, you can obtain a solution of acceptable accuracy.

Although there are many different finite difference schemes available, we consider here only two methods that are chosen for their simplicity. The interested student may read any number of texts on numerical analysis if he or she wishes a wider view of other possible schemes.

Let us focus on second-order differential equations; the solution of higher-order differential equations follows by analog. In the case of second-order ordinary differential equations, the differential equation can be rewritten as

$$x'' = f(x, x', t), \qquad x_0 = x(0), \quad x_0' = x'(0), \tag{2.9.1}$$

where the initial conditions x_0 and x_0' are assumed to be known.

For the present moment, let us treat the second-order ordinary differential equation

$$x'' = f(x, t), \qquad x_0 = x(0), \quad x_0' = x'(0). \tag{2.9.2}$$

The following scheme, known as the *central difference method*, computes the solution from Taylor expansions at x_{i+1} and x_{i-1}:

$$x_{i+1} = x_i + hx_i' + \tfrac{1}{2}h^2 x_i'' + \tfrac{1}{6}h^3 x_i''' + O(h^4) \tag{2.9.3}$$

and

$$x_{i-1} = x_i - hx_i' + \tfrac{1}{2}h^2 x_i'' - \tfrac{1}{6}h^3 x_i''' + O(h^4), \tag{2.9.4}$$

where h denotes the time interval Δt. Subtracting and ignoring higher-order terms, we obtain

$$x_i' = \frac{x_{i+1} - x_{i-1}}{2h}. \tag{2.9.5}$$

Adding Equation 2.9.3 and Equation 2.9.4 yields

$$x_i'' = \frac{x_{i+1} - 2x_i + x_{i-1}}{h^2}. \tag{2.9.6}$$

In both Equation 2.9.5 and Equation 2.9.6 we ignored terms of $O(h^2)$. After substituting into the differential equation, Equation 2.9.2, Equation 2.9.6 can be rearranged to

$$x_{i+1} = 2x_i - x_{i-1} + h^2 f(x_i, t_i), \qquad i \geq 1, \tag{2.9.7}$$

which is known as the *recurrence formula*.

Consider now the situation when $i = 0$. We note that although we have x_0 we do not have x_{-1}. Thus, to start the computation, we need another equation for x_1. This is supplied by Equation 2.9.3, which gives

$$x_1 = x_0 + hx_0' + \tfrac{1}{2}h^2 x_0'' = x_0 + hx_0' + \tfrac{1}{2}h^2 f(x_0, t_0). \qquad (2.9.8)$$

Once we have computed x_1, then we can switch to Equation 2.9.6 for all subsequent calculations.

In this development we have ignored higher-order terms that introduce what is known as *truncation errors*. Other errors, such as *round-off errors*, are introduced due to loss of significant figures. These errors are all related to the time increment h in a rather complicated manner that is investigated in numerical analysis books. In general, better accuracy is obtained by choosing a smaller h, but the number of computations will then increase together with errors.

• **Example 2.9.1**

Let us solve $x'' - 4x = 2t$ subject to $x(0) = x'(0) = 1$. The exact solution is

$$x(t) = \tfrac{7}{8}e^{2t} + \tfrac{1}{8}e^{-2t} - \tfrac{1}{2}t. \qquad (2.9.9)$$

The MATLAB script:

```
clear
% test out different time steps
for i = 1:3
% set up time step increment and number of time steps
  h = 1/10^i; n = 10/h;
% set up initial conditions
  t=zeros(n+1,1); t(1) = 0; x(1) = 1; x_exact(1) = 1;
% define right side of differential equation
  f = inline('4*xx+2*tt','tt','xx');
% set up difference arrays for plotting purposes
  diff = zeros(n,1); t_plot = zeros(n,1);
% compute first time step
  t(2) = t(1) + h; x(2) = x(1) + h + 0.5*h*h*f(t(1),x(1));
  x_exact(2) = (7/8)*exp(2*t(2))+(1/8)*exp(-2*t(2))-t(2)/2;
  t_plot(1) = t(2);
  diff(1)=x(2)-x_exact(2); diff(1)=abs(diff(1)/x_exact(2));
% compute the remaining time steps
    for k = 2:n
    t(k+1) = t(k) + h; t_plot(k) = t(k+1);
    x(k+1) = 2*x(k) - x(k-1) + h*h*f(t(k),x(k));
    x_exact(k+1) = (7/8)*exp(2*t(k+1))+(1/8)*exp(-2*t(k+1)) ...
                - t(k+1)/2;
    diff(k) = x(k+1) - x_exact(k+1);
    diff(k) = abs(diff(k) / x_exact(k+1));
    end
% plot the relative error
  semilogy(t_plot,diff,'-')
```

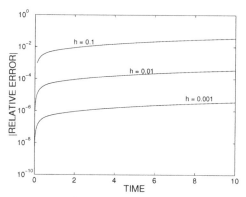

Figure 2.9.1: The numerical solution of $x'' - 4x = 2t$ when $x(0) = x'(0) = 1$ using a simple finite difference approach.

```
hold on
num = 0.2*n;
text(3*i,diff(num),['h = ',num2str(h)],'Fontsize',15,...
   'HorizontalAlignment','right','VerticalAlignment','bottom')
xlabel('TIME','Fontsize',20);
ylabel('|RELATIVE ERROR|','Fontsize',20);
end
```

implements our simple finite difference method of solving a second-order ordinary differential equation. In Figure 2.9.1 we have plotted results for three different values of the time step. As our analysis suggests, the relative error is related to h^2.   □

An alternative method for integrating higher-order ordinary differential equations is Runge-Kutta. It is popular because it is self-starting and the results are very accurate.

For second-order ordinary differential equations, this method first reduces the differential equation into two first-order equations. For example, the differential equation

$$x'' = \frac{f(t) - kx - cx'}{m} = F(x, x', t) \tag{2.9.10}$$

becomes the first-order differential equations

$$x' = y, \qquad y' = F(x, y, t). \tag{2.9.11}$$

The Runge-Kutta procedure can then be applied to each of these equations. Using a fourth-order scheme, the procedure is as follows:

$$x_{i+1} = x_i + \tfrac{1}{6}h(k_1 + 2k_2 + 2k_3 + k_4), \tag{2.9.12}$$

and

$$y_{i+1} = y_i + \tfrac{1}{6}h(K_1 + 2K_2 + 2K_3 + K_4), \tag{2.9.13}$$

where

$$k_1 = y_i, \qquad\qquad K_1 = F(x_i, y_i, t_i), \tag{2.9.14}$$

$$k_2 = y_i + \tfrac{h}{2}K_1, \qquad K_2 = F(x_i + \tfrac{h}{2}k_1, k_2, t_i + \tfrac{h}{2}), \tag{2.9.15}$$

$$k_3 = y_i + \tfrac{h}{2}K_2, \qquad K_3 = F(x_i + \tfrac{h}{2}k_2, k_3, t_i + \tfrac{h}{2}), \tag{2.9.16}$$

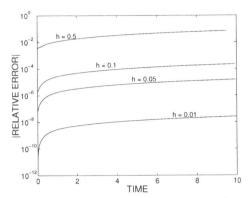

Figure 2.9.2: The numerical solution of $x'' - 4x = 2t$ when $x(0) = x'(0) = 1$ using the Runge-Kutta method.

and

$$k_4 = y_i + K_3 h, \qquad K_4 = F(x_i + h k_3, k_4, t_i + h). \qquad (2.9.17)$$

• **Example 2.9.2**

The MATLAB script:

```
clear
% test out different time steps
for i = 1:4
% set up time step increment and number of time steps
  if i==1 h = 0.50; end; if i==2 h = 0.10; end;
  if i==3 h = 0.05; end; if i==4 h = 0.01; end;
  nn = 10/h;
% set up initial conditions
  t=zeros(n+1,1); t(1) = 0;
  x_rk=zeros(n+1,1); x_rk(1) = 1;
  y_rk=zeros(n+1,1); y_rk(1) = 1;
  x_exact=zeros(n+1,1); x_exact(1) = 1;
% set up difference arrays for plotting purposes
  t_plot = zeros(n,1); diff = zeros(n,1);
% define right side of differential equation
  f = inline('4*xx+2*tt','tt','xx','yy');
  for k = 1:n
    t_local = t(k); x_local = x_rk(k); y_local = y_rk(k);
    k1 = y_local; K1 = f(t_local,x_local,y_local);
    k2 = y_local + h*K1/2;
    K2 = f(t_local + h/2,x_local + h*k1/2,k2);
    k3 = y_local + h*K2/2;
    K3 = f(t_local + h/2,x_local + h*k2/2,k3);
    k4 = y_local + h*K3; K4 = f(t_local + h,x_local + h*k3,k4);
    t(k+1) = t_local + h;
    x_rk(k+1) = x_local + (h/6) * (k1+2*k2+2*k3+k4);
    y_rk(k+1) = y_local + (h/6) * (K1+2*K2+2*K3+K4);
    x_exact(k+1) = (7/8)*exp(2*t(k+1))+(1/8)*exp(-2*t(k+1)) ...
                 - t(k+1)/2;
```

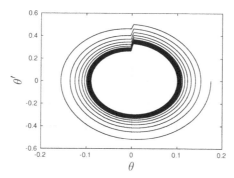

Figure 2.9.3: The phase diagram of the dynamical system given by Equations 2.9.18 and 2.9.19 by modified Euler method. Here the parameters are $g = 9.8$ m/sec^2, $L = 1$ m, $b = 0.22$/sec, $k = 0.02$/sec, and $\Delta t = 0.005$. The initial conditions are $\theta(0) = \pi/18$ and $\theta'(0) = 0$.

```
  t_plot(k) = t(k);
  diff(k) = x_rk(k+1) - x_exact(k+1);
  diff(k) = abs(diff(k) / x_exact(k+1));
 end
% plot the relative errors
 semilogy(t_plot,diff,'-')
 hold on
 xlabel('TIME','Fontsize',20);
 ylabel('|RELATIVE ERROR|','Fontsize',20);
 text(2*i,diff(0.2*n),['h = ',num2str(h)],'Fontsize',15,...
    'HorizontalAlignment','right','VerticalAlignment','bottom')
end
```

was used to resolve Example 2.9.1 using the Runge-Kutta approach. Figure 2.9.2 illustrates the results for time steps of various sizes.

Problems

In previous sections, you found exact solutions to second-order ordinary differential equations. Confirm these earlier results by using MATLAB and the Runge-Kutta scheme to find the numerical solution to the following problems drawn from previous sections.

1. Section 2.1, Problem 1

2. Section 2.1, Problem 5

3. Section 2.4, Problem 1

4. Section 2.4, Problem 5

5. Section 2.6, Problem 1

6. Section 2.6, Problem 5

Project: Pendulum Clock

In his exposition on pendulum clocks, M. Denny[8] modeled the system by the second-order differential equation in time t:

$$\theta'' + b\theta' + \omega_0^2\theta = kf(\theta, \theta'), \tag{2.9.18}$$

---

[8] Denny, M., 2002: The pendulum clock: A venerable dynamical system. *Eur. J. Phys.*, **23**, 449–458.

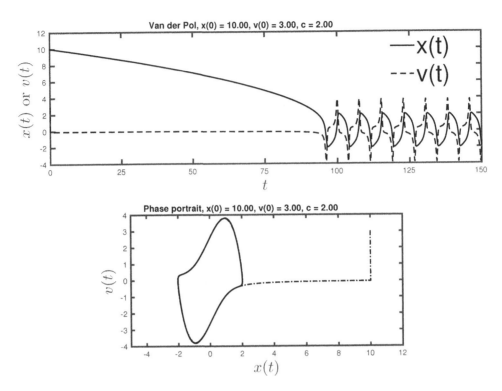

Figure 2.9.4: $x(t)$ and $v(t)$ for van der Pol's oscillator with $A = 0$ when numerically integrated using the Runge-Kutta scheme with a time step of $\Delta t = 0.001$.

where

$$f(\theta, \theta') = \begin{cases} 1/(\Delta t), & |\theta| < \Delta t/2, \quad \theta' > 0; \\ 0, & \text{otherwise;} \end{cases} \qquad (2.9.19)$$

and $\omega^2 = g/L - b^2/4$. Here Δt denotes some arbitrarily small nondimensional time. In Chapter 7 we identify this forcing as the Dirac delta function.

Using the numerical scheme of your choice, develop a MATLAB code to numerically integrate this differential equation. Plot the results as a phase diagram with θ as the abscissa and θ' as the ordinate. What happens with time? What happens as k varies? Figure 2.9.3 illustrates the solution.

Project: Van der Pol Oscillator

In the 1920s and 1930s, Balthazar van der Pol (1889–1959), a Dutch physicist and electrical engineer, studied simple electrical circuits that included vacuum tubes during his employment at Phillips Lab. He modeled the response of the system to the forcing of $A \sin(\omega t)$ by the second-order, nonlinear, ordinary differential equation:

$$\frac{d^2 x(t)}{dt^2} - c[1 - x^2(t)]\frac{dx(t)}{dt} + x(t) = A \sin(\omega t), \qquad c > 0,$$

commonly referred to as van der Pol's equation. Because of its nonlinearity, the general solution must be found numerically. Generally the second-order differential equation is

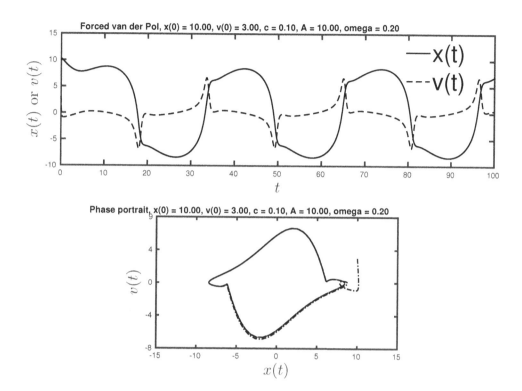

Figure 2.9.5: Same as Figure 2.9.4 except that now $A = 10$ and $\omega = 0.2$.

replaced by two first-order differential equations:

$$\frac{dx(t)}{dt} = v(t), \qquad \frac{dv(t)}{dt} = c[1 - x^2(t)]v(t) - x(t) + A\sin(\omega t),$$

before a particular numerical scheme is employed. The purpose of this project is to analyze van der Pol's equation via numerical integration.

Step 1: Show that our system of first-order equations is equivalent to Van der Pol's equation.

Step 2: Using MATLAB's routine **ode45**, write code which solves van der Pol's equation for given $x(0)$, $v(0)$, c, A, ω and time step Δt.

Step 3: Conduct numerical experiments when $A = 0$ for various values of $c \geq 0$ and discuss differences in the solution as the parameter c varies. In particular, plot $x(t)$ and $v(t)$ as a function of t and plot the corresponding phase diagram.

Step 4: Now choose an A and ω and repeat Step 3.

Further Readings

Boyce, W. E., and R. C. DiPrima, 2004: *Elementary Differential Equations and Boundary Value Problems.* Wiley, 800 pp. Classic textbook.

Ince, E. L., 1956: *Ordinary Differential Equations.* Dover, 558 pp. The source book on ordinary differential equations.

Zill, D. G., and M. R. Cullen, 2008: *Differential Equations with Boundary-Value Problems.* Brooks Cole, 640 pp. Nice undergraduate textbook.

$$\begin{pmatrix} 1 & \gamma_{12} & \cdots & \gamma_{1k} & \gamma_{1k+1} & \cdots & \gamma_{1n} \\ 0 & 1 & \cdots & \gamma_{2k} & \gamma_{2k+1} & \cdots & \gamma_{2n} \\ & & & \vdots & & & \\ 0 & 0 & \cdots & 1 & \gamma_{kk+1} & \cdots & \gamma_{kn} \\ 0 & 0 & \cdots & 0 & 0 & \cdots & 0 \\ & & & \vdots & & & \\ 0 & 0 & \cdots & 0 & 0 & \cdots & 0 \end{pmatrix}$$

Chapter 3

Linear Algebra

Linear algebra was developed for the systematic solving of linear algebraic or differential equations where we have several unknowns. These equations are written and solved using the concepts of *matrices* and *vectors*. In this chapter we explore these concepts and their use in solving linear equations.

3.1 FUNDAMENTALS

Consider the following system of m simultaneous linear equations in n unknowns $x_1, x_2, x_3, \ldots, x_n$:

$$a_{11}x_1 + a_{12}x_2 + \cdots + a_{1n}x_n = b_1,$$

$$a_{21}x_1 + a_{22}x_2 + \cdots + a_{2n}x_n = b_2,$$

$$\vdots \qquad\qquad (3.1.1)$$

$$a_{m1}x_1 + a_{m2}x_2 + \cdots + a_{mn}x_n = b_m,$$

where the coefficients a_{ij} and constants b_j denote known real or complex numbers. The purpose of this chapter is to show how *matrix algebra* can be used to solve these systems by first introducing succinct notation so that we can replace Equation 3.1.1 with rather simple expressions, and then by employing a set of rules to manipulate these expressions. In this section we focus on developing these simple expressions.

One of the fundamental quantities in linear algebra is the *matrix*.[1] A matrix is an ordered rectangular array of numbers or mathematical expressions. We shall use uppercase

---

[1] This term was first used by J. J. Sylvester, 1850: Additions to the articles, "On a new class of theorems," and "On Pascal's theorem." *Philos. Mag., Ser. 4*, **37**, 363–370.

letters to denote them. The $m \times n$ matrix

$$
A = \begin{pmatrix}
a_{11} & a_{12} & a_{13} & \cdots & \cdots & \cdots & a_{1n} \\
a_{21} & a_{22} & a_{23} & \cdots & \cdots & \cdots & a_{2n} \\
\vdots & \vdots & \vdots & \vdots & \vdots & \vdots & \vdots \\
\vdots & \vdots & \vdots & \vdots & a_{ij} & \vdots & \vdots \\
\vdots & \vdots & \vdots & \vdots & \vdots & \vdots & \vdots \\
a_{m1} & a_{m2} & a_{m3} & \vdots & \vdots & \vdots & a_{mn}
\end{pmatrix}
\tag{3.1.2}
$$

has m *rows* and n *columns*. The *order* (or size) of a matrix is determined by the number of rows and columns; Equation 3.1.2 is of order m by n. If $m = n$, the matrix is a *square* matrix; otherwise, A is *rectangular*. The numbers or expressions in the array a_{ij} are the *elements* of A and can be either real or complex. When all of the elements are real, A is a *real matrix*. If some or all of the elements are complex, then A is a *complex matrix*. For a square matrix, the diagonal from the top left corner to the bottom right corner is the *principal diagonal*. The *trace* (often abbreviated to tr) of a square matrix is defined to be the sum of elements on the main diagonal (from the upper left to the lower right).

From the limitless number of possible matrices, certain ones appear with sufficient regularity that they are given special names. A *zero* matrix (sometimes called a *null* matrix) has all of its elements equal to zero. It fulfills the role in matrix algebra that is analogous to that of zero in scalar algebra. The *unit* or *identity* matrix is an $n \times n$ matrix having 1's along its principal diagonal and zero everywhere else. The unit matrix serves essentially the same purpose in matrix algebra as does the number one in scalar algebra. A *symmetric* matrix is one where $a_{ij} = a_{ji}$ for all i and j.

Two matrices A and B are equal if and only if $a_{ij} = b_{ij}$ for all possible i and j and they have the same dimensions.

● **Example 3.1.1**

Examples of zero, identity, and symmetric matrices are

$$
O = \begin{pmatrix} 0 & 0 & 0 \\ 0 & 0 & 0 \\ 0 & 0 & 0 \end{pmatrix}, \quad
I = \begin{pmatrix} 1 & 0 \\ 0 & 1 \end{pmatrix}, \quad \text{and} \quad
A = \begin{pmatrix} 3 & 2 & 4 \\ 2 & 1 & 0 \\ 4 & 0 & 5 \end{pmatrix},
\tag{3.1.3}
$$

respectively.   □

Having defined a matrix, let us explore some of its arithmetic properties. For two matrices A and B with the same dimensions (conformable for addition), the matrix $C = A + B$ contains the elements $c_{ij} = a_{ij} + b_{ij}$. Similarly, $C = A - B$ contains the elements $c_{ij} = a_{ij} - b_{ij}$. Because the order of addition does not matter, addition is *commutative*: $A + B = B + A$.

Consider now a scalar constant k. The product kA is formed by multiplying every element of A by k. Thus the matrix kA has elements ka_{ij}.

So far, the rules for matrix arithmetic conform to their scalar counterparts. However, there are several possible ways of multiplying two matrices together. For example, we might simply multiply together the corresponding elements from each matrix. This is *not* correct and we now derive the *right* rule.

We begin by requiring that the dimensions of A be $m \times n$ while for B they are $n \times p$. That is, the number of columns in A must equal the number of rows in B. The matrices A

and B are then said to be *conformable* for multiplication. If this is true, then $C = AB$ is a matrix $m \times p$, where its elements equal

$$c_{ij} = \sum_{k=1}^{n} a_{ik}\, b_{kj}. \tag{3.1.4}$$

The right side of Equation 3.1.4 is referred to as an *inner product* of the ith row of A and the jth column of B. Although Equation 3.1.4 is the method used with a computer, an easier method for human computation is as a running sum of the products given by successive elements of the ith row of A and the corresponding elements of the jth column of B.

The product AA is usually written A^2; the product AAA, A^3, and so forth.

• **Example 3.1.2**

If

$$A = \begin{pmatrix} -1 & 4 \\ 2 & -3 \end{pmatrix}, \quad \text{and} \quad B = \begin{pmatrix} 1 & 2 \\ 3 & 4 \end{pmatrix}, \tag{3.1.5}$$

then

$$AB = \begin{pmatrix} [(-1)(1) + (4)(3)] & [(-1)(2) + (4)(4)] \\ [(2)(1) + (-3)(3)] & [(2)(2) + (-3)(4)] \end{pmatrix} = \begin{pmatrix} 11 & 14 \\ -7 & -8 \end{pmatrix}. \tag{3.1.6}$$

Checking our results using MATLAB, we have that:

```
>> A = [-1 4; 2 -3];
>> B = [1 2; 3 4];
>> C = A*B
C =
        11   14
        -7   -8
```

Note that there is a tremendous difference between the MATLAB command for matrix multiplication ∗ and element-by-element multiplication .∗. □

Matrix multiplication is associative and distributive with respect to addition:

$$(kA)B = k(AB) = A(kB), \tag{3.1.7}$$

$$A(BC) = (AB)C, \tag{3.1.8}$$

$$(A + B)C = AC + BC, \tag{3.1.9}$$

and

$$C(A + B) = CA + CB. \tag{3.1.10}$$

On the other hand, matrix multiplication is *not commutative*. In general, $AB \neq BA$.

• **Example 3.1.3**

Does $AB = BA$ if

$$A = \begin{pmatrix} 1 & 0 \\ 0 & 0 \end{pmatrix}, \quad \text{and} \quad B = \begin{pmatrix} 1 & 1 \\ 1 & 0 \end{pmatrix}? \tag{3.1.11}$$

Because

$$AB = \begin{pmatrix} 1 & 0 \\ 0 & 0 \end{pmatrix} \begin{pmatrix} 1 & 1 \\ 1 & 0 \end{pmatrix} = \begin{pmatrix} 1 & 1 \\ 0 & 0 \end{pmatrix}, \tag{3.1.12}$$

and

$$BA = \begin{pmatrix} 1 & 1 \\ 1 & 0 \end{pmatrix} \begin{pmatrix} 1 & 0 \\ 0 & 0 \end{pmatrix} = \begin{pmatrix} 1 & 0 \\ 1 & 0 \end{pmatrix}, \tag{3.1.13}$$

$$AB \neq BA. \tag{3.1.14}$$

$\square$

• **Example 3.1.4**

Given

$$A = \begin{pmatrix} 1 & 1 \\ 3 & 3 \end{pmatrix}, \quad \text{and} \quad B = \begin{pmatrix} -1 & 1 \\ 1 & -1 \end{pmatrix}, \tag{3.1.15}$$

find the product AB.

Performing the calculation, we find that

$$AB = \begin{pmatrix} 1 & 1 \\ 3 & 3 \end{pmatrix} \begin{pmatrix} -1 & 1 \\ 1 & -1 \end{pmatrix} = \begin{pmatrix} 0 & 0 \\ 0 & 0 \end{pmatrix}. \tag{3.1.16}$$

The point here is that just because $AB = 0$, this does *not* imply that either A or B equals the zero matrix. $\square$

We cannot properly speak of division when we are dealing with matrices. Nevertheless, a matrix A is said to be *nonsingular* or *invertible* if there exists a matrix B such that $AB = BA = I$. This matrix B is the multiplicative inverse of A or simply the *inverse* of A, written A^{-1}. An $n \times n$ matrix is *singular* if it does not have a multiplicative inverse.

• **Example 3.1.5**

If

$$A = \begin{pmatrix} 1 & 0 & 1 \\ 3 & 3 & 4 \\ 2 & 2 & 3 \end{pmatrix}, \tag{3.1.17}$$

let us verify that its inverse is

$$A^{-1} = \begin{pmatrix} 1 & 2 & -3 \\ -1 & 1 & -1 \\ 0 & -2 & 3 \end{pmatrix}. \tag{3.1.18}$$

We perform the check by finding AA^{-1} or $A^{-1}A$,

$$AA^{-1} = \begin{pmatrix} 1 & 0 & 1 \\ 3 & 3 & 4 \\ 2 & 2 & 3 \end{pmatrix} \begin{pmatrix} 1 & 2 & -3 \\ -1 & 1 & -1 \\ 0 & -2 & 3 \end{pmatrix} = \begin{pmatrix} 1 & 0 & 0 \\ 0 & 1 & 0 \\ 0 & 0 & 1 \end{pmatrix}. \tag{3.1.19}$$

In a later section we will show how to compute the inverse, given A. $\square$

Another matrix operation is transposition. The *transpose* of a matrix A with dimensions $m \times n$ is another matrix, written A^T, where we interchange the rows and columns from A. In MATLAB, A^T is computed by typing `A'`. Clearly, $(A^T)^T = A$ as well as $(A + B)^T = A^T + B^T$, and $(kA)^T = kA^T$. If A is symmetric, then $A^T = A$. Finally, if A and B are conformable for multiplication, then $(AB)^T = B^T A^T$. Note the reversal of order between the two sides. To prove this last result, we first show that the results are true for two 3×3 matrices A and B and then generalize to larger matrices.

Having explored matrix operations, we now introduce two more fundamental quantities: *column vectors* and *row vectors*:

$$\mathbf{x} = \begin{pmatrix} x_1 \\ x_2 \\ \vdots \\ x_m \end{pmatrix}, \qquad \mathbf{y} = \begin{pmatrix} y_1 & y_2 & \cdots & y_n \end{pmatrix}. \tag{3.1.20}$$

We denote row and column vectors by lowercase, boldfaced letters. They are a special class of matrices, namely an m matrix and a $1 \times n$ matrix, respectively. Consequently, they are *not* the same thing, as MATLAB continuously reminds you. The length or *norm* of the vector $\mathbf{x}$ of n elements is

$$||\mathbf{x}|| = \left(\sum_{k=1}^{n} x_k^2 \right)^{1/2}. \tag{3.1.21}$$

● **Example 3.1.6: Vector space R^n**

Why have we called these special matrices vectors? The answer lies in the strong analog between these n-dimensional vectors and the three-dimensional vectors from physics. Both column and row vectors are examples of n-vectors and consist of n components. If the components are real, then the totality of all n-vectors is denoted by R^n and called an R^n *vector space*. An example of an R^2 vector space consists of all two-dimensional vectors with the components $(x_1 \ x_2)$. Vector spaces enjoy the property that the sum (addition) of two vectors in R^n yields another vector in R^n. Another property is that a vector in R^n can be multiplied by a real number (scalar) that merely scales the vector. □

Having introduced some of the basic concepts of linear algebra, we are ready to rewrite Equation 3.1.1 in a canonical form so that we can present techniques for its solution. We begin by writing Equation 3.1.1 as a single column vector:

$$\begin{pmatrix} a_{11}x_1 & + & a_{12}x_2 & + & \cdots & + & a_{1n}x_n \\ a_{21}x_1 & + & a_{22}x_2 & + & \cdots & + & a_{2n}x_n \\ \vdots & & \vdots & & \vdots & & \vdots \\ \vdots & & \vdots & & \vdots & & \vdots \\ a_{m1}x_1 & + & a_{m2}x_2 & + & \cdots & + & a_{mn}x_n \end{pmatrix} = \begin{pmatrix} b_1 \\ b_2 \\ \vdots \\ \vdots \\ b_m \end{pmatrix}. \tag{3.1.22}$$

We now use the multiplication rule to rewrite Equation 3.1.22 as

$$\begin{pmatrix} a_{11} & a_{12} & \cdots & a_{1n} \\ a_{21} & a_{22} & \cdots & a_{2n} \\ \vdots & \vdots & \vdots & \vdots \\ \vdots & \vdots & \vdots & \vdots \\ a_{m1} & a_{m2} & \cdots & a_{mn} \end{pmatrix} \begin{pmatrix} x_1 \\ x_2 \\ \vdots \\ \vdots \\ x_n \end{pmatrix} = \begin{pmatrix} b_1 \\ b_2 \\ \vdots \\ \vdots \\ b_m \end{pmatrix}, \tag{3.1.23}$$

or

$$Ax = b, \tag{3.1.24}$$

where x is the solution vector. If $b = 0$, we have a *homogeneous* set of equations; otherwise, we have a *nonhomogeneous* set. In the next few sections, we will give a number of methods for finding x.

- **Example 3.1.7: Solution of a tridiagonal system**

A common problem in linear algebra involves solving systems such as

$$b_1 y_1 + c_1 y_2 = d_1, \tag{3.1.25}$$
$$a_2 y_1 + b_2 y_2 + c_2 y_3 = d_2, \tag{3.1.26}$$
$$\vdots$$
$$a_{N-1} y_{N-2} + b_{N-1} y_{N-1} + c_{N-1} y_N = d_{N-1}, \tag{3.1.27}$$
$$b_N y_{N-1} + c_N y_N = d_N. \tag{3.1.28}$$

Such systems arise in the numerical solution of ordinary and partial differential equations. We begin by rewriting Equation 3.1.25 through Equation 3.1.28 in the matrix notation:

$$\begin{pmatrix} b_1 & c_1 & 0 & \cdots & 0 & 0 & 0 \\ a_2 & b_2 & c_2 & \cdots & 0 & 0 & 0 \\ 0 & a_3 & b_3 & \cdots & 0 & 0 & 0 \\ \vdots & \vdots & \vdots & \ddots & \vdots & \vdots & \vdots \\ 0 & 0 & 0 & \cdots & a_{N-1} & b_{N-1} & c_{N-1} \\ 0 & 0 & 0 & \cdots & 0 & a_N & b_N \end{pmatrix} \begin{pmatrix} y_1 \\ y_2 \\ y_3 \\ \vdots \\ y_{N-1} \\ y_N \end{pmatrix} = \begin{pmatrix} d_1 \\ d_2 \\ d_3 \\ \vdots \\ d_{N-1} \\ d_N \end{pmatrix}. \tag{3.1.29}$$

The matrix in Equation 3.1.29 is an example of a *banded matrix*: a matrix where all of the elements in each row are zero except for the diagonal element and a limited number on either side of it. In the present case, we have a *tridiagonal* matrix in which only the diagonal element and the elements immediately to its left and right in each row are nonzero.

Consider the nth equation. We can eliminate a_n by multiplying the $(n-1)$th equation by a_n/b_{n-1} and subtracting this new equation from the nth equation. The values of b_n and d_n become

$$b'_n = b_n - a_n c_{n-1}/b_{n-1}, \quad \text{and} \quad d'_n = d_n - a_n d_{n-1}/b_{n-1} \tag{3.1.30}$$

for $n = 2, 3, \ldots, N$. The coefficient c_n is unaffected. Because elements a_1 and c_N are never involved, their values can be anything or they can be left undefined. The new system of equations may be written

$$\begin{pmatrix} b'_1 & c_1 & 0 & \cdots & 0 & 0 & 0 \\ 0 & b'_2 & c_2 & \cdots & 0 & 0 & 0 \\ 0 & 0 & b'_3 & \cdots & 0 & 0 & 0 \\ \vdots & \vdots & \vdots & \ddots & \vdots & \vdots & \vdots \\ 0 & 0 & 0 & \cdots & 0 & b'_{N-1} & c_{N-1} \\ 0 & 0 & 0 & \cdots & 0 & 0 & b'_N \end{pmatrix} \begin{pmatrix} y_1 \\ y_2 \\ y_3 \\ \vdots \\ y_{N-1} \\ y_N \end{pmatrix} = \begin{pmatrix} d'_1 \\ d'_2 \\ d'_3 \\ \vdots \\ d'_{N-1} \\ d'_N \end{pmatrix}. \tag{3.1.31}$$

The matrix in Equation 3.1.31 is in *upper triangular* form because all of the elements below the principal diagonal are zero. This is particularly useful because y_n can be computed by *back substitution*. That is, we first compute y_N. Next, we calculate y_{N-1} in terms of y_N. The solution y_{N-2} can then be computed in terms of y_N and y_{N-1}. We continue this process until we find y_1 in terms of $y_N, y_{N-1}, \ldots, y_2$. In the present case, we have the rather simple:

$$y_N = d'_N/b'_N, \quad \text{and} \quad y_n = (d'_n - c_n d'_{n+1})/b'_n \qquad (3.1.32)$$

for $n = N - 1, N - 2, \ldots, 2, 1$.

As we shall show shortly, this is an example of solving a system of linear equations by Gaussian elimination. For a tridiagonal case, we have the advantage that the solution can be expressed in terms of a recurrence relationship, a very convenient feature from a computational point of view. This algorithm is very robust, being stable[2] as long as $|a_i + c_i| < |b_i|$. By stability, we mean that if we change $\mathbf{b}$ by $\Delta\mathbf{b}$ so that $\mathbf{x}$ changes by $\Delta\mathbf{x}$, then $||\Delta\mathbf{x}|| < M\epsilon$, where $||\Delta\mathbf{b}|| \le \epsilon$, $0 < M < \infty$, for *any* N. Here $||\cdot||$ denotes the norm which is defined by Equation 3.1.21. $\qquad\square$

● **Example 3.1.8: Linear transformation**

Consider a set of linear equations

$$
\begin{aligned}
y_1 &= a_{11}x_1 + a_{12}x_2 + a_{13}x_3 + a_{14}x_4, \\
y_2 &= a_{21}x_1 + a_{22}x_2 + a_{23}x_3 + a_{24}x_4, \\
y_3 &= a_{31}x_1 + a_{31}x_2 + a_{33}x_3 + a_{34}x_4.
\end{aligned}
\qquad (3.1.33)
$$

Each of the right-side expressions is called a *linear combination* of x_1, x_2, x_3, and x_4: a sum where each term consists of a constant times x_i raised to the first power. An expression such as $a_{11}x_1^2 + a_{12}x_2^2 + a_{13}x_3^2 + a_{14}x_4^2$ is an example of a nonlinear combination. Note that we are using 4 values of x_i to find only 3 values of y_i.

If we were given values of x_i, we could determine a set of values for y_i using Equation 3.1.33. Such a set of linear equations that yields values of y_i for given x_i's is called a *linear transform of x into y*. The point here is that given x, the corresponding y will be evaluated.

Matrix notation and multiplication are very convenient in expressing linear transformations. We begin by writing Equation 3.1.33 in matrix format:

$$
\begin{pmatrix} y_1 \\ y_2 \\ y_3 \end{pmatrix} = \begin{pmatrix} a_{11}x_1 + a_{12}x_2 + a_{13}x_3 + a_{14}x_4 \\ a_{21}x_1 + a_{22}x_2 + a_{23}x_3 + a_{24}x_4 \\ a_{31}x_1 + a_{31}x_2 + a_{33}x_3 + a_{34}x_4 \end{pmatrix} = \begin{pmatrix} a_{11} & a_{12} & a_{13} & a_{14} \\ a_{21} & a_{22} & a_{23} & a_{24} \\ a_{31} & a_{32} & a_{33} & a_{34} \end{pmatrix} \begin{pmatrix} x_1 \\ x_2 \\ x_3 \\ x_4 \end{pmatrix},
$$
$$(3.1.34)$$

where we have used the matrix multiplication rule Equation 3.1.4 to replace the middle term with the right term in Equation 3.1.34. Finally, by introducing the (column) vectors $\mathbf{x}$ and $\mathbf{y}$ and the matrix A:

$$
\mathbf{x} = \begin{pmatrix} x_1 \\ x_2 \\ x_3 \\ x_4 \end{pmatrix}, \quad \mathbf{y} = \begin{pmatrix} y_1 \\ y_2 \\ y_3 \end{pmatrix}. \quad \text{and} \quad A = \begin{pmatrix} a_{11} & a_{12} & a_{13} & a_{14} \\ a_{21} & a_{22} & a_{23} & a_{24} \\ a_{31} & a_{32} & a_{33} & a_{34} \end{pmatrix}, \qquad (3.1.35)
$$

---

[2] Torii, T., 1966: Inversion of tridiagonal matrices and the stability of tridiagonal systems of linear systems. *Tech. Rep. Osaka Univ.*, **16**, 403–414.

we have the compact expression $\mathbf{y} = A\mathbf{x}$ for the linear transformation expressed by Equation 3.1.33.

Let us now introduce the notation $T(\mathbf{x}) = A\mathbf{x}$. Then, for any vectors $\mathbf{u}$ and $\mathbf{v}$ in the vector space R^n, we have a linear transformation if we satisfy two conditions: (1) $T(\mathbf{u} + \mathbf{v}) = T(\mathbf{u}) + T(\mathbf{v})$ and (2) $T(k\mathbf{u}) = kT(\mathbf{u})$, where k is a scalar.

Problems

Given $A = \begin{pmatrix} 3 & 4 \\ 1 & 2 \end{pmatrix}$, and $B = \begin{pmatrix} 1 & 1 \\ 2 & 2 \end{pmatrix}$, find

1. $A + B$, $B + A$ 2. $A - B$, $B - A$ 3. $3A - 2B$, $3(2A - B)$

4. $A^T, B^T, (B^T)^T$ 5. $(A + B)^T$, $A^T + B^T$ 6. $B + B^T$, $B - B^T$

7. $AB, A^T B, BA, B^T A$ 8. A^2, B^2 9. BB^T, $B^T B$

10. $A^2 - 3A + I$ 11. $A^3 + 2A$ 12. $A^4 - 4A^2 + 2I$

by hand and using MATLAB.

Can multiplication occur between the following matrices? If so, compute it.

13. $\begin{pmatrix} 3 & 5 & 1 \\ -2 & 1 & 2 \end{pmatrix} \begin{pmatrix} 2 & 1 \\ 4 & 1 \\ 1 & 3 \end{pmatrix}$ 14. $\begin{pmatrix} -2 & 4 \\ -4 & 6 \\ -6 & 1 \end{pmatrix} (1 \quad 2 \quad 3)$

15. $\begin{pmatrix} 1 & 4 & 2 \\ 0 & 0 & 4 \\ 0 & 1 & 2 \end{pmatrix} \begin{pmatrix} 3 & 2 \\ 1 & 1 \\ 2 & 1 \end{pmatrix}$ 16. $\begin{pmatrix} 4 & 6 \\ 1 & 2 \end{pmatrix} \begin{pmatrix} 1 & 3 & 6 \\ 1 & 2 & 5 \end{pmatrix}$

17. $\begin{pmatrix} 6 & 4 & 2 \\ 1 & 2 & 3 \end{pmatrix} \begin{pmatrix} 3 & 1 & 4 \\ 2 & 0 & 6 \end{pmatrix}$ 18. $\begin{pmatrix} -2 & 4 \\ 3 & 9 \end{pmatrix} \begin{pmatrix} 1 & -3 & 7 & 2 \\ -5 & 6 & 1 & 0 \end{pmatrix}$

If $A = \begin{pmatrix} 1 & 1 \\ 1 & 2 \\ 3 & 1 \end{pmatrix}$, verify that

19. $7A = 4A + 3A$,   20. $10A = 5(2A)$,   21. $(A^T)^T = A$

by hand and using MATLAB.

If $A = \begin{pmatrix} 2 & 1 \\ 3 & 1 \end{pmatrix}$, $B = \begin{pmatrix} 1 & -2 \\ 4 & 0 \end{pmatrix}$, and $C = \begin{pmatrix} 1 & 1 \\ 1 & 1 \end{pmatrix}$, verify that

22. $(A + B) + C = A + (B + C)$,   23. $(AB)C = A(BC)$,

24. $A(B + C) = AB + AC$,   25. $(A + B)C = AC + BC$

by hand and using MATLAB.

Verify that the following A^{-1} are indeed the inverse of A:

26. $A = \begin{pmatrix} 3 & -1 \\ -5 & 2 \end{pmatrix}$ $A^{-1} = \begin{pmatrix} 2 & 1 \\ 5 & 3 \end{pmatrix}$ 27. $A = \begin{pmatrix} 0 & 1 & 0 \\ 1 & 0 & 0 \\ 0 & 0 & 1 \end{pmatrix}$ $A^{-1} = \begin{pmatrix} 0 & 1 & 0 \\ 1 & 0 & 0 \\ 0 & 0 & 1 \end{pmatrix}$

by hand and using MATLAB.

Write the following linear systems of equations in matrix form: $A\mathbf{x} = \mathbf{b}$.

28. $x_1 - 2x_2 = 5$ $3x_1 + x_2 = 1$

29. $2x_1 + x_2 + 4x_3 = 2$ $4x_1 + 2x_2 + 5x_3 = 6$ $6x_1 - 3x_2 + 5x_3 = 2$

30. $x_2 + 2x_3 + 3x_4 = 2$ $3x_1 - 4x_3 - 4x_4 = 5$ $x_1 + x_2 + x_3 + x_4 = -3$
                                   $2x_1 - 3x_2 + x_3 - 3x_4 = 7$

31. Given a square matrix A that satisfies the equation $A\mathbf{x} = \mathbf{0}$ and $\mathbf{x} \neq \mathbf{0}$. Show that A does not possess an inverse and is thus singular.

32. If R is a rectangular matrix, prove that $R^T R$ is a symmetric matrix.

3.2 DETERMINANTS

Determinants appear naturally during the solution of simultaneous equations. Consider, for example, two simultaneous equations with two unknowns x_1 and x_2,

$$a_{11}x_1 + a_{12}x_2 = b_1, \tag{3.2.1}$$

and

$$a_{21}x_1 + a_{22}x_2 = b_2. \tag{3.2.2}$$

The solution to these equations for the value of x_1 and x_2 is

$$x_1 = \frac{b_1 a_{22} - a_{12} b_2}{a_{11} a_{22} - a_{12} a_{21}}, \tag{3.2.3}$$

and

$$x_2 = \frac{b_2 a_{11} - a_{21} b_1}{a_{11} a_{22} - a_{12} a_{21}}. \tag{3.2.4}$$

Note that the denominator of Equation 3.2.3 and Equation 3.2.4 is the same. This term, which always appears in the solution of 2×2 systems, is formally given the name of *determinant* and written

$$\det(A) = \begin{vmatrix} a_{11} & a_{12} \\ a_{21} & a_{22} \end{vmatrix} = a_{11} a_{22} - a_{12} a_{21}. \tag{3.2.5}$$

MATLAB provides a simple command `det(A)`, which computes the determinant of A. For example, in the present case,

```
>> A = [2 -1 2; 1 3 2; 5 1 6];
>> det(A)

ans =

    0
```

Although determinants have their origin in the solution of systems of equations, any square array of numbers or expressions possesses a unique determinant, independent of whether it is involved in a system of equations or not. This determinant is evaluated (or expanded) according to a formal rule known as *Laplace's expansion of cofactors.*[3] The

---

[3] Laplace, P. S., 1772: Recherches sur le calcul intégral et sur le système du monde. *Hist. Acad. R. Sci., IIe Partie*, 267–376. *Œuvres*, **8**, pp. 369–501. See Muir, T., 1960: *The Theory of Determinants in the Historical Order of Development, Vol. I, Part 1, General Determinants Up to 1841.* Dover Publishers, pp. 24–33.

process revolves around expanding the determinant using any arbitrary column *or* row of A. If the ith row or jth column is chosen, the determinant is given by

$$\det(A) = a_{i1}A_{i1} + a_{i2}A_{i2} + \cdots + a_{in}A_{in} = a_{1j}A_{1j} + a_{2j}A_{2j} + \cdots + a_{nj}A_{nj}, \qquad (\mathbf{3.2.6})$$

where A_{ij}, the *cofactor* of a_{ij}, equals $(-1)^{i+j}M_{ij}$. The minor M_{ij} is the determinant of the $(n-1) \times (n-1)$ submatrix obtained by deleting row i, column j of A. This rule, of course, was chosen so that determinants are still useful in solving systems of equations.

• **Example 3.2.1**

Let us evaluate

$$\begin{vmatrix} 2 & -1 & 2 \\ 1 & 3 & 2 \\ 5 & 1 & 6 \end{vmatrix}$$

by an expansion in cofactors.

Using the first column,

$$\begin{vmatrix} 2 & -1 & 2 \\ 1 & 3 & 2 \\ 5 & 1 & 6 \end{vmatrix} = 2(-1)^2 \begin{vmatrix} 3 & 2 \\ 1 & 6 \end{vmatrix} + 1(-1)^3 \begin{vmatrix} -1 & 2 \\ 1 & 6 \end{vmatrix} + 5(-1)^4 \begin{vmatrix} -1 & 2 \\ 3 & 2 \end{vmatrix} \qquad (\mathbf{3.2.7})$$

$$= 2(16) - 1(-8) + 5(-8) = 0. \qquad (\mathbf{3.2.8})$$

The greatest source of error is forgetting to take the factor $(-1)^{i+j}$ into account during the expansion. □

Although Laplace's expansion does provide a method for calculating $\det(A)$, the number of calculations equals $n!$. Consequently, for hand calculations, an obvious strategy is to select the column or row that has the greatest number of zeros. An even better strategy would be to manipulate a determinant with the goal of introducing zeros into a particular column or row. In the remaining portion of this section, we show some operations that may be performed on a determinant to introduce the desired zeros. Most of the properties follow from the expansion of determinants by cofactors.

• $\boxed{Rule\ 1}$: For every square matrix A, $\det(A^T) = \det(A)$.

The proof is left as an exercise.

• $\boxed{Rule\ 2}$: If any two rows or columns of A are identical, $\det(A) = 0$.

To see that this is true, consider the following 3×3 matrix:

$$\begin{vmatrix} b_1 & b_1 & c_1 \\ b_2 & b_2 & c_2 \\ b_3 & b_3 & c_3 \end{vmatrix} = c_1(b_2b_3 - b_3b_2) - c_2(b_1b_3 - b_3b_1) + c_3(b_1b_2 - b_2b_1) = 0. \qquad (\mathbf{3.2.9})$$

• $\boxed{Rule\ 3}$: The determinant of a triangular matrix is equal to the product of its diagonal elements.

If A is lower triangular, successive expansions by elements in the first column give

$$\det(A) = \begin{vmatrix} a_{11} & 0 & \cdots & 0 \\ a_{21} & a_{22} & \cdots & 0 \\ \vdots & \vdots & \ddots & \vdots \\ a_{n1} & a_{n2} & \cdots & a_{nn} \end{vmatrix} = a_{11} \begin{vmatrix} a_{22} & \cdots & 0 \\ \vdots & \ddots & \vdots \\ a_{n2} & \cdots & a_{nn} \end{vmatrix} = \cdots = a_{11}a_{22}\cdots a_{nn}. \quad \textbf{(3.2.10)}$$

If A is upper triangular, successive expansions by elements of the first row prove the property.

- $\boxed{\textit{Rule 4}}$: If a square matrix A has either a row or a column of all zeros, then $\det(A) = 0$.

  The proof is left as an exercise.

- $\boxed{\textit{Rule 5}}$: If each element in one row (column) of a determinant is multiplied by a number c, the value of the determinant is multiplied by c.

  Suppose that $|B|$ has been obtained from $|A|$ by multiplying row i (column j) of $|A|$ by c. Upon expanding $|B|$ in terms of row i (column j), each term in the expansion contains c as a factor. Factor out the common c, and the result is just c times the expansion $|A|$ by the same row (column).

- $\boxed{\textit{Rule 6}}$: If each element of a row (or a column) of a determinant can be expressed as a binomial, the determinant can be written as the sum of two determinants.

  To understand this property, consider the following 3×3 determinant:

$$\begin{vmatrix} a_1 + d_1 & b_1 & c_1 \\ a_2 + d_2 & b_2 & c_2 \\ a_3 + d_3 & b_3 & c_3 \end{vmatrix} = \begin{vmatrix} a_1 & b_1 & c_1 \\ a_2 & b_2 & c_2 \\ a_3 & b_3 & c_3 \end{vmatrix} + \begin{vmatrix} d_1 & b_1 & c_1 \\ d_2 & b_2 & c_2 \\ d_3 & b_3 & c_3 \end{vmatrix}. \quad \textbf{(3.2.11)}$$

The proof follows by expanding the determinant by the row (or column) that contains the binomials.

- $\boxed{\textit{Rule 7}}$: If B is a matrix obtained by interchanging any two rows (columns) of a square matrix A, then $\det(B) = -\det(A)$.

  The proof is by induction. It is easily shown for any 2×2 matrix. Assume that this rule holds for any $(n-1) \times (n-1)$ matrix. If A is $n \times n$, then let B be a matrix formed by interchanging rows i and j. Expanding $|B|$ and $|A|$ by a different row, say k, we have that

$$|B| = \sum_{s=1}^{n} (-1)^{k+s} b_{ks} M_{ks}, \quad \text{and} \quad |A| = \sum_{s=1}^{n} (-1)^{k+s} a_{ks} N_{ks}, \quad \textbf{(3.2.12)}$$

where M_{ks} and N_{ks} are the minors formed by deleting row k, column s from $|B|$ and $|A|$, respectively. For $s = 1, 2, \ldots, n$, we obtain N_{ks} and M_{ks} by interchanging rows i and j. By the induction hypothesis and recalling that N_{ks} and M_{ks} are $(n-1) \times (n-1)$ determinants, $N_{ks} = -M_{ks}$ for $s = 1, 2, \ldots, n$. Hence, $|B| = -|A|$. Similar arguments hold if two columns are interchanged.

• $\boxed{Rule\ 8}$: If one row (column) of a square matrix A equals a number c times some other row (column), then $\det(A) = 0$.

Suppose one row of a square matrix A is equal to c times some other row. If $c = 0$, then $|A| = 0$. If $c \neq 0$, then $|A| = c|B|$, where $|B| = 0$ because $|B|$ has two identical rows. A similar argument holds for two columns.

• $\boxed{Rule\ 9}$: The value of $\det(A)$ is unchanged if any arbitrary multiple of any line (row or column) is added to any other line.

To see that this is true, consider the simple example:

$$\begin{vmatrix} a_1 & b_1 & c_1 \\ a_2 & b_2 & c_2 \\ a_3 & b_3 & c_3 \end{vmatrix} + \begin{vmatrix} cb_1 & b_1 & c_1 \\ cb_2 & b_2 & c_2 \\ cb_3 & b_3 & c_3 \end{vmatrix} = \begin{vmatrix} a_1 + cb_1 & b_1 & c_1 \\ a_2 + cb_2 & b_2 & c_2 \\ a_3 + cb_3 & b_3 & c_3 \end{vmatrix}, \qquad (3.2.13)$$

where $c \neq 0$. The first determinant on the left side is our original determinant. In the second determinant, we again expand the first column and find that

$$\begin{vmatrix} cb_1 & b_1 & c_1 \\ cb_2 & b_2 & c_2 \\ cb_3 & b_3 & c_3 \end{vmatrix} = c \begin{vmatrix} b_1 & b_1 & c_1 \\ b_2 & b_2 & c_2 \\ b_3 & b_3 & c_3 \end{vmatrix} = 0. \qquad (3.2.14)$$

• **Example 3.2.2**

Let us evaluate

$$\begin{vmatrix} 1 & 2 & 3 & 4 \\ -1 & 1 & 2 & 3 \\ 1 & -1 & 1 & 2 \\ -1 & 1 & -1 & 5 \end{vmatrix}$$

using a combination of the properties stated above and expansion by cofactors.

By adding or subtracting the first row to the other rows, we have that

$$\begin{vmatrix} 1 & 2 & 3 & 4 \\ -1 & 1 & 2 & 3 \\ 1 & -1 & 1 & 2 \\ -1 & 1 & -1 & 5 \end{vmatrix} = \begin{vmatrix} 1 & 2 & 3 & 4 \\ 0 & 3 & 5 & 7 \\ 0 & -3 & -2 & -2 \\ 0 & 3 & 2 & 9 \end{vmatrix} = \begin{vmatrix} 3 & 5 & 7 \\ -3 & -2 & -2 \\ 3 & 2 & 9 \end{vmatrix} \qquad (3.2.15)$$

$$= \begin{vmatrix} 3 & 5 & 7 \\ 0 & 3 & 5 \\ 0 & -3 & 2 \end{vmatrix} = 3 \begin{vmatrix} 3 & 5 \\ -3 & 2 \end{vmatrix} = 3 \begin{vmatrix} 3 & 5 \\ 0 & 7 \end{vmatrix} = 63. \qquad (3.2.16)$$

Problems

Evaluate the following determinants. Check your answer using MATLAB.

1. $\begin{vmatrix} 3 & 5 \\ -2 & -1 \end{vmatrix}$ 2. $\begin{vmatrix} 5 & -1 \\ -8 & 4 \end{vmatrix}$ 3. $\begin{vmatrix} 3 & 1 & 2 \\ 2 & 4 & 5 \\ 1 & 4 & 5 \end{vmatrix}$ 4. $\begin{vmatrix} 4 & 3 & 0 \\ 3 & 2 & 2 \\ 5 & -2 & -4 \end{vmatrix}$

5. $\begin{vmatrix} 1 & 3 & 2 \\ 4 & 1 & 1 \\ 2 & 1 & 3 \end{vmatrix}$　　6. $\begin{vmatrix} 2 & -1 & 2 \\ 1 & 3 & 3 \\ 5 & 1 & 6 \end{vmatrix}$　　7. $\begin{vmatrix} 2 & 0 & 0 & 1 \\ 0 & 1 & 0 & 0 \\ 1 & 6 & 1 & 0 \\ 1 & 1 & -2 & 3 \end{vmatrix}$　　8. $\begin{vmatrix} 2 & 1 & 2 & 1 \\ 3 & 0 & 2 & 2 \\ -1 & 2 & -1 & 1 \\ -3 & 2 & 3 & 1 \end{vmatrix}$

9. Using the properties of determinants, show that

$$\begin{vmatrix} 1 & 1 & 1 & 1 \\ a & b & c & d \\ a^2 & b^2 & c^2 & d^2 \\ a^3 & b^3 & c^3 & d^3 \end{vmatrix} = (b-a)(c-a)(d-a)(c-b)(d-b)(d-c).$$

This determinant is called *Vandermonde's determinant*.

10. Show that

$$\begin{vmatrix} a & b+c & 1 \\ b & a+c & 1 \\ c & a+b & 1 \end{vmatrix} = 0.$$

11. Show that if all of the elements of a row or column are zero, then $\det(A) = 0$.

12. Prove that $\det(A^T) = \det(A)$.

3.3 CRAMER'S RULE

One of the most popular methods for solving simple systems of linear equations is Cramer's rule.[4] It is very useful for 2×2 systems, acceptable for 3×3 systems, and of doubtful use for 4×4 or larger systems.

Let us have n equations with n unknowns, $A\mathbf{x} = \mathbf{b}$. Cramer's rule states that

$$x_1 = \frac{\det(A_1)}{\det(A)}, \quad x_2 = \frac{\det(A_2)}{\det(A)}, \quad \cdots, \quad x_n = \frac{\det(A_n)}{\det(A)}, \tag{3.3.1}$$

where A_i is a matrix obtained from A by replacing the ith column with $\mathbf{b}$ and n is the number of unknowns and equations. Obviously, $\det(A) \neq 0$ if Cramer's rule is to work.

To prove[5] Cramer's rule, consider

$$x_1 \det(A) = \begin{vmatrix} a_{11}x_1 & a_{12} & a_{13} & \cdots & a_{1n} \\ a_{21}x_1 & a_{22} & a_{23} & \cdots & a_{2n} \\ a_{31}x_1 & a_{32} & a_{33} & \cdots & a_{3n} \\ \vdots & \vdots & \vdots & \ddots & \vdots \\ a_{n1}x_1 & a_{n2} & a_{n3} & \cdots & a_{nn} \end{vmatrix} \tag{3.3.2}$$

---

[4] Cramer, G., 1750: *Introduction à l'analyse des lignes courbes algébriques.* Geneva, p. 657.

[5] First proved by Cauchy, L. A., 1815: Mémoire sur les fonctions quine peuvent obtemir que deux valeurs égales et de signes contraires par suite des transportations opérées entre les variables qúelles renferment. *J. l'École Polytech.*, **10**, 29–112.

by Rule 5 from the previous section. By adding x_2 times the second column to the first column,

$$
x_1 \det(A) = \begin{vmatrix}
a_{11}x_1 + a_{12}x_2 & a_{12} & a_{13} & \cdots & a_{1n} \\
a_{21}x_1 + a_{22}x_2 & a_{22} & a_{23} & \cdots & a_{2n} \\
a_{31}x_1 + a_{32}x_2 & a_{32} & a_{33} & \cdots & a_{3n} \\
\vdots & & \vdots & \vdots & \ddots & \vdots \\
a_{n1}x_1 + a_{n2}x_2 & a_{n2} & a_{n3} & \cdots & a_{nn}
\end{vmatrix}.
\tag{3.3.3}
$$

Multiplying each of the columns by the corresponding x_i and adding it to the first column yields

$$
x_1 \det(A) = \begin{vmatrix}
a_{11}x_1 + a_{12}x_2 + \cdots + a_{1n}x_n & a_{12} & a_{13} & \cdots & a_{1n} \\
a_{21}x_1 + a_{22}x_2 + \cdots + a_{2n}x_n & a_{22} & a_{23} & \cdots & a_{2n} \\
a_{31}x_1 + a_{32}x_2 + \cdots + a_{3n}x_n & a_{32} & a_{33} & \cdots & a_{3n} \\
\vdots & & \vdots & \vdots & \ddots & \vdots \\
a_{n1}x_1 + a_{n2}x_2 + \cdots + a_{nn}x_n & a_{n2} & a_{n3} & \cdots & a_{nn}
\end{vmatrix}.
\tag{3.3.4}
$$

The first column of Equation 3.3.4 equals $A\mathbf{x}$ and we replace it with $\mathbf{b}$. Thus,

$$
x_1 \det(A) = \begin{vmatrix}
b_1 & a_{12} & a_{13} & \cdots & a_{1n} \\
b_2 & a_{22} & a_{23} & \cdots & a_{2n} \\
b_3 & a_{32} & a_{33} & \cdots & a_{3n} \\
\vdots & \vdots & \vdots & \ddots & \vdots \\
b_n & a_{n2} & a_{n3} & \cdots & a_{nn}
\end{vmatrix} = \det(A_1),
\tag{3.3.5}
$$

or

$$
x_1 = \frac{\det(A_1)}{\det(A)}
\tag{3.3.6}
$$

provided $\det(A) \neq 0$. To complete the proof we do exactly the same procedure to the jth column.

• **Example 3.3.1**

Let us solve the following system of equations by Cramer's rule:

$$
2x_1 + x_2 + 2x_3 = -1,
\tag{3.3.7}
$$

$$
x_1 + x_3 = -1,
\tag{3.3.8}
$$

and

$$
-x_1 + 3x_2 - 2x_3 = 7.
\tag{3.3.9}
$$

From the matrix form of the equations,

$$
\begin{pmatrix}
2 & 1 & 2 \\
1 & 0 & 1 \\
-1 & 3 & -2
\end{pmatrix}
\begin{pmatrix}
x_1 \\
x_2 \\
x_3
\end{pmatrix}
=
\begin{pmatrix}
-1 \\
-1 \\
7
\end{pmatrix},
\tag{3.3.10}
$$

we have that

$$
\det(A) = \begin{vmatrix}
2 & 1 & 2 \\
1 & 0 & 1 \\
-1 & 3 & -2
\end{vmatrix} = 1,
\tag{3.3.11}
$$

$$\det(A_1) = \begin{vmatrix} -1 & 1 & 2 \\ -1 & 0 & 1 \\ 7 & 3 & -2 \end{vmatrix} = 2, \tag{3.3.12}$$

$$\det(A_2) = \begin{vmatrix} 2 & -1 & 2 \\ 1 & -1 & 1 \\ -1 & 7 & -2 \end{vmatrix} = 1, \tag{3.3.13}$$

and

$$\det(A_3) = \begin{vmatrix} 2 & 1 & -1 \\ 1 & 0 & -1 \\ -1 & 3 & 7 \end{vmatrix} = -3. \tag{3.3.14}$$

Finally,

$$x_1 = \frac{2}{1} = 2, \quad x_2 = \frac{1}{1} = 1, \quad \text{and} \quad x_3 = \frac{-3}{1} = -3. \tag{3.3.15}$$

You can also use MATLAB to perform Cramer's rule. In the present example, the script is as follows:

```
clear; % clear all previous computations
A = [2 1 2; 1 0 1; -1 3 -2]; % input coefficient matrix
b = [-1 ; -1; 7]; % input right side
A1 = A; A1(:,1) = b; % compute A_1
A2 = A; A2(:,2) = b; % compute A_2
A3 = A; A3(:,3) = b; % compute A_3
% compute solution vector
x = [det(A1), det(A2), det(A3)] / det(A)
```

Problems

Solve the following systems of equations by Cramer's rule:

1. $x_1 + 2x_2 = 3, \qquad 3x_1 + x_2 = 6$

2. $2x_1 + x_2 = -3, \qquad x_1 - x_2 = 1$

3. $2x_1 - 3x_2 = 7, \qquad 4x_1 - 5x_2 = 9$

4. $6x_1 - 5x_2 = 2, \qquad x_1 + 2x_2 = 7$

5. $x_1 + 2x_2 - 2x_3 = 4, \qquad 2x_1 + x_2 + x_3 = -2, \qquad -x_1 + x_2 - x_3 = 2$

6. $2x_1 + 3x_2 - x_3 = -1, \qquad -x_1 - 2x_2 + x_3 = 5, \qquad 3x_1 - x_2 = -2.$

7. $x_1 - 8x_2 + x_3 = 4, \qquad -x_1 + 2x_2 + x_3 = 2, \qquad x_1 - x_2 + 2x_3 = -1.$

8. $2x_1 - 3x_2 + 5x_3 = 7, \qquad -x_1 + 4x_2 - 8x_3 = 9, \qquad 12x_1 + 7x_2 - 13x_3 = 17.$

Check your answer using MATLAB.

3.4 ROW ECHELON FORM AND GAUSSIAN ELIMINATION

So far, we assumed that every system of equations has a unique solution. This is not necessarily true, as the following examples show.

● **Example 3.4.1**

Consider the system
$$x_1 + x_2 = 2 \tag{3.4.1}$$
and
$$2x_1 + 2x_2 = -1. \tag{3.4.2}$$

This system is inconsistent because the second equation does not follow after multiplying the first by 2. Geometrically, Equation 3.4.1 and Equation 3.4.2 are parallel lines; they never intersect to give a unique x_1 and x_2. ☐

● **Example 3.4.2**

Even if a system is consistent, it still may not have a unique solution. For example, the system
$$x_1 + x_2 = 2 \tag{3.4.3}$$
and
$$2x_1 + 2x_2 = 4 \tag{3.4.4}$$

is consistent, with the second equation formed by multiplying the first by 2. However, there are an infinite number of solutions. ☐

Our examples suggest the following:

Theorem: *A system of m linear equations in n unknowns may: (1) have no solution, in which case it is called an* inconsistent *system, or (2) have exactly one solution (called a* unique solution*), or (3) have an infinite number of solutions. In the latter two cases, the system is said to be* consistent.

Before we can prove this theorem at the end of this section, we need to introduce some new concepts.

The first one is equivalent systems. Two systems of equations involving the same variables are *equivalent* if they have the same solution set. Of course, the only reason for introducing equivalent systems is the possibility of transforming one system of linear systems into another that is easier to solve. But what operations are permissible? Also, what is the ultimate goal of our transformation?

From a complete study of possible operations, there are only three operations for transforming one system of linear equations into another. These three *elementary row operations* are

---

(1) interchanging any two rows in the matrix,

(2) multiplying any row by a nonzero scalar, and

(3) adding any arbitrary multiple of any row to any other row.

Armed with our elementary row operations, let us now solve the following set of linear equations:
$$x_1 - 3x_2 + 7x_3 = 2, \tag{3.4.5}$$

$$2x_1 + 4x_2 - 3x_3 = -1, \tag{3.4.6}$$

and

$$-x_1 + 13x_2 - 21x_3 = 2. \tag{3.4.7}$$

We begin by writing Equation 3.4.5 through Equation 3.4.7 in matrix notation:

$$\begin{pmatrix} 1 & -3 & 7 \\ 2 & 4 & -3 \\ -1 & 13 & -21 \end{pmatrix} \begin{pmatrix} x_1 \\ x_2 \\ x_3 \end{pmatrix} = \begin{pmatrix} 2 \\ -1 \\ 2 \end{pmatrix}. \tag{3.4.8}$$

The matrix in Equation 3.4.8 is called the *coefficient matrix* of the system.

We now introduce the concept of the *augmented matrix*: a matrix B composed of A plus the column vector $\mathbf{b}$ or

$$B = \begin{pmatrix} 1 & -3 & 7 & 2 \\ 2 & 4 & -3 & -1 \\ -1 & 13 & -21 & 2 \end{pmatrix}. \tag{3.4.9}$$

We can solve our original system by performing elementary row operations on the augmented matrix. Because x_i functions essentially as a placeholder, we can omit them until the end of the computation.

Returning to the problem, the first row can be used to eliminate the elements in the first column of the remaining rows. For this reason the first row is called the *pivotal* row and the element a_{11} is the *pivot*. By using the third elementary row operation twice (to eliminate the 2 and -1 in the first column), we have the equivalent system

$$B = \begin{pmatrix} 1 & -3 & 7 & 2 \\ 0 & 10 & -17 & -5 \\ 0 & 10 & -14 & 4 \end{pmatrix}. \tag{3.4.10}$$

At this point we choose the second row as our new pivotal row and again apply the third row operation to eliminate the last element in the second column. This yields

$$B = \begin{pmatrix} 1 & -3 & 7 & 2 \\ 0 & 10 & -17 & -5 \\ 0 & 0 & 3 & 9 \end{pmatrix}. \tag{3.4.11}$$

Thus, elementary row operations transformed Equation 3.4.5 through Equation 3.4.7 into the triangular system:

$$x_1 - 3x_2 + 7x_3 = 2, \tag{3.4.12}$$
$$10x_2 - 17x_3 = -5, \tag{3.4.13}$$
$$3x_3 = 9, \tag{3.4.14}$$

which is *equivalent* to the original system. The final solution is obtained by *back substitution*, solving from Equation 3.4.14 back to Equation 3.4.12. In the present case, $x_3 = 3$. Then, $10x_2 = 17(3) - 5$, or $x_2 = 4.6$. Finally, $x_1 = 3x_2 - 7x_3 + 2 = -5.2$.

In general, any $n \times n$ linear system can be reduced to triangular form. This reduction involves $n-1$ steps. In the first step, a pivot element (or simply *pivot*), and thus the pivotal row, is chosen from the nonzero entries in the first column of the matrix. We interchange rows (if necessary) so that the pivotal row is the first row. Multiples of the pivotal row are then subtracted from each of the remaining $n-1$ rows so that there are 0's in the $(2,1), ..., (n,1)$ positions. In the second step, a pivot element (another pivot) is chosen from the nonzero entries in column 2, rows 2 through n, of the matrix. The row containing the pivot is then interchanged with the second row (if necessary) of the matrix and is used as the pivotal row. Multiples of the pivotal row are then subtracted from the remaining $n-2$ rows, eliminating all entries below the diagonal in the second column. The same procedure is repeated for columns 3 through $n-1$. Note that in the second step, row 1 and column 1 remain unchanged, in the third step the first two rows and first two columns remain unchanged, and so on.

If elimination is carried out as described, we arrive at an equivalent upper triangular system after $n-1$ steps. There are two possible outcomes. If all n pivots are nonzero, then the system of equations has a unique solution that we can obtain by performing back substitution. This reduction involves $n-1$ steps. However, the procedure fails if, at any step, all possible choices for a pivot element equal zero. Let us now examine such cases.

Consider now the system

$$x_1 + 2x_2 + x_3 = -1, \tag{3.4.15}$$
$$2x_1 + 4x_2 + 2x_3 = -2, \tag{3.4.16}$$
$$x_1 + 4x_2 + 2x_3 = 2. \tag{3.4.17}$$

Its augmented matrix is

$$B = \begin{pmatrix} 1 & 2 & 1 & -1 \\ 2 & 4 & 2 & -2 \\ 1 & 4 & 2 & 2 \end{pmatrix}. \tag{3.4.18}$$

Choosing the first row as our pivotal row, we find that

$$B = \begin{pmatrix} 1 & 2 & 1 & -1 \\ 0 & 0 & 0 & 0 \\ 0 & 2 & 1 & 3 \end{pmatrix}, \tag{3.4.19}$$

or

$$B = \begin{pmatrix} 1 & 2 & 1 & -1 \\ 0 & 2 & 1 & 3 \\ 0 & 0 & 0 & 0 \end{pmatrix}. \tag{3.4.20}$$

The difficulty here is the presence of the zeros in the third row. Clearly any finite numbers satisfy the equation $0x_1 + 0x_2 + 0x_3 = 0$ and we have an infinite number of solutions. Closer examination of the original system shows an underdetermined system; Equation 3.4.15 and Equation 3.4.16 differ by a multiplicative factor of 2. An important aspect of this problem is the fact that the final augmented matrix is of the form of a staircase or *echelon form* rather than of triangular form.

Let us modify Equation 3.4.15 through Equation 3.4.17 to read

$$x_1 + 2x_2 + x_3 = -1, \tag{3.4.21}$$
$$2x_1 + 4x_2 + 2x_3 = 3, \tag{3.4.22}$$
$$x_1 + 4x_2 + 2x_3 = 2, \tag{3.4.23}$$

then the final augmented matrix is

$$B = \begin{pmatrix} 1 & 2 & 1 & -1 \\ 0 & 2 & 1 & 3 \\ 0 & 0 & 0 & 5 \end{pmatrix}. \tag{3.4.24}$$

We again have a problem with the third row because $0x_1 + 0x_2 + 0x_3 = 5$, which is impossible. There is no solution in this case and we have an *inconsistent system*. Note, once again, that our augmented matrix has a row echelon form rather than a triangular form.

In summary, to include all possible situations in our procedure, we must rewrite the augmented matrix in row echelon form. We have *row echelon form* when:

---

(1) The first nonzero entry in each row is 1.

(2) If row k does not consist entirely of zeros, the number of leading zero entries in row $k + 1$ is greater than the number of leading zero entries in row k.

(3) If there are rows whose entries are all zero, they are below the rows having nonzero entries.

The number of nonzero rows in the row echelon form of a matrix is known as its *rank*. In MATLAB, the rank is easily found using the command `rank( )`. *Gaussian elimination* is the process of using elementary row operations to transform a linear system into one whose augmented matrix is in row echelon form.

• **Example 3.4.3**

Each of the following matrices is *not* of row echelon form because they violate one of the conditions for row echelon form:

$$\begin{pmatrix} 2 & 2 & 3 \\ 0 & 2 & 1 \\ 0 & 0 & 4 \end{pmatrix}, \begin{pmatrix} 0 & 0 & 0 \\ 0 & 2 & 0 \end{pmatrix}, \begin{pmatrix} 0 & 1 \\ 1 & 0 \end{pmatrix}. \tag{3.4.25}$$

□

• **Example 3.4.4**

The following matrices are in row echelon form:

$$\begin{pmatrix} 1 & 2 & 3 \\ 0 & 1 & 1 \\ 0 & 0 & 1 \end{pmatrix}, \begin{pmatrix} 1 & 4 & 6 \\ 0 & 0 & 1 \\ 0 & 0 & 0 \end{pmatrix}, \begin{pmatrix} 1 & 3 & 4 & 0 \\ 0 & 0 & 1 & 3 \\ 0 & 0 & 0 & 0 \end{pmatrix}. \tag{3.4.26}$$

□

• **Example 3.4.5: Gauss-Jordan elimination**

Gaussian elimination can also be used to solve the general problem $AX = B$. A particular popular form is *Gauss-Jordan elimination*. In this particular form of Gaussian elimination, row reduction is continued until the A matrix consists of elements equaling 1 along the principal diagonal and all of the other elements equal zero, yielding a *reduced row echelon*. It is commonly used to find the inverse of a matrix.

Let us find the inverse of the matrix

$$A = \begin{pmatrix} 4 & -2 & 2 \\ -2 & -4 & 4 \\ -4 & 2 & 8 \end{pmatrix} \tag{3.4.27}$$

by Gaussian elimination.

Because the inverse is defined by $AA^{-1} = I$, our augmented matrix is

$$\left(\begin{array}{ccc|ccc} 4 & -2 & 2 & 1 & 0 & 0 \\ -2 & -4 & 4 & 0 & 1 & 0 \\ -4 & 2 & 8 & 0 & 0 & 1 \end{array} \right). \tag{3.4.28}$$

Then, by elementary row operations,

$$\left(\begin{array}{ccc|ccc} 4 & -2 & 2 & 1 & 0 & 0 \\ -2 & -4 & 4 & 0 & 1 & 0 \\ -4 & 2 & 8 & 0 & 0 & 1 \end{array} \right) = \left(\begin{array}{ccc|ccc} -2 & -4 & 4 & 0 & 1 & 0 \\ 4 & -2 & 2 & 1 & 0 & 0 \\ -4 & 2 & 8 & 0 & 0 & 1 \end{array} \right) \tag{3.4.29}$$

$$= \left(\begin{array}{ccc|ccc} -2 & -4 & 4 & 0 & 1 & 0 \\ 4 & -2 & 2 & 1 & 0 & 0 \\ 0 & 0 & 10 & 1 & 0 & 1 \end{array} \right) \tag{3.4.30}$$

$$= \left(\begin{array}{ccc|ccc} -2 & -4 & 4 & 0 & 1 & 0 \\ 0 & -10 & 10 & 1 & 2 & 0 \\ 0 & 0 & 10 & 1 & 0 & 1 \end{array} \right) \tag{3.4.31}$$

$$= \left(\begin{array}{ccc|ccc} -2 & -4 & 4 & 0 & 1 & 0 \\ 0 & -10 & 0 & 0 & 2 & -1 \\ 0 & 0 & 10 & 1 & 0 & 1 \end{array} \right) \tag{3.4.32}$$

$$= \left(\begin{array}{ccc|ccc} -2 & -4 & 0 & -2/5 & 1 & -2/5 \\ 0 & -10 & 0 & 0 & 2 & -1 \\ 0 & 0 & 10 & 1 & 0 & 1 \end{array} \right) \tag{3.4.33}$$

$$= \left(\begin{array}{ccc|ccc} -2 & 0 & 0 & -2/5 & 1/5 & 0 \\ 0 & -10 & 0 & 0 & 2 & -1 \\ 0 & 0 & 10 & 1 & 0 & 1 \end{array} \right) \tag{3.4.34}$$

$$= \left(\begin{array}{ccc|ccc} 1 & 0 & 0 & 1/5 & -1/10 & 0 \\ 0 & 1 & 0 & 0 & -1/5 & 1/10 \\ 0 & 0 & 1 & 1/10 & 0 & 1/10 \end{array} \right). \tag{3.4.35}$$

Thus, the right half of the augmented matrix yields the inverse and it equals

$$A^{-1} = \left(\begin{array}{ccc} 1/5 & -1/10 & 0 \\ 0 & -1/5 & 1/10 \\ 1/10 & 0 & 1/10 \end{array} \right). \tag{3.4.36}$$

MATLAB has the ability of doing Gaussian elimination step by step. We begin by typing:

```
>>% input augmented matrix
>>aug = [4 -2 2 1 0 0 ; -2 -4 4 0 1 0;-4 2 8 0 0 1];
>>rrefmovie(aug);
```

The MATLAB command `rrefmovie(A)` produces the reduced row echelon form of A. Repeated pressing of any key gives the next step in the calculation along with a statement of how it computed the modified augmented matrix. Eventually you obtain

```
A =
        1           0           0          1/5        -1/10          0
```

$$\begin{matrix} 0 & 1 & 0 & 0 & -1/5 & 1/10 \\ 0 & 0 & 1 & 1/10 & 0 & 1/10 \end{matrix}$$

You can read the inverse matrix just as we did earlier. □

Gaussian elimination may be used with overdetermined systems. *Overdetermined systems* are linear systems where there are more equations than unknowns $(m > n)$. These systems are usually (but not always) inconsistent.

• **Example 3.4.6**

Consider the linear system

$$x_1 + x_2 = 1, \tag{3.4.37}$$
$$-x_1 + 2x_2 = -2, \tag{3.4.38}$$
$$x_1 - x_2 = 4. \tag{3.4.39}$$

After several row operations, the augmented matrix

$$\left(\begin{array}{cc|c} 1 & 1 & 1 \\ -1 & 2 & -2 \\ 1 & -1 & 4 \end{array} \right) \tag{3.4.40}$$

becomes

$$\left(\begin{array}{cc|c} 1 & 1 & 1 \\ 0 & 1 & 2 \\ 0 & 0 & -7 \end{array} \right). \tag{3.4.41}$$

From the last row of the augmented matrix, Equation 3.4.41, we see that the system is inconsistent.

If we test this system using MATLAB by typing:

```
>>% input augmented matrix
>>aug = [1 1 1 ; -1 2 -2; 1 -1 4];
>>rrefmovie(aug);
```

eventually you obtain

```
A =
```

$$\begin{matrix} 1 & 0 & 0 \\ 0 & 1 & 0 \\ 0 & 0 & 1 \end{matrix}$$

Although the numbers have changed from our hand calculation, we still have an inconsistent system because $x_1 = x_2 = 0$ does not satisfy $x_1 + x_2 = 1$.

Considering now a slight modification of this system to

$$x_1 + x_2 = 1, \tag{3.4.42}$$
$$-x_1 + 2x_2 = 5, \tag{3.4.43}$$
$$x_1 = -1, \tag{3.4.44}$$

the final form of the augmented matrix is

$$
\left(\begin{array}{cc|c}
1 & 1 & 1 \\
0 & 1 & 2 \\
0 & 0 & 0
\end{array}\right),
\tag{3.4.45}
$$

which has the unique solution $x_1 = -1$ and $x_2 = 2$.

How does MATLAB handle this problem? Typing:

```
>>% input augmented matrix
>>aug = [1 1 1 ; -1 2 5; 1 0 -1];
>>rrefmovie(aug);
```

we eventually obtain

```
A =
          1                 0               -1
          0                 1                2
          0                 0                0
```

This yields $x_1 = -1$ and $x_2 = 2$, as we found by hand.

Finally, by introducing the set:

$$
x_1 + x_2 = 1,
\tag{3.4.46}
$$
$$
2x_1 + 2x_2 = 2,
\tag{3.4.47}
$$
$$
3x_1 + 3x_3 = 3,
\tag{3.4.48}
$$

the final form of the augmented matrix is

$$
\left(\begin{array}{cc|c}
1 & 1 & 1 \\
0 & 0 & 0 \\
0 & 0 & 0
\end{array}\right).
\tag{3.4.49}
$$

There are an infinite number of solutions: $x_1 = 1 - \alpha$, and $x_2 = \alpha$.

Turning to MATLAB, we first type:

```
>>% input augmented matrix
>>aug = [1 1 1 ; 2 2 2; 3 3 3];
>>rrefmovie(aug);
```

and we eventually obtain

```
A =
          1                 1                1
          0                 0                0
          0                 0                0
```

This is the same as Equation 3.4.49 and the final answer is the same.   □

Gaussian elimination can also be employed with underdetermined systems. An *under-determined linear system* is one where there are fewer equations than unknowns ($m < n$). These systems usually have an infinite number of solutions although they can be inconsistent.

• **Example 3.4.7**

Consider the underdetermined system:

$$2x_1 + 2x_2 + x_3 = -1, \tag{3.4.50}$$
$$4x_1 + 4x_2 + 2x_3 = 3. \tag{3.4.51}$$

Its augmented matrix can be transformed into the form:

$$\begin{pmatrix} 2 & 2 & 1 & -1 \\ 0 & 0 & 0 & 5 \end{pmatrix}. \tag{3.4.52}$$

Clearly, this case corresponds to an inconsistent set of equations. On the other hand, if Equation 3.4.51 is changed to

$$4x_1 + 4x_2 + 2x_3 = -2, \tag{3.4.53}$$

then the final form of the augmented matrix is

$$\begin{pmatrix} 2 & 2 & 1 & -1 \\ 0 & 0 & 0 & 0 \end{pmatrix} \tag{3.4.54}$$

and we have an infinite number of solutions, namely $x_3 = \alpha$, $x_2 = \beta$, and $2x_1 = -1-\alpha-2\beta$. We can write this solution as

$$\mathbf{x} = \begin{pmatrix} x_1 \\ x_2 \\ x_3 \end{pmatrix} = \begin{pmatrix} -\frac{1}{2} \\ 0 \\ 0 \end{pmatrix} + \alpha \begin{pmatrix} -\frac{1}{2} \\ 0 \\ 1 \end{pmatrix} + \beta \begin{pmatrix} -1 \\ 1 \\ 0 \end{pmatrix}. \tag{3.4.55}$$

□

Consider now one of the most important classes of linear equations: the homogeneous equations $A\mathbf{x} = \mathbf{0}$. If $\det(A) \neq 0$, then by Cramer's rule $x_1 = x_2 = x_3 = \cdots = x_n = 0$. Thus, the only possibility for a nontrivial solution is $\det(A) = 0$. In this case, A is singular, no inverse exists, and nontrivial solutions exist but they are not unique.

• **Example 3.4.8**

Consider the two homogeneous equations:

$$x_1 + x_2 = 0, \tag{3.4.56}$$
$$x_1 - x_2 = 0. \tag{3.4.57}$$

Note that $\det(A) = -2$. Solving this system yields $x_1 = x_2 = 0$.
On the other hand, if we change the system to

$$x_1 + x_2 = 0, \tag{3.4.58}$$
$$x_1 + x_2 = 0, \tag{3.4.59}$$

which has the $\det(A) = 0$ so that A is singular. Both equations yield $x_1 = -x_2 = \alpha$, any constant. Thus, there is an infinite number of solutions for this set of homogeneous equations.

□

• **Example 3.4.9: Null space**

At this point it is convenient to introduce the concept of null space: For an $m \times n$ matrix A, the *null space* is the set of solutions to the homogeneous system $A\mathbf{x} = \mathbf{0}$.

Consider the matrix

$$A = \begin{pmatrix} 2 & 2 & 1 \\ 4 & 4 & 2 \\ 0 & 0 & 0 \end{pmatrix}. \tag{3.4.60}$$

To find its null space we first set up the augment matrix

$$\left(\begin{array}{ccc|c} 2 & 2 & 1 & 0 \\ 4 & 4 & 2 & 0 \\ 0 & 0 & 0 & 0 \end{array} \right). \tag{3.4.61}$$

Using Gaussian elimination, Equation 3.4.61 can be rewritten

$$\left(\begin{array}{ccc|c} 2 & 2 & 1 & 0 \\ 0 & 0 & 0 & 0 \\ 0 & 0 & 0 & 0 \end{array} \right) \tag{3.4.62}$$

and the solution to our homogeneous equations can be written

$$\mathbf{x} = \begin{pmatrix} x_1 \\ x_2 \\ x_3 \end{pmatrix} = \alpha \begin{pmatrix} -\frac{1}{2} \\ 0 \\ 1 \end{pmatrix} + \beta \begin{pmatrix} -1 \\ 1 \\ 0 \end{pmatrix}. \tag{3.4.63}$$

Thus, the null space for the matrix A consists of two vectors:

$$\begin{pmatrix} -\frac{1}{2} \\ 0 \\ 1 \end{pmatrix} \qquad \text{and} \qquad \begin{pmatrix} -1 \\ 1 \\ 0 \end{pmatrix}.$$

Let us return to Example 3.4.7. There we solved the system $A\mathbf{x} = \mathbf{b}$ where A is given by Equation 3.4.60, $\mathbf{x} = (x_1\ x_2\ x_3)^T$, and $\mathbf{b} = (-1\ 3\ 0)^T$. We found that its general solution, given by Equation 3.4.55, contained two parts: the particular solution $(-\frac{1}{2}\ 0\ 0)^T$ and a general solution consisting of two terms. Each term equaled a free constant times a vector from the null space. □

We close this section by outlining the proof of the theorem, which we introduced at the beginning.

Consider the system $A\mathbf{x} = \mathbf{b}$. By elementary row operations, the first equation in this system can be reduced to

$$x_1 + \alpha_{12}x_2 + \cdots + \alpha_{1n}x_n = \beta_1. \tag{3.4.64}$$

The second equation has the form

$$x_p + \alpha_{2p+1}x_{p+1} + \cdots + \alpha_{2n}x_n = \beta_2, \tag{3.4.65}$$

where $p > 1$. The third equation has the form

$$x_q + \alpha_{3q+1}x_{q+1} + \cdots + \alpha_{3n}x_n = \beta_3, \tag{3.4.66}$$

where $q > p$, and so on. To simplify the notation, we introduce z_i where we choose the first k values so that $z_1 = x_1$, $z_2 = x_p$, $z_3 = x_q$, ... Thus, the question of the existence of solutions depends upon the three integers: m, n, and k. The resulting set of equations has the form:

$$\begin{pmatrix} 1 & \gamma_{12} & \cdots & \gamma_{1k} & \gamma_{1k+1} & \cdots & \gamma_{1n} \\ 0 & 1 & \cdots & \gamma_{2k} & \gamma_{2k+1} & \cdots & \gamma_{2n} \\ & & \vdots & & & & \\ 0 & 0 & \cdots & 1 & \gamma_{kk+1} & \cdots & \gamma_{kn} \\ 0 & 0 & \cdots & 0 & 0 & \cdots & 0 \\ & & \vdots & & & & \\ 0 & 0 & \cdots & 0 & 0 & \cdots & 0 \end{pmatrix} \begin{pmatrix} z_1 \\ z_2 \\ \vdots \\ z_k \\ z_{k+1} \\ \vdots \\ z_n \end{pmatrix} = \begin{pmatrix} \beta_1 \\ \beta_2 \\ \vdots \\ \beta_k \\ \beta_{k+1} \\ \vdots \\ \beta_m \end{pmatrix}. \tag{3.4.67}$$

Note that $\beta_{k+1}, \ldots, \beta_m$ need not be all zeros.

There are three possibilities:

(a) $k < m$ and at least one of the elements $\beta_{k+1}, \ldots, \beta_m$ is nonzero. Suppose that an element β_p is nonzero $(p > k)$. Then the pth equation is

$$0z_1 + 0z_2 + \cdots + 0z_n = \beta_p \neq 0. \tag{3.4.68}$$

However, this is a contradiction and the equations are inconsistent.

(b) $k = n$ and either (i) $k < m$ and all of the elements $\beta_{k+1}, \ldots, \beta_m$ are zero, or (ii) $k = m$. Then the equations have a unique solution that can be obtained by back-substitution.

(c) $k < n$ and either (i) $k < m$ and all of the elements $\beta_{k+1}, \ldots, \beta_m$ are zero, or (ii) $k = m$. Then, arbitrary values can be assigned to the $n - k$ variables $z_{k+1}, \ldots, z_n$. The equations can be solved for $z_1, z_2, \ldots, z_k$ and there is an infinity of solutions.

For homogeneous equations $\mathbf{b} = \mathbf{0}$, all of the β_i are zero. In this case, we have only two cases:

(b') $k = n$, then Equation 3.4.67 has the solution $\mathbf{z} = \mathbf{0}$, which leads to the trivial solution for the original system $A\mathbf{x} = \mathbf{0}$.

(c') $k < n$, the equations possess an infinity of solutions given by assigning arbitrary values to $z_{k+1}, \ldots, z_n$.

Problems

Solve the following systems of linear equations by Gaussian elimination. Check your answer using MATLAB.

1. $2x_1 + x_2 = 4$, $5x_1 - 2x_2 = 1$

2. $x_1 + x_2 = 0$, $3x_1 - 4x_2 = 1$

3. $-x_1 + x_2 + 2x_3 = 0$, $3x_1 + 4x_2 + x_3 = 0$, $-x_1 + x_2 + 2x_3 = 0$

4. $4x_1 + 6x_2 + x_3 = 2$, $2x_1 + x_2 - 4x_3 = 3$, $3x_1 - 2x_2 + 5x_3 = 8$

5. $3x_1 + x_2 - 2x_3 = -3$, $x_1 - x_2 + 2x_3 = -1$, $-4x_1 + 3x_2 - 6x_3 = 4$

6. $x_1 - 3x_2 + 7x_3 = 2$, $2x_1 + 4x_2 - 3x_3 = -1$, $-3x_1 + 7x_2 + 2x_3 = 3$

7. $x_1 - x_2 + 3x_3 = 5$, $2x_1 - 4x_2 + 7x_3 = 7$, $4x_1 - 9x_2 + 2x_3 = -15$

8. $x_1 + x_2 + x_3 + x_4 = -1$, $2x_1 - x_2 + 3x_3 = 1$,
    $2x_2 + 3x_4 = 15$, $-x_1 + 2x_2 + x_4 = -2$

Find the inverse of each of the following matrices by Gauss-Jordan elimination. Check your answers using MATLAB.

9. $\begin{pmatrix} -3 & 5 \\ 2 & 1 \end{pmatrix}$ 10. $\begin{pmatrix} 3 & -1 \\ -5 & 2 \end{pmatrix}$ 11. $\begin{pmatrix} 19 & 2 & -9 \\ -4 & -1 & 2 \\ -2 & 0 & 1 \end{pmatrix}$ 12. $\begin{pmatrix} 1 & 2 & 5 \\ 0 & -1 & 2 \\ 2 & 4 & 11 \end{pmatrix}$

13. Does $(A^2)^{-1} = (A^{-1})^2$? Justify your answer.

Project: Solving Fredholm Integral Equations of the Second Kind

Fredholm integral equations of the second kind and their variants appear in many scientific and engineering applications. In this project you will use matrix methods to solve this equation:

$$u(x) = \int_a^b K(x,t)u(t)\,dt + f(x), \qquad a < x < b,$$

where the kernel $K(x,t)$ is a given real-valued and continuous function and $u(x)$ is the unknown. One method for solving this integral equation replaces the integral with some grid-point representation. The goal of this project is to examine how we can use linear algebra to solve this numerical approximation to this integral equation.

Step 1: Using Simpson's rule, show that our Fredholm equation can be written in the matrix form $(I - KD)\mathbf{u} = \mathbf{f}$, where

$$D = \begin{pmatrix} A_0 & 0 & \cdots & 0 & 0 \\ 0 & A_1 & \cdots & 0 & 0 \\ \vdots & \vdots & \ddots & \vdots & \vdots \\ 0 & 0 & \cdots & A_{n-1} & 0 \\ 0 & 0 & \cdots & 0 & A_n \end{pmatrix},$$

$$K = \begin{pmatrix} K(x_0,x_0) & K(x_0,x_1) & \cdots & K(x_0,x_{n-1}) & K(x_0,x_n) \\ K(x_1,x_0) & K(x_1,x_1) & \cdots & K(x_1,x_{n-1}) & K(x_1,x_n) \\ \vdots & \vdots & \ddots & \vdots & \vdots \\ K(x_{n-1},x_0) & K(x_{n-1},x_1) & \cdots & K(x_{n-1},x_{n-1}) & K(x_{n-1},x_n) \\ K(x_n,x_0) & K(x_n,x_1) & \cdots & K(x_n,x_{n-1}) & K(x_n,x_n) \end{pmatrix},$$

$$\mathbf{u} = [u(x_0), u(x_1), \cdots, u(x_n)]^T, \quad \text{and} \quad \mathbf{f} = [f(x_0), f(x_1), \cdots, f(x_n)]^T,$$

where $A_0 = A_n = h/3$, $A_2 = A_4 = \cdots = A_{n-2} = 2h/3$, $A_1 = A_3 = \cdots = A_{n-1} = 4h/3$, $x_i = ih$, and $h = (b-a)/n$. Here n must be an *even* integer.

Step 2: Use MATLAB to solve our matrix equation to find $\mathbf{u}$. Use the following known solutions:

(a) $K(x,t) = \frac{1}{2}x^2t^2$, $f(x) = 0.9x^2$, $u(x) = x^2$,

(b) $K(x,t) = x^2 e^{t(x-1)}$, $f(x) = x + (1-x)e^x$, $u(x) = e^x$,

(c) $K(x,t) = \frac{1}{3}e^{x-t}$, $f(x) = \frac{2}{3}e^x$, $u(x) = e^x$,

(d) $K(x,t) = -\frac{1}{3}e^{2x-5t/3}$, $f(x) = \exp(2x + \frac{1}{3})$, $u(x) = e^{2x}$,

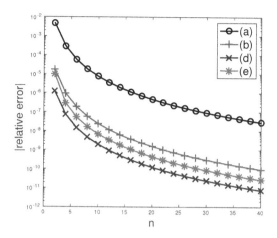

Figure 3.4.1: The absolute value of the relative error of the numerical solution of a Fredholm integral equation of the second kind as a function of n for test problems (a), (b), (d), and (e).

$$(e) \qquad K(x,t) = -x\left(e^{xt} - 1\right), \qquad f(x) = e^x - x, \qquad u(x) = 1,$$
$$(f) \qquad K(x,t) = \tfrac{1}{2}xt, \qquad f(x) = \tfrac{5}{6}x, \qquad u(x) = x,$$

when $0 \le x \le 1$. How does the accuracy of this method vary with n (or h)? What happens when n becomes large? Figure 3.4.1 shows the absolute value of the relative error in the numerical solution at $x = \tfrac{1}{2}$ as a function of n. For test cases (c) and (f) the error was the same order of magnitude as the round-off error.

Project: LU Decomposition

In this section we showed how Gaussian elimination can be used to find solutions to sets of linear equations. A popular alternative involves rewriting the $n \times n$ coefficient matrix:

$$
A = \begin{pmatrix}
a_{11} & a_{12} & a_{13} & \cdots & \cdots & \cdots & a_{1n} \\
a_{21} & a_{22} & a_{23} & \cdots & \cdots & \cdots & a_{2n} \\
\vdots & \vdots & \vdots & \vdots & \vdots & \vdots & \vdots \\
\cdots & \cdots & \cdots & \cdots & a_{ij} & \cdots & \cdots \\
\vdots & \vdots & \vdots & \vdots & \vdots & \vdots & \vdots \\
a_{n1} & a_{n2} & a_{n3} & \cdots & \cdots & \cdots & a_{nn}
\end{pmatrix}
$$

as the product of a lower $n \times n$ triangular matrix:

$$
L = \begin{pmatrix}
\ell_{11} & 0 & 0 & 0 & 0 & 0 & 0 \\
\ell_{21} & \ell_{22} & 0 & 0 & 0 & 0 & 0 \\
\ell_{31} & \ell_{32} & \ell_{33} & 0 & 0 & 0 & 0 \\
\vdots & \vdots & \vdots & \ddots & 0 & 0 & 0 \\
\vdots & \vdots & \vdots & \vdots & \ell_{ij} & 0 & 0 \\
\vdots & \vdots & \vdots & \vdots & \vdots & \ddots & 0 \\
\ell_{n1} & \ell_{n2} & \ell_{n3} & \cdots & \cdots & \cdots & \ell_{nn}
\end{pmatrix}
$$

and an upper $n \times n$ triangular matrix:

$$U = \begin{pmatrix} 1 & u_{12} & u_{13} & u_{14} & u_{15} & \cdots & u_{1n} \\ 0 & 1 & u_{23} & u_{24} & u_{25} & \cdots & u_{2n} \\ 0 & 0 & 1 & u_{34} & u_{35} & \cdots & u_{3n} \\ 0 & 0 & 0 & 1 & u_{45} & \cdots & u_{4n} \\ \vdots & \vdots & \vdots & \vdots & \ddots & \vdots & \vdots \\ 0 & 0 & 0 & 0 & 0 & 1 & u_{n-1n} \\ 0 & 0 & 0 & 0 & 0 & 0 & 1 \end{pmatrix},$$

so that $A = LU$. By simply doing the matrix multiplication, we find the following Crout algorithm to compute ℓ_{ij} and u_{ij}:

$$\ell_{ij} = a_{ij} - \sum_{k=1}^{j-1} \ell_{ik} u_{kj}, \qquad j \le i, \quad i = 1, 2, \ldots, n;$$

and

$$u_{ij} = \left[a_{ij} - \sum_{k=1}^{i-1} \ell_{ik} u_{kj} \right] \bigg/ \ell_{ii}, \qquad i < j, \quad j = 2, 3, \ldots, n.$$

For the special case of $j = 1$, $\ell_{i1} = a_{i1}$; for $i = 1$, $u_{1j} = a_{1j}/\ell_{11} = a_{i1}/a_{11}$. Clearly we could write code to compute L and U given A. However, MATLAB has a subroutine for doing this factorization [L,U] = lu(A).

Note: If you use [L,U] = lu(A), L may not be a lower triangular matrix as shown above but still A = L*U. For example, if A = [4,3;6,3], [L,U] = lu(A) yields the lower triangular matrix L = [0.667,1;1,0], U = [6,3;0,1], and A = L*U. But, L = [1,0;1.5,1] and U = [4,3;0,-1.5] also gives A = LU and L and U do conform to our format. If you want to use MATLAB you must use the decomposition that it gives you.

How does this factorization help us to solve $A\mathbf{x} = \mathbf{b}$? The goal of this project is to answer this question.

Step 1: Show that $L\mathbf{y} = \mathbf{b}$ and $U\mathbf{x} = \mathbf{y}$ can be combined together to yield $A\mathbf{x} = \mathbf{b}$.

Step 2: Write a MATLAB script to solve $A\mathbf{x} = \mathbf{b}$ using LU decomposition. Hint: Use y = L\b and x = U\y rather than the MATLAB function inv.

Step 3: Check your program by resolving Problems 4, 6, 7, and 8.

The principal reason that this scheme is so popular is its economy of storage. The 0's in either L or U are not stored. Furthermore, after the element a_{ij} is used, it never appears again.

Project: QR Decomposition

In the previous project, you discovered that by factoring the matrix A into upper and lower diagonal matrices, we could solve $A\mathbf{x} = \mathbf{b}$. Here we will again factor the matrix A into the product QR, but Q will have the property that $Q^T Q = I$ (orthogonal matrix) and R is an upper triangular matrix.

Step 1: Assuming that we can rewrite $A\mathbf{x} = \mathbf{b}$ as $QR\mathbf{x} = \mathbf{b}$, multiply both sides of this second equation by Q^T and show that you obtain $R\mathbf{x} = Q^T\mathbf{b} = \mathbf{y}$.

Step 2: Show that x_i can be computed from $x_n = y_n/r_{nn}$ and

$$x_i = \left[y_i - \sum_{j=i+1}^{n} r_{ij}y_j \right] \bigg/ r_{ii}, \qquad i = n-1, n-2, \dots, 1.$$

Step 3: Write a MATLAB script to solve $A\mathbf{x} = \mathbf{b}$ using QR decomposition. Note that MATLAB has the subroutine `qr` for QR decomposition of a matrix A.

Step 4: Check your program by resolving Problems 4, 6, 7, and 8.

What advantages does QR decomposition have over LU decomposition? First, solving $A\mathbf{x} = \mathbf{b}$ via $R\mathbf{x} = Q^T\mathbf{b}$ is as well-conditioned as the original problem. Second, QR decomposition finds the least-squares solutions when no exact solution exists. When there are exact solutions, it finds all of them.

3.5 EIGENVALUES AND EIGENVECTORS

One of the classic problems of linear algebra[6] is finding all of the λ's that satisfy the $n \times n$ system

$$A\mathbf{x} = \lambda\mathbf{x}. \tag{3.5.1}$$

The nonzero quantity λ is the *eigenvalue* or *characteristic value* of A. The vector $\mathbf{x}$ is the *eigenvector* or *characteristic vector* belonging to λ. The set of the eigenvalues of A is called the *spectrum* of A. The largest of the absolute values of the eigenvalues of A is called the *spectral radius* of A.

To find λ and $\mathbf{x}$, we first rewrite Equation 3.5.1 as a set of homogeneous equations:

$$(A - \lambda I)\mathbf{x} = \mathbf{0}. \tag{3.5.2}$$

From the theory of linear equations, Equation 3.5.2 has trivial solutions unless its determinant equals zero. On the other hand, if

$$\det(A - \lambda I) = 0, \tag{3.5.3}$$

there is an infinity of solutions.

The expansion of the determinant, Equation 3.5.3, yields an nth-degree polynomial in λ, the *characteristic polynomial*. The roots of the characteristic polynomial are the eigenvalues of A. Because the characteristic polynomial has exactly n roots, A has n eigenvalues, some of which can be repeated (with multiplicity $k \leq n$) and some of which can be complex numbers. For each eigenvalue λ_i, there is a corresponding eigenvector $\mathbf{x}_i$. This eigenvector is the solution of the homogeneous equations $(A - \lambda_i I)\mathbf{x}_i = \mathbf{0}$.

An important property of eigenvectors is their *linear independence* if there are n distinct eigenvalues. Vectors are linearly independent if the equation

$$\alpha_1\mathbf{x}_1 + \alpha_2\mathbf{x}_2 + \cdots + \alpha_n\mathbf{x}_n = \mathbf{0} \tag{3.5.4}$$

---

[6] The standard reference is Wilkinson, J. H., 1965: *The Algebraic Eigenvalue Problem*. Oxford University Press, 662 pp.

can be satisfied only by taking *all* of the coefficients α_n equal to zero.

This concept of linear independence or dependence actually extends to vectors in general, not just eigenvectors. Algebraists would say that our n linearly independent vectors form a *basis* that *spans* a *vector space* V. A vector space is simply a set V of vectors that can be added and scaled. The maximum number of linearly independent vectors in a vector space gives its *dimension* of V. A vector space V can have many different bases, but there are always the same number of basis vectors in each of them.

Returning to the eigenvalue problem, we now show that in the case of n distinct eigenvalues $\lambda_1, \lambda_2, \ldots, \lambda_n$, each eigenvalue λ_i having a corresponding eigenvector $\mathbf{x}_i$, the eigenvectors form a basis. We first write down the linear dependence condition

$$\alpha_1 \mathbf{x}_1 + \alpha_2 \mathbf{x}_2 + \cdots + \alpha_n \mathbf{x}_n = \mathbf{0}. \tag{3.5.5}$$

Premultiplying Equation 3.5.5 by A,

$$\alpha_1 A\mathbf{x}_1 + \alpha_2 A\mathbf{x}_2 + \cdots + \alpha_n A\mathbf{x}_n = \alpha_1 \lambda_1 \mathbf{x}_1 + \alpha_2 \lambda_2 \mathbf{x}_2 + \cdots + \alpha_n \lambda_n \mathbf{x}_n = \mathbf{0}. \tag{3.5.6}$$

Premultiplying Equation 3.5.5 by A^2,

$$\alpha_1 A^2 \mathbf{x}_1 + \alpha_2 A^2 \mathbf{x}_2 + \cdots + \alpha_n A^2 \mathbf{x}_n = \alpha_1 \lambda_1^2 \mathbf{x}_1 + \alpha_2 \lambda_2^2 \mathbf{x}_2 + \cdots + \alpha_n \lambda_n^2 \mathbf{x}_n = \mathbf{0}. \tag{3.5.7}$$

In a similar manner, we obtain the system of equations:

$$\begin{pmatrix} 1 & 1 & \cdots & 1 \\ \lambda_1 & \lambda_2 & \cdots & \lambda_n \\ \lambda_1^2 & \lambda_2^2 & \cdots & \lambda_n^2 \\ \vdots & \vdots & \ddots & \vdots \\ \lambda_1^{n-1} & \lambda_2^{n-1} & \cdots & \lambda_n^{n-1} \end{pmatrix} \begin{pmatrix} \alpha_1 \mathbf{x}_1 \\ \alpha_2 \mathbf{x}_2 \\ \alpha_3 \mathbf{x}_3 \\ \vdots \\ \alpha_n \mathbf{x}_n \end{pmatrix} = \begin{pmatrix} 0 \\ 0 \\ 0 \\ \vdots \\ 0 \end{pmatrix}. \tag{3.5.8}$$

Because

$$\begin{vmatrix} 1 & 1 & \cdots & 1 \\ \lambda_1 & \lambda_2 & \cdots & \lambda_n \\ \lambda_1^2 & \lambda_2^2 & \cdots & \lambda_n^2 \\ \vdots & \vdots & \ddots & \vdots \\ \lambda_1^{n-1} & \lambda_2^{n-1} & \cdots & \lambda_n^{n-1} \end{vmatrix} = \begin{matrix} (\lambda_2 - \lambda_1)(\lambda_3 - \lambda_2)(\lambda_3 - \lambda_1)(\lambda_4 - \lambda_3) \\ (\lambda_4 - \lambda_2) \cdots (\lambda_n - \lambda_1) \neq 0, \end{matrix} \tag{3.5.9}$$

since it is a Vandermonde determinant, $\alpha_1 \mathbf{x}_1 = \alpha_2 \mathbf{x}_2 = \alpha_3 \mathbf{x}_3 = \cdots = \alpha_n \mathbf{x}_n = 0$. Because the eigenvectors are nonzero, $\alpha_1 = \alpha_2 = \alpha_3 = \cdots = \alpha_n = 0$, and the eigenvectors are linearly independent. $\square$

This property of eigenvectors allows us to express any arbitrary vector $\mathbf{x}$ as a linear sum of the eigenvectors $\mathbf{x}_i$, or

$$\mathbf{x} = c_1 \mathbf{x}_1 + c_2 \mathbf{x}_2 + \cdots + c_n \mathbf{x}_n. \tag{3.5.10}$$

We will make good use of this property in Example 8.5.1.

• **Example 3.5.1**

Let us find the eigenvalues and corresponding eigenvectors of the matrix

$$A = \begin{pmatrix} -4 & 2 \\ -1 & -1 \end{pmatrix}. \tag{3.5.11}$$

We begin by setting up the characteristic equation:

$$\det(A - \lambda I) = \begin{vmatrix} -4 - \lambda & 2 \\ -1 & -1 - \lambda \end{vmatrix} = 0. \tag{3.5.12}$$

Expanding the determinant,

$$(-4 - \lambda)(-1 - \lambda) + 2 = \lambda^2 + 5\lambda + 6 = (\lambda + 3)(\lambda + 2) = 0. \tag{3.5.13}$$

Thus, the eigenvalues of the matrix A are $\lambda_1 = -3$, and $\lambda_2 = -2$.

To find the corresponding eigenvectors, we must solve the linear system:

$$\begin{pmatrix} -4 - \lambda & 2 \\ -1 & -1 - \lambda \end{pmatrix} \begin{pmatrix} x_1 \\ x_2 \end{pmatrix} = \begin{pmatrix} 0 \\ 0 \end{pmatrix}. \tag{3.5.14}$$

For example, for $\lambda_1 = -3$,

$$\begin{pmatrix} -1 & 2 \\ -1 & 2 \end{pmatrix} \begin{pmatrix} x_1 \\ x_2 \end{pmatrix} = \begin{pmatrix} 0 \\ 0 \end{pmatrix}, \tag{3.5.15}$$

or

$$x_1 = 2x_2. \tag{3.5.16}$$

Thus, any nonzero multiple of the vector $\begin{pmatrix} 2 \\ 1 \end{pmatrix}$ is an eigenvector belonging to $\lambda_1 = -3$.

Similarly, for $\lambda_2 = -2$, the eigenvector is any nonzero multiple of the vector $\begin{pmatrix} 1 \\ 1 \end{pmatrix}$.

Of course, MATLAB will do all of the computations for you via the command `eig`, which computes the eigenvalues and corresponding eigenvalues. In the present case, you would type:

```
>> A = [-4 2; -1 -1]; % load in array A
>> % find eigenvalues and eigenvectors
>> [eigenvector,eigenvalue] = eig(A)
```

This yields:

```
eigenvector =
            -0.8944              -0.7071
            -0.4472              -0.7071
```

and

```
eigenvalue =
             -3                   0
              0                  -2.
```

The eigenvalues are given as the elements along the principal diagonal of `eigenvalue`. The corresponding vectors are given by the corresponding column of `eigenvector`. As this example shows, these eigenvectors have been normalized so that their norm, Equation 3.1.5, equals one. Also, their sign may be different than any you would choose. We can recover our hand-computed results by dividing the first eigenvector by -0.4472 while in the second case we would divide by -0.7071. Finally, note that the product `eigenvector*eigenvalue*inv(eigenvector)` would yield `A`. □

• **Example 3.5.2**

Let us now find the eigenvalues and corresponding eigenvectors of the matrix

$$A = \begin{pmatrix} -4 & 5 & 5 \\ -5 & 6 & 5 \\ -5 & 5 & 6 \end{pmatrix}. \tag{3.5.17}$$

Setting up the characteristic equation:

$$\det(A - \lambda I) = \begin{vmatrix} -4-\lambda & 5 & 5 \\ -5 & 6-\lambda & 5 \\ -5 & 5 & 6-\lambda \end{vmatrix} = \begin{vmatrix} -4-\lambda & 5 & 5 \\ -5 & 6-\lambda & 5 \\ 0 & \lambda-1 & 1-\lambda \end{vmatrix} \tag{3.5.18}$$

$$= (\lambda-1) \begin{vmatrix} -4-\lambda & 5 & 5 \\ -5 & 6-\lambda & 5 \\ 0 & 1 & -1 \end{vmatrix} = (\lambda-1)^2 \begin{vmatrix} -1 & 1 & 0 \\ -5 & 6-\lambda & 5 \\ 0 & 1 & -1 \end{vmatrix} \tag{3.5.19}$$

$$= (\lambda-1)^2 \begin{vmatrix} -1 & 0 & 0 \\ -5 & 6-\lambda & 0 \\ 0 & 1 & -1 \end{vmatrix} = (\lambda-1)^2(6-\lambda) = 0. \tag{3.5.20}$$

Thus, the eigenvalues of the matrix A are $\lambda_{1,2} = 1$ (twice), and $\lambda_3 = 6$.

To find the corresponding eigenvectors, we must solve the linear system:

$$(-4-\lambda)x_1 + 5x_2 + 5x_3 = 0, \tag{3.5.21}$$

$$-5x_1 + (6-\lambda)x_2 + 5x_3 = 0, \tag{3.5.22}$$

and

$$-5x_1 + 5x_2 + (6-\lambda)x_3 = 0. \tag{3.5.23}$$

For $\lambda_3 = 6$, Equations 3.5.21 through 3.5.23 become

$$-10x_1 + 5x_2 + 5x_3 = 0, \tag{3.5.24}$$

$$-5x_1 + 5x_3 = 0, \tag{3.5.25}$$

and

$$-5x_1 + 5x_2 = 0. \tag{3.5.26}$$

Thus, $x_1 = x_2 = x_3$ and the eigenvector is any nonzero multiple of the vector $\begin{pmatrix} 1 \\ 1 \\ 1 \end{pmatrix}$.

The interesting aspect of this example centers on finding the eigenvector for the eigenvalue $\lambda_{1,2} = 1$. If $\lambda_{1,2} = 1$, then Equations 3.5.21 through 3.5.23 collapse into one equation,

$$-x_1 + x_2 + x_3 = 0, \tag{3.5.27}$$

and we have *two* free parameters at our disposal. Let us take $x_2 = \alpha$, and $x_3 = \beta$. Then the eigenvector equals $\alpha \begin{pmatrix} 1 \\ 1 \\ 0 \end{pmatrix} + \beta \begin{pmatrix} 1 \\ 0 \\ 1 \end{pmatrix}$ for $\lambda_{1,2} = 1$.

In this example, we may associate the eigenvector $\begin{pmatrix} 1 \\ 1 \\ 0 \end{pmatrix}$ with $\lambda_1 = 1$, and $\begin{pmatrix} 1 \\ 0 \\ 1 \end{pmatrix}$ with $\lambda_2 = 1$ so that, along with the eigenvector $\begin{pmatrix} 1 \\ 1 \\ 1 \end{pmatrix}$ with $\lambda_3 = 6$, we still have n *linearly independent* eigenvectors for our 3×3 matrix. However, with repeated eigenvalues this is not always true. For example,

$$A = \begin{pmatrix} 1 & -1 \\ 0 & 1 \end{pmatrix} \qquad (3.5.28)$$

has the repeated eigenvalues $\lambda_{1,2} = 1$. Still, there is only a single eigenvector $\begin{pmatrix} 1 \\ 0 \end{pmatrix}$ for *both* λ_1 and λ_2.

What happens in MATLAB in the present case? Typing in:

```
>> A = [-4 5 5; -5 6 5; -5 5 6]; % load in array A
>> % find eigenvalues and eigenvectors
>> [eigenvector,eigenvalue] = eig(A)
```

we obtain

```
eigenvector =
            -0.8165          0.5774          0.4259
            -0.4082          0.5774         -0.3904
            -0.4082          0.5774          0.8162
```

and

```
eigenvalue =
          1               0               0
          0               6               0
          0               0               1
```

The second eigenvector is clearly the same as the hand-computed one if you normalize it with 0.5774. The equivalence of the first and third eigenvectors is not as clear. However, if you choose $\alpha = \beta = -0.4082$, then the first eigenvector agrees with the hand-computed value. Similarly, taking $\alpha = -0.3904$ and $\beta = 0.8162$ result in agreement with the third MATLAB eigenvector. Finally, note that the product `eigenvector*eigenvalue*inv(eigenvector)` would yield `A`. □

• **Example 3.5.3: Diagonalization of a matrix A with distinct eigenvalues**

Consider an $n \times n$ matrix A which has n distinct eigenvalues $\lambda_1, \lambda_2, \cdots, \lambda_n$ and n corresponding (and distinct) eigenvectors $\mathbf{p}_1, \mathbf{p}_2, \cdots, \mathbf{p}_n$. If we introduce a matrix $P = [\mathbf{p}_1 \mathbf{p}_2 \cdots \mathbf{p}_n]$ (the eigenvectors form the columns of P), called the modal matrix, and recall that $P\mathbf{p}_j = \lambda_j \mathbf{p}_j$, then

$$AP = A[\mathbf{p}_1 \mathbf{p}_2 \cdots \mathbf{p}_n] = [A\mathbf{p}_1 A\mathbf{p}_2 \cdots A\mathbf{p}_n] = [\lambda_1 \mathbf{p}_1 \lambda_2 \mathbf{p}_2 \cdots \lambda_n \mathbf{p}_n]. \qquad (3.5.29)$$

Therefore, $AP = PD$, where

$$D = \begin{pmatrix} \lambda_1 & 0 & \cdots & 0 \\ 0 & \lambda_2 & \cdots & 0 \\ \vdots & \vdots & \ddots & \vdots \\ 0 & 0 & \cdots & \lambda_n \end{pmatrix}. \qquad (3.5.30)$$

Because P has rank n, P^{-1} exists and $D = P^{-1}AP$. Therefore, performing the operation $P^{-1}AP$ is a process by which we can *diagonalize* the matrix A using the eigenvectors of A. Diagonalizable matrices are of interest because diagonal matrices are especially easy to use. Finally, we note that

$$D^2 = DD = P^{-1}APP^{-1}AP = P^{-1}AAP = P^{-1}A^2P. \tag{3.5.31}$$

Repeating this process, we eventually obtain the general result that $D^m = P^{-1}A^mP$.
    To verify $P^{-1}AP = D$, let us use

$$A = \begin{pmatrix} 3 & 4 \\ 1 & 3 \end{pmatrix}. \tag{3.5.32}$$

This matrix has the eigenvalues $\lambda_{1,2} = 1, 5$ with the corresponding eigenvectors $\mathbf{p}_1 = (2 \quad -1)^T$ and $\mathbf{p}_2 = (2 \quad 1)^T$. Therefore,

$$P = \begin{pmatrix} 2 & 2 \\ -1 & 1 \end{pmatrix}, \quad \text{and} \quad P^{-1} = \begin{pmatrix} 1/4 & -1/2 \\ 1/4 & 1/2 \end{pmatrix}. \tag{3.5.33}$$

Therefore,

$$P^{-1}AP = \begin{pmatrix} 1/4 & -1/2 \\ 1/4 & 1/2 \end{pmatrix} \begin{pmatrix} 3 & 4 \\ 1 & 3 \end{pmatrix} \begin{pmatrix} 2 & 2 \\ -1 & 1 \end{pmatrix} = \begin{pmatrix} 1/4 & -1/2 \\ 1/4 & 1/2 \end{pmatrix} \begin{pmatrix} 2 & 10 \\ -1 & 5 \end{pmatrix} = \begin{pmatrix} 1 & 0 \\ 0 & 5 \end{pmatrix}. \tag{3.5.34}$$

Similarly, if we are given D and P for an $n \times n$ matrix A, we can express A as the product PDP^{-1} and $A^m = PD^mP^{-1}$.

Problems

Find the eigenvalues and corresponding eigenvectors for the following matrices. Check your answers using MATLAB.

1. $A = \begin{pmatrix} 3 & 2 \\ 3 & -2 \end{pmatrix}$
2. $A = \begin{pmatrix} 3 & -1 \\ 1 & 1 \end{pmatrix}$
3. $A = \begin{pmatrix} 1 & 3 \\ 2 & 1 \end{pmatrix}$

4. $A = \begin{pmatrix} -5 & 0 \\ 1 & 2 \end{pmatrix}$
5. $A = \begin{pmatrix} 2 & -3 & 1 \\ 1 & -2 & 1 \\ 1 & -3 & 2 \end{pmatrix}$
6. $A = \begin{pmatrix} 0 & 1 & 0 \\ 0 & 0 & 1 \\ 0 & 0 & 0 \end{pmatrix}$

7. $A = \begin{pmatrix} 1 & 1 & 1 \\ 0 & 2 & 1 \\ 0 & 0 & 1 \end{pmatrix}$
8. $A = \begin{pmatrix} 1 & 2 & 1 \\ 0 & 3 & 1 \\ 0 & 5 & -1 \end{pmatrix}$
9. $A = \begin{pmatrix} 4 & -5 & 1 \\ 1 & 0 & -1 \\ 0 & 1 & -1 \end{pmatrix}$

10. $A = \begin{pmatrix} -2 & 0 & 1 \\ 3 & 0 & -1 \\ 0 & 1 & 1 \end{pmatrix}$
11. $A = \begin{pmatrix} 2 & 0 & 0 \\ 1 & 0 & 2 \\ 0 & 0 & 3 \end{pmatrix}$
12. $A = \begin{pmatrix} -3 & 1 & 1 \\ 0 & 0 & 0 \\ 0 & 1 & 0 \end{pmatrix}$

13. Using the formula $A^m = PD^mP^{-1}$, compute A^{23} if $A = \begin{pmatrix} 2 & 3 \\ 3 & 2 \end{pmatrix}$.

Project: Singular Value Decomposition (SVD) and Linear Least Squares

In projects at the end of Section 3.4 we showed two ways that linear equations can be solved by factoring the matrix A in $A\mathbf{x} = \mathbf{b}$ into a product of two matrices. The LU and QR decompositions are *not* the only possible factorization. One popular version rewrites the square matrix A as PDP^{-1}, where D is a diagonal matrix with the n eigenvalues along the principal diagonal and P contains the eigenvectors. MATLAB's routine `eig` yields both D and P via `[P,D] = eig(A)`. See Example 3.5.2, for example. In this project we focus on singular value decomposition, possibly the most important matrix decomposition of them all. It is used in signal processing, statistics, and numerical methods and theory.

Singular value decomposition factors a matrix A of dimension $m \times n$ into the product of three matrices: $A = UDV^T$, where U is an $m \times m$ orthogonal matrix, V is an $n \times n$ orthogonal matrix, and D is an $m \times n$ diagonal matrix. The diagonal entries of D are called the *singular values* of A. The rank of a matrix equals the number of non-zero singular values.

Consider the matrix B given by

$$B = \begin{pmatrix} 45 & -108 & 36 & -45 \\ 21 & -68 & 26 & -33 \\ 72 & -32 & -16 & 24 \\ -56 & 64 & -8 & 8 \\ 50 & -32 & -6 & 10 \end{pmatrix}.$$

Using MATLAB's subroutine `svd`, confirm that it can factor the array B into U, D, and V. Then check that $B = UDV^T$ and find the rank of this matrix.

Let us now seek to find the parameters β and γ so that the line $y = \beta x + \gamma$ gives the best fit to n data points. If there are only two data points, there is no problem because we could immediately find the slope β and the intercept γ. On the other hand, if $n > 2$ the question immediately arises as to the best method for computing β and γ so that a straight line best fits the data. How does singular value decomposition come to the rescue?

To find the answer, we begin by noting that each data point (x_i, y_i) must satisfy the linear equation $\beta x_i + \gamma = y_i$, or

$$\beta \begin{pmatrix} x_1 \\ x_2 \\ \vdots \\ x_n \end{pmatrix} + \gamma \begin{pmatrix} 1 \\ 1 \\ \vdots \\ 1 \end{pmatrix} = \begin{pmatrix} x_1 & 1 \\ x_2 & 1 \\ \vdots & \vdots \\ x_n & 1 \end{pmatrix} \begin{pmatrix} \beta \\ \gamma \end{pmatrix} = A\mathbf{x} = \begin{pmatrix} y_1 \\ y_2 \\ \vdots \\ y_n \end{pmatrix} = \mathbf{y}.$$

We can write this over-determined system of linear equations $A\mathbf{x} = \mathbf{y}$. Ideally we would like to choose our vector $\mathbf{x}$ so that it yields the smallest Euclidean norm of the residual $A\mathbf{x} - \mathbf{y}$; such a solution fits the data in the linear "least-squares sense." Denoting this solution by $\mathbf{x}^*$, we have that $A\mathbf{x}^* \approx \mathbf{y}$. Using singular decomposition we can rewrite this system as $UDV^T\mathbf{x}^* \approx \mathbf{y}$ or $\mathbf{x}^* \approx VD^{-1}U^T\mathbf{y}$.

Step 1: Show that the rank for the matrix A is 2.

Step 2: In Step 1, we found that the rank for the matrix A is 2. What values of $\mathbf{x}^*$ does singular value decomposition give if $\mathbf{y} = (-2\ 4\ 4\ 6\ -4)^T$? This solution equals the least-squares solution of minimum length.

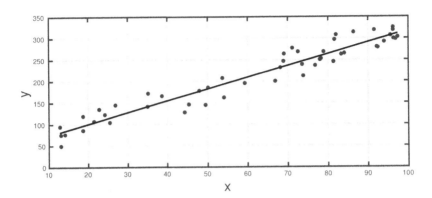

Figure 3.5.2: Using 50 "data points," the singular value decomposition (SVD) provides a least-squares fit to the data.

Step 3: Returning to our original goal of finding the best linear fit to data, create data for your numerical experiment. One way would use the simple line $y = \beta x + \gamma$ (with arbitrarily chosen values of β and γ) to create the initial data and then use the random number generator `rand` to modify this initial data (both x and y) so that both have "noise."

Step 4: Using your data, construct the array A and column vector **y**. Using the MATLAB routine `svd`, find $\mathbf{x}^* = VD^{-1}U^T\mathbf{y}$. An alternative method is to first find `M = transpose(A)*A` followed by `x = inv(M)*transpose(A)*b`. Why does alternative method work?

Step 5: Construct the least-squares fit for the data and plot this curve and your data on the same figure. See Figure 3.5.2.

Step 6: Singular value decomposition is not the only method to solve for **x**. We could also use MATLAB's backslash to find `x = A\y` where MATLAB uses the QR factorization to find the inverse. Compare this method of finding β and γ with your singular value decomposition results.

Step 7: Sometimes a linear fit is not appropriate and we should use a higher-order expansion, for example $y = \delta x^2 + \beta x + \gamma$. Redo Steps 3 to 6 to find δ, β and γ for noisy data where the initial data fits a quadratic.

3.6 SYSTEMS OF LINEAR DIFFERENTIAL EQUATIONS

In this section we show how we may apply the classic algebraic eigenvalue problem to solve a system of ordinary differential equations.

Let us solve the following system:

$$x_1' = x_1 + 3x_2, \tag{3.6.1}$$

and

$$x_2' = 3x_1 + x_2, \tag{3.6.2}$$

where the primes denote the time derivative.

We begin by rewriting Equation 3.6.1 and Equation 3.6.2 in matrix notation:

$$\mathbf{x}' = A\mathbf{x}, \tag{3.6.3}$$

where

$$\mathbf{x} = \begin{pmatrix} x_1 \\ x_2 \end{pmatrix}, \qquad \text{and} \qquad A = \begin{pmatrix} 1 & 3 \\ 3 & 1 \end{pmatrix}. \tag{3.6.4}$$

Note that

$$\begin{pmatrix} x_1' \\ x_2' \end{pmatrix} = \frac{d}{dt} \begin{pmatrix} x_1 \\ x_2 \end{pmatrix} = \mathbf{x}'. \tag{3.6.5}$$

Assuming a solution of the form

$$\mathbf{x} = \mathbf{x}_0 e^{\lambda t}, \qquad \text{where} \qquad \mathbf{x}_0 = \begin{pmatrix} a \\ b \end{pmatrix} \tag{3.6.6}$$

is a constant vector, we substitute Equation 3.6.6 into Equation 3.6.3 and find that

$$\lambda e^{\lambda t} \mathbf{x}_0 = A e^{\lambda t} \mathbf{x}_0. \tag{3.6.7}$$

Because $e^{\lambda t}$ does not generally equal zero, we have that

$$(A - \lambda I)\mathbf{x}_0 = \mathbf{0}, \tag{3.6.8}$$

which we solved in the previous section. This set of homogeneous equations is the *classic eigenvalue problem*. In order for this set not to have trivial solutions,

$$\det(A - \lambda I) = \begin{vmatrix} 1 - \lambda & 3 \\ 3 & 1 - \lambda \end{vmatrix} = 0. \tag{3.6.9}$$

Expanding the determinant,

$$(1 - \lambda)^2 - 9 = 0 \quad \text{or} \quad \lambda = -2, 4. \tag{3.6.10}$$

Thus, we have two real and distinct eigenvalues: $\lambda = -2$ and 4.

We must now find the corresponding $\mathbf{x}_0$ or *eigenvector* for each eigenvalue. From Equation 3.6.8,

$$(1 - \lambda)a + 3b = 0, \tag{3.6.11}$$

and

$$3a + (1 - \lambda)b = 0. \tag{3.6.12}$$

If $\lambda = 4$, these equations are consistent and yield $a = b = c_1$. If $\lambda = -2$, we have that $a = -b = c_2$. Therefore, the general solution in matrix notation is

$$\mathbf{x} = c_1 \begin{pmatrix} 1 \\ 1 \end{pmatrix} e^{4t} + c_2 \begin{pmatrix} 1 \\ -1 \end{pmatrix} e^{-2t}. \tag{3.6.13}$$

To evaluate c_1 and c_2, we must have initial conditions. For example, if $x_1(0) = x_2(0) = 1$, then

$$\begin{pmatrix} 1 \\ 1 \end{pmatrix} = c_1 \begin{pmatrix} 1 \\ 1 \end{pmatrix} + c_2 \begin{pmatrix} 1 \\ -1 \end{pmatrix}. \tag{3.6.14}$$

Solving for c_1 and c_2, $c_1 = 1$, $c_2 = 0$, and the solution with this particular set of initial conditions is

$$\mathbf{x} = \begin{pmatrix} 1 \\ 1 \end{pmatrix} e^{4t}. \tag{3.6.15}$$

• **Example 3.6.1**

Let us solve the following set of linear ordinary differential equations

$$x_1' = -x_2 + x_3, \tag{3.6.16}$$

$$x_2' = 4x_1 - x_2 - 4x_3, \tag{3.6.17}$$

and

$$x_3' = -3x_1 - x_2 + 4x_3; \tag{3.6.18}$$

or in matrix form,

$$\mathbf{x}' = \begin{pmatrix} 0 & -1 & 1 \\ 4 & -1 & -4 \\ -3 & -1 & 4 \end{pmatrix} \mathbf{x}, \qquad \mathbf{x} = \begin{pmatrix} x_1 \\ x_2 \\ x_3 \end{pmatrix}. \tag{3.6.19}$$

Assuming the solution $\mathbf{x} = \mathbf{x}_0 e^{\lambda t}$,

$$\begin{pmatrix} 0 & -1 & 1 \\ 4 & -1 & -4 \\ -3 & -1 & 4 \end{pmatrix} \mathbf{x}_0 = \lambda \mathbf{x}_0, \tag{3.6.20}$$

or

$$\begin{pmatrix} -\lambda & -1 & 1 \\ 4 & -1-\lambda & -4 \\ -3 & -1 & 4-\lambda \end{pmatrix} \mathbf{x}_0 = \mathbf{0}. \tag{3.6.21}$$

For nontrivial solutions,

$$\begin{vmatrix} -\lambda & -1 & 1 \\ 4 & -1-\lambda & -4 \\ -3 & -1 & 4-\lambda \end{vmatrix} = \begin{vmatrix} 0 & 0 & 1 \\ 4-4\lambda & -5-\lambda & -4 \\ -3+4\lambda-\lambda^2 & 3-\lambda & 4-\lambda \end{vmatrix} = 0, \tag{3.6.22}$$

and

$$(\lambda - 1)(\lambda - 3)(\lambda + 1) = 0, \quad \text{or} \quad \lambda = -1, 1, 3. \tag{3.6.23}$$

To determine the eigenvectors, we rewrite Equation 3.6.21 as

$$-\lambda a - b + c = 0, \tag{3.6.24}$$

$$4a - (1+\lambda)b - 4c = 0, \tag{3.6.25}$$

and

$$-3a - b + (4-\lambda)c = 0. \tag{3.6.26}$$

For example, if $\lambda = 1$,

$$-a - b + c = 0, \tag{3.6.27}$$

$$4a - 2b - 4c = 0, \tag{3.6.28}$$

and

$$-3a - b + 3c = 0; \tag{3.6.29}$$

or $a = c$, and $b = 0$. Thus, the eigenvector for $\lambda = 1$ is $\mathbf{x}_0 = \begin{pmatrix} 1 \\ 0 \\ 1 \end{pmatrix}$. Similarly, for $\lambda = -1$,

$\mathbf{x}_0 = \begin{pmatrix} 1 \\ 2 \\ 1 \end{pmatrix}$; and for $\lambda = 3$, $\mathbf{x}_0 = \begin{pmatrix} 1 \\ -1 \\ 2 \end{pmatrix}$. Thus, the most general solution is

$$\mathbf{x} = c_1 \begin{pmatrix} 1 \\ 0 \\ 1 \end{pmatrix} e^t + c_2 \begin{pmatrix} 1 \\ 2 \\ 1 \end{pmatrix} e^{-t} + c_3 \begin{pmatrix} 1 \\ -1 \\ 2 \end{pmatrix} e^{3t}. \tag{3.6.30}$$

$\square$

• **Example 3.6.2**

Let us solve the following set of linear ordinary differential equations:

$$x_1' = x_1 - 2x_2, \tag{3.6.31}$$

and

$$x_2' = 2x_1 - 3x_2; \tag{3.6.32}$$

or in matrix form,

$$\mathbf{x}' = \begin{pmatrix} 1 & -2 \\ 2 & -3 \end{pmatrix} \mathbf{x}, \qquad \mathbf{x} = \begin{pmatrix} x_1 \\ x_2 \end{pmatrix}. \tag{3.6.33}$$

Assuming the solution $\mathbf{x} = \mathbf{x}_0 e^{\lambda t}$,

$$\begin{pmatrix} 1 - \lambda & -2 \\ 2 & -3 - \lambda \end{pmatrix} \mathbf{x}_0 = \mathbf{0}. \tag{3.6.34}$$

For nontrivial solutions,

$$\begin{vmatrix} 1 - \lambda & -2 \\ 2 & -3 - \lambda \end{vmatrix} = (\lambda + 1)^2 = 0. \tag{3.6.35}$$

Thus, we have the solution

$$\mathbf{x} = c_1 \begin{pmatrix} 1 \\ 1 \end{pmatrix} e^{-t}. \tag{3.6.36}$$

The interesting aspect of this example is the single solution that the traditional approach yields because we have repeated roots. To find the second solution, we try the solution

$$\mathbf{x} = \begin{pmatrix} a + ct \\ b + dt \end{pmatrix} e^{-t}. \tag{3.6.37}$$

We guessed Equation 3.6.37 using our knowledge of solutions to differential equations when the characteristic polynomial has repeated roots. Substituting Equation 3.6.37 into Equation 3.6.33, we find that $c = d = 2c_2$, and $a - b = c_2$. Thus, we have one free parameter, which we choose to be b, and set it equal to zero. This is permissible because Equation 3.6.37 can be broken into two terms: $b \begin{pmatrix} 1 \\ 1 \end{pmatrix} e^{-t}$ and $c_2 \begin{pmatrix} 1 + 2t \\ 2t \end{pmatrix} e^{-t}$. The first term can be incorporated into the $c_1 \begin{pmatrix} 1 \\ 1 \end{pmatrix} e^{-t}$ term. Thus, the general solution is

$$\mathbf{x} = c_1 \begin{pmatrix} 1 \\ 1 \end{pmatrix} e^{-t} + c_2 \begin{pmatrix} 1 \\ 0 \end{pmatrix} e^{-t} + 2c_2 \begin{pmatrix} 1 \\ 1 \end{pmatrix} te^{-t}. \tag{3.6.38}$$

$\square$

• **Example 3.6.3**

Let us solve the system of linear differential equations:

$$x_1' = 2x_1 - 3x_2, \tag{3.6.39}$$

and

$$x_2' = 3x_1 + 2x_2; \tag{3.6.40}$$

or in matrix form,

$$\mathbf{x}' = \begin{pmatrix} 2 & -3 \\ 3 & 2 \end{pmatrix} \mathbf{x}, \qquad \mathbf{x} = \begin{pmatrix} x_1 \\ x_2 \end{pmatrix}. \tag{3.6.41}$$

Assuming the solution $\mathbf{x} = \mathbf{x}_0 e^{\lambda t}$,

$$\begin{pmatrix} 2 - \lambda & -3 \\ 3 & 2 - \lambda \end{pmatrix} \mathbf{x}_0 = \mathbf{0}. \tag{3.6.42}$$

For nontrivial solutions,

$$\begin{vmatrix} 2 - \lambda & -3 \\ 3 & 2 - \lambda \end{vmatrix} = (2 - \lambda)^2 + 9 = 0, \tag{3.6.43}$$

and $\lambda = 2 \pm 3i$. If $\mathbf{x}_0 = \begin{pmatrix} a \\ b \end{pmatrix}$, then $b = -ai$ if $\lambda = 2 + 3i$, and $b = ai$ if $\lambda = 2 - 3i$. Thus, the general solution is

$$\mathbf{x} = c_1 \begin{pmatrix} 1 \\ -i \end{pmatrix} e^{2t+3it} + c_2 \begin{pmatrix} 1 \\ i \end{pmatrix} e^{2t-3it}, \tag{3.6.44}$$

where c_1 and c_2 are arbitrary complex constants. Using Euler relationships, we can rewrite Equation 3.6.44 as

$$\mathbf{x} = c_3 \begin{bmatrix} \cos(3t) \\ \sin(3t) \end{bmatrix} e^{2t} + c_4 \begin{bmatrix} \sin(3t) \\ -\cos(3t) \end{bmatrix} e^{2t}, \tag{3.6.45}$$

where $c_3 = c_1 + c_2$ and $c_4 = i(c_1 - c_2)$.

Problems

Find the general solution of the following sets of ordinary differential equations using the matrix technique. You may find the eigenvalues and eigenvectors either by hand or use MATLAB.

1. $x_1' = x_1 + 2x_2$ $\qquad\qquad\qquad\qquad\qquad$ $x_2' = 2x_1 + x_2$

2. $x_1' = x_1 - 4x_2$ $\qquad\qquad\qquad\qquad\qquad$ $x_2' = 3x_1 - 6x_2$

3. $x_1' = x_1 + x_2$ $\qquad\qquad\qquad\qquad\qquad$ $x_2' = 4x_1 + x_2$

4. $x_1' = x_1 + 5x_2$ $\qquad\qquad\qquad\qquad\qquad$ $x_2' = -2x_1 - 6x_2$

5. $x_1' = -\frac{3}{2}x_1 - 2x_2$ $\qquad\qquad\qquad\qquad$ $x_2' = 2x_1 + \frac{5}{2}x_2.$

6. $x_1' = -3x_1 - 2x_2$ $x_2' = 2x_1 + x_2$

7. $x_1' = x_1 - x_2$ $x_2' = x_1 + 3x_2$

8. $x_1' = 3x_1 + 2x_2$ $x_2' = -2x_1 - x_2$

9. $x_1' = -2x_1 - 13x_2$ $x_2' = x_1 + 4x_2$

10. $x_1' = 3x_1 - 2x_2$ $x_2' = 5x_1 - 3x_2$

11. $x_1' = 4x_1 - 2x_2$ $x_2' = 25x_1 - 10x_2$

12. $x_1' = -3x_1 - 4x_2$ $x_2' = 2x_1 + x_2$

13. $x_1' = 3x_1 + 4x_2$ $x_2' = -2x_1 - x_2$

14. $x_1' + 5x_1 + x_2' + 3x_2 = 0$ $2x_1' + x_1 + x_2' + x_2 = 0$

15. $x_1' - x_1 + x_2' - 2x_2 = 0$ $x_1' - 5x_1 + 2x_2' - 7x_2 = 0$

16. $x_1' = x_1 - 2x_2$ $x_2' = 0$ $x_3' = -5x_1 + 7x_3.$

17. $x_1' = 2x_1$ $x_2' = x_1 + 2x_3$ $x_3' = x_3.$

18. $x_1' = 3x_1 - 2x_3$ $x_2' = -x_1 + 2x_2 + x_3$ $x_3' = 4x_1 - 3x_3$

19. $x_1' = 3x_1 - x_3$ $x_2' = -2x_1 + 2x_2 + x_3$ $x_3' = 8x_1 - 3x_3$

3.7 MATRIX EXPONENTIAL

In the previous section we solved initial-value problems involving systems of linear ordinary differential equations via the eigenvalue problem. Here we introduce an alternative method based on the *matrix exponential*, defined by

$$e^{At} = I + At + \tfrac{1}{2!}A^2t^2 + \cdots + \tfrac{1}{k!}A^k t^k + \cdots. \tag{3.7.1}$$

Clearly

$$e^0 = e^{A0} = I, \quad \text{and} \quad \frac{d}{dt}\left(e^{At}\right) = Ae^{At}. \tag{3.7.2}$$

Therefore, using the matrix exponential function, the solution to the system of homogeneous linear first-order differential equations with constant coefficients

$$\mathbf{x}' = A\mathbf{x}, \quad \mathbf{x}(0) = \mathbf{x}_0, \tag{3.7.3}$$

is $\mathbf{x}(t) = e^{At}\mathbf{x}_0$.

The question now arises as to how to compute this matrix exponential. There are several methods. For example, from the concept of diagonalization of a matrix (see Example 3.6.4) we can write $A = PDP^{-1}$, where P are the eigenvectors of A. Then,

$$e^A = \sum_{k=0}^{\infty} \frac{(PDP^{-1})^k}{k!} = \sum_{k=0}^{\infty} P\frac{D^k}{k!}P^{-1} = P\left(\sum_{k=0}^{\infty} \frac{D^k}{k!}\right)P^{-1} = Pe^D P^{-1}, \tag{3.7.4}$$

where

$$
e^D = \begin{pmatrix} e^{\lambda_1} & 0 & \cdots & 0 \\ 0 & e^{\lambda_2} & \cdots & 0 \\ \vdots & \vdots & \vdots & \vdots \\ 0 & 0 & \cdots & e^{\lambda_n} \end{pmatrix}.
\tag{3.7.5}
$$

Because many software packages contain routines for finding eigenvalues and eigenvectors, Equation 3.7.4 provides a convenient method for computing e^A. In the case of MATLAB, we just have to invoke the intrinsic function `expm(·)`.

In this section we focus on a recently developed method by Liz,[7] who improved a method constructed by Leonard.[8] The advantage of this method is that it uses techniques that we have already introduced. We will first state the result and then illustrate its use.

The main result of Liz's analysis is:

Theorem: *Let A be a constant $n \times n$ matrix with characteristic polynomial $p(\lambda) = \lambda^n + c_{n-1}\lambda^{n-1} + \cdots + c_1\lambda + c_0$. Then*

$$
e^{At} = x_1(t)I + x_2(t)A + \cdots + x_n(t)A^{n-1},
\tag{3.7.6}
$$

where

$$
\begin{pmatrix} x_1(t) \\ x_2(t) \\ \vdots \\ x_n(t) \end{pmatrix} = B_0^{-1} \begin{pmatrix} \varphi_1(t) \\ \varphi_2(t) \\ \vdots \\ \varphi_n(t) \end{pmatrix},
\tag{3.7.7}
$$

$$
B_t = \begin{pmatrix} \varphi_1(t) & \varphi_1'(t) & \cdots & \varphi_1^{(n-1)}(t) \\ \varphi_2(t) & \varphi_2'(t) & \cdots & \varphi_2^{(n-1)}(t) \\ \vdots & \vdots & \cdots & \vdots \\ \varphi_n(t) & \varphi_n'(t) & \cdots & \varphi_n^{(n-1)}(t) \end{pmatrix},
\tag{3.7.8}
$$

and $S = \{\varphi_1(t), \varphi_2(t), \ldots, \varphi_n(t)\}$ being a fundamental system of solutions for the homogeneous linear differential equations whose characteristic equation is the characteristic equation of A, $p(\lambda) = 0$. The proof is given in Liz's paper. Note that for this technique to work, $x_1(0) = 1$ and $x_2(0) = x_3(0) = \cdots = x_n(0) = 0$. ☐

● **Example 3.7.1**

Let us illustrate this method of computing the matrix exponential by solving

$$
x' = 2x - y + z,
\tag{3.7.9}
$$

$$
y' = 3y - z,
\tag{3.7.10}
$$

and

$$
z' = 2x + y + 3z.
\tag{3.7.11}
$$

---

[7] Liz, E., 1998: A note on the matrix exponential. *SIAM Rev.*, **40**, 700–702.

[8] Leonard, I. E., 1996: The matrix exponential. *SIAM Rev.*, **38**, 507–512.

The solution to this system of equations is $\mathbf{x} = e^{At}\mathbf{x}_0$, where

$$A = \begin{pmatrix} 2 & -1 & 1 \\ 0 & 3 & -1 \\ 2 & 1 & 3 \end{pmatrix} \qquad \text{and} \qquad \mathbf{x} = \begin{pmatrix} x(t) \\ y(t) \\ z(t) \end{pmatrix}. \tag{3.7.12}$$

The vector $\mathbf{x}_0$ is the value of $\mathbf{x}(t)$ at $t = 0$.

Our first task is to compute the characteristic polynomial $p(\lambda) = 0$. This is simply

$$|\lambda I - A| = \begin{vmatrix} \lambda - 2 & 1 & -1 \\ 0 & \lambda - 3 & 1 \\ -2 & -1 & \lambda - 3 \end{vmatrix} = (\lambda - 2)^2(\lambda - 4) = 0. \tag{3.7.13}$$

Consequently, $\lambda = 2$ twice and $\lambda = 4$, and the fundamental solutions are $S = \{e^{2t}, te^{2t}, e^{4t}\}$. Therefore,

$$B_t = \begin{pmatrix} e^{4t} & 4e^{4t} & 16e^{4t} \\ e^{2t} & 2e^{2t} & 4e^{2t} \\ te^{2t} & e^{2t} + 2te^{2t} & 4e^{2t} + 4te^{2t} \end{pmatrix}, \tag{3.7.14}$$

and

$$A^2 = \begin{pmatrix} 6 & -4 & 6 \\ -2 & 8 & -6 \\ 10 & 4 & 10 \end{pmatrix}, \quad B_0 = \begin{pmatrix} 1 & 4 & 16 \\ 1 & 2 & 4 \\ 0 & 1 & 4 \end{pmatrix}, \quad B_0^{-1} = \begin{pmatrix} -1 & 0 & -4 \\ -1 & 1 & 3 \\ \frac{1}{4} & -\frac{1}{4} & -\frac{1}{2} \end{pmatrix}. \tag{3.7.15}$$

The inverse B_0^{-1} can be found using either Gaussian elimination or MATLAB.

To find $x_1(t)$, $x_2(t)$ and $x_3(t)$, we have from Equation 3.7.7 that

$$\begin{pmatrix} x_1(t) \\ x_2(t) \\ x_3(t) \end{pmatrix} = \begin{pmatrix} -1 & 0 & -4 \\ -1 & 1 & 3 \\ \frac{1}{4} & -\frac{1}{4} & -\frac{1}{2} \end{pmatrix} \begin{pmatrix} e^{4t} \\ e^{2t} \\ te^{2t} \end{pmatrix} = \begin{pmatrix} e^{4t} - 4te^{2t} \\ e^{2t} - e^{4t} + 3te^{2t} \\ \frac{1}{4}e^{4t} - \frac{1}{4}e^{2t} - \frac{1}{2}te^{2t} \end{pmatrix}, \tag{3.7.16}$$

or

$$x_1(t) = e^{4t} - 4te^{2t}, \quad x_2(t) = e^{2t} - e^{4t} + 3te^{2t}, \quad x_3(t) = \frac{1}{4}e^{4t} - \frac{1}{4}e^{2t} - \frac{1}{2}te^{2t}. \tag{3.7.17}$$

Note that $x_1(0) = 1$ while $x_2(0) = x_3(0) = 0$.

Finally, we have that

$$e^{At} = x_1(t) \begin{pmatrix} 1 & 0 & 0 \\ 0 & 1 & 0 \\ 0 & 0 & 1 \end{pmatrix} + x_2(t) \begin{pmatrix} 2 & -1 & 1 \\ 0 & 3 & -1 \\ 2 & 1 & 3 \end{pmatrix} + x_3(t) \begin{pmatrix} 6 & -4 & 6 \\ -2 & 8 & -6 \\ 10 & 4 & 10 \end{pmatrix}. \tag{3.7.18}$$

Substituting for $x_1(t)$, $x_2(t)$ and $x_3(t)$ and simplifying, we finally obtain

$$e^{At} = \frac{1}{2} \begin{pmatrix} e^{4t} + e^{2t} - 2te^{2t} & -2te^{2t} & e^{4t} - e^{2t} \\ e^{2t} - e^{4t} + 2te^{2t} & 2(t+1)e^{2t} & e^{2t} - e^{4t} \\ e^{4t} - e^{2t} + 2te^{2t} & 2te^{2t} & e^{4t} + e^{2t} \end{pmatrix}; \tag{3.7.19}$$

or

$$x_1(t) = \left(\tfrac{1}{2}e^{4t} + \tfrac{1}{2}e^{2t} - te^{2t}\right)x_1(0) - te^{2t}x_2(0) + \left(\tfrac{1}{2}e^{4t} - \tfrac{1}{2}e^{2t}\right)x_3(0), \tag{3.7.20}$$

$$x_2(t) = \left(\tfrac{1}{2}e^{2t} - \tfrac{1}{2}e^{4t} + te^{2t}\right)x_1(0) + (t+1)e^{2t}x_2(0) + \left(\tfrac{1}{2}e^{2t} - \tfrac{1}{2}e^{4t}\right)x_3(0), \tag{3.7.21}$$

and

$$x_3(t) = \left(\tfrac{1}{2}e^{4t} - \tfrac{1}{2}e^{2t} + te^{2t}\right) x_1(0) + te^{2t}x_2(0) + \left(\tfrac{1}{2}e^{4t} + \tfrac{1}{2}e^{2t}\right) x_3(0). \qquad (3.7.22)$$

$\square$

• **Example 3.7.2**

The matrix exponential can also be used to solve systems of first-order, nonhomogeneous linear ordinary differential equations. To illustrate this, consider the following system of linear ordinary differential equations:

$$x' = x - 4y + e^{2t}, \qquad (3.7.23)$$

and

$$y' = x + 5y + t. \qquad (3.7.24)$$

We can rewrite this system as

$$\mathbf{x}' = A\mathbf{x} + \mathbf{b}, \qquad (3.7.25)$$

where

$$A = \begin{pmatrix} 1 & -4 \\ 1 & 5 \end{pmatrix}, \qquad \mathbf{b} = \begin{pmatrix} e^{2t} \\ t \end{pmatrix}, \qquad \text{and} \qquad \mathbf{x} = \begin{pmatrix} x(t) \\ y(t) \end{pmatrix}. \qquad (3.7.26)$$

We leave as an exercise the computation of the matrix exponential and find that

$$e^{At} = \begin{pmatrix} e^{3t} - 2te^{3t} & -4te^{3t} \\ te^{3t} & e^{3t} + 2te^{3t} \end{pmatrix}. \qquad (3.7.27)$$

Clearly the homogeneous solution is $\mathbf{x}_H(t) = e^{At}\mathbf{C}$, where $\mathbf{C}$ is the arbitrary constant that is determined by the initial condition. But how do we find the particular solution, $\mathbf{x}_p(t)$? Let $\mathbf{x}_p(t) = e^{At}\mathbf{y}(t)$. Then

$$\mathbf{x}_p'(t) = Ae^{At}\mathbf{y}(t) + e^{At}\mathbf{y}'(t), \qquad (3.7.28)$$

or

$$\mathbf{x}_p'(t) = A\mathbf{x}_p(t) + e^{At}\mathbf{y}'(t). \qquad (3.7.29)$$

Therefore,

$$e^{At}\mathbf{y}'(t) = \mathbf{b}(t), \qquad \text{or} \qquad \mathbf{y}(t) = e^{-At}\mathbf{b}(t), \qquad (3.7.30)$$

since $\left(e^A\right)^{-1} = e^{-A}$. Integrating both sides of Equation 3.7.29 and multiplying through by e^{At}, we find that

$$\mathbf{x}_p(t) = \int_0^t e^{A(t-s)}\mathbf{b}(s)\,ds = \int_0^t e^{As}\mathbf{b}(t-s)\,ds. \qquad (3.7.31)$$

Returning to our original problem,

$$\int_0^t e^{As}\mathbf{b}(t-s)\,ds = \int_0^t \begin{pmatrix} e^{3s} - 2se^{3s} & -4se^{3s} \\ se^{3s} & e^{3s} + 2se^{3s} \end{pmatrix} \begin{pmatrix} e^{2(t-s)} \\ t-s \end{pmatrix} ds \qquad (3.7.32)$$

$$= \int_0^t \begin{pmatrix} e^{2t}\left(e^s - 2se^s\right) - 4s(t-s)e^{3s} \\ e^{2t}se^s + (t-s)e^{3s} + 2s(t-s)e^{3s} \end{pmatrix} ds \qquad (3.7.33)$$

$$= \begin{pmatrix} \tfrac{89}{27}e^{3t} - 3e^{2t} - \tfrac{22}{9}te^{3t} - \tfrac{4}{9}t - \tfrac{8}{27} \\ -\tfrac{28}{27}e^{3t} + e^{2t} + \tfrac{11}{9}te^{3t} - \tfrac{1}{9}t + \tfrac{1}{27} \end{pmatrix}. \qquad (3.7.34)$$

The final answer consists of the homogeneous solution plus the particular solution.

Problems

Find e^{At} for the following matrices A:

1. $A = \begin{pmatrix} 5 & -3 \\ 1 & 1 \end{pmatrix}$
2. $A = \begin{pmatrix} 1 & 3 \\ 0 & 1 \end{pmatrix}$
3. $A = \begin{pmatrix} 3 & 5 \\ 0 & 3 \end{pmatrix}$

4. $A = \begin{pmatrix} 1 & 1 & 0 \\ 0 & 1 & 1 \\ 0 & 0 & 1 \end{pmatrix}$
5. $A = \begin{pmatrix} 2 & 3 & 4 \\ 0 & 2 & 3 \\ 0 & 0 & 2 \end{pmatrix}$
6. $A = \begin{pmatrix} 1 & 2 & 0 \\ 0 & 1 & 2 \\ 0 & 0 & 1 \end{pmatrix}$

For each of the following A's and $\mathbf{b}$'s, use the matrix exponential to find the general solution for the system of first-order, linear ordinary differential equations $\mathbf{x}' = A\mathbf{x} + \mathbf{b}$:

7. $A = \begin{pmatrix} 3 & -2 \\ 4 & -1 \end{pmatrix}$, $\mathbf{b} = \begin{pmatrix} e^t \\ e^t \end{pmatrix}$
8. $A = \begin{pmatrix} 2 & -1 \\ 3 & -2 \end{pmatrix}$, $\mathbf{b} = \begin{pmatrix} e^t \\ t \end{pmatrix}$

9. $A = \begin{pmatrix} 2 & 1 \\ -4 & 2 \end{pmatrix}$, $\mathbf{b} = \begin{pmatrix} te^{2t} \\ -e^{2t} \end{pmatrix}$
10. $A = \begin{pmatrix} 2 & -5 \\ 1 & -2 \end{pmatrix}$, $\mathbf{b} = \begin{pmatrix} -\cos(t) \\ \sin(t) \end{pmatrix}$

11. $A = \begin{pmatrix} 2 & -1 \\ 5 & -2 \end{pmatrix}$, $\mathbf{b} = \begin{pmatrix} \cos(t) \\ \sin(t) \end{pmatrix}$
12. $A = \begin{pmatrix} 2 & 1 & 1 \\ 1 & 2 & 1 \\ 1 & 1 & 2 \end{pmatrix}$, $\mathbf{b} = \begin{pmatrix} 0 \\ te^t \\ e^t \end{pmatrix}$

13. $A = \begin{pmatrix} 1 & 1 & 1 \\ 0 & 2 & 1 \\ 0 & 0 & 3 \end{pmatrix}$, $\mathbf{b} = \begin{pmatrix} 2t \\ t+2 \\ 3t \end{pmatrix}$
14. $A = \begin{pmatrix} 1 & 0 & 1 \\ 0 & -2 & 0 \\ 4 & 0 & 1 \end{pmatrix}$, $\mathbf{b} = \begin{pmatrix} -3e^t \\ 6e^t \\ -4e^t \end{pmatrix}$

15. $A = \begin{pmatrix} 1 & 0 & 2 \\ 0 & 1 & 0 \\ 1 & 0 & 0 \end{pmatrix}$, $\mathbf{b} = \begin{pmatrix} e^t \\ e^t \\ e^t \end{pmatrix}$
16. $A = \begin{pmatrix} 1 & 1 & 2 \\ -1 & 3 & 4 \\ 0 & 0 & 2 \end{pmatrix}$, $\mathbf{b} = \begin{pmatrix} t \\ 1 \\ e^t \end{pmatrix}$.

Further Readings

Bronson, R., and G. B. Costa, 2007: *Linear Algebra: An Introduction.* Academic Press, 520 pp. Provides a step-by-step explanation of linear algebra.

Davis, H. T., and K. T. Thomson, 2000: *Linear Algebra and Linear Operators in Engineering with Applications in Mathematica.* Academic Press, 547 pp. Advanced textbook designed for first-year graduate students in the physical sciences and engineering.

Hoffman, J., 2001: *Numerical Methods for Engineers and Scientists.* Mc-Graw Hill, 823 pp. A first course in numerical methods that is both lucid and in depth.

Munakata, T., 1979: *Matrices and Linear Programming with Applications.* Holden-Day, 469 pp. Provides the basic concepts with clarity.

Noble, B., and J. W. Daniel, 1977: *Applied Linear Algebra.* Prentice Hall, 494 pp. Excellent conceptual explanations with well-motivated proofs.

Strang, G., 2009: *Linear Algebra and Its Applications.* Academic Press, 584 pp. A good supplement to a formal text that provides intuitive understanding.

Wilkinson, J. H., 1988: *The Algebraic Eigenvalue Problem.* Clarendon Press, 662 pp. The classic source book on the eigenvalue problem

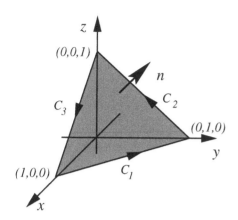

Chapter 4

Vector Calculus

Physicists invented vectors and vector operations to facilitate their mathematical expression of such diverse topics as mechanics and electromagnetism. In this chapter we focus on multivariable differentiations and integrations of vector fields, such as the velocity of a fluid, where the vector field is solely a function of its position.

4.1 REVIEW

The physical sciences and engineering abound with vectors and scalars. *Scalars* are physical quantities that only possess magnitude. Examples include mass, temperature, density, and pressure. *Vectors* are physical quantities that possess both magnitude and direction. Examples include velocity, acceleration, and force. We shall denote vectors by boldfaced letters.

Two vectors are equal if they have the same magnitude and direction. From the limitless number of possible vectors, two special cases are the *zero vector* **0**, which has no magnitude and unspecified direction, and the *unit vector*, which has unit magnitude.

The most convenient method for expressing a vector analytically is in terms of its components. A vector **a** in three-dimensional real space is any order triplet of real numbers (*components*) a_1, a_2, and a_3 such that $\mathbf{a} = a_1\mathbf{i} + a_2\mathbf{j} + a_3\mathbf{k}$, where $a_1\mathbf{i}$, $a_2\mathbf{j}$, and $a_3\mathbf{k}$ are vectors that lie along the coordinate axes and have their origin at a common initial point. The *magnitude*, *length*, or *norm* of a vector **a**, $|\mathbf{a}|$, equals $\sqrt{a_1^2 + a_2^2 + a_3^2}$. A particularly important vector is the *position vector*, defined by $\mathbf{r} = x\mathbf{i} + y\mathbf{j} + z\mathbf{k}$.

As in the case of scalars, certain arithmetic rules hold. Addition and subtraction are very similar to their scalar counterparts:

$$\mathbf{a} + \mathbf{b} = (a_1 + b_1)\mathbf{i} + (a_2 + b_2)\mathbf{j} + (a_3 + b_3)\mathbf{k}, \tag{4.1.1}$$

147

and

$$\mathbf{a} - \mathbf{b} = (a_1 - b_1)\mathbf{i} + (a_2 - b_2)\mathbf{j} + (a_3 - b_3)\mathbf{k}. \tag{4.1.2}$$

In contrast to its scalar counterpart, there are two types of multiplication. The *dot product* is defined as

$$\mathbf{a} \cdot \mathbf{b} = |\mathbf{a}||\mathbf{b}| \cos(\theta) = a_1 b_1 + a_2 b_2 + a_3 b_3, \tag{4.1.3}$$

where θ is the angle between the vector such that $0 \leq \theta \leq \pi$. The dot product yields a scalar answer. A particularly important case is $\mathbf{a} \cdot \mathbf{b} = 0$ with $|\mathbf{a}| \neq 0$, and $|\mathbf{b}| \neq 0$. In this case the vectors are orthogonal (perpendicular) to each other.

The other form of multiplication is the *cross product*, which is defined by $\mathbf{a} \times \mathbf{b} = |\mathbf{a}||\mathbf{b}| \sin(\theta)\mathbf{n}$, where θ is the angle between the vectors such that $0 \leq \theta \leq \pi$, and $\mathbf{n}$ is a unit vector perpendicular to the plane of $\mathbf{a}$ and $\mathbf{b}$, with the direction given by the right-hand rule. A convenient method for computing the cross product from the scalar components of $\mathbf{a}$ and $\mathbf{b}$ is

$$\mathbf{a} \times \mathbf{b} = \begin{vmatrix} \mathbf{i} & \mathbf{j} & \mathbf{k} \\ a_1 & a_2 & a_3 \\ b_1 & b_2 & b_3 \end{vmatrix} = \begin{vmatrix} a_2 & a_3 \\ b_2 & b_3 \end{vmatrix} \mathbf{i} - \begin{vmatrix} a_1 & a_3 \\ b_1 & b_3 \end{vmatrix} \mathbf{j} + \begin{vmatrix} a_1 & a_2 \\ b_1 & b_2 \end{vmatrix} \mathbf{k}. \tag{4.1.4}$$

Two nonzero vectors $\mathbf{a}$ and $\mathbf{b}$ are *parallel* if and only if $\mathbf{a} \times \mathbf{b} = \mathbf{0}$.

Most of the vectors that we will use are vector-valued functions. These functions are vectors that vary either with a single parametric variable t or multiple variables, say x, y, and z.

The most commonly encountered example of a vector-valued function that varies with a single independent variable involves the trajectory of particles. If a *space curve* is parameterized by the equations $x = f(t)$, $y = g(t)$, and $z = h(t)$ with $a \leq t \leq b$, the position vector $\mathbf{r}(t) = f(t)\mathbf{i} + g(t)\mathbf{j} + h(t)\mathbf{k}$ gives the location of a point P as it moves from its initial position to its final position. Furthermore, because the increment quotient $\Delta\mathbf{r}/\Delta t$ is in the direction of a secant line, then the limit of this quotient as $\Delta t \to 0$, $\mathbf{r}'(t)$ gives the tangent (tangent vector) to the curve at P.

• Example 4.1.1: Foucault pendulum

One of the great experiments of mid-nineteenth-century physics was the demonstration by J. B. L. Foucault (1819–1868) in 1851 of the earth's rotation by designing a (spherical) pendulum, supported by a long wire, that essentially swings in an nonaccelerating coordinate system. This problem demonstrates many of the fundamental concepts of vector calculus.

The total force[1] acting on the bob of the pendulum is $\mathbf{F} = \mathbf{T} + m\mathbf{G}$, where $\mathbf{T}$ is the tension in the pendulum and $\mathbf{G}$ is the gravitational attraction per unit mass. Using Newton's second law,

$$\left. \frac{d^2\mathbf{r}}{dt^2} \right|_{\text{inertial}} = \frac{\mathbf{T}}{m} + \mathbf{G}, \tag{4.1.5}$$

where $\mathbf{r}$ is the position vector from a fixed point in an inertial coordinate system to the bob. This system is inconvenient because we live on a rotating coordinate system. Employing

---

[1] See Broxmeyer, C., 1960: Foucault pendulum effect in a Schuler-tuned system. *J. Aerosp. Sci.*, **27**, 343–347.

the conventional geographic coordinate system,[2] Equation 4.1.5 becomes

$$\frac{d^2\mathbf{r}}{dt^2} + 2\mathbf{\Omega} \times \frac{d\mathbf{r}}{dt} + \mathbf{\Omega} \times (\mathbf{\Omega} \times \mathbf{r}) = \frac{\mathbf{T}}{m} + \mathbf{G}, \tag{4.1.6}$$

where $\mathbf{\Omega}$ is the angular rotation vector of the earth and $\mathbf{r}$ now denotes a position vector in the rotating reference system with its origin at the center of the earth and terminal point at the bob. If we define the gravity vector $\mathbf{g} = \mathbf{G} - \mathbf{\Omega} \times (\mathbf{\Omega} \times \mathbf{r})$, then the dynamical equation is

$$\frac{d^2\mathbf{r}}{dt^2} + 2\mathbf{\Omega} \times \frac{d\mathbf{r}}{dt} = \frac{\mathbf{T}}{m} + \mathbf{g}, \tag{4.1.7}$$

where the second term on the left side of Equation 4.1.7 is called the *Coriolis force*.

Because the equation is *linear*, let us break the position vector $\mathbf{r}$ into two separate vectors: $\mathbf{r}_0$ and $\mathbf{r}_1$, where $\mathbf{r} = \mathbf{r}_0 + \mathbf{r}_1$. The vector $\mathbf{r}_0$ extends from the center of the earth to the pendulum's point of support, and $\mathbf{r}_1$ extends from the support point to the bob. Because $\mathbf{r}_0$ is a constant in the geographic system,

$$\frac{d^2\mathbf{r}_1}{dt^2} + 2\mathbf{\Omega} \times \frac{d\mathbf{r}_1}{dt} = \frac{\mathbf{T}}{m} + \mathbf{g}. \tag{4.1.8}$$

If the length of the pendulum is L, then for small oscillations $\mathbf{r}_1 \approx x\mathbf{i} + y\mathbf{j} + L\mathbf{k}$ and the equations of motion are

$$\frac{d^2x}{dt^2} + 2\Omega \sin(\lambda)\frac{dy}{dt} = \frac{T_x}{m}, \tag{4.1.9}$$

$$\frac{d^2y}{dt^2} - 2\Omega \sin(\lambda)\frac{dx}{dt} = \frac{T_y}{m}, \tag{4.1.10}$$

and

$$2\Omega \cos(\lambda)\frac{dy}{dt} - g = \frac{T_z}{m}, \tag{4.1.11}$$

where λ denotes the latitude of the point and Ω is the rotation rate of the earth. The relationships between the components of tension are $T_x = xT_z/L$, and $T_y = yT_z/L$. From Equation 4.1.11,

$$\frac{T_z}{m} + g = 2\Omega \cos(\lambda)\frac{dy}{dt} \approx 0. \tag{4.1.12}$$

Substituting the definitions of T_x, T_y, and Equation 4.1.12 into Equation 4.1.9 and Equation 4.1.10,

$$\frac{d^2x}{dt^2} + \frac{g}{L}x + 2\Omega \sin(\lambda)\frac{dy}{dt} = 0, \tag{4.1.13}$$

and

$$\frac{d^2y}{dt^2} + \frac{g}{L}y - 2\Omega \sin(\lambda)\frac{dx}{dt} = 0. \tag{4.1.14}$$

The approximate solution to these coupled differential equations is

$$x(t) \approx A_0 \cos[\Omega \sin(\lambda)t] \sin\left(\sqrt{g/L}\,t\right), \tag{4.1.15}$$

and

$$y(t) \approx A_0 \sin[\Omega \sin(\lambda)t] \sin\left(\sqrt{g/L}\,t\right), \tag{4.1.16}$$

---

[2] For the derivation, see Marion, J. B., 1965: *Classical Dynamics of Particles and Systems*. Academic Press, Sections 12.2–12.3.

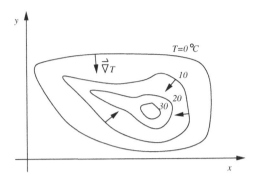

Figure 4.1.1: For a two-dimensional field $T(x, y)$, the gradient is a vector that is perpendicular to the isotherms $T(x, y) =$ constant and points in the direction of most rapidly increasing temperatures.

if $\Omega^2 \ll g/L$. Thus, we have a pendulum that swings with an angular frequency $\sqrt{g/L}$. However, depending upon the *latitude* λ, the direction in which the pendulum swings changes counterclockwise with time, completing a full cycle in $2\pi/[\Omega \sin(\lambda)]$. This result is most clearly seen when $\lambda = \pi/2$ and we are at the North Pole. There, the earth is turning underneath the pendulum. If initially we set the pendulum swinging along the $0°$ longitude, the pendulum will shift with time to longitudes east of the Greenwich median. Eventually, after 24 hours, the process repeats itself.   □

Consider now vector-valued functions that vary with several variables. A *vector function of position* assigns a vector value for every value of x, y, and z within some domain. Examples include the velocity field of a fluid at a given instant:

$$\mathbf{v} = u(x, y, z)\mathbf{i} + v(x, y, z)\mathbf{j} + w(x, y, z)\mathbf{k}. \tag{4.1.17}$$

Another example arises in electromagnetism where electric and magnetic fields often vary as a function of the space coordinates. For us, however, probably the most useful example involves the vector differential operator, *del* or *nabla*,

$$\nabla = \frac{\partial}{\partial x}\mathbf{i} + \frac{\partial}{\partial y}\mathbf{j} + \frac{\partial}{\partial z}\mathbf{k}, \tag{4.1.18}$$

which we apply to the multivariable differentiable scalar function $F(x, y, z)$ to give the *gradient* ∇F.

An important geometric interpretation of the gradient—one which we shall use frequently—is the fact that ∇f is perpendicular (normal) to the level surface at a given point P. To prove this, let the equation $F(x, y, z) = c$ describe a three-dimensional surface. If the differentiable functions $x = f(t)$, $y = g(t)$, and $z = h(t)$ are the parametric equations of a curve on the surface, then the derivative of $F[f(t), g(t), h(t)] = c$ is

$$\frac{\partial F}{\partial x}\frac{dx}{dt} + \frac{\partial F}{\partial y}\frac{dy}{dt} + \frac{\partial F}{\partial z}\frac{dz}{dt} = 0, \tag{4.1.19}$$

or

$$\nabla F \cdot \mathbf{r}' = 0. \tag{4.1.20}$$

When $\mathbf{r}' \neq \mathbf{0}$, the vector ∇F is orthogonal to the tangent vector. Because our argument holds for any differentiable curve that passes through the arbitrary point (x, y, z), then ∇F is normal to the level surface at that point.

Figure 4.1.1 gives a common application of the gradient. Consider a two-dimensional temperature field $T(x, y)$. The level curves $T(x, y) =$ constant are lines that connect points where the temperature is the same (isotherms). The gradient in this case, ∇T, is a vector that is perpendicular or normal to these isotherms and points in the direction of most rapidly increasing temperature.

- **Example 4.1.2**

Let us find the gradient of the function $f(x, y, z) = x^2 z^2 \sin(4y)$.
Using the definition of gradient,

$$\nabla f = \frac{\partial [x^2 z^2 \sin(4y)]}{\partial x} \mathbf{i} + \frac{\partial [x^2 z^2 \sin(4y)]}{\partial y} \mathbf{j} + \frac{\partial [x^2 z^2 \sin(4y)]}{\partial z} \mathbf{k} \qquad (4.1.21)$$

$$= 2xz^2 \sin(4y)\mathbf{i} + 4x^2 z^2 \cos(4y)\mathbf{j} + 2x^2 z \sin(4y)\mathbf{k}. \qquad (4.1.22)$$

$\square$

- **Example 4.1.3**

Let us find the unit normal to the unit sphere at any arbitrary point (x, y, z).
The surface of a unit sphere is defined by the equation $f(x, y, z) = x^2 + y^2 + z^2 = 1$. Therefore, the normal is given by the gradient

$$\mathbf{N} = \nabla f = 2x\mathbf{i} + 2y\mathbf{j} + 2z\mathbf{k}, \qquad (4.1.23)$$

and the unit normal

$$\mathbf{n} = \frac{\nabla f}{|\nabla f|} = \frac{2x\mathbf{i} + 2y\mathbf{j} + 2z\mathbf{k}}{\sqrt{4x^2 + 4y^2 + 4z^2}} = x\mathbf{i} + y\mathbf{j} + z\mathbf{k}, \qquad (4.1.24)$$

because $x^2 + y^2 + z^2 = 1$.

$\square$

- **Example 4.1.4**

In Figure 4.1.2, MATLAB has been used to illustrate the unit normal of the surface $z = 4 - x^2 - y^2$. Here $f(x, y, z) = z + x^2 + y^2 = 4$ so that $\nabla f = 2x\mathbf{i} + 2y\mathbf{j} + \mathbf{k}$. The corresponding script is:

```
clear % clear variables
clf % clear figures
[x,y] = meshgrid(-2:0.5:2); % create the grid
z = 4 - x.^2 - y.^2; % compute surface within domain
% compute the gradient of f(x,y,z) = z + x^2 + y^2 = 4
% the x, y, and z components are u, v, and w
u = 2*x; v = 2*y; w = 1;
% find magnitude of gradient at each point
magnitude = sqrt(u.*u + v.*v + w.*w);
% compute unit gradient vector
u = u./magnitude; v = v./magnitude; w = w./magnitude;
mesh(x,y,z) % plot the surface
axis square
xlabel('x'); ylabel('y')
```

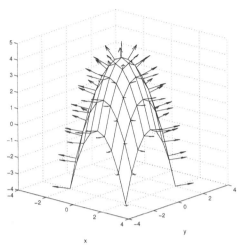

Figure 4.1.2: MATLAB plot of the function $z = 4 - x^2 - y^2$. The arrows give the unit normal to this surface.

```
hold on
% plot the unit gradient vector
quiver3(x,y,z,u,v,w,0)
```
This figure clearly shows that gradient gives a vector which is perpendicular to the surface.   □

A popular method for visualizing a vector field **F** is to draw space curves that are tangent to the vector field at each x, y, z. In fluid mechanics these lines are called *streamlines*, while in physics they are generally called *lines of force* or *flux lines* for an electric, magnetic, or gravitational field. For a fluid with a velocity field that does not vary with time, the streamlines give the paths along which small parcels of the fluid move.

To find the streamlines of a given vector field **F** with components $P(x, y, z)$, $Q(x, y, z)$, and $R(x, y, z)$, we assume that we can parameterize the streamlines in the form $\mathbf{r}(t) = x(t)\mathbf{i} + y(t)\mathbf{j} + z(t)\mathbf{k}$. Then the tangent line is $\mathbf{r}'(t) = x'(t)\mathbf{i} + y'(t)\mathbf{j} + z'(t)\mathbf{k}$. Because the streamline must be parallel to the vector field at any t, $\mathbf{r}'(t) = \lambda\mathbf{F}$, or

$$\frac{dx}{dt} = \lambda P(x, y, z), \quad \frac{dy}{dt} = \lambda Q(x, y, z), \quad \text{and} \quad \frac{dz}{dt} = \lambda R(x, y, z), \qquad (4.1.25)$$

or

$$\frac{dx}{P(x, y, z)} = \frac{dy}{Q(x, y, z)} = \frac{dz}{R(x, y, z)}. \qquad (4.1.26)$$

The solution of this system of differential equations yields the streamlines.

• **Example 4.1.5**

Let us find the streamlines for the vector field $\mathbf{F} = \sec(x)\mathbf{i} - \cot(y)\mathbf{j} + \mathbf{k}$ that passes through the point $(\pi/4, \pi, 1)$. In this particular example, **F** represents a measured or computed fluid's velocity at a particular instant.

From Equation 4.1.26,

$$\frac{dx}{\sec(x)} = -\frac{dy}{\cot(y)} = \frac{dz}{1}. \qquad (4.1.27)$$

This yields two differential equations:

$$\cos(x)\,dx = -\frac{\sin(y)}{\cos(y)}\,dy, \quad \text{and} \quad dz = -\frac{\sin(y)}{\cos(y)}\,dy. \qquad (4.1.28)$$

Integrating these equations gives

$$\sin(x) = \ln|\cos(y)| + c_1, \quad \text{and} \quad z = \ln|\cos(y)| + c_2. \qquad (4.1.29)$$

Substituting for the given point, we finally have that

$$\sin(x) = \ln|\cos(y)| + \sqrt{2}/2, \quad \text{and} \quad z = \ln|\cos(y)| + 1. \qquad (4.1.30)$$

$\square$

- ## Example 4.1.6

Let us find the streamlines for the vector field $\mathbf{F} = \sin(z)\mathbf{j} + e^y\mathbf{k}$ that passes through the point $(2, 0, 0)$.

From Equation 4.1.26,

$$\frac{dx}{0} = \frac{dy}{\sin(z)} = \frac{dz}{e^y}. \qquad (4.1.31)$$

This yields two differential equations:

$$dx = 0, \quad \text{and} \quad \sin(z)\,dz = e^y\,dy. \qquad (4.1.32)$$

Integrating these equations gives

$$x = c_1, \quad \text{and} \quad e^y = -\cos(z) + c_2. \qquad (4.1.33)$$

Substituting for the given point, we finally have that

$$x = 2, \quad \text{and} \quad e^y = 2 - \cos(z). \qquad (4.1.34)$$

Note that Equation 4.1.34 only applies for a certain strip in the yz-plane.

Problems

Given the following vectors $\mathbf{a}$ and $\mathbf{b}$, verify that $\mathbf{a} \cdot (\mathbf{a} \times \mathbf{b}) = 0$, and $\mathbf{b} \cdot (\mathbf{a} \times \mathbf{b}) = 0$:

1. $\mathbf{a} = 4\mathbf{i} - 2\mathbf{j} + 5\mathbf{k}, \quad \mathbf{b} = 3\mathbf{i} + \mathbf{j} - \mathbf{k}$ 2. $\mathbf{a} = \mathbf{i} - 3\mathbf{j} + \mathbf{k}, \quad \mathbf{b} = 2\mathbf{i} + 4\mathbf{k}$

3. $\mathbf{a} = \mathbf{i} + \mathbf{j} + \mathbf{k}, \quad \mathbf{b} = -5\mathbf{i} + 2\mathbf{j} + 3\mathbf{k}$ 4. $\mathbf{a} = 8\mathbf{i} + \mathbf{j} - 6\mathbf{k}, \quad \mathbf{b} = \mathbf{i} - 2\mathbf{j} + 10\mathbf{k}$

5. $\mathbf{a} = 2\mathbf{i} + 7\mathbf{j} - 4\mathbf{k}, \quad \mathbf{b} = \mathbf{i} + \mathbf{j} - \mathbf{k}.$ 6. $\mathbf{a} = -4\mathbf{i} - 2\mathbf{j} + 3\mathbf{k}, \quad \mathbf{b} = 6\mathbf{i} - 2\mathbf{j} + 4\mathbf{k}.$

7. Prove $\mathbf{a} \times (\mathbf{b} \times \mathbf{c}) = (\mathbf{a} \cdot \mathbf{c})\mathbf{b} - (\mathbf{a} \cdot \mathbf{b})\mathbf{c}.$

8. Prove $\mathbf{a} \times (\mathbf{b} \times \mathbf{c}) + \mathbf{b} \times (\mathbf{c} \times \mathbf{a}) + \mathbf{c} \times (\mathbf{a} \times \mathbf{b}) = \mathbf{0}.$

Find the gradient of the following functions:

9. $f(x,y,z) = xy^2/z^3$ 10. $f(x,y,z) = xy\cos(yz)$ 11. $f(x,y,z) = \ln(x^2 + y^2 + z^2)$

12. $f(x,y,z) = x^2y^2(2z+1)^2$ 13. $f(x,y,z) = 2x - y^2 + z^2$ 14. $f(x,y,z) = \cosh(x - y + 2z)$.

Use MATLAB to illustrate the following surfaces as well as the unit normal.

15. $z = 3$ 16. $x^2 + y^2 = 4$ 17. $z = x^2 + y^2$ 18. $z = \sqrt{x^2 + y^2}$

19. $z = y$ 20. $x + y + z = 1$ 21. $z = x^2$ 22. $z^2 = x^2 - y^2$.

Find the streamlines for the following vector fields that pass through the specified point:

23. $\mathbf{F} = \mathbf{i} + \mathbf{j} + \mathbf{k}$; $(0,1,1)$ 24. $\mathbf{F} = 2\mathbf{i} - y^2\mathbf{j} + z\mathbf{k}$; $(1,1,1)$

25. $\mathbf{F} = 3x^2\mathbf{i} - y^2\mathbf{j} + z^2\mathbf{k}$; $(2,1,3)$ 26. $\mathbf{F} = x^2\mathbf{i} + y^2\mathbf{j} - z^3\mathbf{k}$; $(1,1,1)$

27. $\mathbf{F} = (1/x)\mathbf{i} + e^y\mathbf{j} - \mathbf{k}$; $(2,0,4)$ 28. $\mathbf{F} = \cos(y)\mathbf{i} + \sin(x)\mathbf{j}$; $(\pi/2, 0, -4)$.

29. Solve the differential equations, Equation 4.1.13 and Equation 4.1.14, with the initial conditions $x(0) = y(0) = y'(0) = 0$, and $x'(0) = A_0\sqrt{g/L}$ assuming that $\Omega^2 \ll g/L$.

30. If a fluid is bounded by a fixed surface $f(x,y,z) = c$, show that the fluid must satisfy the boundary condition $\mathbf{v} \cdot \nabla f = 0$, where $\mathbf{v}$ is the velocity of the fluid.

31. A sphere of radius a is moving in a fluid with the constant velocity $\mathbf{u}$. Show that the fluid satisfies the boundary condition $(\mathbf{v} - \mathbf{u}) \cdot (\mathbf{r} - \mathbf{u}t) = 0$ at the surface of the sphere, if the center of the sphere coincides with the origin at $t = 0$ and $\mathbf{v}$ denotes the velocity of the fluid.

4.2 DIVERGENCE AND CURL

Consider a vector field $\mathbf{v}$ defined in some region of three-dimensional space. The function $\mathbf{v}(\mathbf{r})$ can be resolved into components along the $\mathbf{i}$, $\mathbf{j}$, and $\mathbf{k}$ directions, or

$$\mathbf{v}(\mathbf{r}) = u(x,y,z)\mathbf{i} + v(x,y,z)\mathbf{j} + w(x,y,z)\mathbf{k}. \tag{4.2.1}$$

If $\mathbf{v}$ is a fluid's velocity field, then we can compute the flow rate through a small (differential) rectangular box defined by increments $(\Delta x, \Delta y, \Delta z)$ centered at the point (x, y, z). See Figure 4.2.1. The flow out from the box through the face with the outwardly pointing normal $\mathbf{n} = -\mathbf{j}$ is

$$\mathbf{v} \cdot (-\mathbf{j}) = -v(x, y - \Delta y/2, z)\Delta x \Delta z, \tag{4.2.2}$$

and the flow through the face with the outwardly pointing normal $\mathbf{n} = \mathbf{j}$ is

$$\mathbf{v} \cdot \mathbf{j} = v(x, y + \Delta y/2, z)\Delta x \Delta z. \tag{4.2.3}$$

The net flow through the two faces is

$$[v(x, y + \Delta y/2, z) - v(x, y - \Delta y/2, z)]\Delta x \Delta z \approx v_y(x, y, z)\Delta x \Delta y \Delta z. \tag{4.2.4}$$

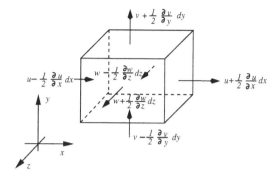

Figure 4.2.1: Divergence of a vector function $\mathbf{v}(x, y, z)$.

A similar analysis of the other faces and combination of the results gives the approximate total flow from the box as

$$[u_x(x, y, z) + v_y(x, y, z) + w_z(x, y, z)]\Delta x \Delta y \Delta z. \tag{4.2.5}$$

Dividing by the volume $\Delta x \Delta y \Delta z$ and taking the limit as the dimensions of the box tend to zero yield $u_x + v_y + w_z$ as the flow out from (x, y, z) per unit volume per unit time. This scalar quantity is called the *divergence* of the vector $\mathbf{v}$:

$$\text{div}(\mathbf{v}) = \nabla \cdot \mathbf{v} = \left(\frac{\partial}{\partial x}\mathbf{i} + \frac{\partial}{\partial y}\mathbf{j} + \frac{\partial}{\partial z}\mathbf{k}\right) \cdot (u\mathbf{i} + v\mathbf{j} + w\mathbf{k}) = u_x + v_y + w_z. \tag{4.2.6}$$

Thus, if the divergence is positive, either the fluid is expanding and its density at the point is falling with time, or the point is a *source* at which fluid is entering the field. When the divergence is negative, either the fluid is contracting and its density is rising at the point, or the point is a negative source or *sink* at which fluid is leaving the field.

   If the divergence of a vector field is zero everywhere within a domain, then the flux entering any element of space exactly balances the flux leaving it and the vector field is called *nondivergent* or *solenoidal* (from a Greek word meaning a tube). For a fluid, if there are no sources or sinks, then its density cannot change.

   Some useful properties of the divergence operator are

$$\nabla \cdot (\mathbf{F} + \mathbf{G}) = \nabla \cdot \mathbf{F} + \nabla \cdot \mathbf{G}, \tag{4.2.7}$$

$$\nabla \cdot (\varphi \mathbf{F}) = \varphi \nabla \cdot \mathbf{F} + \mathbf{F} \cdot \nabla \varphi \tag{4.2.8}$$

and

$$\nabla^2 \varphi = \nabla \cdot \nabla \varphi = \varphi_{xx} + \varphi_{yy} + \varphi_{zz}. \tag{4.2.9}$$

Equation 4.2.9 is very important in physics and is given the special name of the *Laplacian*.[3]

• **Example 4.2.1**

   If $\mathbf{F} = x^2 z \mathbf{i} - 2y^3 z^2 \mathbf{j} + xy^2 z \mathbf{k}$, compute the divergence of $\mathbf{F}$.

---

[3] Some mathematicians write Δ instead of ∇^2.

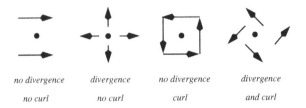

no divergence   divergence   no divergence   divergence

no curl   no curl   curl   and curl

Figure 4.2.2: Examples of vector fields with and without divergence and curl.

$$\nabla \cdot \mathbf{F} = \frac{\partial}{\partial x}\left(x^2 z\right) + \frac{\partial}{\partial y}\left(-2y^3 z^2\right) + \frac{\partial}{\partial z}\left(xy^2 z\right) = 2xz - 6y^2 z^2 + xy^2. \qquad (\mathbf{4.2.10})$$

$\square$

• **Example 4.2.2**

If $\mathbf{r} = x\mathbf{i} + y\mathbf{j} + z\mathbf{k}$, show that $\mathbf{r}/|\mathbf{r}|^3$ is nondivergent.

$$\nabla \cdot \left(\frac{\mathbf{r}}{|\mathbf{r}|^3}\right) = \frac{\partial}{\partial x}\left[\frac{x}{(x^2 + y^2 + z^2)^{3/2}}\right] + \frac{\partial}{\partial y}\left[\frac{y}{(x^2 + y^2 + z^2)^{3/2}}\right]$$

$$+ \frac{\partial}{\partial z}\left[\frac{z}{(x^2 + y^2 + z^2)^{3/2}}\right] \qquad (\mathbf{4.2.11})$$

$$= \frac{3}{(x^2 + y^2 + z^2)^{3/2}} - \frac{3x^2 + 3y^2 + 3z^2}{(x^2 + y^2 + z^2)^{5/2}} = 0. \qquad (\mathbf{4.2.12})$$

$\square$

Another important vector function involving the vector field $\mathbf{v}$ is the curl of $\mathbf{v}$, written curl($\mathbf{v}$) or rot($\mathbf{v}$) in some older textbooks. In fluid flow problems it is proportional to the instantaneous angular velocity of a fluid element. In rectangular coordinates,

$$\text{curl}(\mathbf{v}) = \nabla \times \mathbf{v} = (w_y - v_z)\mathbf{i} + (u_z - w_x)\mathbf{j} + (v_x - u_y)\mathbf{k}, \qquad (\mathbf{4.2.13})$$

where $\mathbf{v} = u\mathbf{i} + v\mathbf{j} + w\mathbf{k}$ as before. However, it is best remembered in the mnemonic form:

$$\nabla \times \mathbf{F} = \begin{vmatrix} \mathbf{i} & \mathbf{j} & \mathbf{k} \\ \frac{\partial}{\partial x} & \frac{\partial}{\partial y} & \frac{\partial}{\partial z} \\ u & v & w \end{vmatrix} = (w_y - v_z)\mathbf{i} + (u_z - w_x)\mathbf{j} + (v_x - u_y)\mathbf{k}. \qquad (\mathbf{4.2.14})$$

If the curl of a vector field is zero everywhere within a region, then the field is *irrotational*.

Figure 4.2.2 illustrates graphically some vector fields that do and do not possess divergence and curl. Let the vectors that are illustrated represent the motion of fluid particles. In the case of divergence only, fluid is streaming from the point at which the density is falling. Alternatively, the point could be a source. In the case where there is only curl, the fluid rotates about the point and the fluid is incompressible. Finally, the point that possesses both divergence and curl is a compressible fluid with rotation.

Some useful computational formulas exist for both the divergence and curl operations:

$$\nabla \times (\mathbf{F} + \mathbf{G}) = \nabla \times \mathbf{F} + \nabla \times \mathbf{G}, \qquad (\mathbf{4.2.15})$$

$$\nabla \times \nabla \varphi = 0, \tag{4.2.16}$$

$$\nabla \cdot \nabla \times \mathbf{F} = 0, \tag{4.2.17}$$

$$\nabla \times (\varphi \mathbf{F}) = \varphi \nabla \times \mathbf{F} + \nabla \varphi \times \mathbf{F}, \tag{4.2.18}$$

$$\nabla(\mathbf{F} \cdot \mathbf{G}) = (\mathbf{F} \cdot \nabla)\mathbf{G} + (\mathbf{G} \cdot \nabla)\mathbf{F} + \mathbf{F} \times (\nabla \times \mathbf{G}) + \mathbf{G} \times (\nabla \times \mathbf{F}), \tag{4.2.19}$$

$$\nabla \times (\mathbf{F} \times \mathbf{G}) = (\mathbf{G} \cdot \nabla)\mathbf{F} - (\mathbf{F} \cdot \nabla)\mathbf{G} + \mathbf{F}(\nabla \cdot \mathbf{G}) - \mathbf{G}(\nabla \cdot \mathbf{F}), \tag{4.2.20}$$

$$\nabla \times (\nabla \times \mathbf{F}) = \nabla(\nabla \cdot \mathbf{F}) - (\nabla \cdot \nabla)\mathbf{F}, \tag{4.2.21}$$

and

$$\nabla \cdot (\mathbf{F} \times \mathbf{G}) = \mathbf{G} \cdot \nabla \times \mathbf{F} - \mathbf{F} \cdot \nabla \times \mathbf{G}. \tag{4.2.22}$$

In this book, the operation $\nabla \mathbf{F}$ is undefined.

- **Example 4.2.3**

  If $\mathbf{F} = xz^3\mathbf{i} - 2x^2yz\mathbf{j} + 2yz^4\mathbf{k}$, compute the curl of $\mathbf{F}$ and verify that $\nabla \cdot \nabla \times \mathbf{F} = 0$.
  From the definition of curl,

$$\nabla \times \mathbf{F} = \begin{vmatrix} \mathbf{i} & \mathbf{j} & \mathbf{k} \\ \frac{\partial}{\partial x} & \frac{\partial}{\partial y} & \frac{\partial}{\partial z} \\ xz^3 & -2x^2yz & 2yz^4 \end{vmatrix} \tag{4.2.23}$$

$$= \left[\frac{\partial}{\partial y}\left(2yz^4\right) - \frac{\partial}{\partial z}\left(-2x^2yz\right)\right]\mathbf{i} - \left[\frac{\partial}{\partial x}\left(2yz^4\right) - \frac{\partial}{\partial z}\left(xz^3\right)\right]\mathbf{j}$$

$$+ \left[\frac{\partial}{\partial x}\left(-2x^2yz\right) - \frac{\partial}{\partial y}\left(xz^3\right)\right]\mathbf{k} \tag{4.2.24}$$

$$= (2z^4 + 2x^2y)\mathbf{i} - (0 - 3xz^2)\mathbf{j} + (-4xyz - 0)\mathbf{k} \tag{4.2.25}$$

$$= (2z^4 + 2x^2y)\mathbf{i} + 3xz^2\mathbf{j} - 4xyz\mathbf{k}. \tag{4.2.26}$$

From the definition of divergence and Equation 4.2.26,

$$\nabla \cdot \nabla \times \mathbf{F} = \frac{\partial}{\partial x}(2z^4 + 2x^2y) + \frac{\partial}{\partial y}(3xz^2) + \frac{\partial}{\partial z}(-4xyz) = 4xy + 0 - 4xy = 0. \tag{4.2.27}$$

$\square$

- **Example 4.2.4: Potential flow theory**

  One of the topics in most elementary fluid mechanics courses is the study of irrotational and nondivergent fluid flows. Because the fluid is irrotational, the velocity vector field $\mathbf{v}$ satisfies $\nabla \times \mathbf{v} = \mathbf{0}$. From Equation 4.2.16 we can introduce a potential φ such that $\mathbf{v} = \nabla \varphi$. Because the flow field is nondivergent, $\nabla \cdot \mathbf{v} = \nabla^2 \varphi = 0$. Thus, the fluid flow can be completely described in terms of solutions to Laplace's equation. This area of fluid mechanics is called *potential flow theory*.

Problems

Compute $\nabla \cdot \mathbf{F}, \nabla \times \mathbf{F}, \nabla \cdot (\nabla \times \mathbf{F})$, and $\nabla(\nabla \cdot \mathbf{F})$, for the following vector fields:

1. $\mathbf{F} = x^2z\mathbf{i} + yz^2\mathbf{j} + xy^2\mathbf{k}$

2. $\mathbf{F} = 4x^2y^2\mathbf{i} + (2x + 2yz)\mathbf{j} + (3z + y^2)\mathbf{k}$

3. $\mathbf{F} = (x - y)^2\mathbf{i} + e^{-xy}\mathbf{j} + xze^{2y}\mathbf{k}$ $\qquad$ 4. $\mathbf{F} = 3xy\mathbf{i} + 2xz^2\mathbf{j} + y^3\mathbf{k}$

5. $\mathbf{F} = 5yz\mathbf{i} + x^2z\mathbf{j} + 3x^3\mathbf{k}$ $\qquad$ 6. $\mathbf{F} = y^3\mathbf{i} + (x^3y^2 - xy)\mathbf{j} - (x^3yz - xz)\mathbf{k}$

7. $\mathbf{F} = xe^{-y}\mathbf{i} + yz^2\mathbf{j} + 3e^{-z}\mathbf{k}$ $\qquad$ 8. $\mathbf{F} = y\ln(x)\mathbf{i} + (2 - 3yz)\mathbf{j} + xyz^3\mathbf{k}$

9. $\mathbf{F} = xyz\mathbf{i} + x^3yze^z\mathbf{j} + xye^z\mathbf{k}$ $\qquad$ 10. $\mathbf{F} = (xy^3 - z^4)\mathbf{i} + 4x^4y^2z\mathbf{j} - y^4z^5\mathbf{k}$

11. $\mathbf{F} = xy^2\mathbf{i} + xyz^2\mathbf{j} + xy\cos(z)\mathbf{k}$ $\qquad$ 12. $\mathbf{F} = xy^2\mathbf{i} + xyz^2\mathbf{j} + xy\sin(z)\mathbf{k}$

13. $\mathbf{F} = xy^2\mathbf{i} + xyz\mathbf{j} + xy\cos(z)\mathbf{k}$ $\qquad$ 14. $\mathbf{F} = \sinh(x)\mathbf{i} + \cosh(y)\mathbf{j} - xyz\mathbf{k}$

15. (a) Assuming continuity of all partial derivatives, show that

$$\nabla \times (\nabla \times \mathbf{F}) = \nabla(\nabla \cdot \mathbf{F}) - \nabla^2\mathbf{F}.$$

(b) Using $\mathbf{F} = 3xy\mathbf{i} + 4yz\mathbf{j} + 2xz\mathbf{k}$, verify the results in part (a).

16. If $\mathbf{E} = \mathbf{E}(x, y, z, t)$ and $\mathbf{B} = \mathbf{B}(x, y, z, t)$ represent the electric and magnetic fields in a vacuum, Maxwell's field equations are:

$$\nabla \cdot \mathbf{E} = 0, \qquad \nabla \times \mathbf{E} = -\frac{1}{c}\frac{\partial \mathbf{B}}{\partial t},$$

$$\nabla \cdot \mathbf{B} = 0, \qquad \nabla \times \mathbf{B} = \frac{1}{c}\frac{\partial \mathbf{E}}{\partial t},$$

where c is the speed of light. Using the results from Problem 15, show that $\mathbf{E}$ and $\mathbf{B}$ satisfy

$$\nabla^2\mathbf{E} = \frac{1}{c^2}\frac{\partial^2\mathbf{E}}{\partial t^2}, \quad \text{and} \quad \nabla^2\mathbf{B} = \frac{1}{c^2}\frac{\partial^2\mathbf{B}}{\partial t^2}.$$

17. If f and g are continuously differentiable scalar fields, show that $\nabla f \times \nabla g$ is solenoidal. Hint: Show that $\nabla f \times \nabla g = \nabla \times (f\nabla g)$.

18. An inviscid (frictionless) fluid in equilibrium obeys the relationship $\nabla p = \rho\mathbf{F}$, where ρ denotes the density of the fluid, p denotes the pressure, and $\mathbf{F}$ denotes the body forces (such as gravity). Show that $\mathbf{F} \cdot \nabla \times \mathbf{F} = 0$.

4.3 LINE INTEGRALS

Line integrals are ubiquitous in physics. In mechanics they are used to compute work. In electricity and magnetism, they provide simple methods for computing the electric and magnetic fields for simple geometries.

The line integral most frequently encountered is an *oriented* one in which the path C is directed and the integrand is the dot product between the vector function $\mathbf{F}(\mathbf{r})$ and the tangent of the path $d\mathbf{r}$. It is usually written in the economical form

$$\int_C \mathbf{F} \cdot d\mathbf{r} = \int_C P(x, y, z)\,dx + Q(x, y, z)\,dy + R(x, y, z)\,dz, \qquad (4.3.1)$$

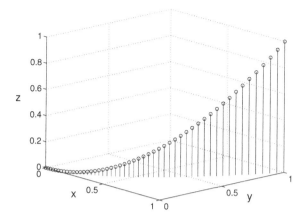

Figure 4.3.1: Diagram for the line integration in Example 4.3.1.

where $\mathbf{F} = P(x, y, z)\mathbf{i} + Q(x, y, z)\mathbf{j} + R(x, y, z)\mathbf{k}$. If the starting and terminal points are the same so that the contour is closed, then this *closed contour integral* will be denoted by $\oint_C$. In the following examples we show how to evaluate the line integrals along various types of curves.

- **Example 4.3.1**

If $\mathbf{F} = (3x^2 + 6y)\mathbf{i} - 14yz\mathbf{j} + 20xz^2\mathbf{k}$, let us evaluate the line integral $\int_C \mathbf{F} \cdot d\mathbf{r}$ along the parametric curves $x(t) = t$, $y(t) = t^2$, and $z(t) = t^3$ from the point $(0, 0, 0)$ to $(1, 1, 1)$. Using the MATLAB commands:

```
>> clear
>> t = 0:0.02:1
>> stem3(t,t.^2,t.^3); xlabel('x','Fontsize',20); ...
   ylabel('y','Fontsize',20); zlabel('z','Fontsize',20);
```

we illustrate these parametric curves in Figure 4.3.1.

We begin by finding the values of t, which give the corresponding endpoints. A quick check shows that $t = 0$ gives $(0, 0, 0)$ while $t = 1$ yields $(1, 1, 1)$. It should be noted that the same value of t must give the correct coordinates in each direction. Failure to do so suggests an error in the parameterization. Therefore,

$$\int_C \mathbf{F} \cdot d\mathbf{r} = \int_0^1 (3t^2 + 6t^2)\, dt - 14t^2(t^3)\, d(t^2) + 20t(t^3)^2 d(t^3) \tag{4.3.2}$$

$$= \int_0^1 9t^2\, dt - 28t^6\, dt + 60t^9\, dt = \left(3t^3 - 4t^7 + 6t^{10}\right)\big|_0^1 = 5. \tag{4.3.3}$$

$\square$

- **Example 4.3.2**

Let us redo the previous example with a contour that consists of three "dog legs," namely straight lines from $(0, 0, 0)$ to $(1, 0, 0)$, from $(1, 0, 0)$ to $(1, 1, 0)$, and from $(1, 1, 0)$ to $(1, 1, 1)$. See Figure 4.3.2.

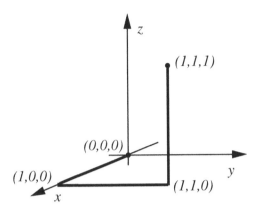

Figure 4.3.2: Diagram for the line integration in Example 4.3.2.

In this particular problem we break the integration down into integrals along each of the legs:

$$\int_C \mathbf{F} \cdot d\mathbf{r} = \int_{C_1} \mathbf{F} \cdot d\mathbf{r} + \int_{C_2} \mathbf{F} \cdot d\mathbf{r} + \int_{C_3} \mathbf{F} \cdot d\mathbf{r}. \tag{4.3.4}$$

For C_1, $y = z = dy = dz = 0$, and

$$\int_{C_1} \mathbf{F} \cdot d\mathbf{r} = \int_0^1 (3x^2 + 6 \cdot 0)\, dx - 14 \cdot 0 \cdot 0 \cdot 0 + 20x \cdot 0^2 \cdot 0 = \int_0^1 3x^2\, dx = 1. \tag{4.3.5}$$

For C_2, $x = 1$ and $z = dx = dz = 0$, so that

$$\int_{C_2} \mathbf{F} \cdot d\mathbf{r} = \int_0^1 (3 \cdot 1^2 + 6y) \cdot 0 - 14y \cdot 0 \cdot dy + 20 \cdot 1 \cdot 0^2 \cdot 0 = 0. \tag{4.3.6}$$

For C_3, $x = y = 1$ and $dx = dy = 0$, so that

$$\int_{C_3} \mathbf{F} \cdot d\mathbf{r} = \int_0^1 (3 \cdot 1^2 + 6 \cdot 1) \cdot 0 - 14 \cdot 1 \cdot z \cdot 0 + 20 \cdot 1 \cdot z^2\, dz = \int_0^1 20z^2\, dz = \tfrac{20}{3}. \tag{4.3.7}$$

Therefore,

$$\int_C \mathbf{F} \cdot d\mathbf{r} = \tfrac{23}{3}. \tag{4.3.8}$$

$\square$

● **Example 4.3.3**

For our third calculation, we redo the first example where the contour is a straight line. The parameterization in this case is $x = y = z = t$ with $0 \le t \le 1$. See Figure 4.3.3. Then,

$$\int_C \mathbf{F} \cdot d\mathbf{r} = \int_0^1 (3t^2 + 6t)\, dt - 14(t)(t)\, dt + 20t(t)^2\, dt \tag{4.3.9}$$

$$= \int_0^1 (3t^2 + 6t - 14t^2 + 20t^3)\, dt = \tfrac{13}{3}. \tag{4.3.10}$$

$\square$

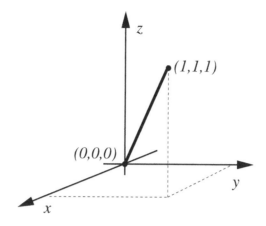

Figure 4.3.3: Diagram for the line integration in Example 4.3.3.

An interesting aspect of these three examples is that, although we used a common vector field and moved from $(0,0,0)$ to $(1,1,1)$ in each case, we obtained a different answer in each case. Thus, for this vector field, the line integral is *path dependent*. This is generally true. In the next section we will meet *conservative vector fields* where the results will be path independent.

• **Example 4.3.4**

If $\mathbf{F} = (x^2 + y^2)\mathbf{i} - 2xy\mathbf{j} + x\mathbf{k}$, let us evaluate $\int_C \mathbf{F} \cdot d\mathbf{r}$ if the contour is that portion of the circle $x^2 + y^2 = a^2$ from the point $(a, 0, 3)$ to $(-a, 0, 3)$. See Figure 4.3.4.

The parametric equations for this example are $x = a\cos(\theta)$, $dx = -a\sin(\theta)\,d\theta$, $y = a\sin(\theta)$, $dy = a\cos(\theta)\,d\theta$, $z = 3$, and $dz = 0$ with $0 \le \theta \le \pi$. Therefore,

$$\int_C \mathbf{F} \cdot d\mathbf{r} = \int_0^\pi [a^2\cos^2(\theta) + a^2\sin^2(\theta)][-a\sin(\theta)\,d\theta]$$

$$- 2a^2\cos(\theta)\sin(\theta)[a\cos(\theta)\,d\theta] + a\cos(\theta)\cdot 0 \qquad (4.3.11)$$

$$= -a^3 \int_0^\pi \sin(\theta)\,d\theta - 2a^3 \int_0^\pi \cos^2(\theta)\sin(\theta)\,d\theta \qquad (4.3.12)$$

$$= a^3\cos(\theta)\big|_0^\pi + \tfrac{2}{3}a^3\cos^3(\theta)\big|_0^\pi = -2a^3 - \tfrac{4}{3}a^3 = -\tfrac{10}{3}a^3. \qquad (4.3.13)$$

□

• **Example 4.3.5: Circulation**

Let $\mathbf{v}(x, y, z)$ denote the velocity at the point (x, y, z) in a moving fluid. If it varies with time, this is the velocity at a particular instant of time. The integral $\oint_C \mathbf{v} \cdot d\mathbf{r}$ around a closed path C is called the *circulation* around that path. The average component of velocity along the path is

$$\overline{v}_s = \frac{\oint_C v_s\, ds}{s} = \frac{\oint_C \mathbf{v} \cdot d\mathbf{r}}{s}, \qquad (4.3.14)$$

where s is the total length of the path. The circulation is thus $\oint_C \mathbf{v} \cdot d\mathbf{r} = \overline{v}_s s$, the product of the length of the path and the average velocity along the path. When the circulation is

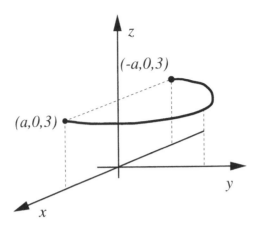

Figure 4.3.4: Diagram for the line integration in Example 4.3.4.

positive, the flow is more in the direction of integration than opposite to it. Circulation is thus an indication, and to some extent a measure, of motion around the path.

Problems

Evaluate $\int_C \mathbf{F} \cdot d\mathbf{r}$ for the following vector fields and curves:

1. $\mathbf{F} = y\sin(\pi z)\mathbf{i} + x^2 e^y \mathbf{j} + 3xz\mathbf{k}$ and C is the curve $x = t$, $y = t^2$, and $z = t^3$ from $(0,0,0)$ to $(1,1,1)$. Use MATLAB to illustrate the parametric curves.

2. $\mathbf{F} = y\mathbf{i} + z\mathbf{j} + x\mathbf{k}$ and C consists of the line segments from $(0,0,0)$ to $(2,3,0)$, and from $(2,3,0)$ to $(2,3,4)$. Use MATLAB to illustrate the parametric curves.

3. $\mathbf{F} = e^x \mathbf{i} + xe^{xy}\mathbf{j} + xye^{xyz}\mathbf{k}$ and C is the curve $x = t$, $y = t^2$, and $z = t^3$ with $0 \le t \le 2$. Use MATLAB to illustrate the parametric curves.

4. $\mathbf{F} = yz\mathbf{i} + xz\mathbf{j} + xy\mathbf{k}$ and C is the curve $x = t^3$, $y = t^2$, and $z = t$ with $1 \le t \le 2$. Use MATLAB to illustrate the parametric curves.

5. $\mathbf{F} = y\mathbf{i} - x\mathbf{j} + 3xy\mathbf{k}$ and C consists of the semicircle $x^2 + y^2 = 4$, $z = 0$, $y > 0$, and the line segment from $(-2,0,0)$ to $(2,0,0)$. Use MATLAB to illustrate the parametric curves.

6. $\mathbf{F} = (x + 2y)\mathbf{i} + (6y - 2x)\mathbf{j}$ and C consists of the sides of the triangle with vertices at $(0,0,0)$, $(1,1,1)$, and $(1,1,0)$. Proceed from $(0,0,0)$ to $(1,1,1)$ to $(1,1,0)$ and back to $(0,0,0)$. Use MATLAB to illustrate the parametric curves.

7. $\mathbf{F} = 2xz\mathbf{i} + 4y^2\mathbf{j} + x^2\mathbf{k}$ and C is taken counterclockwise around the ellipse $x^2/4 + y^2/9 = 1$, $z = 1$. Use MATLAB to illustrate the parametric curves.

8. $\mathbf{F} = 2x\mathbf{i} + y\mathbf{j} + z\mathbf{k}$ and C is the contour $x = t$, $y = \sin(t)$, and $z = \cos(t) + \sin(t)$ with $0 \le t \le 2\pi$. Use MATLAB to illustrate the parametric curves.

9. $\mathbf{F} = (2y^2 + z)\mathbf{i} + 4xy\mathbf{j} + x\mathbf{k}$ and C is the spiral $x = \cos(t)$, $y = \sin(t)$, and $z = t$ with $0 \le t \le 2\pi$ between the points $(1,0,0)$ and $(1,0,2\pi)$. Use MATLAB to illustrate the parametric curves.

10. $\mathbf{F} = x^2\mathbf{i} + y^2\mathbf{j} + (z^2 + 2xy)\mathbf{k}$ and C consists of the edges of the triangle with vertices at $(0,0,0)$, $(1,1,0)$, and $(0,1,0)$. Proceed from $(0,0,0)$ to $(1,1,0)$ to $(0,1,0)$ and back to $(0,0,0)$. Use MATLAB to illustrate the parametric curves.

4.4 THE POTENTIAL FUNCTION

In Section 4.2 we showed that the curl operation applied to a gradient produces the zero vector: $\nabla \times \nabla\varphi = \mathbf{0}$. Consequently, if we have a vector field $\mathbf{F}$ such that $\nabla \times \mathbf{F} \equiv \mathbf{0}$ everywhere, then that vector field is called a *conservative* field and we can compute a potential φ such that $\mathbf{F} = \nabla\varphi$.

• **Example 4.4.1**

Let us show that the vector field $\mathbf{F} = ye^{xy}\cos(z)\mathbf{i} + xe^{xy}\cos(z)\mathbf{j} - e^{xy}\sin(z)\mathbf{k}$ is conservative and then find the corresponding potential function.

To show that the field is conservative, we compute the curl of $\mathbf{F}$ or

$$\nabla \times \mathbf{F} = \begin{vmatrix} \mathbf{i} & \mathbf{j} & \mathbf{k} \\ \frac{\partial}{\partial x} & \frac{\partial}{\partial y} & \frac{\partial}{\partial z} \\ ye^{xy}\cos(z) & xe^{xy}\cos(z) & -e^{xy}\sin(z) \end{vmatrix} = \mathbf{0}. \tag{4.4.1}$$

To find the potential we must solve three partial differential equations:

$$\varphi_x = ye^{xy}\cos(z) = \mathbf{F} \cdot \mathbf{i}, \tag{4.4.2}$$

$$\varphi_y = xe^{xy}\cos(z) = \mathbf{F} \cdot \mathbf{j}, \tag{4.4.3}$$

and

$$\varphi_z = -e^{xy}\sin(z) = \mathbf{F} \cdot \mathbf{k}. \tag{4.4.4}$$

We begin by integrating any one of these three equations. Choosing Equation 4.4.2,

$$\varphi(x,y,z) = e^{xy}\cos(z) + f(y,z). \tag{4.4.5}$$

To find $f(y,z)$ we differentiate Equation 4.4.5 with respect to y and find that

$$\varphi_y = xe^{xy}\cos(z) + f_y(y,z) = xe^{xy}\cos(z) \tag{4.4.6}$$

from Equation 4.4.3. Thus, $f_y = 0$ and $f(y,z)$ can only be a function of z, say $g(z)$. Then,

$$\varphi(x,y,z) = e^{xy}\cos(z) + g(z). \tag{4.4.7}$$

Finally,

$$\varphi_z = -e^{xy}\sin(z) + g'(z) = -e^{xy}\sin(z) \tag{4.4.8}$$

from Equation 4.4.4 and $g'(z) = 0$. Therefore, the potential is

$$\varphi(x,y,z) = e^{xy}\cos(z) + \text{constant}. \tag{4.4.9}$$

$\square$

Potentials can be very useful in computing line integrals, because

$$\int_C \mathbf{F} \cdot d\mathbf{r} = \int_C \varphi_x\,dx + \varphi_y\,dy + \varphi_z\,dz = \int_C d\varphi = \varphi(B) - \varphi(A), \tag{4.4.10}$$

where the point B is the terminal point of the integration while the point A is the starting point. Thus, any path integration between any two points is *path independent*.

Finally, if we close the path so that A and B coincide, then

$$\oint_C \mathbf{F} \cdot d\mathbf{r} = 0. \tag{4.4.11}$$

It should be noted that the converse is *not* true. Just because $\oint_C \mathbf{F} \cdot d\mathbf{r} = 0$, we do not necessarily have a conservative field $\mathbf{F}$.

In summary then, an irrotational vector in a given region has three fundamental properties: (1) its integral around every simply connected circuit is zero, (2) its curl equals zero, and (3) it is the gradient of a scalar function. For continuously differentiable vectors, these properties are equivalent. For vectors that are only piece-wise differentiable, this is not true. Generally the first property is the most fundamental and is taken as the definition of irrotationality.

• **Example 4.4.2**

Using the potential found in Example 4.4.1, let us find the value of the line integral $\int_C \mathbf{F} \cdot d\mathbf{r}$ from the point $(0, 0, 0)$ to $(-1, 2, \pi)$.

From Equation 4.4.9,

$$\int_C \mathbf{F} \cdot d\mathbf{r} = \left[e^{xy} \cos(z) + \text{constant} \right] \Big|_{(0,0,0)}^{(-1,2,\pi)} = -1 - e^{-2}. \tag{4.4.12}$$

Problems

Verify that the following vector fields are conservative and then find the corresponding potential:

1. $\mathbf{F} = 2xy\mathbf{i} + (x^2 + 2yz)\mathbf{j} + (y^2 + 4)\mathbf{k}$ 2. $\mathbf{F} = (2x + 2ze^{2x})\mathbf{i} + (2y - 1)\mathbf{j} + e^{2x}\mathbf{k}$

3. $\mathbf{F} = yz\mathbf{i} + xz\mathbf{j} + xy\mathbf{k}$ 4. $\mathbf{F} = 2x\mathbf{i} + 3y^2\mathbf{j} + 4z^3\mathbf{k}$

5. $\mathbf{F} = (2x + 5)\mathbf{i} + 3y^2\mathbf{j} + (1/z)\mathbf{k}$ 6. $\mathbf{F} = [2x\sin(y) + e^{3z}]\mathbf{i} + x^2\cos(y)\mathbf{j} + (3xe^{3z} + 4)\mathbf{k}$

7. $\mathbf{F} = e^{2z}\mathbf{i} + 3y^2\mathbf{j} + 2xe^{2z}\mathbf{k}$ 8. $\mathbf{F} = y\mathbf{i} + (x + z)\mathbf{j} + y\mathbf{k}$

9. $\mathbf{F} = (z + y)\mathbf{i} + x\mathbf{j} + x\mathbf{k}$ 10. $\mathbf{F} = 2xy\cos(z)\mathbf{i} + x^2\cos(z)\mathbf{j} - x^2y\sin(z)\mathbf{k}$.

4.5 SURFACE INTEGRALS

Surface integrals appear in such diverse fields as electromagnetism and fluid mechanics. For example, if we were oceanographers we might be interested in the rate of volume of seawater through an instrument that has the curved surface S. The volume rate equals $\iint_S \mathbf{v} \cdot \mathbf{n} \, d\sigma$, where $\mathbf{v}$ is the velocity and $\mathbf{n} \, d\sigma$ is an infinitesimally small element on the surface of the instrument. The surface element $\mathbf{n} \, d\sigma$ must have an orientation (given by $\mathbf{n}$) because it makes a considerable difference whether the flow is directly through the surface

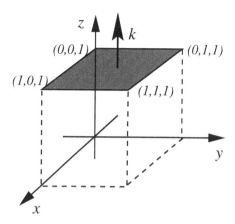

Figure 4.5.1: Diagram for the surface integration in Example 4.5.1.

or at right angles. In the special case when the surface encloses a three-dimensional volume, then we have a *closed surface integral*.

To illustrate the concept of computing a surface integral, we will do three examples with simple geometries. Later we will show how to use surface coordinates to do more complicated geometries.

- **Example 4.5.1**

Let us find the flux out the top of a unit cube if the vector field is $\mathbf{F} = x\mathbf{i} + y\mathbf{j} + z\mathbf{k}$. See Figure 4.5.1.

The top of a unit cube consists of the surface $z = 1$ with $0 \le x \le 1$ and $0 \le y \le 1$. By inspection, the unit normal to this surface is $\mathbf{n} = \mathbf{k}$, or $\mathbf{n} = -\mathbf{k}$. Because we are interested in the flux *out* of the unit cube, $\mathbf{n} = \mathbf{k}$, and

$$\iint_S \mathbf{F} \cdot \mathbf{n} \, d\sigma = \int_0^1 \int_0^1 (x\mathbf{i} + y\mathbf{j} + \mathbf{k}) \cdot \mathbf{k} \, dx \, dy = 1, \qquad (4.5.1)$$

because $z = 1$. $\qquad\qquad\square$

- **Example 4.5.2**

Let us find the flux out of that portion of the cylinder $y^2 + z^2 = 4$ in the first octant bounded by $x = 0$, $x = 3$, $y = 0$, and $z = 0$. The vector field is $\mathbf{F} = x\mathbf{i} + 2z\mathbf{j} + y\mathbf{k}$. See Figure 4.5.2.

Because we are dealing with a cylinder, cylindrical coordinates are appropriate. Let $y = 2\cos(\theta)$, $z = 2\sin(\theta)$, and $x = x$ with $0 \le \theta \le \pi/2$. To find $\mathbf{n}$, we use the gradient in conjunction with the definition of the surface of the cylinder $f(x, y, z) = y^2 + z^2 = 4$. Then,

$$\mathbf{n} = \frac{\nabla f}{|\nabla f|} = \frac{2y\mathbf{j} + 2z\mathbf{k}}{\sqrt{4y^2 + 4z^2}} = \frac{y}{2}\mathbf{j} + \frac{z}{2}\mathbf{k}, \qquad (4.5.2)$$

because $y^2 + z^2 = 4$ along the surface. Since we want the flux *out* of the surface, then $\mathbf{n} = y\mathbf{j}/2 + z\mathbf{k}/2$, whereas the flux *into* the surface would require $\mathbf{n} = -y\mathbf{j}/2 - z\mathbf{k}/2$. Therefore,

$$\mathbf{F} \cdot \mathbf{n} = (x\mathbf{i} + 2z\mathbf{j} + y\mathbf{k}) \cdot \left(\frac{y}{2}\mathbf{j} + \frac{z}{2}\mathbf{k}\right) = \frac{3yz}{2} = 6\cos(\theta)\sin(\theta). \qquad (4.5.3)$$

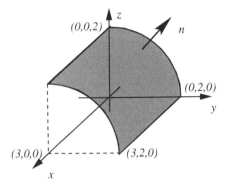

Figure 4.5.2: Diagram for the surface integration in Example 4.5.2.

What is $d\sigma$? Our infinitesimal surface area has a side in the x direction of length dx and a side in the θ direction of length $2\,d\theta$ because the radius equals 2. Therefore, $d\sigma = 2\,dx\,d\theta$.

Bringing all of these elements together,

$$\iint_S \mathbf{F} \cdot \mathbf{n}\,d\sigma = \int_0^3 \int_0^{\pi/2} 12\cos(\theta)\sin(\theta)\,d\theta\,dx = 6\int_0^3 \left[\sin^2(\theta)\big|_0^{\pi/2}\right]dx = 6\int_0^3 dx = 18. \tag{4.5.4}$$

As counterpoint to this example, let us find the flux out of the pie-shaped surface at $x = 3$. In this case, $y = r\cos(\theta)$, $z = r\sin(\theta)$, and

$$\iint_S \mathbf{F} \cdot \mathbf{n}\,d\sigma = \int_0^{\pi/2} \int_0^2 [3\mathbf{i} + 2r\sin(\theta)\mathbf{j} + r\cos(\theta)\mathbf{k}] \cdot \mathbf{i}\,r\,dr\,d\theta = 3\int_0^{\pi/2}\int_0^2 r\,dr\,d\theta = 3\pi. \tag{4.5.5}$$

$\square$

• **Example 4.5.3**

Let us find the flux of the vector field $\mathbf{F} = y^2\mathbf{i} + x^2\mathbf{j} + 5z\mathbf{k}$ out of the hemispheric surface $x^2 + y^2 + z^2 = a^2$, $z > 0$. See Figure 4.5.3.

We begin by finding the outwardly pointing normal. Because the surface is defined by $f(x,y,z) = x^2 + y^2 + z^2 = a^2$,

$$\mathbf{n} = \frac{\nabla f}{|\nabla f|} = \frac{2x\mathbf{i} + 2y\mathbf{j} + 2z\mathbf{k}}{\sqrt{4x^2 + 4y^2 + 4z^2}} = \frac{x}{a}\mathbf{i} + \frac{y}{a}\mathbf{j} + \frac{z}{a}\mathbf{k}, \tag{4.5.6}$$

because $x^2 + y^2 + z^2 = a^2$. This is also the outwardly pointing normal since $\mathbf{n} = \mathbf{r}/a$, where $\mathbf{r}$ is the radial vector.

Using spherical coordinates, $x = a\cos(\varphi)\sin(\theta)$, $y = a\sin(\varphi)\sin(\theta)$, and $z = a\cos(\theta)$, where φ is the angle made by the projection of the point onto the equatorial plane, measured from the x-axis, and θ is the colatitude or "cone angle" measured from the z-axis. To compute $d\sigma$, the infinitesimal length in the θ direction is $a\,d\theta$ while in the φ direction it is $a\sin(\theta)\,d\varphi$, where the $\sin(\theta)$ factor takes into account the convergence of the meridians. Therefore, $d\sigma = a^2\sin(\theta)\,d\theta\,d\varphi$, and

$$\iint_S \mathbf{F} \cdot \mathbf{n}\,d\sigma = \int_0^{2\pi}\int_0^{\pi/2} (y^2\mathbf{i} + x^2\mathbf{j} + 5z\mathbf{k})\left(\frac{x}{a}\mathbf{i} + \frac{y}{a}\mathbf{j} + \frac{z}{a}\mathbf{k}\right)a^2\sin(\theta)\,d\theta\,d\varphi \tag{4.5.7}$$

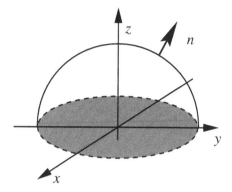

Figure 4.5.3: Diagram for the surface integration in Example 4.5.3.

$$\iint_S \mathbf{F} \cdot \mathbf{n}\, d\sigma = \int_0^{2\pi} \int_0^{\pi/2} \left(\frac{xy^2}{a} + \frac{x^2 y}{a} + \frac{5z^2}{a} \right) a^2 \sin(\theta)\, d\theta\, d\varphi \tag{4.5.8}$$

$$= \int_0^{\pi/2} \int_0^{2\pi} \left[a^4 \cos(\varphi) \sin^2(\varphi) \sin^4(\theta) \right. $$
$$\left. + a^4 \cos^2(\varphi) \sin(\varphi) \sin^4(\theta) + 5a^3 \cos^2(\theta) \sin(\theta) \right] d\varphi\, d\theta \tag{4.5.9}$$

$$= \int_0^{\pi/2} \left[\frac{a^4}{3} \sin^3(\varphi) \Big|_0^{2\pi} \sin^4(\theta) - \frac{a^4}{3} \cos^3(\varphi) \Big|_0^{2\pi} \sin^4(\theta) \right.$$
$$\left. + 5a^3 \cos^2(\theta) \sin(\theta)\varphi \Big|_0^{2\pi} \right] d\theta \tag{4.5.10}$$

$$= 10\pi a^3 \int_0^{\pi/2} \cos^2(\theta) \sin(\theta)\, d\theta = -\frac{10\pi a^3}{3} \cos^3(\theta) \Big|_0^{\pi/2} = \frac{10\pi a^3}{3}. \tag{4.5.11}$$

$\square$

Although these techniques apply for simple geometries such as a cylinder or sphere, we would like a *general* method for treating any arbitrary surface. We begin by noting that a surface is an aggregate of points whose coordinates are functions of two variables. For example, in the previous example, the surface was described by the coordinates φ and θ. Let us denote these surface coordinates in general by u and v. Consequently, on any surface we can re-express x, y, and z in terms of u and v: $x = x(u,v)$, $y = y(u,v)$, and $z = z(u,v)$.

Next, we must find an infinitesimal element of area. The position vector to the surface is $\mathbf{r} = x(u,v)\mathbf{i} + y(u,v)\mathbf{j} + z(u,v)\mathbf{k}$. Therefore, the tangent vectors along $v = $ constant, $\mathbf{r}_u$, and along $u = $ constant, $\mathbf{r}_v$, equal

$$\mathbf{r}_u = x_u \mathbf{i} + y_u \mathbf{j} + z_u \mathbf{k}, \tag{4.5.12}$$

and

$$\mathbf{r}_v = x_v \mathbf{i} + y_v \mathbf{j} + z_v \mathbf{k}. \tag{4.5.13}$$

Consequently, the sides of the infinitesimal area are $\mathbf{r}_u\, du$ and $\mathbf{r}_v\, dv$. Therefore, the vectorial area of the parallelogram that these vectors form is

$$\mathbf{n}\, d\sigma = \mathbf{r}_u \times \mathbf{r}_v\, du\, dv \tag{4.5.14}$$

and is called the *vector element of area* on the surface. Thus, we may convert $\mathbf{F} \cdot \mathbf{n}\, d\sigma$ into an expression involving only u and v and then evaluate the surface integral by integrating

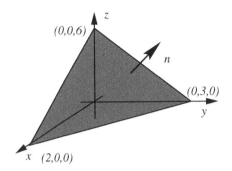

Figure 4.5.4: Diagram for the surface integration in Example 4.5.4.

over the appropriate domain in the uv-plane. Of course, we are in trouble if $\mathbf{r}_u \times \mathbf{r}_v = \mathbf{0}$. Therefore, we only treat regular points where $\mathbf{r}_u \times \mathbf{r}_v \neq \mathbf{0}$. In the next few examples, we show how to use these surface coordinates to evaluate surface integrals.

• **Example 4.5.4**

Let us find the flux of the vector field $\mathbf{F} = x\mathbf{i} + y\mathbf{j} + z\mathbf{k}$ through the top of the plane $3x + 2y + z = 6$, which lies in the first octant. See Figure 4.5.4.

Our parametric equations are $x = u$, $y = v$, and $z = 6 - 3u - 2v$. Therefore,

$$\mathbf{r} = u\mathbf{i} + v\mathbf{j} + (6 - 3u - 2v)\mathbf{k}, \tag{4.5.15}$$

so that

$$\mathbf{r}_u = \mathbf{i} - 3\mathbf{k}, \quad \mathbf{r}_v = \mathbf{j} - 2\mathbf{k}, \tag{4.5.16}$$

and

$$\mathbf{r}_u \times \mathbf{r}_v = 3\mathbf{i} + 2\mathbf{j} + \mathbf{k}. \tag{4.5.17}$$

Bring all of these elements together,

$$\iint_S \mathbf{F} \cdot \mathbf{n} \, d\sigma = \int_0^2 \int_0^{3-3u/2} (3u + 2v + 6 - 3u - 2v) \, dv \, du = 6 \int_0^2 \int_0^{3-3u/2} dv \, du \tag{4.5.18}$$

$$= 6 \int_0^2 (3 - 3u/2) \, du = 6 \left(3u - \tfrac{3}{4}u^2\right)\big|_0^2 = 18. \tag{4.5.19}$$

To set up the limits of integration, we note that the area in u, v space corresponds to the xy-plane. On the xy-plane, $z = 0$ and $3u + 2v = 6$, along with boundaries $u = v = 0$.   □

• **Example 4.5.5**

Let us find the flux of the vector field $\mathbf{F} = x\mathbf{i} + y\mathbf{j} + z\mathbf{k}$ through the top of the surface $z = xy + 1$, which covers the square $0 \leq x \leq 1$, $0 \leq y \leq 1$ in the xy-plane. See Figure 4.5.5.

Our parametric equations are $x = u$, $y = v$, and $z = uv + 1$ with $0 \leq u \leq 1$ and $0 \leq v \leq 1$. Therefore,

$$\mathbf{r} = u\mathbf{i} + v\mathbf{j} + (uv + 1)\mathbf{k}, \tag{4.5.20}$$

so that

$$\mathbf{r}_u = \mathbf{i} + v\mathbf{k}, \qquad \mathbf{r}_v = \mathbf{j} + u\mathbf{k}, \tag{4.5.21}$$

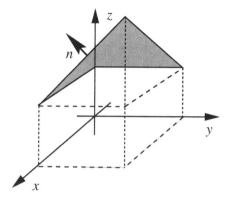

Figure 4.5.5: Diagram for the surface integration in Example 4.5.5.

and

$$\mathbf{r}_u \times \mathbf{r}_v = -v\mathbf{i} - u\mathbf{j} + \mathbf{k}. \tag{4.5.22}$$

Bring all of these elements together,

$$\iint_S \mathbf{F} \cdot \mathbf{n}\, d\sigma = \int_0^1 \int_0^1 [u\mathbf{i} + v\mathbf{j} + (uv+1)\mathbf{k}] \cdot (-v\mathbf{i} - u\mathbf{j} + \mathbf{k})\, du\, dv \tag{4.5.23}$$

$$= \int_0^1 \int_0^1 (1 - uv)\, du\, dv = \int_0^1 \left(u - \tfrac{1}{2}u^2 v\right)\big|_0^1 dv \tag{4.5.24}$$

$$= \int_0^1 \left(1 - \tfrac{1}{2}v\right)\, dv = \left(v - \tfrac{1}{4}v^2\right)\big|_0^1 = \tfrac{3}{4}. \tag{4.5.25}$$

$$\square$$

- **Example 4.5.6**

Let us find the flux of the vector field $\mathbf{F} = 4xz\mathbf{i} + xyz^2\mathbf{j} + 3z\mathbf{k}$ through the exterior surface of the cone $z^2 = x^2 + y^2$ above the xy-plane and below $z = 4$. See Figure 4.5.6.

A natural choice for the surface coordinates is polar coordinates r and θ. Because $x = r\cos(\theta)$ and $y = r\sin(\theta)$, $z = r$. Then,

$$\mathbf{r} = r\cos(\theta)\mathbf{i} + r\sin(\theta)\mathbf{j} + r\mathbf{k} \tag{4.5.26}$$

with $0 \le r \le 4$ and $0 \le \theta \le 2\pi$ so that

$$\mathbf{r}_r = \cos(\theta)\mathbf{i} + \sin(\theta)\mathbf{j} + \mathbf{k} \qquad \mathbf{r}_\theta = -r\sin(\theta)\mathbf{i} + r\cos(\theta)\mathbf{j}, \tag{4.5.27}$$

and

$$\mathbf{r}_r \times \mathbf{r}_\theta = -r\cos(\theta)\mathbf{i} - r\sin(\theta)\mathbf{j} + r\mathbf{k}. \tag{4.5.28}$$

This is the unit area *inside* the cone. Because we want the exterior surface, we must take the negative of Equation 4.5.28. Bring all of these elements together,

$$\iint_S \mathbf{F} \cdot \mathbf{n}\, d\sigma = \int_0^4 \int_0^{2\pi} \big\{ [4r\cos(\theta)]r[r\cos(\theta)]$$

$$+ [r^2 \sin(\theta)\cos(\theta)]r^2[r\sin(\theta)] - 3r^2 \big\}\, d\theta\, dr \tag{4.5.29}$$

$$= \int_0^4 \left\{ 2r^3\left[\theta + \tfrac{1}{2}\sin(2\theta)\right]\big|_0^{2\pi} + r^5 \tfrac{1}{3}\sin^3(\theta)\big|_0^{2\pi} - 3r^2\theta\big|_0^{2\pi} \right\} dr \tag{4.5.30}$$

$$= \int_0^4 \left(4\pi r^3 - 6\pi r^2\right)\, dr = \left(\pi r^4 - 2\pi r^3\right)\big|_0^4 = 128\pi. \tag{4.5.31}$$

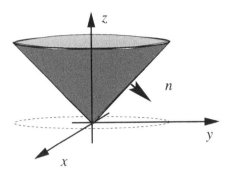

Figure 4.5.6: Diagram for the surface integration in Example 4.5.6.

Problems

Compute the surface integral $\iint_S \mathbf{F} \cdot \mathbf{n}\, d\sigma$ for the following vector fields and surfaces:

1. $\mathbf{F} = x\mathbf{i} - z\mathbf{j} + y\mathbf{k}$ and the surface is the top side of the $z = 1$ plane where $0 \le x \le 1$ and $0 \le y \le 1$.

2. $\mathbf{F} = x\mathbf{i} + y\mathbf{j} + xz\mathbf{k}$ and the surface is the top side of the cylinder $x^2 + y^2 = 9$, $z = 0$, and $z = 1$.

3. $\mathbf{F} = xy\mathbf{i} + z\mathbf{j} + xz\mathbf{k}$ and the surface consists of both exterior *ends* of the cylinder defined by $x^2 + y^2 = 4$, $z = 0$, and $z = 2$.

4. $\mathbf{F} = x\mathbf{i} + z\mathbf{j} + y\mathbf{k}$ and the surface is the lateral and exterior sides of the cylinder defined by $x^2 + y^2 = 4$, $z = -3$, and $z = 3$.

5. $\mathbf{F} = xy\mathbf{i} + z^2\mathbf{j} + y\mathbf{k}$ and the surface is the curved exterior side of the cylinder $y^2 + z^2 = 9$ in the first octant bounded by $x = 0$, $x = 1$, $y = 0$, and $z = 0$.

6. $\mathbf{F} = y\mathbf{j} + z^2\mathbf{k}$ and the surface is the exterior of the semicircular cylinder $y^2 + z^2 = 4$, $z \ge 0$, cut by the planes $x = 0$ and $x = 1$.

7. $\mathbf{F} = z\mathbf{i} + x\mathbf{j} + y\mathbf{k}$ and the surface is the curved exterior side of the cylinder $x^2 + y^2 = 4$ in the first octant cut by the planes $z = 1$ and $z = 2$.

8. $\mathbf{F} = x^2\mathbf{i} - z^2\mathbf{j} + yz\mathbf{k}$ and the surface is the exterior of the hemispheric surface of $x^2 + y^2 + z^2 = 16$ above the plane $z = 2$.

9. $\mathbf{F} = y\mathbf{i} + x\mathbf{j} + y\mathbf{k}$ and the surface is the top of the surface $z = x + 1$, where $-1 \le x \le 1$ and $-1 \le y \le 1$.

10. $\mathbf{F} = z\mathbf{i} + x\mathbf{j} - 3z\mathbf{k}$ and the surface is the top side of the plane $x + y + z = 2a$ that lies above the square $0 \le x \le a$, $0 \le y \le a$ in the xy-plane.

11. $\mathbf{F} = (y^2 + z^2)\mathbf{i} + (x^2 + z^2)\mathbf{j} + (x^2 + y^2)\mathbf{k}$ and the surface is the top side of the surface $z = 1 - x^2$ with $-1 \le x \le 1$ and $-2 \le y \le 2$.

12. $\mathbf{F} = y^2\mathbf{i} + xz\mathbf{j} - \mathbf{k}$ and the surface is the cone $z = \sqrt{x^2 + y^2}$, $0 \le z \le 1$, with the normal pointing away from the z-axis.

13. $\mathbf{F} = y^2\mathbf{i} + x^2\mathbf{j} + 5z\mathbf{k}$ and the surface is the top side of the plane $z = y + 1$, where $-1 \le x \le 1$ and $-1 \le y \le 1$.

14. $\mathbf{F} = -y\mathbf{i} + x\mathbf{j} + z\mathbf{k}$ and the surface is the exterior or bottom side of the paraboloid $z = x^2 + y^2$, where $0 \le z \le 1$.

15. $\mathbf{F} = -y\mathbf{i} + x\mathbf{j} + 6z^2\mathbf{k}$ and the surface is the exterior of the paraboloids $z = 4 - x^2 - y^2$ and $z = x^2 + y^2$.

4.6 GREEN'S LEMMA

Consider a rectangle in the xy-plane that is bounded by the lines $x = a$, $x = b$, $y = c$, and $y = d$. We assume that the boundary of the rectangle is a piece-wise smooth curve that we denote by C. If we have a continuously differentiable vector function $\mathbf{F} = P(x,y)\mathbf{i} + Q(x,y)\mathbf{j}$ at each point of enclosed region R, then

$$\iint_R \frac{\partial Q}{\partial x}\, dA = \int_c^d \left[\int_a^b \frac{\partial Q}{\partial x}\, dx \right] dy = \int_c^d Q(b,y)\, dy - \int_c^d Q(a,y)\, dy = \oint_C Q(x,y)\, dy,$$

$$(4.6.1)$$

where the last integral is a closed line integral counterclockwise around the rectangle because the horizontal sides vanish, since $dy = 0$. By similar arguments,

$$\iint_R \frac{\partial P}{\partial y}\, dA = - \oint_C P(x,y)\, dx \tag{4.6.2}$$

so that

$$\iint_R \left(\frac{\partial Q}{\partial x} - \frac{\partial P}{\partial y} \right) dA = \oint_C P(x,y)\, dx + Q(x,y)\, dy. \tag{4.6.3}$$

This result, often known as *Green's lemma*, may be expressed in vector form as

$$\boxed{\oint_C \mathbf{F} \cdot d\mathbf{r} = \iint_R \nabla \times \mathbf{F} \cdot \mathbf{k}\, dA.} \tag{4.6.4}$$

Although this proof was for a rectangular area, it can be generalized to *any* simply closed region on the xy-plane as follows. Consider an area that is surrounded by simply closed curves. Within the closed contour we can divide the area into an infinite number of infinitesimally small rectangles and apply Equation 4.6.4 to each rectangle. When we sum up all of these rectangles, we find $\iint_R \nabla \times \mathbf{F} \cdot \mathbf{k}\, dA$, where the integration is over the entire surface area. On the other hand, away from the boundary, the line integral along any one edge of a rectangle cancels the line integral along the same edge in a contiguous rectangle.

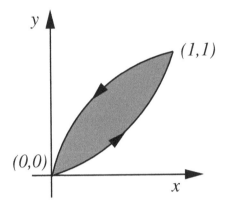

Figure 4.6.1: Diagram for the verification of Green's lemma in Example 4.6.1.

Thus, the only nonvanishing contribution from the line integrals arises from the outside boundary of the domain $\oint_C \mathbf{F} \cdot d\mathbf{r}$.

• **Example 4.6.1**

Let us *verify* Green's lemma using the vector field $\mathbf{F} = (3x^2 - 8y^2)\mathbf{i} + (4y - 6xy)\mathbf{j}$, and the enclosed area lies between the curves $y = \sqrt{x}$ and $y = x^2$. The two curves intersect at $x = 0$ and $x = 1$. See Figure 4.6.1.

We begin with the line integral:

$$
\oint_C \mathbf{F} \cdot d\mathbf{r} = \int_0^1 (3x^2 - 8x^4)\,dx + (4x^2 - 6x^3)(2x\,dx)
$$

$$
+ \int_1^0 (3x^2 - 8x)\,dx + (4x^{1/2} - 6x^{3/2})(\tfrac{1}{2}x^{-1/2}\,dx) \tag{4.6.5}
$$

$$
= \int_0^1 (-20x^4 + 8x^3 + 11x - 2)\,dx = \tfrac{3}{2}. \tag{4.6.6}
$$

In Equation 4.6.6 we used $y = x^2$ in the first integral and $y = \sqrt{x}$ in our return integration. For the areal integration,

$$
\iint_R \nabla \times \mathbf{F} \cdot \mathbf{k}\,dA = \int_0^1 \int_{x^2}^{\sqrt{x}} 10y\,dy\,dx = \int_0^1 5y^2\big|_{x^2}^{\sqrt{x}}\,dx = 5\int_0^1 (x - x^4)\,dx = \tfrac{3}{2} \tag{4.6.7}
$$

and Green's lemma is verified in this particular case.   □

• **Example 4.6.2**

Let us redo Example 4.6.1 except that the closed contour is the triangular region defined by the lines $x = 0$, $y = 0$, and $x + y = 1$.

The line integral is

$$
\oint_C \mathbf{F} \cdot d\mathbf{r} = \int_0^1 (3x^2 - 8 \cdot 0^2)dx + (4 \cdot 0 - 6x \cdot 0) \cdot 0
$$

$$
+ \int_0^1 [3(1-y)^2 - 8y^2](-dy) + [4y - 6(1-y)y]\,dy
$$

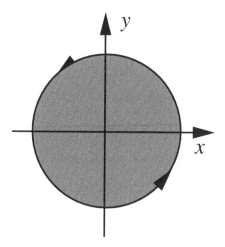

Figure 4.6.2: Diagram for the verification of Green's lemma in Example 4.6.3.

$$+ \int_1^0 (3 \cdot 0^2 - 8y^2) \cdot 0 + (4y - 6 \cdot 0 \cdot y) \, dy \tag{4.6.8}$$

$$= \int_0^1 3x^2 \, dx - \int_0^1 4y \, dy + \int_0^1 (-3 + 4y + 11y^2) \, dy \tag{4.6.9}$$

$$= x^3 \big|_0^1 - 2y^2 \big|_0^1 + \left(-3y + 2y^2 + \tfrac{11}{3} y^3\right) \big|_0^1 = \tfrac{5}{3}. \tag{4.6.10}$$

On the other hand, the areal integration is

$$\iint_R \nabla \times \mathbf{F} \cdot \mathbf{k} \, dA = \int_0^1 \int_0^{1-x} 10y \, dy \, dx = \int_0^1 5y^2 \big|_0^{1-x} \, dx \tag{4.6.11}$$

$$= 5 \int_0^1 (1 - x)^2 \, dx = -\tfrac{5}{3}(1 - x)^3 \big|_0^1 = \tfrac{5}{3} \tag{4.6.12}$$

and Green's lemma is verified in this particular case. □

- **Example 4.6.3**

Let us verify Green's lemma using the vector field $\mathbf{F} = (3x + 4y)\mathbf{i} + (2x - 3y)\mathbf{j}$, and the closed contour is a circle of radius two centered at the origin of the xy-plane. See Figure 4.6.2.

Beginning with the line integration,

$$\oint_C \mathbf{F} \cdot d\mathbf{r} = \int_0^{2\pi} [6 \cos(\theta) + 8 \sin(\theta)][-2 \sin(\theta) \, d\theta] + [4 \cos(\theta) - 6 \sin(\theta)][2 \cos(\theta) \, d\theta]$$
$$\tag{4.6.13}$$

$$= \int_0^{2\pi} [-24 \cos(\theta) \sin(\theta) - 16 \sin^2(\theta) + 8 \cos^2(\theta)] \, d\theta \tag{4.6.14}$$

$$= 12 \cos^2(\theta) \big|_0^{2\pi} - 8 \left[\theta - \tfrac{1}{2} \sin(2\theta)\right] \big|_0^{2\pi} + 4 \left[\theta + \tfrac{1}{2} \sin(2\theta)\right] \big|_0^{2\pi} = -8\pi. \tag{4.6.15}$$

For the areal integration,

$$\iint_R \nabla \times \mathbf{F} \cdot \mathbf{k} \, dA = \int_0^2 \int_0^{2\pi} -2\,r \, d\theta \, dr = -8\pi \tag{4.6.16}$$

and Green's lemma is verified in the special case.

Problems

Verify Green's lemma for the following two-dimensional vector fields and contours:

1. $\mathbf{F} = (x^2 + 4y)\mathbf{i} + (y - x)\mathbf{j}$ and the contour is the square bounded by the lines $x = 0$, $y = 0$, $x = 1$, and $y = 1$.

2. $\mathbf{F} = (x - y)\mathbf{i} + xy\mathbf{j}$ and the contour is the square bounded by the lines $x = 0$, $y = 0$, $x = 1$, and $y = 1$.

3. $\mathbf{F} = -y^2\mathbf{i} + x^2\mathbf{j}$ and the contour is the triangle bounded by the lines $x = 1$, $y = 0$, and $y = x$.

4. $\mathbf{F} = (xy - x^2)\mathbf{i} + x^2y\mathbf{j}$ and the contour is the triangle bounded by the lines $y = 0$, $x = 1$, and $y = x$.

5. $\mathbf{F} = \sin(y)\mathbf{i} + x\cos(y)\mathbf{j}$ and the contour is the triangle bounded by the lines $x + y = 1$, $y - x = 1$, and $y = 0$.

6. $\mathbf{F} = y^2\mathbf{i} + x^2\mathbf{j}$ and the contour is the same contour used in Problem 4.

7. $\mathbf{F} = -y^2\mathbf{i} + x^2\mathbf{j}$ and the contour is the circle $x^2 + y^2 = 4$.

8. $\mathbf{F} = -x^2\mathbf{i} + xy^2\mathbf{j}$ and the contour is the closed circle of radius a.

9. $\mathbf{F} = (6y + x)\mathbf{i} + (y + 2x)\mathbf{j}$ and the contour is the circle $(x - 1)^2 + (y - 2)^2 = 4$.

10. $\mathbf{F} = (x + y)\mathbf{i} + (2x^2 - y^2)\mathbf{j}$ and the contour is the boundary of the region determined by the curves $y = x^2$ and $y = 4$.

11. $\mathbf{F} = 3y\mathbf{i} + 2x\mathbf{j}$ and the contour is the boundary of the region determined by the curves $y = 0$ and $y = \sin(x)$ with $0 \le x \le \pi$.

12. $\mathbf{F} = -16y\mathbf{i} + (4e^y + 3x^2)\mathbf{j}$ and the contour is the pie wedge defined by the lines $y = x$, $y = -x$, $x^2 + y^2 = 4$, and $y > 0$.

4.7 STOKES' THEOREM[4]

In Section 4.2 we introduced the vector quantity $\nabla \times \mathbf{v}$, which gives a measure of the rotation of a parcel of fluid lying within the velocity field $\mathbf{v}$. In this section we show how the curl can be used to simplify the calculation of certain closed line integrals.

---

[4] For the history behind the development of Stokes' theorem, see Katz, V. J., 1979: The history of Stokes' theorem. *Math. Mag.*, **52**, 146–156.

Sir George Gabriel Stokes (1819–1903) was Lucasian Professor of Mathematics at Cambridge University from 1849 until his death. Having learned of an integral theorem from his friend Lord Kelvin, Stokes included it a few years later among his questions on an examination that he wrote for the Smith Prize. It is this integral theorem that we now call Stokes' theorem. (Portrait courtesy of the Royal Society of London.)

This relationship between a closed line integral and a surface integral involving the curl is:

Stokes' Theorem: *The circulation of* $\mathbf{F} = P\mathbf{i} + Q\mathbf{j} + R\mathbf{k}$ *around the closed boundary* C *of an oriented surface* S *in the direction counterclockwise with respect to the surface's unit normal vector* $\mathbf{n}$ *equals the integral of* $\nabla \times \mathbf{F} \cdot \mathbf{n}$ *over* S, *or*

$$\oint_C \mathbf{F} \cdot d\mathbf{r} = \iint_S \nabla \times \mathbf{F} \cdot \mathbf{n} \, d\sigma. \tag{4.7.1}$$

Stokes' theorem requires that all of the functions and derivatives be continuous.

The proof of Stokes' theorem is as follows: Consider a finite surface S whose boundary is the loop C. We divide this surface into a number of small elements $\mathbf{n} \, d\sigma$ and compute the *circulation* $d\Gamma = \oint_L \mathbf{F} \cdot d\mathbf{r}$ around each element. When we add all of the circulations together, the contribution from an integration along a boundary line between two adjoining elements cancels out because the boundary is crossed once in each direction. For this reason, the only contributions that survive are those parts where the element boundaries form part of C. Thus, the sum of all circulations equals $\oint_C \mathbf{F} \cdot d\mathbf{r}$, the circulation around the edge of the whole surface.

Next, let us compute the circulation another way. We begin by finding the Taylor expansion for $P(x, y, z)$ about the arbitrary point (x_0, y_0, z_0):

$$P(x, y, z) = P(x_0, y_0, z_0) + (x - x_0)\frac{\partial P(x_0, y_0, z_0)}{\partial x} + (y - y_0)\frac{\partial P(x_0, y_0, z_0)}{\partial y}$$
$$+ (z - z_0)\frac{\partial P(x_0, y_0, z_0)}{\partial z} + \cdots \qquad (4.7.2)$$

with similar expansions for $Q(x, y, z)$ and $R(x, y, z)$. Then

$$d\Gamma = \oint_L \mathbf{F} \cdot d\mathbf{r} = P(x_0, y_0, z_0) \oint_L dx + \frac{\partial P(x_0, y_0, z_0)}{\partial x} \oint_L (x - x_0)\, dx \qquad (4.7.3)$$
$$+ \frac{\partial P(x_0, y_0, z_0)}{\partial y} \oint_L (y - y_0)\, dy + \cdots + \frac{\partial Q(x_0, y_0, z_0)}{\partial x} \oint_L (x - x_0)\, dy + \cdots,$$

where L denotes some small loop located in the surface S. Note that integrals such as $\oint_L dx$ and $\oint_L (x - x_0)\, dx$ vanish.

If we now require that the loop integrals be in the *clockwise* or *positive* sense so that we preserve the right-hand screw convention, then

$$\mathbf{n} \cdot \mathbf{k}\, \delta\sigma = \oint_L (x - x_0)\, dy = -\oint_L (y - y_0)\, dx, \qquad (4.7.4)$$

$$\mathbf{n} \cdot \mathbf{j}\, \delta\sigma = \oint_L (z - z_0)\, dx = -\oint_L (x - x_0)\, dz, \qquad (4.7.5)$$

$$\mathbf{n} \cdot \mathbf{i}\, \delta\sigma = \oint_L (y - y_0)\, dz = -\oint_L (z - z_0)\, dy, \qquad (4.7.6)$$

and

$$d\Gamma = \left(\frac{\partial R}{\partial y} - \frac{\partial Q}{\partial z}\right) \mathbf{i} \cdot \mathbf{n}\, \delta\sigma + \left(\frac{\partial P}{\partial z} - \frac{\partial R}{\partial x}\right) \mathbf{j} \cdot \mathbf{n}\, \delta\sigma + \left(\frac{\partial Q}{\partial x} - \frac{\partial P}{\partial y}\right) \mathbf{k} \cdot \mathbf{n}\, \delta\sigma = \nabla \times \mathbf{F} \cdot \mathbf{n}\, \delta\sigma.$$
$$(4.7.7)$$

Therefore, the sum of all circulations in the limit when all elements are made infinitesimally small becomes the surface integral $\iint_S \nabla \times \mathbf{F} \cdot \mathbf{n}\, d\sigma$ and Stokes' theorem is proven. $\qquad \square$

In the following examples we first apply Stokes' theorem to a few simple geometries. We then show how to apply this theorem to more complicated surfaces.[5]

● **Example 4.7.1**

Let us verify Stokes' theorem using the vector field $\mathbf{F} = x^2\mathbf{i} + 2x\mathbf{j} + z^2\mathbf{k}$, and the closed curve is a square with vertices at $(0, 0, 3)$, $(1, 0, 3)$, $(1, 1, 3)$, and $(0, 1, 3)$. See Figure 4.7.1.

We begin with the line integral:

$$\oint_C \mathbf{F} \cdot d\mathbf{r} = \int_{C_1} \mathbf{F} \cdot d\mathbf{r} + \int_{C_2} \mathbf{F} \cdot d\mathbf{r} + \int_{C_3} \mathbf{F} \cdot d\mathbf{r} + \int_{C_4} \mathbf{F} \cdot d\mathbf{r}, \qquad (4.7.8)$$

---

[5] Thus, different Stokes for different folks.

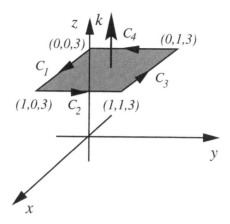

Figure 4.7.1: Diagram for the verification of Stokes' theorem in Example 4.7.1.

where C_1, C_2, C_3, and C_4 represent the four sides of the square. Along C_1, x varies while $y = 0$ and $z = 3$. Therefore,

$$\int_{C_1} \mathbf{F} \cdot d\mathbf{r} = \int_0^1 x^2\, dx + 2x \cdot 0 + 9 \cdot 0 = \tfrac{1}{3}, \qquad (4.7.9)$$

because $dy = dz = 0$, and $z = 3$. Along C_2, y varies with $x = 1$ and $z = 3$. Therefore,

$$\int_{C_2} \mathbf{F} \cdot d\mathbf{r} = \int_0^1 1^2 \cdot 0 + 2 \cdot 1 \cdot dy + 9 \cdot 0 = 2. \qquad (4.7.10)$$

Along C_3, x again varies with $y = 1$ and $z = 3$, and so,

$$\int_{C_3} \mathbf{F} \cdot d\mathbf{r} = \int_1^0 x^2\, dx + 2x \cdot 0 + 9 \cdot 0 = -\tfrac{1}{3}. \qquad (4.7.11)$$

Note how the limits run from 1 to 0 because x is decreasing. Finally, for C_4, y again varies with $x = 0$ and $z = 3$. Hence,

$$\int_{C_4} \mathbf{F} \cdot d\mathbf{r} = \int_1^0 0^2 \cdot 0 + 2 \cdot 0 \cdot dy + 9 \cdot 0 = 0. \qquad (4.7.12)$$

Hence,

$$\oint_C \mathbf{F} \cdot d\mathbf{r} = 2. \qquad (4.7.13)$$

Turning to the other side of the equation,

$$\nabla \times \mathbf{F} = \begin{vmatrix} \mathbf{i} & \mathbf{j} & \mathbf{k} \\ \frac{\partial}{\partial x} & \frac{\partial}{\partial y} & \frac{\partial}{\partial z} \\ x^2 & 2x & z^2 \end{vmatrix} = 2\mathbf{k}. \qquad (4.7.14)$$

Our line integral has been such that the normal vector must be $\mathbf{n} = \mathbf{k}$. Therefore,

$$\iint_S \nabla \times \mathbf{F} \cdot \mathbf{n}\, d\sigma = \int_0^1 \int_0^1 2\mathbf{k} \cdot \mathbf{k}\, dx\, dy = 2 \qquad (4.7.15)$$

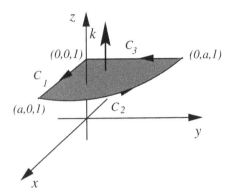

Figure 4.7.2: Diagram for the verification of Stokes' theorem in Example 4.7.2.

and Stokes' theorem is verified for this special case.   □

● **Example 4.7.2**

Let us verify Stokes' theorem using the vector field $\mathbf{F} = (x^2 - y)\mathbf{i} + 4z\mathbf{j} + x^2\mathbf{k}$, where the closed contour consists of the x and y coordinate axes and that portion of the circle $x^2 + y^2 = a^2$ that lies in the first quadrant with $z = 1$. See Figure 4.7.2.

The line integral consists of three parts:

$$\oint_C \mathbf{F} \cdot d\mathbf{r} = \int_{C_1} \mathbf{F} \cdot d\mathbf{r} + \int_{C_2} \mathbf{F} \cdot d\mathbf{r} + \int_{C_3} \mathbf{F} \cdot d\mathbf{r}. \qquad (4.7.16)$$

Along C_1, x varies while $y = 0$ and $z = 1$. Therefore,

$$\int_{C_1} \mathbf{F} \cdot d\mathbf{r} = \int_0^a (x^2 - 0)\,dx + 4 \cdot 1 \cdot 0 + x^2 \cdot 0 = \frac{a^3}{3}. \qquad (4.7.17)$$

Along the circle C_2, we use polar coordinates with $x = a\cos(t)$, $y = a\sin(t)$, and $z = 1$. Therefore,

$$\int_{C_2} \mathbf{F} \cdot d\mathbf{r} = \int_0^{\pi/2} [a^2\cos^2(t) - a\sin(t)][-a\sin(t)\,dt] + 4 \cdot 1 \cdot a\cos(t)\,dt + a^2\cos^2(t) \cdot 0, \qquad (4.7.18)$$

$$= \int_0^{\pi/2} -a^3\cos^2(t)\sin(t)\,dt + a^2\sin^2(t)\,dt + 4a\cos(t)\,dt \qquad (4.7.19)$$

$$= \frac{a^3}{3}\cos^3(t)\Big|_0^{\pi/2} + \frac{a^2}{2}\left[t - \frac{1}{2}\sin(2t)\right]\Big|_0^{\pi/2} + 4a\sin(t)\Big|_0^{\pi/2} \qquad (4.7.20)$$

$$= -\frac{a^3}{3} + \frac{a^2\pi}{4} + 4a, \qquad (4.7.21)$$

because $dx = -a\sin(t)\,dt$, and $dy = a\cos(t)\,dt$. Finally, along C_3, y varies with $x = 0$ and $z = 1$. Therefore,

$$\int_{C_3} \mathbf{F} \cdot d\mathbf{r} = \int_a^0 (0^2 - y) \cdot 0 + 4 \cdot 1 \cdot dy + 0^2 \cdot 0 = -4a, \qquad (4.7.22)$$

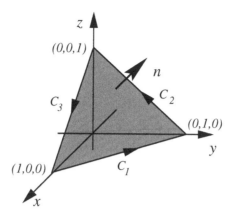

Figure 4.7.3: Diagram for the verification of Stokes' theorem in Example 4.7.3.

so that

$$\oint_C \mathbf{F} \cdot d\mathbf{r} = \frac{a^2 \pi}{4}. \tag{4.7.23}$$

Turning to the other side of the equation,

$$\nabla \times \mathbf{F} = \begin{vmatrix} \mathbf{i} & \mathbf{j} & \mathbf{k} \\ \frac{\partial}{\partial x} & \frac{\partial}{\partial y} & \frac{\partial}{\partial z} \\ x^2 - y & 4z & x^2 \end{vmatrix} = -4\mathbf{i} - 2x\mathbf{j} + \mathbf{k}. \tag{4.7.24}$$

From the path of our line integral, our unit normal vector must be $\mathbf{n} = \mathbf{k}$. Then,

$$\iint_S \nabla \times \mathbf{F} \cdot \mathbf{n}\, d\sigma = \int_0^a \int_0^{\pi/2} [-4\mathbf{i} - 2r\cos(\theta)\mathbf{j} + \mathbf{k}] \cdot \mathbf{k}\, r\, d\theta\, dr = \frac{\pi a^2}{4} \tag{4.7.25}$$

and Stokes' theorem is verified for this case. □

● **Example 4.7.3**

Let us verify Stokes' theorem using the vector field $\mathbf{F} = 2yz\mathbf{i} - (x+3y-2)\mathbf{j} + (x^2+z)\mathbf{k}$, where the closed triangular region is that portion of the plane $x + y + z = 1$ that lies in the first octant.

As shown in Figure 4.7.3, the closed line integration consists of three line integrals:

$$\oint_C \mathbf{F} \cdot d\mathbf{r} = \int_{C_1} \mathbf{F} \cdot d\mathbf{r} + \int_{C_2} \mathbf{F} \cdot d\mathbf{r} + \int_{C_3} \mathbf{F} \cdot d\mathbf{r}. \tag{4.7.26}$$

Along C_1, $z = 0$ and $y = 1 - x$. Therefore, using x as the independent variable,

$$\int_{C_1} \mathbf{F} \cdot d\mathbf{r} = \int_1^0 2(1-x) \cdot 0 \cdot dx - (x+3-3x-2)(-dx) + (x^2+0) \cdot 0 = -x^2\big|_1^0 + x\big|_1^0 = 0. \tag{4.7.27}$$

Along C_2, $x = 0$ and $y = 1 - z$. Thus,

$$\int_{C_2} \mathbf{F} \cdot d\mathbf{r} = \int_0^1 2(1-z)z \cdot 0 - (0+3-3z-2)(-dz) + (0^2+z)\, dz = -\tfrac{3}{2}z^2 + z + \tfrac{1}{2}z^2\big|_0^1 = 0. \tag{4.7.28}$$

Finally, along C_3, $y = 0$ and $z = 1 - x$. Hence,

$$\int_{C_3} \mathbf{F} \cdot d\mathbf{r} = \int_0^1 2 \cdot 0 \cdot (1 - x)\, dx - (x + 0 - 2) \cdot 0 + (x^2 + 1 - x)(-dx) = -\tfrac{1}{3}x^3 - x + \tfrac{1}{2}x^2\big|_0^1 = -\tfrac{5}{6}. \tag{4.7.29}$$

Thus,

$$\oint_C \mathbf{F} \cdot d\mathbf{r} = -\tfrac{5}{6}. \tag{4.7.30}$$

On the other hand,

$$\nabla \times \mathbf{F} = \begin{vmatrix} \mathbf{i} & \mathbf{j} & \mathbf{k} \\ \frac{\partial}{\partial x} & \frac{\partial}{\partial y} & \frac{\partial}{\partial z} \\ 2yz & -x - 3y + 2 & x^2 + z \end{vmatrix} = (-2x + 2y)\mathbf{j} + (-1 - 2z)\mathbf{k}. \tag{4.7.31}$$

To find $\mathbf{n}\, d\sigma$, we use the general coordinate system $x = u$, $y = v$, and $z = 1 - u - v$. Therefore, $\mathbf{r} = u\mathbf{i} + v\mathbf{j} + (1 - u - v)\mathbf{k}$ and

$$\mathbf{r}_u \times \mathbf{r}_v = \begin{vmatrix} \mathbf{i} & \mathbf{j} & \mathbf{k} \\ 1 & 0 & -1 \\ 0 & 1 & -1 \end{vmatrix} = \mathbf{i} + \mathbf{j} + \mathbf{k}. \tag{4.7.32}$$

Thus,

$$\iint_S \nabla \times \mathbf{F} \cdot \mathbf{n}\, d\sigma = \int_0^1 \int_0^{1-u} [(-2u + 2v)\mathbf{j} + (-1 - 2 + 2u + 2v)\mathbf{k}] \cdot [\mathbf{i} + \mathbf{j} + \mathbf{k}]\, dv\, du \tag{4.7.33}$$

$$= \int_0^1 \int_0^{1-u} (4v - 3)\, dv\, du = \int_0^1 [2(1 - u)^2 - 3(1 - u)]\, du \tag{4.7.34}$$

$$= \int_0^1 (-1 - u + 2u^2)\, du = -\tfrac{5}{6} \tag{4.7.35}$$

and Stokes' theorem is verified for this case.

Problems

Verify Stokes' theorem using the following vector fields and surfaces:

1. $\mathbf{F} = 5y\mathbf{i} - 5x\mathbf{j} + 3z\mathbf{k}$ and the surface S is that portion of the plane $z = 1$ with the square at the vertices $(0, 0, 1)$, $(1, 0, 1)$, $(1, 1, 1)$, and $(0, 1, 1)$.

2. $\mathbf{F} = x^2\mathbf{i} + y^2\mathbf{j} + z^2\mathbf{k}$ and the surface S is the rectangular portion of the plane $z = 2$ defined by the corners $(0, 0, 2)$, $(2, 0, 2)$, $(2, 1, 2)$, and $(0, 1, 2)$.

3. $\mathbf{F} = z\mathbf{i} + x\mathbf{j} + y\mathbf{k}$ and the surface S is the triangular portion of the plane $z = 1$ defined by the vertices $(0, 0, 1)$, $(2, 0, 1)$, and $(0, 2, 1)$.

4. $\mathbf{F} = 2z\mathbf{i} - 3x\mathbf{j} + 4y\mathbf{k}$ and the surface S is that portion of the plane $z = 5$ within the cylinder $x^2 + y^2 = 4$.

5. $\mathbf{F} = z\mathbf{i} + x\mathbf{j} + y\mathbf{k}$ and the surface S is that portion of the plane $z = 3$ bounded by the lines $y = 0$, $x = 0$, and $x^2 + y^2 = 4$.

6. $\mathbf{F} = (2z + x)\mathbf{i} + (y - z)\mathbf{j} + (x + y)\mathbf{k}$ and the surface S is the interior of the triangularly shaped plane with vertices at $(1, 0, 0)$, $(0, 1, 0)$, and $(0, 0, 1)$.

7. $\mathbf{F} = z\mathbf{i} + x\mathbf{j} + y\mathbf{k}$ and the surface S is that portion of the plane $2x + y + 2z = 6$ in the first octant.

8. $\mathbf{F} = x\mathbf{i} + xz\mathbf{j} + y\mathbf{k}$ and the surface S is that portion of the paraboloid $z = 9 - x^2 - y^2$ within the cylinder $x^2 + y^2 = 4$.

4.8 DIVERGENCE THEOREM

Although Stokes' theorem is useful in computing closed line integrals, it is usually very difficult to go the other way and convert a surface integral into a closed line integral because the integrand must have a very special form, namely $\nabla \times \mathbf{F} \cdot \mathbf{n}$. In this section we introduce a theorem that allows with equal facility the conversion of a closed surface integral into a volume integral and *vice versa*. Furthermore, if we can convert a given surface integral into a closed one by the introduction of a simple surface (for example, closing a hemispheric surface by adding an equatorial plate), it may be easier to use the divergence theorem and subtract off the contribution from the new surface integral rather than do the original problem.

This relationship between a closed surface integral and a volume integral involving the divergence operator is:

The Divergence or Gauss's Theorem: *Let V be a closed and bounded region in three-dimensional space with a piece-wise smooth boundary S that is oriented outward. Let $\mathbf{F} = P(x, y, z)\mathbf{i} + Q(x, y, z)\mathbf{j} + R(x, y, z)\mathbf{k}$ be a vector field for which P, Q, and R are continuous and have continuous first partial derivatives in a region of three-dimensional space containing V. Then*

$$\oiint_S \mathbf{F} \cdot \mathbf{n}\, d\sigma = \iiint_V \nabla \cdot \mathbf{F}\, dV. \tag{4.8.1}$$

Here, the circle on the double integral signs denotes a closed surface integral.

A nonrigorous proof of Gauss's theorem is as follows. Imagine that our volume V is broken down into small elements $d\tau$ of volume of any shape so long as they include all of the original volume. In general, the surfaces of these elements are composed of common interfaces between adjoining elements. However, for the elements at the periphery of V, part of their surface will be part of the surface S that encloses V. Now $d\Phi = \nabla \cdot \mathbf{F}\, d\tau$ is the net flux of the vector $\mathbf{F}$ out from the element $d\tau$. At the common interface between elements, the flux *out* of one element equals the flux *into* its neighbor. Therefore, the sum of all such terms yields

$$\Phi = \iiint_V \nabla \cdot \mathbf{F}\, d\tau \tag{4.8.2}$$

Carl Friedrich Gauss (1777–1855), the prince of mathematicians, must be on the list of the greatest mathematicians who ever lived. Gauss, a child prodigy, is almost as well known for what he did not publish during his lifetime as for what he did. This is true of Gauss's divergence theorem, which he proved while working on the theory of gravitation. It was only when his notebooks were published in 1898 that his precedence over the published work of Ostrogradsky (1801–1862) was established. (Portrait courtesy of Photo AKG, London, with permission.)

and all of the contributions from these common interfaces cancel; only the contribution from the parts on the outer surface S is left. These contributions, when added together, give $\iint_S \mathbf{F} \cdot \mathbf{n} \, d\sigma$ over S and the proof is completed. $\square$

• **Example 4.8.1**

Let us verify the divergence theorem using the vector field $\mathbf{F} = 4x\mathbf{i} - 2y^2\mathbf{j} + z^2\mathbf{k}$ and the enclosed surface is the cylinder $x^2 + y^2 = 4$, $z = 0$, and $z = 3$. See Figure 4.8.1.

We begin by computing the volume integration. Because

$$\nabla \cdot \mathbf{F} = \frac{\partial(4x)}{\partial x} + \frac{\partial(-2y^2)}{\partial y} + \frac{\partial(z^2)}{\partial z} = 4 - 4y + 2z, \tag{4.8.3}$$

$$\iiint_V \nabla \cdot \mathbf{F} \, dV = \iiint_V (4 - 4y + 2z) \, dV \tag{4.8.4}$$

$$= \int_0^3 \int_0^2 \int_0^{2\pi} [4 - 4r\sin(\theta) + 2z] \, d\theta \, r \, dr \, dz \tag{4.8.5}$$

$$= \int_0^3 \int_0^2 \left[4\theta \Big|_0^{2\pi} + 4r\cos(\theta) \Big|_0^{2\pi} + 2z\theta \Big|_0^{2\pi} \right] r \, dr \, dz \tag{4.8.6}$$

$$= \int_0^3 \int_0^2 (8\pi + 4\pi z) \, r \, dr \, dz = \int_0^3 4\pi(2 + z) \tfrac{1}{2} r^2 \Big|_0^2 \, dz \tag{4.8.7}$$

$$= 4\pi \int_0^3 2(2 + z) \, dz = 8\pi(2z + \tfrac{1}{2}z^2) \Big|_0^3 = 84\pi. \tag{4.8.8}$$

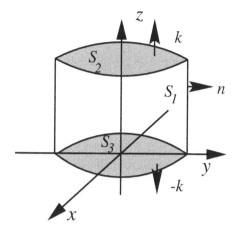

Figure 4.8.1: Diagram for the verification of the divergence theorem in Example 4.8.1.

Turning to the surface integration, we have three surfaces:

$$\oiint_S \mathbf{F} \cdot \mathbf{n} \, d\sigma = \iint_{S_1} \mathbf{F} \cdot \mathbf{n} \, d\sigma + \iint_{S_2} \mathbf{F} \cdot \mathbf{n} \, d\sigma + \iint_{S_3} \mathbf{F} \cdot \mathbf{n} \, d\sigma. \qquad (4.8.9)$$

The first integral is over the exterior to the cylinder. Because the surface is defined by $f(x, y, z) = x^2 + y^2 = 4$,

$$\mathbf{n} = \frac{\nabla f}{|\nabla f|} = \frac{2x\mathbf{i} + 2y\mathbf{j}}{\sqrt{4x^2 + 4y^2}} = \frac{x}{2}\mathbf{i} + \frac{y}{2}\mathbf{j}. \qquad (4.8.10)$$

Therefore,

$$\iint_{S_1} \mathbf{F} \cdot \mathbf{n} \, d\sigma = \iint_{S_1} (2x^2 - y^3) \, d\sigma = \int_0^3 \int_0^{2\pi} \left\{ 2[2\cos(\theta)]^2 - [2\sin(\theta)]^3 \right\} 2 \, d\theta \, dz \qquad (4.8.11)$$

$$= 8 \int_0^3 \int_0^{2\pi} \left\{ \tfrac{1}{2}[1 + \cos(2\theta)] - \sin(\theta) + \cos^2(\theta)\sin(\theta) \right\} 2 \, d\theta \, dz \qquad (4.8.12)$$

$$= 16 \int_0^3 \left[\tfrac{1}{2}\theta + \tfrac{1}{4}\sin(2\theta) + \cos(\theta) - \tfrac{1}{3}\cos^3(\theta) \right] \Big|_0^{2\pi} dz \qquad (4.8.13)$$

$$= 16\pi \int_0^3 dz = 48\pi, \qquad (4.8.14)$$

because $x = 2\cos(\theta)$, $y = 2\sin(\theta)$, and $d\sigma = 2 \, d\theta \, dz$ in cylindrical coordinates.

Along the top of the cylinder, $z = 3$, the outward pointing normal is $\mathbf{n} = \mathbf{k}$, and $d\sigma = r \, dr \, d\theta$. Then,

$$\iint_{S_2} \mathbf{F} \cdot \mathbf{n} \, d\sigma = \iint_{S_2} z^2 \, d\sigma = \int_0^{2\pi} \int_0^2 9 \, r \, dr \, d\theta = 2\pi \times 9 \times 2 = 36\pi. \qquad (4.8.15)$$

However, along the bottom of the cylinder, $z = 0$, the outward pointing normal is $\mathbf{n} = -\mathbf{k}$ and $d\sigma = r \, dr \, d\theta$. Then,

$$\iint_{S_3} \mathbf{F} \cdot \mathbf{n} \, d\sigma = \iint_{S_3} z^2 \, d\sigma = \int_0^{2\pi} \int_0^2 0 \, r \, dr \, d\theta = 0. \qquad (4.8.16)$$

Consequently, the flux out of the entire cylinder is

$$\oiint_S \mathbf{F} \cdot \mathbf{n}\, d\sigma = 48\pi + 36\pi + 0 = 84\pi, \tag{4.8.17}$$

and the divergence theorem is verified for this special case.   □

• **Example 4.8.2**

Let us verify the divergence theorem given the vector field $\mathbf{F} = 3x^2y^2\mathbf{i} + y\mathbf{j} - 6xy^2z\mathbf{k}$ and the volume is the region bounded by the paraboloid $z = x^2 + y^2$, and the plane $z = 2y$. See Figure 4.8.2.

Computing the divergence,

$$\nabla \cdot \mathbf{F} = \frac{\partial(3x^2y^2)}{\partial x} + \frac{\partial(y)}{\partial y} + \frac{\partial(-6xy^2z)}{\partial z} = 6xy^2 + 1 - 6xy^2 = 1. \tag{4.8.18}$$

Then,

$$\iiint_V \nabla \cdot \mathbf{F}\, dV = \iiint_V dV = \int_0^\pi \int_0^{2\sin(\theta)} \int_{r^2}^{2r\sin(\theta)} dz\, r\, dr\, d\theta \tag{4.8.19}$$

$$= \int_0^\pi \int_0^{2\sin(\theta)} [2r\sin(\theta) - r^2]\, r\, dr\, d\theta \tag{4.8.20}$$

$$= \int_0^\pi \left[\frac{2}{3}r^3 \Big|_0^{2\sin(\theta)} \sin(\theta) - \frac{1}{4}r^4 \Big|_0^{2\sin(\theta)} \right] d\theta \tag{4.8.21}$$

$$= \int_0^\pi \left[\frac{16}{3}\sin^4(\theta) - 4\sin^4(\theta) \right] d\theta = \int_0^\pi \frac{4}{3}\sin^4(\theta)\, d\theta \tag{4.8.22}$$

$$= \frac{1}{3} \int_0^\pi [1 - 2\cos(2\theta) + \cos^2(2\theta)]\, d\theta \tag{4.8.23}$$

$$= \frac{1}{3} \left[\theta \Big|_0^\pi - \sin(2\theta) \Big|_0^\pi + \frac{1}{2}\theta \Big|_0^\pi + \frac{1}{8}\sin(4\theta) \Big|_0^\pi \right] = \frac{\pi}{2}. \tag{4.8.24}$$

The limits in the radial direction are given by the intersection of the paraboloid and plane: $r^2 = 2r\sin(\theta)$, or $r = 2\sin(\theta)$, and y is greater than zero.

Turning to the surface integration, we have two surfaces:

$$\oiint_S \mathbf{F} \cdot \mathbf{n}\, d\sigma = \iint_{S_1} \mathbf{F} \cdot \mathbf{n}\, d\sigma + \iint_{S_2} \mathbf{F} \cdot \mathbf{n}\, d\sigma, \tag{4.8.25}$$

where S_1 is the plane $z = 2y$, and S_2 is the paraboloid. For either surface, polar coordinates are best so that $x = r\cos(\theta)$, and $y = r\sin(\theta)$. For the integration over the plane, $z = 2r\sin(\theta)$. Therefore,

$$\mathbf{r} = r\cos(\theta)\mathbf{i} + r\sin(\theta)\mathbf{j} + 2r\sin(\theta)\mathbf{k}, \tag{4.8.26}$$

so that

$$\mathbf{r}_r = \cos(\theta)\mathbf{i} + \sin(\theta)\mathbf{j} + 2\sin(\theta)\mathbf{k}, \tag{4.8.27}$$

and

$$\mathbf{r}_\theta = -r\sin(\theta)\mathbf{i} + r\cos(\theta)\mathbf{j} + 2r\cos(\theta)\mathbf{k}. \tag{4.8.28}$$

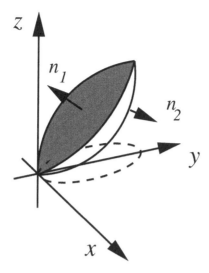

Figure 4.8.2: Diagram for the verification of the divergence theorem in Example 4.8.2. The dashed line denotes the curve $r = 2\sin(\theta)$.

Then,

$$\mathbf{r}_r \times \mathbf{r}_\theta = \begin{vmatrix} \mathbf{i} & \mathbf{j} & \mathbf{k} \\ \cos(\theta) & \sin(\theta) & 2\sin(\theta) \\ -r\sin(\theta) & r\cos(\theta) & 2r\cos(\theta) \end{vmatrix} = -2r\mathbf{j} + r\mathbf{k}. \tag{4.8.29}$$

This is an outwardly pointing normal so that we can immediately set up the surface integral:

$$\iint_{S_1} \mathbf{F} \cdot \mathbf{n}\,d\sigma = \int_0^\pi \int_0^{2\sin(\theta)} \big\{ 3r^4 \cos^2(\theta)\sin^2(\theta)\mathbf{i} + r\sin(\theta)\mathbf{j}$$

$$- 6[2r\sin(\theta)][r\cos(\theta)][r^2\sin^2(\theta)]\mathbf{k} \big\} \cdot (-2r\mathbf{j} + r\mathbf{k})\,dr\,d\theta \tag{4.8.30}$$

$$= \int_0^\pi \int_0^{2\sin(\theta)} \big[-2r^2\sin(\theta) - 12r^5\sin^3(\theta)\cos(\theta) \big]\,dr\,d\theta \tag{4.8.31}$$

$$= \int_0^\pi \left[-\tfrac{2}{3}r^3\big|_0^{2\sin(\theta)}\sin(\theta) - 2r^6\big|_0^{2\sin(\theta)}\sin^3(\theta)\cos(\theta) \right]d\theta \tag{4.8.32}$$

$$= \int_0^\pi \left[-\tfrac{16}{3}\sin^4(\theta) - 128\sin^9(\theta)\cos(\theta) \right]d\theta \tag{4.8.33}$$

$$= -\tfrac{4}{3}\left[\theta\big|_0^\pi - \sin(2\theta)\big|_0^\pi + \tfrac{1}{2}\theta\big|_0^\pi + \tfrac{1}{8}\sin(4\theta)\big|_0^\pi \right] - \tfrac{64}{5}\sin^{10}(\theta)\big|_0^\pi \tag{4.8.34}$$

$$= -2\pi. \tag{4.8.35}$$

For the surface of the paraboloid,

$$\mathbf{r} = r\cos(\theta)\mathbf{i} + r\sin(\theta)\mathbf{j} + r^2\mathbf{k}, \tag{4.8.36}$$

so that

$$\mathbf{r}_r = \cos(\theta)\mathbf{i} + \sin(\theta)\mathbf{j} + 2r\mathbf{k}, \tag{4.8.37}$$

and

$$\mathbf{r}_\theta = -r\sin(\theta)\mathbf{i} + r\cos(\theta)\mathbf{j}. \tag{4.8.38}$$

Then,

$$
\mathbf{r}_r \times \mathbf{r}_\theta = \begin{vmatrix} \mathbf{i} & \mathbf{j} & \mathbf{k} \\ \cos(\theta) & \sin(\theta) & 2r \\ -r\sin(\theta) & r\cos(\theta) & 0 \end{vmatrix} = -2r^2\cos(\theta)\mathbf{i} - 2r^2\sin(\theta)\mathbf{j} + r\mathbf{k}. \qquad (4.8.39)
$$

This is an inwardly pointing normal, so that we must take the negative of it before we do the surface integral. Then,

$$
\iint_{S_2} \mathbf{F} \cdot \mathbf{n}\, d\sigma = \int_0^\pi \int_0^{2\sin(\theta)} \left\{ 3r^4\cos^2(\theta)\sin^2(\theta)\mathbf{i} + r\sin(\theta)\mathbf{j} - 6r^2[r\cos(\theta)][r^2\sin^2(\theta)]\mathbf{k} \right\}
$$
$$
\cdot \left[2r^2\cos(\theta)\mathbf{i} + 2r^2\sin(\theta)\mathbf{j} - r\mathbf{k} \right] dr\, d\theta \qquad (4.8.40)
$$

$$
= \int_0^\pi \int_0^{2\sin(\theta)} \left[6r^6\cos^3(\theta)\sin^2(\theta) + 2r^3\sin^2(\theta) + 6r^6\cos(\theta)\sin^2(\theta) \right] dr\, d\theta \qquad (4.8.41)
$$

$$
= \int_0^\pi \left[\tfrac{6}{7}r^7\big|_0^{2\sin(\theta)}\cos^3(\theta)\sin^2(\theta) + \tfrac{1}{2}r^4\big|_0^{2\sin(\theta)}\sin^2(\theta) \right.
$$
$$
\left. + \tfrac{6}{7}r^7\big|_0^{2\sin(\theta)}\cos(\theta)\sin^2(\theta) \right] d\theta \qquad (4.8.42)
$$

$$
= \int_0^\pi \left\{ \tfrac{768}{7}\sin^9(\theta)[1 - \sin^2(\theta)]\cos(\theta) + 8\sin^6(\theta) + \tfrac{768}{7}\sin^9(\theta)\cos(\theta) \right\} d\theta \qquad (4.8.43)
$$

$$
= \tfrac{1536}{70}\sin^{10}(\theta)\big|_0^\pi - \tfrac{64}{7}\sin^{12}(\theta)\big|_0^\pi + \int_0^\pi [1 - \cos(2\theta)]^3\, d\theta \qquad (4.8.44)
$$

$$
= \int_0^\pi \left\{ 1 - 3\cos(2\theta) + 3\cos^2(2\theta) - \cos(2\theta)[1 - \sin^2(2\theta)] \right\} d\theta \qquad (4.8.45)
$$

$$
= \theta\big|_0^\pi - \tfrac{3}{2}\sin(2\theta)\big|_0^\pi + \tfrac{3}{2}[\theta + \tfrac{1}{4}\sin(4\theta)]\big|_0^\pi - \tfrac{1}{2}\sin(2\theta)\big|_0^\pi + \tfrac{1}{3}\sin^3(2\theta)\big|_0^\pi \qquad (4.8.46)
$$

$$
= \pi + \tfrac{3}{2}\pi = \tfrac{5}{2}\pi. \qquad (4.8.47)
$$

Consequently,

$$
\oiint_S \mathbf{F} \cdot \mathbf{n}\, d\sigma = -2\pi + \tfrac{5}{2}\pi = \tfrac{1}{2}\pi, \qquad (4.8.48)
$$

and the divergence theorem is verified for this special case. □

• **Example 4.8.3: Archimedes' principle**

Consider a solid[6] of volume V and surface S that is immersed in a vessel filled with a fluid of density ρ. The pressure field p in the fluid is a function of the distance from the liquid/air interface and equals

$$
p = p_0 - \rho g z, \qquad (4.8.49)
$$

---

[6] Adapted from Altintas, A., 1990: Archimedes' principle as an application of the divergence theorem. *IEEE Trans. Educ.*, **33**, 222.

where g is the gravitational acceleration, z is the vertical distance measured from the interface (increasing in the $\mathbf{k}$ direction), and p_0 is the constant pressure along the liquid/air interface.

If we define $\mathbf{F} = -p\mathbf{k}$, then $\mathbf{F} \cdot \mathbf{n} \, d\sigma$ is the vertical component of the force on the surface due to the pressure and $\iint_S \mathbf{F} \cdot \mathbf{n} \, d\sigma$ is the total lift. Using the divergence theorem and noting that $\nabla \cdot \mathbf{F} = \rho g$, the total lift also equals

$$\iiint_V \nabla \cdot \mathbf{F} \, dV = \rho g \iiint_V dV = \rho g V, \tag{4.8.50}$$

which is the weight of the displaced liquid. This is *Archimedes' principle*: The buoyant force on a solid immersed in a fluid of constant density equals the weight of the fluid displaced.

□

• Example 4.8.4: Conservation of charge

Let a charge of density ρ flow with an average velocity $\mathbf{v}$. Then the charge crossing the element $d\mathbf{S}$ per unit time is $\rho \mathbf{v} \cdot d\mathbf{S} = \mathbf{J} \cdot d\mathbf{S}$, where $\mathbf{J}$ is defined as the conduction current vector or current density vector. The current across any surface drawn in the medium is $\iint_S \mathbf{J} \cdot d\mathbf{S}$.

The total charge inside the closed surface is $\iiint_V \rho \, dV$. If there are no sources or sinks inside the surface, the rate at which the charge decreases is $-\iiint_V \rho_t \, dV$. Because this change is due to the outward flow of charge,

$$-\iiint_V \frac{\partial \rho}{\partial t} \, dV = \oiint_S \mathbf{J} \cdot d\mathbf{S}. \tag{4.8.51}$$

Applying the divergence theorem,

$$\iiint_V \left(\frac{\partial \rho}{\partial t} + \nabla \cdot \mathbf{J} \right) dV = 0. \tag{4.8.52}$$

Because the result holds true for any arbitrary volume, the integrand must vanish identically and we have the equation of continuity or the *equation of conservation of charge*:

$$\frac{\partial \rho}{\partial t} + \nabla \cdot \mathbf{J} = 0. \tag{4.8.53}$$

Problems

Verify the divergence theorem using the following vector fields and volumes:

1. $\mathbf{F} = x^2 \mathbf{i} + y^2 \mathbf{j} + z^2 \mathbf{k}$ and the volume V is the cube cut from the first octant by the planes $x = 1$, $y = 1$, and $z = 1$.

2. $\mathbf{F} = xy \mathbf{i} + yz \mathbf{j} + xz \mathbf{k}$ and the volume V is the cube bounded by $0 \le x \le 1$, $0 \le y \le 1$, and $0 \le z \le 1$.

3. $\mathbf{F} = (y - x)\mathbf{i} + (z - y)\mathbf{j} + (y - x)\mathbf{k}$ and the volume V is the cube bounded by $-1 \le x \le 1$, $-1 \le y \le 1$, and $-1 \le z \le 1$.

4. $\mathbf{F} = x^2\mathbf{i} + y\mathbf{j} + z\mathbf{k}$ and the volume V is the cylinder defined by the surfaces $x^2 + y^2 = 1$, $z = 0$, and $z = 1$.

5. $\mathbf{F} = x^2\mathbf{i} + y^2\mathbf{j} + z^2\mathbf{k}$ and the volume V is the cylinder defined by the surfaces $x^2 + y^2 = 4$, $z = 0$, and $z = 1$.

6. $\mathbf{F} = y^2\mathbf{i} + xz^3\mathbf{j} + (z-1)^2\mathbf{k}$ and the volume V is the cylinder bounded by the surface $x^2 + y^2 = 4$, and the planes $z = 1$ and $z = 5$.

7. $\mathbf{F} = 6xy\mathbf{i} + 4yz\mathbf{j} + xe^{-y}\mathbf{k}$ and the volume V is that region created by the plane $x + y + z = 1$, and the three coordinate planes.

8. $\mathbf{F} = y\mathbf{i} + xy\mathbf{j} - z\mathbf{k}$ and the volume V is that solid created by the paraboloid $z = x^2 + y^2$ and plane $z = 1$.

Further Readings

Davis, H. F., and A. D. Snider, 1995: *Introduction to Vector Analysis.* Wm. C. Brown Publ., 416 pp. Designed as a reference book for engineering majors.

Kendall, P. C., and D. E. Bourne, 1992: *Vector Analysis and Cartesian Tensors.* Wm. C. Brown, Publ., 304 pp. A clear introduction to the concepts and techniques of vector analysis.

Matthews, P. C., 2005: *Vector Calculus.* Springer, 200 pp. A good book for self-study with complete solutions to the problems.

Schey, H. M., 2005: *Div, Grad, Curl, and All That.* Chapman & Hall, 176 pp. A book to hone your vector calculus skills.

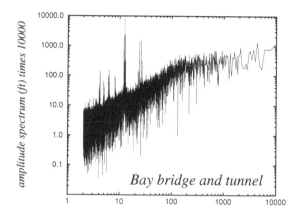

amplitude spectrum (ft) times 10000

Bay bridge and tunnel

Chapter 5
Fourier Series

In differential calculus you developed the technique of representing a function by the Taylor series:

$$f(x) = \sum_{n=0}^{\infty} \frac{f^{(n)}(a)}{n!}(x-a)^n. \tag{5.0.1}$$

A Taylor expansion is particularly useful for approximating functions in an interval near the point of expansion $x = a$. In this chapter we derive an alternate expansion, called the Fourier series, which approximates a function from a sum of sine and cosines:

$$f(x) = \frac{a_0}{2} + \sum_{n=1}^{\infty} a_n \cos\left(\frac{n\pi x}{L}\right) + b_n \sin\left(\frac{n\pi x}{L}\right). \tag{5.0.2}$$

The Fourier series converges to the function *in the mean* over the entire interval $(-L, L)$. This convergence generally improves as more and more sines and cosines (harmonics) are included.

An obvious question is: Why do we need such an expansion? Why use sines and cosines? As we shall see in Chapter 12 we could have used other functions. Although there are several different ways of explaining our choice of Equation 5.0.2, the most direct is its use in the solution of linear partial differential equations in rectangular coordinates in Chapters 8 to 10. There, our need to re-express a function as a sum of sine or cosines will justify our interest.

Since its development in the latter half of the eighteenth and beginning half of the nineteen centuries for the solution of linear partial differential equations, Fourier analysis has taken on a life of its own. For example, it is used to describe physical processes in which events recur in a regular pattern such as a musical note which consists of a simple note, called the fundamental, and a series of auxiliary vibrations, called overtones. For this reason we devote this chapter to the development of Fourier series, its properties, and some of its applications outside of partial differential equations. Later on, when we study the wave, heat and Laplace's equations, we will use them there to solve these equations.

5.1 FOURIER SERIES

One of the crowning glories[1] of nineteenth-century mathematics was the discovery that the infinite series

$$f(t) = \frac{a_0}{2} + \sum_{n=1}^{\infty} a_n \cos\left(\frac{n\pi t}{L}\right) + b_n \sin\left(\frac{n\pi t}{L}\right) \tag{5.1.1}$$

can represent a function $f(t)$ under certain general conditions. This series, called a *Fourier series*, converges to the value of the function $f(t)$ at every point in the interval $[-L, L]$ with the possible exceptions of the points at any discontinuities and the endpoints of the interval. Because each term has a period of $2L$, the sum of the series also has the same period. The *fundamental* of the periodic function $f(t)$ is the $n = 1$ term while the *harmonics* are the remaining terms whose frequencies are integer multiples of the fundamental. We must now find some easy method for computing the coefficients a_n and b_n for a given function $f(t)$. As a first attempt, we integrate Equation 5.1.1 term by term[2] from $-L$ to L. On the right side, all of the integrals multiplied by a_n and b_n vanish because the average of $\cos(n\pi t/L)$ and $\sin(n\pi t/L)$ is zero. Therefore, we are left with

$$a_0 = \frac{1}{L} \int_{-L}^{L} f(t)\, dt. \tag{5.1.2}$$

Consequently a_0 is twice the mean value of $f(t)$ over one period.

We next multiply each side of Equation 5.1.1 by $\cos(m\pi t/L)$, where m is a fixed integer. Integrating from $-L$ to L,

$$\int_{-L}^{L} f(t) \cos\left(\frac{m\pi t}{L}\right) dt = \frac{a_0}{2} \int_{-L}^{L} \cos\left(\frac{m\pi t}{L}\right) dt + \sum_{n=1}^{\infty} a_n \int_{-L}^{L} \cos\left(\frac{n\pi t}{L}\right) \cos\left(\frac{m\pi t}{L}\right) dt$$

$$+ \sum_{n=1}^{\infty} b_n \int_{-L}^{L} \sin\left(\frac{n\pi t}{L}\right) \cos\left(\frac{m\pi t}{L}\right) dt. \tag{5.1.3}$$

---

[1] "Fourier's Theorem ... is not only one of the most beautiful results of modern analysis, but may be said to furnish an indispensable instrument in the treatment of nearly every recondite question in modern physics. To mention only sonorous vibrations, the propagation of electric signals along a telegraph wire, and the conduction of heat by the earth's crust, as subjects in their generality intractable without it, is to give but a feeble idea of its importance." (Quote taken from Thomson, W., and P. G. Tait, 1879: *Treatise on Natural Philosophy, Part 1.* Cambridge University Press, Section 75.)

[2] We assume that the integration of the series can be carried out term by term. This is sometimes difficult to justify but we do it anyway.

The a_0 and b_n terms vanish by direct integration. Finally, all of the a_n integrals vanish when $n \neq m$. Therefore, Equation 5.1.3 simplifies to

$$a_n = \frac{1}{L} \int_{-L}^{L} f(t) \cos\left(\frac{n\pi t}{L}\right) dt, \tag{5.1.4}$$

because $\int_{-L}^{L} \cos^2(n\pi t/L) \, dt = L$. Finally, by multiplying both sides of Equation 5.1.1 by $\sin(m\pi t/L)$ (m is again a fixed integer) and integrating from $-L$ to L,

$$b_n = \frac{1}{L} \int_{-L}^{L} f(t) \sin\left(\frac{n\pi t}{L}\right) dt. \tag{5.1.5}$$

Although Equation 5.1.2, Equation 5.1.4, and Equation 5.1.5 give us a_0, a_n, and b_n for periodic functions over the interval $[-L, L]$, in certain situations it is convenient to use the interval $[\tau, \tau + 2L]$, where τ is any real number. In that case, Equation 5.1.1 still gives the Fourier series of $f(t)$ and

$$
\begin{aligned}
a_0 &= \frac{1}{L} \int_{\tau}^{\tau+2L} f(t) \, dt, \\
a_n &= \frac{1}{L} \int_{\tau}^{\tau+2L} f(t) \cos\left(\frac{n\pi t}{L}\right) dt, \\
b_n &= \frac{1}{L} \int_{\tau}^{\tau+2L} f(t) \sin\left(\frac{n\pi t}{L}\right) dt.
\end{aligned}
\tag{5.1.6}
$$

These results follow when we recall that the function $f(t)$ is a periodic function that extends from minus infinity to plus infinity. The results must remain unchanged, therefore, when we shift from the interval $[-L, L]$ to the new interval $[\tau, \tau + 2L]$.

We now ask the question: what types of functions have Fourier series? Secondly, if a function is discontinuous at a point, what value will the Fourier series give? Dirichlet[3,4] answered these questions in the first half of the nineteenth century. His results may be summarized as follows.

Dirichlet's Theorem: *If for the interval $[-L, L]$ the function $f(t)$ (1) is single-valued, (2) is bounded, (3) has at most a finite number of maxima and minima, and (4) has only a finite number of discontinuities (piecewise continuous), and if (5) $f(t + 2L) = f(t)$ for values of t outside of $[-L, L]$, then*

$$f(t) = \frac{a_0}{2} + \sum_{n=1}^{N} a_n \cos\left(\frac{n\pi t}{L}\right) + b_n \sin\left(\frac{n\pi t}{L}\right) \tag{5.1.7}$$

---

[3] Dirichlet, P. G. L., 1829: Sur la convergence des séries trigonométriques qui servent à représenter une fonction arbitraire entre des limites données. *J. Reine Angew. Math.*, **4**, 157–169.

[4] Dirichlet, P. G. L., 1837: Sur l'usage des intégrales définies dans la sommation des séries finies ou infinies. *J. Reine Angew. Math.*, **17**, 57–67.

A product of the French Revolution, (Jean Baptiste) Joseph Fourier (1768–1830) held positions within the Napoleonic Empire during his early career. After Napoleon's fall from power, Fourier devoted his talents exclusively to science. Although he won the Institut de France prize in 1811 for his work on heat diffusion, criticism of its mathematical rigor and generality led him to publish the classic book *Théorie analytique de la chaleur* in 1823. Within this book he introduced the world to the series that bears his name. (Portrait courtesy of the Archives de l'Académie des sciences, Paris.)

converges to $f(t)$ as $N \to \infty$ at values of t for which $f(t)$ is continuous and to $\frac{1}{2}[f(t^-) + f(t^+)]$ at points of discontinuity. Here $f(t^-)$ equals the value of the function at a point that is located infinitesimally to the left of t while $f(t^+)$ equals the value of the function at a point that is located infinitesimally to the right of t. The coefficients in Equation 5.1.7 are given by Equation 5.1.2, Equation 5.1.4, and Equation 5.1.5. A function $f(t)$ is bounded if the inequality $|f(t)| \leq M$ holds for some constant M for all values of t. Because *Dirichlet's conditions* (1)–(4) are very mild, it is very rare that a convergent Fourier series does not exist for a function that appears in an engineering or scientific problem.   □

- **Example 5.1.1**

Let us find the Fourier series for the function

$$f(t) = \begin{cases} 0, & -\pi < t \leq 0, \\ t, & 0 \leq t < \pi. \end{cases} \qquad (5.1.8)$$

We compute the Fourier coefficients a_n and b_n using Equation 5.1.6 by letting $L = \pi$ and $\tau = -\pi$. We then find that

$$a_0 = \frac{1}{\pi} \int_{-\pi}^{\pi} f(t)\, dt = \frac{1}{\pi} \int_0^{\pi} t\, dt = \frac{\pi}{2}, \qquad (5.1.9)$$

Second to Gauss, Peter Gustav Lejeune Dirichlet (1805–1859) was Germany's leading mathematician during the first half of the nineteenth century. Initially drawn to number theory, his later studies in analysis and applied mathematics led him to consider the convergence of Fourier series. These studies eventually produced the modern concept of a function as a correspondence that associates with each real x in an interval some unique value denoted by $f(x)$. (Taken from the frontispiece of Dirichlet, P. G. L., 1889: *Werke.* Druck und Verlag von Georg Reimer, 644 pp.)

$$a_n = \frac{1}{\pi} \int_0^\pi t \cos(nt)\, dt = \frac{1}{\pi} \left[\frac{t \sin(nt)}{n} + \frac{\cos(nt)}{n^2} \right] \Big|_0^\pi = \frac{\cos(n\pi) - 1}{n^2 \pi} = \frac{(-1)^n - 1}{n^2 \pi}$$

(5.1.10)

because $\cos(n\pi) = (-1)^n$, and

$$b_n = \frac{1}{\pi} \int_0^\pi t \sin(nt)\, dt = \frac{1}{\pi} \left[\frac{-t \cos(nt)}{n} + \frac{\sin(nt)}{n^2} \right] \Big|_0^\pi = -\frac{\cos(n\pi)}{n} = \frac{(-1)^{n+1}}{n} \quad (5.1.11)$$

for $n = 1, 2, 3, \ldots$. Thus, the Fourier series for $f(t)$ is

$$f(t) = \frac{\pi}{4} + \sum_{n=1}^\infty \frac{(-1)^n - 1}{n^2 \pi} \cos(nt) + \frac{(-1)^{n+1}}{n} \sin(nt) \qquad (5.1.12)$$

$$= \frac{\pi}{4} - \frac{2}{\pi} \sum_{m=1}^\infty \frac{\cos[(2m-1)t]}{(2m-1)^2} - \sum_{n=1}^\infty \frac{(-1)^n}{n} \sin(nt). \qquad (5.1.13)$$

We note that at the points $t = \pm(2n-1)\pi$, where $n = 1, 2, 3, \ldots$, the function jumps from zero to π. To what value does the Fourier series converge at these points? From

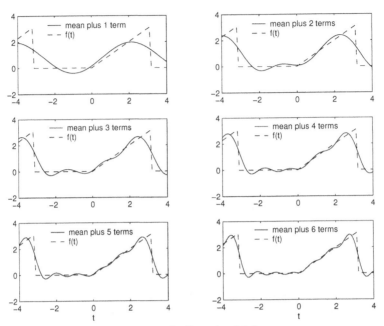

Figure 5.1.1: Partial sum of the Fourier series for Equation 5.1.8.

Dirichlet's theorem, the series converges to the average of the values of the function just to the right and left of the point of discontinuity, i.e., $(\pi + 0)/2 = \pi/2$. At the remaining points the series converges to $f(t)$.

Figure 5.1.1 shows how well Equation 5.1.12 approximates the function by graphing various partial sums of this expansion as we include more and more terms (harmonics). The MATLAB script that created this figure is:

```
clear;
t = [-4:0.1:4]; % create time points in plot
f = zeros(size(t)); % initialize function f(t)
for k = 1:length(t) % construct function f(t)
  if t(k) < 0; f(k) = 0; else f(k) = t(k); end;
  if t(k) < -pi; f(k) = t(k) + 2*pi; end;
  if t(k) > pi ; f(k) = 0; end;
end
% initialize Fourier series with the mean term
fs = (pi/4) * ones(size(t));
clf % clear any figures
for n = 1:6
% create plot of truncated FS with only n harmonic
  fs = fs - (2/pi) * cos((2*n-1)*t) / (2*n-1)^2;
  fs = fs - (-1)^n * sin(n*t) / n;
  subplot(3,2,n), plot(t,fs,t,f,'--')
  if n==1
    legend('mean plus 1 term','f(t)'); legend boxoff;
  else
    legend(['mean plus ',num2str(n),' terms'],'f(t)')
    legend boxoff
```

```
    end
  if n >= 5; xlabel('t'); end;
end
```

As the figure shows, successive corrections are made to the mean value of the series, $\pi/2$. As each harmonic is added, the Fourier series fits the function better in the sense of least squares:

$$\int_{\tau}^{\tau+2L} [f(x) - f_N(x)]^2 \, dx = \text{minimum},\qquad (5.1.14)$$

where $f_N(x)$ is the truncated Fourier series of N terms. Still, near the points of discontinuity $t = \pm\pi$, the Fourier series struggles to express a discontinuity as a sum of continuous sinusoidal functions. Failing to do so, it oscillates above and below the given function. Indeed, even if we could actually compute the series with an infinite number of harmonics, we would still obtain incorrect answers near the discontinuities. We will discuss this Gibbs phenomena at the end of Section 5.2.   □

• **Example 5.1.2**

Let us calculate the Fourier series of the function $f(t) = |t|$, which is defined over the range $-\pi \le t \le \pi$.

From the definition of the Fourier coefficients,

$$a_0 = \frac{1}{\pi}\left[\int_{-\pi}^{0} -t\,dt + \int_{0}^{\pi} t\,dt\right] = \frac{\pi}{2} + \frac{\pi}{2} = \pi,\qquad (5.1.15)$$

$$a_n = \frac{1}{\pi}\left[\int_{-\pi}^{0} -t\cos(nt)\,dt + \int_{0}^{\pi} t\cos(nt)\,dt\right]\qquad (5.1.16)$$

$$= -\left.\frac{nt\sin(nt) + \cos(nt)}{n^2\pi}\right|_{-\pi}^{0} + \left.\frac{nt\sin(nt) + \cos(nt)}{n^2\pi}\right|_{0}^{\pi}\qquad (5.1.17)$$

$$= \frac{2}{n^2\pi}[(-1)^n - 1]\qquad (5.1.18)$$

and

$$b_n = \frac{1}{\pi}\left[\int_{-\pi}^{0} -t\sin(nt)\,dt + \int_{0}^{\pi} t\sin(nt)\,dt\right]\qquad (5.1.19)$$

$$= \left.\frac{nt\cos(nt) - \sin(nt)}{n^2\pi}\right|_{-\pi}^{0} - \left.\frac{nt\cos(nt) - \sin(nt)}{n^2\pi}\right|_{0}^{\pi} = 0\qquad (5.1.20)$$

for $n = 1, 2, 3, \ldots$. Therefore,

$$|t| = \frac{\pi}{2} + \frac{2}{\pi}\sum_{n=1}^{\infty} \frac{[(-1)^n - 1]}{n^2}\cos(nt) = \frac{\pi}{2} - \frac{4}{\pi}\sum_{m=1}^{\infty} \frac{\cos[(2m-1)t]}{(2m-1)^2}\qquad (5.1.21)$$

for $-\pi \le t \le \pi$.

In Figure 5.1.2 we show how well Equation 5.1.21 approximates the function by graphing various partial sums of this expansion. As the figure shows, the Fourier series does very

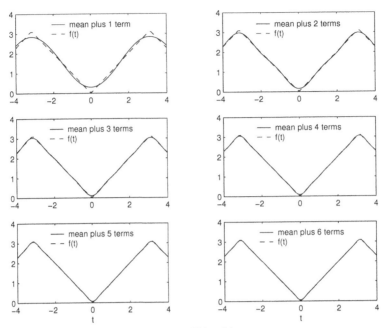

Figure 5.1.2: Partial sum of the Fourier series for $f(t) = |t|$.

well even when we use very few terms. The reason for this rapid convergence is the nature
of the function: it does not possess any jump discontinuities.   □

• **Example 5.1.3**

Sometimes the function $f(t)$ is an even or odd function.[5] Can we use this property to
simplify our work? The answer is yes.

Let $f(t)$ be an even function. Then

$$a_0 = \frac{1}{L} \int_{-L}^{L} f(t)\, dt = \frac{2}{L} \int_{0}^{L} f(t)\, dt, \tag{5.1.22}$$

and

$$a_n = \frac{1}{L} \int_{-L}^{L} f(t) \cos\left(\frac{n\pi t}{L}\right) dt = \frac{2}{L} \int_{0}^{L} f(t) \cos\left(\frac{n\pi t}{L}\right) dt, \tag{5.1.23}$$

whereas

$$b_n = \frac{1}{L} \int_{-L}^{L} f(t) \sin\left(\frac{n\pi t}{L}\right) dt = 0. \tag{5.1.24}$$

Here we used the properties that $\int_{-L}^{L} f_e(x)\, dx = 2 \int_{0}^{L} f_e(x)\, dx$ and $\int_{-L}^{L} f_o(x) dx = 0$. Thus,
if we have an even function, we merely compute a_0 and a_n via Equation 5.1.22 and Equation
5.1.23, and $b_n = 0$. Because the corresponding series contains only cosine terms, it is often
called a *Fourier cosine series*.

---

[5] An even function $f_e(t)$ has the property that $f_e(-t) = f_e(t)$; an odd function $f_o(t)$ has the property
that $f_o(-t) = -f_o(t)$.

Similarly, if $f(t)$ is odd, then

$$a_0 = a_n = 0, \qquad \text{and} \qquad b_n = \frac{2}{L} \int_0^L f(t) \sin\left(\frac{n\pi t}{L}\right) dt. \qquad (5.1.25)$$

Thus, if we have an odd function, we merely compute b_n via Equation 5.1.25 and $a_0 = a_n = 0$. Because the corresponding series contains only sine terms, it is often called a *Fourier sine series*. ☐

• Example 5.1.4

In the case when $f(x)$ consists of a constant and/or trigonometric functions, it is much easier to find the corresponding Fourier series by inspection rather than by using Equation 5.1.6. For example, let us find the Fourier series for $f(x) = \sin^2(x)$ defined over the range $-\pi \le x \le \pi$.

We begin by rewriting $f(x) = \sin^2(x)$ as $f(x) = \frac{1}{2}[1 - \cos(2x)]$. Next, we note that any function defined over the range $-\pi < x < \pi$ has the Fourier series

$$f(x) = \frac{a_0}{2} + \sum_{n=1}^{\infty} a_n \cos(nx) + b_n \sin(nx) \qquad (5.1.26)$$

$$= \frac{a_0}{2} + a_1 \cos(x) + b_1 \sin(x) + a_2 \cos(2x) + b_2 \sin(2x) + \cdots. \qquad (5.1.27)$$

On the other hand,

$$f(x) = \frac{1}{2} - \frac{1}{2}\cos(2x) = \frac{1}{2} + 0\cos(x) + 0\sin(x) - \frac{1}{2}\cos(2x) + 0\sin(2x) + \cdots. \qquad (5.1.28)$$

By inspection, we can immediately write that

$$a_0 = 1, \quad a_1 = b_1 = 0, \quad a_2 = -\tfrac{1}{2}, \quad b_2 = 0, \quad a_n = b_n = 0, \quad n \ge 3. \qquad (5.1.29)$$

Thus, instead of the usual expansion involving an infinite number of sine and cosine terms, our Fourier series contains only two terms and is simply

$$f(x) = \tfrac{1}{2} - \tfrac{1}{2}\cos(2x), \qquad -\pi \le x \le \pi. \qquad (5.1.30)$$

☐

• Example 5.1.5: Quieting snow tires

An application of Fourier series to a problem in industry occurred several years ago, when drivers found that snow tires produced a loud whine[6] on dry pavement. Tire sounds are produced primarily by the dynamic interaction of the tread elements with the road surface.[7] As each tread element passes through the contact patch, it contributes a pulse of acoustic energy to the total sound field radiated by the tire.

---

[6] See Varterasian, J. H., 1969: Math quiets rotating machines. *SAE J.*, **77(10)**, 53.

[7] Willett, P. R., 1975: Tire tread pattern sound generation. *Tire Sci. Tech.*, **3**, 252–266.

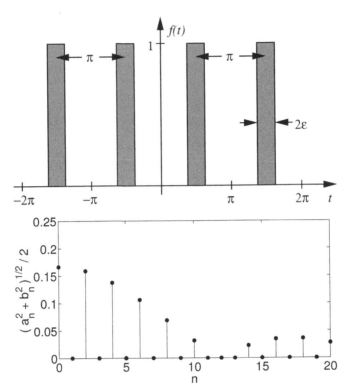

Figure 5.1.3: Temporal spacing (over two periods) and frequency spectrum of uniformly spaced snow tire treads.

For evenly spaced treads we envision that the release of acoustic energy resembles the top of Figure 5.1.3. If we perform a Fourier analysis of this distribution, we find that

$$a_0 = \frac{1}{\pi} \left[\int_{-\pi/2-\epsilon}^{-\pi/2+\epsilon} 1 \, dt + \int_{\pi/2-\epsilon}^{\pi/2+\epsilon} 1 \, dt \right] = \frac{4\epsilon}{\pi}, \tag{5.1.31}$$

where ϵ is half of the width of the tread and

$$a_n = \frac{1}{\pi} \left[\int_{-\pi/2-\epsilon}^{-\pi/2+\epsilon} \cos(nt) \, dt + \int_{\pi/2-\epsilon}^{\pi/2+\epsilon} \cos(nt) \, dt \right] \tag{5.1.32}$$

$$= \frac{1}{n\pi} \left[\sin(nt) \Big|_{-\pi/2-\epsilon}^{-\pi/2+\epsilon} + \sin(nt) \Big|_{\pi/2-\epsilon}^{\pi/2+\epsilon} \right] \tag{5.1.33}$$

$$= \frac{1}{n\pi} \left[\sin\left(-\frac{n\pi}{2} + n\epsilon\right) - \sin\left(-\frac{n\pi}{2} - n\epsilon\right) + \sin\left(\frac{n\pi}{2} + n\epsilon\right) - \sin\left(\frac{n\pi}{2} - n\epsilon\right) \right] \tag{5.1.34}$$

$$= \frac{1}{n\pi} \left[2\cos\left(-\frac{n\pi}{2}\right) + 2\cos\left(\frac{n\pi}{2}\right) \right] \sin(n\epsilon) = \frac{4}{n\pi} \cos\left(\frac{n\pi}{2}\right) \sin(n\epsilon). \tag{5.1.35}$$

Because $f(t)$ is an even function, $b_n = 0$.

The question now arises of how to best illustrate our Fourier coefficients. In Section 5.4 we will show that any harmonic can be represented as a single wave $A_n \cos(n\pi t/L + \varphi_n)$ or $A_n \sin(n\pi t/L + \psi_n)$, where the amplitude $A_n = \sqrt{a_n^2 + b_n^2}$. In the bottom frame of Figure 5.1.3, MATLAB was used to plot this amplitude, usually called the *amplitude* or *frequency spectrum* $\frac{1}{2}\sqrt{a_n^2 + b_n^2}$, as a function of n for an arbitrarily chosen $\epsilon = \pi/12$. Although the value of ϵ will affect the exact shape of the spectrum, the qualitative arguments that we will present remain unchanged. We have added the factor $\frac{1}{2}$ so that our definition of the frequency spectrum is consistent with that for a complex Fourier series stated after Equation 5.5.12. The amplitude spectrum in Figure 5.1.3 shows that the spectrum for periodically placed tire treads has its largest amplitude at small n. This produces one loud tone plus strong harmonic overtones because the fundamental and its overtones are the dominant terms in the Fourier series representation.

Clearly, this loud, monotone whine is undesirable. How might we avoid it? Just as soldiers marching in step produce a loud uniform sound, we suspect that our uniform tread pattern is the problem. Therefore, let us now vary the interval between the treads so that the distance between any tread and its nearest neighbor is not equal, as illustrated in Figure 5.1.4. Again we perform its Fourier analysis and obtain that

$$a_0 = \frac{1}{\pi}\left[\int_{-\pi/2-\epsilon}^{-\pi/2+\epsilon} 1\,dt + \int_{\pi/4-\epsilon}^{\pi/4+\epsilon} 1\,dt\right] = \frac{4\epsilon}{\pi}, \tag{5.1.36}$$

$$a_n = \frac{1}{\pi}\left[\int_{-\pi/2-\epsilon}^{-\pi/2+\epsilon} \cos(nt)\,dt + \int_{\pi/4-\epsilon}^{\pi/4+\epsilon} \cos(nt)\,dt\right] \tag{5.1.37}$$

$$= \frac{1}{n\pi}\sin(nt)\Big|_{-\pi/2-\epsilon}^{-\pi/2+\epsilon} + \frac{1}{n\pi}\sin(nt)\Big|_{\pi/4-\epsilon}^{\pi/4+\epsilon} \tag{5.1.38}$$

$$= -\frac{1}{n\pi}\left[\sin\left(\frac{n\pi}{2} - n\epsilon\right) - \sin\left(\frac{n\pi}{2} + n\epsilon\right)\right] + \frac{1}{n\pi}\left[\sin\left(\frac{n\pi}{4} + n\epsilon\right) - \sin\left(\frac{n\pi}{4} - n\epsilon\right)\right] \tag{5.1.39}$$

$$a_n = \frac{2}{n\pi}\left[\cos\left(\frac{n\pi}{2}\right) + \cos\left(\frac{n\pi}{4}\right)\right]\sin(n\epsilon), \tag{5.1.40}$$

and

$$b_n = \frac{1}{\pi}\left[\int_{-\pi/2-\epsilon}^{-\pi/2+\epsilon} \sin(nt)\,dt + \int_{\pi/4-\epsilon}^{\pi/4+\epsilon} \sin(nt)\,dt\right] \tag{5.1.41}$$

$$= -\frac{1}{n\pi}\left[\cos\left(\frac{n\pi}{2} - n\epsilon\right) - \cos\left(\frac{n\pi}{2} + n\epsilon\right)\right] - \frac{1}{n\pi}\left[\cos\left(\frac{n\pi}{4} + n\epsilon\right) - \cos\left(\frac{n\pi}{4} - n\epsilon\right)\right] \tag{5.1.42}$$

$$= \frac{2}{n\pi}\left[\sin\left(\frac{n\pi}{4}\right) - \sin\left(\frac{n\pi}{2}\right)\right]\sin(n\epsilon). \tag{5.1.43}$$

The MATLAB script is:

```
epsilon = pi/12; % set up parameter for fs coefficient
n = 1:20; % number of harmonics
arg1 = (pi/2)*n; arg2 = (pi/4)*n; arg3 = epsilon*n;
% compute the Fourier coefficient a_n
an = (cos(arg1) + cos(arg2)).*sin(arg3);
an = (2/pi) * an./n;
% compute the Fourier coefficient b_n
bn = (sin(arg2) - sin(arg1)).*sin(arg3);
bn = (2/pi) * bn./n;
% compute the magnitude
cn = 0.5 * sqrt(an.*an + bn.*bn);
```

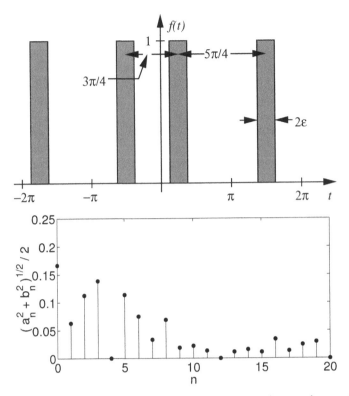

Figure 5.1.4: Temporal spacing and frequency spectrum of nonuniformly spaced snow tire treads.

```
% add in the a_0 term
cn = [2*epsilon/pi,cn];
n = [0,n];
clf % clear any figures
axes('FontSize',20) % set font size
stem(n,cn,'filled') % plot spectrum
set (gca,'PlotBoxAspectRatio',[8 4 1]) % set aspect ratio
xlabel('n') % label x-axis
ylabel('( a_n^2 + b_n^2 )^{1/2}/2') % label y-axis,
```

was used to compute the amplitude of each harmonic as a function of n and the results were plotted. See Figure 5.1.4. The important point is that our new choice for the spacing of the treads has reduced or eliminated some of the harmonics compared to the case of equally spaced treads. On the negative side, we have excited some of the harmonics that were previously absent. However, the net effect is advantageous because the treads produce less noise at more frequencies rather than a lot of noise at a few select frequencies.

If we were to extend this technique so that the treads occurred at completely random positions, then the treads would produce very little noise at many frequencies and the total noise would be comparable to that generated by other sources within the car. To find the distribution of treads with the whitest noise[8] is a process of trial and error. Assuming a distribution, we can perform a Fourier analysis to obtain its frequency spectrum. If annoying peaks are present in the spectrum, we can then adjust the elements in the tread

---

[8] White noise is sound that is analogous to white light in that it is uniformly distributed throughout the complete audible sound spectrum.

distribution that may contribute to the peak and analyze the revised distribution. You are finished when no peaks appear.

Problems

Find the Fourier series for the following functions. Using MATLAB, plot the Fourier spectrum. Then plot various partial sums and compare them against the exact function.

1. $f(t) = \begin{cases} 1, & -\pi < t < 0 \\ 0, & 0 < t < \pi \end{cases}$

2. $f(t) = \begin{cases} t, & -\pi < t \le 0 \\ 0, & 0 \le t < \pi \end{cases}$

3. $f(t) = \begin{cases} -\pi, & -\pi < t < 0 \\ t, & 0 < t < \pi \end{cases}$

4. $f(t) = \begin{cases} 1/2 + t, & -1 \le t \le 0 \\ 1/2 - t, & 0 \le t \le 1 \end{cases}$

5. $f(t) = \begin{cases} 0, & -\pi \le t \le 0 \\ t, & 0 \le t \le \pi/2 \\ \pi - t, & \pi/2 \le t \le \pi \end{cases}$

6. $f(t) = \begin{cases} 0, & -\pi \le t \le -\pi/2 \\ \sin(2t), & -\pi/2 \le t \le \pi/2 \\ 0, & \pi/2 \le t \le \pi \end{cases}$

7. $f(t) = e^{at}, \qquad -L < t < L$

8. $f(t) = t + t^2, \qquad -L < t < L$

9. $f(t) = \begin{cases} 0, & -\pi \le t \le 0 \\ \sin(t), & 0 \le t \le \pi \end{cases}$

10. $f(t) = \begin{cases} t, & -\frac{1}{2} \le t \le \frac{1}{2} \\ 1 - t, & \frac{1}{2} \le t \le \frac{3}{2} \end{cases}$

11. $f(t) = \begin{cases} 0, & -a < t < 0 \\ 2t, & 0 < t < a \end{cases}$

12. $f(t) = \begin{cases} 0, & -\pi < t \le 0 \\ t^2, & 0 \le t < \pi \end{cases}$

13. $f(t) = (\pi - t)/2, \quad 0 < t < 2$

14. $f(t) = t \cos(\pi t/L), \ -L < t < L$

15. $f(t) = \sinh[a\,(\pi/2 - |t|)], \ -\pi \le t \le \pi$

16. $f(t) = \begin{cases} t(2L - t), & 0 \le t \le 2L \\ t^2 - 6Lt + 8L^2, & 2L \le t \le 4L \end{cases}$

17. Given the Fourier series

$$f(x) = \sum_{n=1}^{\infty} \frac{\sin(n^2)}{n^\alpha} \sin(n\pi x), \qquad 0 < x < 1,$$

use MATLAB to plot this Fourier series for $\frac{1}{2} < \alpha < 2$ as a three-dimensional plot with axes x, $f(x)$ and α. What happens as α crosses the line $\alpha = 1$?

Project: Design Your Own Snow Tire, Part I

In the example on quieting snow tires, there are at least three important parameters: the thickness of the tread 2ϵ, the time at which the first tread hits the road $t = -a$, and the time at which the second thread hits the road $t = b$, where $\epsilon < a, b < \pi - \epsilon$. The values that I used were arbitrary. In this project you will redo the calculations to see the sensitivity of the results various values of a, b and ϵ.

Step 1: Recompute the Fourier coefficients where one of the treads strikes the road between $-a - \epsilon$ and $-a + \epsilon$ and the second strikes the road between $b - \epsilon$ and $b + \epsilon$. What is the effect of a thicker tread on the spectrum?

Step 2: In the text we examined the case of $a = -\pi/2$ and $b = \pi/4$. What happens if $a = -3\pi/2$ and $b = \pi/4$? Explain your results.

Step 3: What happens to the spectrum when the distance between the treads lies at some other arbitrary distance between 0 and π?

5.2 PROPERTIES OF FOURIER SERIES

In the previous section we introduced the Fourier series and showed how to compute one given the function $f(t)$. In this section we examine some particular properties of these series.

> Differentiation of a Fourier series

In certain instances we only have the Fourier series representation of a function $f(t)$. Can we find the derivative or the integral of $f(t)$ merely by differentiating or integrating the Fourier series term by term? Is this permitted? Let us consider the case of differentiation first.

Consider a function $f(t)$ of period $2L$, which has the derivative $f'(t)$. Let us assume that we can expand $f'(t)$ as a Fourier series. This implies that $f'(t)$ is continuous except for a finite number of discontinuities and $f(t)$ is continuous over an interval that starts at $t = \tau$ and ends at $t = \tau + 2L$. Then

$$f'(t) = \frac{a_0'}{2} + \sum_{n=1}^{\infty} a_n' \cos\left(\frac{n\pi t}{L}\right) + b_n' \sin\left(\frac{n\pi t}{L}\right), \tag{5.2.1}$$

where we denoted the Fourier coefficients of $f'(t)$ with a prime. Computing the Fourier coefficients,

$$a_0' = \frac{1}{L} \int_\tau^{\tau+2L} f'(t)\, dt = \frac{1}{L}\left[f(\tau + 2L) - f(\tau)\right] = 0, \tag{5.2.2}$$

if $f(\tau + 2L) = f(\tau)$. Similarly, by integrating by parts,

$$a_n' = \frac{1}{L} \int_\tau^{\tau+2L} f'(t) \cos\left(\frac{n\pi t}{L}\right) dt \tag{5.2.3}$$

$$= \frac{1}{L}\left[f(t) \cos\left(\frac{n\pi t}{L}\right)\right]\Bigg|_\tau^{\tau+2L} + \frac{n\pi}{L^2} \int_\tau^{\tau+2L} f(t) \sin\left(\frac{n\pi t}{L}\right) dt \tag{5.2.4}$$

$$= \frac{n\pi b_n}{L}, \tag{5.2.5}$$

and

$$b'_n = \frac{1}{L} \int_\tau^{\tau+2L} f'(t) \sin\left(\frac{n\pi t}{L}\right) dt \tag{5.2.6}$$

$$= \frac{1}{L}\left[f(t) \sin\left(\frac{n\pi t}{L}\right) \right]\Big|_\tau^{\tau+2L} - \frac{n\pi}{L^2} \int_\tau^{\tau+2L} f(t) \cos\left(\frac{n\pi t}{L}\right) dt \tag{5.2.7}$$

$$= -\frac{n\pi a_n}{L}. \tag{5.2.8}$$

If we have a function $f(t)$ whose derivative $f'(t)$ is continuous except for a finite number of discontinuities and $f(\tau) = f(\tau + 2L)$, then

$$f'(t) = \sum_{n=1}^\infty \frac{n\pi}{L}\left[b_n \cos\left(\frac{n\pi t}{L}\right) - a_n \sin\left(\frac{n\pi t}{L}\right) \right]. \tag{5.2.9}$$

That is, the derivative of $f(t)$ is given by a term-by-term differentiation of the Fourier series of $f(t)$.

- **Example 5.2.1**

    The Fourier series for the periodic function

$$f(t) = \begin{cases} 0, & -\pi \le t \le 0, \\ t, & 0 \le t \le \pi/2, \\ \pi - t, & \pi/2 \le t \le \pi, \end{cases} \qquad f(t) = f(t + 2\pi), \tag{5.2.10}$$

is

$$f(t) = \frac{\pi}{8} - \frac{1}{\pi}\sum_{n=1}^\infty \frac{\cos[2(2n-1)t]}{(2n-1)^2} - \frac{2}{\pi}\sum_{n=1}^\infty \frac{(-1)^n}{(2n-1)^2} \sin[(2n-1)t]. \tag{5.2.11}$$

Because $f(t)$ is continuous over the entire interval $(-\pi, \pi)$ and $f(-\pi) = f(\pi) = 0$, we can find $f'(t)$ by taking the derivative of Equation 5.2.11 term by term:

$$f'(t) = \frac{2}{\pi}\sum_{n=1}^\infty \frac{\sin[2(2n-1)t]}{2n-1} - \frac{2}{\pi}\sum_{n=1}^\infty \frac{(-1)^n}{2n-1} \cos[(2n-1)t]. \tag{5.2.12}$$

This is the same Fourier series that we would obtain by computing the Fourier series for

$$f'(t) = \begin{cases} 0, & -\pi < t < 0, \\ 1, & 0 < t < \pi/2, \\ -1, & \pi/2 < t < \pi. \end{cases} \tag{5.2.13}$$

$\square$

Integration of a Fourier series

To determine whether we can find the integral of $f(t)$ by term-by-term integration of its Fourier series, consider a form of the antiderivative of $f(t)$:

$$F(t) = \int_0^t \left[f(\tau) - \frac{a_0}{2} \right] d\tau. \tag{5.2.14}$$

Now

$$F(t + 2L) = \int_0^t \left[f(\tau) - \frac{a_0}{2} \right] d\tau + \int_t^{t+2L} \left[f(\tau) - \frac{a_0}{2} \right] d\tau \qquad (5.2.15)$$

$$= F(t) + \int_{-L}^L \left[f(\tau) - \frac{a_0}{2} \right] d\tau \qquad (5.2.16)$$

$$= F(t) + \int_{-L}^L f(\tau) \, d\tau - La_0 = F(t), \qquad (5.2.17)$$

so that $F(t)$ has a period of $2L$. As a result we may expand $F(t)$ as the Fourier series

$$F(t) = \frac{A_0}{2} + \sum_{n=1}^{\infty} A_n \cos\left(\frac{n\pi t}{L} \right) + B_n \sin\left(\frac{n\pi t}{L} \right). \qquad (5.2.18)$$

For A_n,

$$A_n = \frac{1}{L} \int_{-L}^L F(t) \cos\left(\frac{n\pi t}{L} \right) dt \qquad (5.2.19)$$

$$= \frac{1}{L} \left[F(t) \frac{\sin(n\pi t/L)}{n\pi/L} \right] \Big|_{-L}^{L} - \frac{1}{n\pi} \int_{-L}^L \left[f(t) - \frac{a_0}{2} \right] \sin\left(\frac{n\pi t}{L} \right) dt \qquad (5.2.20)$$

$$= -\frac{b_n}{n\pi/L}. \qquad (5.2.21)$$

Similarly,

$$B_n = \frac{a_n}{n\pi/L}. \qquad (5.2.22)$$

Therefore,

$$\int_0^t f(\tau) \, d\tau = \frac{a_0 t}{2} + \frac{A_0}{2} + \sum_{n=1}^{\infty} \frac{a_n \sin(n\pi t/L) - b_n \cos(n\pi t/L)}{n\pi/L}. \qquad (5.2.23)$$

This is identical to a term-by-term integration of the Fourier series for $f(t)$. Thus, we can always find the integral of $f(t)$ by a term-by-term integration of its Fourier series.

• **Example 5.2.2**

The Fourier series for $f(t) = t$ for $-\pi < t < \pi$ is

$$f(t) = -2 \sum_{n=1}^{\infty} \frac{(-1)^n}{n} \sin(nt). \qquad (5.2.24)$$

To find the Fourier series for $f(t) = t^2$, we integrate Equation 5.2.24 term by term and find that

$$\frac{\tau^2}{2} \Big|_0^t = 2 \sum_{n=1}^{\infty} \frac{(-1)^n}{n^2} \cos(nt) - 2 \sum_{n=1}^{\infty} \frac{(-1)^n}{n^2}. \qquad (5.2.25)$$

But $\sum_{n=1}^{\infty} (-1)^n/n^2 = -\pi^2/12$. Substituting and multiplying by 2, we obtain the final result that

$$t^2 = \frac{\pi^2}{3} + 4 \sum_{n=1}^{\infty} \frac{(-1)^n}{n^2} \cos(nt). \qquad (5.2.26)$$

$\square$

Parseval's equality

One of the fundamental quantities in engineering is power. The *power content* of a periodic signal $f(t)$ of period $2L$ is $\int_{\tau}^{\tau+2L} f^2(t)\, dt/L$. This mathematical definition mirrors the power dissipation $I^2 R$ that occurs in a resistor of resistance R where I is the root mean square (RMS) of the current. We would like to compute this power content as simply as possible given the coefficients of its Fourier series.

Assume that $f(t)$ has the Fourier series

$$f(t) = \frac{a_0}{2} + \sum_{n=1}^{\infty} a_n \cos\left(\frac{n\pi t}{L}\right) + b_n \sin\left(\frac{n\pi t}{L}\right). \qquad (5.2.27)$$

Then,

$$\frac{1}{L} \int_{\tau}^{\tau+2L} f^2(t)\, dt = \frac{a_0}{2L} \int_{\tau}^{\tau+2L} f(t)\, dt + \sum_{n=1}^{\infty} \frac{a_n}{L} \int_{\tau}^{\tau+2L} f(t) \cos\left(\frac{n\pi t}{L}\right) dt$$

$$+ \sum_{n=1}^{\infty} \frac{b_n}{L} \int_{\tau}^{\tau+2L} f(t) \sin\left(\frac{n\pi t}{L}\right) dt \qquad (5.2.28)$$

$$= \frac{a_0^2}{2} + \sum_{n=1}^{\infty} (a_n^2 + b_n^2). \qquad (5.2.29)$$

Equation 5.2.29 is *Parseval's equality*.[9] It allows us to sum squares of Fourier coefficients (which we have already computed) rather than performing the integration $\int_{\tau}^{\tau+2L} f^2(t)\, dt$ analytically or numerically.

- **Example 5.2.3**

The Fourier series for $f(t) = t^2$ over the interval $[-\pi, \pi]$ is

$$t^2 = \frac{\pi^2}{3} + 4 \sum_{n=1}^{\infty} \frac{(-1)^n}{n^2} \cos(nt). \qquad (5.2.30)$$

---

[9] Parseval, M.-A., 1805: Mémoire sur les séries et sur l'intégration complète d'une équation aux différences partielles linéaires du second ordre, à coefficients constants. *Mémoires présentés a l'Institut des sciences, lettres et arts, par divers savans, et lus dans ses assemblées: Sciences mathématiques et Physiques*, **1**, 638–648.

Then, by Parseval's equality,

$$\frac{1}{\pi} \int_{-\pi}^{\pi} t^4 \, dt = \frac{2t^5}{5\pi} \bigg|_0^\pi = \frac{4\pi^4}{18} + 16 \sum_{n=1}^{\infty} \frac{1}{n^4}, \quad \text{or} \quad \left(\frac{2}{5} - \frac{4}{18} \right) \pi^4 = 16 \sum_{n=1}^{\infty} \frac{1}{n^4}. \quad (5.2.31)$$

Consequently,

$$\frac{\pi^4}{90} = \sum_{n=1}^{\infty} \frac{1}{n^4}. \quad (5.2.32)$$

$\square$

Gibbs phenomena

In the actual application of Fourier series, we cannot sum an infinite number of terms but must be content with N terms. If we denote this partial sum of the Fourier series by $S_N(t)$, we have from the definition of the Fourier series:

$$S_N(t) = \tfrac{1}{2}a_0 + \sum_{n=1}^{N} a_n \cos(nt) + b_n \sin(nt) \quad (5.2.33)$$

$$= \frac{1}{2\pi} \int_0^{2\pi} f(x) \, dx + \frac{1}{\pi} \int_0^{2\pi} f(x) \left[\sum_{n=1}^{N} \cos(nt)\cos(nx) + \sin(nt)\sin(nx) \right] dx \quad (5.2.34)$$

$$= \frac{1}{\pi} \int_0^{2\pi} f(x) \left\{ \frac{1}{2} + \sum_{n=1}^{N} \cos[n(t-x)] \right\} dx \quad (5.2.35)$$

$$= \frac{1}{2\pi} \int_0^{2\pi} f(x) \frac{\sin[(N+\tfrac{1}{2})(x-t)]}{\sin[\tfrac{1}{2}(x-t)]} \, dx. \quad (5.2.36)$$

The quantity $\sin[(N+\tfrac{1}{2})(x-t)]/\sin[\tfrac{1}{2}(x-t)]$ is called a *scanning function*. Over the range $0 \le x \le 2\pi$ it has a very large peak at $x = t$ where the amplitude equals $2N+1$. See Figure 5.2.1. On either side of this peak there are oscillations that decrease rapidly with distance from the peak. As $N \to \infty$, the scanning function becomes essentially a long narrow slit corresponding to the area under the large peak at $x = t$. If we neglect for the moment the small area under the minor ripples adjacent to this slit, then the integral, Equation 5.2.36, essentially equals $f(t)$ times the area of the slit divided by 2π. If $1/2\pi$ times the area of the slit equals unity, then the value of $S_N(t) \approx f(t)$ to a good approximation for large N.

For relatively small values of N, the scanning function deviates considerably from its ideal form, and the partial sum $S_N(t)$ only crudely approximates $f(t)$. As the partial sum includes more terms and N becomes relatively large, the form of the scanning function improves and so does the agreement between $S_N(t)$ and $f(t)$. The improvement in the scanning function is due to the large hump becoming taller and narrower. At the same time, the adjacent ripples become more numerous as well as narrower in the same proportion as the large hump does.

The reason why $S_N(t)$ and $f(t)$ will never become identical, even in the limit of $N \to \infty$, is the presence of the positive and negative side lobes near the large peak. Because

$$\frac{\sin[(N+\tfrac{1}{2})(x-t)]}{\sin[\tfrac{1}{2}(x-t)]} = 1 + 2 \sum_{n=1}^{N} \cos[n(t-x)], \quad (5.2.37)$$

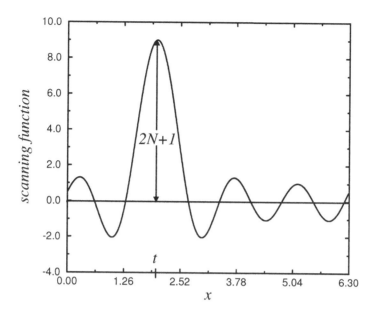

Figure 5.2.1: The scanning function over $0 \le x \le 2\pi$ for $N = 5$.

an integration of the scanning function over the interval 0 to 2π shows that the total area under the scanning function equals 2π. However, from Figure 5.2.1 the net area contributed by the ripples is numerically negative so that the area under the large peak must exceed 2π if the total area equals 2π. Although the exact value depends upon N, it is important to note that this excess does not become zero as $N \to \infty$.

Thus, the presence of these negative side lobes explains the departure of our scanning function from the idealized slit of area 2π. To illustrate this departure, consider the function:

$$f(t) = \begin{cases} 1, & 0 < t < \pi, \\ -1, & \pi < t < 2\pi. \end{cases} \qquad (5.2.38)$$

Then,

$$S_N(t) = \frac{1}{2\pi} \int_0^\pi \frac{\sin[(N+\frac{1}{2})(x-t)]}{\sin[\frac{1}{2}(x-t)]} \, dx - \frac{1}{2\pi} \int_\pi^{2\pi} \frac{\sin[(N+\frac{1}{2})(x-t)]}{\sin[\frac{1}{2}(x-t)]} \, dx \qquad (5.2.39)$$

$$= \frac{1}{2\pi} \int_0^\pi \left\{ \frac{\sin[(N+\frac{1}{2})(x-t)]}{\sin[\frac{1}{2}(x-t)]} \, dx + \frac{\sin[(N+\frac{1}{2})(x+t)]}{\sin[\frac{1}{2}(x+t)]} \, dx \right\} \qquad (5.2.40)$$

$$= \frac{1}{2\pi} \int_{-t}^{\pi-t} \frac{\sin[(N+\frac{1}{2})\theta]}{\sin(\frac{1}{2}\theta)} \, d\theta - \frac{1}{2\pi} \int_t^{\pi+t} \frac{\sin[(N+\frac{1}{2})\theta]}{\sin(\frac{1}{2}\theta)} \, d\theta. \qquad (5.2.41)$$

The first integral in Equation 5.2.41 gives the contribution to $S_N(t)$ from the jump discontinuity at $t = 0$ while the second integral gives the contribution from $t = \pi$. In Figure 5.2.2 we have plotted $S_N(t)$ when $N = 27$ and $N = 81$. Residual discrepancies remain even for very large values of N. Indeed, as N increases, this figure changes only in that the ripples in the vicinity of the discontinuity of $f(t)$ proportionally increase their rate of oscillation as a function of t while their relative magnitude remains the same. As $N \to \infty$ these ripples compress into a single vertical line at the point of discontinuity. True, these oscillations occupy smaller and smaller spaces but they still remain. Thus, we can never

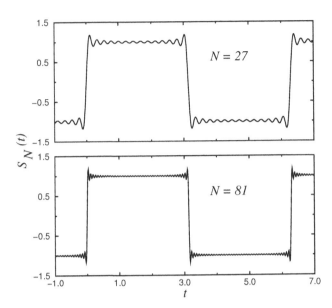

Figure 5.2.2: The finite Fourier series representation $S_N(t)$ for the function, Equation 5.2.38, for the range $-1 \leq t \leq 7$ for $N = 27$ and $N = 81$.

approximate a function in the vicinity of a discontinuity by a finite Fourier series without suffering from this over- and undershooting of the series. This peculiarity of Fourier series is called the *Gibbs phenomena*.[10] Gibbs phenomena can only be eliminated by removing the discontinuity.[11]

Problems

Additional Fourier series representations can be generated by differentiating or integrating known Fourier series. Work out the following two examples.

1. Given

$$\frac{\pi^2 - 2\pi x}{8} = \sum_{n=0}^{\infty} \frac{\cos[(2n+1)x]}{(2n+1)^2}, \qquad 0 \leq x \leq \pi,$$

obtain

$$\frac{\pi^2 x - \pi x^2}{8} = \sum_{n=0}^{\infty} \frac{\sin[(2n+1)x]}{(2n+1)^3}, \qquad 0 \leq x \leq \pi,$$

by term-by-term integration. Could we go the other way, i.e., take the derivative of the second equation to obtain the first? Explain.

2. Given

$$\frac{\pi^2 - 3x^2}{12} = \sum_{n=1}^{\infty} (-1)^{n+1} \frac{\cos(nx)}{n^2}, \qquad -\pi \leq x \leq \pi,$$

[10] Gibbs, J. W., 1898: Fourier's series. *Nature*, **59**, 200; Gibbs, J. W., 1899: Fourier's series. *Nature*, **59**, 606. For the historical development, see Hewitt, E., and R. E. Hewitt, 1979: The Gibbs-Wilbraham phenomenon: An episode in Fourier analysis. *Arch. Hist. Exact Sci.*, **21**, 129–160.

[11] For a particularly clever method for improving the convergence of a trigonometric series, see Kantorovich, L. V., and V. I. Krylov, 1964: *Approximate Methods of Higher Analysis*. Interscience, pp. 77–88.

obtain

$$\frac{\pi^2 x - x^3}{12} = \sum_{n=1}^{\infty} (-1)^{n+1} \frac{\sin(nx)}{n^3}, \qquad -\pi \le x \le \pi,$$

by term-by-term integration. Could we go the other way, i.e., take the derivative of the second equation to obtain the first? Explain.

3. Consider the function

$$f(x) = \begin{cases} 0, & -\pi < x < 0, \\ 1, & 0 < x < \pi, \end{cases}$$

with $f(x + 2\pi) = f(x)$.

(a) Find the Fourier series for $f(x)$ and show that

$$f(x) = \frac{1}{2} + \sum_{n=1}^{\infty} \frac{(-1)^n - 1}{n\pi} \sin(nx).$$

(b) Because

$$g(x) = \int_{-\pi}^{x} f(t)\, dt = \begin{cases} 0, & -\pi \le x \le 0, \\ x, & 0 \le x \le \pi, \end{cases}$$

show the Fourier series for $g(x)$ is

$$g(x) = \frac{\pi}{4} + \sum_{n=1}^{\infty} \frac{(-1)^n - 1}{n^2 \pi} \cos(nx) - \frac{(-1)^n}{n} \sin(nx).$$

(c) Using the results from part (a), integrate it term-by-term and show that you obtain part (b).

4. Consider the function

$$f(x) = \begin{cases} 0, & -\pi \le x \le 0, \\ x, & 0 \le x \le \pi, \end{cases}$$

with $f(x + 2\pi) = f(x)$.

(a) Find the Fourier series for $f(x)$ and show that

$$f(x) = \frac{\pi}{4} + \sum_{n=1}^{\infty} \frac{(-1)^n - 1}{\pi n^2} \cos(nx) - \frac{(-1)^n}{n} \sin(nx).$$

(b) Because

$$g(x) = \int_{-\pi}^{x} f(t)\, dt = \begin{cases} 0, & -\pi \le x \le 0, \\ x^2/2, & 0 \le x \le \pi, \end{cases}$$

show that Fourier series for $g(x)$ is

$$g(x) = \frac{\pi^2}{12} + \sum_{n=1}^{\infty} \left[\frac{(-1)^n \pi}{2n} + \frac{(-1)^n - 1}{\pi n^3} \right] \sin(nx) + \frac{(-1)^n}{n^2} \cos(nx).$$

(c) Using the results from part (a), integrate it term-by-term and show that you obtain part (b).

5. Consider the function $f(x) = |x|, -1 < x < 1$, with $f(x+1) = f(x)$.

(a) Find the Fourier series for $f(x)$ and show that

$$f(x) = \frac{1}{2} + \frac{2}{\pi^2} \sum_{n=1}^{\infty} \frac{(-1)^n - 1}{n^2} \cos(n\pi x).$$

(b) Because

$$g(x) = \int_{-1}^{x} f(t)\, dt = \begin{cases} \frac{1}{2} - x^2/2, & -1 \le x \le 0, \\ \frac{1}{2} + x^2/2, & 0 \le x \le 1, \end{cases}$$

show that the Fourier series for $g(x)$ is

$$g(x) = \frac{1}{2} + \sum_{n=1}^{\infty} \left[2\frac{(-1)^n - 1}{n^3 \pi^3} - \frac{(-1)^n}{n\pi} \right] \sin(n\pi x).$$

(c) Using the results from part (a), integrate it term-by-term and show that you obtain part (b).

6. Consider the function $f(x) = 1 + x, -1 \le x \le 1$, with $f(x+1) = f(x)$.

(a) Find the Fourier series for $f(x)$ and show that

$$f(x) = 1 - \frac{2}{\pi} \sum_{n=1}^{\infty} \frac{(-1)^n}{n} \sin(n\pi x).$$

(b) Because

$$g(x) = \int_{-1}^{x} f(t)\, dt = (x+1)^2/2,$$

show that the Fourier series for $g(x)$ is

$$g(x) = \frac{2}{3} + \frac{2}{\pi^2} \sum_{n=1}^{\infty} \frac{(-1)^n}{n^2} \cos(n\pi x) - \frac{2}{\pi} \sum_{n=1}^{\infty} \frac{(-1)^n}{n} \sin(n\pi x).$$

(c) Using the results from part (a), integrate it term-by-term and show that you obtain part (b).

7. (a) Show that the Fourier series for the odd function:

$$f(t) = \begin{cases} 2t + t^2, & -2 \le t \le 0, \\ 2t - t^2, & 0 \le t \le 2, \end{cases} \quad \text{is} \quad f(t) = \frac{32}{\pi^3} \sum_{n=1}^{\infty} \frac{1}{(2n-1)^3} \sin\left[\frac{(2n-1)\pi t}{2} \right].$$

(b) Use Parseval's equality to show that

$$\frac{\pi^6}{960} = \sum_{n=1}^{\infty} \frac{1}{(2n-1)^6}.$$

This series converges very rapidly to $\pi^6/960$ and provides a convenient method for computing π^6.

5.3 HALF-RANGE EXPANSIONS

In certain applications, we will find that we need a Fourier series representation for a function $f(x)$ that applies over the interval $(0, L)$ rather than $(-L, L)$. Because we are completely free to define the function over the interval $(-L, 0)$, it is simplest to have a series that consists only of sines or cosines. In this section we shall show how we can obtain these so-called *half-range expansions*.

Recall in Example 5.1.3 how we saw that if $f(x)$ is an even function, then $b_n = 0$ for all n. Similarly, if $f(x)$ is an odd function, then $a_0 = a_n = 0$ for all n. We now use these results to find a Fourier half-range expansion by extending the function defined over the interval $(0, L)$ as either an even or odd function into the interval $(-L, 0)$. If we extend $f(x)$ as an even function, we will get a half-range cosine series; if we extend $f(x)$ as an odd function, we obtain a half-range sine series.

It is important to remember that half-range expansions are a special case of the general Fourier series. For any $f(x)$ we can construct either a Fourier sine or cosine series over the interval $(-L, L)$. Both of these series will give the correct answer over the interval of $(0, L)$. Which one we choose to use depends upon whether we wish to deal with a cosine or sine series.

● **Example 5.3.1**

Let us find the half-range sine expansion of

$$f(x) = 1, \qquad 0 < x < \pi. \tag{5.3.1}$$

We begin by defining the periodic odd function

$$\widetilde{f}(x) = \begin{cases} -1, & -\pi < x < 0, \\ 1, & 0 < x < \pi, \end{cases} \tag{5.3.2}$$

with $\widetilde{f}(x + 2\pi) = \widetilde{f}(x)$. Because $\widetilde{f}(x)$ is odd, $a_0 = a_n = 0$ and

$$b_n = \frac{2}{\pi} \int_0^\pi 1 \sin(nx)\, dx = -\frac{2}{n\pi} \cos(nx)\Big|_0^\pi = -\frac{2}{n\pi} [\cos(n\pi) - 1] = -\frac{2}{n\pi} [(-1)^n - 1]. \tag{5.3.3}$$

The Fourier half-range sine series expansion of $f(x)$ is therefore

$$f(x) = \frac{2}{\pi} \sum_{n=1}^{\infty} \frac{[1 - (-1)^n]}{n} \sin(nx) = \frac{4}{\pi} \sum_{m=1}^{\infty} \frac{\sin[(2m-1)x]}{2m-1}. \tag{5.3.4}$$

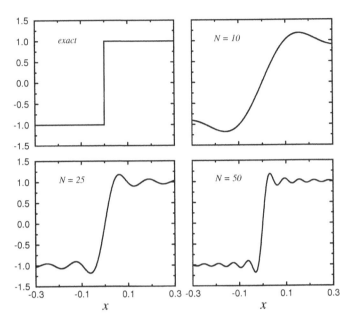

Figure 5.3.1: Partial sum of N terms in the Fourier half-range sine representation of a square wave.

As a counterpoint, let us find the half-range cosine expansion of $f(x) = 1$, $0 < x < \pi$. Now, we have that $b_n = 0$,

$$a_0 = \frac{2}{\pi} \int_0^\pi 1\, dx = 2, \tag{5.3.5}$$

and

$$a_n = \frac{2}{\pi} \int_0^\pi \cos(nx)\, dx = \frac{2}{n\pi} \sin(nx)\Big|_0^\pi = 0. \tag{5.3.6}$$

Thus, the Fourier half-range cosine expansion equals the single term:

$$f(x) = 1, \qquad 0 < x < \pi. \tag{5.3.7}$$

This is perfectly reasonable. To form a half-range cosine expansion we extend $f(x)$ as an even function into the interval $(-\pi, 0)$. In this case, we would obtain $\widetilde{f}(x) = 1$ for $-\pi < x < \pi$. Finally, we note that the Fourier series of a constant is simply that constant.

In practice it is impossible to sum Equation 5.3.4 exactly and we actually sum only the first N terms. Figure 5.3.1 illustrates $f(x)$ when this Fourier series contains N terms. As seen from the figure, the truncated series tries to achieve the infinite slope at $x = 0$, but in the attempt, it *overshoots* the discontinuity by a certain amount (in this particular case, by 17.9%). This is another example of the Gibbs phenomena. Increasing the number of terms does not remove this peculiarity, even if we could take an infinite number of terms; it merely shifts it nearer to the discontinuity. □

• Example 5.3.2: Inertial supercharging of an engine

An important aspect of designing any gasoline engine involves the motion of the fuel, air, and exhaust gas mixture through the engine. Ordinarily, an engineer would consider the motion as steady flow; but in the case of a four-stroke, single-cylinder gasoline engine, the closing of the intake valve interrupts the steady flow of the gasoline-air mixture for nearly three quarters of the engine cycle. This periodic interruption sets up standing waves in the intake pipe—waves that can build up an appreciable pressure amplitude just outside the input value.

When one of the harmonics of the engine frequency equals one of the resonance frequencies of the intake pipe, then the pressure fluctuations at the valve will be large. If the intake valve closes during that portion of the cycle when the pressure is less than average, then the waves will reduce the power output. However, if the intake valve closes when the pressure is greater than atmospheric, then the waves will have a supercharging effect and will produce an increase of power. This effect is called *inertia supercharging*.

While studying this problem, Morse et al.[12] found it necessary to express the velocity of the air-gas mixture in the valve, given by

$$f(t) = \begin{cases} 0, & -\pi < \omega t < -\pi/4, \\ \pi \cos(2\omega t)/2, & -\pi/4 < \omega t < \pi/4, \\ 0, & \pi/4 < \omega t < \pi, \end{cases} \tag{5.3.8}$$

in terms of a Fourier expansion. The advantage of working with the Fourier series rather than the function itself lies in the ability to write the velocity as a periodic forcing function that highlights the various harmonics that might resonate with the structure comprising the fuel line.

Clearly $f(t)$ is an even function and its Fourier representation will be a cosine series. In this problem, $\tau = -\pi/\omega$, and $L = \pi/\omega$. Therefore,

$$a_0 = \frac{2\omega}{\pi} \int_{-\pi/4\omega}^{\pi/4\omega} \frac{\pi}{2} \cos(2\omega t)\, dt = \tfrac{1}{2} \sin(2\omega t)\big|_{-\pi/4\omega}^{\pi/4\omega} = 1, \tag{5.3.9}$$

and

$$a_n = \frac{2\omega}{\pi} \int_{-\pi/4\omega}^{\pi/4\omega} \frac{\pi}{2} \cos(2\omega t) \cos\left(\frac{n\pi t}{\pi/\omega}\right) dt \tag{5.3.10}$$

$$= \frac{\omega}{2} \int_{-\pi/4\omega}^{\pi/4\omega} \{\cos[(n+2)\omega t] + \cos[(n-2)\omega t]\}\, dt \tag{5.3.11}$$

$$= \begin{cases} \left.\frac{\sin[(n+2)\omega t]}{2(n+2)} + \frac{\sin[(n-2)\omega t]}{2(n-2)}\right|_{-\pi/4\omega}^{\pi/4\omega}, & n \neq 2, \\ \left.\frac{\omega t}{2} + \frac{\sin(4\omega t)}{4}\right|_{-\pi/4\omega}^{\pi/4\omega}, & n = 2, \end{cases} \tag{5.3.12}$$

$$= \begin{cases} -\frac{4}{n^2-4} \cos\left(\frac{n\pi}{4}\right), & n \neq 2, \\ \frac{\pi}{4}, & n = 2. \end{cases} \tag{5.3.13}$$

Plotting these Fourier coefficients using the MATLAB script:

```
for m = 1:21;
  n = m-1; % compute the indices for the harmonic
% compute the Fourier coefficients a_n
  if n == 2; an(m) = pi/4; else;
```

[12] Morse, P. M., R. H. Boden, and H. Schecter, 1938: Acoustic vibrations and internal combustion engine performance. I. Standing waves in the intake pipe system. *J. Appl. Phys.*, **9**, 16–23.

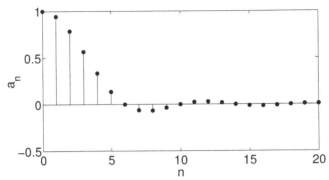

Figure 5.3.2: The spectral coefficients of the Fourier cosine series of the function given by Equation 5.3.9.

```
   an(m) = 4.*cos(pi*n/4)/(4-n*n); end;
end
nn=0:20; % create indices for x-axis
fzero=zeros(size(nn)); % create the zero line
clf % clear any figures
axes('FontSize',20) % set font size
stem(nn,an,'filled') % plot spectrum
hold on
plot(nn,fzero,'-') % plot the zero line
set (gca,'PlotBoxAspectRatio',[8 4 1]) % set aspect ratio
xlabel('n') % label x-axis
ylabel('a_n') % label y-axis,
```

we see that these Fourier coefficients become small rapidly (see Figure 5.3.2). For that reason, Morse et al. showed that there are only about three resonances where the acoustic properties of the intake pipe can enhance engine performance. These peaks occur when $q = 30c/NL = 3, 4$, or 5, where c is the velocity of sound in the air-gas mixture, L is the effective length of the intake pipe, and N is the engine speed in rpm. See Figure 5.3.3. Subsequent experiments[13] verified these results.

Such analyses are valuable to automotive engineers. Engineers are always seeking ways to optimize a system with little or no additional cost. Our analysis shows that by tuning the length of the intake pipe so that it falls on one of the resonance peaks, we could obtain higher performance from the engine with little or no extra work. Of course, the problem is that no car always performs at some optimal condition.

Problems

Find the Fourier cosine and sine series for the following functions. Then, use MATLAB to plot the Fourier coefficients.

1. $f(t) = t, \qquad 0 < t < \pi$

2. $f(t) = \pi - t, \qquad 0 < t < \pi$

3. $f(t) = t(a - t), \qquad 0 < t < a$

4. $f(t) = e^{kt}, \qquad 0 < t < a$

---

[13] Boden, R. H., and H. Schecter, 1944: Dynamics of the inlet system of a four-stroke engine. *NACA Tech. Note 935.*

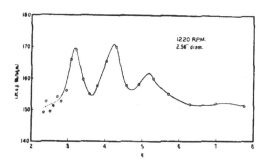

Figure 5.3.3: Experimental verification of inertial supercharging within a gasoline engine resulting from the resonance of the air-gas mixture and the intake pipe system. The peaks correspond to the $n = 3, 4,$ and 5 harmonics of the Fourier representation, Equation 5.3.13, and the parameter q is defined in the text. (From Morse, P., R. H. Boden, and H. Schecter, 1938: Acoustic vibrations and internal combustion engine performance. *J. Appl. Phys.*, **9**, 17 with permission.)

5. $f(t) = \begin{cases} t, & 0 \le t \le \frac{1}{2} \\ 1 - t, & \frac{1}{2} \le t \le 1 \end{cases}$

6. $f(t) = \begin{cases} t, & 0 < t \le 1 \\ 1, & 1 \le t < 2 \end{cases}$

7. $f(t) = \pi^2 - t^2, \quad 0 < t < \pi$

8. $f(t) = \begin{cases} 0, & 0 < t < \frac{a}{2} \\ 1, & \frac{a}{2} < t < a \end{cases}$

9. $f(t) = \begin{cases} 0, & 0 < t \le \frac{a}{3} \\ t - \frac{a}{3}, & \frac{a}{3} \le t \le \frac{2a}{3} \\ \frac{a}{3}, & \frac{2a}{3} \le t < a \end{cases}$

10. $f(t) = \begin{cases} 0, & 0 < t < \frac{a}{4} \\ 1, & \frac{a}{4} < t < \frac{3a}{4} \\ 0, & \frac{3a}{4} < t < a \end{cases}$

11. $f(t) = \begin{cases} \frac{1}{2}, & 0 < t < \frac{a}{2} \\ 1, & \frac{a}{2} < t < a \end{cases}$

12. $f(t) = \begin{cases} \frac{2t}{a}, & 0 < t \le \frac{a}{2} \\ \frac{3a - 2t}{2a}, & \frac{a}{2} \le t < a \end{cases}$

13. $f(t) = \begin{cases} t, & 0 < t \le \frac{a}{2} \\ \frac{a}{2}, & \frac{a}{2} \le t < a \end{cases}$

14. $f(t) = \frac{a - t}{a}, \quad 0 < t < a$

15. Using the relationships[14] that

$$\int_0^1 \frac{\cos(ax)}{\sqrt{1 - x^2}}\, dx = \frac{\pi}{2} J_0(a), \text{ and } \int_0^u (u^2 - x^2)^{\nu - \frac{1}{2}} \cos(ax)\, dx = \frac{\sqrt{\pi}}{2} \left(\frac{2u}{a} \right)^\nu \Gamma\!\left(\nu + \tfrac{1}{2} \right) J_\nu(au),$$

---

[14] Gradshteyn, I. S., and I. M. Ryzhik, 1965: *Table of Integrals, Series, and Products.* Academic Press, Section 3.753, Formula 2 and Section 3.771, Formula 8.

with $a > 0$, $u > 0$, $\Re(\nu) > -\frac{1}{2}$, obtain the following half-range expansions:

$$\frac{1}{\sqrt{1-x^2}} = \frac{\pi}{2} + \pi \sum_{n=1}^{\infty} J_0(n\pi) \cos(n\pi x), \qquad 0 < x < 1,$$

and

$$\sqrt{1-x^2} = 2 \sum_{n=1}^{\infty} \frac{J_1[(2n-1)\pi/2]}{2n-1} \cos[(2n-1)\pi x/2], \qquad 0 < x < 1.$$

Here $J_\nu(\cdot)$ denotes the Bessel function of the first kind and order ν (see Section 11.5) and $\Gamma(\cdot)$ is the gamma function.[15]

16. The function

$$f(t) = 1 - (1+a)\frac{t}{\pi} + (a-1)\frac{t^2}{\pi^2} + (a+1)\frac{t^3}{\pi^3} - a\frac{t^4}{\pi^4}, \qquad 0 < t < \pi,$$

is a curve fit to the observed pressure trace of an explosion wave in the atmosphere. Because the observed transmission of atmospheric waves depends on the five-fourths power of the frequency, Reed[16] had to re-express this curve fit as a Fourier sine series before he could use the transmission law. He found that

$$f(t) = \frac{1}{\pi} \sum_{n=1}^{\infty} \frac{1}{n} \left[1 - \frac{3(a-1)}{2\pi^2 n^2} \right] \sin(2nt)$$

$$+ \frac{1}{\pi} \sum_{n=1}^{\infty} \frac{2}{2n-1} \left[1 + \frac{2(a-1)}{\pi^2(2n-1)^2} - \frac{48a}{\pi^4(2n-1)^4} \right] \sin[(2n-1)t].$$

Confirm his result. Figure 5.3.4 illustrates these results for $a = 5$.

5.4 FOURIER SERIES WITH PHASE ANGLES

Sometimes it is desirable to rewrite a general Fourier series as a purely cosine or purely sine series with a phase angle. Engineers often speak of some quantity leading or lagging another quantity. Re-expressing a Fourier series in terms of amplitude and phase provides a convenient method for determining these phase relationships.

Suppose, for example, that we have a function $f(t)$ of period $2L$, given in the interval $[-L, L]$, whose Fourier series expansion is

$$f(t) = \frac{a_0}{2} + \sum_{n=1}^{\infty} a_n \cos\left(\frac{n\pi t}{L}\right) + b_n \sin\left(\frac{n\pi t}{L}\right). \tag{5.4.1}$$

We wish to replace Equation 5.4.1 by the series:

$$f(t) = \frac{a_0}{2} + \sum_{n=1}^{\infty} B_n \sin\left(\frac{n\pi t}{L} + \varphi_n\right). \tag{5.4.2}$$

[15] Gradshteyn and Ryzhik, op. cit., Section 6.41.

[16] Reed, J. W., 1977: Atmospheric attenuation of explosion waves. *J. Acoust. Soc. Am.*, **61**, 39–47.

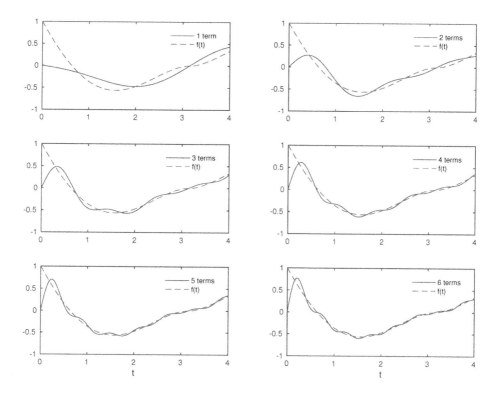

Figure 5.3.4: Partial sum of the Fourier sine series given in Problem 16 when $a = 5$.

To do this we note that

$$B_n \sin\left(\frac{n\pi t}{L} + \varphi_n\right) = a_n \cos\left(\frac{n\pi t}{L}\right) + b_n \sin\left(\frac{n\pi t}{L}\right) \tag{5.4.3}$$

$$= B_n \sin\left(\frac{n\pi t}{L}\right) \cos(\varphi_n) + B_n \sin(\varphi_n) \cos\left(\frac{n\pi t}{L}\right). \tag{5.4.4}$$

We equate coefficients of $\sin(n\pi t/L)$ and $\cos(n\pi t/L)$ on both sides and obtain

$$a_n = B_n \sin(\varphi_n), \qquad \text{and} \qquad b_n = B_n \cos(\varphi_n). \tag{5.4.5}$$

Hence, upon squaring and adding,

$$B_n = \sqrt{a_n^2 + b_n^2}, \tag{5.4.6}$$

while taking the ratio gives

$$\varphi_n = \tan^{-1}(a_n/b_n). \tag{5.4.7}$$

Similarly, we could rewrite Equation 5.4.1 as

$$f(t) = \frac{a_0}{2} + \sum_{n=1}^{\infty} A_n \cos\left(\frac{n\pi t}{L} + \varphi_n\right), \tag{5.4.8}$$

where

$$A_n = \sqrt{a_n^2 + b_n^2}, \qquad \text{and} \qquad \varphi_n = \tan^{-1}(-b_n/a_n), \tag{5.4.9}$$

and

$$a_n = A_n \cos(\varphi_n), \qquad \text{and} \qquad b_n = -A_n \sin(\varphi_n). \tag{5.4.10}$$

In both cases, we must be careful in computing φ_n because there are two possible values of φ_n that satisfy Equation 5.4.7 or Equation 5.4.9. These angles φ_n must give the correct a_n and b_n using either Equation 5.4.5 or Equation 5.4.10.

• **Example 5.4.1**

The Fourier series for $f(t) = e^t$ over the interval $-L < t < L$ is

$$f(t) = \frac{\sinh(aL)}{aL} + 2\sinh(aL) \sum_{n=1}^{\infty} \frac{aL(-1)^n}{a^2 L^2 + n^2 \pi^2} \cos\left(\frac{n\pi t}{L}\right)$$

$$- 2\sinh(aL) \sum_{n=1}^{\infty} \frac{n\pi(-1)^n}{a^2 L^2 + n^2 \pi^2} \sin\left(\frac{n\pi t}{L}\right). \tag{5.4.11}$$

Let us rewrite Equation 5.4.11 as a Fourier series with a phase angle. Regardless of whether we want the new series to contain $\cos(n\pi t/L + \varphi_n)$ or $\sin(n\pi t/L + \varphi_n)$, the amplitude A_n or B_n is the same in both series:

$$A_n = B_n = \sqrt{a_n^2 + b_n^2} = \frac{2\sinh(aL)}{\sqrt{a^2 L^2 + n^2 \pi^2}}. \tag{5.4.12}$$

If we want our Fourier series to read

$$f(t) = \frac{\sinh(aL)}{aL} + 2\sinh(aL) \sum_{n=1}^{\infty} \frac{\cos(n\pi t/L + \varphi_n)}{\sqrt{a^2 L^2 + n^2 \pi^2}}, \tag{5.4.13}$$

then

$$\varphi_n = \tan^{-1}\left(-\frac{b_n}{a_n}\right) = \tan^{-1}\left(\frac{n\pi}{aL}\right), \tag{5.4.14}$$

where φ_n lies in the first quadrant if n is even and in the third quadrant if n is odd. This ensures that the sign from the $(-1)^n$ is correct.

Figure 5.4.1 illustrates the amplitudes and phases for Equation 5.4.13 for the case when $aL = 2$. The phase angle was computed using the MATLAB command `tan2(-b_n,a_n)`, where `a_n = aL*(-1)^n` and `b_n = n*pi*(-1)^n`. Because we have an amplitude and phase for each discrete value of n, these amplitude and phase plots, taken together, are called the *line specta* for our function $f(t)$.

On the other hand, if we prefer

$$f(t) = \frac{\sinh(aL)}{aL} + 2\sinh(aL) \sum_{n=1}^{\infty} \frac{\sin(n\pi t/L + \varphi_n)}{\sqrt{a^2 L^2 + n^2 \pi^2}}, \tag{5.4.15}$$

then

$$\varphi_n = \tan^{-1}\left(\frac{a_n}{b_n}\right) = -\tan^{-1}\left(\frac{aL}{n\pi}\right), \tag{5.4.16}$$

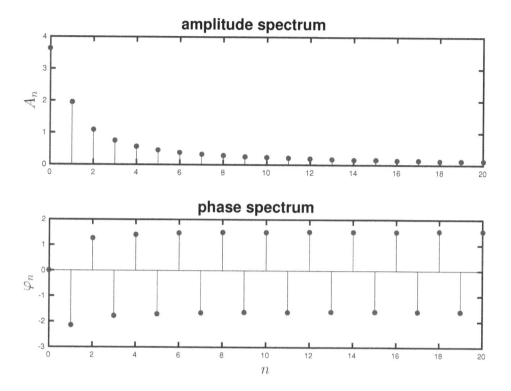

Figure 5.4.1: The amplitude and phase spectra for the Fourier series given by Equation 5.4.13 with $aL = 2$.

where φ_n lies in the fourth quadrant if n is odd and in the second quadrant if n is even. Again, we could plot the amplitude and phase spectrum for each harmonic. The amplitude spectrum is the same as shown in Figure 5.4.1 but the phase spectrum changes to values near π for n even and slightly less than zero for n odd. Here we computed the phase angle using the MATLAB command `tan2(a_n,b_n)`.

Problems

For each of the following Fourier series, (1) plot the Fourier series over the specified range, (2) write the series in both the cosine and sine phase angle form, and (3) plot both the amplitude and phase spectra.

1. $f(t) == \dfrac{8}{\pi} \displaystyle\sum_{n=1}^{\infty} \dfrac{n}{4n^2 - 1} \sin(2n\pi t)$, $-1 \le t \le 2$,

2. $f(t) = \dfrac{1}{2} + \dfrac{2}{\pi} \displaystyle\sum_{n=1}^{\infty} \dfrac{\sin[(2n-1)\pi t]}{2n-1}$, $-2\pi \le t \le 4\pi$,

3. $f(t) = \dfrac{3}{2} + \dfrac{2}{\pi} \displaystyle\sum_{n=1}^{\infty} \dfrac{(-1)^n}{2n-1} \cos\left[\dfrac{(2n-1)\pi t}{2}\right]$, $-4 \le t \le 8$,

4. $f(t) = -2 \displaystyle\sum_{n=1}^{\infty} \dfrac{(-1)^n}{n} \sin(nt)$, $-2\pi \le t \le 4\pi$,

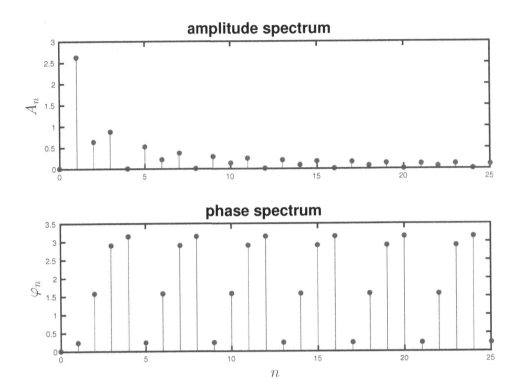

Figure 5.4.2: The same as Figure 5.4.1 except for the Fourier series given in Problem 7.

5. $f(t) = \dfrac{\pi}{2} - \dfrac{4}{\pi} \sum_{n=1}^{\infty} \dfrac{\cos[(2n-1)t]}{(2n-1)^2}$, $\qquad -2\pi \leq t < 4\pi$,

6. $f(t) = 16 + \dfrac{48}{\pi^2} \sum_{n=1}^{\infty} \dfrac{1}{n^2} \cos\left(\dfrac{n\pi t}{2}\right) - \dfrac{48}{\pi} \sum_{n=1}^{\infty} \dfrac{1}{n} \sin\left(\dfrac{n\pi t}{2}\right)$, $\qquad -4 \leq t \leq 8$,

7. $f(t) = \dfrac{8}{\pi} \sum_{n=1}^{\infty} \dfrac{1}{n} \sin\left(\dfrac{n\pi}{2}\right) \cos(2nt) - \dfrac{2}{\pi} \sum_{n=1}^{\infty} \dfrac{1}{n}\left[1 - \cos\left(\dfrac{n\pi}{2}\right)\right] \sin(2nt)$, $\qquad -4 \leq t \leq 8$,

8. $f(t) = L + \dfrac{2L}{\pi^2} \sum_{n=1}^{\infty} \dfrac{[(-1)^n - 1]}{n^2} \cos\left(\dfrac{n\pi x}{L}\right) - \dfrac{L}{\pi} \sum_{n=1}^{\infty} \dfrac{[1 + (-1)^n]}{n} \sin\left(\dfrac{n\pi x}{L}\right)$, $-2L \leq x \leq 4L$.

5.5 COMPLEX FOURIER SERIES

So far in our discussion, we expressed Fourier series in terms of sines and cosines. We are now ready to re-express a Fourier series as a series of complex exponentials. There are two reasons for this. First, in certain engineering and scientific applications of Fourier series, the expansion of a function in terms of complex exponentials results in coefficients of considerable simplicity and clarity. Second, these complex Fourier series point the way to the development of the Fourier transform in the next chapter.

We begin by introducing the variable $\omega_n = n\pi/L$, where $n = 0, \pm 1, \pm 2, \ldots$ Using Euler's formula we can replace the sine and cosine in the Fourier series by exponentials and find that

$$f(t) = \frac{a_0}{2} + \sum_{n=1}^{\infty} \frac{a_n}{2}\left(e^{i\omega_n t} + e^{-i\omega_n t}\right) + \frac{b_n}{2i}\left(e^{i\omega_n t} - e^{-i\omega_n t}\right) \qquad (5.5.1)$$

$$= \frac{a_0}{2} + \sum_{n=1}^{\infty} \left(\frac{a_n}{2} - \frac{b_n i}{2} \right) e^{i\omega_n t} + \left(\frac{a_n}{2} + \frac{b_n i}{2} \right) e^{-i\omega_n t}. \tag{5.5.2}$$

If we define $c_n = \frac{1}{2}(a_n - ib_n)$, then

$$c_n = \frac{1}{2}(a_n - ib_n) = \frac{1}{2L} \int_\tau^{\tau+2L} f(t)[\cos(\omega_n t) - i\sin(\omega_n t)]\, dt = \frac{1}{2L} \int_\tau^{\tau+2L} f(t)e^{-i\omega_n t} dt. \tag{5.5.3}$$

Similarly, the complex conjugate of c_n, c_n^*, equals

$$c_n^* = \frac{1}{2}(a_n + ib_n) = \frac{1}{2L} \int_\tau^{\tau+2L} f(t)e^{i\omega_n t} dt. \tag{5.5.4}$$

To simplify Equation 5.5.2 we note that

$$\omega_{-n} = \frac{(-n)\pi}{L} = -\frac{n\pi}{L} = -\omega_n, \tag{5.5.5}$$

which yields the result that

$$c_{-n} = \frac{1}{2L} \int_\tau^{\tau+2L} f(t)e^{-i\omega_{-n} t} dt = \frac{1}{2L} \int_\tau^{\tau+2L} f(t)e^{i\omega_n t} dt = c_n^* \tag{5.5.6}$$

so that we can write Equation 5.5.2 as

$$f(t) = \frac{a_0}{2} + \sum_{n=1}^{\infty} c_n e^{i\omega_n t} + c_n^* e^{-i\omega_n t} = \frac{a_0}{2} + \sum_{n=1}^{\infty} c_n e^{i\omega_n t} + c_{-n} e^{-i\omega_n t}. \tag{5.5.7}$$

Letting $n = -m$ in the second summation on the right side of Equation 5.5.7,

$$\sum_{n=1}^{\infty} c_{-n} e^{-i\omega_n t} = \sum_{m=-1}^{-\infty} c_m e^{-i\omega_{-m} t} = \sum_{m=-\infty}^{-1} c_m e^{i\omega_m t} = \sum_{n=-\infty}^{-1} c_n e^{i\omega_n t}, \tag{5.5.8}$$

where we introduced $m = n$ into the last summation in Equation 5.5.8. Therefore,

$$f(t) = \frac{a_0}{2} + \sum_{n=1}^{\infty} c_n e^{i\omega_n t} + \sum_{n=-\infty}^{-1} c_n e^{i\omega_n t}. \tag{5.5.9}$$

On the other hand,

$$\frac{a_0}{2} = \frac{1}{2L} \int_\tau^{\tau+2L} f(t)\, dt = c_0 = c_0 e^{i\omega_0 t}, \tag{5.5.10}$$

because $\omega_0 = 0\pi/L = 0$. Thus, our final result is

$$\boxed{f(t) = \sum_{n=-\infty}^{\infty} c_n e^{i\omega_n t},} \tag{5.5.11}$$

where

$$
c_n = \frac{1}{2L} \int_{\tau}^{\tau+2L} f(t) e^{-i\omega_n t}\, dt \tag{5.5.12}
$$

and $n = 0, \pm 1, \pm 2, \dots$. Note that even though c_n is generally complex, the summation Equation 5.5.11 always gives a *real-valued* function $f(t)$.

Just as we can represent the function $f(t)$ graphically by a plot of t against $f(t)$, we can plot c_n as a function of n, commonly called the frequency *spectrum*. Because c_n is generally complex, it is necessary to make two plots. Typically the plotted quantities are the amplitude spectrum $|c_n|$ and the phase spectrum φ_n, where φ_n is the phase of c_n. However, we could just as well plot the real and imaginary parts of c_n. Because n is an integer, these plots consist merely of a series of vertical lines representing the ordinates of the quantity $|c_n|$ or φ_n for each n. For this reason we refer to these plots as the *line spectra*.

Because $2c_n = a_n - ib_n$, the coefficients c_n for an even function will be purely real; the coefficients c_n for an odd function are purely imaginary. It is important to note that we lose the advantage of even and odd functions in the sense that we cannot just integrate over the interval 0 to L and then double the result. In the present case we have a line integral of a complex function along the real axis.

• **Example 5.5.1**

Let us find the complex Fourier series for

$$
f(t) = \begin{cases} 1, & 0 < t < \pi, \\ -1, & -\pi < t < 0, \end{cases} \tag{5.5.13}
$$

which has the periodicity $f(t + 2\pi) = f(t)$.

With $L = \pi$ and $\tau = -\pi$, $\omega_n = n\pi/L = n$. Therefore,

$$
c_n = \frac{1}{2\pi} \int_{-\pi}^{0} (-1) e^{-int}\, dt + \frac{1}{2\pi} \int_{0}^{\pi} (1) e^{-int}\, dt \tag{5.5.14}
$$

$$
= \frac{1}{2n\pi i} e^{-int} \Big|_{-\pi}^{0} - \frac{1}{2n\pi i} e^{-int} \Big|_{0}^{\pi} \tag{5.5.15}
$$

$$
= -\frac{i}{2n\pi} \left(1 - e^{n\pi i}\right) + \frac{i}{2n\pi} \left(e^{-n\pi i} - 1\right), \tag{5.5.16}
$$

if $n \neq 0$. Because $e^{n\pi i} = \cos(n\pi) + i\sin(n\pi) = (-1)^n$ and $e^{-n\pi i} = \cos(-n\pi) + i\sin(-n\pi) = (-1)^n$, then

$$
c_n = -\frac{i}{n\pi}[1 - (-1)^n] = \begin{cases} 0, & n \text{ even}, \\ -\frac{2i}{n\pi}, & n \text{ odd}, \end{cases} \tag{5.5.17}
$$

with

$$
f(t) = \sum_{n=-\infty}^{\infty} c_n e^{int}. \tag{5.5.18}
$$

In this particular problem we must treat the case $n = 0$ specially because Equation 5.5.15 is undefined for $n = 0$. In that case,

$$c_0 = \frac{1}{2\pi} \int_{-\pi}^{0} (-1)\, dt + \frac{1}{2\pi} \int_{0}^{\pi} (1)\, dt = \frac{1}{2\pi}(-t)\Big|_{-\pi}^{0} + \frac{1}{2\pi}(t)\Big|_{0}^{\pi} = 0. \qquad (5.5.19)$$

Because $c_0 = 0$, we can write the expansion:

$$f(t) = -\frac{2i}{\pi} \sum_{m=-\infty}^{\infty} \frac{e^{(2m-1)it}}{2m - 1}, \qquad (5.5.20)$$

since we can write all odd integers as $2m - 1$, where $m = 0, \pm1, \pm2, \pm3, \ldots$. Using the MATLAB script:

```
max = 31; % total number of harmonics
mid = (max+1)/2; % in the array, location of c_0
for m = 1:max;
  n = m - mid; % compute value of harmonic
% compute complex Fourier coefficient c_n = (cnr,cni)
  if mod(n,2) == 0; cnr(m) = 0; cni(m) = 0; else;
  cnr(m) = 0; cni(m) = - 2/(pi*n); end;
end
nn=(1-mid):(max-mid); % create indices for x-axis
fzero=zeros(size(nn)); % create the zero line
clf % clear any figures
amplitude = sqrt(cnr.*cnr+cni.*cni);
phase = atan2(cni,cnr);
% plot amplitude of c_n
subplot(2,1,1), stem(nn,amplitude,'filled')
% label amplitude plot
text(6,0.75,'amplitude','FontSize',20)
subplot(2,1,2), stem(nn,phase,'filled') % plot phases of c_n
text(7,1,'phase','FontSize',20) % label phase plot
xlabel('n','Fontsize',20) % label x-axis,
```

we plot the amplitude and phase spectra for the function, Equation 5.5.13, as a function of n in Figure 5.5.1. □

• **Example 5.5.2**

The concept of Fourier series can be generalized to multivariable functions. Consider the function $f(x, y)$ defined over $0 < x < L$ and $0 < y < H$. Taking y constant, we have that

$$c_n(y) = \frac{1}{L} \int_{0}^{L} f(x, y)e^{-i\xi_n x}\, dx, \qquad \xi_n = \frac{2\pi n}{L}. \qquad (5.5.21)$$

Similarly, holding ξ_n constant,

$$c_{nm} = \frac{1}{H} \int_{0}^{H} c_n(y)e^{-i\eta_m y}\, dy, \qquad \eta_m = \frac{2\pi m}{H}. \qquad (5.5.22)$$

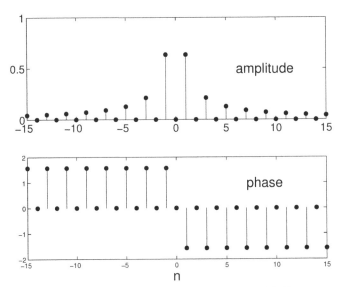

Figure 5.5.1: Amplitude and phase spectra for the function, Equation 5.5.13.

Therefore, the (complex) Fourier coefficient for the two-dimensional function $f(x, y)$ is

$$c_{nm} = \frac{1}{LH} \int_0^L \int_0^H f(x, y) e^{-i(\xi_n x + \eta_m y)} \, dx \, dy, \qquad (5.5.23)$$

assuming that the integral exists.

To recover $f(x, y)$ given c_{nm}, we reverse the process of deriving c_{nm}. Starting with

$$c_n(y) = \sum_{m=-\infty}^{\infty} c_{nm} e^{i\eta_m y}, \qquad (5.5.24)$$

we find that

$$f(x, y) = \sum_{n=-\infty}^{\infty} c_n(y) e^{i\xi_n x}. \qquad (5.5.25)$$

Therefore,

$$f(x, y) = \sum_{n=-\infty}^{\infty} \sum_{m=-\infty}^{\infty} c_{nm} e^{i(\xi_n x + \eta_m y)}. \qquad (5.5.26)$$

Problems

Find the complex Fourier series for the following functions. Then use MATLAB to plot the corresponding spectra.

1. $f(t) = |t|, \quad -\pi \le t \le \pi$

2. $f(t) = e^t, \quad 0 < t < 2$

3. $f(t) = t, \quad 0 < t < 2$

4. $f(t) = t^2, \quad -\pi \le t \le \pi$

5. $f(t) = \begin{cases} 0, & -\pi/2 < t < 0 \\ 1, & 0 < t < \pi/2 \end{cases}$

6. $f(t) = t, \quad -1 < t < 1$

7. $f(t) = 1 - t, \quad 0 \le t < 6$

8. $f(t) = \cos(t), \quad 0 < t < 1$

5.6 THE USE OF FOURIER SERIES IN THE SOLUTION OF ORDINARY DIFFERENTIAL EQUATIONS

An important application of Fourier series is the solution of ordinary differential equations. Structural engineers especially use this technique because the occupants of buildings and bridges often subject these structures to forces that are periodic in nature.[17]

● **Example 5.6.1**

Let us find the general solution to the ordinary differential equation

$$y'' + 9y = f(t), \tag{5.6.1}$$

where the forcing is

$$f(t) = |t|, \qquad -\pi \le t \le \pi, \qquad f(t + 2\pi) = f(t). \tag{5.6.2}$$

This equation represents an oscillator forced by a driver whose displacement is the saw-tooth function.

We begin by replacing the function $f(t)$ by its Fourier series representation because the forcing function is periodic. The advantage of expressing $f(t)$ as a Fourier series is its validity for any time t. The alternative would be to construct a solution over each interval $n\pi < t < (n+1)\pi$ and then piece together the final solution assuming that the solution and its first derivative are continuous at each junction $t = n\pi$. Because the function is an even function, all of the sine terms vanish and the Fourier series is

$$|t| = \frac{\pi}{2} - \frac{4}{\pi} \sum_{n=1}^{\infty} \frac{\cos[(2n-1)t]}{(2n-1)^2}. \tag{5.6.3}$$

Next, we note that the general solution consists of the complementary solution, which equals

$$y_H(t) = A\cos(3t) + B\sin(3t), \tag{5.6.4}$$

and the particular solution $y_p(t)$, which satisfies the differential equation

$$y_p'' + 9y_p = \frac{\pi}{2} - \frac{4}{\pi} \sum_{n=1}^{\infty} \frac{\cos[(2n-1)t]}{(2n-1)^2}. \tag{5.6.5}$$

To determine this particular solution, we write Equation 5.6.5 as

$$y_p'' + 9y_p = \frac{\pi}{2} - \frac{4}{\pi}\cos(t) - \frac{4}{9\pi}\cos(3t) - \frac{4}{25\pi}\cos(5t) - \cdots. \tag{5.6.6}$$

By the method of undetermined coefficients, we guess the particular solution:

$$y_p(t) = \frac{a_0}{2} + a_1\cos(t) + b_1\sin(t) + a_2\cos(3t) + b_2\sin(3t) + \cdots \tag{5.6.7}$$

---

[17] Timoshenko, S. P., 1943: Theory of suspension bridges. Part II. *J. Franklin Inst.*, **235**, 327–349; Inglis, C. E., 1934: *A Mathematical Treatise on Vibrations in Railway Bridges*. Cambridge University Press, 203 pp.

or

$$y_p(t) = \tfrac{1}{2}a_0 + \sum_{n=1}^{\infty} a_n \cos[(2n-1)t] + b_n \sin[(2n-1)t]. \qquad (5.6.8)$$

Because

$$y_p''(t) = \sum_{n=1}^{\infty} -(2n-1)^2 \{a_n \cos[(2n-1)t] + b_n \sin[(2n-1)t]\}, \qquad (5.6.9)$$

$$\sum_{n=1}^{\infty} -(2n-1)^2 \{a_n \cos[(2n-1)t] + b_n \sin[(2n-1)t]\} \qquad (5.6.10)$$

$$+ \tfrac{9}{2}a_0 + 9\sum_{n=1}^{\infty} a_n \cos[(2n-1)t] + b_n \sin[(2n-1)t] = \frac{\pi}{2} - \frac{4}{\pi}\sum_{n=1}^{\infty} \frac{\cos[(2n-1)t]}{(2n-1)^2},$$

or

$$\frac{9a_0}{2} - \frac{\pi}{2} + \sum_{n=1}^{\infty} \left\{ [9-(2n-1)^2]a_n + \frac{4}{\pi(2n-1)^2} \right\} \cos[(2n-1)t]$$

$$+ \sum_{n=1}^{\infty} [9-(2n-1)^2]b_n \sin[(2n-1)t] = 0. \qquad (5.6.11)$$

Because Equation 5.6.11 must hold true for any time, each harmonic must vanish separately and

$$a_0 = \frac{\pi}{9}, \qquad a_n = -\frac{4}{\pi(2n-1)^2[9-(2n-1)^2]} \qquad (5.6.12)$$

and $b_n = 0$. All of the coefficients a_n are finite except for $n=2$, where a_2 becomes undefined. This coefficient is undefined because the harmonic $\cos(3t)$ in the forcing function resonates with the natural mode of the system.

Let us review our analysis to date. We found that each harmonic in the forcing function yields a corresponding harmonic in the particular solution, Equation 5.6.8. The only difficulty arises with the harmonic $n=2$. Although our particular solution is not correct because it contains $\cos(3t)$, we suspect that if we remove that term, then the remaining harmonic solutions are correct. The problem is linear, and difficulties with one harmonic term should not affect other harmonics. But how shall we deal with the $\cos(3t)$ term in the forcing function? Let us denote that particular solution by $Y(t)$ and modify our particular solution as follows:

$$y_p(t) = \tfrac{1}{2}a_0 + a_1 \cos(t) + Y(t) + a_3 \cos(5t) + \cdots. \qquad (5.6.13)$$

Substituting this solution into the differential equation and simplifying, everything cancels except

$$Y'' + 9Y = -\frac{4}{9\pi} \cos(3t). \qquad (5.6.14)$$

The solution of this equation by the method of undetermined coefficients is

$$Y(t) = -\frac{2}{27\pi} t \sin(3t). \qquad (5.6.15)$$

This term, called a *secular term*, is the most important one in the solution. While the other terms merely represent simple oscillatory motion, the term $t\sin(3t)$ grows linearly with time and eventually becomes the dominant term in the series. Consequently, the general solution equals the complementary plus the particular solution, or

$$y(t) = A\cos(3t) + B\sin(3t) + \frac{\pi}{18} - \frac{2}{27\pi}t\sin(3t) - \frac{4}{\pi}\sum_{\substack{n=1 \\ n\neq 2}}^{\infty}\frac{\cos[(2n-1)t]}{(2n-1)^2[9-(2n-1)^2]}. \quad (5.6.16)$$

□

• **Example 5.6.2**

Let us redo the previous problem only using complex Fourier series. That is, let us find the general solution to the ordinary differential equation

$$y'' + 9y = \frac{\pi}{2} - \frac{2}{\pi}\sum_{n=-\infty}^{\infty}\frac{e^{i(2n-1)t}}{(2n-1)^2}. \quad (5.6.17)$$

From the method of undetermined coefficients we guess the particular solution for Equation 5.6.17 to be

$$y_p(t) = c_0 + \sum_{n=-\infty}^{\infty}c_n e^{i(2n-1)t}. \quad (5.6.18)$$

Then

$$y_p''(t) = \sum_{n=-\infty}^{\infty}-(2n-1)^2 c_n e^{i(2n-1)t}. \quad (5.6.19)$$

Substituting Equation 5.6.18 and Equation 5.6.19 into Equation 5.6.17,

$$9c_0 + \sum_{n=-\infty}^{\infty}[9-(2n-1)^2]c_n e^{i(2n-1)t} = \frac{\pi}{2} - \frac{2}{\pi}\sum_{n=-\infty}^{\infty}\frac{e^{i(2n-1)t}}{(2n-1)^2}. \quad (5.6.20)$$

Because Equation 5.6.20 must be true for any t,

$$c_0 = \frac{\pi}{18}, \quad \text{and} \quad c_n = \frac{2}{\pi(2n-1)^2[(2n-1)^2-9]}. \quad (5.6.21)$$

Therefore,

$$y_p(t) = \frac{\pi}{18} + \frac{2}{\pi}\sum_{n=-\infty}^{\infty}\frac{e^{i(2n-1)t}}{(2n-1)^2[(2n-1)^2-9]}e^{i(2n-1)t}. \quad (5.6.22)$$

However, there is a problem when $n = -1$ and $n = 2$. Therefore, we modify Equation 5.6.22 to read

$$y_p(t) = \frac{\pi}{18} + c_2 te^{3it} + c_{-1}te^{-3it} + \frac{2}{\pi}\sum_{\substack{n=-\infty \\ n\neq-1,2}}^{\infty}\frac{e^{i(2n-1)t}}{(2n-1)^2[(2n-1)^2-9]}e^{i(2n-1)t}. \quad (5.6.23)$$

Introducing Equation 5.6.23 into Equation 5.6.17 and simplifying,

$$c_2 = -\frac{1}{27\pi i}, \quad \text{and} \quad c_{-1} = -\frac{1}{27\pi i}. \quad (5.6.24)$$

The general solution is then

$$
y(t) = Ae^{3it} + Be^{-3it} + \frac{\pi}{18} - \frac{te^{3it}}{27\pi i} + \frac{te^{-3it}}{27\pi i} + \frac{2}{\pi} \sum_{\substack{n=-\infty \\ n\neq -1,2}}^{\infty} \frac{e^{i(2n-1)t}}{(2n-1)^2[(2n-1)^2-9]}. \quad (5.6.25)
$$

The first two terms on the right side of Equation 5.6.25 represent the complementary solution. Although this expansion is equivalent to Equation 5.6.16, we have all of the advantages of dealing with exponentials rather than sines and cosines. These advantages include ease of differentiation and integration, and writing the series in terms of amplitude and phase.

$\square$

• **Example 5.6.3: Temperature within a spinning satellite**

In the design of artificial satellites, it is important to determine the temperature distribution on the spacecraft's surface. An interesting special case is the temperature fluctuation in the skin due to the spinning of the vehicle. If the craft is thin-walled so that there is no radial dependence, Hrycak[18] showed that he could approximate the nondimensional temperature field at the equator of the rotating satellite by

$$
\frac{d^2T}{d\eta^2} + b\frac{dT}{d\eta} - c\left(T - \frac{3}{4}\right) = -\frac{\pi c}{4}\frac{F(\eta) + \beta/4}{1 + \pi\beta/4}, \quad (5.6.26)
$$

where

$$
b = 4\pi^2 r^2 f/a, \quad c = \frac{16\pi S}{\gamma T_\infty}\left(1 + \frac{\pi\beta}{4}\right), \quad T_\infty = \left(\frac{S}{\pi\sigma\epsilon}\right)^{1/4}\left(\frac{1 + \pi\beta/4}{1 + \beta}\right)^{1/4}, \quad (5.6.27)
$$

$$
F(\eta) = \begin{cases} \cos(2\pi\eta), & 0 \leq \eta \leq \frac{1}{4}, \\ 0, & \frac{1}{4} \leq \eta \leq \frac{3}{4}, \\ \cos(2\pi\eta), & \frac{3}{4} \leq \eta \leq 1, \end{cases} \quad (5.6.28)
$$

a is the thermal diffusivity of the shell, f is the rate of spin, r is the radius of the spacecraft, S is the net direct solar heating, β is the ratio of the emissivity of the interior shell to the emissivity of the exterior surface, ϵ is the overall emissivity of the exterior surface, γ is the satellite's skin conductance, and σ is the Stefan-Boltzmann constant. The independent variable η is the longitude along the equator with the effect of rotation subtracted out ($2\pi\eta = \varphi - 2\pi ft$). The reference temperature T_∞ equals the temperature that the spacecraft would have if it spun with infinite angular speed so that the solar heating would be uniform around the craft. We nondimensionalized the temperature with respect to T_∞.

We begin by introducing the new variables

$$
y = T - \frac{3}{4} - \frac{\pi\beta}{16 + 4\pi\beta}, \quad \nu_0 = \frac{2\pi^2 r^2 f}{a\rho_0}, \quad A_0 = -\frac{\pi\rho^2}{4 + \pi\beta} \quad (5.6.29)
$$

and $\rho_0^2 = c$ so that Equation 5.6.26 becomes

$$
\frac{d^2y}{d\eta^2} + 2\rho_0\nu_0\frac{dy}{d\eta} - \rho_0^2 y = A_0 F(\eta). \quad (5.6.30)
$$

---

[18] From Hrycak, P., 1963: Temperature distribution in a spinning spherical space vehicle. *AIAA J.*, **1**, 96–99. Reprinted with permission of the American Institute of Aeronautics and Astronautics.

Next, we expand $F(\eta)$ as a Fourier series because it is a periodic function of period 1. Since it is an even function,

$$f(\eta) = \tfrac{1}{2}a_0 + \sum_{n=1}^{\infty} a_n \cos(2n\pi\eta), \tag{5.6.31}$$

where

$$a_0 = \frac{1}{1/2} \int_0^{1/4} \cos(2\pi x)\, dx + \frac{1}{1/2} \int_{3/4}^{1} \cos(2\pi x)\, dx = \frac{2}{\pi}, \tag{5.6.32}$$

$$a_1 = \frac{1}{1/2} \int_0^{1/4} \cos^2(2\pi x)\, dx + \frac{1}{1/2} \int_{3/4}^{1} \cos^2(2\pi x)\, dx = \frac{1}{2} \tag{5.6.33}$$

and

$$a_n = \frac{1}{1/2} \int_0^{1/4} \cos(2\pi x) \cos(2n\pi x)\, dx + \frac{1}{1/2} \int_{3/4}^{1} \cos(2\pi x) \cos(2n\pi x)\, dx \tag{5.6.34}$$

$$= -\frac{2(-1)^n}{\pi(n^2-1)} \cos\left(\frac{n\pi}{2}\right), \tag{5.6.35}$$

if $n \geq 2$. Therefore,

$$f(\eta) = \frac{1}{\pi} + \frac{1}{2}\cos(2\pi\eta) - \frac{2}{\pi}\sum_{n=1}^{\infty} \frac{(-1)^n}{4n^2-1}\cos(4n\pi\eta). \tag{5.6.36}$$

From the method of undetermined coefficients, the particular solution is

$$y_p(\eta) = \tfrac{1}{2}a_0 + a_1\cos(2\pi\eta) + b_1\sin(2\pi\eta) + \sum_{n=1}^{\infty} a_{2n}\cos(4n\pi\eta) + b_{2n}\sin(4n\pi\eta), \tag{5.6.37}$$

which yields

$$y_p'(\eta) = -2\pi a_1\sin(2\pi\eta) + 2\pi b_1\cos(2\pi\eta) + \sum_{n=1}^{\infty}[-4n\pi a_{2n}\sin(4n\pi\eta) + 4n\pi b_{2n}\cos(4n\pi\eta)], \tag{5.6.38}$$

and

$$y_p''(\eta) = -4\pi^2[a_1\cos(2\pi\eta) + b_1\sin(2\pi\eta)] - \sum_{n=1}^{\infty} 16n^2\pi^2[a_{2n}\cos(4n\pi\eta) + b_{2n}\sin(4n\pi\eta)]. \tag{5.6.39}$$

Substituting into Equation 5.6.30,

$$-\frac{1}{2}\rho_0^2 a_0 - \frac{A_0}{\pi} + \left(-4\pi^2 a_1 + 4\pi\rho_0\nu_0 b_1 - \rho_0^2 a_1 - \frac{A_0}{2}\right)\cos(2\pi\eta)$$

$$+ \left(-4\pi^2 b_1 - 4\pi\rho_0\nu_0 a_1 - \rho_0^2 b_1\right)\sin(2\pi\eta)$$

$$+ \sum_{n=1}^{\infty}\left[-16n^2\pi^2 a_{2n} + 8n\pi\rho_0\nu_0 b_{2n} - \rho_0^2 a_{2n} + \frac{2A_0(-1)^n}{\pi(4n^2-1)}\right]\cos(4n\pi\eta)$$

$$+ \sum_{n=1}^{\infty}\left(-16n^2\pi^2 b_{2n} - 8n\pi\rho_0\nu_0 a_{2n} - \rho_0^2 b_{2n}\right)\sin(4n\pi\eta) = 0. \tag{5.6.40}$$

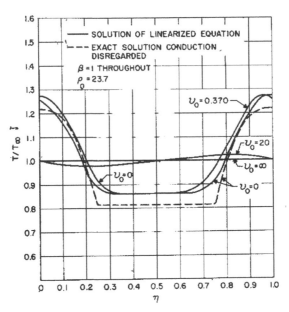

Figure 5.6.1: Temperature distribution along the equator of a spinning spherical satellite. (From Hrycak, P., 1963: Temperature distribution in a spinning spherical space vehicle. *AIAA J.*, **1**, 97. ©1963 AIAA, reprinted with permission.)

To satisfy Equation 5.6.40 for any η, we set

$$a_0 = -\frac{2A_0}{\pi\rho_0^2}, \quad -(4\pi^2 + \rho_0^2)a_1 + 4\pi\rho_0\nu_0 b_1 = \frac{A_0}{2}, \quad 4\pi\rho_0\nu_0 a_1 + (4\pi^2 + \rho_0^2)b_1 = 0, \quad (\mathbf{5.6.41})$$

$$(16n^2\pi^2 + \rho_0^2)a_{2n} - 8n\pi\rho_0\nu_0 b_{2n} = \frac{2A_0(-1)^n}{\pi(4n^2 - 1)}, \quad \text{and} \quad 8n\pi\rho_0\nu_0 a_{2n} + (16n^2\pi^2 + \rho_0^2)b_{2n} = 0,$$

$$(\mathbf{5.6.42})$$

or

$$[16\pi^2\rho_0^2\nu_0^2 + (4\pi^2 + \rho_0^2)^2]a_1 = -\frac{(4\pi^2 + \rho_0^2)A_0}{2}, \quad [16\pi^2\rho_0^2\nu_0^2 + (4\pi^2 + \rho_0^2)^2]b_1 = 2\pi\rho_0\nu_0 A_0,$$

$$(\mathbf{5.6.43})$$

$$[64n^2\pi^2\rho_0^2\nu_0^2 + (16n^2\pi^2 + \rho_0^2)^2]a_{2n} = \frac{2A_0(-1)^n(16n^2\pi^2 + \rho_0^2)}{\pi(4n^2 - 1)}, \quad (\mathbf{5.6.44})$$

and

$$[64n^2\pi^2\rho_0^2\nu_0^2 + (16n^2\pi^2 + \rho_0^2)^2]b_{2n} = -\frac{16(-1)^n\rho_0\nu_0 nA_0}{4n^2 - 1}. \quad (\mathbf{5.6.45})$$

Substituting for a_0, a_1, b_1, a_{2n}, and b_{2n}, the particular solution is

$$\begin{aligned} y_p(\eta) = & -\frac{A_0}{\pi\rho_0^2} - \frac{(4\pi^2 + \rho_0^2)A_0\cos(2\pi\eta)}{2[(4\pi^2 + \rho_0^2)^2 + 16\pi^2\rho_0^2\nu_0^2]} + \frac{2\pi\rho_0\nu_0 A_0\sin(2\pi\eta)}{(4\pi^2 + \rho_0^2)^2 + 16\pi^2\rho_0^2\nu_0^2} \\ & + \frac{2A_0}{\pi}\sum_{n=1}^{\infty}\frac{(-1)^n(16n^2\pi^2 + \rho_0^2)\cos(2n\pi\eta)}{(4n^2 - 1)[64n^2\pi^2\rho_0^2\nu_0^2 + (16n^2\pi^2 + \rho_0^2)^2]} \\ & - 16\rho_0\nu_0 A_0\sum_{n=1}^{\infty}\frac{(-1)^n n\sin(2n\pi\eta)}{(4n^2 - 1)[64n^2\pi^2\rho_0^2\nu_0^2 + (16n^2\pi^2 + \rho_0^2)^2]}. \end{aligned} \quad (\mathbf{5.6.46})$$

Figure 5.6.1 is from Hrycak's paper and shows the variation of the nondimensional temperature as a function of η for the spinning rate ν_0. The other parameters are typical of a satellite with aluminum skin and fully covered with glass-protected solar cells. As a check on the solution, we show the temperature field (the dashed line) of a nonrotating satellite where we neglect the effects of conduction and only radiation occurs. The difference between the $\nu_0 = 0$ solid and dashed lines arises primarily due to the *linearization* of the nonlinear radiation boundary condition during the derivation of the governing equations.

Problems

Solve the following ordinary differential equations by Fourier series if the forcing is given by the periodic function

$$f(t) = \begin{cases} 1, & 0 < t < \pi, \\ 0, & \pi < t < 2\pi, \end{cases}$$

and $f(t) = f(t + 2\pi)$:

1. $y'' - y = f(t)$,   2. $y'' + y = f(t)$,   3. $y'' - 3y' + 2y = f(t)$.

4. Solve the boundary-value problem:

$$y'' + 2y = 3x, \qquad 0 < x < 1,$$

subject to the boundary conditions $y(0) = y(1) = 0$. Solve it two ways: (1) using techniques from Chapter 2 and (2) using Fourier sine series.

5. Given

$$y' + ky = \sum_{n=-\infty}^{\infty} c_n e^{inx}, \qquad -\pi < x < \pi,$$

find its solution in terms of a complex Fourier series.

Solve the following ordinary differential equations by *complex* Fourier series if the forcing is given by the periodic function

$$f(t) = |t|, \qquad -\pi \leq t \leq \pi,$$

and $f(t) = f(t + 2\pi)$:

6. $y'' - y = f(t)$,   7. $y'' + 4y = f(t)$.

8. An object radiating into its nocturnal surrounding has a temperature $y(t)$ governed by the equation[19]

$$\frac{dy}{dt} + ay = A_0 + \sum_{n=1}^{\infty} A_n \cos(n\omega t) + B_n \sin(n\omega t),$$

where the constant a is the heat loss coefficient and the Fourier series describes the temporal variation of the atmospheric air temperature and the effective sky temperature. If $y(0) = T_0$, find $y(t)$.

---

[19] See Sodha, M. S., 1982: Transient radiative cooling. *Solar Energy*, **28**, 541.

9. The equation that governs the charge q on the capacitor of an LRC electrical circuit is

$$q'' + 2\alpha q' + \omega^2 q = \omega^2 E,$$

where $\alpha = R/(2L)$, $\omega^2 = 1/(LC)$, R denotes resistance, C denotes capacitance, L denotes the inductance, and E is the electromotive force driving the circuit. If E is given by

$$E = \sum_{n=-\infty}^{\infty} \varphi_n e^{in\omega_0 t},$$

find $q(t)$.

10. Use Fourier series to find the *particular* solution[20] of the ordinary differential equation

$$y''(x) + k^2 y(x) = -k^2 V_L \begin{cases} 1, & 0 < x < \lambda/4, \\ -1, & \lambda/4 < x < \lambda/2, \end{cases}$$

where k, V_L and λ are constants. Extend the forcing function as an even function into the interval $(-\lambda/2, 0)$.

5.7 FINITE FOURIER SERIES

In many applications we must construct a Fourier series from values given by data or a graph. Unlike the situation with analytic formulas where we have an infinite number of data points and, consequently, an infinite number of terms in the Fourier series, the Fourier series contains a finite number of sines and cosines where the number of coefficients equals the number of data points.

Assuming that these series are useful, the next question is how do we find the Fourier coefficients? We could compute them by numerically integrating Equation 5.1.6. However, the results would suffer from the truncation errors that afflict all numerical schemes. On the other hand, we can avoid this problem if we again employ the orthogonality properties of sines and cosines, now in their discrete form. Just as in the case of conventional Fourier series, we can use these properties to derive formulas for computing the Fourier coefficients. These results will be *exact* except for roundoff errors.

We start by deriving some preliminary results. Let us define $x_m = mP/(2N)$. Then, if k is an integer,

$$\sum_{m=0}^{2N-1} \exp\left(\frac{2\pi i k x_m}{P}\right) = \sum_{m=0}^{2N-1} \exp\left(\frac{km\pi i}{N}\right) = \sum_{m=0}^{2N-1} r^m = \begin{cases} \frac{1-r^{2N}}{1-r} = 0, & r \neq 1, \\ 2N, & r = 1, \end{cases} \quad (5.7.1)$$

because $r^{2N} = \exp(2\pi ki) = 1$ if $r \neq 1$. If $r = 1$, then the sum consists of $2N$ terms, each of which equals one. The condition $r = 1$ corresponds to $k = 0, \pm 2N, \pm 4N, \ldots$. Taking the real and imaginary part of Equation 5.7.1,

$$\sum_{m=0}^{2N-1} \cos\left(\frac{2\pi k x_m}{P}\right) = \begin{cases} 0, & k \neq 0, \pm 2N, \pm 4N, \ldots, \\ 2N, & k = 0, \pm 2N, \pm 4N, \ldots, \end{cases} \quad (5.7.2)$$

---

[20] See Chabert, P., J. L. Raimbault, J. M. Rax, and M. A. Lieberman, 2004: Self-consistent nonlinear transmission line model of standing wave effects in a capacitive discharge. *Phys. Plasmas*, **11**, 1775–1785.

and

$$\sum_{m=0}^{2N-1} \sin\left(\frac{2\pi k x_m}{P}\right) = 0 \qquad (5.7.3)$$

for all k.

Consider now the following sum:

$$\sum_{m=0}^{2N-1} \cos\left(\frac{2\pi k x_m}{P}\right)\cos\left(\frac{2\pi j x_m}{P}\right) = \frac{1}{2}\sum_{m=0}^{2N-1}\left\{\cos\left[\frac{2\pi(k+j)x_m}{P}\right] + \cos\left[\frac{2\pi(k-j)x_m}{P}\right]\right\}$$

$$(5.7.4)$$

$$= \begin{cases} 0, & |k-j| \text{ and } |k+m| \neq 0, 2N, 4N, \ldots, \\ N, & |k-j| \text{ or } |k+m| \neq 0, 2N, 4N, \ldots, \\ 2N, & |k-j| \text{ and } |k+m| = 0, 2N, 4N, \ldots. \end{cases} \quad (5.7.5)$$

Let us simplify the right side of Equation 5.7.5 by restricting ourselves to $k+j$ lying between 0 to $2N$. This is permissible because of the periodic nature of this equation. If $k+j = 0$, $k = j = 0$; if $k+j = 2N$, $k = j = N$. In either case, $k-j = 0$ and the right side of Equation 5.7.5 equals $2N$. Consider now the case $k \neq j$. Then $k+j \neq 0$ or $2N$ and $k-j \neq 0$ or $2N$. The right side of this equation must equal 0. Finally, if $k = j \neq 0$ or N, then $k+j \neq 0$ or $2N$ but $k-j = 0$ and the right side of this equation equals N. In summary,

$$\sum_{m=0}^{2N-1} \cos\left(\frac{2\pi k x_m}{P}\right)\cos\left(\frac{2\pi j x_m}{P}\right) = \begin{cases} 0, & k \neq j \\ N, & k = j \neq 0, N \\ 2N, & k = j = 0, N. \end{cases} \quad (5.7.6)$$

In a similar manner,

$$\sum_{m=0}^{2N-1} \cos\left(\frac{2\pi k x_m}{P}\right)\sin\left(\frac{2\pi j x_m}{P}\right) = 0 \qquad (5.7.7)$$

for all k and j and

$$\sum_{m=0}^{2N-1} \sin\left(\frac{2\pi k x_m}{P}\right)\sin\left(\frac{2\pi j x_m}{P}\right) = \begin{cases} 0, & k \neq j \\ N, & k = j \neq 0, N, \\ 0, & k = j = 0, N. \end{cases} \quad (5.7.8)$$

Armed with these equations we are ready to find the coefficients A_n and B_n of the finite Fourier series,

$$f(x) = \frac{A_0}{2} + \sum_{k=1}^{N-1}\left[A_k \cos\left(\frac{2\pi k x}{P}\right) + B_k \sin\left(\frac{2\pi k x}{P}\right)\right] + \frac{A_N}{2}\cos\left(\frac{2\pi N x}{P}\right), \quad (5.7.9)$$

where we have $2N$ data points and now define P as the period of the function.

To find A_k we proceed as before and multiply Equation 5.7.9 by $\cos(2\pi j x /P)$ (j may take on values from 0 to N) and sum from 0 to $2N-1$. At the point $x = x_m$,

$$\sum_{m=0}^{2N-1} f(x_m)\cos\left(\frac{2\pi j}{P}x_m\right) = \frac{A_0}{2}\sum_{m=0}^{2N-1}\cos\left(\frac{2\pi j}{P}x_m\right)$$

$$+ \sum_{k=1}^{N-1} A_k \sum_{m=0}^{2N-1}\cos\left(\frac{2\pi k}{P}x_m\right)\cos\left(\frac{2\pi j}{P}x_m\right)$$

$$+ \sum_{k=1}^{N-1} B_k \sum_{m=0}^{2N-1}\sin\left(\frac{2\pi k}{P}x_m\right)\cos\left(\frac{2\pi j}{P}x_m\right)$$

$$+ \frac{A_N}{2}\sum_{m=0}^{2N-1}\cos\left(\frac{2\pi N}{P}x_m\right)\cos\left(\frac{2\pi j}{P}x_m\right). \quad (5.7.10)$$

If $j \neq 0$ or N, then the first summation on the right side vanishes by Equation 5.7.2, the third by Equation 5.7.8, and the fourth by Equation 5.7.6. The second summation does *not* vanish if $k = j$ and equals N. Similar considerations lead to the formulas for the calculation of A_k and B_k:

$$A_k = \frac{1}{N} \sum_{m=0}^{2N-1} f_m \cos\left(\frac{\pi k m}{N}\right), \qquad k = 0, 1, 2, \ldots, N, \tag{5.7.11}$$

and

$$B_k = \frac{1}{N} \sum_{m=0}^{2N-1} f_m \sin\left(\frac{\pi k m}{N}\right), \qquad k = 1, 2, \ldots, N-1, \tag{5.7.12}$$

where f_m is the value of the function at the mth data point.

It is important to note that $2N$ data points yield $2N$ Fourier coefficients A_k and B_k. As a consequence, our sampling frequency will always limit the amount of information, whether in the form of data points or Fourier coefficients. It might be argued that from the Fourier series representation of $f(t)$ we could find the value of $f(t)$ for any given t, which is more than we can do with the data alone. This is not true. Although we can calculate $f(t)$ at any t using the finite Fourier series, the values may or may not be correct since the constraint on the finite Fourier series is that the series must fit the data in a least-squared sense. Despite the limitations imposed by only having a finite number of Fourier coefficients, the Fourier analysis of finite data sets yields valuable physical insights into the processes governing many physical systems.

In the case when we have $2N + 1$ (odd numbered) data points, a similar derivation gives

$$f_j = \frac{A_0}{2} + \sum_{n=1}^{N} A_n \cos\left(\frac{2\pi n j}{2N+1}\right) + B_n \sin\left(\frac{2\pi n j}{2N+1}\right), \tag{5.7.13}$$

where $j = 0, 1, 2, \ldots, 2N$, A_n and B_n are given by

$$A_0 = \frac{2}{2N+1} \sum_{j=0}^{2N} f_j, \tag{5.7.14}$$

$$A_m = \frac{2}{2N+1} \sum_{j=0}^{2N} f_j \cos\left(\frac{2\pi m j}{2N+1}\right), \qquad m = 1, 2, \ldots, N, \tag{5.7.15}$$

$$B_m = \frac{2}{2N+1} \sum_{j=0}^{2N} f_j \sin\left(\frac{2\pi m j}{2N+1}\right), \qquad m = 1, 2, \ldots, N, \tag{5.7.16}$$

and f_j is the value of the function at the jth data point.

Finally, following Section 5.5, we can re-express the finite Fourier series in complex form. For N data points the complex Fourier series representation is

$$f_n = \frac{1}{N} \sum_{k=0}^{N-1} C_k \exp\left(2\pi i k n / N\right), \qquad n = 0, 1, 2, \ldots, N-1, \tag{5.7.17}$$

where the complex Fourier coefficient C_k is given by

$$C_k = \sum_{n=0}^{N-1} f_n \exp\left(-2\pi i n k / N\right), \qquad k = 0, 1, 2, \ldots, N-1. \tag{5.7.18}$$

Table 5.7.1: The Depth of Water in the Harbor at Buffalo, NY (Minus the Low-Water Datum of 568.8 ft) on the 15th Day of Each Month during 1977

mo	n	depth	mo	n	depth	mo	n	depth
Jan	1	1.61	May	5	3.16	Sep	9	2.42
Feb	2	1.57	Jun	6	2.95	Oct	10	2.95
Mar	3	2.01	Jul	7	3.10	Nov	11	2.74
Apr	4	2.68	Aug	8	2.90	Dec	12	2.63

• **Example 5.7.1**

Consider the simple case where we have 5 data points: $f_0 = 1$, $f_1 = 1$, $f_2 = 1$, $f_3 = 0$, and $f_4 = 0$. Using Equations 5.7.14 through 5.7.16 with $N = 2$, we find that

$$A_0 = (2/5)(x_0 + x_1 + x_2 + x_3 + x_4), \tag{5.7.19}$$

$$A_k = (2/5)[1 + \cos(2\pi k/5) + \cos(4\pi k/5), \qquad k = 1, 2, \tag{5.7.20}$$

and

$$B_k = (2/5)[\sin(2\pi k/5) + \sin(4\pi k/5), \qquad k = 1, 2. \tag{5.7.21}$$

These equations yield $A_0 = 1.2$, $A_1 = A_2 = 0.2$, $B_1 = 0.61584$, and $B_2 = -0.14531$. When these Fourier coefficients were substituted into Equation 5.7.13, they gave back the original f_j.

On the other hand, Equation 5.7.18 yields the complex Fourier coefficients

$$C_k = \exp(-2\pi i k/5)\,[1 + 2\cos(2\pi k/5)], \qquad k = 0, 1, 2, 3, 4, \tag{5.7.22}$$

or $C_0 = 3$, $C_1 = 0.5 - 1.53884i$, $C_2 = 0.5 + 0.36327i$, $C_3 = 0.5 - 0.36327i$, and $C_4 = 0.5 + 1.53884i$. Upon substituting these complex Fourier coefficients into Equation 5.7.17 we obtain the original data.

We could have also used MATLAB to compute the complex Fourier coefficients C_0 through C_4. By typing `fft([1 1 1 0 0],5)` we would have obtained the desired quantities.

□

• **Example 5.7.2: Water depth at Buffalo, NY**

Each entry[21] in Table 5.7.1 gives the observed depth of water at Buffalo, NY (minus the low-water datum of 568.6 ft) on the 15th of the corresponding month during 1977. Assuming that the water level is a periodic function of 1 year, and that we took the observations at equal intervals, let us construct a finite Fourier series from these data. This corresponds to computing the Fourier coefficients $A_0, A_1, \ldots, A_6, B_1, \ldots, B_5$, which give the mean level and harmonic fluctuations of the depth of water, the harmonics having the periods 12 months, 6 months, 4 months, and so forth.

---

[21] National Ocean Survey, 1977: *Great Lakes Water Level, 1977, Daily and Monthly Average Water Surface Elevations*. National Oceanic and Atmospheric Administration.

In this problem, $N = 6$. Each data point gives the water depth for each month. From Equation 5.7.11 and Equation 5.7.12,

$$A_k = \frac{1}{6} \sum_{m=0}^{11} f(x_m) \cos\left(\frac{mk\pi}{6}\right), \quad k = 0, 1, 2, 3, 4, 5, 6, \tag{5.7.23}$$

and

$$B_k = \frac{1}{6} \sum_{m=0}^{11} f(x_m) \sin\left(\frac{mk\pi}{6}\right), \quad k = 1, 2, 3, 4, 5. \tag{5.7.24}$$

Substituting the data into these equations yields

A_0	= twice the mean level			= +5.120 ft
A_1	= harmonic component with a period of	12	mo	= −0.566 ft
B_1	= harmonic component with a period of	12	mo	= −0.128 ft
A_2	= harmonic component with a period of	6	mo	= −0.177 ft
B_2	= harmonic component with a period of	6	mo	= −0.372 ft
A_3	= harmonic component with a period of	4	mo	= −0.110 ft
B_3	= harmonic component with a period of	4	mo	= −0.123 ft
A_4	= harmonic component with a period of	3	mo	= +0.025 ft
B_4	= harmonic component with a period of	3	mo	= +0.052 ft
A_5	= harmonic component with a period of	2.4	mo	= −0.079 ft
B_5	= harmonic component with a period of	2.4	mo	= −0.131 ft
A_6	= harmonic component with a period of	2	mo	= −0.107 ft

Figure 5.7.1 is a plot of our results using Equation 5.7.9. Note that when we include all of the harmonic terms, the finite Fourier series fits the data points exactly. The values given by the series at points between the data points may be right or they may not. To illustrate this, we also plotted the values for the first of each month. Sometimes the values given by the Fourier series and these intermediate data points are quite different.

Let us now examine our results in terms of various physical processes. In the long run the depth of water in the harbor at Buffalo, NY depends upon the three-way balance between precipitation, evaporation, and inflow-outflow of any rivers. Because the inflow and outflow of the rivers depends strongly upon precipitation, and evaporation is of secondary importance, the water level should correlate with the precipitation rate. It is well known that more precipitation falls during the warmer months rather than the colder months. The large amplitude of the Fourier coefficient A_1 and B_1, corresponding to the annual cycle ($k = 1$), reflects this.

Another important term in the harmonic analysis corresponds to the semiannual cycle ($k = 2$). During the winter months around Lake Ontario, precipitation falls as snow. Therefore, the inflow from rivers is greatly reduced. When spring comes, the snow and ice melt and a jump in the water level occurs. Because the second harmonic gives periodic variations associated with seasonal variations, this harmonic is absolutely necessary if we want to get the correct answer while the higher harmonics do not represent any specific physical process.   □

• Example 5.7.3: Numerical computation of Fourier coefficients

At the beginning of this chapter, we showed how you could compute the Fourier coefficients a_0, a_n, and b_n from Equation 5.1.6 given a function $f(t)$. All of this assumed that you

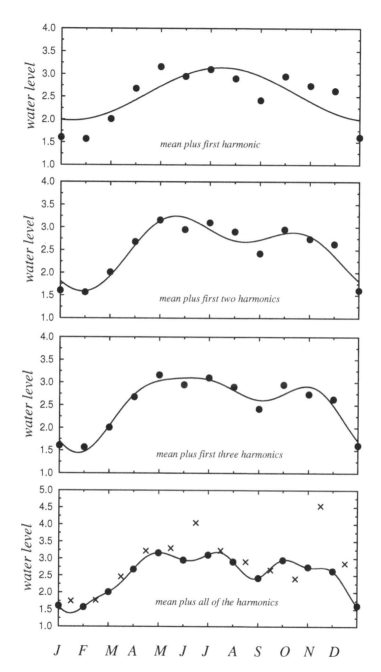

Figure 5.7.1: Partial sums of the finite Fourier series for the depth of water in the harbor of Buffalo, NY during 1977. Circles indicate observations on the 15th of the month; crosses are observations on the first.

could carry out the integrations. What do you do if you cannot perform the integrations? The obvious solution is perform it numerically. In this section we showed that the best approximation to Equation 5.1.6 is given by Equation 5.7.11 and Equation 5.7.12. In the case when we have $f(t)$ this is still true but we may choose N as large as necessary to obtain the desired number of Fourier coefficients.

To illustrate this we have redone Example 5.1.1 and plotted the exact (analytic) and

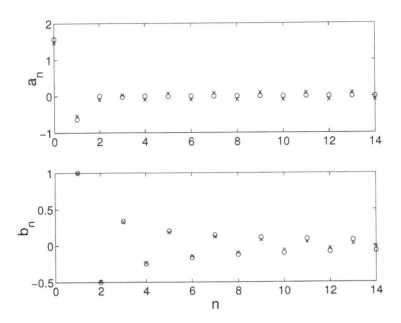

Figure 5.7.2: The computation of Fourier coefficients using a finite Fourier series when $f(t)$ is given by Equation 5.1.8. The circles give a_n and b_n as computed from Equation 5.1.9, Equation 5.1.10, and Equation 5.1.11. The crosses give the corresponding Fourier coefficients given by the finite Fourier series with $N = 15$.

numerically computed Fourier coefficients in Figure 5.7.2. This figure was created using the MATLAB script:

```
clear;
N = 15, M = 2*N; dt = 2*pi/M; % number of points in interval
% create time points assuming x(t) = x(t+period)
t = [-pi:dt:pi-dt];
%
f = zeros(size(t)); % initialize function f(t)
for k = 1:length(t) % construct function f(t)
   if t(k) < 0; f(k) = 0; else f(k) = t(k); end; end;
%
% compute Fourier coefficients using fast Fourier transform
%
fourier = fft(f) / N;
a_0_comp = real(fourier(1)); sign = 1;
for n = 2:N;
   a_n_comp(n-1) = - sign * real(fourier(n));
   b_n_comp(n-1) = sign * imag(fourier(n));
   sign = - sign;
end
%
% plot comparisons
%
NN = linspace(0,N-1,N);
exact_coeff(1) = pi/2;
numer_coeff(1) = a_0_comp;
```

```
for n = 1:N-1;
  exact_coeff(n+1) = ((-1)^n-1) / (pi*(2*n-1)^2);
  numer_coeff(n+1) = a_n_comp(n);
end;
subplot(2,1,1),plot(NN,exact_coeff,'o',NN,numer_coeff,'kx')
ylabel('a_n','Fontsize',20)
clear exact_coeff numer_coeff
NN = linspace(1,N-1,N-1);
for n = 1:N-1;
  exact_coeff(n) = -(-1)^n/n; numer_coeff(n) = b_n_comp(n);
end;
subplot(2,1,2), plot(NN,exact_coeff,'o',NN,numer_coeff,'kx')
xlabel('n','Fontsize',20); ylabel('b_n','Fontsize',20);
```

It shows that relatively few data points can yield quite reasonable answers.

Let us examine this script a little closer. One of the first things that you will note is that there is no explicit reference to Equation 5.7.11 and Equation 5.7.12. How did we get the correct answer?

Although we could have coded Equation 5.7.11 and Equation 5.7.12, no one does that anymore. In the 1960s, J. W. Cooley and J. W. Tukey[22] devised an incredibly clever method of performing these calculations. This method, commonly called a fast Fourier transform or FFT, is so popular that all computational packages contain it as an intrinsic function and MATLAB is no exception, calling it `fft`. This is what has been used here.

Although we now have an `fft` to compute the coefficients, this routine does not directly give the coefficients a_n and b_n but rather some mysterious (complex) number that is related to $a_n + ib_n$. This is a common problem in using a package's FFT rather than your own and why the script divides by N and we keep changing the sign. The best method for discovering how to extract the coefficients a_n and b_n is to test it with a dataset created by a simple, finite series such as

$$f(x) = 20 + \cos(t) + 3\sin(t) + 6\cos(2t) - 20\sin(2t) - 10\cos(3t) - 30\sin(3t). \quad (\mathbf{5.7.25})$$

If the code is correct, it must give back the coefficient in Equation 5.7.25 to within round-off. Otherwise, something is wrong.

Finally, most FFTs assume that the dataset will start repeating after the final data point. Therefore, when reading in the dataset, the point corresponding to $x = L$ must be excluded. □

● **Example 5.7.4: Aliasing**

In the previous example, we could only resolve phenomena with a period of 2 months or greater although we had data for each of the 12 months. This is an example of *Nyquist's sampling criteria:*[23] At least two samples are required to resolve the highest frequency in a periodically sampled record.

Figure 5.7.3 will help explain this phenomenon. In case (a) we have quite a few data points over one cycle. Thus, our picture, constructed from data, is fairly good. In case (b),

---

[22] Cooley, J. W., and J. W. Tukey, 1965: An algorithm for machine calculation of complex Fourier series. *Math. Comput.*, **19**, 297–301.

[23] Nyquist, H., 1928: Certain topics in telegraph transmission theory. *AIEE Trans.*, **47**, 617–644.

real world   instrument data points

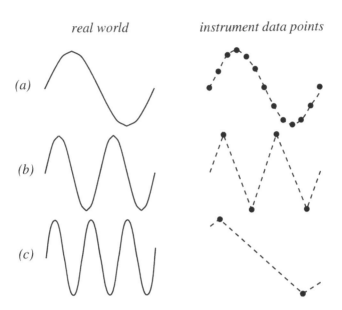

Figure 5.7.3: The effect of sampling in the representation of periodic functions.

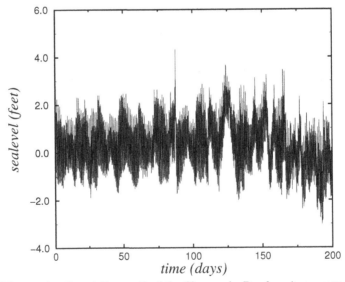

Figure 5.7.4: The sea elevation at the mouth of the Chesapeake Bay from its average depth as a function of time after 1 July 1985.

we took only samples at the ridges and troughs of the wave. Although our picture of the real phenomenon is poor, at least we know that there is a wave. From this picture we see that even if we are lucky enough to take our observations at the ridges and troughs of a wave, we need at least two data points per cycle (one for the ridge, the other for the trough) to resolve the highest-frequency wave.

In case (c) we have made a big mistake. We have taken a wave of frequency N Hz and misrepresented it as a wave of frequency $N/2$ Hz. This misrepresentation of a high-frequency wave by a lower-frequency wave is called *aliasing*. It arises because we are sampling a continuous signal at equal intervals. By comparing cases (b) and (c), we see that there is

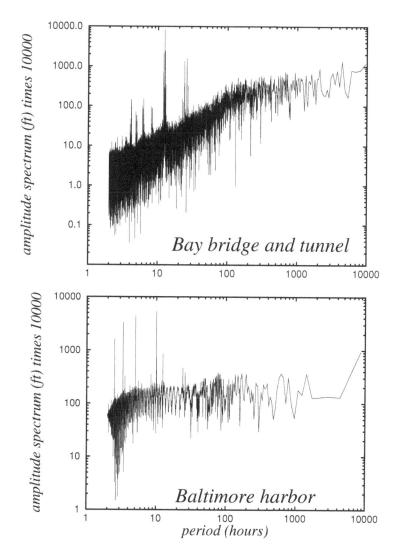

Figure 5.7.5: The amplitude of the Fourier coefficients for the sea elevation at the Chesapeake Bay bridge and tunnel (top) and Baltimore harbor (bottom) as a function of period.

a cutoff between aliased and nonaliased frequencies. This frequency is called the *Nyquist* or *folding* frequency. It corresponds to the highest frequency resolved by our finite Fourier analysis.

Because most periodic functions require an infinite number of harmonics for their representation, aliasing of signals is a common problem. Thus the question is not "can I avoid aliasing?" but "can I live with it?" Quite often, we can construct our experiments to say yes. An example where aliasing is unavoidable occurs in a Western at the movies when we see the rapidly rotating spokes of the stagecoach's wheel. A movie is a sampling of continuous motion where we present the data as a succession of pictures. As such, a film aliases the high rate of revolution of the stagecoach's wheel in such a manner so that it appears to be stationary or rotating very slowly. ☐

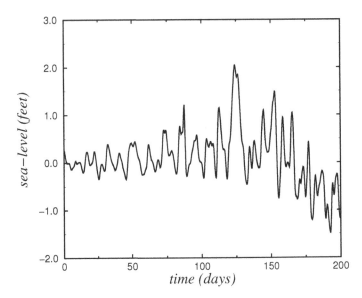

Figure 5.7.6: Same as Figure 5.7.4 but with the tides removed.

• Example 5.7.5: Spectrum of the Chesapeake Bay

For our final example, we perform a Fourier analysis of hourly sea-level measurements taken at the mouth of the Chesapeake Bay during the 2000 days from 9 April 1985 to 29 June 1990. Figure 5.7.4 shows 200 days of this record, starting from 1 July 1985. As this figure shows, the measurements contain a wide range of oscillations. In particular, note the large peak near day 90 that corresponds to the passage of Hurricane Gloria during the early hours of 27 September 1985.

Utilizing the entire 2000 days, we plotted the amplitude of the Fourier coefficients as a function of period in Figure 5.7.5. We see a general rise of the amplitude as the period increases. Especially noteworthy are the sharp peaks near periods of 12 and 24 hours. The largest peak is at 12.417 hours and corresponds to the semidiurnal tide. Thus, our Fourier analysis shows that the dominant oscillations at the mouth of the Chesapeake Bay are the tides. A similar situation occurs in Baltimore harbor. Furthermore, with this spectral information we could predict high and low tides very accurately.

Although the tides are of great interest to some, they are a nuisance to others because they mask other physical processes that might be occurring. For that reason we would like to remove them from the tidal gauge history and see what is left. One way would be to zero out the Fourier coefficients corresponding to the tidal components and then plot the resulting Fourier series. Another method is to replace each hourly report with an average of hourly reports that occurred 24 hours ahead of and behind a particular report. We construct this average in such a manner that waves with periods of the tides sum to zero.[24] Such a *filter* is a popular method for eliminating unwanted waves from a record. Filters play an important role in the analysis of data. We plotted the filtered sea level data in Figure 5.7.6. Note that summertime (0–50 days) produces little variation in the sea level compared to wintertime (100–150 days) when intense coastal storms occur.

---

[24] See Godin, G., 1972: *The Analysis of Tides*. University of Toronto Press, Section 2.1.

Problems

By hand, find the finite Fourier series for the following pieces of data:

1. $f(0) = 0$, $f(1) = 1$, $f(2) = 2$, $f(3) = 3$, and $N = 2$.

2. $f(0) = 1$, $f(1) = 1$, $f(2) = -1$, $f(3) = -1$, and $N = 2$.

Project: Construction of a Finite Fourier Series Using Linear Algebra

In Example 5.1.1 we show that the function $f(t)$ given by Equation 5.1.8 can be re-expressed

$$f(t) = \frac{a_0}{2} + \sum_{n=1}^{\infty} a_n \cos(nt) + b_n \sin(nt), \qquad -\pi < t < \pi,$$

if

$$a_0 = \frac{\pi}{2}, \qquad a_n = \frac{(-1)^n - 1}{n^2 \pi}, \qquad \text{and} \qquad b_n = \frac{(-1)^{n+1}}{n}.$$

There we stated the Fourier series fits $f(t)$ in a "least squares sense." In Section 5.7 we showed that we could also approximate $f(t)$ with the finite Fourier series

$$f(t) = \tfrac{1}{2}A_0 + \sum_{k=1}^{M-1} A_k \cos(kt) + B_k \sin(kt) + \tfrac{1}{2}A_M \cos(Mt),$$

if we sample $f(t)$ at $t_m = (2m+1-M)\pi/M$, where $m = 0, 1, 2, \ldots, M-1$ and M is an even integer. Then we would use Equation 5.7.11 and Equation 5.7.12 to compute A_k and B_k. Because MATLAB solves linear equations in a least-squares sense, this suggests that we could use MATLAB as an alternative method for finding a finite Fourier series approximation.

Let us assume that

$$f(t) = \frac{A_0}{2} + \sum_{n=1}^{N} A_n \cos(nt) + B_n \sin(nt).$$

Then sampling $f(t)$ at the temporal points $t_m = -\pi + (2m-1)\pi/M$, we obtain the following system of linear equations:

$$\frac{A_0}{2} + \sum_{n=1}^{N} A_n \cos(nt_m) + B_n \sin(nt_m) = f(t_m),$$

where $m = 1, 2, \ldots, M$. Write a MATLAB program using `linsolve` that solves this system for given N and M and compare your results with the exact answers a_0, a_n and b_n for various N and M. Consider the case when $M > 2N+1$ (overspecified system), $M < 2N+1$ (underspecified system), and $M = 2N+1$ (equal number of unknowns and equations). Does this method yield any good results? If so, under which conditions?

Project: Using MATLAB's Fast Fourier Transform

For each function defined in Problems 1 to 16 at the end of Section 5.1, use MATLAB's function `fft` to compute the Fourier coefficients. Depending upon the number of data

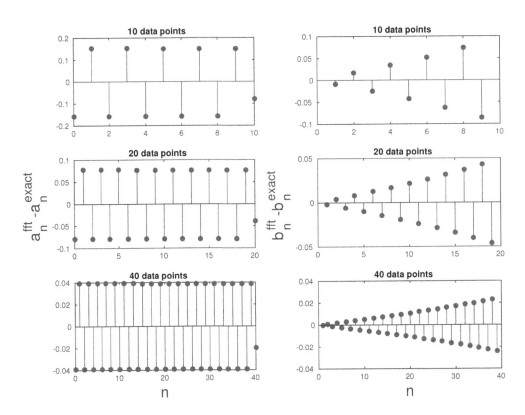

Figure 5.7.7: Given the time function Equation 5.1.8 from Example 5.1.1, this figure gives the difference between the Fourier coefficients given by MATLAB's function `fft` and the exact Fourier coefficients given by Equation 5.1.10 and Equation 5.1.11 as a function of harmonic.

points that define the function, how do the Fourier coefficients that you compute from the `fft` compare with the ones that you found exactly?

Project: Design Your Own Snow Tire, Part II

In Part I of this project you examined the spectrum of a simple model for a snow tire and you found the effect of tread thickness and spacing on the spectrum of the noise produced as the treads struck the pavement. Clearly, if you were to introduce a more complex and perhaps more realistic model, the calculations would become very cumbersome. But now, capitalizing on the work of the previous project, you can use the fast Fourier transform to compute the spectrum once you decide on a tread pattern. That is the object of this project.

Step 1: For $-\pi < t < \pi$ and assuming that this pattern repeats with a period of 2π, devise a function $f(t)$, where it equals 1 when the tread is striking the roadway and 0 when it is not.

Step 2: Use the code that you developed in the previous project to find the amplitude of the Fourier coefficients as a function of the harmonic. Test out your code by recovering the results from Example 5.1.3.

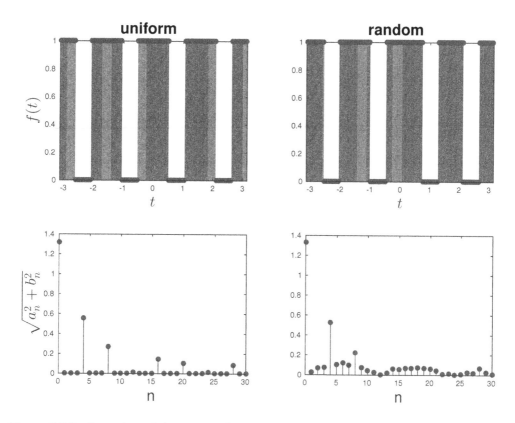

Figure 5.7.8: Comparison of the spectrum between two types of tires. On the left the gap between the treads is uniform. On the right the gaps have been randomly placed.

Step 3: Our tire pattern of two thin treads is not very realistic. Modify your code so that it now has 4 thick treads separated by thin gaps. Compare the spectrum when the gaps are uniformly spaced and one where they are randomly spaced.

Project: Spectrum of the Earth's Orography

Table 5.7.2 gives the orographic height of the earth's surface used in an atmospheric general circulation model (GCM) at a resolution of 2.5° longitude along the latitude belts of 28°S, 36°N, and 66°N. In this project you will find the spectrum of this orographic field along the various latitude belts.

Step 1: Write a MATLAB script that reads in the data and find A_n and B_n and then construct the amplitude spectra for this data.

Step 2: Construct several spectra by using every data point, every other data point, etc. How do the magnitudes of the Fourier coefficient change? You might like to read about *leakage* from a book on harmonic analysis.[25]

Step 3: Compare and contrast the spectra from the various latitude belts. How do the magnitudes of the Fourier coefficients decrease with n? Why are there these differences?

---

[25] For example, Bloomfield, P., 1976: *Fourier Analysis of Time Series: An Introduction.* John Wiley & Sons, 258 pp.

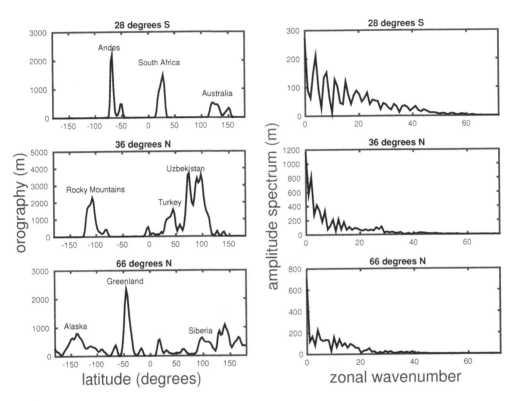

Figure 5.7.9: The orography of the earth and its spectrum in meters along three latitude belts using a topography dataset with a resolution of 2.5° longitude.

Step 4: You may have noted that some of the heights are negative, even in the middle of the ocean! Take the original data (for any latitude belt) and zero out all of the negative heights. Find the spectra for this new data set. How have the spectra changed? Is there a reason why the negative heights were introduced?

Further Readings

Carslaw, H. S., 1950: *An Introduction to the Theory of Fourier's Series and Integrals*. Dover, 368 pp. A classic treatment of the Fourier technique.

Tolstov, Georgi P., 1976: *Fourier Series*. Dover, 336 pp. This book covers the basic theory of Fourier series and its use in mathematical physics.

Table 5.7.2: Orographic Heights (in m) Times the Gravitational Acceleration Constant ($g = 9.81$ m/s^2) along Three Latitude Belts. The data is also given in the file `orographic.txt` on the CRC website.

Longitude	28°S	36°N	66°N	Longitude	28°S	36°N	66°N
−180.0	4.	3.	2532.	−70.0	19317.	−8.	1830.
−177.5	1.	−2.	1665.	−67.5	21681.	0.	3000.
−175.0	1.	2.	1432.	−65.0	9222.	−2.	3668.
−172.5	1.	−3.	1213.	−62.5	1949.	−2.	2147.
−170.0	1.	1.	501.	−60.0	774.	0.	391.
−167.5	1.	−3.	367.	−57.5	955.	5.	−77.
−165.0	1.	1.	963.	−55.0	2268.	6.	601.
−162.5	0.	0.	1814.	−52.5	4636.	−1.	3266.
−160.0	−1.	6.	2562.	−50.0	4621.	2.	9128.
−157.5	0.	1.	3150.	−47.5	1300.	−4.	17808.
−155.0	0.	3.	4008.	−45.0	−91.	1.	22960.
−152.5	1.	−2.	4980.	−42.5	57.	−1.	20559.
−150.0	−1.	4.	6011.	−40.0	−25.	4.	14296.
−147.5	6.	−1.	6273.	−37.5	13.	−1.	9783.
−145.0	14.	3.	5928.	−35.0	−10.	6.	5969.
−142.5	6.	−1.	6509.	−32.5	8.	2.	1972.
−140.0	−2.	6.	7865.	−30.0	−4.	22.	640.
−137.5	0.	3.	7752.	−27.5	6.	33.	379.
−135.0	−2.	5.	6817.	−25.0	−2.	39.	286.
−132.5	1.	−2.	6272.	−22.5	3.	2.	981.
−130.0	−2.	0.	5582.	−20.0	−3.	11.	1971.
−127.5	0.	5.	4412.	−17.5	1.	−6.	2576.
−125.0	−2.	423.	3206.	−15.0	−1.	19.	1692.
−122.5	1.	3688.	2653.	−12.5	0.	−18.	357.
−120.0	−3.	10919.	2702.	−10.0	−1.	490.	−21.
−117.5	2.	16148.	3062.	−7.5	0.	2164.	−5.
−115.0	−3.	17624.	3344.	−5.0	1.	4728.	−10.
−112.5	7.	18132.	3444.	−2.5	0.	5347.	0.
−110.0	12.	19511.	3262.	0.0	4.	2667.	−6.
−107.5	9.	22619.	3001.	2.5	−5.	1213.	−1.
−105.0	−5.	20273.	2931.	5.0	7.	1612.	−31.
−102.5	3.	12914.	2633.	7.5	−13.	1744.	−58.
−100.0	−5.	7434.	1933.	10.0	28.	1153.	381.
−97.5	6.	4311.	1473.	12.5	107.	838.	2472.
−95.0	−8.	2933.	1689.	15.0	2208.	1313.	5263.
−92.5	8.	2404.	2318.	17.5	6566.	862.	5646.
−90.0	−12.	1721.	2285.	20.0	9091.	1509.	3672.
−87.5	18.	1681.	1561.	22.5	10690.	2483.	1628.
−85.0	−23.	2666.	1199.	25.0	12715.	1697.	889.
−82.5	36.	4047.	737.	27.5	14583.	3377.	1366.
−80.0	−64.	3938.	185.	30.0	11351.	7682.	1857.
−77.5	138.	1669.	71.	32.5	3370.	9663.	1534.
−75.0	−363.	236.	160.	35.0	15.	10197.	993.
−72.5	4692.	31.	823.	37.5	49.	10792.	863.

Table 5.7.2, contd.: Orographic Heights (in m) Times the Gravitational Acceleration Constant ($g = 9.81$ m/s^2) along Three Latitude Belts

Longitude	28°S	36°N	66°N	Longitude	28°S	36°N	66°N
40.0	−31.	11322.	756.	110.0	−17.	12639.	4674.
42.5	20.	13321.	620.	112.5	302.	10543.	4435.
45.0	−17.	15414.	626.	115.0	1874.	4967.	3646.
47.5	−19.	12873.	836.	117.5	4005.	1119.	2655.
50.0	−18.	6114.	1029.	120.0	4989.	696.	2065.
52.5	6.	2962.	946.	122.5	4887.	475.	1583.
55.0	−2.	4913.	828.	125.0	4445.	1631.	3072.
57.5	3.	6600.	1247.	127.5	4362.	2933.	7290.
60.0	−3.	4885.	2091.	130.0	4368.	1329.	8541.
62.5	2.	3380.	2276.	132.5	3485.	88.	7078.
65.0	−1.	5842.	1870.	135.0	1921.	598.	7322.
67.5	2.	12106.	1215.	137.5	670.	1983.	9445.
70.0	0.	23032.	680.	140.0	666.	2511.	10692.
72.5	2.	35376.	531.	142.5	1275.	866.	9280.
75.0	−1.	36415.	539.	145.0	1865.	13.	8372.
77.5	1.	26544.	579.	147.5	2452.	11.	6624.
80.0	0.	19363.	554.	150.0	3160.	−4.	3617.
82.5	1.	17915.	632.	152.5	2676.	−1.	2717.
85.0	−2.	22260.	791.	155.0	697.	0.	3474.
87.5	−1.	30442.	1455.	157.5	−67.	−3.	4337.
90.0	−3.	33601.	3194.	160.0	25.	3.	4824.
92.5	−1.	30873.	4878.	162.5	−12.	−1.	5525.
95.0	0.	31865.	5903.	165.0	10.	4.	6323.
97.5	0.	35538.	6222.	167.5	−5.	−2.	5899.
100.0	−2.	31985.	5523.	170.0	0.	1.	4330.
102.5	0.	23246.	4823.	172.5	0.	−4.	3338.
105.0	−4.	17363.	4689.	175.0	4.	3.	3408.
107.5	2.	14315.	4698.	177.5	3.	−1.	3407.

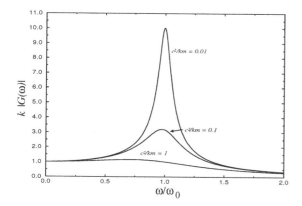

Chapter 6
The Fourier Transform

In the previous chapter you learned how to expand a periodic function in terms of an infinite sum of sines and cosines. However, most functions encountered in engineering are aperiodic. As we shall see, the extension of Fourier series to these functions leads to the Fourier transform.

6.1 FOURIER TRANSFORMS

The Fourier transform is the natural extension of Fourier series to a function $f(t)$ of infinite period. To show this, consider a periodic function $f(t)$ of period $2T$ that satisfies the so-called Dirichlet's conditions.[1] If the integral $\int_a^b |f(t)|\, dt$ exists, this function has the complex Fourier series

$$f(t) = \sum_{n=-\infty}^{\infty} c_n e^{in\pi t/T}, \tag{6.1.1}$$

where

$$c_n = \frac{1}{2T} \int_{-T}^{T} f(t)e^{-in\pi t/T} dt. \tag{6.1.2}$$

Equation 6.1.1 applies only if $f(t)$ is continuous at t; if $f(t)$ suffers from a jump discontinuity at t, then the left side of Equation 6.1.1 equals $\frac{1}{2}[f(t^+)+f(t^-)]$, where $f(t^+) = \lim_{x \to t^+} f(x)$ and $f(t^-) = \lim_{x \to t^-} f(x)$. Substituting Equation 6.1.2 into Equation 6.1.1,

$$f(t) = \frac{1}{2T} \sum_{n=-\infty}^{\infty} e^{in\pi t/T} \int_{-T}^{T} f(x)e^{-in\pi x/T} dx. \tag{6.1.3}$$

---

[1] A function $f(t)$ satisfies Dirichlet's conditions in the interval (a, b) if (1) it is bounded in (a, b), and (2) it has at most a finite number of discontinuities and a finite number of maxima and minima in that interval.

Let us now introduce the notation $\omega_n = n\pi/T$ so that $\Delta\omega_n = \omega_{n+1} - \omega_n = \pi/T$. Then,

$$f(t) = \frac{1}{2\pi} \sum_{n=-\infty}^{\infty} F(\omega_n)e^{i\omega_n t}\Delta\omega_n, \tag{6.1.4}$$

where

$$F(\omega_n) = \int_{-T}^{T} f(x)e^{-i\omega_n x}dx. \tag{6.1.5}$$

As $T \to \infty$, ω_n approaches a continuous variable ω, and $\Delta\omega_n$ may be interpreted as the infinitesimal $d\omega$. Therefore, ignoring any possible difficulties,[2]

$$f(t) = \frac{1}{2\pi} \int_{-\infty}^{\infty} F(\omega)e^{i\omega t}d\omega, \tag{6.1.6}$$

and

$$F(\omega) = \int_{-\infty}^{\infty} f(t)e^{-i\omega t}dt. \tag{6.1.7}$$

Equation 6.1.7 is the *Fourier transform* of $f(t)$ while Equation 6.1.6 is the *inverse Fourier transform* that converts a Fourier transform back to $f(t)$. Alternatively, we may combine Equation 6.1.6 and Equation 6.1.7 to yield the equivalent real form

$$f(t) = \frac{1}{\pi} \int_0^{\infty} \left\{ \int_{-\infty}^{\infty} f(x)\cos[\omega(t-x)]\,dx \right\} d\omega. \tag{6.1.8}$$

Hamming[3] suggested the following analog in understanding the Fourier transform. Let us imagine that $f(t)$ is a light beam. Then the Fourier transform, like a glass prism, breaks up the function into its component frequencies ω, each of intensity $F(\omega)$. In optics, the various frequencies are called colors; by analogy the Fourier transform gives us the color spectrum of a function. On the other hand, the inverse Fourier transform blends a function's spectrum to give back the original function.

Most signals encountered in practice have Fourier transforms because they are absolutely integrable, since they are bounded and of finite duration. However, there are some notable exceptions. Examples include the trigonometric functions sine and cosine.

---

[2] For a rigorous derivation, see Titchmarsh, E. C., 1948: *Introduction to the Theory of Fourier Integrals.* Oxford University Press, Chapter 1.

[3] Hamming, R. W., 1977: *Digital Filters.* Prentice-Hall, p. 136.

• **Example 6.1.1**

Let us find the Fourier transform for

$$f(t) = \begin{cases} 1, & |t| < a, \\ 0, & |t| > a. \end{cases} \qquad (6.1.9)$$

From the definition of the Fourier transform,

$$F(\omega) = \int_{-\infty}^{-a} 0\, e^{-i\omega t}\, dt + \int_{-a}^{a} 1\, e^{-i\omega t}\, dt + \int_{a}^{\infty} 0\, e^{-i\omega t}\, dt \qquad (6.1.10)$$

$$= \frac{e^{\omega a i} - e^{-\omega a i}}{\omega i} = \frac{2\sin(\omega a)}{\omega} = 2a\,\mathrm{sinc}(\omega a), \qquad (6.1.11)$$

where $\mathrm{sinc}(x) = \sin(x)/x$ is the *sinc function*.

From the definition of the inverse Fourier transform,

$$f(t) = \frac{1}{\pi} \int_{-\infty}^{\infty} \frac{\sin(\omega a)}{\omega} e^{i\omega t}\, d\omega = \begin{cases} 1, & |t| < a, \\ 0, & |t| > a. \end{cases} \qquad (6.1.12)$$

An important question is what value does $f(t)$ converge to in the limit as $t \to a$ and $t \to -a$? Because Fourier transforms are an extension of Fourier series, the behavior at a jump is the same as that for a Fourier series. For that reason, $f(a) = \frac{1}{2}[f(a^+) + f(a^-)] = \frac{1}{2}$ and $f(-a) = \frac{1}{2}[f(-a^+) + f(-a^-)] = \frac{1}{2}$. ☐

Although our previous example does not show it, the Fourier transform is, in general, a complex function. The most common method of displaying it is to plot its amplitude and phase on two separate graphs for all values of ω. Another problem here is the ratio of $0/0$ when $\omega = 0$. Applying L'Hôpital's rule, we find that $F(0) = 2$. Thus, we can plot the amplitude and phase of $F(\omega)$ using the MATLAB script:

```
clear; % clear all previous computations
omegan = [-20:0.01:-0.01]; % set up negative frequencies
omegap = [0.01:0.01:20]; % set up positive frequencies
% compute Fourier transform for negative frequencies
f_omegan = 2.*sin(omegan)./omegan;
% compute Fourier transform for positive frequencies
f_omegap = 2.*sin(omegap)./omegap;
% concatenate all of the frequencies
omega = [omegan,0,omegap];
% bring together the Fourier transforms found
%      at positive and negative frequencies
f_omega = [f_omegan,2,f_omegap];
amplitude = abs(f_omega); % compute the amplitude
phase = atan2(0,f_omega); % compute the phase
clf; % clear all previous figures
% plot frequency spectrum
subplot(2,1,1), plot(omega,amplitude)
% label amplitude plot
ylabel('|F(\omega)|/a','FontSize',15)
subplot(2,1,2), plot(omega,phase) % plot phase of transform
```

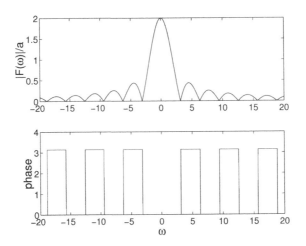

Figure 6.1.1: Graph of the Fourier transform for Equation 6.1.9.

```
ylabel('phase','FontSize',15) % label amplitude plot
xlabel('\omega','FontSize',15) % label x-axis.
```

Figure 6.1.1 shows the output from the MATLAB script. Of these two quantities, the amplitude is by far the more popular one and is given the special name of *frequency spectrum*.

- **Example 6.1.2: Dirac delta function**

Of the many functions that have a Fourier transform, a particularly important one is the *(Dirac) delta function*.[4] For example, in Section 6.6 we will use it to solve differential equations. We *define* it as the inverse of the Fourier transform $F(\omega) = 1$. Therefore,

$$\delta(t) = \frac{1}{2\pi} \int_{-\infty}^{\infty} e^{i\omega t} d\omega. \tag{6.1.13}$$

$\square$

To give some insight into the nature of the delta function, consider another band-limited transform

$$F_{\Omega}(\omega) = \begin{cases} 1, & |\omega| < \Omega, \\ 0, & |\omega| > \Omega, \end{cases} \tag{6.1.14}$$

where Ω is real and positive. Then,

$$f_{\Omega}(t) = \frac{1}{2\pi} \int_{-\Omega}^{\Omega} e^{i\omega t} d\omega = \frac{\Omega}{\pi} \frac{\sin(\Omega t)}{\Omega t}. \tag{6.1.15}$$

Figure 6.1.2 illustrates $f_{\Omega}(t)$ for a large value of Ω. We observe that as $\Omega \to \infty$, $f_{\Omega}(t)$ becomes very large near $t = 0$ as well as very narrow. On the other hand, $f_{\Omega}(t)$ rapidly approaches zero as $|t|$ increases. Therefore, the delta function is given by the limit

$$\delta(t) = \lim_{\Omega \to \infty} \frac{\sin(\Omega t)}{\pi t} = \begin{cases} \infty, & t = 0, \\ 0, & t \neq 0. \end{cases} \tag{6.1.16}$$

---

[4] Dirac, P. A. M., 1947: *The Principles of Quantum Mechanics*. Oxford University Press, Section 15.

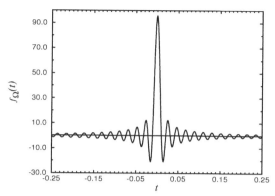

Figure 6.1.2: Graph of the function given in Equation 6.1.15 for $\Omega = 300$.

Because the Fourier transform of the delta function equals one,

$$\int_{-\infty}^{\infty} \delta(t)e^{-i\omega t}\,dt = 1. \tag{6.1.17}$$

Since Equation 6.1.17 must hold for any ω, we take $\omega = 0$ and find that

$$\int_{-\infty}^{\infty} \delta(t)\,dt = 1. \tag{6.1.18}$$

Thus, the area under the delta function equals unity. Taking Equation 6.1.16 into account, we can also write Equation 6.1.18 as

$$\int_{-a}^{b} \delta(t)\,dt = 1, \qquad a, b > 0. \tag{6.1.19}$$

Finally, from the law of the mean of integrals, we have the *sifting property* that

$$\int_{a}^{b} f(t)\delta(t - t_0)\,dt = f(t_0), \tag{6.1.20}$$

if $a < t_0 < b$. This property is given its name because $\delta(t - t_0)$ acts as a sieve, selecting from all possible values of $f(t)$ its value at $t = t_0$.

We can also use several other functions with equal validity to represent the delta function. These include the limiting case of the following rectangular or triangular distributions:

$$\delta(t) = \lim_{\epsilon \to 0} \begin{cases} \frac{1}{\epsilon}, & |t| < \frac{\epsilon}{2}, \\ 0, & |t| > \frac{\epsilon}{2}, \end{cases} \quad \text{or} \quad \delta(t) = \lim_{\epsilon \to 0} \begin{cases} \frac{1}{\epsilon}\left(1 - \frac{|t|}{\epsilon}\right), & |t| < \epsilon, \\ 0, & |t| > \epsilon, \end{cases} \tag{6.1.21}$$

and the Gaussian function:

$$\delta(t) = \lim_{\epsilon \to 0} \frac{\exp(-\pi t^2/\epsilon)}{\sqrt{\epsilon}}. \tag{6.1.22}$$

Note that the delta function is an even function. $\qquad\qquad\qquad\qquad\qquad\qquad\qquad$ $\square$

The Fourier Transforms of Some Commonly Encountered Functions

	$f(t)$, $|t| < \infty$	$F(\omega)$
1.	$e^{-at}H(t), \quad a > 0$	$\dfrac{1}{a + \omega i}$
2.	$e^{at}H(-t), \quad a > 0$	$\dfrac{1}{a - \omega i}$
3.	$te^{-at}H(t), \quad a > 0$	$\dfrac{1}{(a + \omega i)^2}$
4.	$te^{at}H(-t), \quad a > 0$	$\dfrac{-1}{(a - \omega i)^2}$
5.	$t^n e^{-at}H(t), \ \Re(a) > 0, \ n = 1, 2, \ldots$	$\dfrac{n!}{(a + \omega i)^{n+1}}$
6.	$e^{-a|t|}, \quad a > 0$	$\dfrac{2a}{\omega^2 + a^2}$
7.	$te^{-a|t|}, \quad a > 0$	$\dfrac{-4a\omega i}{(\omega^2 + a^2)^2}$
8.	$\dfrac{1}{1 + a^2 t^2}$	$\dfrac{\pi}{|a|} e^{-|\omega/a|}$
9.	$\dfrac{\cos(at)}{1 + t^2}$	$\dfrac{\pi}{2}\left(e^{-|\omega - a|} + e^{-|\omega + a|}\right)$
10.	$\dfrac{\sin(at)}{1 + t^2}$	$\dfrac{\pi}{2i}\left(e^{-|\omega - a|} - e^{-|\omega + a|}\right)$
11.	$\begin{cases} 1, & |t| < a \\ 0, & |t| > a \end{cases}$	$\dfrac{2\sin(\omega a)}{\omega}$
12.	$\dfrac{\sin(at)}{at}$	$\begin{cases} \pi/a, & |\omega| < a \\ 0, & |\omega| > a \end{cases}$
13.	$e^{-at^2}, \quad a > 0$	$\sqrt{\dfrac{\pi}{a}}\exp\left(-\dfrac{\omega^2}{4a}\right)$

Note: The Heaviside step function $H(t)$ is defined by Equation 6.1.35.

• Example 6.1.3: Fourier transform of Bessel functions

In Section 12.2 we will introduce *Bessel functions of the first kind of order n*, $J_n(t)$; they are useful in solving partial differential equations. For the present we can consider them tabulated functions, similar to sine and cosine, that are computed by power series. In the case of $J_0(t)$ it could also be computed via the definite integral:[5]

$$J_0(t) = \frac{1}{\pi} \int_0^\pi \cos[t\,\sin(\varphi)]\,d\varphi. \tag{6.1.23}$$

---

[5] Watson, G. N., 1966: *A Treatise on the Theory of Bessel Functions*. Cambridge University Press, p. 176, Equation (4) with $\nu = 0$.

If we introduce

$$\varphi = \begin{cases} \arcsin(\omega), & 0 \leq \varphi \leq \pi/2, \\ \pi - \arcsin(\omega), & \pi/2 < \varphi < \pi, \end{cases} \tag{6.1.24}$$

we can rewrite Equation 6.1.23 as

$$J_0(t) = \frac{2}{\pi} \int_0^1 \frac{\cos(t\omega)}{\sqrt{1-\omega^2}} \, d\omega = \frac{1}{\pi} \int_{-1}^1 \frac{\cos(t\omega)}{\sqrt{1-\omega^2}} \, d\omega = \frac{1}{\pi} \int_{-1}^1 \frac{1}{\sqrt{1-\omega^2}} e^{it\omega} \, d\omega \tag{6.1.25}$$

$$= \frac{1}{2\pi} \int_{-\infty}^{\infty} H(1-|\omega|) \frac{2}{\sqrt{1-\omega^2}} e^{it\omega} \, d\omega. \tag{6.1.26}$$

From the form of Equation 6.1.26, we recognize that $J_0(t)$ is the inverse Fourier transform of $2H(1-|\omega|)/\sqrt{1-\omega^2}$. Note that the Fourier transform is nonzero if $|\omega| < 1$. This is an example of a *bandlimited Fourier transform*.

For the general case of $J_n(t)$, H. O. Bèca[6] proved that

$$\mathcal{F}[J_{2m}(\omega)] = \frac{2}{\sqrt{1-\omega^2}} \cos[2m \arcsin(\omega)] H(1-|\omega|), \tag{6.1.27}$$

and

$$\mathcal{F}[J_{2m+1}(\omega)] = \frac{2i}{\sqrt{1-\omega^2}} \sin[(2m+1) \arcsin(\omega)] H(1-|\omega|), \tag{6.1.28}$$

where $m = 0, \pm 1, \pm 2, \ldots$. $\qquad\qquad\square$

- **Example 6.1.4: Multiple Fourier transforms**

The concept of Fourier transforms can be extended to multivariable functions. Consider a two-dimensional function $f(x,y)$. Then, holding y constant,

$$G(\xi, y) = \int_{-\infty}^{\infty} f(x,y) e^{-i\xi x} \, dx. \tag{6.1.29}$$

Then, holding ξ constant,

$$F(\xi, \eta) = \int_{-\infty}^{\infty} G(\xi, y) e^{-i\eta y} \, dy. \tag{6.1.30}$$

Therefore, the double Fourier transform of $f(x,y)$ is

$$F(\xi, \eta) = \int_{-\infty}^{\infty} \int_{-\infty}^{\infty} f(x,y) e^{-i(\xi x + \eta y)} \, dx \, dy, \tag{6.1.31}$$

assuming that the integral exists.

In a similar manner, we can compute $f(x,y)$ given $F(\xi, \eta)$ by reversing the process. Starting with

$$G(\xi, y) = \frac{1}{2\pi} \int_{-\infty}^{\infty} F(\xi, \eta) e^{i\eta y} \, d\eta, \tag{6.1.32}$$

---

[6] Bèca, H. O., 1980: An orthogonal set based on Bessel functions of the first kind. *Univ. Beograd. Publ. Elektrotehn. Fak., Ser. Mat. Fiz.*, No. 695, 85–90.

followed by

$$f(x,y) = \frac{1}{2\pi} \int_{-\infty}^{\infty} G(\xi, y)\, e^{i\xi x}\, d\xi, \tag{6.1.33}$$

we find that

$$f(x,y) = \frac{1}{4\pi^2} \int_{-\infty}^{\infty} \int_{-\infty}^{\infty} F(\xi, \eta)\, e^{i(\xi x + \eta y)}\, d\xi\, d\eta. \tag{6.1.34}$$

$\square$

• **Example 6.1.5: Computation of Fourier transforms using** MATLAB

The Heaviside (unit) step function is a piecewise continuous function defined by

$$H(t-a) = \begin{cases} 1, & t > a, \\ 0, & t < a, \end{cases} \tag{6.1.35}$$

where $a \geq 0$. We will have much to say about this very useful function in the chapter on Laplace transforms. Presently we will use it to express functions whose definition changes over different ranges of t. For example, the "top hat" function Equation 6.1.9 can be rewritten $f(t) = H(t+a) - H(t-a)$. We can see that this is correct by considering various ranges of t. For example, if $t < -a$, both step functions equal zero and $f(t) = 0$. On the other hand, if $t > a$, both step functions equal one and again $f(t) = 0$. Finally, for $-a < t < a$, the first step function equals one while the second one equals zero. In this case, $f(t) = 1$. Therefore, $f(t) = H(t+a) - H(t-a)$ is equivalent to Equation 6.1.9.

This ability to rewrite functions in terms of the step function is crucial if you want to use MATLAB to compute the Fourier transform via the MATLAB routine `fourier`. For example, how would we compute the Fourier transform of the signum function? The MATLAB commands:

```
>> syms omega t; syms a positive
>> fourier('Heaviside(t+a)-Heaviside(t-a)',t,omega)
>> simplify(ans)
```

yields

```
ans =
2*sin(a*omega)/omega
```

the correct answer. $\square$

• **Example 6.1.6: Numerical computation of Fourier transforms using** MATLAB

As our table on Fourier transform suggests, there are relatively few functions that possess a Fourier transform and even fewer for which we can write it down. For that reason we must numerically integrate Equation 6.1.7. The idea here is that we replace the integration from $-\infty$ to ∞ with one from $-T$ to T and then use the trapezoid rule. We must choose T large enough so that it captures the behavior of $f(t)$.

To illustrate this technique using MATLAB, let us use the function $f(t) = te^{-t}H(t)$ which has the transform $F(\omega) = 1/(1 + \omega i)^2$. We begin by tabulating the function for times:

```
N = 20; dt = 0.1; t = [-N/2:N/2]*dt;

for n = 1:N+1
```

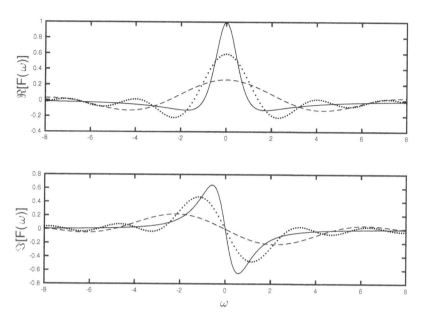

Figure 6.1.3: The real and imaginary part of the Fourier transform $F(\omega)$ for $f(t) = te^{-t}H(t)$ computed by numerically integrating the definition for the Fourier transform, Equation 6.1.7. The upper and lower limits have been replaced by $\pm T$, respectively. The solid line gives the exact result, the dashed line $T = 1$, and the dotted line $T = 2$. The function $f(t)$ was sampled at the interval $\Delta t = 0.1$.

```
if t(n) < 0
f(n) = 0;
else
f(n) = t(n) * exp(-t(n));
end

end
```

Having computed `f(n)`, we do the numerical integral by the trapezoid rule using MATLAB's procedure `trapz`:

```
k = 0;
for omega = -8:0.1:8
k = k+1;
Omega(k) = omega;
F(k) = trapz(t,f.*exp(-i*omega*t)); % the Fourier transform
end

FR = real(F); % ℜ[F(ω)]
FI = imag(F); % ℑ[F(ω)]
```

Figure 5.1.3 compares results from the numerical integration when $T = 1$ (dashed line) and $T = 2$ (dotted line) with the exact answer (solid line). Because the integrand rapidly approaches zero for increasing t the results are rather good for a relatively small T. The optimal choice for T is dictated by the behavior of $f(t)$ with t.

An alternative method of computing the Fourier transform uses the fast Fourier transform. Recall that we derived the Fourier transform by considering a periodic function $f(t)$ defined on the interval $(-T, T)$. The Fourier transform equals the Fourier coefficients in the

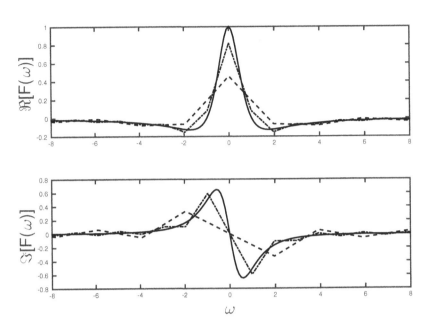

Figure 6.1.4: The real and imaginary part of the Fourier transform $F(\omega)$ for $f(t) = te^{-t}H(t)$ computed using the fast Fourier transform. The solid line gives the exact result, the dashed line $T = 1$, and the dotted line $T = 2$. The function $f(t)$ was sampled at the interval $\Delta t = 0.1$.

limit $T \to \infty$. As we showed in Section 5.7 we can compute these Fourier coefficients very efficiently using the fast Fourier transform.

To illustrate this technique, we again will compute the Fourier transform for $f(t) = te^{-t}H(t)$. Our first step is to compute $f(t)$ over the interval $[0, 2T]$ using the periodicity condition that $f(t + 2T) = f(t)$. The corresponding MATLAB code to create $\mathbf{f(n)}$ for $T = 1.6$ is:

```
N = 32; dt = 0.1; t = [0:N-1]*dt;
x = zeros(1,N);
for n = 1:N/2
x(n) = t(n) * exp(-t(n));
end
```

Next, we use MATLAB's intrinsic function `fft` to compute the Fourier coefficients over the interval $[0, 2\Omega)$, where $\Omega = 1/(2\Delta t)$ and Δt equals the time interval that we have sampled $f(t)$. The Fourier transform is simply `X = fft(x) * dt`.

Because we generally prefer to have the Fourier transform over the interval $[-\Omega, \Omega]$, we use MATLAB's intrinsic function `fftshift` to shift X to that interval. The MATLAB code is:

```
M = length(X)
fshift = [-M/2:M/2-1] / (M*dt); % frequency of F(ω)
yshift = fftshift(X); % F(ω)
Omega1 = 2*pi*fshift; % ω
```

Figure 5.1.4 presents the results when $T = 1.6$ (dashed line) and $T = 3.2$ (dotted line) with the exact answer (solid line).

Problems

1. (a) Show that the Fourier transform of

$$f(t) = e^{-a|t|}, \qquad a > 0, \qquad \text{is} \qquad F(\omega) = \frac{2a}{\omega^2 + a^2}.$$

Using MATLAB, plot the amplitude and phase spectra for this transform.

(b) Use MATLAB's `fourier` to find $F(\omega)$.

2. (a) Show that the Fourier transform of

$$f(t) = te^{-a|t|}, \quad a > 0, \qquad \text{is} \qquad F(\omega) = -\frac{4a\omega i}{(\omega^2 + a^2)^2}.$$

Using MATLAB, plot the amplitude and phase spectra for this transform.

(b) Use MATLAB's `fourier` to find $F(\omega)$.

3. (a) Show that the Fourier transform of

$$f(t) = e^{-at^2}, \quad a > 0, \qquad \text{is} \qquad F(\omega) = \sqrt{\frac{\pi}{a}} \exp\left(-\frac{\omega^2}{4a}\right).$$

Using MATLAB, plot the amplitude and phase spectra for this transform.

(b) Use MATLAB's `fourier` to find $F(\omega)$.

4. (a) Show that the Fourier transform of

$$f(t) = \begin{cases} e^{2t}, & t < 0, \\ e^{-t}, & t > 0, \end{cases} \qquad \text{is} \qquad F(\omega) = \frac{3}{(2 - i\omega)(1 + i\omega)}.$$

Using MATLAB, plot the amplitude and phase spectra for this transform.

(b) Rewrite $f(t)$ in terms of step functions. Then use MATLAB's `fourier` to find $F(\omega)$.

5. (a) Show that the Fourier transform of

$$f(t) = \begin{cases} e^{-(1+i)t}, & t > 0, \\ -e^{(1-i)t}, & t < 0, \end{cases} \qquad \text{is} \qquad F(\omega) = \frac{-2i(\omega + 1)}{(\omega + 1)^2 + 1}.$$

Using MATLAB, plot the amplitude and phase spectra for this transform.

(b) Rewrite $f(t)$ in terms of step functions. Then use MATLAB's `fourier` to find $F(\omega)$.

6. (a) Show that the Fourier transform of

$$f(t) = \begin{cases} \cos(at), & |t| < 1, \\ 0, & |t| > 1, \end{cases} \qquad \text{is} \qquad F(\omega) = \frac{\sin(\omega - a)}{\omega - a} + \frac{\sin(\omega + a)}{\omega + a}.$$

Using MATLAB, plot the amplitude and phase spectra for this transform.

(b) Rewrite $f(t)$ in terms of step functions. Then use MATLAB's `fourier` to find $F(\omega)$.

7. (a) Show that the Fourier transform of

$$f(t) = \begin{cases} \sin(t), & 0 \le t < 1, \\ 0, & \text{otherwise}, \end{cases}$$

is

$$F(\omega) = -\frac{1}{2} \left[\frac{1 - \cos(\omega - 1)}{\omega - 1} + \frac{\cos(\omega + 1) - 1}{\omega + 1} \right] - \frac{i}{2} \left[\frac{\sin(\omega - 1)}{\omega - 1} - \frac{\sin(\omega + 1)}{\omega + 1} \right].$$

Using MATLAB, plot the amplitude and phase spectra for this transform.

(b) Rewrite $f(t)$ in terms of step functions. Then use MATLAB's `fourier` to find $F(\omega)$.

8. (a) Show that the Fourier transform of

$$f(t) = \begin{cases} t/a, & |t| < a, \\ 0, & |t| > a, \end{cases} \qquad \text{is} \qquad F(\omega) = \frac{2i \cos(\omega a)}{\omega} - \frac{2i \sin(\omega a)}{\omega^2 a}.$$

Using MATLAB, plot the amplitude and phase spectra for this transform.

(b) Rewrite $f(t)$ in terms of step functions. Then use MATLAB's `fourier` to find $F(\omega)$.

9. (a) Show that the Fourier transform of

$$f(t) = \begin{cases} (t/a)^2, & |t| < a, \\ 0, & |t| > a, \end{cases} \qquad \text{is} \qquad F(\omega) = \frac{4 \cos(\omega a)}{\omega^2 a} - \frac{4 \sin(\omega a)}{\omega^3 a^2} + \frac{2 \sin(\omega a)}{\omega}.$$

Using MATLAB, plot the amplitude and phase spectra for this transform.

(b) Rewrite $f(t)$ in terms of step functions. Then use MATLAB's `fourier` to find $F(\omega)$.

10. (a) Show that the Fourier transform of

$$f(t) = \begin{cases} 1 - t/\tau, & 0 \le t < 2\tau, \\ 0, & \text{otherwise}, \end{cases} \qquad \text{is} \qquad F(\omega) = \frac{2e^{-i\omega\tau}}{i\omega} \left[\frac{\sin(\omega\tau)}{\omega\tau} - \cos(\omega\tau) \right].$$

Using MATLAB, plot the amplitude and phase spectra for this transform.

(b) Rewrite $f(t)$ in terms of step functions. Then use MATLAB's `fourier` to find $F(\omega)$.

11. (a) Show that the Fourier transform of

$$f(t) = \begin{cases} 1 - (t/a)^2, & |t| \le a, \\ 0, & |t| \ge a, \end{cases} \qquad \text{and} \qquad F(\omega) = \frac{4 \sin(\omega a) - 4a\omega \cos(\omega a)}{a^2 \omega^3}.$$

Using MATLAB, plot the amplitude and phase spectra for this transform.

(b) Rewrite $f(t)$ in terms of step functions. Then use MATLAB's `fourier` to find $F(\omega)$.

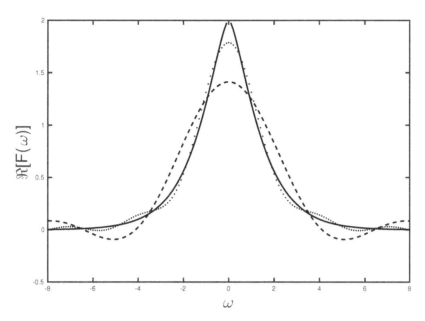

Figure 6.1.5: The real part of the Fourier transform $F(\omega)$ for $f(t) = 1/(t^2 + 1)^{3/2}$ using the trapezoidal rule to numerically integrate the definition for the Fourier transform, Equation 6.1.7. The imaginary part equals zero. The solid line gives the exact result from Problem 12, the dashed line $T = 1$, and the dotted line $T = 2$. The function $f(t)$ was sampled at the interval $\Delta t = 0.1$.

12. The integral representation[7] of the modified Bessel function $K_\nu(\cdot)$ is

$$K_\nu\left(a|\omega|\right) = \frac{\Gamma\left(\nu + \frac{1}{2}\right)(2a)^\nu}{|\omega|^\nu \Gamma\left(\frac{1}{2}\right)} \int_0^\infty \frac{\cos(\omega t)}{(t^2 + a^2)^{\nu + 1/2}}\, dt,$$

where $\Gamma(\cdot)$ is the gamma function, $\nu \geq 0$ and $a > 0$. Use this relationship to show that

$$\mathcal{F}\left[\frac{1}{(t^2 + a^2)^{\nu + 1/2}}\right] = \frac{2|\omega|^\nu \Gamma\left(\frac{1}{2}\right) K_\nu\left(a|\omega|\right)}{\Gamma\left(\nu + \frac{1}{2}\right)(2a)^\nu}.$$

Use MATLAB to verify your result by numerical integration.

13. Show that

$$\mathcal{F}[H(a - |t|)] = \frac{2\sin(\omega a)}{\omega}, \qquad a > 0.$$

14. Show that

$$\int_a^b \tau\,\delta(t - \tau)\,d\tau = t\left[H(t - a) - H(t - b)\right].$$

Hint: Use integration by parts.

15. For the real function $f(t)$ with Fourier transform $F(\omega)$, prove that $|F(\omega)| = |F(-\omega)|$ and the phase of $F(\omega)$ is an odd function of ω.

---

[7] Watson, G. N., 1966: *A Treatise on the Theory of Bessel Functions*. Cambridge University Press, p. 185.

6.2 FOURIER TRANSFORMS CONTAINING THE DELTA FUNCTION

In the previous section we stressed the fact that such simple functions as cosine and sine are not absolutely integrable. Does this mean that these functions do not possess a Fourier transform? In this section we shall show that certain functions can still have a Fourier transform even though we cannot compute them directly.

The reason why we can find the Fourier transform of certain functions that are not absolutely integrable lies with the introduction of the delta function because

$$\int_{-\infty}^{\infty} \delta(\omega - \omega_0)e^{it\omega}\, d\omega = e^{i\omega_0 t} \tag{6.2.1}$$

for all t. Thus, the inverse of the Fourier transform $\delta(\omega - \omega_0)$ is the complex exponential $e^{i\omega_0 t}/2\pi$ or

$$\mathcal{F}\left(e^{i\omega_0 t}\right) = 2\pi\delta(\omega - \omega_0). \tag{6.2.2}$$

This immediately yields the result that

$$\mathcal{F}(1) = 2\pi\delta(\omega), \tag{6.2.3}$$

if we set $\omega_0 = 0$. Thus, the Fourier transform of 1 is an impulse at $\omega = 0$ with weight 2π. Because the Fourier transform equals zero for all $\omega \neq 0$, $f(t) = 1$ does not contain a nonzero frequency and is consequently a DC signal.

Another set of transforms arises from Euler's formula because we have that

$$\mathcal{F}[\sin(\omega_0 t)] = \left[\mathcal{F}\left(e^{i\omega_0 t}\right) - \mathcal{F}\left(e^{-i\omega_0 t}\right)\right]/(2i) \tag{6.2.4}$$

$$= \pi\left[\delta(\omega - \omega_0) - \delta(\omega + \omega_0)\right]/i \tag{6.2.5}$$

$$= -\pi i\delta(\omega - \omega_0) + \pi i\delta(\omega + \omega_0) \tag{6.2.6}$$

and

$$\mathcal{F}[\cos(\omega_0 t)] = \tfrac{1}{2}\left[\mathcal{F}\left(e^{i\omega_0 t}\right) + \mathcal{F}\left(e^{-i\omega_0 t}\right)\right] = \pi\left[\delta(\omega - \omega_0) + \delta(\omega + \omega_0)\right]. \tag{6.2.7}$$

Note that although the amplitude spectra of $\sin(\omega_0 t)$ and $\cos(\omega_0 t)$ are the same, their phase spectra are different.

Let us consider the Fourier transform of any arbitrary periodic function. Recall that any such function $f(t)$ with period $2L$ can be rewritten as the complex Fourier series

$$f(t) = \sum_{n=-\infty}^{\infty} c_n e^{in\omega_0 t}, \tag{6.2.8}$$

where $\omega_0 = \pi/L$. The Fourier transform of $f(t)$ is

$$F(\omega) = \mathcal{F}[f(t)] = \sum_{n=-\infty}^{\infty} 2\pi c_n \delta(\omega - n\omega_0). \tag{6.2.9}$$

Therefore, the Fourier transform of any arbitrary periodic function is a sequence of impulses with weight $2\pi c_n$ located at $\omega = n\omega_0$ with $n = 0, \pm 1, \pm 2, \ldots$. Thus, the Fourier series and transform of a periodic function are closely related.

• **Example 6.2.1: Fourier transform of the sign function**

Consider the sign function

$$\operatorname{sgn}(t) = \begin{cases} 1, & t > 0, \\ 0, & t = 0, \\ -1, & t < 0. \end{cases} \tag{6.2.10}$$

The function is not absolutely integrable. However, let us approximate it by $e^{-\epsilon|t|}\operatorname{sgn}(t)$, where ϵ is a small positive number. This new function is absolutely integrable and we have that

$$\mathcal{F}[\operatorname{sgn}(t)] = \lim_{\epsilon \to 0} \left[-\int_{-\infty}^{0} e^{\epsilon t} e^{-i\omega t}\, dt + \int_{0}^{\infty} e^{-\epsilon t} e^{-i\omega t}\, dt \right] = \lim_{\epsilon \to 0} \left(\frac{-1}{\epsilon - i\omega} + \frac{1}{\epsilon + i\omega} \right). \tag{6.2.11}$$

If $\omega \neq 0$, Equation 6.2.11 equals $2/i\omega$. If $\omega = 0$, Equation 6.2.11 equals 0 because

$$\lim_{\epsilon \to 0} \left(\frac{-1}{\epsilon} + \frac{1}{\epsilon} \right) = 0. \tag{6.2.12}$$

Thus, we conclude that

$$\mathcal{F}[\operatorname{sgn}(t)] = \begin{cases} 2/i\omega, & \omega \neq 0, \\ 0, & \omega = 0. \end{cases} \tag{6.2.13}$$

$\square$

• **Example 6.2.2: Fourier transform of the step function**

An important function in transform methods is the *(Heaviside) step function*

$$H(t) = \begin{cases} 1, & t > 0, \\ 0, & t < 0. \end{cases} \tag{6.2.14}$$

In terms of the sign function it can be written

$$H(t) = \tfrac{1}{2} + \tfrac{1}{2}\operatorname{sgn}(t). \tag{6.2.15}$$

Because the Fourier transforms of 1 and $\operatorname{sgn}(t)$ are $2\pi\delta(\omega)$ and $2/i\omega$, respectively, we have that

$$\mathcal{F}[H(t)] = \pi\delta(\omega) + \frac{1}{i\omega}. \tag{6.2.16}$$

These transforms are used in engineering but the presence of the delta function requires extra care to ensure their proper use.

Problems

1. Show that the Fourier transform of a constant K is $2\pi K\delta(\omega)$.

2. Verify that

$$\mathcal{F}[\sin(\omega_0 t)H(t)] = \frac{\omega_0}{\omega_0^2 - \omega^2} + \frac{\pi i}{2}[\delta(\omega + \omega_0) - \delta(\omega - \omega_0)].$$

3. Verify that

$$\mathcal{F}[\cos(\omega_0 t) H(t)] = \frac{i\omega}{\omega_0^2 - \omega^2} + \frac{\pi}{2}[\delta(\omega + \omega_0) + \delta(\omega - \omega_0)].$$

4. Using the definition of Fourier transforms and Equation 6.2.16, show that

$$\int_0^\infty e^{-i\omega t}\, dt = \pi\delta(\omega) - \frac{i}{\omega}, \qquad \text{or} \qquad \int_0^\infty e^{i\omega t}\, dt = \pi\delta(\omega) + \frac{i}{\omega}.$$

5. Following Example 6.2.1, show that

$$\mathcal{F}[\text{sgn}(t)\sin(\omega_0 t)] = \frac{2\omega_0}{\omega_0^2 - \omega^2}, \qquad \text{and} \qquad \mathcal{F}[\text{sgn}(t)\cos(\omega_0 t)] = \frac{2\omega i}{\omega_0^2 - \omega^2}.$$

6.3 PROPERTIES OF FOURIER TRANSFORMS

In principle we can compute any Fourier transform from its definition. However, it is far more efficient to derive some simple relationships that relate transforms to each other. This is the purpose of this section.

> ### Linearity

If $f(t)$ and $g(t)$ are functions with Fourier transforms $F(\omega)$ and $G(\omega)$, respectively, then

$$\mathcal{F}[c_1 f(t) + c_2 g(t)] = c_1 F(\omega) + c_2 G(\omega), \tag{6.3.1}$$

where c_1 and c_2 are (real or complex) constants.

This result follows from the integral definition

$$\mathcal{F}[c_1 f(t) + c_2 g(t)] = \int_{-\infty}^\infty [c_1 f(t) + c_2 g(t)] e^{-i\omega t}\, dt \tag{6.3.2}$$

$$= c_1 \int_{-\infty}^\infty f(t) e^{-i\omega t}\, dt + c_2 \int_{-\infty}^\infty g(t) e^{-i\omega t}\, dt \tag{6.3.3}$$

$$= c_1 F(\omega) + c_2 G(\omega). \tag{6.3.4}$$

> ### Time shifting

If $f(t)$ is a function with a Fourier transform $F(\omega)$, then $\mathcal{F}[f(t - \tau)] = e^{-i\omega\tau} F(\omega)$.
This follows from the definition of the Fourier transform

$$\mathcal{F}[f(t - \tau)] = \int_{-\infty}^\infty f(t - \tau) e^{-i\omega t}\, dt = \int_{-\infty}^\infty f(x) e^{-i\omega(x+\tau)}\, dx \tag{6.3.5}$$

$$= e^{-i\omega\tau} \int_{-\infty}^\infty f(x) e^{-i\omega x}\, dx = e^{-i\omega\tau} F(\omega). \tag{6.3.6}$$

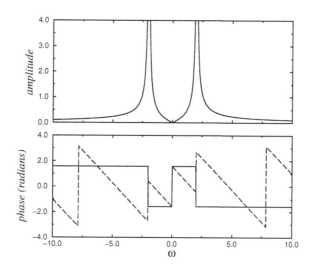

Figure 6.3.1: The amplitude and phase spectra of the Fourier transform for $\cos(2t) \, H(t)$ (solid line) and $\cos[2(t-1)]H(t-1)$ (dashed line). The amplitude becomes infinite at $\omega = \pm 2$.

• **Example 6.3.1**

The Fourier transform of $f(t) = \cos(at)H(t)$ is $F(\omega) = i\omega/(a^2 - \omega^2) + \pi[\delta(\omega + a) + \delta(\omega - a)]/2$. Therefore,

$$\mathcal{F}\{\cos[a(t-k)]H(t-k)\} = e^{-ik\omega}\mathcal{F}[\cos(at)H(t)], \qquad (6.3.7)$$

or

$$\mathcal{F}\{\cos[a(t-k)]H(t-k)\} = \frac{i\omega e^{-ik\omega}}{a^2 - \omega^2} + \frac{\pi}{2}e^{-ik\omega}[\delta(\omega + a) + \delta(\omega - a)]. \qquad (6.3.8)$$

In Figure 6.3.1 we present the amplitude and phase spectra for $\cos(2t) \, H(t)$ (the solid line) while the dashed line gives these spectra for $\cos[2(t-1)]H(t-1)$. This figure shows that the amplitude spectra are identical (why?) while the phase spectra are considerably different.

□

Scaling factor

Let $f(t)$ be a function with a Fourier transform $F(\omega)$ and k be a real, nonzero constant. Then $\mathcal{F}[f(kt)] = F(\omega/k)/|k|$.

From the definition of the Fourier transform:

$$\mathcal{F}[f(kt)] = \int_{-\infty}^{\infty} f(kt)e^{-i\omega t}dt = \frac{1}{|k|}\int_{-\infty}^{\infty} f(x)e^{-i(\omega/k)x}dx = \frac{1}{|k|}F\left(\frac{\omega}{k}\right). \qquad (6.3.9)$$

• **Example 6.3.2**

The Fourier transform of $f(t) = e^{-t}H(t)$ is $F(\omega) = 1/(1 + \omega i)$. Therefore, the Fourier transform for $f(at) = e^{-at}H(t)$, $a > 0$, is

$$\mathcal{F}[f(at)] = \left(\frac{1}{a}\right)\left(\frac{1}{1 + i\omega/a}\right) = \frac{1}{a + \omega i}. \qquad (6.3.10)$$

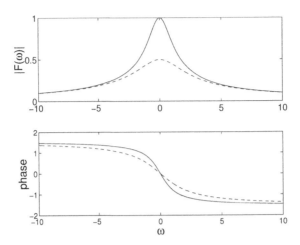

Figure 6.3.2: The amplitude and phase spectra of the Fourier transform for $e^{-t}H(t)$ (solid line) and $e^{-2t}H(t)$ (dashed line).

To illustrate this scaling property we use the MATLAB script:

```
clear; % clear all previous computations
omega = [-10:0.01:10]; % set up frequencies
% real part of transform with a = 1
f1r_omega = 1./(1+omega.*omega);
% imaginary part of transform with a = 1
f1i_omega = - omega./(1+omega.*omega);
% real part of transform with a = 2
f2r_omega = 2./(4+omega.*omega);
% imaginary part of transform with a = 2
f2i_omega = - omega./(4+omega.*omega);
% compute the amplitude of the first transform
ampl1 = sqrt(f1r_omega.*f1r_omega + f1i_omega.*f1i_omega);
% compute the amplitude of the second transform
ampl2 = sqrt(f2r_omega.*f2r_omega + f2i_omega.*f2i_omega);
% compute phase of first transform
phase1 = atan2(f1i_omega,f1r_omega);
% compute phase of second transform
phase2 = atan2(f2i_omega,f2r_omega);
clf; % clear all previous figures
% plot amplitudes of Fourier transforms
subplot(2,1,1), plot(omega,ampl1,omega,ampl2,'--')
ylabel('|F(\omega)|','FontSize',15) % label amplitude plot
% plot phases of Fourier transforms
subplot(2,1,2), plot(omega,phase1,omega,phase2,'--')
ylabel('phase','FontSize',15) % label amplitude plot
xlabel('\omega','FontSize',15) % label x-axis
```

to plot the amplitude and phase when $a = 1$ and $a = 2$. Figure 6.3.2 shows the results from the MATLAB script: The amplitude spectra decreased by a factor of two for $e^{-2t}H(t)$ compared to $e^{-t}H(t)$ while the differences in the phase are smaller.   □

Symmetry

If the function $f(t)$ has the Fourier transform $F(\omega)$, then $\mathcal{F}[F(t)] = 2\pi f(-\omega)$.
From the definition of the inverse Fourier transform,

$$f(t) = \frac{1}{2\pi} \int_{-\infty}^{\infty} F(\omega)e^{i\omega t}d\omega = \frac{1}{2\pi} \int_{-\infty}^{\infty} F(x)e^{ixt}dx. \qquad (6.3.11)$$

Then

$$2\pi f(-\omega) = \int_{-\infty}^{\infty} F(x)e^{-i\omega x}dx = \int_{-\infty}^{\infty} F(t)e^{-i\omega t}dt = \mathcal{F}[F(t)]. \qquad (6.3.12)$$

• **Example 6.3.3**

The Fourier transform of $1/(1+t^2)$ is $\pi e^{-|\omega|}$. Therefore,

$$\mathcal{F}\left(\pi e^{-|t|}\right) = \frac{2\pi}{1+\omega^2} \qquad \text{or} \qquad \mathcal{F}\left(e^{-|t|}\right) = \frac{2}{1+\omega^2}. \qquad (6.3.13)$$

$\square$

Derivatives of functions

Let $f^{(k)}(t), k = 0, 1, 2, \ldots, n-1$, be continuous and $f^{(n)}(t)$ be piecewise continuous. Let
$|f^{(k)}(t)| \le Ke^{-bt}, b > 0, 0 \le t < \infty; |f^{(k)}(t)| \le Me^{at}, a > 0, -\infty < t \le 0, k = 0, 1, \ldots, n$.
Then, $\mathcal{F}[f^{(n)}(t)] = (i\omega)^n F(\omega)$.
We begin by noting that if the transform $\mathcal{F}[f'(t)]$ exists, then

$$\mathcal{F}[f'(t)] = \int_{-\infty}^{\infty} f'(t)e^{-i\omega t}dt = \int_{-\infty}^{\infty} f'(t)e^{\omega_i t}[\cos(\omega_r t) - i\sin(\omega_r t)]\,dt \qquad (6.3.14)$$

$$= (-\omega_i + i\omega_r) \int_{-\infty}^{\infty} f(t)e^{\omega_i t}[\cos(\omega_r t) - i\sin(\omega_r t)]\,dt \qquad (6.3.15)$$

$$= i\omega \int_{-\infty}^{\infty} f(t)e^{-i\omega t}dt = i\omega F(\omega). \qquad (6.3.16)$$

Finally,

$$\mathcal{F}[f^{(n)}(t)] = i\omega \mathcal{F}[f^{(n-1)}(t)] = (i\omega)^2 \mathcal{F}[f^{(n-2)}(t)] = \cdots = (i\omega)^n F(\omega). \qquad (6.3.17)$$

• **Example 6.3.4**

The Fourier transform of $f(t) = 1/(1+t^2)$ is $F(\omega) = \pi e^{-|\omega|}$. Therefore,

$$\mathcal{F}\left[-\frac{2t}{(1+t^2)^2}\right] = i\omega\pi e^{-|\omega|}, \qquad \text{or} \qquad \mathcal{F}\left[\frac{t}{(1+t^2)^2}\right] = -\frac{i\omega\pi}{2}e^{-|\omega|}. \qquad (6.3.18)$$

$\square$

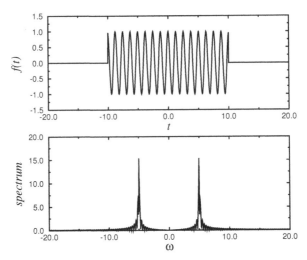

Figure 6.3.3: The (amplitude) spectrum of a rectangular pulse Equation 6.1.9 with a half width $a = 10$ that has been modulated with $\cos(5t)$.

> Modulation

In communications a popular method of transmitting information is by *amplitude modulation* (AM). In this process the signal is carried according to the expression $f(t)e^{i\omega_0 t}$, where ω_0 is the *carrier frequency* and $f(t)$ is an arbitrary function of time whose amplitude spectrum peaks at some frequency that is usually small compared to ω_0. We now show that the Fourier transform of $f(t)e^{i\omega_0 t}$ is $F(\omega - \omega_0)$, where $F(\omega)$ is the Fourier transform of $f(t)$.

We begin by using the definition of the Fourier transform, or

$$\mathcal{F}[f(t)e^{i\omega_0 t}] = \int_{-\infty}^{\infty} f(t)e^{i\omega_0 t}e^{-i\omega t}dt = \int_{-\infty}^{\infty} f(t)e^{-i(\omega-\omega_0)t}dt = F(\omega - \omega_0). \qquad \textbf{(6.3.19)}$$

Therefore, if we have the spectrum of a particular function $f(t)$, then the Fourier transform of the modulated function $f(t)e^{i\omega_0 t}$ is the same as that for $f(t)$ except that it is now centered on the frequency ω_0 rather than on the zero frequency.

• **Example 6.3.5**

Let us determine the Fourier transform of a square pulse modulated by a cosine wave as shown in Figures 6.3.3 and 6.3.4. Because $\cos(\omega_0 t) = \frac{1}{2}[e^{i\omega_0 t} + e^{-i\omega_0 t}]$ and the Fourier transform of a square pulse is $F(\omega) = 2\sin(\omega a)/\omega$,

$$\mathcal{F}[f(t)\cos(\omega_0 t)] = \frac{\sin[(\omega - \omega_0)a]}{\omega - \omega_0} + \frac{\sin[(\omega + \omega_0)a]}{\omega + \omega_0}. \qquad \textbf{(6.3.20)}$$

Therefore, the Fourier transform of the modulated pulse equals one half of the sum of the Fourier transform of the pulse centered on ω_0 and $-\omega_0$. See Figures 6.3.3 and 6.3.4.

In many practical situations, $\omega_0 \gg \pi/a$. In this case we may treat each term as completely independent from the other; the contribution from the peak at $\omega = \omega_0$ has a negligible effect on the peak at $\omega = -\omega_0$.   □

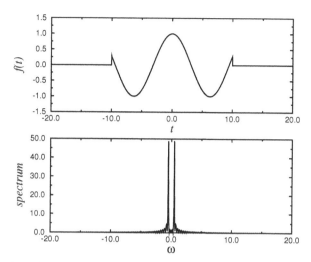

Figure 6.3.4: The (amplitude) spectrum of a rectangular pulse Equation 6.1.9 with a half width $a = 10$ that has been modulated with $\cos(t/2)$.

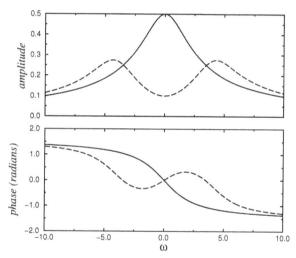

Figure 6.3.5: The amplitude and phase spectra of the Fourier transform for $e^{-2t}\,H(t)$ (solid line) and $e^{-2t}\cos(4t)H(t)$ (dashed line).

- **Example 6.3.6**

The Fourier transform of $f(t) = e^{-bt}H(t)$ is $F(\omega) = 1/(b + i\omega)$. Therefore,

$$\mathcal{F}[e^{-bt}\cos(at)H(t)] = \tfrac{1}{2}\mathcal{F}\left(e^{iat}e^{-bt} + e^{-iat}e^{-bt}\right) \qquad (6.3.21)$$

$$= \frac{1}{2}\left(\left.\frac{1}{b + i\omega'}\right|_{\omega'=\omega-a} + \left.\frac{1}{b + i\omega'}\right|_{\omega'=\omega+a}\right) \qquad (6.3.22)$$

$$= \frac{1}{2}\left[\frac{1}{(b + i\omega) - ai} + \frac{1}{(b + i\omega) + ai}\right] \qquad (6.3.23)$$

$$= \frac{b + i\omega}{(b + i\omega)^2 + a^2}. \qquad (6.3.24)$$

We illustrate this result using $e^{-2t}H(t)$ and $e^{-2t}\cos(4t)H(t)$ in Figure 6.3.5. $\qquad\qquad\square$

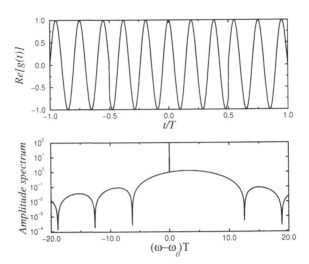

Figure 6.3.6: The (amplitude) spectrum $|G(\omega)|/T$ of a frequency-modulated signal (shown top) when $\omega_1 T = 2\pi$ and $\omega_0 T = 10\pi$. The transform becomes undefined at $\omega = \omega_0$.

• **Example 6.3.7: Frequency modulation**

In contrast to amplitude modulation, *frequency modulation* (FM) transmits information by instantaneous variations of the carrier frequency. It can be expressed mathematically as $\exp\left[i\int_{-\infty}^{t} f(\tau)\,d\tau + iC\right] e^{i\omega_0 t}$, where C is a constant. To illustrate this concept, let us find the Fourier transform of a simple frequency modulation

$$f(t) = \begin{cases} \omega_1, & |t| < T/2, \\ 0, & |t| > T/2, \end{cases} \qquad (6.3.25)$$

and $C = -\omega_1 T/2$. In this case, the signal in the time domain is

$$g(t) = \exp\left[i\int_{-\infty}^{t} f(\tau)\,d\tau + iC\right] e^{i\omega_0 t} = \begin{cases} e^{-i\omega_1 T/2} e^{i\omega_0 t}, & t < -T/2, \\ e^{i\omega_1 t} e^{i\omega_0 t}, & -T/2 < t < T/2, \\ e^{i\omega_1 T/2} e^{i\omega_0 t}, & T/2 < t. \end{cases} \quad (6.3.26)$$

We illustrate this signal in Figures 6.3.6 and 6.3.7.

The Fourier transform of the signal $G(\omega)$ equals

$$G(\omega) = e^{-i\omega_1 T/2} \int_{-\infty}^{-T/2} e^{i(\omega_0 - \omega)t}\,dt + \int_{-T/2}^{T/2} e^{i(\omega_0 + \omega_1 - \omega)t}\,dt + e^{i\omega_1 T/2} \int_{T/2}^{\infty} e^{i(\omega_0 - \omega)t}\,dt$$

$$(6.3.27)$$

$$= e^{-i\omega_1 T/2} \int_{-\infty}^{0} e^{i(\omega_0 - \omega)t}\,dt + e^{i\omega_1 T/2} \int_{0}^{\infty} e^{i(\omega_0 - \omega)t}\,dt - e^{-i\omega_1 T/2} \int_{-T/2}^{0} e^{i(\omega_0 - \omega)t}\,dt$$

$$+ \int_{-T/2}^{T/2} e^{i(\omega_0 + \omega_1 - \omega)t}\,dt - e^{i\omega_1 T/2} \int_{0}^{T/2} e^{i(\omega_0 - \omega)t}\,dt. \qquad (6.3.28)$$

Applying the fact that

$$\int_{0}^{\infty} e^{\pm i\alpha t}\,dt = \pi\delta(\alpha) \pm \frac{i}{\alpha}, \qquad (6.3.29)$$

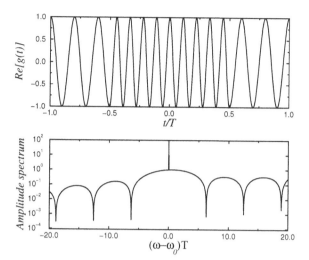

Figure 6.3.7: The (amplitude) spectrum $|G(\omega)|/T$ of a frequency-modulated signal (shown top) when $\omega_1 T = 8\pi$ and $\omega_0 T = 10\pi$. The transform becomes undefined at $\omega = \omega_0$.

$$G(\omega) = \pi\delta(\omega - \omega_0)\left[e^{i\omega_1 T/2} + e^{-i\omega_1 T/2}\right] + \frac{\left[e^{i(\omega_0 + \omega_1 - \omega)T/2} - e^{-i(\omega_0 + \omega_1 - \omega)T/2}\right]}{i(\omega_0 + \omega_1 - \omega)}$$

$$- \frac{\left[e^{i(\omega_0 + \omega_1 - \omega)T/2} - e^{-i(\omega_0 + \omega_1 - \omega)T/2}\right]}{i(\omega_0 - \omega)} \tag{6.3.30}$$

$$= 2\pi\delta(\omega - \omega_0)\cos(\omega_1 T/2) + \frac{2\omega_1 \sin[(\omega - \omega_0 - \omega_1)T/2]}{(\omega - \omega_0)(\omega - \omega_0 - \omega_1)}. \tag{6.3.31}$$

Figures 6.3.6 and 6.3.7 illustrate the amplitude spectrum for various parameters. In general, the transform is not symmetric, with an increasing number of humped curves as $\omega_1 T$ increases. $\qquad\square$

Parseval's equality

In applying Fourier methods to practical problems we may encounter a situation where we are interested in computing the energy of a system. Energy is usually expressed by the integral $\int_{-\infty}^{\infty} |f(t)|^2\, dt$. Can we compute this integral if we only have the Fourier transform of $F(\omega)$?

From the definition of the inverse Fourier transform

$$f(t) = \frac{1}{2\pi}\int_{-\infty}^{\infty} F(\omega)e^{i\omega t}\, d\omega, \tag{6.3.32}$$

we have that

$$\int_{-\infty}^{\infty} |f(t)|^2\, dt = \frac{1}{2\pi}\int_{-\infty}^{\infty} f(t)\left[\int_{-\infty}^{\infty} F(\omega)e^{i\omega t}\, d\omega\right] dt. \tag{6.3.33}$$

Interchanging the order of integration on the right side of Equation 6.3.33,

$$\int_{-\infty}^{\infty} |f(t)|^2\, dt = \frac{1}{2\pi}\int_{-\infty}^{\infty} F(\omega)\left[\int_{-\infty}^{\infty} f(t)e^{i\omega t}\, dt\right] d\omega. \tag{6.3.34}$$

However,

$$F^*(\omega) = \int_{-\infty}^{\infty} f(t)e^{i\omega t}dt. \tag{6.3.35}$$

Therefore,

$$\int_{-\infty}^{\infty} |f(t)|^2 \, dt = \frac{1}{2\pi} \int_{-\infty}^{\infty} |F(\omega)|^2 \, d\omega. \tag{6.3.36}$$

This is *Parseval's equality*[8] as it applies to Fourier transforms. The quantity $|F(\omega)|^2$ is called the *power spectrum*.

• Example 6.3.8

In Example 6.1.1, we showed that the Fourier transform for a unit rectangular pulse between $-a < t < a$ is $2\sin(\omega a)/\omega$. Therefore, by Parseval's equality,

$$\frac{2}{\pi} \int_{-\infty}^{\infty} \frac{\sin^2(\omega a)}{\omega^2} \, d\omega = \int_{-a}^{a} 1^2 \, dt = 2a, \qquad \text{or} \qquad \int_{-\infty}^{\infty} \frac{\sin^2(\omega a)}{\omega^2} \, d\omega = \pi a. \tag{6.3.37}$$

$\square$

Poisson's summation formula

If $f(x)$ is integrable over $(-\infty, \infty)$, there exists a relationship between the function and its Fourier transform, commonly called *Poisson's summation formula.*[9]

We begin by inventing a periodic function $g(x)$ defined by

$$g(x) = \sum_{k=-\infty}^{\infty} f(x + 2\pi k). \tag{6.3.38}$$

Because $g(x)$ is a periodic function of 2π, it can be represented by the complex Fourier series:

$$g(x) = \sum_{n=-\infty}^{\infty} c_n e^{inx}, \qquad \text{or} \qquad g(0) = \sum_{k=-\infty}^{\infty} f(2\pi k) = \sum_{n=-\infty}^{\infty} c_n. \tag{6.3.39}$$

Computing c_n, we find that

$$c_n = \frac{1}{2\pi} \int_{-\pi}^{\pi} g(x)e^{-inx} \, dx = \frac{1}{2\pi} \int_{-\pi}^{\pi} \sum_{k=-\infty}^{\infty} f(x + 2k\pi)e^{-inx} \, dx \tag{6.3.40}$$

$$= \frac{1}{2\pi} \sum_{k=-\infty}^{\infty} \int_{-\pi}^{\pi} f(x + 2k\pi)e^{-inx} \, dx = \frac{1}{2\pi} \int_{-\infty}^{\infty} f(x)e^{-inx} \, dx = \frac{F(n)}{2\pi}, \tag{6.3.41}$$

---

[8] Apparently first derived by Rayleigh, J. W., 1889: On the character of the complete radiation at a given temperature. *Philos. Mag., Ser. 5*, **27**, 460–469.

[9] Poisson, S. D., 1823: Suite du mémoire sur les intégrales définies et sur la sommation des séries. *J. École Polytech.*, **19**, 404–509. See page 451.

Some General Properties of Fourier Transforms

	function, $f(t)$	Fourier transform, $F(\omega)$
1. Linearity	$c_1 f(t) + c_2 g(t)$	$c_1 F(\omega) + c_2 G(\omega)$
2. Complex conjugate	$f^*(t)$	$F^*(-\omega)$
3. Scaling	$f(\alpha t)$	$F(\omega/\alpha)/|\alpha|$
4. Time shift	$f(t \pm \tau)$	$e^{\pm i\omega\tau} F(\omega)$
5. Frequency translation	$e^{i\omega_0 t} f(t)$	$F(\omega - \omega_0)$
6. Duality-time frequency	$F(t)$	$2\pi f(-\omega)$
7. Modulation	$\cos(\omega_0 t) f(t)$	$\frac{1}{2}[F(\omega + \omega_0) + F(\omega - \omega_0)]$
8. Time differentiation	$f^{(n)}(t)$	$(i\omega)^n F(\omega)$
9. Frequency differentiation	$t^n f(t)$	$i^n F^{(n)}(\omega)$
10. Time integration	$\displaystyle\int_{-\infty}^{t} f(\tau)\,d\tau$	$F(\omega)/(\omega i) + \pi F(0)\delta(\omega)$
11. Reversal	$f(-t)$	$F(-\omega)$ or $F^*(\omega)$
12. Convolution in t	$f_1(t) * f_2(t)$	$F_1(\omega) F_2(\omega)$
13. Convolution in ω	$f_1(t) f_2(t)$	$F_1(\omega) * F_2(\omega)/(2\pi)$

where $F(\omega)$ is the Fourier transform of $f(x)$. Substituting Equation 6.3.41 into the right side of Equation 6.3.39, we obtain

$$\sum_{k=-\infty}^{\infty} f(2\pi k) = \frac{1}{2\pi} \sum_{n=-\infty}^{\infty} F(n) \tag{6.3.42}$$

or

$$\sum_{k=-\infty}^{\infty} f(\alpha k) = \frac{1}{\alpha} \sum_{n=-\infty}^{\infty} F\left(\frac{2\pi n}{\alpha}\right). \tag{6.3.43}$$

• **Example 6.3.9**

One of the popular uses of Poisson's summation formula is the evaluation of infinite series. For example, let $f(x) = 1/(a^2 + x^2)$ with a real and nonzero. Then, $F(\omega) = \pi e^{-|a\omega|}/|a|$ and

$$\sum_{k=-\infty}^{\infty} \frac{1}{a^2 + (2\pi k)^2} = \frac{1}{2} \sum_{n=-\infty}^{\infty} \frac{1}{|a|} e^{-|a|n|} = \frac{1}{2|a|} \left(1 + 2 \sum_{n=1}^{\infty} e^{-|a|n} \right) \tag{6.3.44}$$

$$= \frac{1}{2|a|} \left(-1 + \frac{2}{1 - e^{-|a|}} \right) = \frac{1}{2|a|} \coth\left(\frac{|a|}{2} \right). \tag{6.3.45}$$

Problems

1. Find the Fourier transform of $1/(1 + a^2 t^2)$, where a is real, given that $\mathcal{F}[1/(1 + t^2)] = \pi e^{-|\omega|}$.

2. Find the Fourier transform of $J_0(at)$, where a is real, given that $\mathcal{F}[J_0(t)] = 2H(1 - |\omega|)/\sqrt{1 - \omega^2}$.

3. Find the Fourier transform of $2[H(t - 3) - H(t - 11)]$, given Equation 6.2.16. Check your answer by performing a direct calculation using the definition of the Fourier transform.

4. Find the Fourier transform of $\cos(at)/(1 + t^2)$, where a is real, given that $\mathcal{F}[1/(1 + t^2)] = \pi e^{-|\omega|}$.

5. Use the fact that $\mathcal{F}[e^{-at}H(t)] = 1/(a + i\omega)$ with $a > 0$ and Parseval's equality to show that

$$\int_{-\infty}^{\infty} \frac{dx}{x^2 + a^2} = \frac{\pi}{a}.$$

6. Use the fact that $\mathcal{F}[1/(1 + t^2)] = \pi e^{-|\omega|}$ and Parseval's equality to show that

$$\int_{-\infty}^{\infty} \frac{dx}{(x^2 + 1)^2} = \frac{\pi}{2}.$$

7. Use the function $f(t) = e^{-at} \sin(bt) H(t)$ with $a > 0$ and Parseval's equality to show that

$$2 \int_0^{\infty} \frac{dx}{(x^2 + a^2 - b^2)^2 + 4a^2 b^2} = \int_{-\infty}^{\infty} \frac{dx}{(x^2 + a^2 - b^2)^2 + 4a^2 b^2} = \frac{\pi}{2a(a^2 + b^2)}.$$

8. Using the modulation property and $\mathcal{F}[e^{-bt}H(t)] = 1/(b + i\omega)$, show that

$$\mathcal{F}\left[e^{-bt} \sin(at) H(t) \right] = \frac{a}{(b + i\omega)^2 + a^2}.$$

Use MATLAB to plot and compare the amplitude and phase spectra for $e^{-t} H(t)$ and $e^{-t} \sin(2t) H(t)$.

9. Use Poisson's summation formula with $f(t) = e^{-|t|}$ to show that

$$\sum_{n=-\infty}^{\infty} \frac{1}{n^2+1} = \pi \frac{1+e^{-2\pi}}{1-e^{-2\pi}}.$$

10. Use Poisson's summation formula to prove[10] that

$$\sum_{n=-\infty}^{\infty} e^{-a(n+c)^2+2b(n+c)} = \sqrt{\frac{\pi}{a}} e^{b^2/a} \sum_{n=-\infty}^{\infty} e^{-n^2\pi^2/a-2n\pi i(b/a-c)}.$$

11. Use Poisson's summation formula to prove that

$$\sum_{n=-\infty}^{\infty} e^{-ianT} = \frac{2\pi}{T} \sum_{n=-\infty}^{\infty} \delta\left(\frac{2\pi n}{T} - a\right),$$

where $\delta(\cdot)$ is the Dirac delta function.

12. Prove the two-dimensional form[11] of Poisson's summation formula:

$$\sum_{k_1=-\infty}^{\infty} \sum_{k_2=-\infty}^{\infty} f(\alpha_1 k_1, \alpha_2 k_2) = \frac{1}{\alpha_1\alpha_2} \sum_{n_1=-\infty}^{\infty} \sum_{n_2=-\infty}^{\infty} F\left(\frac{2\pi n_1}{\alpha_1}, \frac{2\pi n_2}{\alpha_2}\right),$$

where

$$F(\omega_1, \omega_2) = \int_{-\infty}^{\infty} \int_{-\infty}^{\infty} f(x, y) e^{-i\omega_1 x - i\omega_2 y} \, dx \, dy.$$

6.4 INVERSION OF FOURIER TRANSFORMS

Having focused on the Fourier transform in the previous sections, we now consider the inverse Fourier transform. Recall that the improper integral, Equation 6.1.6, defines the inverse. Consequently, one method of inversion is direct integration.

• **Example 6.4.1**

Let us find the inverse of $F(\omega) = \pi e^{-|\omega|}$.
From the definition of the inverse Fourier transform,

$$f(t) = \frac{1}{2\pi} \int_{-\infty}^{\infty} \pi e^{-|\omega|} e^{i\omega t} d\omega = \frac{1}{2} \int_{-\infty}^{0} e^{(1+it)\omega} d\omega + \frac{1}{2} \int_{0}^{\infty} e^{(-1+it)\omega} d\omega \qquad (6.4.1)$$

$$= \frac{1}{2}\left[\frac{e^{(1+it)\omega}}{1+it}\Big|_{-\infty}^{0} + \frac{e^{(-1+it)\omega}}{-1+it}\Big|_{0}^{\infty}\right] = \frac{1}{2}\left[\frac{1}{1+it} - \frac{1}{-1+it}\right] = \frac{1}{1+t^2}. \qquad (6.4.2)$$

---

[10] First proved by Ewald, P. P., 1921: Die Berechnung optischer und elektrostatischer Gitterpotentiale. *Ann. Phys., 4te Folge*, **64**, 253–287.

[11] Lucas, S. K., R. Sipcic, and H. A. Stone, 1997: An integral equation solution for the steady-state current at a periodic array of surface microelectrodes. *SIAM J. Appl. Math.*, **57**, 1615–1638.

An alternative to direct integration is the MATLAB function `ifourier`. For example, to invert $F(\omega) = \pi e^{-|\omega|}$, we type in the commands:

```
>> syms pi omega t
>> ifourier('pi*exp(-abs(omega))',omega,t)
```

This yields

```
ans =
1/(1+t^2)
```
□

Another method for inverting Fourier transforms is rewriting the Fourier transform using partial fractions so that we can use transform tables. The following example illustrates this technique.

• **Example 6.4.2**

Let us invert the transform

$$F(\omega) = \frac{1}{(1+i\omega)(1-2i\omega)^2}. \qquad (6.4.3)$$

We begin by rewriting Equation 6.4.3 as

$$F(\omega) = \frac{1}{9}\left[\frac{1}{1+i\omega} + \frac{2}{1-2i\omega} + \frac{6}{(1-2i\omega)^2}\right] = \frac{1}{9(1+i\omega)} + \frac{1}{9(\frac{1}{2}-i\omega)} + \frac{1}{6(\frac{1}{2}-i\omega)^2}. \qquad (6.4.4)$$

Using a table of Fourier transforms (see Section 6.1), we invert Equation 6.4.4 term by term and find that

$$f(t) = \tfrac{1}{9}e^{-t}H(t) + \tfrac{1}{9}e^{t/2}H(-t) - \tfrac{1}{6}te^{t/2}H(-t). \qquad (6.4.5)$$

To check our answer, we type the following commands into MATLAB:

```
>> syms omega t
>> ifourier(1/((1+i*omega)*(1-2*i*omega)^2),omega,t)
```

which yields

```
ans =
1/9*exp(-t)*Heaviside(t)-1/6*exp(1/2*t)*t*Heaviside(-t)
    +1/9*exp(1/2*t)*Heaviside(-t)
```
□

• **Example 6.4.3: Numerical computation of the inverse Fourier transforms using** MATLAB

In Example 6.1.6 we showed how we can compute the Fourier transform of a temporal function $f(t)$ by sampling at intervals of Δt. We performed the computations two ways: the trapezoidal rule and the fast Fourier transform. In this example we show how to use these two methods to numerically compute *inverse* Fourier transforms. In particular, we use the Fourier transform pair:

$$\mathcal{F}[J_0(t)] = \frac{2}{\sqrt{1-\omega^2}}H(1-|\omega|). \qquad (6.4.6)$$

Turning to the trapezoidal methods first, we approximate the integral by replacing the limits from $-\infty$ to ∞ to one from $-L$ to L:

$$f(t) \approx \frac{1}{2\pi}\int_{-L}^{L} F(\omega)e^{it\omega}\,d\omega, \qquad (6.4.7)$$

where L is a "large number" that will be specified later. The MATLAB code that yields $f(t)$ is:

```
clear

N = 40; d_omega = 0.2; omega = [-N/2:N/2]*d_omega; % L = (N/2)*d_omega

%%%%%%%%%%%%%%%%%%%%%%%%%%%%%%%%%%%%%%%%%%%%%%%%%%%%%%%%%%%%%%%%%%%%%%%%%%%%%
%                   Compute the Fourier transform X(omega)
%%%%%%%%%%%%%%%%%%%%%%%%%%%%%%%%%%%%%%%%%%%%%%%%%%%%%%%%%%%%%%%%%%%%%%%%%%%%%

temp = 2 ./ sqrt(1-omega.*omega);
for n = 1:length(omega)
if abs(omega(n)) < 1
X(n) = temp(n); % load in transform
else
X(n) = 0;
end; end

%%%%%%%%%%%%%%%%%%%%%%%%%%%%%%%%%%%%%%%%%%%%%%%%%%%%%%%%%%%%%%%%%%%%%%%%%%%%%
%                   Perform the integration in Equation 6.4.7
%%%%%%%%%%%%%%%%%%%%%%%%%%%%%%%%%%%%%%%%%%%%%%%%%%%%%%%%%%%%%%%%%%%%%%%%%%%%%

k = 0;
for t = -40:dt:40
k = k+1;
tt(k) = t;
xx(k) = trapz(omega,X.*exp(i*omega*t));
end

x = real(xx) / (2*pi) % finally divide by 2π
```

Figure 6.4.1 illustrates the inversion of the Fourier transform using the trapezoidal rule for various $\Delta\omega$'s but the same L. Our code is not optimal because we have included ω's where $F(\omega) = 0$. However, I designed the code so that it could be used for any arbitrary Fourier transform.

Looking at the figure closely, we note that numerical inversion lies between $-T$ and T, where $T = 2\pi/\Delta\omega$ and $\Delta\omega$ is the size of our sampling of the transform $F(\omega)$. It is also most accurate for small $|t|$. An interesting artifact is the repetition of the numerical inversion in the intervals $(-3T, -T)$ and $(T, 3T)$ in the top frame. Here we are reminded that our numerical inverse is a Fourier series approximation of the Fourier transform. Only in the limit of $\Delta\omega \to 0$ would we obtain the exact result.

Let us now return to Equation 6.4.7 and apply the first-order extended rectangular rule to it. This yields

$$f(t) \approx \frac{\Delta\omega}{2\pi} \sum_{m=-M}^{M-1} F(\omega_m)e^{it\omega_m}, \tag{6.4.8}$$

where $\omega_m = m\Delta\omega$ and $\Delta\omega = L/M$. Although we could perform the summation in Equation 6.4.8 for a given time t, let us examine the case when $t = t_n = \pi n/L$,

$$f_n = f(t_n) \approx \frac{\Delta\omega}{2\pi} \sum_{m=-M}^{M-1} F(\omega_m)e^{\pi imn/M}, \tag{6.4.9}$$

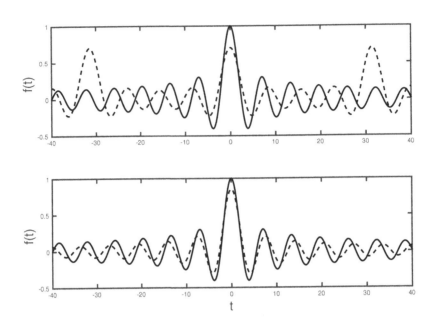

Figure 6.4.1: The numerical inversion of the Fourier transform $F(\omega) = 2H(1 - |\omega|)/\sqrt{1 - \omega^2}$ using the trapezoidal rule. In both frames the solid lines gives the exact inverse. In the top frame $N = 40$ and $\Delta\omega = 0.2$ ($L = 4$) while in the lower frame $N = 160$ and $\Delta\omega = 0.05$ ($L = 4$).

or

$$f_n \approx \frac{\Delta\omega}{2\pi} \sum_{m=-M}^{-1} F(\omega_m)e^{\pi imn/M} + \frac{\Delta\omega}{2\pi} \sum_{m=0}^{M-1} F(\omega_m)e^{\pi imn/M}. \tag{6.4.10}$$

Invoking the periodicity condition $F(\omega_m) = F(\omega_{m+2M})$,

$$f_n \approx \frac{L}{\pi} \left[\frac{1}{2M} \sum_{m=0}^{2M-1} F(\omega_m)e^{\pi imn/M} \right]. \tag{6.4.11}$$

The quantity within the square brackets is the inverse of the discrete Fourier transform. We see now why we choose to evaluate Equation 6.4.8 at $t = t_n$; it can be rapidly computed. Equation 6.4.11 then yields the inverse for times $0 \le t_n \le (2M - 1)\pi/L$. For other times outside of this range, the periodicity condition gives $f_n = f_{n\pm 2M}$.

Returning to our test transform, we can program Equation 6.4.11 using MATLAB, our inversion scheme is:

```
clear; L = 20; M = 64; d_omega = L / M; % d_omega = Δω
for m = 1:2*M
omega = -L + (m-1) * d_omega; % compute ω_m
if (abs(omega(m)) < 1)
temp = 2 / sqrt(1-omega(m)*omega(m));
else
temp = 0;
end
```

```
%%%%%%%%%%%%%%%%%%%%%%%%%%%%%%%%%%%%%%%%%%%%%%%%%%%%%%%%%%%%%%%%%%%%%%
Load in the transform
%%%%%%%%%%%%%%%%%%%%%%%%%%%%%%%%%%%%%%%%%%%%%%%%%%%%%%%%%%%%%%%%%%%%%%

if (omega(m) < 0)
F(m+M) = temp;
else
F(m-M) = temp;
end
end


%%%%%%%%%%%%%%%%%%%%%%%%%%%%%%%%%%%%%%%%%%%%%%%%%%%%%%%%%%%%%%%%%%%%%%
% compute inverse of discrete Fourier transform
%%%%%%%%%%%%%%%%%%%%%%%%%%%%%%%%%%%%%%%%%%%%%%%%%%%%%%%%%%%%%%%%%%%%%%
f = ifft(F,2*M); % compute inverse of discrete Fourier transform
```

We have special code for `F(m)` so that `F(m)` runs from $m = 1$ to $m = 2M$. To correctly order the computed values of f_n so that they run from $t = -M\pi/L$ to $(M-1)\pi/L$, we need

```
for m = 1:2*M
time(m) = pi * (m-1-M) / L;
if (time(m) >= 0)
f_n(m) = L * real(f(m-M)) / pi;
else
f_n(m) = L * real(f(m+M)) / pi;
end; end
```

Figure 6.4.2 shows the results when $L = 20$ for two different values of M. From this figure, the results are rather good even when the resolution is $\Delta\omega = 0.3125$. The best results occur at small $|t|$.

Problems

1. Use direct integration to find the inverse of the Fourier transform $F(\omega) = i\omega\pi e^{-|\omega|}/2$. Check your answer using MATLAB.

Use partial fractions to invert the following Fourier transforms:

2. $\dfrac{1}{(1+i\omega)(1+2i\omega)}$ 3. $\dfrac{1}{(1+i\omega)(1-i\omega)}$

4. $\dfrac{i\omega}{(1+i\omega)(1+2i\omega)}$ 5. $\dfrac{1}{(1+i\omega)(1+2i\omega)^2}$

Then check your answer using MATLAB.

6.5 CONVOLUTION

The most important property of Fourier transforms is convolution. We shall use it extensively in the solution of differential equations and the design of filters because it yields in time or space the effect of multiplying two transforms together.

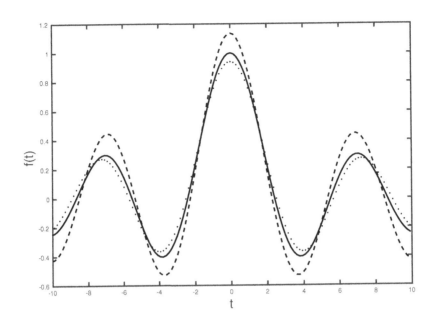

Figure 6.4.2: The numerical inversion of the Fourier transform $F(\omega)$ using Equation 6.4.11 with $L = 20$. The solid line gives the exact inverse while the dashed line corresponds to $M = 64$ while the dotted line gives the solution when $M = 128$.

The convolution operation is

$$f(t) * g(t) = \int_{-\infty}^{\infty} f(x)g(t-x)\,dx = \int_{-\infty}^{\infty} f(t-x)g(x)\,dx. \tag{6.5.1}$$

Then,

$$\mathcal{F}[f(t) * g(t)] = \int_{-\infty}^{\infty} f(x)e^{-i\omega x}\left[\int_{-\infty}^{\infty} g(t-x)e^{-i\omega(t-x)}dt\right]dx \tag{6.5.2}$$

$$= \int_{-\infty}^{\infty} f(x)G(\omega)e^{-i\omega x}dx = F(\omega)G(\omega). \tag{6.5.3}$$

Thus, the Fourier transform of the convolution of two functions equals the product of the Fourier transforms of each of the functions.

• **Example 6.5.1**

Let us verify the convolution theorem using the functions $f(t) = H(t+a) - H(t-a)$ and $g(t) = e^{-t}H(t)$, where $a > 0$.

The convolution of $f(t)$ with $g(t)$ is

$$f(t) * g(t) = \int_{-\infty}^{\infty} e^{-(t-x)}H(t-x)\left[H(x+a) - H(x-a)\right]dx = e^{-t}\int_{-a}^{a} e^{x}H(t-x)\,dx. \tag{6.5.4}$$

If $t < -a$, then the integrand of Equation 6.5.4 is always zero and $f(t) * g(t) = 0$. If $t > a$,

$$f(t) * g(t) = e^{-t} \int_{-a}^{a} e^x dx = e^{-(t-a)} - e^{-(t+a)}. \qquad (6.5.5)$$

Finally, for $-a < t < a$,

$$f(t) * g(t) = e^{-t} \int_{-a}^{t} e^x dx = 1 - e^{-(t+a)}. \qquad (6.5.6)$$

In summary,

$$f(t) * g(t) = \begin{cases} 0, & t \leq -a, \\ 1 - e^{-(t+a)}, & -a \leq t \leq a, \\ e^{-(t-a)} - e^{-(t+a)}, & a \leq t. \end{cases} \qquad (6.5.7)$$

$\square$

As an alternative to examining various cases involving the value of t, we could have used MATLAB to evaluate Equation 6.5.4. The MATLAB instructions are as follows:

```
>> syms f t x
>> syms a positive
>> f = 'exp(x-t)*Heaviside(t-x)*(Heaviside(x+a)-Heaviside(x-a))'
>> int(f,x,-inf,inf)
```

This yields

```
ans =
Heaviside(t+a)-Heaviside(t+a)*exp(-a-t)
   -Heaviside(t-a)+Heaviside(t-a)*exp(a-t)
```

The Fourier transform of $f(t) * g(t)$ is

$$\mathcal{F}[f(t) * g(t)] = \int_{-a}^{a} \left[1 - e^{-(t+a)}\right] e^{-i\omega t} dt + \int_{a}^{\infty} \left[e^{-(t-a)} - e^{-(t+a)}\right] e^{-i\omega t} dt \qquad (6.5.8)$$

$$= \frac{2\sin(\omega a)}{\omega} - \frac{2i\sin(\omega a)}{1 + \omega i} = \frac{2\sin(\omega a)}{\omega} \left(\frac{1}{1 + \omega i}\right) = F(\omega)G(\omega) \qquad (6.5.9)$$

and the convolution theorem is true for this special case. The Fourier transform Equation 6.5.9 could also be obtained by substituting our earlier MATLAB result into `fourier` and then using `simplify(ans)`.

• **Example 6.5.2**

Let us consider the convolution of $f(t) = f_+(t)H(t)$ with $g(t) = g_+ H(t)$. Note that both of the functions are nonzero only for $t > 0$.

From the definition of convolution,

$$f(t)*g(t) = \int_{-\infty}^{\infty} f_+(t-x)H(t-x)g_+(x)H(x)\,dx = \int_{0}^{\infty} f_+(t-x)H(t-x)g_+(x)\,dx. \qquad (6.5.10)$$

For $t < 0$, the integrand is always zero and $f(t) * g(t) = 0$. For $t > 0$,

$$f(t) * g(t) = \int_{0}^{t} f_+(t - x)g_+(x)\,dx. \qquad (6.5.11)$$

Therefore, in general,

$$f(t) * g(t) = \left[\int_0^t f_+(t-x)g_+(x)\,dx \right] H(t). \tag{6.5.12}$$

This is the definition of convolution that we will use for Laplace transforms where all of the functions equal zero for $t < 0$. □

The convolution operation also applies to Fourier transforms, in what is commonly known as *frequency convolution*. We now prove that

$$\mathcal{F}[f(t)g(t)] = \frac{F(\omega) * G(\omega)}{2\pi}, \tag{6.5.13}$$

where

$$F(\omega) * G(\omega) = \int_{-\infty}^{\infty} F(\tau)G(\omega - \tau)\,d\tau, \tag{6.5.14}$$

where $F(\omega)$ and $G(\omega)$ are the Fourier transforms of $f(t)$ and $g(t)$, respectively.

Proof: Starting with

$$f(t) = \frac{1}{2\pi} \int_{-\infty}^{\infty} F(\tau)e^{i\tau t}\,d\tau, \tag{6.5.15}$$

we can multiply the inverse of $F(\tau)$ by $g(t)$ so that we obtain

$$f(t)g(t) = \frac{1}{2\pi} \int_{-\infty}^{\infty} F(\tau)g(t)e^{i\tau t}\,d\tau. \tag{6.5.16}$$

Then, taking the Fourier transform of Equation 6.5.16, we find that

$$\mathcal{F}[f(t)g(t)] = \int_{-\infty}^{\infty} \left[\frac{1}{2\pi} \int_{-\infty}^{\infty} F(\tau)g(t)e^{i\tau t}\,d\tau \right] e^{-i\omega t}\,dt \tag{6.5.17}$$

$$= \frac{1}{2\pi} \int_{-\infty}^{\infty} F(\tau) \left[\int_{-\infty}^{\infty} g(t)e^{-i(\omega-\tau)t}\,dt \right] d\tau \tag{6.5.18}$$

$$= \frac{1}{2\pi} \int_{-\infty}^{\infty} F(\tau)G(\omega - \tau)\,d\tau = \frac{F(\omega) * G(\omega)}{2\pi}. \tag{6.5.19}$$

Thus, the multiplication of two functions in the time domain is equivalent to the convolution of their spectral densities in the frequency domain. □

• Example 6.5.3

Let us find the convolution of $H(t)$ with $f(t)$. This convolution is

$$H(t) * f(t) = \int_{-\infty}^{\infty} f(\tau)H(t - \tau)\,d\tau = \int_{-\infty}^{t} f(\tau)\,d\tau. \tag{6.5.20}$$

Hence,

$$\mathcal{F}\left[\int_{-\infty}^{t} f(\tau)\,d\tau \right] = F(\omega)\mathcal{F}[H(t)] = F(\omega)\left[\frac{1}{\omega i} + \pi\delta(\omega) \right] = \frac{F(\omega)}{\omega i} + \pi F(0)\delta(\omega), \tag{6.5.21}$$

where we have used Equation 6.2.16 and the property that $f(t)\delta(t) = f(0)\delta(t)$.

Problems

1. Show that $e^{-t}H(t) * e^{-t}H(t) = te^{-t}H(t)$. Then verify your result using MATLAB.

2. Show that $e^{-t}H(t) * e^{t}H(-t) = \frac{1}{2}e^{-|t|}$. Then verify your result using MATLAB.

3. Show that $e^{-t}H(t) * e^{-2t}H(t) = \left(e^{-t} - e^{-2t}\right)H(t)$. Then verify your result using MATLAB.

4. Show that

$$e^{t}H(-t) * [H(t) - H(t-2)] = \begin{cases} e^{t} - e^{t-2}, & t \le 0, \\ 1 - e^{t-2}, & 0 \le t \le 2, \\ 0, & 2 \le t. \end{cases}$$

Then verify your result using MATLAB.

5. Show that

$$[H(t) - H(t-2)] * [H(t) - H(t-2)] = \begin{cases} 0, & t \le 0, \\ t, & 0 \le t \le 2, \\ 4-t, & 2 \le t \le 4, \\ 0, & 4 \le t. \end{cases}$$

Then try and verify your result using MATLAB. What do you have to do to make it work?

6. Show that $e^{-|t|} * e^{-|t|} = (1 + |t|)e^{-|t|}$.

7. Prove that the convolution of two Dirac delta functions is a Dirac delta function.

6.6 THE SOLUTION OF ORDINARY DIFFERENTIAL EQUATIONS BY FOURIER TRANSFORMS

We may use Fourier transforms to solve ordinary differential equations. However, this method gives only the particular solution and we must find the complementary solution separately.

Consider the differential equation

$$y' + y = \tfrac{1}{2}e^{-|t|}, \qquad -\infty < t < \infty. \tag{6.6.1}$$

Taking the Fourier transform of both sides of Equation 6.6.1,

$$i\omega Y(\omega) + Y(\omega) = \frac{1}{\omega^2 + 1}, \tag{6.6.2}$$

where we used the derivative rule, Equation 6.3.17, to obtain the transform of y' and $Y(\omega) = \mathcal{F}[y(t)]$. Therefore,

$$Y(\omega) = \frac{1}{(\omega^2 + 1)(1 + \omega i)}. \tag{6.6.3}$$

Applying the inversion integral to Equation 6.6.3,

$$y(t) = \frac{1}{2\pi} \int_{-\infty}^{\infty} \frac{e^{it\omega}}{(\omega^2 + 1)(1 + \omega i)} \, d\omega = \frac{1}{2\pi} \int_{-\infty}^{\infty} \frac{e^{it\omega}}{(\omega + i)(\omega - i)(1 + \omega i)} \, d\omega. \qquad (6.6.4)$$

Using partial fractions we find that

$$\frac{1}{(\omega + i)(\omega - i)(1 + \omega i)} = \frac{1}{4(1 - \omega i)} + \frac{1}{2(1 + \omega i)^2} + \frac{1}{4(1 + \omega i)}. \qquad (6.6.5)$$

Using tables or MATLAB, term-by-term inversion yields

$$y(t) = \tfrac{1}{4}e^{t}H(-t) + \tfrac{1}{2}te^{-t}H(t) + \tfrac{1}{4}e^{-t}H(t), \qquad (6.6.6)$$

which can be written more compactly as

$$y(t) = \tfrac{1}{4}e^{-|t|} + \tfrac{1}{2}te^{-t}H(t). \qquad (6.6.7)$$

Note that we only found the particular or forced solution to Equation 6.6.1. The most general solution therefore requires that we add the complementary solution Ae^{-t}, yielding

$$y(t) = Ae^{-t} + \tfrac{1}{4}e^{-|t|} + \tfrac{1}{2}te^{-t}H(t). \qquad (6.6.8)$$

The arbitrary constant A would be determined by the initial condition, which we have not specified.

We could also have solved this problem using MATLAB. The MATLAB script is:

```
clear
% define symbolic variables
syms omega t Y
% take Fourier transform of left side of differential equation
LHS = fourier(diff(sym('y(t)'))+sym('y(t)'),t,omega);
% take Fourier transform of right side of differential equation
RHS = fourier(1/2*exp(-abs(t)),t,omega);
% set Y for Fourier transform of y
%      and introduce initial conditions
newLHS = subs(LHS,'fourier(y(t),t,omega)',Y);
% solve for Y
Y = solve(newLHS-RHS,Y);
% invert Fourier transform and find y(t)
y = ifourier(Y,omega,t)
```

yields

```
y =
1/4*exp(t)*Heaviside(-t)+1/2*exp(-t)*t*Heaviside(t)
     +1/4*exp(-t)*Heaviside(t)
```

which is equivalent to Equation 6.6.7.

Consider now a more general problem of

$$y' + y = f(t), \qquad -\infty < t < \infty, \qquad (6.6.9)$$

where we assume that $f(t)$ has the Fourier transform $F(\omega)$. Then the Fourier-transformed solution to Equation 6.6.9 is

$$Y(\omega) = \frac{1}{1 + \omega i} F(\omega) = G(\omega) F(\omega) \qquad \text{or} \qquad y(t) = g(t) * f(t), \tag{6.6.10}$$

where $g(t) = \mathcal{F}^{-1}[1/(1 + \omega i)] = e^{-t} H(t)$. Thus, we can obtain our solution in one of two ways. First, we can take the Fourier transform of $f(t)$, multiply this transform by $G(\omega)$, and finally compute the inverse. The second method requires a convolution of $f(t)$ with $g(t)$. Which method is easiest depends upon $f(t)$ and $g(t)$.

In summary, we can use Fourier transforms to find particular solutions to differential equations. The complete solution consists of this particular solution plus any homogeneous solution that we need to satisfy the initial conditions.

Problems

Find the particular solutions for the following differential equations. Verify your solution using MATLAB.

1. $y'' + 3y' + 2y = e^{-t} H(t)$ 2. $y'' + 4y' + 4y = \frac{1}{2} e^{-|t|}$ 3. $y'' - 4y' + 4y = e^{-t} H(t)$

4. Assuming that the function $g(x)$ possesses a Fourier transform, find the particular solution for the differential equation:

$$y'' - y = -g(x).$$

Step 1: Take the Fourier transform of both sides of the differential equation and show that

$$Y(\omega) = \frac{G(\omega)}{\omega^2 + 1},$$

where $Y(\omega)$ and $G(\omega)$ are the Fourier transforms of $y(x)$ and $g(x)$, respectively.

Step 2: Using tables, show that

$$y(x) = e^{-!x!} * g(x) = e^{-x} \int_{-\infty}^{x} e^{t} g(t)\, dt + e^{x} \int_{x}^{\infty} e^{-t} g(t)\, dt.$$

5. (a) Consider the function $f(x) = ie^{-i\lambda|x|}/(2\lambda)$, where λ is a complex number with a positive real part and a negative imaginary part. Using direct integration, show that the Fourier transform of $f(x)$, $F(\omega)$, is $F(\omega) = 1/(\lambda^2 - \omega^2)$.
(b) Using the results from part (a) and the table given in Section 6.1, find the particular solution to $y^{iv} - \lambda^4 y = \delta(x)$, where λ has a positive real part and a negative imaginary part.

6.7 THE SOLUTION OF LAPLACE'S EQUATION ON THE UPPER HALF-PLANE

In this section we shall use Fourier integrals and convolution to find the solution of Laplace's equation (see Chapter 10) on the upper half-plane $y > 0$. We require that the solution remains bounded over the entire domain and specify it along the x-axis, $u(x, 0) =$

$f(x)$. Under these conditions, we can take the Fourier transform of Laplace's equation and find that

$$\int_{-\infty}^{\infty} \frac{\partial^2 u}{\partial x^2} e^{-i\omega x}\, dx + \int_{-\infty}^{\infty} \frac{\partial^2 u}{\partial y^2} e^{-i\omega x}\, dx = 0. \qquad (6.7.1)$$

If everything is sufficiently differentiable, we may successively integrate by parts the first integral in Equation 6.7.1, which yields

$$\int_{-\infty}^{\infty} \frac{\partial^2 u}{\partial x^2} e^{-i\omega x}\, dx = \frac{\partial u}{\partial x} e^{-i\omega x}\Big|_{-\infty}^{\infty} + i\omega \int_{-\infty}^{\infty} \frac{\partial u}{\partial x} e^{-i\omega x}\, dx \qquad (6.7.2)$$

$$= i\omega\, u(x,y)e^{-i\omega x}\Big|_{-\infty}^{\infty} - \omega^2 \int_{-\infty}^{\infty} u(x,y)e^{-i\omega x}\, dx \qquad (6.7.3)$$

$$= -\omega^2 U(\omega, y), \qquad (6.7.4)$$

where

$$U(\omega, y) = \int_{-\infty}^{\infty} u(x,y)e^{-i\omega x}\, dx. \qquad (6.7.5)$$

The second integral becomes

$$\int_{-\infty}^{\infty} \frac{\partial^2 u}{\partial y^2} e^{-i\omega x}\, dx = \frac{d^2}{dy^2}\left[\int_{-\infty}^{\infty} u(x,y)e^{-i\omega x}\, dx\right] = \frac{d^2 U(\omega, y)}{dy^2}, \qquad (6.7.6)$$

along with the boundary condition that

$$F(\omega) = U(\omega, 0) = \int_{-\infty}^{\infty} f(x)e^{-i\omega x}\, dx. \qquad (6.7.7)$$

Consequently, we reduced Laplace's equation, a partial differential equation, to an ordinary differential equation in y, where ω is merely a parameter:

$$\frac{d^2 U(\omega, y)}{dy^2} - \omega^2 U(\omega, y) = 0, \qquad (6.7.8)$$

with the boundary condition $U(\omega, 0) = F(\omega)$. The solution to Equation 6.7.8 is

$$U(\omega, y) = A(\omega)e^{|\omega|y} + B(\omega)e^{-|\omega|y}, \quad 0 \le y. \qquad (6.7.9)$$

We must discard the $e^{|\omega|y}$ term because it becomes unbounded as we go to infinity along the y-axis. The boundary condition results in $B(\omega) = F(\omega)$. Hence,

$$U(\omega, y) = F(\omega)e^{-|\omega|y}. \qquad (6.7.10)$$

The inverse of the Fourier transform $e^{-|\omega|y}$ equals

$$\frac{1}{2\pi}\int_{-\infty}^{\infty} e^{-|\omega|y}e^{i\omega x}\, d\omega = \frac{1}{2\pi}\int_{-\infty}^{0} e^{\omega y}e^{i\omega x}\, d\omega + \frac{1}{2\pi}\int_{0}^{\infty} e^{-\omega y}e^{i\omega x}\, d\omega \qquad (6.7.11)$$

$$= \frac{1}{2\pi}\int_{0}^{\infty} e^{-\omega y}e^{-i\omega x}\, d\omega + \frac{1}{2\pi}\int_{0}^{\infty} e^{-\omega y}e^{i\omega x}\, d\omega \qquad (6.7.12)$$

$$= \frac{1}{\pi}\int_{0}^{\infty} e^{-\omega y}\cos(\omega x)\, d\omega \qquad (6.7.13)$$

$$= \frac{1}{\pi}\left\{\frac{\exp(-\omega y)}{x^2 + y^2}\left[-y\cos(\omega x) + x\sin(\omega x)\right]\right\}\Big|_{0}^{\infty} \qquad (6.7.14)$$

$$= \frac{1}{\pi}\frac{y}{x^2 + y^2}. \qquad (6.7.15)$$

Furthermore, because Equation 6.7.10 is a convolution of two Fourier transforms, its inverse is

$$u(x, y) = \frac{1}{\pi} \int_{-\infty}^{\infty} \frac{y f(t)}{(x-t)^2 + y^2} \, dt. \tag{6.7.16}$$

Equation 6.7.16 is *Poisson's integral formula*[12] for the half-plane $y > 0$ or *Schwarz' integral formula*.[13]

- **Example 6.7.1**

As an example, let $u(x, 0) = 1$ if $|x| < 1$ and $u(x, 0) = 0$ otherwise. Then,

$$u(x, y) = \frac{1}{\pi} \int_{-1}^{1} \frac{y}{(x-t)^2 + y^2} \, dt = \frac{1}{\pi} \left[\tan^{-1}\left(\frac{1-x}{y}\right) + \tan^{-1}\left(\frac{1+x}{y}\right) \right]. \tag{6.7.17}$$

Problems

Find the solution to Laplace's equation $u_{xx} + u_{yy} = 0$ in the upper half-plane for the following boundary conditions:

1. $u(x, 0) = \begin{cases} 1, & 0 < x < 1, \\ 0, & \text{otherwise.} \end{cases}$

2. $u(x, 0) = \begin{cases} 1, & x > 0, \\ -1, & x < 0. \end{cases}$

3. $u(x, 0) = \begin{cases} T_0, & x < 0, \\ 0, & x > 0. \end{cases}$

4. $u(x, 0) = \begin{cases} 2T_0, & x < -1, \\ T_0, & -1 < x < 1, \\ 0, & 1 < x. \end{cases}$

5. $u(x, 0) = \begin{cases} T_0, & -1 < x < 0, \\ T_0 + (T_1 - T_0)x, & 0 < x < 1, \\ 0, & \text{otherwise.} \end{cases}$

6. $u(x, 0) = \begin{cases} T_0, & x < a_1, \\ T_1, & a_1 < x < a_2, \\ T_2, & a_2 < x < a_3, \\ \vdots & \vdots \\ T_n, & a_n < x. \end{cases}$

6.8 THE SOLUTION OF THE HEAT EQUATION

We now consider the problem of one-dimensional heat flow (see Chapter 9) in a rod of infinite length with insulated sides. Although there are no boundary conditions because the slab is of infinite extent, we do require that the solution remains bounded as we go to either positive or negative infinity. The initial temperature within the rod is $u(x, 0) = f(x)$.

---

[12] Poisson, S. D., 1823: Suite du mémoire sur les intégrales définies et sur la sommation des séries. *J. École Polytech.*, **19**, 404–509. See pg. 462.

[13] Schwarz, H. A., 1870: Über die Integration der partiellen Differentialgleichung $\partial^2 u/\partial x^2 + \partial^2 u/\partial y^2 = 0$ für die Fläche eines Kreises. *Vierteljahrsschr. Naturforsch. Ges. Zürich*, **15**, 113–128.

Employing the product solution technique of separation of variables (see Section 9.3), we begin by assuming that $u(x,t) = X(x)T(t)$ with

$$T' + a^2\lambda T = 0, \tag{6.8.1}$$

and

$$X'' + \lambda X = 0. \tag{6.8.2}$$

Solutions to Equation 6.8.1 and Equation 6.8.2, which remain finite over the entire x-domain, are

$$X(x) = E\cos(kx) + F\sin(kx), \tag{6.8.3}$$

and

$$T(t) = C\exp(-k^2 a^2 t). \tag{6.8.4}$$

Because we do not have any boundary conditions, we must include *all* possible values of k. Thus, when we sum all of the product solutions according to the principle of linear superposition, we obtain the integral

$$u(x,t) = \int_0^\infty [A(k)\cos(kx) + B(k)\sin(kx)]e^{-k^2 a^2 t}\,dk. \tag{6.8.5}$$

We can satisfy the initial condition by choosing

$$A(k) = \frac{1}{\pi}\int_{-\infty}^\infty f(x)\cos(kx)\,dx, \tag{6.8.6}$$

and

$$B(k) = \frac{1}{\pi}\int_{-\infty}^\infty f(x)\sin(kx)\,dx, \tag{6.8.7}$$

because the initial condition has the form of a Fourier integral

$$f(x) = \int_0^\infty [A(k)\cos(kx) + B(k)\sin(kx)]\,dk, \tag{6.8.8}$$

when $t = 0$.

Several important results follow by rewriting Equation 6.8.8 as

$$u(x,t) = \frac{1}{\pi}\int_0^\infty \left[\int_{-\infty}^\infty f(\xi)\cos(k\xi)\cos(kx)\,d\xi + \int_{-\infty}^\infty f(\xi)\sin(k\xi)\sin(kx)\,d\xi\right]e^{-k^2 a^2 t}\,dk. \tag{6.8.9}$$

Combining terms,

$$u(x,t) = \frac{1}{\pi}\int_0^\infty \left\{\int_{-\infty}^\infty f(\xi)[\cos(k\xi)\cos(kx) + \sin(k\xi)\sin(kx)]\,d\xi\right\}e^{-k^2 a^2 t}\,dk \tag{6.8.10}$$

$$= \frac{1}{\pi}\int_0^\infty \left[\int_{-\infty}^\infty f(\xi)\cos[k(\xi - x)]\,d\xi\right]e^{-k^2 a^2 t}\,dk. \tag{6.8.11}$$

Reversing the order of integration,

$$u(x,t) = \frac{1}{\pi}\int_{-\infty}^\infty f(\xi)\left[\int_0^\infty \cos[k(\xi - x)]e^{-k^2 a^2 t}\,dk\right]\,d\xi. \tag{6.8.12}$$

The inner integral is called the *source function*. We may compute its value through an integration on the complex plane; it equals

$$\int_0^\infty \cos[k(\xi - x)] \exp(-k^2 a^2 t)\, dk = \left(\frac{\pi}{4a^2 t}\right)^{1/2} \exp\left[-\frac{(\xi - x)^2}{4a^2 t}\right], \qquad (6.8.13)$$

if $0 < t$. This gives the final form for the temperature distribution:

$$u(x,t) = \frac{1}{\sqrt{4a^2 \pi t}} \int_{-\infty}^\infty f(\xi) \exp\left[-\frac{(\xi - x)^2}{4a^2 t}\right] d\xi. \qquad (6.8.14)$$

• Example 6.8.1

Let us find the temperature field if the initial distribution is

$$u(x,0) = \begin{cases} T_0, & x > 0, \\ -T_0, & x < 0. \end{cases} \qquad (6.8.15)$$

Then

$$u(x,t) = \frac{T_0}{\sqrt{4a^2 \pi t}} \left\{ \int_0^\infty \exp\left[-\frac{(\xi - x)^2}{4a^2 t}\right] d\xi - \int_{-\infty}^0 \exp\left[-\frac{(\xi - x)^2}{4a^2 t}\right] d\xi \right\} \qquad (6.8.16)$$

$$= \frac{T_0}{\sqrt{\pi}} \left[\int_{-x/2a\sqrt{t}}^\infty e^{-\tau^2}\, d\tau - \int_{x/2a\sqrt{t}}^\infty e^{-\tau^2}\, d\tau \right] = \frac{T_0}{\sqrt{\pi}} \int_{-x/2a\sqrt{t}}^{x/2a\sqrt{t}} e^{-\tau^2}\, d\tau \qquad (6.8.17)$$

$$= \frac{2T_0}{\sqrt{\pi}} \int_0^{x/2a\sqrt{t}} e^{-\tau^2}\, d\tau = T_0\, \mathrm{erf}\left(\frac{x}{2a\sqrt{t}}\right), \qquad (6.8.18)$$

where $\mathrm{erf}(\cdot)$ is the error function. □

• Example 6.8.2: Kelvin's estimate of the age of the earth

In the middle of the nineteenth century, Lord Kelvin[14] estimated the age of the earth using the observed vertical temperature gradient at the earth's surface. He hypothesized that the earth was initially formed at a uniform high temperature T_0 and that its surface was subsequently maintained at the lower temperature of T_S. Assuming that most of the heat conduction occurred near the earth's surface, he reasoned that he could neglect the curvature of the earth, consider the earth's surface planar, and employ our one-dimensional heat conduction model in the vertical direction to compute the observed heat flux.

Following Kelvin, we model the earth's surface as a flat plane with an infinitely deep earth below ($z > 0$). Initially the earth has the temperature T_0. Suddenly we drop the temperature at the surface to T_S. We wish to find the heat flux across the boundary at $z = 0$ from the earth into an infinitely deep atmosphere.

The first step is to redefine our temperature scale $v(z,t) = u(z,t) + T_S$, where $v(z,t)$ is the observed temperature so that $u(0,t) = 0$ at the surface. Next, in order to use Equation

---

[14] Thomson, W., 1863: On the secular cooling of the earth. *Philos. Mag.*, Ser. 4, **25**, 157–170.

6.8.14, we must define our initial state for $z < 0$. To maintain the temperature $u(0,t) = 0$, the initial temperature field $f(z)$ must be an odd function, or

$$f(z) = \begin{cases} T_0 - T_S, & z > 0, \\ T_S - T_0, & z < 0. \end{cases} \tag{6.8.19}$$

From Equation 6.8.14,

$$u(z,t) = \frac{T_0 - T_S}{\sqrt{4a^2\pi t}} \left\{ \int_0^\infty \exp\left[-\frac{(\xi - z)^2}{4a^2 t}\right] d\xi - \int_{-\infty}^0 \exp\left[-\frac{(\xi - z)^2}{4a^2 t}\right] d\xi \right\} \tag{6.8.20}$$

$$= (T_0 - T_S)\,\mathrm{erf}\left(\frac{z}{2a\sqrt{t}}\right), \tag{6.8.21}$$

following the work in the previous example.

The heat flux q at the surface $z = 0$ is obtained by differentiating Equation 6.8.21 according to Fourier's law and evaluating the result at $z = 0$:

$$q = -\kappa \frac{\partial v}{\partial z}\bigg|_{z=0} = \frac{\kappa(T_S - T_0)}{a\sqrt{\pi t}}. \tag{6.8.22}$$

The surface heat flux is infinite at $t = 0$ because of the sudden application of the temperature T_S at $t = 0$. After that time, the heat flux decreases with time. Consequently, the time t at which we have the temperature gradient $\partial v(0,t)/\partial z$ is

$$t = \frac{(T_0 - T_S)^2}{\pi a^2 [\partial v(0,t)/\partial z]^2}. \tag{6.8.23}$$

For the present near-surface thermal gradient of 25 K/km, $T_0 - T_S = 2000$ K, and $a^2 = 1$ mm^2/s, the age of the earth from Equation 6.8.23 is 65 million years.

Although Kelvin realized that this was a very rough estimate, his calculation showed that the earth had a finite age. This was in direct contradiction to the contemporary geological principle of *uniformitarianism*: that the earth's surface and upper crust had remained unchanged in temperature and other physical quantities for millions and millions of years. The resulting debate would rage throughout the latter half of the nineteenth century and feature such luminaries as Kelvin, Charles Darwin, Thomas Huxley, and Oliver Heaviside.[15] Eventually Kelvin's arguments would prevail and uniformitarianism would fade into history.

Today, Kelvin's estimate is of academic interest because of the discovery of radioactivity at the turn of the twentieth century. During the first half of the twentieth century, geologists assumed that the radioactivity was uniformly distributed around the globe and restricted to the upper few tens of kilometers of the crust. Using this model they would then use observed heat fluxes to compute the distribution of radioactivity within the solid earth.[16] Now we know that the interior of the earth is quite dynamic; the oceans and continents are mobile and interconnected according to the theory of plate tectonics. However, geophysicists still use measured surface heat fluxes to infer the interior[17] of the earth. □

---

[15] See Burchfield, J. D., 1975: *Lord Kelvin and the Age of the Earth*. Science History Publ., 260 pp.

[16] See Slichter, L. B., 1941: Cooling of the earth. *Bull. Geol. Soc. Am.*, **52**, 561–600.

[17] Sclater, J. G., C. Jaupart, and D. Galson, 1980: The heat flow through oceanic and continental crust and the heat loss of the earth. *Rev. Geophys. Space Phys.*, **18**, 269–311.

• **Example 6.8.3**

So far we have shown how a simple application of separation of variables and the Fourier transform yields solutions to the heat equation over the semi-infinite interval $(0, \infty)$ via Equation 6.8.5. Can we still use this technique for more complicated versions of the heat equation? The answer is yes but the procedure is more complicated. We illustrate it by solving[18]

$$\frac{\partial u}{\partial t} = \alpha \frac{\partial^3 u}{\partial t \partial x^2} + a^2 \frac{\partial^2 u}{\partial x^2}, \qquad 0 < x < \infty, \quad 0 < t, \tag{6.8.24}$$

subject to the boundary conditions

$$u(0, t) = f(t), \quad \lim_{x \to 0} |u(x, t)| < \infty, \qquad 0 < t, \tag{6.8.25}$$

$$\lim_{x \to \infty} u(x, t) \to 0, \; \lim_{x \to \infty} u_x(x, t) \to 0, \qquad 0 < t, \tag{6.8.26}$$

and the initial condition

$$u(x, 0) = 0, \qquad 0 < x < \infty. \tag{6.8.27}$$

We begin by multiplying Equation 6.8.24 by $\sin(kx)$ and integrating over x from 0 to ∞:

$$\alpha \int_0^\infty u_{txx} \sin(kx) \, dx + a^2 \int_0^\infty u_{xx} \sin(kx) \, dx = \int_0^\infty u_t \sin(kx) \, dx. \tag{6.8.28}$$

Next, we integrate by parts. For example,

$$\int_0^\infty u_{xx} \sin(kx) \, dx = u_x \sin(kx) \Big|_0^\infty - k \int_0^\infty u_x \cos(kx) \, dx \tag{6.8.29}$$

$$= -k \int_0^\infty u_x \cos(kx) \, dx \tag{6.8.30}$$

$$= -ku(x, t) \cos(kx) \Big|_0^\infty - k^2 \int_0^\infty u(x, t) \sin(kx) \, dx \tag{6.8.31}$$

$$= kf(t) - k^2 U(k, t), \tag{6.8.32}$$

where

$$U(k, t) = \int_0^\infty u(x, t) \sin(kx) \, dx, \tag{6.8.33}$$

and the boundary conditions have been used to simplify Equation 6.8.29 and Equation 6.8.31. Equation 6.8.33 is the definition of the *Fourier sine transform*. It and its mathematical cousin, the *Fourier cosine transform* $\int_0^\infty u(x, t) \cos(kx) \, dx$, are analogous to the half-range sine and cosine expansions that appear in solving the heat equation over the finite interval $(0, L)$. The difference here is that our range runs from 0 to ∞.

Applying the same technique to the other terms, we obtain

$$\alpha[kf'(t) - k^2 U'(k, t)] + a^2[kf(t) - k^2 U(k, t)] = U'(k, t) \tag{6.8.34}$$

---

[18] See Fetecău, C., and J. Zierep, 2001: On a class of exact solutions of the equations of motion of a second grade fluid. *Acta Mech.*, **150**, 135–138.

with $U(k,0) = 0$, where the primes denote differentiation with respect to time. Solving Equation 6.8.34,

$$e^{a^2 k^2 t/(1+\alpha k^2)} U(k,t) = \frac{\alpha k}{1+\alpha k^2} \int_0^t f'(\tau) e^{a^2 k^2 \tau/(1+\alpha k^2)} d\tau + \frac{a^2 k}{1+\alpha k^2} \int_0^t f(\tau) e^{a^2 k^2 \tau/(1+\alpha k^2)} d\tau. \tag{6.8.35}$$

Using integration by parts on the second integral in Equation 6.8.35, we find that

$$U(k,t) = \frac{1}{k} \left[f(t) - f(0) e^{-a^2 k^2 t/(1+\alpha k^2)} - \frac{1}{1+\alpha k^2} \int_0^t f'(\tau) e^{-a^2 k^2 (t-\tau)/(1+\alpha k^2)} d\tau \right]. \tag{6.8.36}$$

Because

$$u(x,t) = \frac{2}{\pi} \int_0^\infty U(k,t) \sin(kx)\, dk, \tag{6.8.37}$$

$$u(x,t) = \frac{2}{\pi} f(t) \int_0^\infty \frac{\sin(kx)}{k} dk$$
$$- \frac{2}{\pi} \int_0^\infty \frac{\sin(kx)}{k} e^{-a^2 k^2 t/(1+\alpha k^2)}\, dk \left[f(0) + \frac{1}{1+\alpha k^2} \int_0^t f'(\tau) e^{a^2 k^2 \tau/(1+\alpha k^2)}\, d\tau \right] \tag{6.8.38}$$

$$= f(t) - \frac{2}{\pi} \int_0^\infty \frac{\sin(kx)}{k} e^{-a^2 k^2 t/(1+\alpha k^2)} dk \left[f(0) + \frac{1}{1+\alpha k^2} \int_0^t f'(\tau) e^{a^2 k^2 \tau/(1+\alpha k^2)} d\tau \right]. \tag{6.8.39}$$

Problems

For Problems 1–4, find the solution of the heat equation

$$\frac{\partial u}{\partial t} = a^2 \frac{\partial^2 u}{\partial x^2}, \qquad -\infty < x < \infty, \quad 0 < t,$$

subject to the stated initial conditions.

1. $u(x,0) = \begin{cases} 1, & |x| < b, \\ 0, & |x| > b. \end{cases}$ 2. $u(x,0) = e^{-b|x|}$

3. $u(x,0) = \begin{cases} 0, & -\infty < x < 0, \\ T_0, & 0 < x < b, \\ 0, & b < x < \infty. \end{cases}$ 4. $u(x,0) = \delta(x)$

Lovering[19] has applied the solution to Problem 1 to cases involving the cooling of lava.

5. Solve the spherically symmetric equation of diffusion,[20]

$$\frac{\partial u}{\partial t} = a^2 \left(\frac{\partial^2 u}{\partial r^2} + \frac{2}{r} \frac{\partial u}{\partial r} \right), \qquad 0 \le r < \infty, \quad 0 < t,$$

[19] Lovering, T. S., 1935: Theory of heat conduction applied to geological problems. *Bull. Geol. Soc. Am.*, **46**, 69–94.

[20] See Shklovskii, I. S., and V. G. Kurt, 1960: Determination of atmospheric density at a height of 430 km by means of the diffusion of sodium vapors. *Am. Rocket Soc. J.*, **30**, 662–667.

with $u(r, 0) = u_0(r)$.

Step 1: Assuming $v(r, t) = r \, u(r, t)$, show that the problem can be recast as

$$\frac{\partial v}{\partial t} = a^2 \frac{\partial^2 v}{\partial r^2}, \qquad 0 \le r < \infty, \quad 0 < t,$$

with $v(r, 0) = r \, u_0(r)$.

Step 2: Using Equation 6.8.14, show that the general solution is

$$u(r, t) = \frac{1}{2ar\sqrt{\pi t}} \int_0^\infty u_0(\rho) \left\{ \exp\left[-\frac{(r - \rho)^2}{4a^2 t} \right] - \exp\left[-\frac{(r + \rho)^2}{4a^2 t} \right] \right\} \rho \, d\rho.$$

Hint: What is the constraint on Equation 6.8.14 so that the solution remains radially symmetric?

Step 3: For the initial concentration of

$$u_0(r) = \begin{cases} N_0, & 0 \le r < r_0, \\ 0, & r_0 < r, \end{cases}$$

show that

$$u(r, t) = \tfrac{1}{2} N_0 \left[\operatorname{erf}\left(\frac{r_0 - r}{2a\sqrt{t}} \right) + \operatorname{erf}\left(\frac{r_0 + r}{2a\sqrt{t}} \right) + \frac{2a\sqrt{t}}{r\sqrt{\pi}} \left\{ \exp\left[-\frac{(r_0 + r)^2}{4a^2 t} \right] - \exp\left[-\frac{(r_0 - r)^2}{4a^2 t} \right] \right\} \right],$$

where $\operatorname{erf}(\cdot)$ is the error function.

Further Readings

Bracewell, R. N., 2000: *The Fourier Transform and Its Applications.* McGraw-Hill Book Co., 616 pp. This book presents the theory as well as a wealth of applications.

Körner, T. W., 1988: *Fourier Analysis.* Cambridge University Press, 591 pp. Presents several interesting applications.

Sneddon, I. N., 1995: *Fourier Transforms.* Dover, 542 pp. A wonderful book that illustrates the use of Fourier and Bessel transforms in solving a wealth of problems taken from the sciences and engineering.

Titchmarsh, E. C., 1948: *Introduction to the Theory of Fourier Integrals.* Oxford University Press, 391. A source book on the theory of Fourier integrals until 1950.

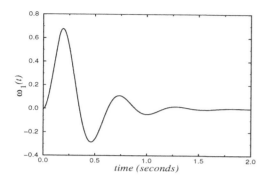

Chapter 7
The Laplace Transform

The previous chapter introduced the concept of the Fourier integral. If the function is nonzero only when $t > 0$, a similar transform, the *Laplace transform*,[1] exists. It is particularly useful in solving initial-value problems involving linear, constant coefficient, ordinary, and partial differential equations. The present chapter develops the general properties and techniques of Laplace transforms.

7.1 DEFINITION AND ELEMENTARY PROPERTIES

Consider a function $f(t)$ such that $f(t) = 0$ for $t < 0$. Then the *Laplace integral*:

$$\mathcal{L}[f(t)] = F(s) = \int_0^\infty f(t)e^{-st}\,dt \tag{7.1.1}$$

defines the Laplace transform of $f(t)$, which we shall write $\mathcal{L}[f(t)]$ or $F(s)$. The Laplace transform converts a function of t into a function of the transform variable s.

Not all functions have a Laplace transform because the integral, Equation 7.1.1, may fail to exist. For example, the function may have infinite discontinuities. For this reason, $f(t) = \tan(t)$ does *not* have a Laplace transform. We can avoid this difficulty by requiring that $f(t)$ be *piece-wise continuous*. That is, we can divide a finite range into a finite number of intervals in such a manner that $f(t)$ is continuous inside each interval and approaches finite values as we approach either end of any interval from the interior.

---

[1] The standard reference for Laplace transforms is Doetsch, G., 1950: *Handbuch der Laplace-Transformation. Band 1. Theorie der Laplace-Transformation.* Birkhäuser Verlag, 581 pp.; Doetsch, G., 1955: *Handbuch der Laplace-Transformation. Band 2. Anwendungen der Laplace-Transformation. 1. Abteilung.* Birkhäuser Verlag, 433 pp.; Doetsch, G., 1956: *Handbuch der Laplace-Transformation. Band 3. Anwendungen der Laplace-Transformation. 2. Abteilung.* Birkhäuser Verlag, 298 pp.

Another unacceptable function is $f(t) = 1/t$ because the integral Equation 7.1.1 fails to exist. This leads to the requirement that the product $t^n|f(t)|$ is bounded near $t = 0$ for some number $n < 1$.

Finally, $|f(t)|$ cannot grow too rapidly or it could overwhelm the e^{-st} term. To express this, we introduce the concept of functions of *exponential order*. By exponential order we mean that there exist some constants, M and k, for which $|f(t)| \leq Me^{kt}$ for all $t > 0$. Then, the Laplace transform of $f(t)$ exists if s, or just the real part of s, is greater than k.

In summary, the Laplace transform of $f(t)$ exists, for sufficiently large s, provided $f(t)$ satisfies the following conditions:

- $f(t) = 0$ for $t < 0$,
- $f(t)$ is continuous or piece-wise continuous in every interval,
- $t^n|f(t)| < \infty$ as $t \to 0$ for some number n, where $n < 1$,
- $e^{-s_0 t}|f(t)| < \infty$ as $t \to \infty$, for some number s_0. The quantity s_0 is called the *abscissa of convergence*.

- **Example 7.1.1**

Let us find the Laplace transform of 1, e^{at}, $\sin(at)$, $\cos(at)$, and t^n from the definition of the Laplace transform. From Equation 7.1.1, direct integration yields

$$\mathcal{L}(1) = \int_0^\infty e^{-st}dt = -\left.\frac{e^{-st}}{s}\right|_0^\infty = \frac{1}{s}, \qquad s > 0, \tag{7.1.2}$$

$$\mathcal{L}(e^{at}) = \int_0^\infty e^{at}e^{-st}dt = \int_0^\infty e^{-(s-a)t}dt \tag{7.1.3}$$

$$= -\left.\frac{e^{-(s-a)t}}{s-a}\right|_0^\infty = \frac{1}{s-a}, \qquad s > a, \tag{7.1.4}$$

$$\mathcal{L}[\sin(at)] = \int_0^\infty \sin(at)e^{-st}dt = -\left.\frac{e^{-st}}{s^2+a^2}[s\sin(at)+a\cos(at)]\right|_0^\infty \tag{7.1.5}$$

$$= \frac{a}{s^2+a^2}, \qquad s > 0, \tag{7.1.6}$$

$$\mathcal{L}[\cos(at)] = \int_0^\infty \cos(at)e^{-st}dt = \left.\frac{e^{-st}}{s^2+a^2}[-s\cos(at)+a\sin(at)]\right|_0^\infty \tag{7.1.7}$$

$$= \frac{s}{s^2+a^2}, \qquad s > 0, \tag{7.1.8}$$

and

$$\mathcal{L}(t^n) = \int_0^\infty t^n e^{-st}dt = n!e^{-st}\sum_{m=0}^n \left.\frac{t^{n-m}}{(n-m)!s^{m+1}}\right|_0^\infty = \frac{n!}{s^{n+1}}, \quad s > 0, \tag{7.1.9}$$

where n is a positive integer.

MATLAB provides the routine `laplace` to compute the Laplace transform for a given function. For example,

```
>> syms a n s t
>> laplace(1,t,s)
ans =
1/s
>> laplace(exp(a*t),t,s)
ans =
1/(s-a)
>> laplace(sin(a*t),t,s)
ans =
a/(s^2+a^2)
>> laplace(cos(a*t),t,s)
ans =
s/(s^2+a^2)
>> laplace(t^5,t,s)
ans =
120/s^6
```

$\square$

The Laplace transform inherits two important properties from its integral definition. First, the transform of a sum equals the sum of the transforms, or

$$\mathcal{L}[c_1 f(t) + c_2 g(t)] = c_1 \mathcal{L}[f(t)] + c_2 \mathcal{L}[g(t)]. \qquad (7.1.10)$$

This linearity property holds with complex numbers and functions as well.

• **Example 7.1.2**

Success with Laplace transforms often rests with the ability to manipulate a given transform into a form that you can invert by inspection. Consider the following examples.

Given $F(s) = 4/s^3$, then

$$F(s) = 2 \times \frac{2}{s^3}, \qquad \text{and} \qquad f(t) = 2t^2 \qquad (7.1.11)$$

from Equation 7.1.9.

Given

$$F(s) = \frac{s+2}{s^2+1} = \frac{s}{s^2+1} + \frac{2}{s^2+1}, \qquad (7.1.12)$$

then

$$f(t) = \cos(t) + 2\sin(t) \qquad (7.1.13)$$

by Equation 7.1.6, Equation 7.1.8, and Equation 7.1.10.

Because

$$F(s) = \frac{1}{s(s-1)} = \frac{1}{s-1} - \frac{1}{s} \qquad (7.1.14)$$

by partial fractions, then

$$f(t) = e^t - 1 \qquad (7.1.15)$$

by Equation 7.1.2, Equation 7.1.4, and Equation 7.1.10.

MATLAB also provides the routine `ilaplace` to compute the inverse Laplace transform for a given function. For example,

```
>> syms s t
>> ilaplace(4/s^3,s,t)
ans =
2*t^2
>> ilaplace((s+2)/(s^2+1),s,t)
ans =
cos(t)+2*sin(t)
>> ilaplace(1/(s*(s-1)),s,t)
ans =
-1+exp(t)
```
□

The second important property deals with derivatives. Suppose $f(t)$ is continuous and has a piece-wise continuous derivative $f'(t)$. Then

$$\mathcal{L}[f'(t)] = \int_0^\infty f'(t)e^{-st}dt = e^{-st}f(t)\big|_0^\infty + s\int_0^\infty f(t)e^{-st}dt \qquad (7.1.16)$$

by integration by parts. If $f(t)$ is of exponential order, $e^{-st}f(t)$ tends to zero as $t \to \infty$, for large enough s, so that $\mathcal{L}[f'(t)] = sF(s) - f(0)$. Similarly, if $f(t)$ and $f'(t)$ are continuous, $f''(t)$ is piece-wise continuous, and all three functions are of exponential order, then

$$\mathcal{L}[f''(t)] = s\mathcal{L}[f'(t)] - f'(0) = s^2F(s) - sf(0) - f'(0). \qquad (7.1.17)$$

In general,

$$\boxed{\mathcal{L}[f^{(n)}(t)] = s^nF(s) - s^{n-1}f(0) - \cdots - sf^{(n-2)}(0) - f^{(n-1)}(0)} \qquad (7.1.18)$$

on the assumption that $f(t)$ and its first $n-1$ derivatives are continuous, $f^{(n)}(t)$ is piece-wise continuous, and all are of exponential order so that the Laplace transform exists.

The converse of Equation 7.1.18 is also of some importance. If

$$u(t) = \int_0^t f(\tau)\,d\tau, \qquad (7.1.19)$$

then

$$\mathcal{L}[u(t)] = \int_0^\infty e^{-st}\left[\int_0^t f(\tau)\,d\tau\right]dt = -\frac{e^{-st}}{s}\int_0^t f(\tau)\,d\tau\bigg|_0^\infty + \frac{1}{s}\int_0^\infty f(t)e^{-st}dt,$$
$$(7.1.20)$$

and

$$\boxed{\mathcal{L}\left[\int_0^t f(\tau)\,d\tau\right] = F(s)/s,} \qquad (7.1.21)$$

where $u(0) = 0$.

Problems

Using the definition of the Laplace transform, find the Laplace transform of the following functions. For Problems 1–4, check your answers using MATLAB.

1. $f(t) = \cosh(at)$

2. $f(t) = \cos^2(at)$

3. $f(t) = (t+1)^2$

4. $f(t) = (t+1)e^{-at}$

5. $f(t) = \begin{cases} e^t, & 0 < t < 2 \\ 0, & 2 < t \end{cases}$

6. $f(t) = \begin{cases} \sin(t), & 0 \le t \le \pi \\ 0, & \pi \le t \end{cases}$

Using your knowledge of the transform for 1, e^{at}, $\sin(at)$, $\cos(at)$, and t^n, find the Laplace transform of

7. $f(t) = 2\sin(t) - \cos(2t) + \cos(3) - t$

8. $f(t) = t - 2 + e^{-5t} - \sin(5t) + \cos(2)$.

Find the inverse of the following transforms. Verify your result using MATLAB.

9. $F(s) = 1/(s+3)$

10. $F(s) = 1/s^4$

11. $F(s) = 1/(s^2+9)$

12. $F(s) = (2s+3)/(s^2+9)$

13. $F(s) = 2/(s^2+1) - 15/s^3 + 2/(s+1) - 6s/(s^2+4)$

14. $F(s) = 3/s + 15/s^3 + (s+5)/(s^2+1) - 6/(s-2)$.

15. Verify the derivative rule for Laplace transforms using the function $f(t) = \sin(at)$.

16. Show that $\mathcal{L}[f(at)] = F(s/a)/a$, where $F(s) = \mathcal{L}[f(t)]$.

17. Using the trigonometric identity $\sin^2(x) = [1 - \cos(2x)]/2$, find the Laplace transform of $f(t) = \sin^2[\pi t/(2T)]$.

7.2 THE HEAVISIDE STEP AND DIRAC DELTA FUNCTIONS

Change can occur abruptly. We throw a switch and electricity suddenly flows. In this section we introduce two functions, the Heaviside step and Dirac delta, that will give us the ability to construct complicated discontinuous functions to express these changes.

> Heaviside step function

We define the *Heaviside step function* as

$$H(t - a) = \begin{cases} 1, & t > a, \\ 0, & t < a, \end{cases} \tag{7.2.1}$$

Largely a self-educated man, Oliver Heaviside (1850–1925) lived the life of a recluse. It was during his studies of the implications of Maxwell's theory of electricity and magnetism that he re-invented Laplace transforms. Initially rejected, it would require the work of Bromwich to justify its use. (Portrait courtesy of the Institution of Engineering and Technology Archives.)

where $a \geq 0$. From this definition,

$$\mathcal{L}[H(t-a)] = \int_a^\infty e^{-st} dt = \frac{e^{-as}}{s}, \qquad s > 0. \tag{7.2.2}$$

Note that this transform is identical to that for $f(t) = 1$ if $a = 0$. This should not surprise us. As pointed out earlier, the function $f(t)$ is zero for all $t < 0$ by definition. Thus, when dealing with Laplace transforms, $f(t) = 1$ and $H(t)$ are identical. Generally we will take 1 rather than $H(t)$ as the inverse of $1/s$. The Heaviside step function is essentially a bookkeeping device that gives us the ability to "switch on" and "switch off" a given function. For example, if we want a function $f(t)$ to become nonzero at time $t = a$, we represent this process by the product $f(t)H(t-a)$. On the other hand, if we only want the function to be "turned on" when $a < t < b$, the desired expression is then $f(t)[H(t-a) - H(t-b)]$. For $t < a$, both step functions in the brackets have the value of zero. For $a < t < b$, the first step function has the value of unity and the second step function has the value of zero, so that we have $f(t)$. For $t > b$, both step functions equal unity so that their difference is zero.

• **Example 7.2.1**

Quite often we need to express the graphical representation of a function by a mathematical equation. We can conveniently do this through the use of step functions in a two-step procedure. The following example illustrates this procedure.

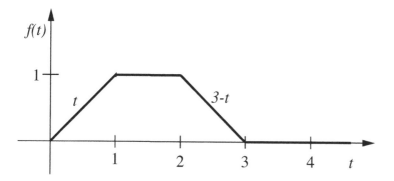

Figure 7.2.1: Graphical representation of Equation 7.2.5.

Consider Figure 7.2.1. We would like to express this graph in terms of Heaviside step functions. We begin by introducing step functions at each point where there is a kink (discontinuity in the first derivative) or jump in the graph—in the present case at $t = 0$, $t = 1$, $t = 2$, and $t = 3$. These are the points of abrupt change. Thus,

$$f(t) = a_0(t)H(t) + a_1(t)H(t-1) + a_2(t)H(t-2) + a_3(t)H(t-3), \qquad (7.2.3)$$

where the coefficients $a_0(t), a_1(t), \ldots$ are yet to be determined. Proceeding from left to right in Figure 7.2.1, the coefficient of each step function equals the mathematical expression that we want after the kink or jump minus the expression before the kink or jump. As each Heaviside turns on, we need to add in the new t behavior and subtract out the old t behavior. Thus, in the present example,

$$f(t) = (t-0)H(t) + (1-t)H(t-1) + [(3-t)-1]H(t-2) + [0-(3-t)]H(t-3) \quad (7.2.4)$$

or

$$f(t) = tH(t) - (t-1)H(t-1) - (t-2)H(t-2) + (t-3)H(t-3). \qquad (7.2.5)$$

We can easily find the Laplace transform of Equation 7.2.5 by the "second shifting" theorem introduced in the next section. □

• Example 7.2.2

Laplace transforms are particularly useful in solving initial-value problems involving linear, constant coefficient, ordinary differential equations where the nonhomogeneous term is discontinuous. As we shall show in the next section, we must first rewrite the nonhomogeneous term using the Heaviside step function before we can use Laplace transforms. For example, given the nonhomogeneous ordinary differential equation:

$$y'' + 3y' + 2y = \begin{cases} t, & 0 < t < 1, \\ 0, & 1 < t, \end{cases} \qquad (7.2.6)$$

we can rewrite the right side of Equation 7.2.6 as

$$y'' + 3y' + 2y = t - tH(t-1) = t - (t-1)H(t-1) - H(t-1). \qquad (7.2.7)$$

In Section 7.7 we will show how to solve this type of ordinary differential equation using Laplace transforms. □

The Laplace Transforms of Some Commonly Encountered Functions

	$f(t), \ t \geq 0$	$F(s)$
1.	1	$\dfrac{1}{s}$
2.	e^{-at}	$\dfrac{1}{s+a}$
3.	$\frac{1}{a}(1 - e^{-at})$	$\dfrac{1}{s(s+a)}$
4.	$\frac{1}{a-b}(e^{-bt} - e^{-at})$	$\dfrac{1}{(s+a)(s+b)}$
5.	$\frac{1}{b-a}(be^{-bt} - ae^{-at})$	$\dfrac{s}{(s+a)(s+b)}$
6.	$\sin(at)$	$\dfrac{a}{s^2 + a^2}$
7.	$\cos(at)$	$\dfrac{s}{s^2 + a^2}$
8.	$\sinh(at)$	$\dfrac{a}{s^2 - a^2}$
9.	$\cosh(at)$	$\dfrac{s}{s^2 - a^2}$
10.	$t\sin(at)$	$\dfrac{2as}{(s^2 + a^2)^2}$
11.	$1 - \cos(at)$	$\dfrac{a^2}{s(s^2 + a^2)}$
12.	$at - \sin(at)$	$\dfrac{a^3}{s^2(s^2 + a^2)}$
13.	$t\cos(at)$	$\dfrac{s^2 - a^2}{(s^2 + a^2)^2}$
14.	$\sin(at) - at\cos(at)$	$\dfrac{2a^3}{(s^2 + a^2)^2}$
15.	$t\sinh(at)$	$\dfrac{2as}{(s^2 - a^2)^2}$
16.	$t\cosh(at)$	$\dfrac{s^2 + a^2}{(s^2 - a^2)^2}$
17.	$at\cosh(at) - \sinh(at)$	$\dfrac{2a^3}{(s^2 - a^2)^2}$
18.	$e^{-bt}\sin(at)$	$\dfrac{a}{(s+b)^2 + a^2}$
19.	$e^{-bt}\cos(at)$	$\dfrac{s+b}{(s+b)^2 + a^2}$
20.	$(1 + a^2t^2)\sin(at) - at\cos(at)$	$\dfrac{8a^3 s^2}{(s^2 + a^2)^3}$

The Laplace Transforms of Some Commonly Encountered Functions (Continued)

	$f(t), \ t \geq 0$	$F(s)$
21.	$\sin(at)\cosh(at) - \cos(at)\sinh(at)$	$\dfrac{4a^3}{s^4 + 4a^4}$
22.	$\sin(at)\sinh(at)$	$\dfrac{2a^2 s}{s^4 + 4a^4}$
23.	$\sinh(at) - \sin(at)$	$\dfrac{2a^3}{s^4 - a^4}$
24.	$\cosh(at) - \cos(at)$	$\dfrac{2a^2 s}{s^4 - a^4}$
25.	$\dfrac{a\sin(at) - b\sin(bt)}{a^2 - b^2}, a^2 \neq b^2$	$\dfrac{s^2}{(s^2 + a^2)(s^2 + b^2)}$
26.	$\dfrac{b\sin(at) - a\sin(bt)}{ab(b^2 - a^2)}, a^2 \neq b^2$	$\dfrac{1}{(s^2 + a^2)(s^2 + b^2)}$
27.	$\dfrac{\cos(at) - \cos(bt)}{b^2 - a^2}, a^2 \neq b^2$	$\dfrac{s}{(s^2 + a^2)(s^2 + b^2)}$
28.	$t^n, n \geq 0$	$\dfrac{n!}{s^{n+1}}$
29.	$\dfrac{t^{n-1}e^{-at}}{(n-1)!}, n > 0$	$\dfrac{1}{(s+a)^n}$
30.	$\dfrac{(n-1) - at}{(n-1)!}t^{n-2}e^{-at}, n > 1$	$\dfrac{s}{(s+a)^n}$
31.	$t^n e^{-at}, n \geq 0$	$\dfrac{n!}{(s+a)^{n+1}}$
32.	$\dfrac{2^n t^{n-(1/2)}}{1 \cdot 3 \cdot 5 \cdots (2n-1)\sqrt{\pi}}, n \geq 1$	$s^{-[n+(1/2)]}$
33.	$J_0(at)$	$\dfrac{1}{\sqrt{s^2 + a^2}}$
34.	$I_0(at)$	$\dfrac{1}{\sqrt{s^2 - a^2}}$
35.	$\dfrac{1}{\sqrt{a}}\operatorname{erf}(\sqrt{at})$	$\dfrac{1}{s\sqrt{s+a}}$
36.	$\dfrac{1}{\sqrt{\pi t}}e^{-at} + \sqrt{a}\operatorname{erf}(\sqrt{at})$	$\dfrac{\sqrt{s+a}}{s}$
37.	$\dfrac{1}{\sqrt{\pi t}} - ae^{a^2 t}\operatorname{erfc}(a\sqrt{t})$	$\dfrac{1}{a + \sqrt{s}}$
38.	$e^{at}\operatorname{erfc}(\sqrt{at})$	$\dfrac{1}{s + \sqrt{as}}$
39.	$\dfrac{1}{2\sqrt{\pi t^3}}\left(e^{bt} - e^{at}\right)$	$\sqrt{s-a} - \sqrt{s-b}$

The Laplace Transforms of Some Commonly Encountered Functions (Continued)

	$f(t), \ t \geq 0$	$F(s)$
40.	$\dfrac{1}{\sqrt{\pi t}} + a e^{a^2 t} \mathrm{erf}(a\sqrt{t})$	$\dfrac{\sqrt{s}}{s - a^2}$
41.	$\dfrac{1}{\sqrt{\pi t}} e^{at}(1 + 2at)$	$\dfrac{s}{(s - a)\sqrt{s - a}}$
42.	$\dfrac{1}{a} e^{a^2 t} \mathrm{erf}(a\sqrt{t})$	$\dfrac{1}{(s - a^2)\sqrt{s}}$
43.	$\sqrt{\dfrac{a}{\pi t^3}} e^{-a/t}, a > 0$	$e^{-2\sqrt{as}}$
44.	$\dfrac{1}{\sqrt{\pi t}} e^{-a/t}, a \geq 0$	$\dfrac{1}{\sqrt{s}} e^{-2\sqrt{as}}$
45.	$\mathrm{erfc}\left(\sqrt{\dfrac{a}{t}}\right), a \geq 0$	$\dfrac{1}{s} e^{-2\sqrt{as}}$
46.	$2\sqrt{\dfrac{t}{\pi}} \exp\left(-\dfrac{a^2}{4t}\right) - a\,\mathrm{erfc}\left(\dfrac{a}{2\sqrt{t}}\right), \ a \geq 0$	$\dfrac{e^{-a\sqrt{s}}}{s\sqrt{s}}$
47.	$-e^{b^2 t + ab}\mathrm{erfc}\left(b\sqrt{t} + \dfrac{a}{2\sqrt{t}}\right) + \mathrm{erfc}\left(\dfrac{a}{2\sqrt{t}}\right), a \geq 0$	$\dfrac{b e^{-a\sqrt{s}}}{s(b + \sqrt{s})}$
48.	$e^{ab} e^{b^2 t}\mathrm{erfc}\left(b\sqrt{t} + \dfrac{a}{2\sqrt{t}}\right), a \geq 0$	$\dfrac{e^{-a\sqrt{s}}}{\sqrt{s}\,(b + \sqrt{s})}$

Notes:

$$\text{Error function: } \mathrm{erf}(x) = \frac{2}{\sqrt{\pi}} \int_0^x e^{-y^2}\,dy$$

$$\text{Complementary error function: } \mathrm{erfc}(x) = 1 - \mathrm{erf}(x)$$

Dirac delta function

The second special function is the *Dirac delta function* or *impulse function*. We define it by

$$\delta(t - a) = \begin{cases} \infty, & t = a, \\ 0, & t \neq a, \end{cases} \qquad \int_0^\infty \delta(t - a)\,dt = 1, \qquad (7.2.8)$$

where $a \geq 0$.

A popular way of visualizing the delta function is as a very narrow rectangular pulse:

$$\delta(t - a) = \lim_{\epsilon \to 0} \begin{cases} 1/\epsilon, & 0 < |t - a| < \epsilon/2, \\ 0, & |t - a| > \epsilon/2, \end{cases} \qquad (7.2.9)$$

where $\epsilon > 0$ is some small number and $a > 0$. See Figure 7.2.2. This pulse has a width ϵ, height $1/\epsilon$, and its center at $t = a$ so that its area is unity. Now, as this pulse shrinks in width ($\epsilon \to 0$), its height increases so that it remains centered at $t = a$ and its area equals unity. If we continue this process, always keeping the area unity and the pulse

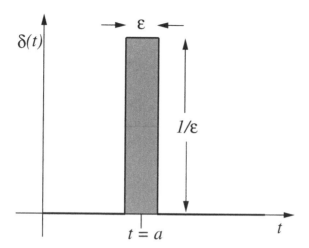

Figure 7.2.2: The Dirac delta function.

symmetric about $t = a$, eventually we obtain an extremely narrow, very large amplitude pulse at $t = a$. If we proceed to the limit, where the width approaches zero and the height approaches infinity (but still with unit area), we obtain the delta function $\delta(t - a)$.

The delta function was introduced earlier during our study of Fourier transforms. So what is the difference between the delta function introduced then and the delta function now? Simply put, the delta function can now only be used on the interval $[0, \infty)$. Outside of that, we shall use it very much as we did with Fourier transforms.

Using Equation 7.2.9, the Laplace transform of the delta function is

$$\mathcal{L}[\delta(t - a)] = \int_0^\infty \delta(t - a)e^{-st}\,dt = \lim_{\epsilon \to 0} \frac{1}{\epsilon} \int_{a-\epsilon/2}^{a+\epsilon/2} e^{-st}\,dt \tag{7.2.10}$$

$$= \lim_{\epsilon \to 0} \frac{1}{\epsilon s}\left(e^{-as+\epsilon s/2} - e^{-as-\epsilon s/2}\right) \tag{7.2.11}$$

$$= \lim_{\epsilon \to 0} \frac{e^{-as}}{\epsilon s}\left(1 + \frac{\epsilon s}{2} + \frac{\epsilon^2 s^2}{8} + \cdots - 1 + \frac{\epsilon s}{2} - \frac{\epsilon^2 s^2}{8} + \cdots\right) \tag{7.2.12}$$

$$= e^{-as}. \tag{7.2.13}$$

In the special case when $a = 0$, $\mathcal{L}[\delta(t)] = 1$, a property that we will use in Section 7.7. Note that this is exactly the result that we obtained for the Fourier transform of the delta function.

If we integrate the impulse function,

$$\int_0^t \delta(\tau - a)\,d\tau = \begin{cases} 0, & t < a, \\ 1, & t > a, \end{cases} \tag{7.2.14}$$

according to whether the impulse does or does not come within the range of integration. This integral gives a result that is precisely the definition of the Heaviside step function, so that we can rewrite Equation 7.2.14:

$$\int_0^t \delta(\tau - a)\,d\tau = H(t - a). \tag{7.2.15}$$

Consequently, the delta function behaves like the derivative of the step function, or

$$\frac{d}{dt}\left[H(t-a)\right] = \delta(t-a). \tag{7.2.16}$$

Because the conventional derivative does not exist at a point of discontinuity, we can only make sense of Equation 7.2.16 if we extend the definition of the derivative. Here we extended the definition formally, but a richer and deeper understanding arises from the theory of generalized functions.[2]

• **Example 7.2.3**

Let us find the (generalized) derivative of

$$f(t) = 3t^2 \left[H(t) - H(t-1)\right]. \tag{7.2.17}$$

Proceeding formally,

$$f'(t) = 6t\left[H(t) - H(t-1)\right] + 3t^2\left[\delta(t) - \delta(t-1)\right] \tag{7.2.18}$$
$$= 6t\left[H(t) - H(t-1)\right] + 0 - 3\delta(t-1) \tag{7.2.19}$$
$$= 6t\left[H(t) - H(t-1)\right] - 3\delta(t-1), \tag{7.2.20}$$

because $f(t)\delta(t-t_0) = f(t_0)\delta(t-t_0)$. □

• **Example 7.2.4**

MATLAB also includes the step and Dirac delta functions among its intrinsic functions. There are two types of step functions. In symbolic calculations, the function is `Heaviside` while `step function` is used in numerical calculations. For example, the Laplace transform of Equation 7.2.5 is:

```
>>syms s,t
>>laplace('t-(t-1)*Heaviside(t-1)-(t-2)*Heaviside(t-2)'...
    '+(t-3)*Heaviside(t-3)',t,s)
ans =
1/s^2-exp(-s)/s^2-exp(-2*s)/s^2+exp(-3*s)/s^2
```

In a similar manner, the symbolic function for the Dirac delta function is `Dirac`. Therefore, the Laplace transform of $(t-1)\delta(t-2)$ is:

```
>>syms s,t
>>laplace('(t-1)*Dirac(t-2)',t,s)
ans =
exp(-2*s)
```

---

[2] The generalization of the definition of a function so that it can express in a mathematically correct form such idealized concepts as the density of a material point, a point charge or point dipole, the space charge of a simple or double layer, the intensity of an instantaneous source, etc.

Problems

Sketch the following functions and express them in terms of the Heaviside step functions:

1. $f(t) = \begin{cases} 0, & 0 \le t \le 2 \\ t - 2, & 2 \le t < 3 \\ 0, & 3 < t \end{cases}$

2. $f(t) = \begin{cases} 0, & 0 < t < a \\ 1, & a < t < 2a \\ -1, & 2a < t < 3a \\ 0, & 3a < t \end{cases}$

Rewrite the following nonhomogeneous ordinary differential equations using the Heaviside step functions:

3. $y'' + 3y' + 2y = \begin{cases} 0, & 0 < t < 1 \\ 1, & 1 < t \end{cases}$

4. $y'' + 4y = \begin{cases} 0, & 0 < t < 4 \\ 3, & 4 < t \end{cases}$

5. $y'' + 4y' + 4y = \begin{cases} 0, & 0 < t < 2 \\ t, & 2 < t \end{cases}$

6. $y'' + 3y' + 2y = \begin{cases} 0, & 0 < t < 1 \\ e^t, & 1 < t \end{cases}$

7. $y'' - 3y' + 2y = \begin{cases} 0, & 0 < t < 2 \\ e^{-t}, & 2 < t \end{cases}$

8. $y'' - 3y' + 2y = \begin{cases} 0, & 0 < t < 1 \\ t^2, & 1 < t \end{cases}$

9. $y'' + y = \begin{cases} \sin(t), & 0 \le t \le \pi \\ 0, & \pi \le t \end{cases}$

10. $y'' + 3y' + 2y = \begin{cases} t, & 0 \le t \le a \\ ae^{-(t-a)}, & a \le t \end{cases}$

7.3 SOME USEFUL THEOREMS

Although at first sight there would appear to be a bewildering number of transforms to either memorize or tabulate, there are several useful theorems, that can extend the applicability of a given transform.

First shifting theorem

Consider the transform of the function $e^{-at} f(t)$, where a is any real number. Then, by definition,

$$\mathcal{L}\left[e^{-at} f(t)\right] = \int_0^\infty e^{-st} e^{-at} f(t)\, dt = \int_0^\infty e^{-(s+a)t} f(t)\, dt, \qquad (7.3.1)$$

or

$$\mathcal{L}\left[e^{-at} f(t)\right] = F(s+a). \qquad (7.3.2)$$

That is, if $F(s)$ is the transform of $f(t)$ and a is a constant, then $F(s+a)$ is the transform of $e^{-at} f(t)$.

• Example 7.3.1

Let us find the Laplace transform of $f(t) = e^{-at} \sin(bt)$. Because the Laplace transform of $\sin(bt)$ is $b/(s^2 + b^2)$,

$$\mathcal{L}\left[e^{-at} \sin(bt)\right] = \frac{b}{(s+a)^2 + b^2}, \qquad (7.3.3)$$

where we simply replaced s by $s + a$ in the transform for $\sin(bt)$.   □

• **Example 7.3.2**

Let us find the inverse of the Laplace transform

$$F(s) = \frac{s+2}{s^2 + 6s + 1}. \tag{7.3.4}$$

Rearranging terms,

$$F(s) = \frac{s+2}{s^2 + 6s + 1} = \frac{s+2}{(s+3)^2 - 8} = \frac{s+3}{(s+3)^2 - 8} - \frac{1}{2\sqrt{2}}\frac{2\sqrt{2}}{(s+3)^2 - 8}. \tag{7.3.5}$$

Immediately, from the first shifting theorem,

$$f(t) = e^{-3t}\cosh\left(2\sqrt{2}t\right) - \frac{e^{-3t}}{2\sqrt{2}}\sinh\left(2\sqrt{2}t\right). \tag{7.3.6}$$

   □

Second shifting theorem

The *second shifting theorem* states that if $F(s)$ is the transform of $f(t)$, then $e^{-bs}F(s)$ is the transform of $f(t-b)H(t-b)$, where b is real and positive. To show this, consider the Laplace transform of $f(t-b)H(t-b)$. Then, from the definition,

$$\mathcal{L}[f(t-b)H(t-b)] = \int_0^\infty f(t-b)H(t-b)e^{-st}\,dt \tag{7.3.7}$$

$$= \int_b^\infty f(t-b)e^{-st}\,dt = \int_0^\infty e^{-bs}e^{-sx}f(x)\,dx \tag{7.3.8}$$

$$= e^{-bs}\int_0^\infty e^{-sx}f(x)\,dx, \tag{7.3.9}$$

or

$$\mathcal{L}[f(t-b)H(t-b)] = e^{-bs}F(s), \tag{7.3.10}$$

where we set $x = t - b$. This theorem is of fundamental importance because it allows us to write down the transforms for "delayed" time functions. That is, functions that "turn on" b units after the initial time.

• **Example 7.3.3**

Let us find the inverse of the transform $(1 - e^{-s})/s$. Since

$$\frac{1 - e^{-s}}{s} = \frac{1}{s} - \frac{e^{-s}}{s}, \tag{7.3.11}$$

$$\mathcal{L}^{-1}\left(\frac{1}{s} - \frac{e^{-s}}{s}\right) = \mathcal{L}^{-1}\left(\frac{1}{s}\right) - \mathcal{L}^{-1}\left(\frac{e^{-s}}{s}\right) = H(t) - H(t-1), \tag{7.3.12}$$

because $\mathcal{L}^{-1}(1/s) = f(t) = 1$, and $f(t-1) = 1$. □

- **Example 7.3.4**

    Let us find the Laplace transform of $f(t) = (t^2 - 1)H(t - 1)$.
    We begin by noting that

$$(t^2 - 1)H(t - 1) = [(t - 1 + 1)^2 - 1]H(t - 1) \tag{7.3.13}$$
$$= [(t - 1)^2 + 2(t - 1)]H(t - 1) \tag{7.3.14}$$
$$= (t - 1)^2 H(t - 1) + 2(t - 1)H(t - 1). \tag{7.3.15}$$

A direct application of the second shifting theorem then leads to

$$\mathcal{L}[(t^2 - 1)H(t - 1)] = \frac{2e^{-s}}{s^3} + \frac{2e^{-s}}{s^2}. \tag{7.3.16}$$

□

- **Example 7.3.5**

    In Example 7.2.2 we discussed the use of Laplace transforms in solving ordinary differential equations. One further step along the road consists of finding $Y(s) = \mathcal{L}[y(t)]$. Now that we have the second shifting theorem, let us do this.
    Continuing Example 7.2.2 with $y(0) = 0$ and $y'(0) = 1$, let us take the Laplace transform of Equation 7.2.7. Employing the second shifting theorem and Equation 7.1.18, we find that

$$s^2 Y(s) - sy(0) - y'(0) + 3sY(s) - 3y(0) + 2Y(s) = \frac{1}{s^2} - \frac{e^{-s}}{s^2} - \frac{e^{-s}}{s}. \tag{7.3.17}$$

Substituting in the initial conditions and solving for $Y(s)$, we finally obtain

$$Y(s) = \frac{1}{(s + 2)(s + 1)} + \frac{1}{s^2(s + 2)(s + 1)} + \frac{e^{-s}}{s^2(s + 2)(s + 1)} + \frac{e^{-s}}{s(s + 2)(s + 1)}. \tag{7.3.18}$$

□

> **Laplace transform of $t^n f(t)$**

    In addition to the shifting theorems, there are two other particularly useful theorems that involve the derivative and integral of the transform $F(s)$. For example, if we write

$$F(s) = \mathcal{L}[f(t)] = \int_0^\infty f(t)e^{-st}dt \tag{7.3.19}$$

and differentiate with respect to s, then

$$F'(s) = \int_0^\infty -tf(t)e^{-st}dt = -\mathcal{L}[tf(t)]. \tag{7.3.20}$$

In general, we have that

$$F^{(n)}(s) = (-1)^n \mathcal{L}[t^n f(t)].$$ (7.3.21)

Laplace transform of $f(t)/t$

Consider the following integration of the Laplace transform $F(s)$:

$$\int_s^\infty F(z)\,dz = \int_s^\infty \left[\int_0^\infty f(t) e^{-zt} dt \right] dz.$$ (7.3.22)

Upon interchanging the order of integration, we find that

$$\int_s^\infty F(z)\,dz = \int_0^\infty f(t) \left[\int_s^\infty e^{-zt} dz \right] dt = -\int_0^\infty f(t) \left. \frac{e^{-zt}}{t} \right|_s^\infty dt = \int_0^\infty \frac{f(t)}{t} e^{-st} dt.$$ (7.3.23)

Therefore,

$$\int_s^\infty F(z)\,dz = \mathcal{L}\left[\frac{f(t)}{t} \right].$$ (7.3.24)

• Example 7.3.6

Let us find the transform of $t\sin(at)$. From Equation 7.3.20,

$$\mathcal{L}[t\sin(at)] = -\frac{d}{ds}\left\{ \mathcal{L}[\sin(at)] \right\} = -\frac{d}{ds}\left[\frac{a}{s^2 + a^2} \right] = \frac{2as}{(s^2 + a^2)^2}.$$ (7.3.25)

□

• Example 7.3.7

Let us find the transform of $[1 - \cos(at)]/t$. To solve this problem, we apply Equation 7.3.24 and find that

$$\mathcal{L}\left[\frac{1 - \cos(at)}{t} \right] = \int_s^\infty \mathcal{L}[1 - \cos(at)] \Big|_{s=z} dz = \int_s^\infty \left(\frac{1}{z} - \frac{z}{z^2 + a^2} \right) dz$$ (7.3.26)

$$= \ln(z) - \tfrac{1}{2}\ln(z^2 + a^2) \Big|_s^\infty = \ln\left(\frac{z}{\sqrt{z^2 + a^2}} \right) \Big|_s^\infty$$ (7.3.27)

$$= \ln(1) - \ln\left(\frac{s}{\sqrt{s^2 + a^2}} \right) = -\ln\left(\frac{s}{\sqrt{s^2 + a^2}} \right).$$ (7.3.28)

□

Initial-value theorem

Let $f(t)$ and $f'(t)$ possess Laplace transforms. Then, from the definition of the Laplace transform,

$$\int_0^\infty f'(t) e^{-st}\,dt = sF(s) - f(0).$$ (7.3.29)

Because s is a parameter in Equation 7.3.29 and the existence of the integral is implied by the derivative rule, we can let $s \to \infty$ before we integrate. In that case, the left side of Equation 7.3.29 vanishes to zero, which leads to

$$\lim_{s \to \infty} sF(s) = f(0). \tag{7.3.30}$$

This is the *initial-value theorem*.

- **Example 7.3.8**

Let us verify the initial-value theorem using $f(t) = e^{3t}$. Because $F(s) = 1/(s-3)$, $\lim_{s \to \infty} s/(s-3) = 1$. This agrees with $f(0) = 1$.

In the common case when the Laplace transform is a ratio of two polynomials, we can use MATLAB to find the initial value. This consists of two steps. First, we construct $sF(s)$ by creating vectors that describe the numerator and denominator of $sF(s)$ and then evaluate the numerator and denominator using very large values of s. For example, in the previous example,

```
>>num = [1 0];
>>den = [1 -3];
>>initialvalue = polyval(num,1e20) / polyval(den,1e20)
initialvalue =
    1
```

$\square$

Final-value theorem

Let $f(t)$ and $f'(t)$ possess Laplace transforms. Then, in the limit of $s \to 0$, Equation 7.3.29 becomes

$$\int_0^\infty f'(t)\,dt = \lim_{t \to \infty} \int_0^t f'(\tau)\,d\tau = \lim_{t \to \infty} f(t) - f(0) = \lim_{s \to 0} sF(s) - f(0). \tag{7.3.31}$$

Because $f(0)$ is not a function of t or s, the quantity $f(0)$ cancels from Equation 7.3.31, leaving

$$\lim_{t \to \infty} f(t) = \lim_{s \to 0} sF(s). \tag{7.3.32}$$

Equation 7.3.32 is the *final-value theorem*. It should be noted that this theorem assumes that $\lim_{t \to \infty} f(t)$ exists. For example, it does not apply to sinusoidal functions.

In the case when $f(t)$ is a periodic function with a period T, Gluskin[3] showed that

$$\lim_{s \to 0} sF(s) = \frac{1}{T} \int_0^T f(t)\,dt. \tag{7.3.33}$$

---

[3] Gluskin, E., 2003: Let us teach this generalization of the final-value theorem. *Eur. J. Phys.*, **24**, 591–597.

• Example 7.3.9

Let us verify the final-value theorem using $f(t) = t$. Because $F(s) = 1/s^2$,

$$\lim_{s \to 0} sF(s) = \lim_{s \to 0} 1/s = \infty. \qquad (7.3.34)$$

The limit of $f(t)$ as $t \to \infty$ is also undefined.

Just as we can use MATLAB to find the initial value of a Laplace transform in the case when $F(s)$ is a ratio of two polynomials, we can do the same here for the final value. Again we define vectors num and den that give $sF(s)$ and then evaluate them at $s = 0$. Using the previous example, the MATLAB commands are:

```
>>num = [0 1 0];
>>den = [1 0 0];
>>finalvalue = polyval(num,0) / polyval(den,0)
Warning:  Divide by zero.
finalvalue =
    NaN
```

This agrees with the result from a hand calculation and shows what happens when the denominator has a zero.   □

• Example 7.3.10

Looking ahead, we will shortly need to find the Laplace transform of $y(t)$, which is defined by a differential equation. For example, we will want $Y(s)$ where $y(t)$ is governed by

$$y'' + 2y' + 2y = \cos(t) + \delta(t - \pi/2), \qquad y(0) = y'(0) = 0. \qquad (7.3.35)$$

Applying Laplace transforms to both sides of Equation 7.3.35, we have that

$$\mathcal{L}(y'') + 2\mathcal{L}(y') + 2\mathcal{L}(y) = \mathcal{L}[\cos(t)] + \mathcal{L}[\delta(t - \pi/2)], \qquad (7.3.36)$$

or

$$s^2 Y(s) - sy(0) - y'(0) + 2sY(s) - 2y(0) + 2Y(s) = \frac{s}{s^2 + 1} + e^{-s\pi/2}. \qquad (7.3.37)$$

Substituting for $y(0)$ and $y'(0)$ and solving for $Y(s)$, we find that

$$Y(s) = \frac{s}{(s^2 + 1)(s^2 + 2s + 2)} + \frac{e^{-s\pi/2}}{s^2 + 2s + 2}. \qquad (7.3.38)$$

Presently this is as far as we can go.

How would we use MATLAB to find $Y(s)$? The following MATLAB script shows you how:

```
clear
% define symbolic variables
syms pi s t Y
% take Laplace transform of left side of differential equation
LHS = laplace(diff(diff(sym('y(t)')))+2*diff(sym('y(t)'))...
    +2*sym('y(t)'));
% take Laplace transform of right side of differential equation
```

```
RHS = laplace(cos(t)+'Dirac(t-pi/2)',t,s);
% set Y for Laplace transform of y
%       and introduce initial conditions
newLHS = subs(LHS,'laplace(y(t),t,s)','y(0)','D(y)(0)',Y,0,0);
% solve for Y
Y = solve(newLHS-RHS,Y)
```

It yields

```
Y =
(s+exp(-1/2*pi*s)*s^2+exp(-1/2*pi*s))/(s^4+3*s^2+2*s^3+2*s+2)
```

Problems

Find the Laplace transform of the following functions and then check your work using MATLAB.

1. $f(t) = e^{-t}\sin(2t)$

2. $f(t) = e^{-2t}\cos(2t)$

3. $f(t) = t^2 H(t-1)$

4. $f(t) = e^{2t} H(t-3)$

5. $f(t) = te^t + \sin(3t)e^t + \cos(5t)e^{2t}$

6. $f(t) = t^4 e^{-2t} + \sin(3t)e^t + \cos(4t)e^{2t}$

7. $f(t) = t^2 e^{-t} + \sin(2t)e^t + \cos(3t)e^{-3t}$

8. $f(t) = t^2 H(t-1) + e^t H(t-2)$

9. $f(t) = (t^2 + 2)H(t-1) + H(t-2)$

10. $f(t) = (t+1)^2 H(t-1) + e^t H(t-2)$

11. $f(t) = te^{-3t}\sin(2t)$

12. $f(t) = e^t[1 - \cosh(t)]$

13. $f(t) = \begin{cases} \sin(t), & 0 \le t \le \pi \\ 0, & \pi \le t \end{cases}$

14. $f(t) = \begin{cases} t, & 0 \le t \le 2 \\ 2, & 2 \le t \end{cases}$

Find the inverse of the following Laplace transforms by hand and using MATLAB:

15. $F(s) = \dfrac{1}{(s+2)^4}$

16. $F(s) = \dfrac{s}{(s+2)^4}$

17. $F(s) = \dfrac{s}{s^2 + 2s + 2}$

18. $F(s) = \dfrac{s+3}{s^2 + 2s + 2}$

19. $F(s) = \dfrac{s}{(s+1)^3} + \dfrac{s+1}{s^2 + 2s + 2}$

20. $F(s) = \dfrac{s}{(s+2)^2} + \dfrac{s+2}{s^2 + 2s + 2}$

21. $F(s) = \dfrac{s}{(s+2)^3} + \dfrac{s+4}{s^2 + 4s + 5}$

22. $F(s) = \dfrac{e^{-3s}}{s-1}$

23. $F(s) = \dfrac{e^{-2s}}{(s+1)^2}$

24. $F(s) = \dfrac{s\,e^{-s}}{s^2 + 2s + 2}$

25. $F(s) = \dfrac{e^{-4s}}{s^2 + 4s + 5}$

26. $F(s) = \dfrac{s\,e^{-s}}{s^2 + 4} + \dfrac{e^{-3s}}{(s-2)^4}$

27. $F(s) = \dfrac{e^{-s}}{s^2 + 4} + \dfrac{(s-1)\,e^{-3s}}{s^4}$

28. $F(s) = \dfrac{(s+1)\,e^{-s}}{s^2 + 4} + \dfrac{e^{-3s}}{s^4}$

29. Find the Laplace transform of $f(t) = te^t[H(t-1)-H(t-2)]$ by using (a) the definition of the Laplace transform, and (b) a joint application of the first and second shifting theorems.

30. Write the function

$$f(t) = \begin{cases} t, & 0 < t < a, \\ 0, & a < t, \end{cases}$$

in terms of Heaviside's step functions. Then find its transform using (a) the definition of the Laplace transform, and (b) the second shifting theorem.

In Problems 31–34, write the function $f(t)$ in terms of the Heaviside step functions and then find its transform using the second shifting theorem. Check your answer using MATLAB.

31. $f(t) = \begin{cases} t/2, & 0 \le t < 2 \\ 0, & 2 < t \end{cases}$

32. $f(t) = \begin{cases} t, & 0 \le t \le 1 \\ 1, & 1 \le t < 2 \\ 0, & 2 < t \end{cases}$

33. $f(t) = \begin{cases} t, & 0 \le t \le 2 \\ 4 - t, & 2 \le t \le 4 \\ 0, & 4 \le t \end{cases}$

34. $f(t) = \begin{cases} 0, & 0 \le t \le 1 \\ t - 1, & 1 \le t \le 2 \\ 1, & 2 \le t < 3 \\ 0, & 3 < t \end{cases}$

Find $Y(s)$ for the following ordinary differential equations and then use MATLAB to check your work.

35. $y'' + 3y' + 2y = H(t - 1);$ $y(0) = y'(0) = 0$

36. $y'' + 4y = 3H(t - 4);$ $y(0) = 1,\ y'(0) = 0$

37. $y'' + 4y' + 4y = tH(t - 2);$ $y(0) = 0,\ y'(0) = 2$

38. $y'' + 3y' + 2y = e^t H(t - 1);$ $y(0) = y'(0) = 0$

39. $y'' - 3y' + 2y = e^{-t} H(t - 2);$ $y(0) = 2,\ y'(0) = 0$

40. $y'' - 3y' + 2y = t^2 H(t - 1);$ $y(0) = 0,\ y'(0) = 5$

41. $y'' + y = \sin(t)[1 - H(t - \pi)];$ $y(0) = y'(0) = 0$

42. $y'' + 3y' + 2y = t + \left[ae^{-(t-a)} - t\right] H(t - a);$ $y(0) = y'(0) = 0.$

For each of the following functions, find its value at $t = 0$. Then check your answer using the initial-value theorem by hand and using MATLAB.

43. $f(t) = t$

44. $f(t) = \cos(at)$

45. $f(t) = te^{-t}$

46. $f(t) = e^t \sin(3t)$

For each of the following Laplace transforms, state whether you can or cannot apply the final-value theorem. If you can, find the final value by hand and using MATLAB. Check your result by finding the inverse and finding the limit as $t \to \infty$.

47. $F(s) = \dfrac{1}{s - 1}$

48. $F(s) = \dfrac{1}{s}$

49. $F(s) = \dfrac{1}{s+1}$ 50. $F(s) = \dfrac{s}{s^2+1}$

51. $F(s) = \dfrac{2}{s(s^2+3s+2)}$ 52. $F(s) = \dfrac{2}{s(s^2-3s+2)}$

53. Using the fact that

$$e^{-c\xi} = 1 - c\int_0^{\xi} e^{-c\eta}\, d\eta,$$

show[4] that

$$\frac{1}{s}\exp\left(-\frac{asx}{s+b}\right) = \frac{1}{s} - \int_0^{ax} \frac{e^{-\eta}}{s+b}\exp\left(\frac{b\eta}{s+b}\right) d\eta$$

if $x > 0$. Therefore, using the fact[5] that

$$\mathcal{L}^{-1}\left(s^{-\nu-1}e^{\alpha/s}\right) = \left(\frac{t}{\alpha}\right)^{\nu/2} I_\nu\left(2\sqrt{\alpha t}\right), \qquad \Re(\nu) > -1,$$

and the first shifting theorem, show

$$\mathcal{L}^{-1}\left[\frac{1}{s}\exp\left(-\frac{asx}{s+b}\right)\right] = 1 - e^{-bt}\int_0^{ax} e^{-\eta} I_0\left(2\sqrt{2t\eta}\right) d\eta,$$

where $I_\nu(\cdot)$ is a modified Bessel function of the first kind and order ν introduced in Section 12.2.

7.4 THE LAPLACE TRANSFORM OF A PERIODIC FUNCTION

Periodic functions frequently occur in engineering problems and we shall now show how to calculate their transform. They possess the property that $f(t+T) = f(t)$ for $t > 0$ and equal zero for $t < 0$, where T is the period of the function.

For convenience, let us define a function $x(t)$ that equals zero except over the interval $(0, T)$ where it equals $f(t)$:

$$x(t) = \begin{cases} f(t), & 0 < t < T, \\ 0, & T < t. \end{cases} \tag{7.4.1}$$

By definition,

$$F(s) = \int_0^\infty f(t)e^{-st}dt = \int_0^T f(t)e^{-st}dt + \int_T^{2T} f(t)e^{-st}dt + \cdots + \int_{kT}^{(k+1)T} f(t)e^{-st}dt + \cdots. \tag{7.4.2}$$

Now let $z = t - kT$, where $k = 0, 1, 2, \ldots$, in the kth integral and $F(s)$ becomes

$$F(s) = \int_0^T f(z)e^{-sz}\, dz + \int_0^T f(z+T)e^{-s(z+T)}\, dz + \cdots + \int_0^T f(z+kT)e^{-s(z+kT)}\, dz + \cdots. \tag{7.4.3}$$

[4] Liaw, C. H., J. S. P. Wang, R. A. Greenkorn, and K. C. Chao, 1979: Kinetics of fixed-bed absorption: A new solution. *AICHE J.*, **25**, 376–381.

[5] Watson, E. J., 1981: *Laplace Transforms and Applications.* Van Nostrand Reinhold Co., p. 195.

However,
$$x(z) = f(z) = f(z+T) = \ldots = f(z+kT) = \ldots, \qquad (7.4.4)$$

because the range of integration in each integral is from 0 to T. Thus, $F(s)$ becomes

$$F(s) = \int_0^T x(z)e^{-sz}\,dz + e^{-sT}\int_0^T x(z)e^{-sz}\,dz + \cdots + e^{-ksT}\int_0^T x(z)e^{-sz}\,dz + \cdots \quad (7.4.5)$$

or
$$F(s) = \left(1 + e^{-sT} + e^{-2sT} + \cdots + e^{-ksT} + \cdots\right)X(s). \qquad (7.4.6)$$

The first term on the right side of Equation 7.4.6 is a geometric series with common ratio e^{-sT}. If $|e^{-sT}| < 1$, then the series converges and

$$F(s) = \frac{X(s)}{1 - e^{-sT}}. \qquad (7.4.7)$$

• **Example 7.4.1**

Let us find the Laplace transform of the square wave with period T:

$$f(t) = \begin{cases} h, & 0 < t < T/2, \\ -h, & T/2 < t < T. \end{cases} \qquad (7.4.8)$$

By definition $x(t)$ is

$$x(t) = \begin{cases} h, & 0 < t < T/2, \\ -h, & T/2 < t < T, \\ 0, & T < t. \end{cases} \qquad (7.4.9)$$

Then

$$X(s) = \int_0^\infty x(t)e^{-st}\,dt = \int_0^{T/2} h\,e^{-st}\,dt + \int_{T/2}^T (-h)\,e^{-st}\,dt \qquad (7.4.10)$$

$$= \frac{h}{s}\left(1 - 2e^{-sT/2} + e^{-sT}\right) = \frac{h}{s}\left(1 - e^{-sT/2}\right)^2, \qquad (7.4.11)$$

and

$$F(s) = \frac{h\left(1 - e^{-sT/2}\right)^2}{s\left(1 - e^{-sT}\right)} = \frac{h\left(1 - e^{-sT/2}\right)}{s\left(1 + e^{-sT/2}\right)}. \qquad (7.4.12)$$

If we multiply numerator and denominator by $\exp(sT/4)$ and recall that $\tanh(u) = (e^u - e^{-u})/(e^u + e^{-u})$, we have that

$$F(s) = \frac{h}{s}\tanh\left(\frac{sT}{4}\right). \qquad (7.4.13)$$

$\square$

• **Example 7.4.2**

Let us find the Laplace transform of the periodic function

$$f(t) = \begin{cases} \sin(2\pi t/T), & 0 \le t \le T/2, \\ 0, & T/2 \le t \le T. \end{cases} \qquad (7.4.14)$$

By definition $x(t)$ is

$$x(t) = \begin{cases} \sin(2\pi t/T), & 0 \le t \le T/2, \\ 0, & T/2 \le t. \end{cases} \qquad (7.4.15)$$

Then

$$X(s) = \int_0^{T/2} \sin\left(\frac{2\pi t}{T}\right) e^{-st}\, dt = \frac{2\pi T}{s^2 T^2 + 4\pi^2}\left(1 + e^{-sT/2}\right). \qquad (7.4.16)$$

Hence,

$$F(s) = \frac{X(s)}{1 - e^{-sT}} = \frac{2\pi T}{s^2 T^2 + 4\pi^2} \times \frac{1 + e^{-sT/2}}{1 - e^{-sT}} = \frac{2\pi T}{s^2 T^2 + 4\pi^2} \times \frac{1}{1 - e^{-sT/2}}. \qquad (7.4.17)$$

Problems

Find the Laplace transform for the following periodic functions:

1. $f(t) = \sin(t)$, $\quad 0 \le t \le \pi$, $\qquad f(t) = f(t + \pi)$

2. $f(t) = \begin{cases} \sin(t), & 0 \le t \le \pi, \\ 0, & \pi \le t \le 2\pi, \end{cases} \qquad f(t) = f(t + 2\pi)$

3. $f(t) = \begin{cases} t, & 0 \le t < a, \\ 0, & a < t \le 2a, \end{cases} \qquad f(t) = f(t + 2a)$

4. $f(t) = \begin{cases} 1, & 0 < t < a, \\ 0, & a < t < 2a, \\ -1, & 2a < t < 3a, \\ 0, & 3a < t < 4a, \end{cases} \qquad f(t) = f(t + 4a)$

7.5 INVERSION BY PARTIAL FRACTIONS: HEAVISIDE'S EXPANSION THEOREM

In the previous sections, we devoted our efforts to calculating the Laplace transform of a given function. Obviously, we must have a method for going the other way. Given a transform, we must find the corresponding function. This is often a very formidable task. In the next few sections we shall present some general techniques for the inversion of a Laplace transform.

The first technique involves transforms that we can express as the ratio of two polynomials: $F(s) = q(s)/p(s)$. We shall assume that the order of $q(s)$ is *less* than $p(s)$ and we have divided out any common factor between them. In principle we know that $p(s)$ has n zeros, where n is the order of the $p(s)$ polynomial. Some of the zeros may be complex, some of them may be real, and some of them may be duplicates of other zeros. In the case when $p(s)$ has n simple, nonrepeating roots (zeros), a simple method exists for inverting the transform.

We want to rewrite $F(s)$ in the form:

$$F(s) = \frac{a_1}{s - s_1} + \frac{a_2}{s - s_2} + \cdots + \frac{a_n}{s - s_n} = \frac{q(s)}{p(s)}, \tag{7.5.1}$$

where $s_1, s_2, \ldots, s_n$ are the n simple zeros of $p(s)$. We now multiply both sides of Equation 7.5.1 by $s - s_1$ so that

$$\frac{(s - s_1)q(s)}{p(s)} = a_1 + \frac{(s - s_1)a_2}{s - s_2} + \cdots + \frac{(s - s_1)a_n}{s - s_n}. \tag{7.5.2}$$

If we set $s = s_1$, the right side of Equation 7.5.2 becomes simply a_1. The left side takes the form $0/0$ and there are two cases. If $p(s) = (s - s_1)g(s)$, then $a_1 = q(s_1)/g(s_1)$. If we cannot explicitly factor out $s - s_1$, l'Hôpital's rule gives

$$a_1 = \lim_{s \to s_1} \frac{(s - s_1)q(s)}{p(s)} = \lim_{s \to s_1} \frac{(s - s_1)q'(s) + q(s)}{p'(s)} = \frac{q(s_1)}{p'(s_1)}. \tag{7.5.3}$$

In a similar manner, we can compute all of the coefficients a_k, where $k = 1, 2, \ldots, n$. Therefore,

$$\mathcal{L}^{-1}[F(s)] = \mathcal{L}^{-1}\left[\frac{q(s)}{p(s)}\right] = \mathcal{L}^{-1}\left(\frac{a_1}{s - s_1} + \frac{a_2}{s - s_2} + \cdots + \frac{a_n}{s - s_n}\right) \tag{7.5.4}$$

$$= a_1 e^{s_1 t} + a_2 e^{s_2 t} + \cdots + a_n e^{s_n t}. \tag{7.5.5}$$

This is *Heaviside's expansion theorem*, applicable when $p(s)$ has only simple zeros.

• Example 7.5.1

Let us invert the transform $s/[(s + 2)(s^2 + 1)]$. It has three simple zeros at $s = -2$ and $s = \pm i$ in the denominator. From our earlier discussion, $q(s) = s$, $p(s) = (s + 2)(s^2 + 1)$, and $p'(s) = 3s^2 + 4s + 1$. Therefore,

$$\mathcal{L}^{-1}\left[\frac{s}{(s + 2)(s^2 + 1)}\right] = \frac{-2}{12 - 8 + 1}e^{-2t} + \frac{i}{-3 + 4i + 1}e^{it} + \frac{-i}{-3 - 4i + 1}e^{-it} \tag{7.5.6}$$

$$= -\frac{2}{5}e^{-2t} + \frac{i}{-2 + 4i}e^{it} - \frac{i}{-2 - 4i}e^{-it} \tag{7.5.7}$$

$$= -\frac{2}{5}e^{-2t} + i\frac{-2 - 4i}{4 + 16}e^{it} - i\frac{-2 + 4i}{4 + 16}e^{-it} \tag{7.5.8}$$

$$= -\frac{2}{5}e^{-2t} + \frac{1}{5}\sin(t) + \frac{2}{5}\cos(t), \tag{7.5.9}$$

where we used $\sin(t) = \frac{1}{2i}(e^{it} - e^{-it})$, and $\cos(t) = \frac{1}{2}(e^{it} + e^{-it})$.   □

• Example 7.5.2

Let us invert the transform $1/[(s - 1)(s - 2)(s - 3)]$. There are three simple zeros in the denominator: $s_1 = 1$, $s_2 = 2$, and $s_3 = 3$. In this case, the easiest method for computing a_1, a_2, and a_3 is

$$a_1 = \lim_{s \to 1} \frac{s - 1}{(s - 1)(s - 2)(s - 3)} = \frac{1}{2}, \tag{7.5.10}$$

$$a_2 = \lim_{s \to 2} \frac{s-2}{(s-1)(s-2)(s-3)} = -1, \quad (7.5.11)$$

and

$$a_3 = \lim_{s \to 3} \frac{s-3}{(s-1)(s-2)(s-3)} = \frac{1}{2}. \quad (7.5.12)$$

Therefore,

$$\mathcal{L}^{-1}\left[\frac{1}{(s-1)(s-2)(s-3)}\right] = \mathcal{L}^{-1}\left[\frac{a_1}{s-1} + \frac{a_2}{s-2} + \frac{a_3}{s-3}\right] = \tfrac{1}{2}e^t - e^{2t} + \tfrac{1}{2}e^{3t}. \quad (7.5.13)$$

□

Note that for inverting transforms of the form $F(s)e^{-as}$ with $a > 0$, you should use Heaviside's expansion theorem to first invert $F(s)$ and then apply the second shifting theorem.

Let us now find the expansion when we have multiple roots, namely

$$F(s) = \frac{q(s)}{p(s)} = \frac{q(s)}{(s-s_1)^{m_1}(s-s_2)^{m_2}\cdots(s-s_n)^{m_n}}, \quad (7.5.14)$$

where the order of the denominator, $m_1 + m_2 + \cdots + m_n$, is greater than that for the numerator. Once again we eliminated any common factor between the numerator and denominator. Now we can write $F(s)$ as

$$F(s) = \sum_{k=1}^{n}\sum_{j=1}^{m_k} \frac{a_{kj}}{(s-s_k)^{m_k-j+1}}. \quad (7.5.15)$$

Multiplying Equation 7.5.15 by $(s-s_k)^{m_k}$,

$$\frac{(s-s_k)^{m_k}q(s)}{p(s)} = a_{k1} + a_{k2}(s-s_k) + \cdots + a_{km_k}(s-s_k)^{m_k-1}$$
$$+ (s-s_k)^{m_k}\left[\frac{a_{11}}{(s-s_1)^{m_1}} + \cdots + \frac{a_{nm_n}}{s-s_n}\right], \quad (7.5.16)$$

where we grouped together into the square-bracketed term all of the terms except for those with a_{kj} coefficients. Taking the limit as $s \to s_k$,

$$a_{k1} = \lim_{s \to s_k} \frac{(s-s_k)^{m_k}q(s)}{p(s)}. \quad (7.5.17)$$

Let us now take the derivative of Equation 7.5.16,

$$\frac{d}{ds}\left[\frac{(s-s_k)^{m_k}q(s)}{p(s)}\right] = a_{k2} + 2a_{k3}(s-s_k) + \cdots + (m_k-1)a_{km_k}(s-s_k)^{m_k-2}$$
$$+ \frac{d}{ds}\left\{(s-s_k)^{m_k}\left[\frac{a_{11}}{(s-s_1)^{m_1}} + \cdots + \frac{a_{nm_n}}{s-s_n}\right]\right\}. \quad (7.5.18)$$

Taking the limit as $s \to s_k$,

$$a_{k2} = \lim_{s \to s_k} \frac{d}{ds}\left[\frac{(s-s_k)^{m_k}q(s)}{p(s)}\right]. \quad (7.5.19)$$

In general,

$$a_{kj} = \lim_{s \to s_k} \frac{1}{(j-1)!} \frac{d^{j-1}}{ds^{j-1}} \left[\frac{(s-s_k)^{m_k} q(s)}{p(s)} \right], \tag{7.5.20}$$

and by direct inversion,

$$f(t) = \sum_{k=1}^{n} \sum_{j=1}^{m_k} \frac{a_{kj}}{(m_k - j)!} t^{m_k - j} e^{s_k t}. \tag{7.5.21}$$

• Example 7.5.3

Let us find the inverse of

$$F(s) = \frac{s}{(s+2)^2 (s^2 + 1)}. \tag{7.5.22}$$

We first note that the denominator has simple zeros at $s = \pm i$ and a repeated root at $s = -2$. Therefore,

$$F(s) = \frac{A}{s-i} + \frac{B}{s+i} + \frac{C}{s+2} + \frac{D}{(s+2)^2}, \tag{7.5.23}$$

where

$$A = \lim_{s \to i} (s-i) F(s) = \tfrac{1}{6+8i}, \tag{7.5.24}$$

$$B = \lim_{s \to -i} (s+i) F(s) = \tfrac{1}{6-8i}, \tag{7.5.25}$$

$$C = \lim_{s \to -2} \frac{d}{ds} \left[(s+2)^2 F(s) \right] = \lim_{s \to -2} \frac{d}{ds} \left(\frac{s}{s^2+1} \right) = -\tfrac{3}{25}, \tag{7.5.26}$$

and

$$D = \lim_{s \to -2} (s+2)^2 F(s) = -\tfrac{2}{5}. \tag{7.5.27}$$

Thus,

$$f(t) = \tfrac{1}{6+8i} e^{it} + \tfrac{1}{6-8i} e^{-it} - \tfrac{3}{25} e^{-2t} - \tfrac{2}{5} t e^{-2t} = \tfrac{3}{25} \cos(t) + \tfrac{4}{25} \sin(t) - \tfrac{3}{25} e^{-2t} - \tfrac{10}{25} t e^{-2t}. \tag{7.5.28}$$

$\square$

Let us now find the inverse of

$$F(s) = \frac{cs + (ca - \omega d)}{(s+a)^2 + \omega^2} = \frac{cs + (ca - \omega d)}{(s+a-\omega i)(s+a+\omega i)} \tag{7.5.29}$$

by Heaviside's expansion theorem. Then

$$F(s) = \frac{c+di}{2(s+a-\omega i)} + \frac{c-di}{2(s+a+\omega i)} = \frac{\sqrt{c^2+d^2} e^{\theta i}}{2(s+a-\omega i)} + \frac{\sqrt{c^2+d^2} e^{-\theta i}}{2(s+a+\omega i)}, \tag{7.5.30}$$

where $\theta = \tan^{-1}(d/c)$. Note that we must choose θ so that it gives the correct sign for c and d.

Taking the inverse of Equation 7.5.30,

$$f(t) = \tfrac{1}{2}\sqrt{c^2 + d^2}\, e^{-at+\omega ti+\theta i} + \tfrac{1}{2}\sqrt{c^2 + d^2}\, e^{-at-\omega ti-\theta i} = \sqrt{c^2 + d^2}\, e^{-at}\cos(\omega t + \theta). \quad (\mathbf{7.5.31})$$

Equation 7.5.31 is the amplitude/phase form of the inverse of Equation 7.5.29. It is partic-ularly popular with electrical engineers.

• Example 7.5.4

Let us express the inverse of

$$F(s) = \frac{8s - 3}{s^2 + 4s + 13} \qquad (\mathbf{7.5.32})$$

in the amplitude/phase form.

Starting with

$$F(s) = \frac{8s - 3}{(s + 2 - 3i)(s + 2 + 3i)} = \frac{4 + 19i/6}{s + 2 - 3i} + \frac{4 - 19i/6}{s + 2 + 3i} \qquad (\mathbf{7.5.33})$$

$$= \frac{5.1017 e^{38.3675°i}}{s + 2 - 3i} + \frac{5.1017 e^{-38.3675°i}}{s + 2 + 3i}, \qquad (\mathbf{7.5.34})$$

or

$$f(t) = 5.1017 e^{-2t+3it+38.3675°i} + 5.1017 e^{-2t-3it-38.3675°i} \qquad (\mathbf{7.5.35})$$

$$= 10.2034 e^{-2t}\cos(3t + 38.3675°). \qquad (\mathbf{7.5.36})$$

□

• Example 7.5.5: The design of film projectors

For our final example we anticipate future work. The primary use of Laplace transforms is the solution of differential equations. In this example we illustrate this technique that includes Heaviside's expansion theorem in the form of amplitude and phase.

This problem[6] arose in the design of projectors for motion pictures. An early problem was ensuring that the speed at which the film passed the electric eye remained essentially constant; otherwise, a frequency modulation of the reproduced sound resulted. Figure 7.5.1(A) shows a diagram of the projector. Many will remember this design from their days as a school projectionist. In this section we shall show that this particular design filters out variations in the film speed caused by irregularities either in the driving-gear trains or in the engagement of the sprocket teeth with the holes in the film.

Let us now focus on the film head—a hollow drum of small moment of inertia J_1. See Figure 7.5.1(B). Within it there is a concentric inner flywheel of moment of inertia J_2, where $J_2 \gg J_1$. The remainder of the space within the drum is filled with oil. The inner flywheel rotates on precision ball bearings on the drum shaft. The only coupling between the drum and flywheel is through fluid friction and the very small friction in the ball bearings.

---

[6] Cook, E. D., 1935: The technical aspects of the high-fidelity reproducer. *J. Soc. Motion Pict. Eng.*, **25**, 289–312.

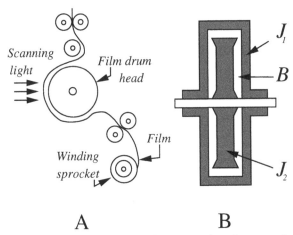

$$A \qquad\qquad\qquad B$$

Figure 7.5.1: (A) The schematic for the scanning light in a motion-picture projector and (B) interior of the film drum head.

The flection of the film-loops between the drum head and idler pulleys provides the spring restoring force for the system as the film runs rapidly through the system.

From Figure 7.5.1 the dynamical equations governing the outer case and inner flywheel are (1) the rate of change of the outer casing of the film head equals the frictional torque given to the casing from the inner flywheel plus the restoring torque due to the flection of the film, and (2) the rate of change of the inner flywheel equals the negative of the frictional torque given to the outer casing by the inner flywheel.

Assuming that the frictional torque between the two flywheels is proportional to the difference in their angular velocities, the frictional torque given to the casing from the inner flywheel is $B(\omega_2 - \omega_1)$, where B is the frictional resistance, ω_1 and ω_2 are the deviations of the drum and inner flywheel from their normal angular velocities, respectively. If r is the ratio of the diameter of the winding sprocket to the diameter of the drum, the restoring torque due to the flection of the film and its corresponding angular twist equals $K \int_0^t (r\omega_0 - \omega_1)\, d\tau$, where K is the rotational stiffness and ω_0 is the deviation of the winding sprocket from its normal angular velocity. The quantity $r\omega_0$ gives the angular velocity at which the film is running through the projector because the winding sprocket is the mechanism that pulls the film. Consequently, the equations governing this mechanical system are

$$J_1 \frac{d\omega_1}{dt} = K \int_0^t (r\omega_0 - \omega_1)\, d\tau + B(\omega_2 - \omega_1), \qquad (7.5.37)$$

and

$$J_2 \frac{d\omega_2}{dt} = -B(\omega_2 - \omega_1). \qquad (7.5.38)$$

With the winding sprocket, the drum, and the flywheel running at their normal uniform angular velocities, let us assume that the winding sprocket introduces a disturbance equivalent to a unit increase in its angular velocity for 0.15 second, followed by the resumption of its normal velocity. It is assumed that the film in contact with the drum cannot slip. The initial conditions are $\omega_1(0) = \omega_2(0) = 0$.

Taking the Laplace transform of Equation 7.5.37 and Equation 7.5.38 and using Equation 7.1.18,

$$\left(J_1 s + B + \frac{K}{s} \right) \Omega_1(s) - B\Omega_2(s) = \frac{rK}{s} \Omega_0(s) = rK \mathcal{L}\left[\int_0^t \omega_0(\tau)\, d\tau \right], \qquad (7.5.39)$$

and

$$-B\Omega_1(s) + (J_2 s + B)\Omega_2(s) = 0. \tag{7.5.40}$$

The solution of Equation 7.5.39 and Equation 7.5.40 for $\Omega_1(s)$ is

$$\Omega_1(s) = \frac{rK}{J_1} \frac{(s + a_0)\Omega_0(s)}{s^3 + b_2 s^2 + b_1 s + b_0}, \tag{7.5.41}$$

where typical values[7] are

$$\frac{rK}{J_1} = 90.8, a_0 = \frac{B}{J_2} = 1.47, b_0 = \frac{BK}{J_1 J_2} = 231, b_1 = \frac{K}{J_1} = 157, \text{and } b_2 = \frac{B(J_1 + J_2)}{J_1 J_2} = 8.20. \tag{7.5.42}$$

The transform $\Omega_1(s)$ has three simple zeros in the denominator located at $s_1 = -1.58$, $s_2 = -3.32 + 11.6i$, and $s_3 = -3.32 - 11.6i$.

Because the sprocket angular velocity deviation $\omega_0(t)$ is a pulse of unit amplitude and 0.15 second duration, we express it as the difference of two Heaviside step functions

$$\omega_0(t) = H(t) - H(t - 0.15). \tag{7.5.43}$$

Its Laplace transform is

$$\Omega_0(s) = \frac{1}{s} - \frac{1}{s}e^{-0.15s} \tag{7.5.44}$$

so that Equation 7.5.41 becomes

$$\Omega_1(s) = \frac{rK}{J_1} \frac{(s + a_0)}{s(s - s_1)(s - s_2)(s - s_3)} \left(1 - e^{-0.15s}\right). \tag{7.5.45}$$

The inversion of Equation 7.5.45 follows directly from the second shifting theorem and Heaviside's expansion theorem, or

$$\omega_1(t) = K_0 + K_1 e^{s_1 t} + K_2 e^{s_2 t} + K_3 e^{s_3 t} \tag{7.5.46}$$
$$- [K_0 + K_1 e^{s_1(t-0.15)} + K_2 e^{s_2(t-0.15)} + K_3 e^{s_3(t-0.15)}]H(t - 0.16),$$

where

$$K_0 = \frac{rK}{J_1} \left.\frac{s + a_0}{(s - s_1)(s - s_2)(s - s_3)}\right|_{s=0} = 0.578, \tag{7.5.47}$$

$$K_1 = \frac{rK}{J_1} \left.\frac{s + a_0}{s(s - s_2)(s - s_3)}\right|_{s=s_1} = 0.046, \tag{7.5.48}$$

$$K_2 = \frac{rK}{J_1} \left.\frac{s + a_0}{s(s - s_1)(s - s_3)}\right|_{s=s_2} = 0.326 e^{165°i}, \tag{7.5.49}$$

and

$$K_3 = \frac{rK}{J_1} \left.\frac{s + a_0}{s(s - s_1)(s - s_2)}\right|_{s=s_3} = 0.326 e^{-165°i}. \tag{7.5.50}$$

---

[7] $J_1 = 1.84 \times 10^4$ dyne cm sec^2 per radian, $J_2 = 8.43 \times 10^4$ dyne cm sec^2 per radian, $B = 12.4 \times 10^4$ dyne cm sec per radian, $K = 2.89 \times 10^6$ dyne cm per radian, and $r = 0.578$.

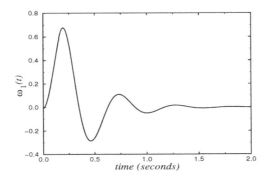

Figure 7.5.2: The deviation $\omega_1(t)$ of a film drum head from its uniform angular velocity when the sprocket angular velocity is perturbed by a unit amount for the duration of 0.15 second.

Using Euler's identity $\cos(t) = (e^{it} + e^{-it})/2$, we can write Equation 7.5.46 as

$$
\begin{aligned}
\omega_1(t) = {}& 0.578 + 0.046e^{-1.58t} + 0.652e^{-3.32t}\cos(11.6t + 165°) \\
& - \{0.578 + 0.046e^{-1.58(t-0.15)} + 0.652e^{-3.32(t-0.15)} \\
& \times \cos[11.6(t - 0.15) + 165°]\}H(t - 0.15).
\end{aligned}
\tag{7.5.51}
$$

Equation 7.5.51 is plotted in Figure 7.5.2. Note that fluctuations in $\omega_1(t)$ are damped out by the particular design of this film projector. Because this mechanical device dampens unwanted fluctuations (or noise) in the motion-picture projector, this particular device is an example of a *mechanical filter*.

Problems

Use Heaviside's expansion theorem to find the inverse of the following Laplace transforms:

1. $F(s) = \dfrac{1}{s^2 + 3s + 2}$

2. $F(s) = \dfrac{s + 3}{(s + 4)(s - 2)}$

3. $F(s) = \dfrac{s - 4}{(s + 2)(s + 1)(s - 3)}$

4. $F(s) = \dfrac{s - 3}{(s^2 + 4)(s + 1)}$

Find the inverse of the following transforms and express them in amplitude/phase form:

5. $F(s) = \dfrac{1}{s^2 + 4s + 5}$

6. $F(s) = \dfrac{1}{s^2 + 6s + 13}$

7. $F(s) = \dfrac{2s - 5}{s^2 + 16}$

8. $F(s) = \dfrac{1}{s(s^2 + 2s + 2)}$

9. $F(s) = \dfrac{s + 2}{s(s^2 + 4)}$

10. $F(s) = \dfrac{2s^2 - 5s + 5}{s(s^2 - 2s + 5)}$

7.6 CONVOLUTION

In this section we turn to a fundamental concept in Laplace transforms: convolution. We shall restrict ourselves to its use in finding the inverse of a transform when that transform consists of the *product* of two simpler transforms. In subsequent sections we will use it to solve ordinary differential equations.

We begin by formally introducing the mathematical operation of the *convolution product*

$$f(t) * g(t) = \int_0^t f(t-x)g(x)\,dx = \int_0^t f(x)g(t-x)\,dx. \qquad (7.6.1)$$

In most cases the operations required by Equation 7.6.1 are straightforward.

• **Example 7.6.1**

Let us find the convolution between $\cos(t)$ and $\sin(t)$.

$$\cos(t) * \sin(t) = \int_0^t \sin(t-x)\cos(x)\,dx = \tfrac{1}{2}\int_0^t [\sin(t) + \sin(t-2x)]\,dx \qquad (7.6.2)$$

$$= \tfrac{1}{2}\int_0^t \sin(t)\,dx + \tfrac{1}{2}\int_0^t \sin(t-2x)\,dx \qquad (7.6.3)$$

$$= \tfrac{1}{2}\sin(t)\,x\big|_0^t + \tfrac{1}{4}\cos(t-2x)\big|_0^t = \tfrac{1}{2}t\sin(t). \qquad (7.6.4)$$

□

• **Example 7.6.2**

Similarly, the convolution between t^2 and $\sin(t)$ is

$$t^2 * \sin(t) = \int_0^t (t-x)^2 \sin(x)\,dx \qquad (7.6.5)$$

$$= -(t-x)^2 \cos(x)\big|_0^t - 2\int_0^t (t-x)\cos(x)\,dx \qquad (7.6.6)$$

$$= t^2 - 2(t-x)\sin(x)\big|_0^t - 2\int_0^t \sin(x)\,dx \qquad (7.6.7)$$

$$= t^2 + 2\cos(t) - 2 \qquad (7.6.8)$$

by integration by parts.

□

• **Example 7.6.3**

Consider now the convolution between e^t and the discontinuous function $H(t-1) - H(t-2)$:

$$e^t * [H(t-1) - H(t-2)] = \int_0^t e^{t-x}[H(x-1) - H(x-2)]\,dx \qquad (7.6.9)$$

$$= e^t \int_0^t e^{-x}[H(x-1) - H(x-2)]\,dx. \qquad (7.6.10)$$

In order to evaluate the integral, Equation 7.6.10, we must examine various cases. If $t < 1$, then both of the step functions equal zero and the convolution equals zero. However, when $1 < t < 2$, the first step function equals one while the second equals zero as the dummy variable x runs between 1 and t. Therefore,

$$e^t * [H(t-1) - H(t-2)] = e^t \int_1^t e^{-x}\,dx = e^{t-1} - 1, \qquad (7.6.11)$$

because the portion of the integral from zero to one equals zero. Finally, when $t > 2$, the integrand is only nonzero for that portion of the integration when $1 < x < 2$. Consequently,

$$e^t * [H(t-1) - H(t-2)] = e^t \int_1^2 e^{-x} dx = e^{t-1} - e^{t-2}. \tag{7.6.12}$$

Thus, the convolution of e^t with the pulse $H(t-1) - H(t-2)$ is

$$e^t * [H(t-1) - H(t-2)] = \begin{cases} 0, & 0 \le t \le 1, \\ e^{t-1} - 1, & 1 \le t \le 2, \\ e^{t-1} - e^{t-2}, & 2 \le t. \end{cases} \tag{7.6.13}$$

MATLAB can also be used to find the convolution of two functions. For example, in the present case the commands are:

```
syms x t positive
int('exp(t-x)*(Heaviside(x-1)-Heaviside(x-2))',x,0,t)
```
yield
```
ans =
-Heaviside(t-1)+Heaviside(t-1)*exp(t-1)+Heaviside(t-2)
    -Heaviside(t-2)*exp(t-2)
```
□

The reason why we introduced convolution stems from the following fundamental theorem (often called *Borel's theorem*[8]). If

$$w(t) = u(t) * v(t), \qquad \text{then} \qquad W(s) = U(s)V(s). \tag{7.6.14}$$

In other words, we can invert a complicated transform by convoluting the inverses to two simpler functions. The proof is as follows:

$$W(s) = \int_0^\infty \left[\int_0^t u(x)v(t-x)\,dx \right] e^{-st} dt \tag{7.6.15}$$

$$= \int_0^\infty \left[\int_x^\infty u(x)v(t-x)e^{-st} dt \right] dx \tag{7.6.16}$$

$$= \int_0^\infty u(x) \left[\int_0^\infty v(r)e^{-s(r+x)} dr \right] dx \tag{7.6.17}$$

$$= \left[\int_0^\infty u(x)e^{-sx} dx \right] \left[\int_0^\infty v(r)e^{-sr} dr \right] = U(s)V(s), \tag{7.6.18}$$

where $t = r + x$. □

• **Example 7.6.4**

Let us find the inverse of the transform

$$\frac{s}{(s^2+1)^2} = \frac{s}{s^2+1} \times \frac{1}{s^2+1} = \mathcal{L}[\cos(t)]\mathcal{L}[\sin(t)] = \mathcal{L}[\cos(t) * \sin(t)] = \mathcal{L}[\tfrac{1}{2}t\sin(t)] \tag{7.6.19}$$

from Example 7.6.1. □

---

[8] Borel, É., 1901: *Leçons sur les séries divergentes.* Gauthier-Villars, p. 104.

• **Example 7.6.5**

Let us find the inverse of the transform

$$\frac{1}{(s^2 + a^2)^2} = \frac{1}{a^2}\left(\frac{a}{s^2 + a^2} \times \frac{a}{s^2 + a^2}\right) = \frac{1}{a^2}\mathcal{L}[\sin(at)]\mathcal{L}[\sin(at)]. \quad (7.6.20)$$

Therefore,

$$\mathcal{L}^{-1}\left[\frac{1}{(s^2 + a^2)^2}\right] = \frac{1}{a^2}\int_0^t \sin[a(t-x)]\sin(ax)\,dx \quad (7.6.21)$$

$$= \frac{1}{2a^2}\int_0^t \cos[a(t-2x)]\,dx - \frac{1}{2a^2}\int_0^t \cos(at)\,dx \quad (7.6.22)$$

$$= -\frac{1}{4a^3}\sin[a(t-2x)]\Big|_0^t - \frac{1}{2a^2}\cos(at)\,x\Big|_0^t \quad (7.6.23)$$

$$= \frac{1}{2a^3}[\sin(at) - at\cos(at)]. \quad (7.6.24)$$

$\square$

• **Example 7.6.6**

Let us use the results from Example 7.6.3 to verify the convolution theorem.

We begin by rewriting Equation 7.6.13 in terms of the Heaviside step functions. Using the method outline in Example 7.2.1,

$$f(t) * g(t) = \left(e^{t-1} - 1\right)H(t-1) + \left(1 - e^{t-2}\right)H(t-2). \quad (7.6.25)$$

Employing the second shifting theorem,

$$\mathcal{L}[f * g] = \frac{e^{-s}}{s-1} - \frac{e^{-s}}{s} + \frac{e^{-2s}}{s} - \frac{e^{-2s}}{s-1} \quad (7.6.26)$$

$$= \frac{e^{-s}}{s(s-1)} - \frac{e^{-2s}}{s(s-1)} = \frac{1}{s-1}\left(\frac{e^{-s}}{s} - \frac{e^{-2s}}{s}\right) \quad (7.6.27)$$

$$= \mathcal{L}[e^t]\mathcal{L}[H(t-1) - H(t-2)] \quad (7.6.28)$$

and the convolution theorem holds true. If we had not rewritten Equation 7.6.13 in terms of step functions, we could still have found $\mathcal{L}[f * g]$ from the definition of the Laplace transform.

Problems

Verify the following convolutions and then show that the convolution theorem is true. Use MATLAB to check your answer.

1. $1 * 1 = t$

2. $1 * \cos(at) = \sin(at)/a$

3. $1 * e^t = e^t - 1$

4. $t * t = t^3/6$

5. $t * \sin(t) = t - \sin(t)$

6. $t * e^t = e^t - t - 1$

7. $t^2 * \sin(at) = \dfrac{t^2}{a} - \dfrac{4}{a^3}\sin^2\left(\dfrac{at}{2}\right)$ 8. $t * H(t-1) = \frac{1}{2}(t-1)^2 H(t-1)$

9. $H(t-a) * H(t-b) = (t-a-b)H(t-a-b)$

10. $t * [H(t) - H(t-2)] = \dfrac{t^2}{2} - \dfrac{(t-2)^2}{2} H(t-2)$

Use the convolution theorem to invert the following functions:

11. $F(s) = \dfrac{1}{s^2(s-1)}$ 12. $F(s) = \dfrac{1}{s^2(s+a)^2}$

13. Given[9]

$$\mathcal{L}^{-1}\left[\dfrac{e^{\alpha/s}}{s}\right] = I_0\left(2\sqrt{\alpha t}\right), \qquad \alpha > 0,$$

show that

$$\dfrac{e^{a/s}}{s-1} = \left(1 + \dfrac{1}{s-1}\right)\dfrac{e^{a/s}}{s}, \qquad a > 0,$$

and

$$\mathcal{L}^{-1}\left(\dfrac{e^{a/s}}{s-1}\right) = [\delta(t) + e^t] * I_0\left(2\sqrt{at}\right) = e^t + e^t \int_0^{\sqrt{4at}} \exp\left(-\dfrac{\tau^2}{4a}\right) I_1(\tau)\, d\tau,$$

where $I_n(\cdot)$ denotes a modified Bessel function of the first kind and order n from Section 12.2. There we showed that $I_0'(\tau) = I_1(\tau)$.

14. Using the fact that

$$\mathcal{L}^{-1}[F(s)G(s)] = \int_0^t g(\tau)f(t-\tau)\, d\tau,$$

and given that

$$\mathcal{L}\left[\dfrac{H(t-a)}{\sqrt{\pi t}}\right] = \dfrac{\mathrm{erfc}(\sqrt{as})}{\sqrt{s}}, \qquad \text{and} \qquad \mathcal{L}\left(\dfrac{1}{\sqrt{\pi t}}\right) = \dfrac{1}{\sqrt{s}},$$

show[10] that

$$\mathcal{L}^{-1}\left[\dfrac{\mathrm{erfc}(\sqrt{as})}{s}\right] = \dfrac{1}{2} + \dfrac{1}{\pi}\arcsin\left(1 - \dfrac{2a}{t}\right),$$

where $\mathrm{erfc}(\cdot)$ is the complementary error function.

15. Prove that the convolution of two Dirac delta functions is a Dirac delta function.

---

[9] Watson, op. cit., p. 195.

[10] DeChant, L. J., 2004: An analytical solution for unconfined, unsteady, inviscid jets; with applications to penetration problem debris cloud formation. *Comput. Math. Applic.*, **48**, 201–213.

7.7 SOLUTION OF LINEAR DIFFERENTIAL EQUATIONS WITH CONSTANT COEFFICIENTS

For the engineer, as it was for Oliver Heaviside, the primary use of Laplace transforms is the solution of ordinary, constant coefficient, linear differential equations. These equations are important not only because they appear in many engineering problems, but also because they may serve as approximations, even if locally, to ordinary differential equations with nonconstant coefficients or to nonlinear ordinary differential equations.

For all of these reasons, we wish to solve the *initial-value problem*

$$\frac{d^n y}{dt^n} + a_1 \frac{d^{n-1} y}{dt^{n-1}} + \cdots + a_{n-1} \frac{dy}{dt} + a_n y = f(t), \quad t > 0, \tag{7.7.1}$$

by Laplace transforms, where $a_1, a_2, \ldots$ are constants and we know the value of $y, y', \ldots, y^{(n-1)}$ at $t = 0$. The procedure is as follows. Applying the derivative rule Equation 7.1.18 to Equation 7.7.1, we reduce the *differential* equation to an *algebraic* one involving the constants $a_1, a_2, \ldots, a_n$, the parameter s, the Laplace transform of $f(t)$, and the values of the initial conditions. We then solve for the Laplace transform of $y(t)$, $Y(s)$. Finally, we apply one of the many techniques of inverting a Laplace transform to find $y(t)$.

Similar considerations hold with *systems* of ordinary differential equations. The Laplace transform of the system of ordinary differential equations results in an algebraic set of equations containing $Y_1(s), Y_2(s), \ldots, Y_n(s)$. By some method we solve this set of equations and invert each transform $Y_1(s), Y_2(s), \ldots, Y_n(s)$ in turn to give $y_1(t), y_2(t), \ldots, y_n(t)$.

The following examples will illustrate the details of the process.

- **Example 7.7.1**

Let us solve the ordinary differential equation

$$y'' + 2y' = 8t, \tag{7.7.2}$$

subject to the initial conditions that $y'(0) = y(0) = 0$. Taking the Laplace transform of both sides of Equation 7.7.2,

$$\mathcal{L}(y'') + 2\mathcal{L}(y') = 8\mathcal{L}(t), \tag{7.7.3}$$

or

$$s^2 Y(s) - sy(0) - y'(0) + 2sY(s) - 2y(0) = \frac{8}{s^2}, \tag{7.7.4}$$

where $Y(s) = \mathcal{L}[y(t)]$. Substituting the initial conditions into Equation 7.7.4 and solving for $Y(s)$,

$$Y(s) = \frac{8}{s^3(s+2)} = \frac{A}{s^3} + \frac{B}{s^2} + \frac{C}{s} + \frac{D}{s+2} = \frac{(s+2)A + s(s+2)B + s^2(s+2)C + s^3 D}{s^3(s+2)}. \tag{7.7.5}$$

Matching powers of s in the numerators of Equation 7.7.5, $C + D = 0$, $B + 2C = 0$, $A + 2B = 0$, and $2A = 8$ or $A = 4$, $B = -2$, $C = 1$, and $D = -1$. Therefore,

$$Y(s) = \frac{4}{s^3} - \frac{2}{s^2} + \frac{1}{s} - \frac{1}{s+2}. \tag{7.7.6}$$

Finally, performing term-by-term inversion of Equation 7.7.6, the final solution is

$$y(t) = 2t^2 - 2t + 1 - e^{-2t}. \tag{7.7.7}$$

We could have done the same operations using the symbolic toolbox with MATLAB. The MATLAB script:

```
clear
% define symbolic variables
syms s t Y
% take Laplace transform of left side of differential equation
LHS = laplace(diff(diff(sym('y(t)')))+2*diff(sym('y(t)')));
% take Laplace transform of right side of differential equation
RHS = laplace(8*t);
% set Y for Laplace transform of y
%     and introduce initial conditions
newLHS = subs(LHS,'laplace(y(t),t,s)','y(0)','D(y)(0)',Y,0,0);
% solve for Y
Y = solve(newLHS-RHS,Y);
% invert Laplace transform and find y(t)
y = ilaplace(Y,s,t)
```

yields the result:

```
y =
1-exp(-2*t)-2*t+2*t^2
```

which agrees with Equation 7.7.7.   □

• **Example 7.7.2**

Let us solve the ordinary differential equation

$$y'' + y = H(t) - H(t-1) \tag{7.7.8}$$

with the initial conditions that $y'(0) = y(0) = 0$. Taking the Laplace transform of both sides of Equation 7.7.8,

$$s^2 Y(s) - sy(0) - y'(0) + Y(s) = \frac{1}{s} - \frac{e^{-s}}{s}, \tag{7.7.9}$$

where $Y(s) = \mathcal{L}[y(t)]$. Substituting the initial conditions into Equation 7.7.9 and solving for $Y(s)$,

$$Y(s) = \left(\frac{1}{s} - \frac{s}{s^2+1}\right) - \left(\frac{1}{s} - \frac{s}{s^2+1}\right)e^{-s}. \tag{7.7.10}$$

Using the second shifting theorem, the final solution is

$$y(t) = 1 - \cos(t) - [1 - \cos(t-1)]H(t-1). \tag{7.7.11}$$

We can check our results using the MATLAB script:

```
clear
% define symbolic variables
syms s t Y
% take Laplace transform of left side of differential equation
LHS = laplace(diff(diff(sym('y(t)')))+sym('y(t)'));
```

```
% take Laplace transform of right side of differential equation
RHS = laplace('Heaviside(t) - Heaviside(t-1)',t,s);
% set Y for Laplace transform of y
%      and introduce initial conditions
newLHS = subs(LHS,'laplace(y(t),t,s)','y(0)','D(y)(0)',Y,0,0);
% solve for Y
Y = solve(newLHS-RHS,Y);
% invert Laplace transform and find y(t)
y = ilaplace(Y,s,t)
```

which yields

```
y =
1-cos(t)-Heaviside(t-1)+Heaviside(t-1)*cos(t-1)
```

□

• **Example 7.7.3**

Let us solve the ordinary differential equation

$$y'' + 2y' + y = f(t) \tag{7.7.12}$$

with the initial conditions that $y'(0) = y(0) = 0$, where $f(t)$ is an unknown function whose Laplace transform exists. Taking the Laplace transform of both sides of Equation 7.7.12,

$$s^2 Y(s) - sy(0) - y'(0) + 2sY(s) - 2y(0) + Y(s) = F(s), \tag{7.7.13}$$

where $Y(s) = \mathcal{L}[y(t)]$. Substituting the initial conditions into Equation 7.7.13 and solving for $Y(s)$,

$$Y(s) = \frac{1}{(s+1)^2} F(s). \tag{7.7.14}$$

We wrote Equation 7.7.14 in this form because the transform $Y(s)$ equals the product of two transforms $1/(s+1)^2$ and $F(s)$. Therefore, by the convolution theorem we can immediately write

$$y(t) = te^{-t} * f(t) = \int_0^t xe^{-x} f(t-x)\, dx. \tag{7.7.15}$$

Without knowing $f(t)$, this is as far as we can go. □

• **Example 7.7.4: Forced harmonic oscillator**

Let us solve the *simple harmonic oscillator* forced by a harmonic forcing

$$y'' + \omega^2 y = \cos(\omega t), \tag{7.7.16}$$

subject to the initial conditions that $y'(0) = y(0) = 0$. Although the complete solution could be found by summing the complementary solution and a particular solution obtained, say, from the method of undetermined coefficients, we now illustrate how we can use Laplace transforms to solve this problem.

Taking the Laplace transform of both sides of Example 7.7.16, substituting in the initial conditions, and solving for $Y(s)$,

$$Y(s) = \frac{s}{(s^2 + \omega^2)^2}, \tag{7.7.17}$$

and

$$y(t) = \frac{1}{\omega}\sin(\omega t) * \cos(\omega t) = \frac{t}{2\omega}\sin(\omega t). \qquad (7.7.18)$$

Equation 7.7.18 gives an oscillation that grows linearly with time although the forcing function is simply periodic. Why does this occur? Recall that our simple harmonic oscillator has the natural frequency ω. But that is exactly the frequency at which we drive the system. Consequently, our choice of forcing has resulted in *resonance*, where energy continuously feeds into the oscillator. $\qquad \square$

• **Example 7.7.5**

Let us solve the *system* of ordinary differential equations:

$$2x' + y = \cos(t), \qquad (7.7.19)$$

and

$$y' - 2x = \sin(t), \qquad (7.7.20)$$

subject to the initial conditions that $x(0) = 0$, and $y(0) = 1$. Taking the Laplace transform of Equation 7.7.19 and Equation 7.7.20,

$$2sX(s) + Y(s) = \frac{s}{s^2 + 1}, \qquad (7.7.21)$$

and

$$-2X(s) + sY(s) = 1 + \frac{1}{s^2 + 1}, \qquad (7.7.22)$$

after introducing the initial conditions. Solving for $X(s)$ and $Y(s)$,

$$X(s) = -\frac{1}{(s^2 + 1)^2}, \qquad (7.7.23)$$

and

$$Y(s) = \frac{s}{s^2 + 1} + \frac{2s}{(s^2 + 1)^2}. \qquad (7.7.24)$$

Taking the inverse of Equation 7.7.23 and Equation 7.7.24 term by term,

$$x(t) = \tfrac{1}{2}[t\cos(t) - \sin(t)], \qquad (7.7.25)$$

and

$$y(t) = t\sin(t) + \cos(t). \qquad (7.7.26)$$

The MATLAB script:

```
clear
% define symbolic variables
syms s t X Y
% take Laplace transform of left side of differential equations
LHS1 = laplace(2*diff(sym('x(t)'))+sym('y(t)'));
LHS2 = laplace(diff(sym('y(t)'))-2*sym('x(t)'));
% take Laplace transform of right side of differential equations
RHS1 = laplace(cos(t)); RHS2 = laplace(sin(t));
% set X and Y for Laplace transforms of x and y
```

```
%        and introduce initial conditions
newLHS1 = subs(LHS1,'laplace(x(t),t,s)','laplace(y(t),t,s)',...
       'x(0)','y(0)',X,Y,0,1);
newLHS2 = subs(LHS2,'laplace(x(t),t,s)','laplace(y(t),t,s)',...
       'x(0)','y(0)',X,Y,0,1);
% solve for X and Y
[X,Y] = solve(newLHS1-RHS1,newLHS2-RHS2,X,Y);
% invert Laplace transform and find x(t) and y(t)
x = ilaplace(X,s,t); y = ilaplace(Y,s,t)
```

uses the symbolic toolbox to solve Equation 7.7.19 and Equation 7.7.20. MATLAB finally gives:

```
x =
1/2*t*cos(t)-1/2*sin(t)
y =
t*sin(t)+cos(t)
```

□

• Example 7.7.6

Let us determine the displacement of a mass m attached to a spring and excited by the driving force

$$F(t) = mA\left(1 - \frac{t}{T}\right)e^{-t/T}. \qquad (7.7.27)$$

The dynamical equation governing this system is

$$y'' + \omega^2 y = A\left(1 - \frac{t}{T}\right)e^{-t/T}, \qquad (7.7.28)$$

where $\omega^2 = k/m$ and k is the spring constant. Assuming that the system is initially at rest, the Laplace transform of the dynamical system is

$$(s^2 + \omega^2)Y(s) = \frac{A}{s + 1/T} - \frac{A}{T(s + 1/T)^2}, \qquad (7.7.29)$$

or

$$Y(s) = \frac{A}{(s^2 + \omega^2)(s + 1/T)} - \frac{A}{T(s^2 + \omega^2)(s + 1/T)^2}. \qquad (7.7.30)$$

Partial fractions yield

$$Y(s) = \frac{A}{\omega^2 + 1/T^2}\left(\frac{1}{s + 1/T} - \frac{s - 1/T}{s^2 + \omega^2}\right) - \frac{A}{T(\omega^2 + 1/T^2)^2}$$
$$\times \left[\frac{1/T^2 - \omega^2}{s^2 + \omega^2} - \frac{2s/T}{s^2 + \omega^2} + \frac{\omega^2 + 1/T^2}{(s + 1/T)^2} + \frac{2/T}{s + 1/T}\right]. \qquad (7.7.31)$$

Inverting Equation 7.7.31 term by term,

$$y(t) = \frac{AT^2}{1 + \omega^2 T^2}\left[e^{-t/T} - \cos(\omega t) + \frac{\sin(\omega t)}{\omega T}\right] - \frac{AT^2}{(1 + \omega^2 T^2)^2}\left\{(1 - \omega^2 T^2)\frac{\sin(\omega t)}{\omega T}\right.$$
$$\left. + 2\left[e^{-t/T} - \cos(\omega t)\right] + (1 + \omega^2 T^2)(t/T)e^{-t/T}\right\}. \qquad (7.7.32)$$

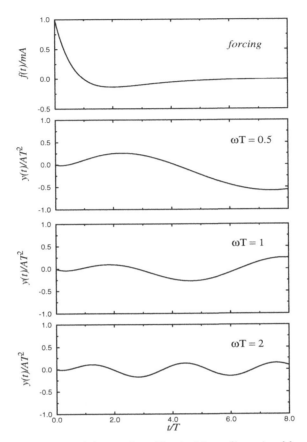

Figure 7.7.1: Displacement of a simple harmonic oscillator with nondimensional frequency ωT as a function of time t/T. The top frame shows the forcing function.

The solution to this problem consists of two parts. The exponential terms result from the forcing and will die away with time. This is the *transient* portion of the solution. The sinusoidal terms are those natural oscillations that are necessary so that the solution satisfies the initial conditions. They are the *steady-state* portion of the solution and endure forever. Figure 7.7.1 illustrates the solution when $\omega T = 0.1$, 1, and 2. Note that the displacement decreases in magnitude as the nondimensional frequency of the oscillator increases.   □

• **Example 7.7.7**

Let us solve the equation
$$y'' + 16y = \delta(t - \pi/4) \tag{7.7.33}$$
with the initial conditions that $y(0) = 1$, and $y'(0) = 0$.

Taking the Laplace transform of Equation 7.7.33 and inserting the initial conditions,
$$(s^2 + 16)Y(s) = s + e^{-s\pi/4}, \tag{7.7.34}$$
or
$$Y(s) = \frac{s}{s^2 + 16} + \frac{e^{-s\pi/4}}{s^2 + 16}. \tag{7.7.35}$$
Applying the second shifting theorem,
$$y(t) = \cos(4t) + \tfrac{1}{4}\sin[4(t - \pi/4)]H(t - \pi/4) = \cos(4t) - \tfrac{1}{4}\sin(4t)H(t - \pi/4). \tag{7.7.36}$$

We can check our results using the MATLAB script:

```
clear
% define symbolic variables
syms pi s t Y
% take Laplace transform of left side of differential equation
LHS = laplace(diff(diff(sym('y(t)')))+16*sym('y(t)'));
% take Laplace transform of right side of differential equation
RHS = laplace('Dirac(t-pi/4)',t,s);
% set Y for Laplace transform of y
%     and introduce initial conditions
newLHS = subs(LHS,'laplace(y(t),t,s)','y(0)','D(y)(0)',Y,1,0);
% solve for Y
Y = solve(newLHS-RHS,Y);
% invert Laplace transform and find y(t)
y = ilaplace(Y,s,t)
```

which yields

```
y =
cos(4*t)-1/4*Heaviside(t-1/4*pi)*sin(4*t)
```

We can also verify that Equation 7.7.36 is the solution to our initial-value problem by computing the (generalized) derivative of Equation 7.7.36, or

$$y'(t) = -4\sin(4t) - \cos(4t)H(t - \pi/4) - \tfrac{1}{4}\sin(4t)\delta(t - \pi/4) \qquad (\mathbf{7.7.37})$$

$$= -4\sin(4t) - \cos(4t)H(t - \pi/4) - \tfrac{1}{4}\sin(\pi)\delta(t - \pi/4) \qquad (\mathbf{7.7.38})$$

$$= -4\sin(4t) - \cos(4t)H(t - \pi/4), \qquad (\mathbf{7.7.39})$$

since $f(t)\delta(t - t_0) = f(t_0)\delta(t - t_0)$. Similarly,

$$y''(t) = -16\cos(4t) + 4\sin(4t)H(t - \pi/4) - \cos(4t)\delta(t - \pi/4) \qquad (\mathbf{7.7.40})$$

$$= -16\cos(4t) + 4\sin(4t)H(t - \pi/4) - \cos(\pi)\delta(t - \pi/4) \qquad (\mathbf{7.7.41})$$

$$= -16\cos(4t) + 4\sin(4t)H(t - \pi/4) + \delta(t - \pi/4). \qquad (\mathbf{7.7.42})$$

Substituting Equation 7.7.36 and Equation 7.7.42 into Equation 7.7.33 completes the verification. A quick check of $y(0)$ and $y'(0)$ also shows that we have the correct solution. □

• Example 7.7.8: Oscillations in electric circuits

During the middle of the nineteenth century, Lord Kelvin[11] analyzed the LCR electrical circuit shown in Figure 7.7.2, which contains resistance R, capacitance C, and inductance L. For reasons that we shall shortly show, this LCR circuit has become one of the quintessential circuits for electrical engineers. In this example, we shall solve the problem by Laplace transforms.

Because we can add the potential differences across the elements, the equation governing the LCR circuit is

$$L\frac{dI}{dt} + RI + \frac{1}{C}\int_0^t I\,d\tau = E(t), \qquad (\mathbf{7.7.43})$$

---

[11] Thomson, W., 1853: On transient electric currents. *Philos. Mag.*, *Ser. 4*, **5**, 393–405.

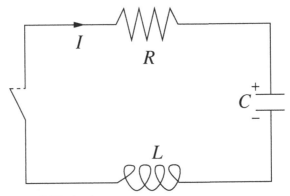

Figure 7.7.2: Schematic of a LCR circuit.

where I denotes the current in the circuit. Let us solve Equation 7.7.43 when we close the circuit and the initial conditions are $I(0) = 0$ and $Q(0) = -Q_0$. Taking the Laplace transform of Equation 7.7.43,

$$\left(Ls + R + \frac{1}{Cs}\right)\overline{I}(s) = LI(0) - \frac{Q(0)}{Cs}. \tag{7.7.44}$$

Solving for $\overline{I}(s)$,

$$\overline{I}(s) = \frac{Q_0}{Cs(Ls + R + 1/Cs)} = \frac{\omega_0^2 Q_0}{s^2 + 2\alpha s + \omega_0^2} = \frac{\omega_0^2 Q_0}{(s+\alpha)^2 + \omega_0^2 - \alpha^2}, \tag{7.7.45}$$

where $\alpha = R/(2L)$, and $\omega_0^2 = 1/(LC)$. From the first shifting theorem,

$$I(t) = \frac{\omega_0^2 Q_0}{\omega} e^{-\alpha t} \sin(\omega t), \tag{7.7.46}$$

where $\omega^2 = \omega_0^2 - \alpha^2 > 0$. The quantity ω is the natural frequency of the circuit, which is lower than the free frequency ω_0 of a circuit formed by a condenser and coil. Most importantly, the solution decays in amplitude with time.

Although Kelvin's solution was of academic interest when he originally published it, this radically changed with the advent of radio telegraphy[12] because the LCR circuit described the fundamental physical properties of wireless transmitters and receivers.[13] The inescapable conclusion from numerous analyses was that no matter how cleverly the receiver was designed, eventually the resistance in the circuit would dampen the electrical oscillations and thus limit the strength of the received signal.

This technical problem was overcome by Armstrong,[14] who invented an electrical circuit that used De Forest's audion (the first vacuum tube) for generating electrical oscillations and for amplifying externally impressed oscillations by "regenerative action." The effect of adding the "thermionic amplifier" is seen by again considering the LRC circuit as shown in Figure 7.7.3 with the modification suggested by Armstrong.[15]

---

[12] Stone, J S., 1914: The resistance of the spark and its effect on the oscillations of electrical oscillators. *Proc. IRE*, **2**, 307–324.

[13] See Hogan, J. L., 1916: Physical aspects of radio telegraphy. *Proc. IRE*, **4**, 397–420.

[14] Armstrong, E. H., 1915: Some recent developments in the audion receiver. *Proc. IRE*, **3**, 215–247.

[15] See Ballantine, S., 1919: The operational characteristics of thermionic amplifiers. *Proc. IRE*, **7**, 129–161.

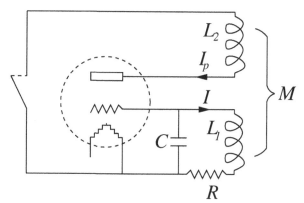

Figure 7.7.3: Schematic of an LCR circuit with the addition of a thermionic amplifier. (From Ballantine, S., 1919: The operational characteristics of thermionic amplifiers. *Proc. IRE*, **7**, 155.)

The governing equations of this new circuit are

$$L_1 \frac{dI}{dt} + RI + \frac{1}{C} \int_0^t I \, d\tau + M \frac{dI_p}{dt} = 0, \tag{7.7.47}$$

and

$$L_2 \frac{dI_p}{dt} + R_0 I_p + M \frac{dI}{dt} + \frac{\mu}{C} \int_0^t I \, d\tau = 0, \tag{7.7.48}$$

where the plate circuit has the current I_p, the resistance R_0, the inductance L_2, and the electromotive force (emf) of $\mu \int_0^t I \, d\tau / C$. The mutual inductance between the two circuits is given by M. Taking the Laplace transform of Equation 7.7.47 and Equation 7.7.48,

$$L_1 s \overline{I}(s) + R \overline{I}(s) + \frac{\overline{I}(s)}{sC} + Ms \overline{I}_p(s) = \frac{Q_0}{sC}, \tag{7.7.49}$$

and

$$L_2 s \overline{I}_p(s) + R_0 \overline{I}_p(s) + Ms \overline{I}(s) + \frac{\mu}{sC} \overline{I}(s) = 0. \tag{7.7.50}$$

Eliminating $\overline{I}_p(s)$ between Equation 7.7.49 and Equation 7.7.50 and solving for $\overline{I}(s)$,

$$\overline{I}(s) = \frac{(L_2 s + R_0) Q_0}{(L_1 L_2 - M^2) C s^3 + (RL_2 + R_0 L_1) C s^2 + (L_2 + CRR_0 - \mu M) s + R_0}. \tag{7.7.51}$$

For high-frequency radio circuits, we can approximate the roots of the denominator of Equation 7.7.51 as

$$s_1 \approx -\frac{R_0}{L_2 + CRR_0 - \mu M}, \tag{7.7.52}$$

and

$$s_{2,3} \approx \frac{R_0}{2(L_2 + CRR_0 - \mu M)} - \frac{R_0 L_1 + RL_2}{2(L_1 L_2 - M^2)} \pm i\omega. \tag{7.7.53}$$

In the limit of M and R_0 vanishing, we recover our previous result for the LRC circuit. However, in reality, R_0 is very large and our solution has three terms. The term associated with s_1 is a rapidly decaying transient while the s_2 and s_3 roots yield oscillatory solutions with a *slight* amount of damping. Thus, our analysis shows that in the ordinary regenerative

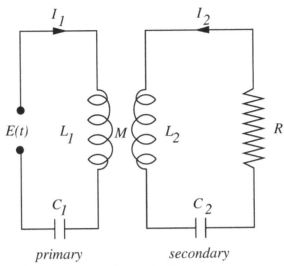

Figure 7.7.4: Schematic of a resonance transformer circuit.

circuit, the tube effectively introduces sufficient "negative" resistance so that the resultant positive resistance of the equivalent LCR circuit is relatively low, and the response of an applied signal voltage at the resonant frequency of the circuit is therefore relatively great. Later, Armstrong[16] extended his work on regeneration by introducing an electrical circuit—the superregenerative circuit—where the regeneration is made large enough so that the resultant resistance is negative, and self-sustained oscillations can occur.[17] It was this circuit[18] that led to the explosive development of radio in the 1920s and 1930s.   □

• **Example 7.7.9: Resonance transformer circuit**

One of the fundamental electrical circuits of early radio telegraphy[19] is the resonance transformer circuit shown in Figure 7.7.4. Its development gave transmitters and receivers the ability to tune to each other.

The governing equations follow from Kirchhoff's law and are

$$L_1 \frac{dI_1}{dt} + M \frac{dI_2}{dt} + \frac{1}{C_1} \int_0^t I_1 \, d\tau = E(t), \tag{7.7.54}$$

and

$$M \frac{dI_1}{dt} + L_2 \frac{dI_2}{dt} + RI_2 + \frac{1}{C_2} \int_0^t I_2 \, d\tau = 0. \tag{7.7.55}$$

Let us examine the oscillations generated if initially the system has no currents or charges and the forcing function is $E(t) = \delta(t)$.

---

[16] Armstrong, E. H., 1922: Some recent developments of regenerative circuits. *Proc. IRE*, **10**, 244–260.

[17] See Frink, F. W., 1938: The basic principles of superregenerative reception. *Proc. IRE*, **26**, 76–106.

[18] Lewis, T., 1991: *Empire of the Air: The Men Who Made Radio.* HarperCollins Publishers, 421 pp.

[19] Fleming, J. A., 1919: *The Principles of Electric Wave Telegraphy and Telephony.* Longmans, Green, 911 pp.

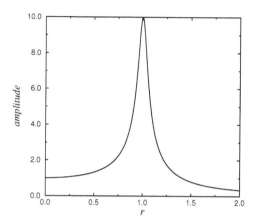

Figure 7.7.5: The resonance curve $1/\sqrt{(r^2 - 1)^2 + 0.01}$ for a resonance transformer circuit with $r = \omega_2/\omega_1$.

Taking the Laplace transform of Equation 7.7.54 and Equation 7.7.55,

$$L_1 s \bar{I}_1 + M s \bar{I}_2 + \frac{\bar{I}_1}{sC_1} = 1, \qquad (7.7.56)$$

and

$$M s \bar{I}_1 + L_2 s \bar{I}_2 + R \bar{I}_2 + \frac{\bar{I}_2}{sC_2} = 0. \qquad (7.7.57)$$

Because the current in the second circuit is of greater interest, we solve for $\bar{I}_2$ and find that

$$\bar{I}_2(s) = -\frac{M s^3}{L_1 L_2 [(1 - k^2)s^4 + 2\alpha \omega_2^2 s^3 + (\omega_1^2 + \omega_2^2)s^2 + 2\alpha \omega_1^2 s + \omega_1^2 \omega_2^2]}, \qquad (7.7.58)$$

where $\alpha = R/(2L_2)$, $\omega_1^2 = 1/(L_1 C_1)$, $\omega_2^2 = 1/(L_2 C_2)$, and $k^2 = M^2/(L_1 L_2)$, the so-called coefficient of coupling.

We can obtain analytic solutions if we assume that the coupling is weak ($k^2 \ll 1$). Equation 7.7.58 becomes

$$\bar{I}_2 = -\frac{M s^3}{L_1 L_2 (s^2 + \omega_1^2)(s^2 + 2\alpha s + \omega_2^2)}. \qquad (7.7.59)$$

Using partial fractions and inverting term by term, we find that

$$I_2(t) = \frac{M}{L_1 L_2} \left[\frac{2\alpha \omega_1^3 \sin(\omega_1 t)}{(\omega_2^2 - \omega_1^2)^2 + 4\alpha^2 \omega_1^2} + \frac{\omega_1^2(\omega_2^2 - \omega_1^2)\cos(\omega_1 t)}{(\omega_2^2 - \omega_1^2)^2 + 4\alpha^2 \omega_1^2} \right. \qquad (7.7.60)$$
$$\left. + \frac{\alpha \omega_2^4 - 3\alpha \omega_1^2 \omega_2^2 + 4\alpha^3 \omega_1^2}{(\omega_2^2 - \omega_1^2)^2 + 4\alpha^2 \omega_1^2} e^{-\alpha t} \frac{\sin(\omega t)}{\omega} - \frac{\omega_2^2(\omega_2^2 - \omega_1^2) + 4\alpha^2 \omega_1^2}{(\omega_2^2 - \omega_1^2)^2 + 4\alpha^2 \omega_1^2} e^{-\alpha t} \cos(\omega t) \right],$$

where $\omega^2 = \omega_2^2 - \alpha^2$.

The exponentially damped solutions will eventually disappear, leaving only the steady-state oscillations that vibrate with the angular frequency ω_1, the natural frequency of the primary circuit. If we rewrite this steady-state solution in amplitude/phase form, the amplitude is

$$\frac{M}{L_1 L_2 \sqrt{(r^2 - 1)^2 + 4\alpha^2/\omega_1^2}}, \qquad (7.7.61)$$

where $r = \omega_2/\omega_1$. As Figure 7.7.5 shows, as r increases from zero to two, the amplitude rises until a very sharp peak occurs at $r = 1$ and then decreases just as rapidly as we approach $r = 2$. Thus, the resonance transformer circuit provides a convenient way to tune a transmitter or receiver to the frequency ω_1. $\square$

• Example 7.7.10: Delay differential equation

Laplace transforms provide a valuable tool in solving a general class of ordinary differential equations called *delay differential equations*. These equations arise in such diverse fields as chemical kinetics[20] and population dynamics.[21]

To illustrate the technique,[22] consider the differential equation

$$x' = -ax(t-1) \tag{7.7.62}$$

with $x(t) = 1 - at$ for $0 < t < 1$. Clearly, $x(0) = 1$.

Multiplying Equation 7.7.62 by e^{-st} and integrating from 1 to ∞,

$$\int_1^\infty x'(t)e^{-st}\,dt = -a\int_1^\infty x(t-1)e^{-st}\,dt \tag{7.7.63}$$

$$\int_0^\infty x'(t)e^{-st}\,dt - \int_0^1 x'(t)e^{-st}\,dt = -a\int_0^\infty x(\tau)e^{-s(\tau+1)}\,d\tau \tag{7.7.64}$$

$$sX(s) - 1 + a\int_0^1 e^{-st}\,dt = -ae^{-s}X(s) \tag{7.7.65}$$

$$sX(s) - 1 - \frac{a}{s}\,e^{-st}\Big|_0^1 = -ae^{-s}X(s) \tag{7.7.66}$$

since $x'(t) = -a$ for $0 < t < 1$. Solving for $X(s)$,

$$X(s) = (1 + ae^{-s}/s - a/s)/[s(1 + ae^{-s}/s)]. \tag{7.7.67}$$

To facilitate the inversion of Equation 7.7.67, we expand its denominator in terms of a geometric series and find that

$$X(s) = \sum_{n=0}^\infty (-a)^n e^{-ns}/s^{n+1} + \sum_{n=0}^\infty (-a)^{n+1} e^{-ns}/s^{n+2} - \sum_{n=0}^\infty (-a)^{n+1} e^{-(n+1)s}/s^{n+2}. \tag{7.7.68}$$

The first and third sums cancel, except for the $n = 0$ term in the first sum. Therefore,

$$X(s) = \frac{1}{s} + \sum_{n=0}^\infty (-a)^{n+1} e^{-ns}/s^{n+2} \tag{7.7.69}$$

---

[20] See Roussel, M. R., 1996: The use of delay differential equations in chemical kinetics. *J. Phys. Chem.*, **100**, 8323–8330.

[21] See the first chapter of MacDonald, N., 1989: *Biological Delay Systems: Linear Stability Theory.* Cambridge University Press, 235 pp.

[22] See Epstein, I. R., 1990: Differential delay equations in chemical kinetics: Some simple linear model systems. *J. Chem. Phys.*, **92**, 1702–1712.

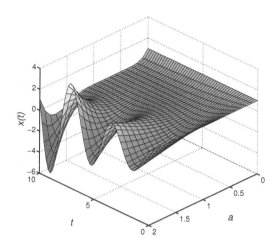

Figure 7.7.6: The solution to the delay differential equation, Equation 7.7.62, at various times t and values of a.

and

$$x(t) = 1 + \sum_{n=0}^{\infty} \frac{(-a)^{n+1}}{(n+1)!} H(t-n)(t-n)^{n+1}. \tag{7.7.70}$$

Figure 7.7.6 illustrates Equation 7.7.70 as a function of time for various values of a. For $0 < a < e^{-1}$, $x(t)$ decays monotonically from 1 to an asymptotic limit of zero. For $e^{-1} < a < \pi/2$, the solution is a damped oscillatory function. If $\pi/2 < a$, then $x(t)$ is oscillatory with an exponentially increasing envelope. When $a = \pi/2$, $x(t)$ oscillates periodically. □

• **Example 7.7.11**

Laplace transforms can sometimes be used to solve ordinary differential equations where the coefficients are powers of t. To illustrate this, let us solve

$$y'' + 2ty' - 4y = 0, \qquad y(0) = 1, \qquad \lim_{t \to \infty} y(t) \to 0. \tag{7.7.71}$$

We begin by taking the Laplace transform of Equation 7.7.71 and find that

$$s^2 Y(s) - s y(0) - y'(0) - 2\frac{d}{ds}[sY(s) - y(0)] - 4Y(s) = 0. \tag{7.7.72}$$

An interesting aspect of this problem is the fact that we do not know $y'(0)$. To circumvent this difficulty, let us temporarily set $y'(0) = -A$ so that Equation 7.7.72 becomes

$$\frac{dY}{ds} + \left(\frac{3}{s} - \frac{s}{2}\right) Y = \frac{A}{2s} - \frac{1}{2}. \tag{7.7.73}$$

Later on, we will find A.

Equation 7.7.73 is a first-order, linear, ordinary differential equation with s as its independent variable. To find $Y(s)$, we use the standard technique of multiplying it by its integrating factor, here $\mu(s) = s^3 e^{-s^2/4}$, and rewriting it as

$$\frac{d}{ds}\left[s^3 e^{-s^2/4} Y(s)\right] = \frac{1}{2}As^2 e^{-s^2/4} - \frac{1}{2}s^3 e^{-s^2/4}. \tag{7.7.74}$$

Advanced Engineering Mathematics with MATLAB

Integrating Equation 7.7.74 from s to ∞, we obtain

$$s^3 e^{-s^2/4} Y(s) = (s^2 + 4)e^{-s^2/4} - A\left[se^{-s^2/4} + \sqrt{\pi}\,\mathrm{erfc}(s/2)\right], \qquad (\textbf{7.7.75})$$

or

$$Y(s) = \frac{4}{s^3} + \frac{1}{s} - \frac{A}{s^2} - \frac{A\sqrt{\pi}}{s^3} e^{s^2/4}\,\mathrm{erfc}(s/2). \qquad (\textbf{7.7.76})$$

We must now evaluate A. From the final-value theorem, $\lim_{t\to\infty} y(t) = \lim_{s\to 0} sY(s) = 0$. Therefore, multiplying Equation 7.7.76 by s and using the expansion for the complementary error function for small s, we have that

$$sY(s) = \frac{4}{s^2} + 1 - \frac{A}{s} - \frac{A\sqrt{\pi}}{s^2}\left[1 + \frac{s^2}{4} - \frac{s}{\sqrt{\pi}} + \cdots\right]. \qquad (\textbf{7.7.77})$$

In order that $\lim_{s\to 0} sY(s) = 0$, $A = 4/\sqrt{\pi}$. Therefore,

$$Y(s) = \frac{4}{s^3} + \frac{1}{s} - \frac{4}{\sqrt{\pi}\,s^2} - \frac{4}{s^3} e^{s^2/4}\,\mathrm{erfc}(s/2). \qquad (\textbf{7.7.78})$$

The final step is to invert Equation 7.7.78. Applying tables and the convolution theorem,

$$y(t) = 2t^2 + 1 - \frac{4t}{\sqrt{\pi}} - \frac{4}{\sqrt{\pi}} \int_0^t (t-x)^2 e^{-x^2}\,dx = (2t^2 + 1)[1 - \mathrm{erf}(t)] - \frac{2t}{\sqrt{\pi}} e^{-t^2}. \qquad (\textbf{7.7.79})$$

Problems

Solve the following ordinary differential equations by Laplace transforms. Then use MATLAB to verify your solution.

1. $y' - 2y = 1 - t; \quad y(0) = 1$

2. $y'' - 4y' + 3y = e^t; \quad y(0) = 0, y'(0) = 0$

3. $y'' - 4y' + 3y = e^{2t}; \quad y(0) = 0, y'(0) = 1$

4. $y'' - 6y' + 8y = e^t; \quad y(0) = 3, y'(0) = 9$

5. $y'' + 4y' + 3y = e^{-t}; \quad y(0) = 1, y'(0) = 1$

6. $y'' + y = t; \quad y(0) = 1, y'(0) = 0$

7. $y'' + 4y' + 3y = e^t; \quad y(0) = 0, y'(0) = 2$

8. $y'' - 4y' + 5y = 0; \quad y(0) = 2, y'(0) = 4$

9. $y' + y = tH(t - 1); \quad y(0) = 0$

10. $y'' + 3y' + 2y = H(t - 1); \quad y(0) = 0, y'(0) = 1$

11. $y'' - 3y' + 2y = H(t - 1); \quad y(0) = 0, y'(0) = 1$

12. $y'' + 4y = 3H(t - 4); \quad y(0) = 1, y'(0) = 0$

13. $y'' + 4y' + 4y = 4H(t - 2); \quad y(0) = 0, y'(0) = 0$

14. $y'' + 3y' + 2y = e^{t-1}H(t - 1); \quad y(0) = 0, y'(0) = 1$

15. $y'' - 3y' + 2y = e^{-(t-2)}H(t - 2); \quad y(0) = 0, y'(0) = 0$

16. $y'' - 3y' + 2y = H(t - 1) - H(t - 2); \quad y(0) = 0, y'(0) = 0$

17. $y'' + y = 1 - H(t - T); \quad y(0) = 0, y'(0) = 0$

18. $y'' + y = \begin{cases} \sin(t), & 0 \le t \le \pi, \\ 0, & \pi \le t; \end{cases} \quad y(0) = 0, y'(0) = 0$

19. $y'' + 3y' + 2y = \begin{cases} t, & 0 \le t \le a, \\ ae^{-(t-a)}, & a \le t; \end{cases} \quad y(0) = 0, y'(0) = 0$

20. $y'' + \omega^2 y = \begin{cases} t/a, & 0 \le t \le a, \\ 1 - (t - a)/(b - a), & a \le t \le b, \\ 0, & b \le t; \end{cases} \quad y(0) = 0, y'(0) = 0$

21. $y'' - 2y' + y = 3\delta(t - 2); \quad y(0) = 0, y'(0) = 1$

22. $y'' - 5y' + 4y = \delta(t - 1); \quad y(0) = 0, y'(0) = 0$

23. $y'' + 5y' + 6y = 3\delta(t - 2) - 4\delta(t - 5); \quad y(0) = y'(0) = 0$

24. $y'' + \omega y' = A\delta(t - \tau) - BH(t - \tau); \quad y(0) = y'(0) = 0$

25. $x' - 2x + y = 0, \quad y' - 3x - 4y = 0; \quad x(0) = 1, y(0) = 0$

26. $x' - 2y' = 1, \quad x' + y - x = 0; \quad x(0) = y(0) = 0$

27. $x' + 2x - y' = 0, \quad x' + y + x = t^2; \quad x(0) = y(0) = 0$

28. $x' + 3x - y = 1, \quad x' + y' + 3x = 0; \quad x(0) = 2, y(0) = 0$

29. Forster, Escobal, and Lieske[23] used Laplace transforms to solve the linearized equations of motion of a vehicle in a gravitational field created by two other bodies. A simplified form of this problem involves solving the following system of ordinary differential equations:

$$x'' - 2y' = F_1 + x + 2y, \qquad 2x' + y'' = F_2 + 2x + 3y,$$

subject to the initial conditions that $x(0) = y(0) = x'(0) = y'(0) = 0$. Find the solution to this system.

---

[23] Forster, K., P. R. Escobal, and H. A. Lieske, 1968: Motion of a vehicle in the transition region of the three-body problem. *Astronaut. Acta,* **14,** 1–10.

Following Example 7.7.11, find the solution for the following ordinary differential equations:

30. $y'' + 2ty' - 8y = 0, \qquad y(0) = 1, \quad y'(0) = 0$

Step 1: Show that the Laplace transform for this differential equation is $2sY'(s) + (10 - s^2)Y(s) = -s$.

Step 2: Solve this first-order ordinary differential equation and show that $Y(s) = 1/s + 8/s^3 + 32/s^5 + Ae^{s^2/4}/s^5$.

Step 3: Invert $Y(s)$ and show that the solution is $y(t) = 1 + 4t^2 + 4t^4/3$.

31. $y'' - ty' + 2y = 0, \qquad y(0) = -1, \quad y'(0) = 0$

Step 1: Show that the Laplace transform for this differential equation is $sY'(s) + (s^2 + 3)Y(s) = -s$.

Step 2: Solve the first-order ordinary differential equation and show that $Y(s) = (A - 2)e^{-s^2/2}/s^3 + 2/s^3 - 1/s$.

Step 3: Invert $Y(s)$ and show that the solution is $y(t) = t^2 - 1$.

32. $ty'' - (2 - t)y' - y = 0$

Step 1: Show that the Laplace transform for this differential equation is $s(s + 1)Y'(s) + 2(2s + 1)Y(s) = 3y(0)$.

Step 2: Solve the first-order ordinary differential equation and show that $Y(s) = y(0)/(s + 1) + y(0)/[2(s + 1)^2] + A/[s^2(s + 1)^2]$.

Step 3: Invert $Y(s)$ and show that the general solution is $y(t) = C_1(t + 2)e^{-t} + C_2(t - 2)$.

33. $ty'' - 2(a + bt)y' + b(2a + bt)y = 0, \qquad a \geq 0$

Step 1: Show that the Laplace transform for this differential equation is $(s - b)^2Y'(s) + 2(1 + a)(s - b)Y(s) = (1 + 2a)y(0)$.

Step 2: Solve the first-order ordinary differential equation and show that $Y(s) = y(0)/(s - b) + A/(s - b)^{2+2a}$.

Step 3: Invert $Y(s)$ and show that the general solution is $y(t) = C_1e^{bt} + C_2t^{2a+1}e^{bt}$.

Further Readings

Churchill, R. V., 1972: *Operational Mathematics*. McGraw-Hill Book Co., 481 pp. A classic textbook on Laplace transforms.

Doetsch, G., 1950: *Handbuch der Laplace-Transformation. Band 1. Theorie der Laplace-Transformation*. Birkhäuser Verlag, 581 pp.; Doetsch, G., 1955: *Handbuch der Laplace-Transformation. Band 2. Anwendungen der Laplace-Transformation. 1. Abteilung*. Birkhäuser Verlag, 433 pp.; Doetsch, G., 1956: *Handbuch der Laplace-Transformation. Band 3. Anwendungen der Laplace-Transformation. 2. Abteilung*. Birkhäuser Verlag, 298 pp. One of the standard reference books on Laplace transforms.

LePage, W. R., 1980: *Complex Variables and the Laplace Transform for Engineers*. Dover, 483 pp. Laplace transforms approached from complex variable theory and a few applications.

McLachlan, N. W., 1944: *Complex Variables & Operational Calculus with Technical Applications.* Cambridge Press, 355 pp. An early book on Laplace transforms from the view of complex variables.

Thomson, W. T., 1960: *Laplace Transformation.* Prentice-Hall, 255 pp. Presents Laplace transforms in the engineering applications that gave them birth.

Watson, E. J., 1981: *Laplace Transforms and Applications.* Van Nostrand Reinhold, 205 pp. A brief and very complete guide to Laplace transforms.

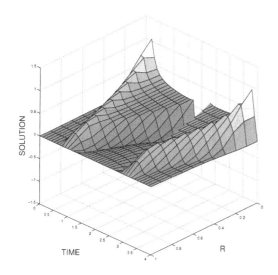

Chapter 8

The Wave Equation

With this chapter we begin our study of *linear* partial differential equations. By definition, partial differential equations are those differential equations which contain partial derivatives. A simple example is

$$\frac{\partial u}{\partial t} = u \frac{\partial u}{\partial x}, \tag{8.0.1}$$

where the solution $u(x,t)$ depends upon distance x and time t. It is also a nonlinear partial differential equation because of the $u\,u_x$ term.

Of all of the possible partial differential equations that one might conceive, we shall study a very special class: linear partial differential equations. By linear we mean that

$$L(c_1 u_1 + c_2 u_2) = c_1 L(u_1) + c_2 L(u_2) \tag{8.0.2}$$

for any two functions u_1 and u_2, where c_1 and c_2 are arbitrary constants. For example, because

$$\frac{\partial^2}{\partial t^2} \left[c_1 u_1 + c_2 u_2 \right] = c_1 \frac{\partial^2 u_1}{\partial t^2} + c_2 \frac{\partial^2 u_2}{\partial t^2}, \tag{8.0.3}$$

and

$$\frac{\partial^2}{\partial x^2} \left[c_1 u_1 + c_2 u_2 \right] = c_1 \frac{\partial^2 u_1}{\partial x^2} + c_2 \frac{\partial^2 u_2}{\partial x^2}, \tag{8.0.4}$$

347

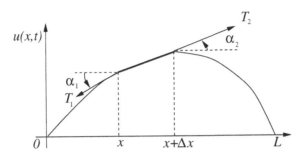

Figure 8.1.1: The vibrating string.

the operator

$$L(\cdot) = \frac{\partial^2(\cdot)}{\partial t^2} - c^2\frac{\partial^2(\cdot)}{\partial x^2} \tag{8.0.5}$$

is also a linear operator and the equation

$$\frac{\partial^2 u}{\partial t^2} - c^2\frac{\partial^2 u}{\partial x^2} = 0 \tag{8.0.6}$$

is an example of a linear partial differential equation, as are

$$\frac{\partial u}{\partial t} - a^2\frac{\partial^2 u}{\partial x^2} = 0, \qquad \text{and} \qquad \frac{\partial^2 u}{\partial x^2} + \frac{\partial^2 u}{\partial y^2} = 0. \tag{8.0.7}$$

Equations 8.0.6 and 8.0.7 are all examples of *homogeneous* linear partial differential equations since they can be written $L(u) = 0$. On the other hand, if $L(u) = f(x,t)$, we have a *nonhomogeneous* linear partial differential equation.

The concepts of linearity and homogeneity also play an important role in the boundary conditions—conditions that the solution of the partial differential equation must satisfy along a specific boundary. Examples of linear boundary conditions are $u(0,t) = f(t)$, $u_x(L,t) = g(t)$, $u_x(0,t) = 0$, and $u_x(L,t) = u(L,t) - h(t)$. Only $u_x(0,t) = 0$ is an example of a homogeneous boundary condition; the other boundary conditions are examples of non-homogeneous boundary conditions. On the other hand, $u_x(0,t) = [u(0,t)]^2$ is an example a nonlinear boundary condition.

In this chapter we will study problems associated with Equation 8.0.6 where c is a constant. This equation, called the *wave equation*, serves as the prototype for a wider class of *hyperbolic equations*

$$a(x,t)\frac{\partial^2 u}{\partial x^2} + b(x,t)\frac{\partial^2 u}{\partial x \partial t} + c(x,t)\frac{\partial^2 u}{\partial t^2} = f\left(x,t,u,\frac{\partial u}{\partial x},\frac{\partial u}{\partial t}\right), \tag{8.0.9}$$

where $b^2 > 4ac$. It arises in the study of many important physical problems involving wave propagation, such as the transverse vibrations of an elastic string and the longitudinal vibrations or torsional oscillations of a rod.

8.1 THE VIBRATING STRING

The motion of a string of length L and constant density ρ (mass per unit *length*) is a simple example of a physical system described by the wave equation. See Figure 8.1.1. Assuming that the equilibrium position of the string and the interval $[0,L]$ along the x-axis coincide, the equation of motion that describes the vertical displacement $u(x,t)$ of the string follows by considering a short piece whose ends are at x and $x + \Delta x$ and applying Newton's second law.

If we assume that the string is perfectly flexible and offers no resistance to bending, Figure 8.1.1 shows the forces on an element of the string. Applying Newton's second law in the x-direction, the sum of forces equals

$$-T(x)\cos(\alpha_1) + T(x + \Delta x)\cos(\alpha_2), \tag{8.1.1}$$

where $T(x)$ denotes the tensile force. If we assume that a point on the string moves only in the vertical direction, the sum of forces in Equation 8.1.1 equals zero and the horizontal component of tension is constant:

$$-T(x)\cos(\alpha_1) + T(x + \Delta x)\cos(\alpha_2) = 0, \tag{8.1.2}$$

and

$$T(x)\cos(\alpha_1) = T(x + \Delta x)\cos(\alpha_2) = T, \text{ a constant.} \tag{8.1.3}$$

If gravity is the only external force, Newton's law in the vertical direction gives

$$-T(x)\sin(\alpha_1) + T(x + \Delta x)\sin(\alpha_2) - mg = m\frac{\partial^2 u}{\partial t^2}, \tag{8.1.4}$$

where u_{tt} is the acceleration. Because

$$T(x) = \frac{T}{\cos(\alpha_1)}, \quad \text{and} \quad T(x + \Delta x) = \frac{T}{\cos(\alpha_2)}, \tag{8.1.5}$$

then

$$-T\tan(\alpha_1) + T\tan(\alpha_2) - \rho g \Delta x = \rho \Delta x \frac{\partial^2 u}{\partial t^2}. \tag{8.1.6}$$

The quantities $\tan(\alpha_1)$ and $\tan(\alpha_2)$ equal the slope of the string at x and $x + \Delta x$, respectively; that is,

$$\tan(\alpha_1) = \frac{\partial u(x,t)}{\partial x}, \quad \text{and} \quad \tan(\alpha_2) = \frac{\partial u(x + \Delta x, t)}{\partial x}. \tag{8.1.7}$$

Substituting Equation 8.1.7 into Equation 8.1.6,

$$T\left[\frac{\partial u(x + \Delta x, t)}{\partial x} - \frac{\partial u(x,t)}{\partial x}\right] = \rho \Delta x \left(\frac{\partial^2 u}{\partial t^2} + g\right). \tag{8.1.8}$$

After dividing through by Δx, we have a difference quotient on the left:

$$\frac{T}{\Delta x}\left[\frac{\partial u(x + \Delta x, t)}{\partial x} - \frac{\partial u(x,t)}{\partial x}\right] = \rho\left(\frac{\partial^2 u}{\partial t^2} + g\right). \tag{8.1.9}$$

In the limit as $\Delta x \to 0$, this difference quotient becomes a partial derivative with respect to x, leaving Newton's second law in the form

$$T\frac{\partial^2 u}{\partial x^2} = \rho\frac{\partial^2 u}{\partial t^2} + \rho g, \tag{8.1.10}$$

or

$$\frac{\partial^2 u}{\partial x^2} = \frac{1}{c^2}\frac{\partial^2 u}{\partial t^2} + \frac{g}{c^2}, \tag{8.1.11}$$

where $c^2 = T/\rho$. Because u_{tt} is generally much larger than g, we can neglect the last term, giving the equation of the vibrating string as

$$\frac{\partial^2 u}{\partial x^2} = \frac{1}{c^2}\frac{\partial^2 u}{\partial t^2}. \tag{8.1.12}$$

Equation 8.1.12 is the one-dimensional *wave equation*.

As a second example[1] we derive the threadline equation, which describes how a thread composed of yarn vibrates as we draw it between two eyelets spaced a distance L apart. We assume that the tension in the thread is constant, the vibrations are small, the thread is perfectly flexible, the effects of gravity and air drag are negligible, and the mass of the thread per unit length is constant. Unlike the vibrating string between two fixed ends, we draw the threadline through the eyelets at a speed V so that a segment of thread experiences motion in both the x and y directions as it vibrates about its equilibrium position. The eyelets may move in the vertical direction.

From Newton's second law,

$$\frac{d}{dt}\left(m\frac{dy}{dt}\right) = \sum \text{ forces}, \tag{8.1.13}$$

where m is the mass of the thread. But

$$\frac{dy}{dt} = \frac{\partial y}{\partial t} + \frac{dx}{dt}\frac{\partial y}{\partial x}. \tag{8.1.14}$$

Because $dx/dt = V$,

$$\frac{dy}{dt} = \frac{\partial y}{\partial t} + V\frac{\partial y}{\partial x}, \tag{8.1.15}$$

and

$$\frac{d}{dt}\left(m\frac{dy}{dt}\right) = \frac{\partial}{\partial t}\left[m\left(\frac{\partial y}{\partial t} + V\frac{\partial y}{\partial x}\right)\right] + V\frac{\partial}{\partial x}\left[m\left(\frac{\partial y}{\partial t} + V\frac{\partial y}{\partial x}\right)\right]. \tag{8.1.16}$$

Because both m and V are constant, it follows that

$$\frac{d}{dt}\left(m\frac{dy}{dt}\right) = m\frac{\partial^2 y}{\partial t^2} + 2mV\frac{\partial^2 y}{\partial x\partial t} + mV^2\frac{\partial^2 y}{\partial x^2}. \tag{8.1.17}$$

The sum of the forces again equals

$$T\frac{\partial^2 y}{\partial x^2}\Delta x \tag{8.1.18}$$

so that the threadline equation is

$$T\frac{\partial^2 y}{\partial x^2}\Delta x = m\frac{\partial^2 y}{\partial t^2} + 2mV\frac{\partial^2 y}{\partial x\partial t} + mV^2\frac{\partial^2 y}{\partial x^2}, \tag{8.1.19}$$

or

$$\frac{\partial^2 y}{\partial t^2} + 2V\frac{\partial^2 y}{\partial x\partial t} + \left(V^2 - \frac{T}{\rho}\right)\frac{\partial^2 y}{\partial x^2} = 0, \tag{8.1.20}$$

---

[1] See Swope, R. D., and W. F. Ames, 1963: Vibrations of a moving threadline. *J. Franklin Inst.*, **275**, 36–55.

where ρ is the density of the thread. Although Equation 8.1.20 is *not* the classic wave equation given in Equation 8.1.12, it is an example of a hyperbolic equation. As we shall see, the solutions to hyperbolic equations share the same behavior, namely, wave-like motion.

8.2 INITIAL CONDITIONS: CAUCHY PROBLEM

Any mathematical model of a physical process must include not only the governing differential equation but also any conditions that are imposed on the solution. For example, in time-dependent problems the solution must conform to the initial condition of the modeled process. Finding those solutions that satisfy the initial conditions (initial data) is called the *Cauchy problem*.

In the case of partial differential equations with second-order derivatives in time, such as the wave equation, we correctly pose the *Cauchy boundary condition* if we specify the value of the solution $u(x, t_0) = f(t)$ *and* its time derivative $u_t(x, t_0) = g(t)$ at some initial time t_0, usually taken to be $t_0 = 0$. The functions $f(t)$ and $g(t)$ are called the *Cauchy data*. We require two conditions involving time because the differential equation has two time derivatives.

In addition to the initial conditions, we must specify boundary conditions in the spatial direction. For example, we may require that the end of the string be fixed. In the next chapter, we discuss boundary conditions in greater depth. However, one boundary condition that is uniquely associated with the wave equation on an open domain is the *radiation condition*. It requires that the waves radiate off to infinity and remain finite as they propagate there.

In summary, Cauchy boundary conditions, along with the appropriate spatial boundary conditions, uniquely determine the solution to the wave equation; any additional information is extraneous. Having developed the differential equation and initial conditions necessary to solve the wave equation, let us now turn to the actual methods used to solve this equation.

8.3 SEPARATION OF VARIABLES

Separation of variables is the most popular method for solving the wave equation. Despite its current widespread use, its initial application to the vibrating string problem was controversial because of the use of a half-range Fourier sine series to represent the initial conditions. On one side, Daniel Bernoulli claimed (in 1775) that he could represent any general initial condition with this technique. To d'Alembert and Euler, however, the half-range Fourier sine series, with its period of $2L$, could not possibly represent any arbitrary function.[2] However, by 1807 Bernoulli was proven correct by the use of separation of variables in the heat conduction problem and it rapidly grew in acceptance.[3] In the following examples we show how to apply this method.

Separation of variables consists of four distinct steps that convert a second-order partial differential equation into two ordinary differential equations. First, we *assume* that the solution equals the product $X(x)T(t)$. Direct substitution into the partial differential

[2] See Hobson, E. W., 1957: *The Theory of Functions of a Real Variable and the Theory of Fourier's Series, Vol. 2*. Dover Publishers, Sections 312–314.

[3] Lützen, J., 1984: Sturm and Liouville's work on ordinary linear differential equations. The emergence of Sturm-Liouville theory. *Arch. Hist. Exact Sci.*, **29**, 317.

equation and boundary conditions yields two ordinary differential equations and the corresponding boundary conditions. Step two involves solving a boundary-value problem. In step three we find the corresponding time dependence. Finally we construct the complete solution as a sum of all product solutions. Upon applying the initial conditions, we have a Fourier half-range expansion and must compute the Fourier coefficients. The substitution of these coefficients into the summation yields the complete solution.

● **Example 8.3.1**

Let us solve the wave equation for the special case when we clamp the string at $x = 0$ and $x = L$. Mathematically, we find the solution to the wave equation

$$\frac{\partial^2 u}{\partial t^2} = c^2 \frac{\partial^2 u}{\partial x^2}, \quad 0 < x < L, \quad 0 < t, \tag{8.3.1}$$

which satisfies the initial conditions

$$u(x, 0) = f(x), \quad \frac{\partial u(x, 0)}{\partial t} = g(x), \quad 0 < x < L, \tag{8.3.2}$$

and the boundary conditions

$$u(0, t) = u(L, t) = 0, \quad 0 < t. \tag{8.3.3}$$

For the present, we leave the Cauchy data quite arbitrary.

We begin by assuming that the solution $u(x, t)$ equals the product $X(x)T(t)$. (Here T no longer denotes tension.) Because

$$\frac{\partial^2 u}{\partial t^2} = X(x)T''(t), \tag{8.3.4}$$

and

$$\frac{\partial^2 u}{\partial x^2} = X''(x)T(t), \tag{8.3.5}$$

the wave equation becomes

$$c^2 X''T = T''X, \tag{8.3.6}$$

or

$$\frac{X''}{X} = \frac{T''}{c^2 T}, \tag{8.3.7}$$

after dividing through by $c^2 X(x)T(t)$. Because the left side of Equation 8.3.7 depends only on x and the right side depends only on t, both sides must equal a constant. We write this separation constant $-\lambda$ and separate Equation 8.3.7 into two ordinary differential equations:

$$T'' + c^2 \lambda T = 0, \quad 0 < t, \tag{8.3.8}$$

and

$$X'' + \lambda X = 0, \quad 0 < x < L. \tag{8.3.9}$$

We now rewrite the boundary conditions in terms of $X(x)$ by noting that the boundary conditions become

$$u(0, t) = X(0)T(t) = 0, \tag{8.3.10}$$

and

$$u(L,t) = X(L)T(t) = 0 \qquad (8.3.11)$$

for $0 < t$. If we were to choose $T(t) = 0$, then we would have a trivial solution for $u(x,t)$. Consequently,

$$X(0) = X(L) = 0. \qquad (8.3.12)$$

This concludes the first step.

In the second step we consider three possible values for λ: $\lambda < 0$, $\lambda = 0$, and $\lambda > 0$. Turning first to $\lambda < 0$, we set $\lambda = -m^2$ so that square roots of λ will not appear later on and m is real. The general solution of Equation 8.3.9 is

$$X(x) = A \cosh(mx) + B \sinh(mx). \qquad (8.3.13)$$

Because $X(0) = 0$, $A = 0$. On the other hand, $X(L) = B \sinh(mL) = 0$. The function $\sinh(mL)$ does not equal zero since $mL \neq 0$ (recall $m > 0$). Thus, $B = 0$ and we have trivial solutions for a positive separation constant.

If $\lambda = 0$, the general solution now becomes

$$X(x) = C + Dx. \qquad (8.3.14)$$

The condition $X(0) = 0$ yields $C = 0$ while $X(L) = 0$ yields $DL = 0$ or $D = 0$. Hence, we have a trivial solution for the $\lambda = 0$ separation constant.

If $\lambda = k^2 > 0$, the general solution to Equation 8.3.9 is

$$X(x) = E \cos(kx) + F \sin(kx). \qquad (8.3.15)$$

The condition $X(0) = 0$ results in $E = 0$. On the other hand, $X(L) = F \sin(kL) = 0$. If we wish to avoid a trivial solution in this case ($F \neq 0$), $\sin(kL) = 0$, or $k_n = n\pi/L$, and $\lambda_n = n^2\pi^2/L^2$. The x-dependence equals $X_n(x) = F_n \sin(n\pi x/L)$. We added the n subscript to k and λ to indicate that these quantities depend on n. This concludes the second step.

Turning to Equation 8.3.8 for the third step, the solution to the $T(t)$ equation is

$$T_n(t) = G_n \cos(k_n ct) + H_n \sin(k_n ct), \qquad (8.3.16)$$

where G_n and H_n are arbitrary constants. For each $n = 1, 2, 3, \ldots$, a particular solution that satisfies the wave equation and prescribed boundary conditions is

$$u_n(x,t) = F_n \sin\left(\frac{n\pi x}{L}\right)\left[G_n \cos\left(\frac{n\pi ct}{L}\right) + H_n \sin\left(\frac{n\pi ct}{L}\right)\right], \qquad (8.3.17)$$

or

$$u_n(x,t) = \sin\left(\frac{n\pi x}{L}\right)\left[A_n \cos\left(\frac{n\pi ct}{L}\right) + B_n \sin\left(\frac{n\pi ct}{L}\right)\right], \qquad (8.3.18)$$

where $A_n = F_n G_n$ and $B_n = F_n H_n$. This concludes the third step.

For any choice of A_n and B_n, Equation 8.3.18 is a solution of the partial differential equation, Equation 8.3.1, also satisfying the boundary conditions, Equation 8.3.3. At this point we must introduce the powerful *principle of linear superposition*.

Principle of Linear Superposition

Let $u_1, u_2, \ldots, u_n$ satisfy a linear homogeneous equation $L(u_i) = 0$, then any arbitrary linear combination of $u = c_1 u_1 + c_2 u_2 + \cdots + c_n u_n$ also satisfies the same linear homogeneous equation.

Proof: We will outline the proof for just two solutions u_1 and u_2 and it can easily be generalized to n solutions. If u_1 and u_2 are two homogeneous solutions of a linear homogeneous equation, then $L(u_1) = 0$ and $L(u_2) = 0$. From the properties of linear operators, $L(c_1 u_1 + c_2 u_2) = c_1 L(u_1) + c_2 L(u_2) = 0$. Thus $u = c_1 u_1 + c_2 u_2$ satisfies the linear homogeneous equation $L(u) = 0$ if u_1 and u_2 satisfy the *same* linear homogeneous equation. $\quad\square$

Returning to our original problem, our solution is

$$u(x,t) = \sum_{n=1}^{\infty} \sin\left(\frac{n\pi x}{L}\right) \left[A_n \cos\left(\frac{n\pi ct}{L}\right) + B_n \sin\left(\frac{n\pi ct}{L}\right) \right]. \tag{8.3.19}$$

We need no new constants because A_n and B_n are still arbitrary.

Our method of using particular solutions to build up the general solution illustrates the powerful *principle of linear superposition*, which is applicable to any *linear* system. This principle states that if u_1 and u_2 are any solutions of a linear homogeneous partial differential equation in any region, then $u = c_1 u_1 + c_2 u_2$ is also a solution of that equation in that region, where c_1 and c_2 are any constants. We can generalize this to an infinite sum. It is extremely important because it allows us to construct general solutions to partial differential equations from particular solutions to the same problem.

Our fourth and final task remains to determine A_n and B_n. At $t = 0$,

$$u(x,0) = \sum_{n=1}^{\infty} A_n \sin\left(\frac{n\pi x}{L}\right) = f(x), \tag{8.3.20}$$

and

$$u_t(x,0) = \sum_{n=1}^{\infty} \frac{n\pi c}{L} B_n \sin\left(\frac{n\pi x}{L}\right) = g(x). \tag{8.3.21}$$

Both of these series are Fourier half-range sine expansions over the interval $(0, L)$. Applying the results from Section 5.3,

$$A_n = \frac{2}{L} \int_0^L f(x) \sin\left(\frac{n\pi x}{L}\right) dx, \tag{8.3.22}$$

and

$$\frac{n\pi c}{L} B_n = \frac{2}{L} \int_0^L g(x) \sin\left(\frac{n\pi x}{L}\right) dx, \tag{8.3.23}$$

or

$$B_n = \frac{2}{n\pi c} \int_0^L g(x) \sin\left(\frac{n\pi x}{L}\right) dx. \tag{8.3.24}$$

At this point we might ask ourselves whether the Fourier series solution to the wave equation always converges. For the case $g(x) = 0$, Carslaw[4] showed that if the initial position of the string forms a curve so that $f(x)$ or the slope $f'(x)$ is continuous between $x = 0$ and $x = L$, then the series converges uniformly.

As an example, let us take the initial conditions

$$f(x) = \begin{cases} 0, & 0 < x \le L/4, \\ 4h\left(\frac{x}{L} - \frac{1}{4}\right), & L/4 \le x \le L/2, \\ 4h\left(\frac{3}{4} - \frac{x}{L}\right), & L/2 \le x \le 3L/4, \\ 0, & 3L/4 \le x < L, \end{cases} \tag{8.3.25}$$

and

$$g(x) = 0, \qquad 0 < x < L. \tag{8.3.26}$$

In this particular example, $B_n = 0$ for all n because $g(x) = 0$. On the other hand,

$$A_n = \frac{8h}{L} \int_{L/4}^{L/2} \left(\frac{x}{L} - \frac{1}{4}\right) \sin\left(\frac{n\pi x}{L}\right) dx + \frac{8h}{L} \int_{L/2}^{3L/4} \left(\frac{3}{4} - \frac{x}{L}\right) \sin\left(\frac{n\pi x}{L}\right) dx \tag{8.3.27}$$

$$= \frac{8h}{n^2\pi^2}\left[2\sin\left(\frac{n\pi}{2}\right) - \sin\left(\frac{3n\pi}{4}\right) - \sin\left(\frac{n\pi}{4}\right)\right] \tag{8.3.28}$$

$$= \frac{8h}{n^2\pi^2}\left[2\sin\left(\frac{n\pi}{2}\right) - 2\sin\left(\frac{n\pi}{2}\right)\cos\left(\frac{n\pi}{4}\right)\right] \tag{8.3.29}$$

$$= \frac{16h}{n^2\pi^2}\sin\left(\frac{n\pi}{2}\right)\left[1 - \cos\left(\frac{n\pi}{4}\right)\right] = \frac{32h}{n^2\pi^2}\sin\left(\frac{n\pi}{2}\right)\sin^2\left(\frac{n\pi}{8}\right), \tag{8.3.30}$$

because $\sin(A) + \sin(B) = 2\sin[\frac{1}{2}(A+B)]\cos[\frac{1}{2}(A-B)]$, and $1 - \cos(2A) = 2\sin^2(A)$. Therefore,

$$u(x,t) = \frac{32h}{\pi^2}\sum_{n=1}^{\infty}\sin\left(\frac{n\pi}{2}\right)\sin^2\left(\frac{n\pi}{8}\right)\frac{1}{n^2}\sin\left(\frac{n\pi x}{L}\right)\cos\left(\frac{n\pi ct}{L}\right). \tag{8.3.31}$$

Because $\sin(n\pi/2)$ vanishes for n even, so does A_n. If Equation 8.3.31 were evaluated on a computer, considerable time and effort would be wasted. Consequently it is preferable to rewrite this equation so that we eliminate these vanishing terms. The most convenient method introduces the general expression $n = 2m - 1$ for any odd integer, where $m = 1, 2, 3, \ldots$, and notes that $\sin[(2m-1)\pi/2] = (-1)^{m+1}$. Therefore, Equation 8.3.31 becomes

$$u(x,t) = \frac{32h}{\pi^2}\sum_{m=1}^{\infty}\frac{(-1)^{m+1}}{(2m-1)^2}\sin^2\left[\frac{(2m-1)\pi}{8}\right]\sin\left[\frac{(2m-1)\pi x}{L}\right]\cos\left[\frac{(2m-1)\pi ct}{L}\right]. \tag{8.3.32}$$

Although we completely solved the problem, it is useful to rewrite Equation 8.3.32 as

$$u(x,t) = \frac{1}{2}\sum_{n=1}^{\infty}A_n\left\{\sin\left[\frac{n\pi}{L}(x - ct)\right] + \sin\left[\frac{n\pi}{L}(x + ct)\right]\right\} \tag{8.3.33}$$

---

[4] Carslaw, H. S., 1902: Note on the use of Fourier's series in the problem of the transverse vibrations of strings. *Proc. Edinburgh Math. Soc., Ser. 1*, **20**, 23–28.

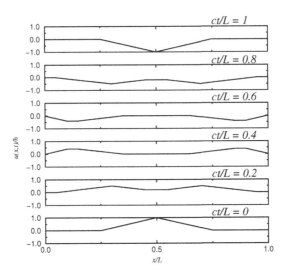

Figure 8.3.1: The vibration of a string $u(x,t)/h$ at various positions x/L at the times $ct/L = 0$, 0.2, 0.4, 0.6, 0.8, and 1. For times $1 < ct/L < 2$ the pictures appear in reverse time order.

through the application of the trigonometric identity $\sin(A)\cos(B) = \frac{1}{2}\sin(A - B) + \frac{1}{2}\sin(A + B)$. From general physics we find expressions like $\sin[k_n(x - ct)]$ or $\sin(kx - \omega t)$ arising in studies of simple wave motions. The quantity $\sin(kx - \omega t)$ is the mathematical description of a propagating wave in the sense that we must move to the right at the speed c if we wish to keep in the same position relative to the nearest crest and trough. The quantities k, ω, and c are the wavenumber, frequency, and phase speed or wave velocity, respectively. The relationship $\omega = kc$ holds between the frequency and phase speed.

It may seem paradoxical that we are talking about traveling waves in a problem dealing with waves confined on a string of length L. In the next section we will show that this is a fundamental property of solutions to the wave equation. For domains of infinite extent, the solution consists of two traveling waves; one runs out to positive infinity while the other runs out to negative infinity. Here we are dealing with standing waves because at the same time that a wave is propagating to the right, its mirror image is running to the left so that there is no resultant progressive wave motion.

Figures 8.3.1 and 8.3.2 illustrate our solution. Figure 8.3.1 gives various cross sections. The single large peak at $t = 0$ breaks into two smaller peaks that race towards the two ends. At each end, they reflect and turn upside down as they propagate back towards $x = L/2$ at $ct/L = 1$. This large, negative peak at $x = L/2$ again breaks apart, with the two smaller peaks propagating towards the endpoints. They reflect and again become positive peaks as they propagate back to $x = L/2$ at $ct/L = 2$. After that time, the whole process repeats itself.

MATLAB can be used to examine the solution in its totality. The script

```
% set parameters for the calculation
clear; M = 50; dx = 0.02; dt = 0.02;
% compute Fourier coefficients
sign = 32;
for m = 1:M
  temp1 = (2*m-1)*pi; temp2 = sin(temp1/8);
  a(m) = sign * temp2 * temp2 / (temp1 * temp1);
  sign = -sign;
end
```

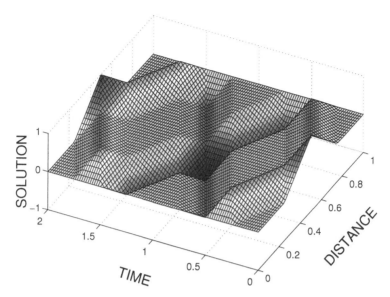

Figure 8.3.2: Two-dimensional plot of the vibration of a string $u(x,t)/h$ at various times ct/L and positions x/L.

```
% compute grid and initialize solution
X = [0:dx:1]; T = [0:dt:2];
u = zeros(length(T),length(X));
XX = repmat(X,[length(T) 1]);
TT = repmat(T',[1 length(X)]);
% compute solution from Equation 8.3.32
for m = 1:M
  temp1 = (2*m-1)*pi;
  u = u + a(m) .* sin(temp1*XX) .* cos(temp1*TT);
end
% plot space/time picture of the solution
surf(XX,TT,u)
xlabel('DISTANCE','Fontsize',20); ylabel('TIME','Fontsize',20)
zlabel('SOLUTION','Fontsize',20)
```

gives a three-dimensional view of Equation 8.3.32. The solution can be viewed in many different prospects using the interactive capacity of MATLAB.

An important dimension to the vibrating string problem is the fact that the wavenumber k_n is not a free parameter but has been restricted to the values of $n\pi/L$. This restriction on wavenumber is common in wave problems dealing with limited domains (for example, a building, ship, lake, or planet) and these oscillations are given the special name of *normal modes* or *natural vibrations*.

In our problem of the vibrating string, all of the components propagate with the same phase speed. That is, all of the waves, regardless of wavenumber k_n, move the characteristic distance $c\Delta t$ or $-c\Delta t$ after the time interval Δt elapsed. In the next example we will see that this is not always true. $\quad\square$

• Example 8.3.2: Dispersion

In the preceding example, the solution to the vibrating string problem consisted of two simple waves, each propagating with a phase speed c to the right and left. In many problems where the equations of motion are a little more complicated than Equation 8.3.1, all of the harmonics no longer propagate with the same phase speed but at a speed that depends upon the wavenumber. In such systems the phase relation varies between the harmonics and these systems are referred to as *dispersive*.

A modification of the vibrating string problem provides a simple illustration. We now subject each element of the string to an additional applied force that is proportional to its displacement:

$$\frac{\partial^2 u}{\partial t^2} = c^2 \frac{\partial^2 u}{\partial x^2} - hu, \quad 0 < x < L, \quad 0 < t, \tag{8.3.34}$$

where $h > 0$ is constant. For example, if we embed the string in a thin sheet of rubber, then in addition to the restoring force due to tension, there is a resistive force on each portion of the string as the rubber seeks to rebound. From its use in the quantum mechanics of "scalar" mesons, Equation 8.3.34 is often referred to as the *Klein-Gordon* equation.

We shall again look for particular solutions of the form $u(x,t) = X(x)T(t)$. This time, however,

$$XT'' - c^2 X''T + hXT = 0, \tag{8.3.35}$$

or

$$\frac{T''}{c^2 T} + \frac{h}{c^2} = \frac{X''}{X} = -\lambda, \tag{8.3.36}$$

which leads to two ordinary differential equations

$$X'' + \lambda X = 0, \tag{8.3.37}$$

and

$$T'' + (\lambda c^2 + h)T = 0. \tag{8.3.38}$$

If we attach the string at $x = 0$ and $x = L$, the $X(x)$ solution is

$$X_n(x) = \sin\left(\frac{n\pi x}{L}\right) \tag{8.3.39}$$

with $k_n = n\pi/L$, and $\lambda_n = n^2\pi^2/L^2$. On the other hand, the $T(t)$ solution becomes

$$T_n(t) = A_n \cos\left(\sqrt{k_n^2 c^2 + h}\, t\right) + B_n \sin\left(\sqrt{k_n^2 c^2 + h}\, t\right), \tag{8.3.40}$$

so that the product solution is

$$u_n(x,t) = \sin\left(\frac{n\pi x}{L}\right)\left[A_n \cos\left(\sqrt{k_n^2 c^2 + h}\, t\right) + B_n \sin\left(\sqrt{k_n^2 c^2 + h}\, t\right)\right]. \tag{8.3.41}$$

Finally, the general solution becomes

$$u(x,t) = \sum_{n=1}^{\infty} \sin\left(\frac{n\pi x}{L}\right)\left[A_n \cos\left(\sqrt{k_n^2 c^2 + h}\, t\right) + B_n \sin\left(\sqrt{k_n^2 c^2 + h}\, t\right)\right] \tag{8.3.42}$$

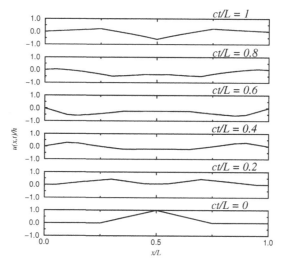

Figure 8.3.3: The vibration of a string $u(x,t)/h$ embedded in a thin sheet of rubber at various positions x/L at the times $ct/L = 0$, 0.2, 0.4, 0.6, 0.8, and 1 for $hL^2/c^2 = 10$. The same parameters were used as in Figure 8.3.1.

from the principle of linear superposition. Let us consider the case when $B_n = 0$. Then we can write Equation 8.3.42 as

$$u(x,t) = \sum_{n=1}^{\infty} \frac{A_n}{2}\left[\sin\left(k_n x + \sqrt{k_n^2 c^2 + h}\, t\right) + \sin\left(k_n x - \sqrt{k_n^2 c^2 + h}\, t\right)\right]. \qquad (8.3.43)$$

Comparing our results with Equation 8.3.33, the distance that a particular mode k_n moves during the time interval Δt depends not only upon external parameters such as h, the tension and density of the string, but also upon its wavenumber (or equivalently, wavelength). Furthermore, the frequency of a particular harmonic is larger than that when $h = 0$. This result is not surprising, because the added stiffness of the medium should increase the natural frequencies.

The importance of dispersion lies in the fact that if the solution $u(x,t)$ is a superposition of progressive waves in the same direction, then the phase relationship between the different harmonics changes with time. Because most signals consist of an infinite series of these progressive waves, dispersion causes the signal to become garbled. We show this by comparing the solution, Equation 8.3.42 given in Figures 8.3.3 and 8.3.4 for the initial conditions, Equation 8.3.25 and Equation 8.3.26, with $hL^2/c^2 = 10$, to the results given in Figures 8.3.1 and 8.3.2. In the case of Figure 8.3.4, the MATLAB script line
```
u = u + a(m) .* sin(temp1*XX) .* cos(temp1*TT);
```
has been replaced with
```
temp2 = temp1 * sqrt(1 + H/(temp1*temp1));
u = u + a(m) .* sin(temp1*XX) .* cos(temp2*TT);
```
where H = 10 is defined earlier in the script. Note how garbled the picture becomes at $ct/L = 2$ in Figure 8.3.4 compared to the nondispersive solution at the same time in Figure 8.3.2. $\qquad\square$

• **Example 8.3.3: Damped wave equation**

In the previous example a slight modification of the wave equation resulted in a wave solution where each Fourier harmonic propagates with its own particular phase speed. In

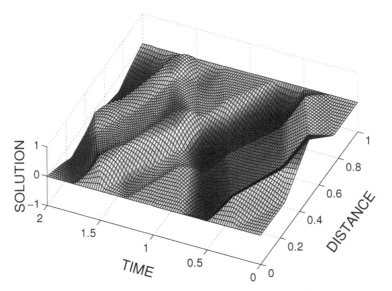

Figure 8.3.4: The two-dimensional plot of the vibration of a string $u(x,t)/h$ embedded in a thin sheet of rubber at various times ct/L and positions x/L for $hL^2/c^2 = 10$.

this example we introduce a modification of the wave equation that results not only in dispersive waves but also in the exponential decay of the amplitude as the wave propagates.

So far we have neglected the reaction of the surrounding medium (air or water, for example) on the motion of the string. For small-amplitude motions this reaction opposes the motion of each element of the string and is proportional to the element's velocity. The equation of motion, when we account for the tension and friction in the medium but not its stiffness or internal friction, is

$$\frac{\partial^2 u}{\partial t^2} + 2h\frac{\partial u}{\partial t} = c^2\frac{\partial^2 u}{\partial x^2}, \quad 0 < x < L, \quad 0 < t. \tag{8.3.44}$$

Because Equation 8.3.44 first arose in the mathematical description of the telegraph,[5] it is generally known as the *equation of telegraphy*. The effect of friction is, of course, to damp out the free vibration.

Let us assume a solution of the form $u(x,t) = X(x)T(t)$ and separate the variables to obtain the two ordinary differential equations:

$$X'' + \lambda X = 0, \tag{8.3.45}$$

and

$$T'' + 2hT' + \lambda c^2 T = 0 \tag{8.3.46}$$

with $X(0) = X(L) = 0$. Friction does not affect the shape of the normal modes; they are still

$$X_n(x) = \sin\left(\frac{n\pi x}{L}\right) \tag{8.3.47}$$

with $k_n = n\pi/L$ and $\lambda_n = n^2\pi^2/L^2$.

---

[5] The first published solution was by Kirchhoff, G., 1857: Über die Bewegung der Electrität in Drähten. *Ann. Phys. Chem.*, **100**, 193–217. English translation: Kirchhoff, G., 1857: On the motion of electricity in wires. *Philos. Mag., Ser. 4*, **13**, 393–412.

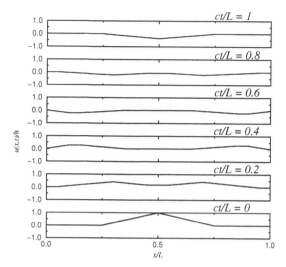

Figure 8.3.5: The vibration of a string $u(x,t)/h$ with frictional dissipation at various positions x/L at the times $ct/L = 0$, 0.2, 0.4, 0.6, 0.8, and 1 for $hL/c = 1$. The same parameters were used as in Figure 8.3.1.

The solution for the $T(t)$ equation is

$$T_n(t) = e^{-ht}\left[A_n \cos\left(\sqrt{k_n^2 c^2 - h^2}\, t\right) + B_n \sin\left(\sqrt{k_n^2 c^2 - h^2}\, t\right)\right] \qquad (8.3.48)$$

with the condition that $k_n c > h$. If we violate this condition, the solutions are two exponentially decaying functions in time. Because most physical problems usually fulfill this condition, we concentrate on this solution.

From the principle of linear superposition, the general solution is

$$u(x,t) = e^{-ht}\sum_{n=1}^{\infty}\sin\left(\frac{n\pi x}{L}\right)\left[A_n \cos\left(\sqrt{k_n^2 c^2 - h^2}\, t\right) + B_n \sin\left(\sqrt{k_n^2 c^2 - h^2}\, t\right)\right], \qquad (8.3.49)$$

where $\pi c > hL$. From Equation 8.3.49 we see two important effects. First, the presence of friction slows all of the harmonics. Furthermore, friction dampens all of the harmonics. Figures 8.3.5 and 8.3.6 illustrate the solution using the initial conditions given by Equation 8.3.25 and Equation 8.3.26 with $hL/c = 1$. In the case of Figure 8.3.6, the script line that produced Figure 8.3.2:

```
u = u + a(m) .* sin(temp1*XX) .* cos(temp1*TT);
```
has been replaced with
```
temp2 = temp1 * sqrt(1 - (H*H)/(temp1*temp1));
u = u + a(m) .* exp(-H*TT) .* sin(temp1*XX) .* cos(temp2*TT);
```
where H = 1 is defined earlier in the script. Because this is a rather large coefficient of friction, Figures 8.3.5 and 8.3.6 exhibit rapid damping as well as dispersion.

This damping and dispersion of waves also occurs in solutions of the equation of telegraphy where the solutions are progressive waves. Because early telegraph lines were short, time delay effects were negligible. However, when engineers laid the first transoceanic cables in the 1850s, the time delay became seconds and differences in the velocity of propagation of different frequencies, as predicted by Equation 8.3.49, became noticeable to the operators. Table 8.3.1 gives the transmission rate for various transatlantic submarine telegraph lines. As it shows, increases in the transmission rates during the nineteenth century were due primarily to improvements in terminal technology.

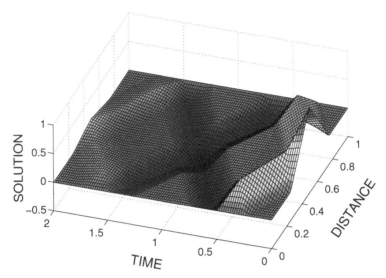

Figure 8.3.6: The vibration of a string $u(x,t)/h$ with frictional dissipation at various times ct/L and positions x/L for $hL/c = 1$.

When they instituted long-distance telephony just before the turn of the twentieth century, this difference in velocity between frequencies should have limited the circuits to a few tens of miles.[6] However, in 1899, Prof. Michael Pupin at Columbia University showed that by adding inductors ("loading coils") to the line at regular intervals, the velocities at the different frequencies could be equalized.[7] Heaviside[8] and the French engineer Vaschy[9] made similar suggestions in the nineteenth century. Thus, adding resistance and inductance, which would seem to make things worse, actually made possible long-distance telephony. Today you can see these loading coils as you drive along the street; they are the black cylinders, approximately one between each pair of telephone poles, spliced into the telephone cable. The loading of long submarine telegraph cables had to wait for the development of permalloy and mu-metal materials of high magnetic induction.

Problems

Solve the wave equation $u_{tt} = c^2 u_{xx}$, $0 < x < L$, $0 < t$, subject to the boundary conditions that $u(0,t) = u(L,t) = 0, 0 < t$, and the following initial conditions for $0 < x < L$. Use MATLAB to illustrate your solution.

1. $u(x,0) = 0, \quad u_t(x,0) = 1$

2. $u(x,0) = 1, \quad u_t(x,0) = 0$

---

[6] Rayleigh, J. W., 1884: On telephoning through a cable. *Brit. Assoc. Rep.*, 632–633; Jordan, D. W., 1982: The adoption of self-induction by telephony, 1886–1889. *Ann. Sci.*, **39**, 433–461.

[7] There is considerable controversy concerning who is exactly the inventor. See Brittain, J. E., 1970: The introduction of the loading coil: George A. Campbell and Michael I. Pupin. *Tech. Culture*, **11**, 36–57.

[8] First published 3 June 1887. Reprinted in Heaviside, O., 1970: *Electrical Papers, Vol. 2.* Chelsea Publishing, pp. 119–124.

[9] See Devaux-Charbonnel, X. G. F., 1917: La contribution des ingénieurs français à la téléphonie à grande distance par câbles souterrains: Vaschy et Barbarat. *Rev. Gén. Électr.*, **2**, 288–295.

Table 8.3.1: Technological Innovation on Transatlantic Telegraph Cables

Year	Technological Innovation	Performance (words/min)
1857–58	Mirror galvanometer	3–7
1870	Condensers	12
1872	Siphon recorder	17
1879	Duplex	24
1894	Larger diameter cable	72–90
1915–20	Brown drum repeater and Heurtley magnifier	100
1923–28	Magnetically loaded lines	300–320
1928–32	Electronic signal shaping amplifiers and time division multiplexing	480
1950	Repeaters on the continental shelf	100–300
1956	Repeater telephone cables	21600

From Coates, V. T., and B. Finn, 1979: *A Retrospective Technology Assessment: Submarine Telegraphy. The Transatlantic Cable of 1866.* San Francisco Press, Inc., 268 pp.

3. $u(x,0) = \begin{cases} 3hx/2L, & 0 < x < 2L/3, \\ 3h(L-x)/L, & 2L/3 < x < L, \end{cases}$ $u_t(x,0) = 0$

4. $u(x,0) = [3\sin(\pi x/L) - \sin(3\pi x/L)]/4, \quad u_t(x,0) = 0,$

5. $u(x,0) = \sin(\pi x/L), \quad u_t(x,0) = \begin{cases} 0, & 0 < x < L/4 \\ a, & L/4 < x < 3L/4 \\ 0, & 3L/4 < x < L \end{cases}$

6. $u(x,0) = 0, \quad u_t(x,0) = \begin{cases} ax/L, & 0 < x < L/2 \\ a(L-x)/L, & L/2 < x < L \end{cases}$

7. $u(x,0) = \begin{cases} x, & 0 < x < L/2, \\ L-x, & L/2 < x < L, \end{cases}$ $u_t(x,0) = 0$

8. Solve the wave equation

$$\frac{\partial^2 u}{\partial t^2} = c^2 \frac{\partial^2 u}{\partial x^2}, \qquad 0 < x < \pi, \qquad 0 < t,$$

subject to the boundary conditions $u_x(0,t) = u_x(\pi,t) = 0$, $0 < t$, and the initial conditions $u(x,0) = 0$, $u_t(x,0) = 1 + \cos^3(x)$, $0 < x < \pi$. Hint: You must include the separation constant of zero.

9. One of the classic applications of the wave equation has been the explanation of the acoustic properties of string instruments. Usually we excite a string in one of three ways: by plucking (as in the harp, zither, etc.), by striking with a hammer (piano), or by bowing (violin, violoncello, etc.). In all of these cases, the governing partial differential equation is

$$\frac{\partial^2 u}{\partial t^2} = c^2 \frac{\partial^2 u}{\partial x^2}, \qquad 0 < x < L, \quad 0 < t,$$

with the boundary conditions $u(0,t) = u(L,t) = 0, 0 < t$. For each of the following methods of exciting a string instrument, find the complete solution to the problem:

(a) Plucked string

For the initial conditions:

$$u(x,0) = \begin{cases} \beta x/a, & 0 < x < a, \\ \beta(L-x)/(L-a), & a < x < L, \end{cases} \qquad u_t(x,0) = 0, \quad 0 < x < L,$$

show that

$$u(x,t) = \frac{2\beta L^2}{\pi^2 a(L-a)} \sum_{n=1}^{\infty} \frac{1}{n^2} \sin\left(\frac{n\pi a}{L}\right) \sin\left(\frac{n\pi x}{L}\right) \cos\left(\frac{n\pi ct}{L}\right).$$

We note that the harmonics are absent where $\sin(n\pi a/L) = 0$. Thus, if we pluck the string at the center, all of the harmonics of even order are absent. Furthermore, the intensity of the successive harmonics varies as n^{-2}. The higher harmonics (overtones) are therefore relatively feeble compared to the $n = 1$ term (the fundamental).

(b) String excited by impact

The effect of the impact of a hammer depends upon the manner and duration of the contact, and is more difficult to estimate. However, as a first estimate, let

$$u(x,0) = 0, \quad 0 < x < L, \qquad u_t(x,0) = \begin{cases} \mu, & a - \epsilon < x < a + \epsilon, \\ 0, & \text{otherwise}, \end{cases}$$

where $\epsilon \ll 1$. Show that the solution in this case is

$$u(x,t) = \frac{4\mu L}{\pi^2 c} \sum_{n=1}^{\infty} \frac{1}{n^2} \sin\left(\frac{n\pi\epsilon}{L}\right) \sin\left(\frac{n\pi a}{L}\right) \sin\left(\frac{n\pi x}{L}\right) \sin\left(\frac{n\pi ct}{L}\right).$$

As in part (a), the nth mode is absent if the origin is at a node. The intensity of the overtones is now of the same order of magnitude; higher harmonics (overtones) are relatively more in evidence than in part (a).

(c) Bowed violin string

The theory of the vibration of a string when excited by bowing is poorly understood. The bow drags the string for a time until the string springs back. After a while the process repeats. It can be shown[10] that the proper initial conditions are

$$u(x,0) = 0, \quad u_t(x,0) = 4\beta c(L-x)/L^2, \qquad 0 < x < L,$$

where β is the maximum displacement. Show that the solution is now

$$u(x,t) = \frac{8\beta}{\pi^2} \sum_{n=1}^{\infty} \frac{1}{n^2} \sin\left(\frac{n\pi x}{L}\right) \sin\left(\frac{n\pi ct}{L}\right).$$

---

[10] See Lamb, H., 1960: *The Dynamical Theory of Sound*. Dover Publishers, Section 27.

Although largely self-educated in mathematics, Jean Le Rond d'Alembert (1717–1783) gained equal fame as a mathematician and *philosophe* of the continental Enlightenment. By the middle of the eighteenth century, he stood with such leading European mathematicians and mathematical physicists as Clairaut, D. Bernoulli, and Euler. Today we best remember him for his work in fluid dynamics and applying partial differential equations to problems in physics. (Portrait courtesy of the Archives de l'Académie des sciences, Paris.)

8.4 D'ALEMBERT'S FORMULA

In the previous section we sought solutions to the homogeneous wave equation in the form of a product $X(x)T(t)$. For the one-dimensional wave equation there is a more general method for constructing the solution, published by d'Alembert[11] in 1747.

Let us determine a solution to the homogeneous wave equation

$$\frac{\partial^2 u}{\partial t^2} = c^2 \frac{\partial^2 u}{\partial x^2}, \quad -\infty < x < \infty, \quad 0 < t, \tag{8.4.1}$$

which satisfies the initial conditions

$$u(x,0) = f(x), \quad \frac{\partial u(x,0)}{\partial t} = g(x), \quad -\infty < x < \infty. \tag{8.4.2}$$

We begin by introducing two new variables ξ, η defined by $\xi = x + ct$, and $\eta = x - ct$, and set $u(x,t) = w(\xi, \eta)$. The variables ξ and η are called the *characteristics* of the wave equation. Using the chain rule,

$$\frac{\partial}{\partial x} = \frac{\partial \xi}{\partial x}\frac{\partial}{\partial \xi} + \frac{\partial \eta}{\partial x}\frac{\partial}{\partial \eta} = \frac{\partial}{\partial \xi} + \frac{\partial}{\partial \eta} \tag{8.4.3}$$

---

[11] D'Alembert, J., 1747: Recherches sur la courbe que forme une corde tendüe mise en vibration. *Hist. Acad. R. Sci. Belles Lett., Berlin*, 214–219.

$$\frac{\partial}{\partial t} = \frac{\partial \xi}{\partial t} \frac{\partial}{\partial \xi} + \frac{\partial \eta}{\partial t} \frac{\partial}{\partial \eta} = c \frac{\partial}{\partial \xi} - c \frac{\partial}{\partial \eta} \qquad (8.4.4)$$

$$\frac{\partial^2}{\partial x^2} = \frac{\partial \xi}{\partial x} \frac{\partial}{\partial \xi} \left(\frac{\partial}{\partial \xi} + \frac{\partial}{\partial \eta} \right) + \frac{\partial \eta}{\partial x} \frac{\partial}{\partial \eta} \left(\frac{\partial}{\partial \xi} + \frac{\partial}{\partial \eta} \right) \qquad (8.4.5)$$

$$= \frac{\partial^2}{\partial \xi^2} + 2 \frac{\partial^2}{\partial \xi \partial \eta} + \frac{\partial^2}{\partial \eta^2}, \qquad (8.4.6)$$

and similarly

$$\frac{\partial^2}{\partial t^2} = c^2 \left(\frac{\partial^2}{\partial \xi^2} - 2 \frac{\partial^2}{\partial \xi \partial \eta} + \frac{\partial^2}{\partial \eta^2} \right), \qquad (8.4.7)$$

so that the wave equation becomes

$$\frac{\partial^2 w}{\partial \xi \partial \eta} = 0. \qquad (8.4.8)$$

The general solution of Equation 8.4.8 is

$$w(\xi, \eta) = F(\xi) + G(\eta). \qquad (8.4.9)$$

Thus, the general solution of Equation 8.4.1 is of the form

$$u(x, t) = F(x + ct) + G(x - ct), \qquad (8.4.10)$$

where F and G are arbitrary functions of one variable and are assumed to be twice differentiable. Setting $t = 0$ in Equation 8.4.10 and using the initial condition that $u(x, 0) = f(x)$,

$$F(x) + G(x) = f(x). \qquad (8.4.11)$$

The partial derivative of Equation 8.4.10 with respect to t yields

$$\frac{\partial u(x, t)}{\partial t} = cF'(x + ct) - cG'(x - ct). \qquad (8.4.12)$$

Here primes denote differentiation with respect to the argument of the function. If we set $t = 0$ in Equation 8.4.12 and apply the initial condition that $u_t(x, 0) = g(x)$,

$$cF'(x) - cG'(x) = g(x). \qquad (8.4.13)$$

Integrating Equation 8.4.13 from 0 to any point x gives

$$F(x) - G(x) = \frac{1}{c} \int_0^x g(\tau) \, d\tau + C, \qquad (8.4.14)$$

where C is the constant of integration. Combining this result with Equation 8.4.11,

$$F(x) = \frac{f(x)}{2} + \frac{1}{2c} \int_0^x g(\tau) \, d\tau + \frac{C}{2}, \qquad (8.4.15)$$

and

$$G(x) = \frac{g(x)}{2} - \frac{1}{2c} \int_0^x g(\tau) \, d\tau - \frac{C}{2}. \qquad (8.4.16)$$

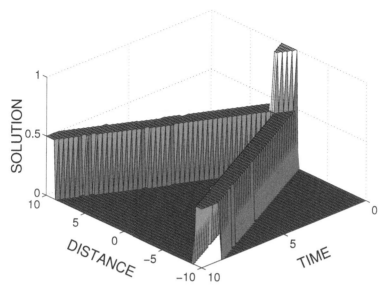

Figure 8.4.1: D'Alembert's solution, Equation 8.4.18, to the wave equation.

If we replace the variable x in the expression for F and G by $x+ct$ and $x-ct$, respectively, and substitute the results into Equation 8.4.10, we finally arrive at the formula

$$u(x,t) = \frac{f(x+ct) + f(x-ct)}{2} + \frac{1}{2c} \int_{x-ct}^{x+ct} g(\tau)\,d\tau. \qquad (\mathbf{8.4.17})$$

This is known as *d'Alembert's formula* for the solution of the wave equation, Equation 8.4.1, subject to the initial conditions, Equation 8.4.2. It gives a *representation* of the solution in terms of *known* initial conditions.

- **Example 8.4.1**

To illustrate d'Alembert's formula, let us find the solution to the wave equation, Equation 8.4.1, satisfying the initial conditions $u(x,0) = H(x+1) - H(x-1)$ and $u_t(x,0) = 0$, $-\infty < x < \infty$. By d'Alembert's formula, Equation 8.4.17,

$$u(x,t) = \tfrac{1}{2}\left[H(x+ct+1) + H(x-ct+1) - H(x+ct-1) - H(x-ct-1)\right]. \qquad (\mathbf{8.4.18})$$

We illustrate this solution in Figure 8.4.1 generated by the MATLAB script:

```
% set mesh size for solution
clear; dx = 0.1; dt = 0.1;
% compute grid
X=[-10:dx:10]; T = [0:dt:10];
for j=1:length(T); t = T(j);
for i=1:length(X); x = X(i);
% compute characteristics
  characteristic_1 = x + t; characteristic_2 = x - t;
% compute solution
  XX(i,j) = x; TT(i,j) = t;
  u(i,j) = 0.5*(stepfun(characteristic_1,-1)+stepfun(characteristic_2,-1)...
```

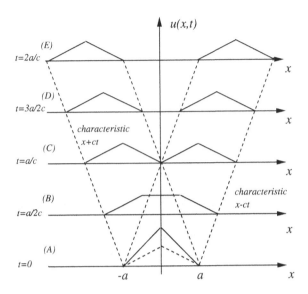

Figure 8.4.2: The propagation of waves due to an initial displacement according to d'Alembert's formula.

```
                    -stepfun(characteristic_1, 1)-stepfun(characteristic_2, 1));
end; end
surf(XX,TT,u); colormap autumn;
xlabel('DISTANCE','Fontsize',20); ylabel('TIME','Fontsize',20)
zlabel('SOLUTION','Fontsize',20)
```

In this figure, you can clearly see the characteristics as they emanate from the discontinuities at $x = \pm 1$.   □

• **Example 8.4.2**

Let us find the solution to the wave equation, Equation 8.4.1, when $u(x,0) = 0$, and $u_t(x,0) = \sin(2x)$, $-\infty < x < \infty$. By d'Alembert's formula, the solution is

$$u(x,t) = \frac{1}{2c} \int_{x-ct}^{x+ct} \sin(2\tau)\, d\tau = \frac{\sin(2x)\sin(2ct)}{2}. \qquad (8.4.19)$$

In addition to providing a method of solving the wave equation, d'Alembert's solution can also provide physical insight into the vibration of a string. Consider the case when we release a string with zero velocity after giving it an initial displacement of $f(x)$. According to Equation 8.4.17, the displacement at a point x at any time t is

$$u(x,t) = \frac{f(x+ct) + f(x-ct)}{2}. \qquad (8.4.20)$$

Because the function $f(x-ct)$ is the same as the function of $f(x)$ translated to the right by a distance equal to ct, $f(x-ct)$ represents a wave of form $f(x)$ traveling to the right with the velocity c, a forward wave. Similarly, we can interpret the function $f(x+ct)$ as representing a wave with the shape $f(x)$ traveling to the left with the velocity c, a backward wave. Thus, the solution, Equation 8.4.17, is a superposition of forward and backward waves traveling with the same velocity c and having the shape of the initial profile $f(x)$ with half of the amplitude. Clearly the characteristics $x + ct$ and $x - ct$ give the propagation paths along which the waveform $f(x)$ propagates.   □

• **Example 8.4.3**

To illustrate our physical interpretation of d'Alembert's solution, suppose that the string has an initial displacement defined by

$$f(x) = \begin{cases} a - |x|, & -a \le x \le a, \\ 0, & \text{otherwise.} \end{cases} \tag{8.4.21}$$

In Figure 8.4.2(A) the forward and backward waves, indicated by the dashed line, coincide at $t = 0$. As time advances, both waves move in opposite directions. In particular, at $t = a/(2c)$, they moved through a distance $a/2$, resulting in the displacement of the string shown in Figure 8.4.2(B). Eventually, at $t = a/c$, the forward and backward waves completely separate. Finally, Figures 8.4.2(D) and 8.4.3(E) show how the waves radiate off to infinity at the speed of c. Note that at each point the string returns to its original position of rest after the passage of each wave.

Consider now the opposite situation when $u(x,0) = 0$, and $u_t(x,0) = g(x)$. The displacement is

$$u(x,t) = \frac{1}{2c} \int_{x-ct}^{x+ct} g(\tau) \, d\tau. \tag{8.4.22}$$

If we introduce the function

$$\varphi(x) = \frac{1}{2c} \int_0^x g(\tau) \, d\tau, \tag{8.4.23}$$

then we can write Equation 8.4.22 as

$$u(x,t) = \varphi(x + ct) - \varphi(x - ct), \tag{8.4.24}$$

which again shows that the solution is a superposition of a forward wave $-\varphi(x - ct)$ and a backward wave $\varphi(x + ct)$ traveling with the same velocity c. The function φ, which we compute from Equation 8.4.23 and the initial velocity $g(x)$, determines the exact form of these waves. □

• **Example 8.4.4: Vibration of a moving threadline**

The characterization and analysis of the oscillations of a string or yarn have an important application in the textile industry because they describe the way that yarn winds on a bobbin.[12] As we showed in Section 8.1, the governing equation, the "threadline equation," is

$$\frac{\partial^2 u}{\partial t^2} + \alpha \frac{\partial^2 u}{\partial x \partial t} + \beta \frac{\partial^2 u}{\partial x^2} = 0, \tag{8.4.25}$$

where $\alpha = 2V$, $\beta = V^2 - gT/\rho$, V is the windup velocity, g is the gravitational attraction, T is the tension in the yarn, and ρ is the density of the yarn. We now introduce the characteristics $\xi = x + \lambda_1 t$, and $\eta = x + \lambda_2 t$, where λ_1 and λ_2 are yet undetermined. Upon substituting ξ and η into Equation 8.4.25,

$$(\lambda_1^2 + 2V\lambda_1 + V^2 - gT/\rho)u_{\xi\xi} + (\lambda_2^2 + 2V\lambda_2 + V^2 - gT/\rho)u_{\eta\eta}$$
$$+ [2V^2 - 2gT/\rho + 2V(\lambda_1 + \lambda_2) + 2\lambda_1\lambda_2]u_{\xi\eta} = 0. \tag{8.4.26}$$

---

[12] See Swope, R. D., and W. F. Ames, 1963: Vibrations of a moving threadline. *J. Franklin Inst.*, **275**, 36–55.

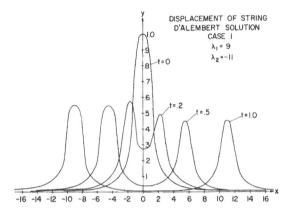

Figure 8.4.3: Displacement of an infinite, moving threadline when $c = 10$, and $V = 1$.

If we choose λ_1 and λ_2 to be roots of the equation

$$\lambda^2 + 2V\lambda + V^2 - gT/\rho = 0, \tag{8.4.27}$$

Equation 8.4.26 reduces to the simple form

$$u_{\xi\eta} = 0, \tag{8.4.28}$$

which has the general solution

$$u(x,t) = F(\xi) + G(\eta) = F(x + \lambda_1 t) + G(x + \lambda_2 t). \tag{8.4.29}$$

Solving Equation 8.4.27 yields

$$\lambda_1 = c - V, \qquad \text{and} \qquad \lambda_2 = -c - V, \tag{8.4.30}$$

where $c = \sqrt{gT/\rho}$. If the initial conditions are

$$u(x,0) = f(x), \qquad \text{and} \qquad u_t(x,0) = g(x), \tag{8.4.31}$$

then

$$u(x,t) = \frac{1}{2c}\left[\lambda_1 f(x + \lambda_2 t) - \lambda_2 f(x + \lambda_1 t) + \int_{x+\lambda_2 t}^{x+\lambda_1 t} g(\tau)\, d\tau\right]. \tag{8.4.32}$$

Because λ_1 does not generally equal λ_2, the two waves that constitute the motion of the string move with different speeds and have different shapes and forms. For example, if

$$f(x) = \frac{1}{x^2 + 1}, \qquad \text{and} \quad g(x) = 0, \tag{8.4.33}$$

$$u(x,t) = \frac{1}{2c}\left\{\frac{c - V}{1 + [x - (c + V)t]^2} + \frac{c + V}{1 + [x - (c - V)t]^2}\right\}. \tag{8.4.34}$$

Figures 8.4.3 and 8.4.4 illustrate this solution for several different parameters.

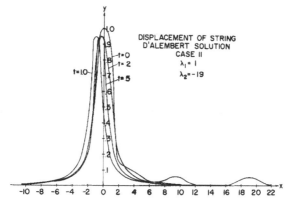

Figure 8.4.4: Displacement of an infinite, moving threadline when $c = 11$, and $V = 10$.

Problems

Use d'Alembert's formula to solve the wave equation, Equation 8.4.1, for the following initial conditions defined for $|x| < \infty$. Then illustrate your solution using MATLAB.

1. $u(x, 0) = 2\sin(x)\cos(x)$ $\qquad\qquad$ $u_t(x, 0) = \cos(x)$

2. $u(x, 0) = x\sin(x)$ $\qquad\qquad$ $u_t(x, 0) = \cos(2x)$

3. $u(x, 0) = 1/(x^2 + 1)$ $\qquad\qquad$ $u_t(x, 0) = e^x$

4. $u(x, 0) = e^{-x}$ $\qquad\qquad$ $u_t(x, 0) = 1/(x^2 + 1)$

5. $u(x, 0) = \cos(\pi x/2)$ $\qquad\qquad$ $u_t(x, 0) = \sinh(ax)$

6. $u(x, 0) = \sin(3x)$ $\qquad\qquad$ $u_t(x, 0) = \sin(2x) - \sin(x)$

7. Assuming that the functions F and G are differentiable, show by direct substitution that

$$u(x, t) = EF(x + ct) - EG(x - ct) - \tfrac{1}{8}kc^2t^2 + \tfrac{3}{8}kx^2,$$

and

$$v(x, t) = cF(x + ct) + cG(x - ct) - \frac{kc^2xt}{4E}$$

are the d'Alembert solutions to the hyperbolic system

$$\frac{\partial u}{\partial t} = E\frac{\partial v}{\partial x}, \qquad \frac{\partial u}{\partial x} = \rho\frac{\partial v}{\partial t} + kx, \qquad -\infty < x < \infty, \quad 0 < t,$$

where $c^2 = E/\rho$ and E, k, and ρ are constants.

8. D'Alembert's solution can also be used in problems over the limited domain $0 < x < L$. To illustrate this, let us solve the wave equation, Equation 8.4.1, with the initial conditions $u(x, 0) = 0$, $u_t(x, 0) = V_{max}(1 - x/L)$, $0 < x < L$, and the boundary conditions $u(0, t) = u(L, t) = 0$, $0 < t$.

Step 1: Show that the solution to this problem is

$$u(x, t) = \tfrac{1}{2}[V_0(x + ct) - V_0(x - ct)],$$

where
$$V_0(\chi) = \frac{1}{c}\int_0^\chi u_t(\xi,0)\,d\xi = \frac{V_{max}\chi}{c}\left(1 - \frac{\chi}{2L}\right), \qquad 0 < \chi < L,$$

along with the periodicity conditions $V_0(\chi) = V_0(-\chi)$, and $V_0(L+\chi) = V_0(L-\chi)$ to take care of those cases when the argument of $V_0(\cdot)$ is outside of $(0,L)$. Hint: Substitute the solution into the boundary conditions.

Step 2: Show that at any point x within the interval $(0,L)$, the solution repeats with a period of $2L/c$ if $ct > 2L$. Therefore, if we know the behavior of the solution for the time interval $0 < ct < 2L$, we know the behavior for any other time.

Step 3: Show that the solution at any point x within the interval $(0,L)$ and time $t + L/c$, where $0 < ct < L$, is the mirror image (about $u = 0$) of the solution at the point $L - x$ and time t, where $0 < ct < L$.

Step 4: Show that the maximum value of $u(x,t)$ occurs at $x = ct$, where $0 < x < L$ and when $0 < ct < L$. At that point,

$$u_{max} = \frac{V_{max}x}{c}\left(1 - \frac{x}{L}\right),$$

where u_{max} equals the largest magnitude of $u(x,t)$ for any time t. Plot u_{max} as a function x and show that it is a parabola. Hint: Find the maximum value of $u(x,t)$ when $0 < x \le ct$ and $ct \le x < L$ with $0 < x + ct < L$ or $L < x + ct < 2L$.

8.5 NUMERICAL SOLUTION OF THE WAVE EQUATION

In addition to separation of variables, linear wave equations can also be solved using the method of characteristics and transform methods. However, when these analytic techniques fail or we have a nonlinear wave equation, we must resort to numerical techniques.

In contrast to the continuous solutions, finite difference methods, a type of numerical solution technique, give discrete numerical values at a specific location (x_m, t_n), called a *grid point*. These numerical values represent a numerical approximation of the continuous solution over the region $(x_m - \Delta x/2, x_m + \Delta x/2)$ and $(t_n - \Delta t/2, t_n + \Delta t/2)$, where Δx and Δt are the distance and time intervals between grid points, respectively. Clearly, in the limit of $\Delta x, \Delta t \to 0$, we recover the continuous solution. However, practical considerations such as computer memory or execution time often require that Δx and Δt, although small, are not negligibly small.

The first task in the numerical solution of a partial differential equation is the replacement of its continuous derivatives with finite differences. The most popular approach employs Taylor expansions. If we focus on the x-derivative, then the value of the solution at $u[(m+1)\Delta x, n\Delta t]$ in terms of the solution at $(m\Delta x, n\Delta t)$ is

$$u[(m+1)\Delta x, n\Delta t] = u(x_m, t_n) + \frac{\Delta x}{1!}\frac{\partial u(x_m, t_n)}{\partial x} + \frac{(\Delta x)^2}{2!}\frac{\partial^2 u(x_m, t_n)}{\partial x^2}$$
$$+ \frac{(\Delta x)^3}{3!}\frac{\partial^3 u(x_m, t_n)}{\partial x^3} + \frac{(\Delta x)^4}{4!}\frac{\partial^4 u(x_m, t_n)}{\partial x^4} + \cdots \qquad (8.5.1)$$
$$= u(x_m, t_n) + \Delta x\frac{\partial u(x_m, t_n)}{\partial x} + O[(\Delta x)^2], \qquad (8.5.2)$$

where $O[(\Delta x)^2]$ gives a measure of the magnitude of neglected terms.[13]

---

[13] The symbol O is a mathematical notation indicating relative magnitude of terms, namely that $f(\epsilon) = O(\epsilon^n)$ provided $\lim_{\epsilon \to 0} |f(\epsilon)/\epsilon^n| < \infty$. For example, as $\epsilon \to 0$, $\sin(\epsilon) = O(\epsilon)$, $\sin(\epsilon^2) = O(\epsilon^2)$, and $\cos(\epsilon) = O(1)$.

From Equation 8.5.2, one possible approximation for u_x is

$$\frac{\partial u(x_m, t_n)}{\partial x} = \frac{u_{m+1}^n - u_m^n}{\Delta x} + O(\Delta x), \tag{8.5.3}$$

where we use the standard notation that $u_m^n = u(x_m, t_n)$. This is an example of a *one-sided finite difference* approximation of the partial derivative u_x. The error in using this approximation grows as Δx.

Another possible approximation for the derivative arises from using $u(m\Delta x, n\Delta t)$ and $u[(m-1)\Delta x, n\Delta t]$. From the Taylor expansion:

$$u[(m-1)\Delta x, n\Delta t] = u(x_m, t_n) - \frac{\Delta x}{1!}\frac{\partial u(x_m, t_n)}{\partial x} + \frac{(\Delta x)^2}{2!}\frac{\partial^2 u(x_m, t_n)}{\partial x^2}$$
$$- \frac{(\Delta x)^3}{3!}\frac{\partial^3 u(x_m, t_n)}{\partial x^3} + \frac{(\Delta x)^4}{4!}\frac{\partial^4 u(x_m, t_n)}{\partial x^4} - \cdots, \tag{8.5.4}$$

we can also obtain the one-sided difference formula

$$\frac{u(x_m, t_n)}{\partial x} = \frac{u_m^n - u_{m-1}^n}{\Delta x} + O(\Delta x). \tag{8.5.5}$$

A third possibility arises from subtracting Equation 8.5.4 from Equation 8.5.1:

$$u_{m+1}^n - u_{m-1}^n = 2\Delta x\frac{\partial u(x_m, t_n)}{\partial x} + O[(\Delta x)^3], \tag{8.5.6}$$

or

$$\frac{\partial u(x_m, t_n)}{\partial x} = \frac{u_{m+1}^n - u_{m-1}^n}{2\Delta x} + O[(\Delta x)^2]. \tag{8.5.7}$$

Thus, the choice of the finite differencing scheme can produce profound differences in the accuracy of the results. In the present case, *centered finite differences* can yield results that are markedly better than using one-sided differences.

To solve the wave equation, we need to approximate u_{xx}. If we add Equation 8.5.1 and Equation 8.5.4,

$$u_{m+1}^n + u_{m-1}^n = 2u_m^n + \frac{\partial^2 u(x_m, t_n)}{\partial x^2}(\Delta x)^2 + O[(\Delta x)^4], \tag{8.5.8}$$

or

$$\frac{\partial^2 u(x_m, t_n)}{\partial x^2} = \frac{u_{m+1}^n - 2u_m^n + u_{m-1}^n}{(\Delta x)^2} + O[(\Delta x)^2]. \tag{8.5.9}$$

Similar considerations hold for the time derivative. Thus, by neglecting errors of $O[(\Delta x)^2]$ and $O[(\Delta t)^2]$, we may approximate the wave equation by

$$\frac{u_m^{n+1} - 2u_m^n + u_m^{n-1}}{(\Delta t)^2} = c^2\frac{u_{m+1}^n - 2u_m^n + u_{m-1}^n}{(\Delta x)^2}. \tag{8.5.10}$$

Because the wave equation represents evolutionary change of some quantity, Equation 8.5.10 is generally used as a predictive equation where we forecast u_m^{n+1} by

$$u_m^{n+1} = 2u_m^n - u_m^{n-1} + \left(\frac{c\Delta t}{\Delta x}\right)^2\left(u_{m+1}^n - 2u_m^n + u_{m-1}^n\right). \tag{8.5.11}$$

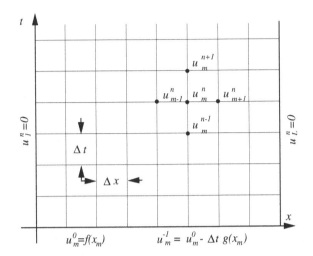

Figure 8.5.1: Schematic of the numerical solution of the wave equation with fixed endpoints.

Figure 8.5.1 illustrates this numerical scheme.

The greatest challenge in using Equation 8.5.11 occurs with the very first prediction. When $n = 0$, clearly u_{m+1}^0, u_m^0, and u_{m-1}^0 are specified from the initial condition $u(m\Delta x, 0) = f(x_m)$. But what about u_m^{-1}? Recall that we still have $u_t(x,0) = g(x)$. If we use the backward difference formula, Equation 8.5.5,

$$\frac{u_m^0 - u_m^{-1}}{\Delta t} = g(x_m). \tag{8.5.12}$$

Solving for u_m^{-1},

$$u_m^{-1} = u_m^0 - \Delta t g(x_m). \tag{8.5.13}$$

One disadvantage of using the backward finite-difference formula is the larger error associated with this term compared to those associated with the finite-differenced form of the wave equation. In the case of the barotropic vorticity equation, a partial differential equation with wave-like solutions, this inconsistency eventually leads to a separation of solution between adjacent time levels.[14] This difficulty is avoided by stopping after a certain number of time steps, averaging the solution, and starting again.

A better solution for computing that first time step employs the centered difference form

$$\frac{u_m^1 - u_m^{-1}}{2\Delta t} = g(x_m), \tag{8.5.14}$$

along with the wave equation

$$\frac{u_m^1 - 2u_m^0 + u_m^{-1}}{(\Delta t)^2} = c^2 \frac{u_{m+1}^0 - 2u_m^0 + u_{m-1}^0}{(\Delta x)^2}, \tag{8.5.15}$$

so that

$$u_m^1 = \left(\frac{c\Delta t}{\Delta x}\right)^2 \frac{f(x_{m+1}) + f(x_{m-1})}{2} + \left[1 - \left(\frac{c\Delta t}{\Delta x}\right)^2\right] f(x_m) + \Delta t g(x_m). \tag{8.5.16}$$

---

[14] Gates, W. L., 1959: On the truncation error, stability, and convergence of difference solutions of the barotropic vorticity equation. *J. Meteorol.*, **16**, 556–568. See Section 4.

Although it appears that we are ready to start calculating, we need to check whether our numerical scheme possesses three properties: convergence, stability, and consistency. By *consistency* we mean that the difference equations approach the differential equation as $\Delta x, \Delta t \to 0$. To prove consistency, we first write u_{m+1}^n, u_{m-1}^n, u_m^{n-1}, and u_m^{n+1} in terms of $u(x,t)$ and its derivatives evaluated at (x_m, t_n). From Taylor expansions,

$$u_{m+1}^n = u_m^n + \Delta x \frac{\partial u}{\partial x}\Big|_n^m + \tfrac{1}{2}(\Delta x)^2 \frac{\partial^2 u}{\partial x^2}\Big|_n^m + \tfrac{1}{6}(\Delta x)^3 \frac{\partial^3 u}{\partial x^3}\Big|_n^m + \cdots, \tag{8.5.17}$$

$$u_{m-1}^n = u_m^n - \Delta x \frac{\partial u}{\partial x}\Big|_n^m + \tfrac{1}{2}(\Delta x)^2 \frac{\partial^2 u}{\partial x^2}\Big|_n^m - \tfrac{1}{6}(\Delta x)^3 \frac{\partial^3 u}{\partial x^3}\Big|_n^m + \cdots, \tag{8.5.18}$$

$$u_m^{n+1} = u_m^n + \Delta t \frac{\partial u}{\partial t}\Big|_n^m + \tfrac{1}{2}(\Delta t)^2 \frac{\partial^2 u}{\partial t^2}\Big|_n^m + \tfrac{1}{6}(\Delta t)^3 \frac{\partial^3 u}{\partial t^3}\Big|_n^m + \cdots, \tag{8.5.19}$$

and

$$u_m^{n-1} = u_m^n - \Delta t \frac{\partial u}{\partial t}\Big|_n^m + \tfrac{1}{2}(\Delta t)^2 \frac{\partial^2 u}{\partial t^2}\Big|_n^m - \tfrac{1}{6}(\Delta t)^3 \frac{\partial^3 u}{\partial t^3}\Big|_n^m + \cdots. \tag{8.5.20}$$

Substituting Equations 8.5.17 through 8.5.20 into Equation 8.5.10, we obtain

$$\frac{u_m^{n+1} - 2u_m^n + u_m^{n-1}}{(\Delta t)^2} - c^2 \frac{u_{m+1}^n - 2u_m^n + u_{m-1}^n}{(\Delta x)^2} \tag{8.5.21}$$
$$= \left(\frac{\partial^2 u}{\partial t^2} - c^2 \frac{\partial^2 u}{\partial x^2}\right)\Big|_n^m + \tfrac{1}{12}(\Delta t)^2 \frac{\partial^4 u}{\partial t^4}\Big|_n^m - \tfrac{1}{12}(c\Delta x)^2 \frac{\partial^4 u}{\partial x^4}\Big|_n^m + \cdots.$$

The first term on the right side of Equation 8.5.21 vanishes because $u(x,t)$ satisfies the wave equation. As $\Delta x \to 0$, $\Delta t \to 0$, the remaining terms on the right side of Equation 8.5.21 tend to zero and Equation 8.5.10 is a consistent finite difference approximation of the wave equation.

Stability is another question. Under certain conditions the small errors inherent in fixed precision arithmetic (round off) can grow for certain choices of Δx and Δt. During the 1920s the mathematicians Courant, Friedrichs, and Lewy[15] found that if $c\Delta t/\Delta x > 1$, then our scheme is unstable. This CFL criterion has its origin in the fact that if $c\Delta t > \Delta x$, then we are asking signals in the numerical scheme to travel faster than their real-world counterparts and this unrealistic expectation leads to instability!

One method of determining *stability*, commonly called the von Neumann method,[16] involves examining solutions to Equation 8.5.11 that have the form

$$u_m^n = e^{im\theta} e^{in\lambda}, \tag{8.5.22}$$

where θ is an arbitrary real number and λ is a yet undetermined complex number. Our choice of Equation 8.5.22 is motivated by the fact that the initial condition u_m^0 can be represented by a Fourier series where a typical term behaves as $e^{im\theta}$.

[15] Courant, R., K. O. Friedrichs, and H. Lewy, 1928: Über die partiellen Differenzengleichungen der mathematischen Physik. *Math. Annalen*, **100**, 32–74. Translated into English in *IBM J. Res. Dev.*, **11**, 215–234.

[16] After its inventor, J. von Neumann. See O'Brien, G. G., M. A. Hyman, and S. Kaplan, 1950: A study of the numerical solution of partial differential equations. *J. Math. Phys. (Cambridge, MA)*, **29**, 223–251.

If we substitute Equation 8.5.22 into Equation 8.5.10 and divide out the common factor $e^{im\theta}e^{in\lambda}$, we have that

$$\frac{e^{i\lambda} - 2 + e^{-i\lambda}}{(\Delta t)^2} = c^2 \frac{e^{i\theta} - 2 + e^{-i\theta}}{(\Delta x)^2}, \tag{8.5.23}$$

or

$$\sin^2\left(\frac{\lambda}{2}\right) = \left(\frac{c\Delta t}{\Delta x}\right)^2 \sin^2\left(\frac{\theta}{2}\right). \tag{8.5.24}$$

The behavior of u_m^n is determined by the values of λ given by Equation 8.5.24. If $c\Delta t/\Delta x \le 1$, then λ is real and u_m^n is bounded for all θ as $n \to \infty$. If $c\Delta t/\Delta x > 1$, then it is possible to find a value of θ such that the right side of Equation 8.5.24 exceeds unity and the corresponding values of λ occur as complex conjugate pairs. The λ with the negative imaginary part produces a solution with exponential growth because $n = t_n/\Delta t \to \infty$ as $\Delta t \to 0$ for a fixed t_n and $c\Delta t/\Delta x$. Thus, the value of u_m^n becomes infinitely large, even though the initial data may be arbitrarily small.

• **Example 8.5.1**

Let us examine the stability question using tools from linear algebra (see Section 3.5).

Consider the explicit scheme for the numerical integration of the wave equation, Equation 8.5.11. We can rewrite that single equation as the coupled difference equations:

$$u_m^{n+1} = 2(1 - r^2)u_m^n + r^2(u_{m+1}^n + u_{m-1}^n) - v_m^n, \tag{8.5.25}$$

and

$$v_m^{n+1} = u_m^n, \tag{8.5.26}$$

where $r = c\Delta t/\Delta x$. Let $u_{m+1}^n = e^{i\beta\Delta x}u_m^n$, and $u_{m-1}^n = e^{-i\beta\Delta x}u_m^n$, where β is real. Then Equation 8.5.25 and Equation 8.5.26 become

$$u_m^{n+1} = 2\left[1 - 2r^2\sin^2\left(\frac{\beta\Delta x}{2}\right)\right]u_m^n - v_m^n, \tag{8.5.27}$$

and

$$v_m^{n+1} = u_m^n, \tag{8.5.28}$$

or in the matrix form

$$\mathbf{u}_m^{n+1} = \begin{pmatrix} 2\left[1 - 2r^2\sin^2\left(\frac{\beta\Delta x}{2}\right)\right] & -1 \\ 1 & 0 \end{pmatrix} \mathbf{u}_m^n, \tag{8.5.29}$$

where $\mathbf{u}_m^n = \begin{pmatrix} u_m^n \\ v_m^n \end{pmatrix}$. The eigenvalues λ of this *amplification matrix* are given by

$$\lambda^2 - 2\left[1 - 2r^2\sin^2\left(\frac{\beta\Delta x}{2}\right)\right]\lambda + 1 = 0, \tag{8.5.30}$$

or

$$\lambda_{1,2} = 1 - 2r^2\sin^2\left(\frac{\beta\Delta x}{2}\right) \pm 2r\sin\left(\frac{\beta\Delta x}{2}\right)\sqrt{r^2\sin^2\left(\frac{\beta\Delta x}{2}\right) - 1}. \tag{8.5.31}$$

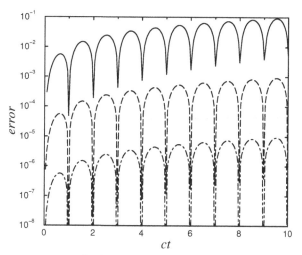

Figure 8.5.2: The growth of error $||e_n||$ as a function of ct for various resolutions. For the top line, $\Delta x = 0.1$; for the middle line, $\Delta x = 0.01$; and for the bottom line, $\Delta x = 0.001$.

Because each successive time step consists of multiplying the solution from the previous time step by the amplification matrix, the solution is stable only if u_m^n remains bounded. This occurs only if all of the eigenvalues have a magnitude less than or equal to one, because

$$\mathbf{u}_m^n = \sum_k c_k A^n \mathbf{x}_k = \sum_k c_k \lambda_k^n \mathbf{x}_k, \tag{8.5.32}$$

where A denotes the amplification matrix and $\mathbf{x}_k$ denotes the eigenvectors corresponding to the eigenvalues λ_k. Equation 8.5.32 follows from our ability to express any initial condition in terms of an eigenvector expansion

$$\mathbf{u}_m^0 = \sum_k c_k \mathbf{x}_k. \tag{8.5.33}$$

In our particular example, two cases arise. If $r^2 \sin^2(\beta \Delta x/2) \leq 1$,

$$\lambda_{1,2} = 1 - 2r^2 \sin^2\left(\frac{\beta \Delta x}{2}\right) \pm 2ri \sin\left(\frac{\beta \Delta x}{2}\right) \sqrt{1 - r^2 \sin^2\left(\frac{\beta \Delta x}{2}\right)} \tag{8.5.34}$$

and $|\lambda_{1,2}| = 1$. On the other hand, if $r^2 \sin^2(\beta \Delta x/2) > 1$, $|\lambda_{1,2}| > 1$. Thus, we have stability only if $c\Delta t/\Delta x \leq 1$. $\square$

Finally, we must check for convergence. A numerical scheme is *convergent* if the numerical solution approaches the continuous solution as $\Delta x, \Delta t \to 0$. The general procedure for proving convergence involves the evolution of the error term e_m^n, which gives the difference between the true solution $u(x_m, t_n)$ and the finite difference solution u_m^n. From Equation 8.5.21,

$$e_m^{n+1} = \left(\frac{c\Delta t}{\Delta x}\right)^2 \left(e_{m+1}^n + e_{m-1}^n\right) + 2\left[1 - \left(\frac{c\Delta t}{\Delta x}\right)^2\right] e_m^n - e_m^{n-1} + O[(\Delta t)^4] + O[(\Delta x)^2(\Delta t)^2].$$

$$\tag{8.5.35}$$

Let us apply Equation 8.5.25 to work backwards from the point (x_m, t_n) by changing n to $n-1$. The nonvanishing terms in e_m^n reduce to a sum of $n+1$ values on the line $n=1$ plus $\frac{1}{2}(n+1)n$ terms of the form $A(\Delta x)^4$. If we define the max norm $||e_n|| = \max_m |e_m^n|$, then

$$||e_n|| \leq nB(\Delta x)^3 + \tfrac{1}{2}(n+1)nA(\Delta x)^4. \qquad (8.5.36)$$

Because $n\Delta x \leq ct_n$, Equation 8.5.26 simplifies to

$$||e_n|| \leq ct_n B(\Delta x)^2 + \tfrac{1}{2}c^2 t_n^2 A(\Delta x)^2. \qquad (8.5.37)$$

Thus, the error tends to zero as $\Delta x \to 0$, verifying convergence. We illustrate Equation 8.5.37 by using the finite difference equation, Equation 8.5.11, to compute $||e_n||$ during a numerical experiment that used $c\Delta t/\Delta x = 0.5$, $f(x) = \sin(\pi x)$, and $g(x) = 0$; $||e_n||$ is plotted in Figure 8.5.2. Note how each increase of resolution by 10 results in a drop in the error by 100.

In the following examples we apply our scheme to solve a few simple initial and boundary conditions:

• **Example 8.5.2**

For our first example, we resolve Equation 8.3.1 through Equation 8.3.3 and Equation 8.3.25 and Equation 8.3.26 numerically using MATLAB. The MATLAB code is:

```
clear
coeff = 0.5; coeffsq = coeff * coeff % coeff = cΔt/Δx
dx = 0.04; dt = coeff * dx; N = 100; x = 0:dx:1;
M = 1/dx + 1; % M = number of spatial grid points
% introduce the initial conditions via F and G
F = zeros(M,1); G = zeros(M,1);
for m = 1:M
  if x(m) >= 0.25 & x(m) <= 0.5
      F(m) = 4 * x(m) - 1; end
  if x(m) >= 0.5 & x(m) <= 0.75
      F(m) = 3 - 4 * x(m); end; end
% at t = 0, the solution is:
  tplot(1) = 0; u = zeros(M,N+1); u(1:M,1) = F(1:M);
% at t = Δt, the solution is given by Equation 8.5.16
  tplot(2) = dt;
  for m = 2:M-1
    u(m,2) = 0.5*coeffsq*(F(m+1)+F(m-1)) + (1-coeffsq)*F(m)+dt*G(m);
  end
% in general, the solution is given by Equation 8.5.11
  for n = 2:N
    tplot(n+1) = dt * n;
  for m = 2:M-1
    u(m,n+1) = 2*u(m,n)-u(m,n-1) + coeffsq*(u(m+1,n)-2*u(m,n)+u(m-1,n));
end; end
X = x' * ones(1,length(tplot)); T = ones(M,1) * tplot;
surf(X,T,u)
xlabel('DISTANCE','Fontsize',20); ylabel('TIME','Fontsize',20)
zlabel('SOLUTION','Fontsize',20)
```

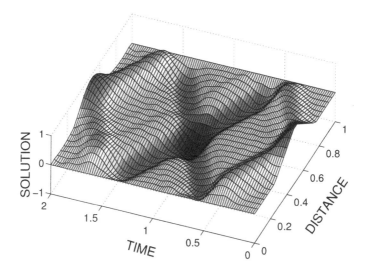

Figure 8.5.3: The numerical solution $u(x,t)/h$ of the wave equation with $c\Delta t/\Delta x = \frac{1}{2}$ using Equation 8.5.11 at various positions $x' = x/L$ and times $t' = ct/L$. The exact solution is plotted in Figure 8.3.2.

Overall, the numerical solution shown in Figure 8.5.3 approximates the exact or analytic solution well. However, we note small-scale noise in the numerical solution at later times. Why does this occur? Recall that the exact solution could be written as an infinite sum of sines in the x dimension. Each successive harmonic adds a contribution from waves of shorter and shorter wavelength. In the case of the numerical solution, the longer-wavelength harmonics are well represented by the numerical scheme because there are many grid points available to resolve a given wavelength. As the wavelengths become shorter, the higher harmonics are poorly resolved by the numerical scheme, move at incorrect phase speeds, and their misplacement (dispersion) creates the small-scale noise that you observe rather than giving the sharp angular features of the exact solution. The only method for avoiding this problem is to devise schemes that minimize dispersion. □

• **Example 8.5.3**

Let us redo Example 8.5.2 except that we introduce the boundary condition that $u_x(L, t) = 0$. This corresponds to a string where we fix the left end and allow the right end to freely move up and down. This requires a new difference condition along the right boundary. If we employ centered differencing,

$$\frac{u_{L+1}^n - u_{L-1}^n}{2\Delta x} = 0, \tag{8.5.38}$$

and

$$u_L^{n+1} = 2u_L^n - u_L^{n-1} + \left(\frac{c\Delta t}{\Delta x}\right)^2 \left(u_{L+1}^n - 2u_L^n + u_{L-1}^n\right). \tag{8.5.39}$$

Eliminating u_{L+1}^n between Equation 8.5.38 and Equation 8.5.39,

$$u_L^{n+1} = 2u_L^n - u_L^{n-1} + \left(\frac{c\Delta t}{\Delta x}\right)^2 \left(2u_{L-1}^n - 2u_L^n\right). \tag{8.5.40}$$

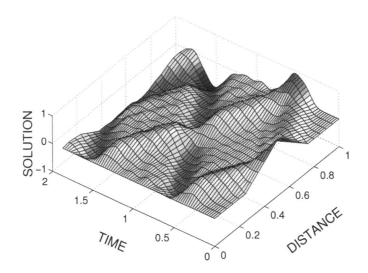

Figure 8.5.4: The numerical solution $u(x,t)/h$ of the wave equation when the right end moves freely with $c\Delta t/\Delta x = \frac{1}{2}$ using Equation 8.5.11 and Equation 8.5.40 at various positions $x' = x/L$ and times $t' = ct/L$.

For the special case of $n = 1$, Equation 8.5.40 becomes

$$u_L^1 = f(x_L) + \left(\frac{c\Delta t}{\Delta x}\right)^2 [f(x_{L-1}) - f(x_L)] + \Delta t f(x_L). \qquad (8.5.41)$$

The MATLAB code used to numerically solve the wave equation with a Neumann boundary condition is very similar to the one used in the previous example that we must add the line

```
u(M,2) = coeffsq * F(M-1) + (1-coeffsq) * F(M) + dt*G(M);
```

after

```
for m = 2:M-1
  u(m,2) = 0.5*coeffsq*(F(m+1)+F(m-1)) + (1-coeffsq)*F(m)+dt*G(m);
end
```

and

```
  u(M,n+1) = 2*u(M,n)-u(M,n-1) + 2*coeffsq*(u(M-1,n)-u(M,n));
```

after

```
for m = 2:M-1
  u(m,n+1) = 2*u(m,n)-u(m,n-1) + coeffsq*(u(m+1,n)-2*u(m,n)+u(m-1,n));
end
```

Figure 8.5.4 shows the results. The numerical solution agrees well with the exact solution

$$u(x,t) = \frac{32h}{\pi^2} \sum_{n=1}^{\infty} \frac{1}{(2n-1)^2} \sin\left[\frac{(2n-1)\pi x}{2L}\right] \cos\left[\frac{(2n-1)\pi ct}{2L}\right]$$

$$\times \left\{2\sin\left[\frac{(2n-1)\pi}{4}\right] - \sin\left[\frac{3(2n-1)\pi}{8}\right] - \sin\left[\frac{(2n-1)\pi}{8}\right]\right\}. \qquad (8.5.42)$$

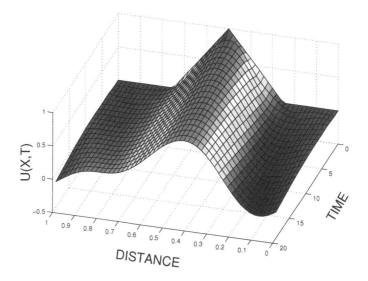

Figure 8.5.5: The numerical solution $u(x,t)$ of the first-order hyperbolic partial differential equation $u_t + u_x = 0$ using the Lax-Wendroff formula as observed at $t = 0, 1, 2, \ldots, 20$. The initial conditions are given by Equation 8.3.25 with $h = 1$, $\Delta t / \Delta x = \frac{2}{3}$, and $\Delta x = 0.02$.

The results are also consistent with those presented in Example 8.5.1, especially with regard to small-scale noise due to dispersion.

Project: Numerical Solution of First-Order Hyperbolic Equations

The equation $u_t + u_x = 0$ is the simplest possible hyperbolic partial differential equation. Indeed, the classic wave equation consists of a system of these equations: $u_t + cv_x = 0$, and $v_t + cu_x = 0$. In this project you will examine several numerical schemes for solving such a partial differential equation using MATLAB.

Step 1: One of the simplest numerical schemes is the forward-in-time, centered-in-space of

$$\frac{u_m^{n+1} - u_m^n}{\Delta t} + \frac{u_{m+1}^n - u_{m-1}^n}{2\Delta x} = 0.$$

Use von Neumann's stability analysis to show that this scheme is *always* unstable.

Step 2: The most widely used method for numerically integrating first-order hyperbolic equations is the *Lax-Wendroff* method:[17]

$$u_m^{n+1} = u_m^n - \frac{\Delta t}{2\Delta x} \left(u_{m+1}^n - u_{m-1}^n \right) + \frac{(\Delta t)^2}{2(\Delta x)^2} \left(u_{m+1}^n - 2u_m^n + u_{m-1}^n \right).$$

This method introduces errors of $O[(\Delta t)^2]$ and $O[(\Delta x)^2]$. Show that this scheme is stable if it satisfies the CFL criteria of $\Delta t / \Delta x \leq 1$.

Using the initial condition given by Equation 8.3.25, write a MATLAB code that uses this scheme to numerically integrate $u_t + u_x = 0$. Plot the results for various $\Delta t / \Delta x$ over

[17] Lax, P. D., and B. Wendroff, 1960: Systems of conservative laws. *Comm. Pure Appl. Math.*, **13**, 217–237.

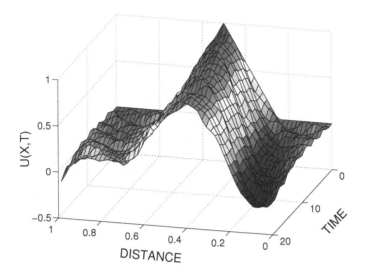

Figure 8.5.6: Same as Figure 8.5.5 except that the centered-in-time, centered-in-space scheme was used.

the interval $0 \le x \le 1$ given the *periodic* boundary conditions of $u(0,t) = u(1,t)$ for the temporal interval $0 \le t \le 20$. See Figure 8.5.5. Discuss the strengths and weaknesses of the scheme with respect to dissipation or damping of the numerical solution and preserving the phase of the solution. Most numerical methods books discuss this.[18]

Step 3: Another simple scheme is the centered-in-time, centered-in-space of

$$\frac{u_m^{n+1} - u_m^{n-1}}{2\Delta t} + \frac{u_{m+1}^n - u_{m-1}^n}{2\Delta x} = 0.$$

This method introduces errors of $O[(\Delta t)^2]$ and $O[(\Delta x)^2]$.

Repeat the analysis from Step 1 for this scheme. One of the difficulties is taking the first time step. Use the scheme in Step 1 to take this first time step. See Figure 8.5.6.

Project: Implicit Schemes for Solving the Wave Equation

Although the simple explicit time differencing scheme of Equation 8.5.10 is very easy to program, the restriction on the size of the time step, $c\Delta t/\Delta x \le 1$, can be computationally expense when Δx is small. In this project we will explore several implicit schemes for the numerical solution of the wave equation. Implicit schemes are useful because they are unconditionally stable for any $\Delta t/\Delta x$.

The wave equation problem that we will solve is:

$$\frac{\partial^2 u}{\partial t^2} = \frac{\partial^2 u}{\partial x^2}, \qquad 0 < x < 1, \quad 0 < t, \tag{8.5.43}$$

which satisfies the initial conditions

$$u(x,0) = \begin{cases} x/a, & 0 < x < a \\ (1-x)/(1-a), & a < x < 1, \end{cases} \tag{8.5.44}$$

---

[18] For example, Lapidus, L., and G. F. Pinder, 1982: *Numerical Solution of Partial Differential Equations in Science and Engineering*. John Wiley & Sons, 677 pp.

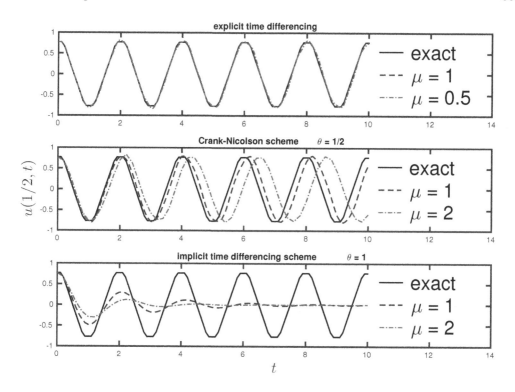

Figure 8.5.7: The numerical solution of the wave equation problem given by Equations 8.5.43 through 8.5.46 using various numerical schemes. The parameters are $a = 0.35$, $\mu = \Delta t / \Delta x$, and $\mu = \Delta t / \Delta x$.

and

$$\frac{\partial u(x,0)}{\partial t} = 0, \quad 0 < x < 1, \tag{8.5.45}$$

and the boundary conditions

$$u(0,t) = u(1,t) = 0, \qquad 0 < t, \tag{8.5.46}$$

where $0 < a < 1$. A knowledge of matrix algebra is required (see Sections 3.1 and 3.4).

We begin by writing the finite differenced wave equation as

$$\frac{u_i^{n+1} - 2u_i^n + u_i^{n-1}}{(\Delta t)^2} = (1-\theta)\frac{u_{i+1}^{n-1} - 2u_{i-1}^{n-1} + u_{i-1}^{n-1}}{(\Delta x)^2} + \theta\frac{u_{i+1}^{n+1} - 2u_{i-1}^{n+1} + u_{i-1}^{n+1}}{(\Delta x)^2}, \tag{8.5.47}$$

or

$$-\theta\mu^2 u_{i-1}^{n+1} + (1 + 2\theta\mu^2)u_i^{n+1} - \theta\mu^2 u_{i+1}^{n+1} \tag{8.5.48}$$
$$= 2u_i^n + (1-\theta)\mu^2 u_{i-1}^{n-1} - [1 + 2(1-\theta)\mu^2]u_i^{n-1} + (1-\theta)\mu^2 u_{i+1}^{n-1},$$

where $0 \leq n$, $0 \leq \theta \leq 1$, and $\mu = \Delta t / \Delta x$.

Care must be taken for the first time step when $n = 0$ because 8.5.48 requires u_i^{-1}, an unknown. From the initial condition $u_t(x,0) = 0$, however, we know that $u_i^{-1} = u_i^1$ (for all i). Consequently,

$$-\mu^2 u_{i-1}^1 + (2 + 2\mu^2)u_i^1 - \mu^2 u_{i+1}^1 = 2u_i^0. \tag{8.5.49}$$

For $n > 1$ we use Equation 8.5.48 because everything on the right side is known. Both Equations 8.5.48 and 8.5.49 must be solved as a system of linear equations.

Step 1: Using separation of variables, show that the solution to Equations 8.5.43 through 8.5.46 is

$$u(x,t) = \frac{2}{a(1-a)\pi^2} \sum_{n=1}^{\infty} \frac{\sin(n\pi a)}{n^2} \sin(n\pi x) \cos(n\pi t).$$

Step 2: Using the simple explicit time differencing scheme of Equation 8.5.10, numerically solve our problem for various values of μ.

Step 3: Write a MATLAB program to solve Equation 8.5.48 for a given value of θ and μ. This will require solving a tridiagonal matrix.

Step 4: When we set $\theta = 1/2$ in Equation 8.5.48, we are using the Crank-Nicolson method to solve the wave equation. Find solutions for various values of μ and compare your answer to the exact solution.

Step 5: When we set $\theta = 1$ in Equation 8.5.48, we are solving this equation by an implicit scheme. Repeat Step 4 and compare your solution against the exact solution.

Step 6: The case when $\theta = 0$ yields an explicit time differencing scheme. For various values of μ, how does this new scheme compare with your numerical solution that you found in Step 2?

Project: Wave Propagation in a Nonhomogeneous Medium

So far everything that we have done has been for a medium that is homogeneous and the phase speed is constant. In this project you will numerically integrate the wave equation where the phase speed c_i differs in two different regions. Different phase speeds can arise in a domain because it possesses different materials with different physical properties in different areas. Such problems are important in many disciplines from antenna theory to seismology.

The objective of this project is to find the numerical solution to the wave equation

$$\frac{\partial^2 g}{\partial t^2} = c_i^2 \frac{\partial^2 g}{\partial x^2} + \delta(x - \xi)\delta(t - \tau), \qquad 0 < x < L, \quad 0 < t,$$

which satisfies the initial conditions $g(x,0) = g_t(x,0) = 0$ for $0 < x < L$, and the boundary conditions $g(0,t) = g(L,t) = 0$ for $0 < t$.

This problem enjoys several differences from previous problems. First, there will be different phase speeds: $c_1 = 2$ for $0 < x < 1$ and $c_2 = 1$ for $1 < x < L$. In order to highlight the discontinuity in the phase speed, we will take L sufficiently large so that we have no reflections of that boundary. Second, we have introduced a source term at the point (ξ, τ). This source term is unique in the sense that it is an impulse forcing at the position $x = \xi$ at time $t = \tau$. The solution of a differential equation whose initial conditions reflect a system initially at rest and which is then subjected to an impulse forcing at a given time and position is called a *Green's function*.

Step 1: Using simple centered time and spatial differencing, write a MATLAB code to numerically solve this problem.

There are two special concerns here. How do you handle the change in c_i? The simplest way is to introduce an array that depends on location that gives the correct value of c_i. The second and more difficult question is how do you model the delta function? Equations 6.1.21 and 6.1.22 provide some simple expressions for small but finite ϵ. You should try them and see what works best.

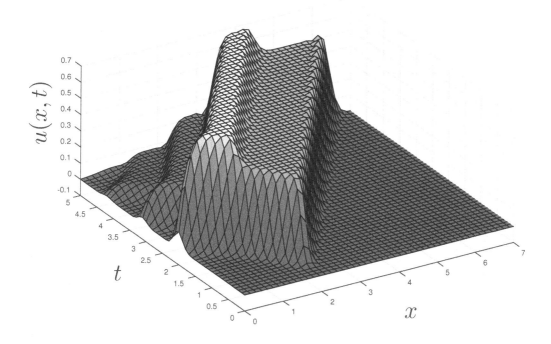

Figure 8.5.8: The numerically computed Green's function associated with a source function located at $\xi = 2$ and $\tau = 0.5$. There are two regions with different phase speeds $c_1 = 2$ for $0 < x < 1$ and $c_2 = 1$ for $1 < x < L$. The other parameters are $\Delta x = 0.05$, $\Delta t = 0.005$, and $L = 20$. The delta function is approximated by a Gaussian.

Step 2: Once you have written and debugged your code, try it out on a situation with uniform phase speed $c_1 = c_2 = 1$ and see what you find. Does it conform to your expectation?

Step 3: You are ready to introduce the discontinuity in phase speed. Redo Step 2 but with different phase speeds in regions $0 < x < 1$ and $1 < x < L$.

Step 4: Plot your results. You should see that, depending upon the values of the phase speeds, some of the wave motion is transmitted and reflected at the interface (as well as at the boundary $x = 0$). Give the history of the wave from the source as it first encounters the interface and then the wall and then the interface again.

Further Readings

King, G. C., 2009: *Vibrations and Waves*. Wiley, 228 pp. This book emphasizes the physical principles, rather than the mathematics.

Koshlyakov, N. S., M. M. Smirnov, and E. B. Gliner, 1964: *Differential Equations of Mathematical Physics*. North-Holland Publishing, 701 pp. See Part I. Detailed presentations of solution techniques.

Morse, P. M., and H. Feshback, 1953: *Methods of Theoretical Physics*. McGraw-Hill Book Co., 997 pp. Chapter 11 is devoted to solving the wave equation.

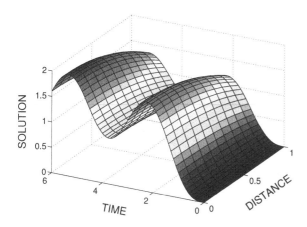

Chapter 9
The Heat Equation

In this chapter we deal with the linear parabolic differential equation

$$\frac{\partial u}{\partial t} = a^2 \frac{\partial^2 u}{\partial x^2} \qquad (9.0.1)$$

in the two independent variables x and t. This equation, known as the one-dimensional heat equation, serves as the prototype for a wider class of *parabolic equations*

$$a(x,t)\frac{\partial^2 u}{\partial x^2} + b(x,t)\frac{\partial^2 u}{\partial x \partial t} + c(x,t)\frac{\partial^2 u}{\partial t^2} = f\left(x,t,u,\frac{\partial u}{\partial x},\frac{\partial u}{\partial t}\right), \qquad (9.0.2)$$

where $b^2 = 4ac$. It arises in the study of heat conduction in solids as well as in a variety of diffusive phenomena. The heat equation is similar to the wave equation in that it is also an equation of evolution. However, the heat equation is not "conservative" because if we reverse the sign of t, we obtain a different solution. This reflects the presence of entropy, which must always increase during heat conduction.

9.1 DERIVATION OF THE HEAT EQUATION

To derive the heat equation, consider a heat-conducting homogeneous rod, extending from $x = 0$ to $x = L$ along the x-axis (see Figure 9.1.1). The rod has uniform cross section A and constant density ρ, is insulated laterally so that heat flows only in the x-direction, and is sufficiently thin so that the temperature at all points on a cross section is constant. Let $u(x,t)$ denote the temperature of the cross section at the point x at any instant of time t, and let c denote the specific heat of the rod (the amount of heat required to raise the

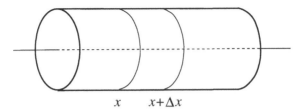

Figure 9.1.1: Heat conduction in a thin bar.

temperature of a unit mass of the rod by a degree). In the segment of the rod between the cross section at x and the cross section at $x + \Delta x$, the amount of heat is

$$Q(t) = \int_x^{x+\Delta x} c\rho A u(s,t)\, ds. \tag{9.1.1}$$

On the other hand, the rate at which heat flows into the segment across the cross section at x is proportional to the cross section and the gradient of the temperature at the cross section (Fourier's law of heat conduction):

$$-\kappa A \frac{\partial u(x,t)}{\partial x}, \tag{9.1.2}$$

where κ denotes the thermal conductivity of the rod. The sign in Equation 9.1.2 indicates that heat flows in the direction of decreasing temperature. Similarly, the rate at which heat flows out of the segment through the cross section at $x + \Delta x$ equals

$$-\kappa A \frac{\partial u(x+\Delta x,t)}{\partial x}. \tag{9.1.3}$$

The difference between the amount of heat that flows in through the cross section at x and the amount of heat that flows out through the cross section at $x+\Delta x$ must equal the change in the heat content of the segment $x \le s \le x + \Delta x$. Hence, by subtracting Equation 9.1.3 from Equation 9.1.2 and equating the result to the time derivative of Equation 9.1.1,

$$\frac{\partial Q}{\partial t} = \int_x^{x+\Delta x} c\rho A \frac{\partial u(s,t)}{\partial t}\, ds = \kappa A \left[\frac{\partial u(x+\Delta x,t)}{\partial x} - \frac{\partial u(x,t)}{\partial x} \right]. \tag{9.1.4}$$

Assuming that the integrand in Equation 9.1.4 is a continuous function of s, then by the mean value theorem for integrals,

$$\int_x^{x+\Delta x} \frac{\partial u(s,t)}{\partial t}\, ds = \frac{\partial u(\xi,t)}{\partial t} \Delta x, \qquad x < \xi < x + \Delta x, \tag{9.1.5}$$

so that Equation 9.1.4 becomes

$$c\rho \Delta x \frac{\partial u(\xi,t)}{\partial t} = \kappa \left[\frac{\partial u(x+\Delta x,t)}{\partial x} - \frac{\partial u(x,t)}{\partial x} \right]. \tag{9.1.6}$$

Dividing both sides of Equation 9.1.6 by $c\rho\Delta x$ and taking the limit as $\Delta x \to 0$,

$$\frac{\partial u(x,t)}{\partial t} = a^2 \frac{\partial^2 u(x,t)}{\partial x^2} \tag{9.1.7}$$

with $a^2 = \kappa/(c\rho)$. Equation 9.1.7 is called the one-dimensional *heat equation*. The constant a^2 is called the *diffusivity* within the solid.

If an external source supplies heat to the rod at a rate $f(x,t)$ per unit volume per unit time, we must add the term $\int_x^{x+\Delta x} f(s,t)\,ds$ to the time derivative term of Equation 9.1.4. Thus, in the limit $\Delta x \to 0$,

$$\frac{\partial u(x,t)}{\partial t} - a^2 \frac{\partial^2 u(x,t)}{\partial x^2} = F(x,t), \tag{9.1.8}$$

where $F(x,t) = f(x,t)/(c\rho)$ is the source density. This equation is called the *nonhomogeneous heat equation*.

9.2 INITIAL AND BOUNDARY CONDITIONS

In the case of heat conduction in a thin rod, the temperature function $u(x,t)$ must satisfy not only the heat equation, Equation 9.1.7, but also how the two ends of the rod exchange heat energy with the surrounding medium. If (1) there is no heat source, (2) the function $f(x), 0 < x < L$ describes the temperature in the rod at $t = 0$, and (3) we maintain both ends at zero temperature for all time, then the partial differential equation

$$\frac{\partial u}{\partial t} = a^2 \frac{\partial^2 u}{\partial x^2}, \qquad 0 < x < L, \quad 0 < t, \tag{9.2.1}$$

describes the temperature distribution $u(x,t)$ in the rod at any later time $0 < t$ subject to the conditions

$$u(x,0) = f(x), \qquad 0 < x < L, \tag{9.2.2}$$

and

$$u(0,t) = u(L,t) = 0, \qquad 0 < t. \tag{9.2.3}$$

Equations 9.2.2 and 9.2.3 describe the *initial-boundary-value problem* for this particular heat conduction problem; Equation 9.2.3 is the boundary condition while Equation 9.2.2 gives the initial condition. Note that in the case of the heat equation, the problem only demands the initial value of $u(x,t)$ and not $u_t(x,0)$, as with the wave equation.

Historically most linear boundary conditions have been classified in one of three ways. The condition, Equation 9.2.3, is an example of a *Dirichlet problem*[1] or *condition of the first kind*. This type of boundary condition gives the value of the solution (which is not necessarily equal to zero) along a boundary.

The next simplest condition involves derivatives. If we insulate both ends of the rod so that no heat flows from the ends, then according to Equation 8.1.2 the boundary condition assumes the form

$$\frac{\partial u(0,t)}{\partial x} = \frac{\partial u(L,t)}{\partial x} = 0, \qquad 0 < t. \tag{9.2.4}$$

This is an example of a *Neumann problem*[2] or *condition of the second kind*. This type of boundary condition specifies the value of the normal derivative (which may not be equal to zero) of the solution along the boundary.

---

[1] Dirichlet, P. G. L., 1850: Über einen neuen Ausdruck zur Bestimmung der Dichtigkeit einer unendlich dünnen Kugelschale, wenn der Werth des Potentials derselben in jedem Punkte ihrer Oberfläche gegeben ist. *Abh. Königlich. Preuss. Akad. Wiss.*, 99–116.

[2] Neumann, C. G., 1877: *Untersuchungen über das Logarithmische und Newton'sche Potential.*

Finally, if there is radiation of heat from the ends of the rod into the surrounding medium, we shall show that the boundary condition is of the form

$$\frac{\partial u(0,t)}{\partial x} - hu(0,t) = \text{a constant},\qquad(\mathbf{9.2.5})$$

and

$$\frac{\partial u(L,t)}{\partial x} + hu(L,t) = \text{another constant}\qquad(\mathbf{9.2.6})$$

for $0 < t$, where h is a positive constant. This is an example of a *condition of the third kind* or *Robin problem*[3] and is a linear combination of Dirichlet and Neumann conditions. To solve these problems you must understand the Sturm-Liouville problem which is presented in Chapter 11. For this reason we will deal with the Robin boundary conditions there.

9.3 SEPARATION OF VARIABLES

As with the wave equation, the most popular and widely used technique for solving the heat equation is separation of variables. Its success depends on our ability to express the solution $u(x,t)$ as the product $X(x)T(t)$. If we cannot achieve this separation, then the technique must be abandoned for others. In the following examples we show how to apply this technique.

• **Example 9.3.1**

Let us find the solution to the homogeneous heat equation

$$\frac{\partial u}{\partial t} = a^2\frac{\partial^2 u}{\partial x^2},\qquad 0 < x < L,\quad 0 < t,\qquad(\mathbf{9.3.1})$$

which satisfies the initial condition

$$u(x,0) = f(x),\quad 0 < x < L,\qquad(\mathbf{9.3.2})$$

and the boundary conditions

$$u(0,t) = u(L,t) = 0,\quad 0 < t.\qquad(\mathbf{9.3.3})$$

This system of equations models heat conduction in a thin metallic bar where both ends are held at the constant temperature of zero and the bar initially has the temperature $f(x)$.

We shall solve this problem by the method of separation of variables. Accordingly, we seek particular solutions of Equation 9.3.1 of the form

$$u(x,t) = X(x)T(t),\qquad(\mathbf{9.3.4})$$

which satisfy the boundary conditions, Equation 9.3.3. Because

$$\frac{\partial u}{\partial t} = X(x)T'(t),\qquad(\mathbf{9.3.5})$$

---

[3] Robin, G., 1886: Sur la distribution de l'électricité à la surface des conducteurs fermés et des conducteurs ouverts. *Ann. Sci. l'Ecole Norm. Sup., Ser. 3*, **3**, S1–S58.

and

$$\frac{\partial^2 u}{\partial x^2} = X''(x)T(t), \tag{9.3.6}$$

Equation 9.3.1 becomes

$$T'(t)X(x) = a^2 X''(x)T(t). \tag{9.3.7}$$

Dividing both sides of Equation 9.3.7 by $a^2 X(x)T(t)$ gives

$$\frac{T'}{a^2 T} = \frac{X''}{X} = -\lambda, \tag{9.3.8}$$

where $-\lambda$ is the separation constant. Equation 9.3.8 immediately yields two ordinary differential equations:

$$X'' + \lambda X = 0, \tag{9.3.9}$$

and

$$T' + a^2 \lambda T = 0 \tag{9.3.10}$$

for the functions $X(x)$ and $T(t)$, respectively.

We now rewrite the boundary conditions in terms of $X(x)$ by noting that the boundary conditions are $u(0,t) = X(0)T(t) = 0$, and $u(L,t) = X(L)T(t) = 0$ for $0 < t$. If we were to choose $T(t) = 0$, then we would have a trivial solution for $u(x,t)$. Consequently, $X(0) = X(L) = 0$.

We now solve Equation 9.3.9. There are three possible cases: $\lambda = -m^2$, $\lambda = 0$, and $\lambda = k^2$. If $\lambda = -m^2 < 0$, then we must solve the boundary-value problem

$$X'' - m^2 X = 0, \qquad X(0) = X(L) = 0. \tag{9.3.11}$$

The general solution to Equation 9.3.11 is

$$X(x) = A \cosh(mx) + B \sinh(mx). \tag{9.3.12}$$

Because $X(0) = 0$, it follows that $A = 0$. The condition $X(L) = 0$ yields $B \sinh(mL) = 0$. Since $\sinh(mL) \neq 0$, $B = 0$, and we have a trivial solution for $\lambda < 0$.

If $\lambda = 0$, the corresponding boundary-value problem is

$$X''(x) = 0, \qquad X(0) = X(L) = 0. \tag{9.3.13}$$

The general solution is

$$X(x) = C + Dx. \tag{9.3.14}$$

From $X(0) = 0$, we have that $C = 0$. From $X(L) = 0$, $DL = 0$, or $D = 0$. Again, we obtain a trivial solution.

Finally, we assume that $\lambda = k^2 > 0$. The corresponding boundary-value problem is

$$X'' + k^2 X = 0, \qquad X(0) = X(L) = 0. \tag{9.3.15}$$

The general solution to Equation 9.3.15 is

$$X(x) = E \cos(kx) + F \sin(kx). \tag{9.3.16}$$

Because $X(0) = 0$, it follows that $E = 0$; from $X(L) = 0$, we obtain $F\sin(kL) = 0$. For a nontrivial solution, $F \neq 0$ and $\sin(kL) = 0$. This implies that $k_n L = n\pi$, where $n = 1, 2, 3, \ldots$. In summary, the x-dependence of the solution is

$$X_n(x) = F_n \sin\left(\frac{n\pi x}{L}\right), \tag{9.3.17}$$

where $\lambda_n = n^2\pi^2/L^2$.

Turning to the time dependence, we use $\lambda_n = n^2\pi^2/L^2$ in Equation 9.3.10

$$T_n' + \frac{a^2 n^2 \pi^2}{L^2} T_n = 0. \tag{9.3.18}$$

The corresponding general solution is

$$T_n(t) = G_n \exp\left(-\frac{a^2 n^2 \pi^2}{L^2} t\right). \tag{9.3.19}$$

Thus, the functions

$$u_n(x,t) = B_n \sin\left(\frac{n\pi x}{L}\right) \exp\left(-\frac{a^2 n^2 \pi^2}{L^2} t\right), \quad n = 1, 2, 3, \ldots, \tag{9.3.20}$$

where $B_n = F_n G_n$, are particular solutions of Equation 9.3.1 and satisfy the homogeneous boundary conditions, Equation 9.3.3.

Having found particular solutions to our problem, the most general solution equals a linear sum of these particular solutions:

$$u(x,t) = \sum_{n=1}^{\infty} B_n \sin\left(\frac{n\pi x}{L}\right) \exp\left(-\frac{a^2 n^2 \pi^2}{L^2} t\right). \tag{9.3.21}$$

The coefficient B_n is chosen so that Equation 9.3.21 yields the initial condition, Equation 9.3.2, if $t = 0$. Thus, setting $t = 0$ in Equation 9.3.21, we see from Equation 9.3.2 that the coefficients B_n must satisfy the relationship

$$f(x) = \sum_{n=1}^{\infty} B_n \sin\left(\frac{n\pi x}{L}\right), \quad 0 < x < L. \tag{9.3.22}$$

This is precisely a Fourier half-range sine series for $f(x)$ on the interval $(0, L)$. Therefore, the formula

$$B_n = \frac{2}{L} \int_0^L f(x) \sin\left(\frac{n\pi x}{L}\right) dx, \quad n = 1, 2, 3, \ldots \tag{9.3.23}$$

gives the coefficients B_n. For example, if $L = \pi$ and $u(x,0) = x(\pi - x)$, then

$$B_n = \frac{2}{\pi} \int_0^\pi x(\pi - x) \sin(nx)\, dx = 2\int_0^\pi x\sin(nx)\, dx - \frac{2}{\pi}\int_0^\pi x^2 \sin(nx)\, dx = 4\frac{1 - (-1)^n}{n^3 \pi}. \tag{9.3.24}$$

Hence,

$$u(x,t) = \frac{8}{\pi} \sum_{n=1}^{\infty} \frac{\sin[(2n-1)x]}{(2n-1)^3} e^{-(2n-1)^2 a^2 t}. \tag{9.3.25}$$

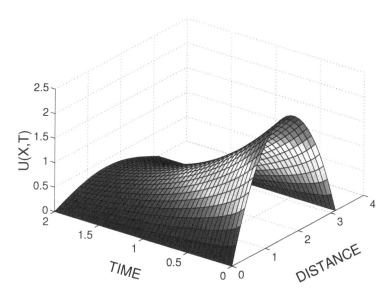

Figure 9.3.1: The temperature $u(x,t)$ within a thin bar as a function of position x and time $a^2 t$ when we maintain both ends at zero and the initial temperature equals $x(\pi - x)$.

Figure 9.3.1 illustrates Equation 9.3.25 for various times. It was created using the MATLAB script:

```
clear
M = 20; dx = pi/25; dt = 0.05;
% compute grid and initialize solution
X = [0:dx:pi]; T = [0:dt:2];
u = zeros(length(T),length(X));
XX = repmat(X,[length(T) 1]); TT = repmat(T',[1 length(X)]);
% compute solution from Equation 9.3.25
for m = 1:M
  temp1 = 2*m-1; coeff = 8 / (pi * temp1 * temp1 * temp1);
  u = u + coeff * sin(temp1*XX) .* exp(-temp1 * temp1 * TT);
end
surf(XX,TT,u)
xlabel('DISTANCE','Fontsize',20); ylabel('TIME','Fontsize',20)
zlabel('U(X,T)','Fontsize',20)
```

Note that both ends of the bar satisfy the boundary conditions, namely that the temperature equals zero. As time increases, heat flows out from the center of the bar to both ends where it is removed. This process is reflected in the collapse of the original parabolic shape of the temperature profile toward zero as time increases.   □

• **Example 9.3.2**

A slight variation on Example 9.3.1 is

$$\frac{\partial u}{\partial t} = a^2 \frac{\partial^2 u}{\partial x^2}, \qquad 0 < x < L, \quad 0 < t, \tag{9.3.26}$$

where

$$u(x,0) = u(0,t) = 0, \quad \text{and} \quad u(L,t) = \theta. \tag{9.3.27}$$

We begin by blindly employing the technique of separation of variables. Once again, we obtain the ordinary differential equations, Equation 9.3.9 and Equation 9.3.10. The initial and boundary conditions become, however,

$$X(0) = T(0) = 0, \tag{9.3.28}$$

and

$$X(L)T(t) = \theta. \tag{9.3.29}$$

Although Equation 9.3.28 is acceptable, Equation 9.3.29 gives us an impossible condition because $T(t)$ cannot be constant. If it were, it would have to equal zero by Equation 9.3.28.

To find a way around this difficulty, suppose that we want the solution to our problem at a time long after $t = 0$. From experience, we know that heat conduction with time-independent boundary conditions eventually results in an evolution from the initial condition to some time-independent (steady-state) equilibrium. If we denote this steady-state solution by $w(x)$, it must satisfy the heat equation

$$a^2 w''(x) = 0, \tag{9.3.30}$$

and the boundary conditions

$$w(0) = 0, \quad \text{and} \quad w(L) = \theta. \tag{9.3.31}$$

We can integrate Equation 9.3.30 immediately to give

$$w(x) = A + Bx, \tag{9.3.32}$$

and the boundary condition, Equation 9.3.31, results in

$$w(x) = \frac{\theta x}{L}. \tag{9.3.33}$$

Clearly, Equation 9.3.33 cannot hope to satisfy the initial condition; that was never expected of it. However, if we add a time-varying (transient) solution $v(x,t)$ to $w(x)$ so that

$$u(x,t) = w(x) + v(x,t), \tag{9.3.34}$$

we could satisfy the initial condition if

$$v(x,0) = u(x,0) - w(x), \tag{9.3.35}$$

and $v(x,t)$ tends to zero as $t \to \infty$. Furthermore, because $w''(x) = w(0) = 0$, and $w(L) = \theta$,

$$\frac{\partial v}{\partial t} = a^2 \frac{\partial^2 v}{\partial x^2}, \qquad 0 < x < L, \quad 0 < t, \tag{9.3.36}$$

with the boundary conditions

$$v(0,t) = v(L,t) = 0, \quad 0 < t. \tag{9.3.37}$$

We can solve Equation 9.3.35, Equation 9.3.36, and Equation 9.3.37 by separation of variables; we did it in Example 9.3.1. However, in place of $f(x)$ we now have $u(x, 0) - w(x)$, or $-w(x)$ because $u(x, 0) = 0$. Therefore, the solution $v(x, t)$ is

$$v(x, t) = \sum_{n=1}^{\infty} B_n \sin\left(\frac{n\pi x}{L}\right) \exp\left(-\frac{a^2 n^2 \pi^2}{L^2} t\right) \tag{9.3.38}$$

with

$$B_n = \frac{2}{L} \int_0^L -w(x) \sin\left(\frac{n\pi x}{L}\right) dx = \frac{2}{L} \int_0^L -\frac{\theta x}{L} \sin\left(\frac{n\pi x}{L}\right) dx \tag{9.3.39}$$

$$= -\frac{2\theta}{L^2} \left[\frac{L^2}{n^2 \pi^2} \sin\left(\frac{n\pi x}{L}\right) - \frac{xL}{n\pi} \cos\left(\frac{n\pi x}{L}\right)\right]_0^L = (-1)^n \frac{2\theta}{n\pi}. \tag{9.3.40}$$

Thus, the entire solution is

$$u(x, t) = \frac{\theta x}{L} + \frac{2\theta}{\pi} \sum_{n=1}^{\infty} \frac{(-1)^n}{n} \sin\left(\frac{n\pi x}{L}\right) \exp\left(-\frac{a^2 n^2 \pi^2}{L^2} t\right). \tag{9.3.41}$$

The quantity $a^2 t / L^2$ is the *Fourier number*.

Figure 9.3.2 illustrates our solution and was created with the MATLAB script:

```
clear
M = 1000; dx = 0.01; dt = 0.01;
% compute grid and initialize solution
X = [0:dx:1]; T = [0:dt:0.2];
XX = repmat(X,[length(T) 1]); TT = repmat(T',[1 length(X)]);
u = XX;
% compute solution from Equation 9.3.41
sign = -2/pi;
for m = 1:M
  coeff = sign/m;
  u = u + coeff * sin((m*pi)*XX) .* exp(-(m*m*pi*pi) * TT);
  sign = -sign;
end
surf(XX,TT,u); axis([0 1 0 0.2 0 1]);
xlabel('DISTANCE','Fontsize',20); ylabel('TIME','Fontsize',20)
zlabel('SOLUTION','Fontsize',20)
```

Clearly it satisfies the boundary conditions. Initially, heat flows rapidly from right to left. As time increases, the rate of heat transfer decreases until the final equilibrium (steady-state) is established and no more heat flows. ☐

• Example 9.3.3

Let us find the solution to the heat equation

$$\frac{\partial u}{\partial t} = a^2 \frac{\partial^2 u}{\partial x^2}, \qquad 0 < x < L, \quad 0 < t, \tag{9.3.42}$$

subject to the Neumann boundary conditions

$$\frac{\partial u(0, t)}{\partial x} = \frac{\partial u(L, t)}{\partial x} = 0, \qquad 0 < t, \tag{9.3.43}$$

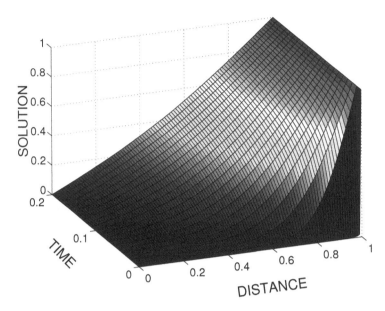

Figure 9.3.2: The temperature $u(x,t)/\theta$ within a thin bar as a function of position x/L and time $a^2 t/L^2$ with the left end held at a temperature of zero and right end held at a temperature θ while the initial temperature of the bar is zero.

and the initial condition that

$$u(x,0) = x, \qquad 0 < x < L. \tag{9.3.44}$$

We have now insulated *both* ends of the bar.

Assuming that $u(x,t) = X(x)T(t)$,

$$\frac{T'}{a^2 T} = \frac{X''}{X} = -k^2, \tag{9.3.45}$$

where we have presently assumed that the separation constant is negative. The Neumann conditions give $u_x(0,t) = X'(0)T(t) = 0$, and $u_x(L,t) = X'(L)T(t) = 0$ so that $X'(0) = X'(L) = 0$.

The boundary-value problem

$$X'' + k^2 X = 0, \tag{9.3.46}$$

and

$$X'(0) = X'(L) = 0 \tag{9.3.47}$$

gives the x-dependence. The solution to Equation 9.3.46 is

$$X_n(x) = \cos\left(\frac{n\pi x}{L}\right), \tag{9.3.48}$$

where $k_n = n\pi/L$ and $n = 1, 2, 3, \ldots$.

The corresponding temporal part equals the solution of

$$T_n' + a^2 k_n^2 T_n = T_n' + \frac{a^2 n^2 \pi^2}{L^2} T_n = 0, \tag{9.3.49}$$

which is

$$T_n(t) = A_n \exp\left(-\frac{a^2 n^2 \pi^2}{L^2} t\right). \tag{9.3.50}$$

Thus, the product solution given by a negative separation constant is

$$u_n(x,t) = X_n(x) T_n(t) = A_n \cos\left(\frac{n\pi x}{L}\right) \exp\left(-\frac{a^2 n^2 \pi^2}{L^2} t\right). \tag{9.3.51}$$

Unlike our previous problems, there is a nontrivial solution for a separation constant that equals zero. In this instance, the x-dependence equals

$$X(x) = Ax + B. \tag{9.3.52}$$

The boundary conditions $X'(0) = X'(L) = 0$ force A to be zero but B is completely free. Consequently, the x-dependence here is

$$X_0(x) = 1. \tag{9.3.53}$$

Because $T_0'(t) = 0$ in this case, the temporal part equals a constant that we shall take to be $A_0/2$. Therefore, the product solution corresponding to the zero separation constant is

$$u_0(x,t) = X_0(x) T_0(t) = A_0/2. \tag{9.3.54}$$

The most general solution to our problem equals the sum of all of the possible solutions:

$$u(x,t) = \frac{A_0}{2} + \sum_{n=1}^{\infty} A_n \cos\left(\frac{n\pi x}{L}\right) \exp\left(-\frac{a^2 n^2 \pi^2}{L^2} t\right). \tag{9.3.55}$$

Upon substituting $t = 0$ into Equation 9.3.55, we can determine A_n because

$$u(x,0) = x = \frac{A_0}{2} + \sum_{n=1}^{\infty} A_n \cos\left(\frac{n\pi x}{L}\right) \tag{9.3.56}$$

is merely a half-range Fourier cosine expansion of the function x over the interval $(0, L)$. From Equation 5.1.22 and Equation 5.1.23,

$$A_0 = \frac{2}{L} \int_0^L x \, dx = L, \tag{9.3.57}$$

and

$$A_n = \frac{2}{L} \int_0^L x \cos\left(\frac{n\pi x}{L}\right) dx = \frac{2}{L} \left[\frac{L^2}{n^2 \pi^2} \cos\left(\frac{n\pi x}{L}\right) + \frac{xL}{n\pi} \sin\left(\frac{n\pi x}{L}\right)\right]_0^L = \frac{2L}{n^2 \pi^2} \left[(-1)^n - 1\right]. \tag{9.3.58}$$

The complete solution is

$$u(x,t) = \frac{L}{2} - \frac{4L}{\pi^2} \sum_{m=1}^{\infty} \frac{1}{(2m-1)^2} \cos\left[\frac{(2m-1)\pi x}{L}\right] \exp\left[-\frac{a^2(2m-1)^2 \pi^2}{L^2} t\right], \tag{9.3.59}$$

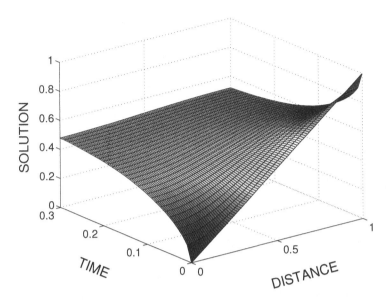

Figure 9.3.3: The temperature $u(x,t)/L$ within a thin bar as a function of position x/L and time a^2t/L^2 when we insulate both ends. The initial temperature of the bar is x.

because all of the even harmonics vanish and we may rewrite the odd harmonics using $n = 2m - 1$, where $m = 1, 2, 3, 4, \ldots$.

Figure 9.3.3 illustrates Equation 9.3.59 for various positions and times. It was generated using the MATLAB script:

```
clear
M = 100; dx = 0.01; dt = 0.01;
% compute grid and initialize solution
X = [0:dx:1]; T = [0:dt:0.3];
u = zeros(length(T),length(X)); u = 0.5;
XX = repmat(X,[length(T) 1]); TT = repmat(T',[1 length(X)]);
% compute solution from Equation 9.3.59
for m = 1:M
  temp1 = (2*m-1) * pi;
  coeff = 4 / (temp1*temp1);
  u = u - coeff * cos(temp1*XX) .* exp(-temp1 * temp1 * TT);
end
surf(XX,TT,u); axis([0 1 0 0.3 0 1]);
xlabel('DISTANCE','Fontsize',20); ylabel('TIME','Fontsize',20)
zlabel('SOLUTION','Fontsize',20)
```

The physical interpretation is quite simple. Since heat cannot flow in or out of the rod because of the insulation, it can only redistribute itself. Thus, heat flows from the warm right end to the cooler left end. Eventually the temperature achieves a steady state when the temperature is uniform throughout the bar.   □

• **Example 9.3.4: Refrigeration of apples**

Some decades ago, shiploads of apples, going from Australia to England, deteriorated from a disease called "brown heart," which occurred under insufficient cooling conditions. Apples, when placed on shipboard, are usually warm and must be cooled to be carried in

Figure 9.3.4: An apple suffering from "brown heart."

cold storage. They also generate heat by their respiration. It was suspected that this heat generation effectively counteracted the refrigeration of the apples, resulting in the "brown heart."

This was the problem that induced Awbery[4] to study the heat distribution within a sphere in which heat is being generated. He first assumed that the apples are initially at a uniform temperature. We can take this temperature to be zero by the appropriate choice of temperature scale. At time $t = 0$, the skins of the apples assume the temperature θ immediately when we introduce them into the hold.

Because of the spherical geometry, the nonhomogeneous heat equation becomes

$$\frac{1}{a^2}\frac{\partial u}{\partial t} = \frac{1}{r^2}\frac{\partial}{\partial r}\left(r^2\frac{\partial u}{\partial r}\right) + \frac{G}{\kappa}, \qquad 0 \le r < b, \quad 0 < t, \qquad (9.3.60)$$

where a^2 is the thermal diffusivity, b is the radius of the apple, κ is the thermal conductivity, and G is the heating rate (per unit time per unit volume).

If we try to use separation of variables on Equation 9.3.60, we find that it does not work because of the G/κ term. To circumvent this difficulty, we ask the simpler question of what happens after a very long time. We anticipate that a balance will eventually be established where conduction transports the heat produced within the apple to the surface of the apple where the surroundings absorb it. Consequently, just as we introduced a steady-state solution in Example 9.3.2, we again anticipate a steady-state solution $w(r)$ where the heat conduction removes the heat generated within the apples. The ordinary differential equation

$$\frac{1}{r^2}\frac{d}{dr}\left(r^2\frac{dw}{dr}\right) = -\frac{G}{\kappa} \qquad (9.3.61)$$

gives the steady state. Furthermore, just as we introduced a transient solution that allowed our solution to satisfy the initial condition, we must also have one here, and the governing

---

[4] Awbery, J. H., 1927: The flow of heat in a body generating heat. *Philos. Mag.*, Ser. 7, **4**, 629–638.

equation is

$$\frac{\partial v}{\partial t} = \frac{a^2}{r^2} \frac{\partial}{\partial r}\left(r^2 \frac{\partial v}{\partial r}\right) = \frac{a^2}{r} \frac{\partial^2(r\,v)}{\partial r^2}. \tag{9.3.62}$$

Solving Equation 9.3.61 first,

$$w(r) = C + \frac{D}{r} - \frac{Gr^2}{6\kappa}. \tag{9.3.63}$$

The constant D equals zero because the solution must be finite at $r = 0$. Since the steady-state solution must satisfy the boundary condition $w(b) = \theta$,

$$C = \theta + \frac{Gb^2}{6\kappa}. \tag{9.3.64}$$

Turning to the transient problem, we introduce a new dependent variable $y(r,t) = rv(r,t)$. This new dependent variable allows us to replace Equation 9.3.62 with

$$\frac{\partial y}{\partial t} = a^2 \frac{\partial^2 y}{\partial r^2}, \tag{9.3.65}$$

which we can solve. If we assume that $y(r,t) = R(r)T(t)$ and we only have a negative separation constant, the $R(r)$ equation becomes

$$\frac{d^2 R}{dr^2} + k^2 R = 0, \tag{9.3.66}$$

which has the solution

$$R(r) = A\cos(kr) + B\sin(kr). \tag{9.3.67}$$

The constant A equals zero because the solution, Equation 9.3.67, must vanish at $r = 0$ so that $v(0,t)$ remains finite. However, because $\theta = w(b) + v(b,t)$ for all time and $v(b,t) = R(b)T(t)/b = 0$, then $R(b) = 0$. Consequently, $k_n = n\pi/b$, and

$$v_n(r,t) = \frac{B_n}{r} \sin\left(\frac{n\pi r}{b}\right) \exp\left(-\frac{n^2\pi^2 a^2 t}{b^2}\right). \tag{9.3.68}$$

Superposition gives the total solution, which equals

$$u(r,t) = \theta + \frac{G}{6\kappa}(b^2 - r^2) + \sum_{n=1}^{\infty} \frac{B_n}{r} \sin\left(\frac{n\pi r}{b}\right) \exp\left(-\frac{n^2\pi^2 a^2 t}{b^2}\right). \tag{9.3.69}$$

Finally, we determine the coefficients B_n by the initial condition that $u(r,0) = 0$. Therefore,

$$B_n = -\frac{2}{b}\int_0^b r\left[\theta + \frac{G}{6\kappa}(b^2 - r^2)\right]\sin\left(\frac{n\pi r}{b}\right)dr = \frac{2\theta b}{n\pi}(-1)^n + \frac{2G}{\kappa}\left(\frac{b}{n\pi}\right)^3(-1)^n. \tag{9.3.70}$$

The complete solution is

$$u(r,t) = \theta + \frac{2\theta b}{r\pi}\sum_{n=1}^{\infty}\frac{(-1)^n}{n}\sin\left(\frac{n\pi r}{b}\right)\exp\left(-\frac{n^2\pi^2 a^2 t}{b^2}\right) \tag{9.3.71}$$

$$+ \frac{G}{6\kappa}(b^2 - r^2) + \frac{2Gb^3}{r\kappa\pi^3}\sum_{n=1}^{\infty}\frac{(-1)^n}{n^3}\sin\left(\frac{n\pi r}{b}\right)\exp\left(-\frac{n^2\pi^2 a^2 t}{b^2}\right).$$

The first line of Equation 9.3.71 gives the temperature distribution due to the imposition of the temperature θ on the surface of the apple while the second line gives the rise in the temperature due to the interior heating.

Returning to our original problem of whether the interior heating is strong enough to counteract the cooling by refrigeration, we merely use the second line of Equation 9.3.71 to find how much the temperature deviates from what we normally expect. Because the highest temperature exists at the center of each apple, its value there is the only one of interest in this problem. Assuming $b = 4$ cm as the radius of the apple, $a^2 G/\kappa = 1.33 \times 10^{-5}$ °C/s, and $a^2 = 1.55 \times 10^{-3}$ cm^2/s, the temperature effect of the heat generation is very small, only 0.0232 °C when, after about 2 hours, the temperatures within the apples reach equilibrium. Thus, we must conclude that heat generation within the apples is not the cause of brown heart.

We now know that brown heart results from an excessive concentration of carbon dioxide and an insufficient amount of oxygen in the storage hold.[5] Presumably this atmosphere affects the metabolic activities that are occurring in the apple[6] and leads to low-temperature breakdown.

Problems

For Problems 1–5, solve the heat equation $u_t = a^2 u_{xx}$, $0 < x < \pi$, $0 < t$, subject to the boundary conditions that $u(0,t) = u(\pi,t) = 0$, $0 < t$, and the following initial conditions for $0 < x < \pi$. Then plot your results using MATLAB.

1. $u(x,0) = A$, a constant

2. $u(x,0) = \sin^3(x) = [3\sin(x) - \sin(3x)]/4$

3. $u(x,0) = x$

4. $u(x,0) = \pi - x$

5. $u(x,0) = \begin{cases} x, & 0 < x < \pi/2, \\ \pi - x, & \pi/2 < x < \pi. \end{cases}$

For Problems 6–10, solve the heat equation $u_t = a^2 u_{xx}$, $0 < x < \pi$, $0 < t$, subject to the boundary conditions that $u_x(0,t) = u_x(\pi,t) = 0$, $0 < t$, and the following initial conditions for $0 < x < \pi$. Then plot your results using MATLAB.

6. $u(x,0) = 1$

7. $u(x,0) = x$

8. $u(x,0) = \cos^2(x) = [1 + \cos(2x)]/2$

9. $u(x,0) = \pi - x$

---

[5] Thornton, N. C., 1931: The effect of carbon dioxide on fruits and vegetables in storage. *Contrib. Boyce Thompson Inst.*, **3**, 219–244.

[6] Fidler, J. C., and C. J. North, 1968: The effect of conditions of storage on the respiration of apples. IV. Changes in concentration of possible substrates of respiration, as related to production of carbon dioxide and uptake of oxygen by apples at low temperatures. *J. Hortic. Sci.*, **43**, 429–439.

10. $u(x,0) = \begin{cases} T_0, & 0 < x < \pi/2, \\ T_1, & \pi/2 < x < \pi. \end{cases}$

For Problems 11 and 12, solve the heat equation $u_t = a^2 u_{xx}$, $0 < x < \pi$, $0 < t$, subject to the following boundary conditions and initial condition. Then plot your results using MATLAB.

11. $u(0,t) = u(\pi,t) = T_0$, $0 < t$; $u(x,0) = T_1 \neq T_0$, $0 < x < \pi$

12. $u(0,t) = 0, u(\pi,t) = T_0$, $0 < t$; $u(x,0) = T_0$, $0 < x < \pi$

13. The linearized Boussinesq equation[7]

$$\frac{\partial u}{\partial t} = \frac{\partial^2 u}{\partial x^2}, \qquad 0 < x < L, \quad 0 < t,$$

governs the height of the water table $u(x,t)$ above some reference point, where a^2 is the product of the storage coefficient times the hydraulic coefficient divided by the aquifer thickness. A typical value of a^2 is 10 m^2/min. Consider the problem of a strip of land of width L that separates two reservoirs of depth h_1. Initially the height of the water table would be h_1. Suddenly we lower the reservoir on the right $x = L$ to a depth h_2 [$u(0,t) = h_1$, $u(L,t) = h_2$, and $u(x,0) = h_1$]. Find the height of the water table at any position x within the aquifer and any time $t > 0$.

14. The equation

$$\frac{\partial u}{\partial t} = \frac{\partial^2 u}{\partial x^2}, \qquad 0 < x < L, \quad 0 < t,$$

governs the height of the water table $u(x,t)$. Consider the problem[8] of a piece of land that suddenly has two drains placed at the points $x = 0$ and $x = L$ so that $u(0,t) = u(L,t) = 0$. If the water table initially has the profile $u(x,0) = 8H(L^3 x - 3L^2 x^2 + 4Lx^3 - 2x^4)/L^4$, find the height of the water table at any point within the aquifer and any time $t > 0$.

15. Solve the nonhomogeneous heat equation

$$\frac{\partial u}{\partial t} - \frac{\partial^2 u}{\partial x^2} = -1, \qquad 0 < x < 1, \quad 0 < t,$$

subject to the boundary conditions $u_x(0,t) = u_x(1,t) = 0$, $0 < t$, and the initial condition $u(x,0) = \frac{1}{2}(1 - x^2)$, $0 < x < 1$. Hint: Note that any function of time satisfies the boundary conditions.

16. Solve the nonhomogeneous heat equation

$$\frac{\partial u}{\partial t} - a^2 \frac{\partial^2 u}{\partial x^2} = A\cos(\omega t), \qquad 0 < x < \pi, \quad 0 < t,$$

---

[7] See, for example, Van Schilfgaarde, J., 1970: Theory of flow to drains. *Advances in Hydroscience*, No. 6, Academic Press, 81–85.

[8] For a similar problem, see Dumm, L. D., 1954: New formula for determining depth and spacing of subsurface drains in irrigated lands. *Agric. Eng.*, **35**, 726–730.

subject to the boundary conditions $u_x(0,t) = u_x(\pi, t) = 0$, $0 < t$, and the initial condition $u(x,0) = f(x)$, $0 < x < \pi$. Hint: Note that any function of time satisfies the boundary conditions.

17. Solve the nonhomogeneous heat equation

$$\frac{\partial u}{\partial t} - \frac{\partial^2 u}{\partial x^2} = \begin{cases} x, & 0 < x \le \pi/2, \\ \pi - x, & \pi/2 \le x < \pi, \end{cases} \quad 0 < t,$$

subject to the boundary conditions $u(0,t) = u(\pi, t) = 0$, $0 < t$, and the initial condition $u(x,0) = 0$, $0 < x < \pi$. Hint: Represent the forcing function as a half-range Fourier sine expansion over the interval $(0, \pi)$.

18. A uniform conducting rod of length L and thermometric diffusivity a^2 is initially at temperature zero. We supply heat uniformly throughout the rod so that the heat conduction equation is

$$a^2 \frac{\partial^2 u}{\partial x^2} = \frac{\partial u}{\partial t} - P, \quad 0 < x < L, \quad 0 < t,$$

where P is the rate at which the temperature would rise if there was no conduction. If we maintain the ends of the rod at the temperature of zero, find the temperature at any position and subsequent time. How would the solution change if the boundary conditions became $u(0,t) = u(L,t) = A \ne 0$, $0 < t$, and the initial condition reads $u(x,0) = A$, $0 < x < L$?

19. Find the solution of

$$\frac{\partial u}{\partial t} = \frac{\partial^2 u}{\partial x^2} - u, \quad 0 < x < L, \quad 0 < t,$$

with the boundary conditions $u(0,t) = 1$, and $u(L,t) = 0$, $0 < t$, and the initial condition $u(x,0) = 0$, $0 < x < L$.

20. Solve the heat equation in spherical coordinates

$$\frac{\partial u}{\partial t} = \frac{a^2}{r^2} \frac{\partial}{\partial r} \left(r^2 \frac{\partial u}{\partial r} \right) = \frac{a^2}{r} \frac{\partial^2 (ru)}{\partial r^2}, \quad 0 \le r < 1, \quad 0 < t,$$

subject to the boundary conditions $\lim_{r \to 0} |u(r,t)| < \infty$, and $u(1,t) = 0$, $0 < t$, and the initial condition $u(r,0) = 1$, $0 \le r < 1$. Hint: Introduce a new independent variable $v(r,t) = r\,u(r,t)$.

21. In their study of heat conduction within a thermocouple through which a steady current flows, Reich and Madigan[9] solved the following nonhomogeneous heat conduction problem:

$$\frac{\partial u}{\partial t} - a^2 \frac{\partial^2 u}{\partial x^2} = J - P\,\delta(x - b), \quad 0 < x < L, \quad 0 < t, \quad 0 < b < L,$$

---

[9] Reich, A. D., and J. R. Madigan, 1961: Transient response of a thermocouple circuit under steady currents. *J. Appl. Phys.*, **32**, 294–301.

where J represents the Joule heating generated by the steady current and the P term represents the heat loss from Peltier cooling.[10] Find $u(x,t)$ if both ends are kept at zero $[u(0,t) = u(L,t) = 0]$ and initially the temperature is zero $[u(x,0) = 0]$. The interesting aspect of this problem is the presence of the delta function $\delta(\cdot)$.

Step 1: Assuming that $u(x,t)$ equals the sum of a steady-state solution $w(x)$ and a transient solution $v(x,t)$, show that the steady-state solution is governed by

$$a^2 \frac{d^2 w}{dx^2} = P\,\delta(x - b) - J, \qquad w(0) = w(L) = 0.$$

Step 2: Show that the steady-state solution is

$$w(x) = \begin{cases} Jx(L - x)/2a^2 + Ax, & 0 < x < b, \\ Jx(L - x)/2a^2 + B(L - x), & b < x < L. \end{cases}$$

Step 3: The temperature must be continuous at $x = b$; otherwise, we would have infinite heat conduction there. Use this condition to show that $Ab = B(L - b)$.

Step 4: To find a second relationship between A and B, integrate the steady-state differential equation across the interface at $x = b$ and show that

$$\lim_{\epsilon \to 0} a^2 \left. \frac{dw}{dx} \right|_{b-\epsilon}^{b+\epsilon} = P.$$

Step 5: Using the result from Step 4, show that $A + B = -P/a^2$, and

$$w(x) = \begin{cases} Jx(L - x)/2a^2 - Px(L - b)/a^2 L, & 0 < x < b, \\ Jx(L - x)/2a^2 - Pb(L - x)/a^2 L, & b < x < L. \end{cases}$$

Step 6: Re-express $w(x)$ as a half-range Fourier sine expansion and show that

$$w(x) = \frac{4JL^2}{a^2 \pi^3} \sum_{m=1}^{\infty} \frac{\sin[(2m - 1)\pi x/L]}{(2m - 1)^3} - \frac{2LP}{a^2 \pi^2} \sum_{n=1}^{\infty} \frac{\sin(n\pi b/L)\sin(n\pi x/L)}{n^2}.$$

Step 7: Use separation of variables to find the transient solution by solving

$$\frac{\partial v}{\partial t} = a^2 \frac{\partial^2 v}{\partial x^2}, \qquad 0 < x < L, \quad 0 < t,$$

subject to the boundary conditions $v(0,t) = v(L,t) = 0$, $0 < t$, and the initial condition $v(x,0) = -w(x), 0 < x < L$.

---

[10] In 1834 Jean Charles Athanase Peltier (1785–1845) discovered that there is a heating or cooling effect, quite apart from ordinary resistance heating, whenever an electric current flows through the junction between two different metals.

Step 8: Add the steady-state and transient solutions together and show that

$$u(x,t) = \frac{4JL^2}{a^2\pi^3} \sum_{m=1}^{\infty} \frac{\sin[(2m-1)\pi x/L]}{(2m-1)^3} \left[1 - e^{-a^2(2m-1)^2\pi^2 t/L^2}\right]$$
$$- \frac{2LP}{a^2\pi^2} \sum_{n=1}^{\infty} \frac{\sin(n\pi b/L)\sin(n\pi x/L)}{n^2} \left[1 - e^{-a^2 n^2\pi^2 t/L^2}\right].$$

9.4 THE SUPERPOSITION INTEGRAL

Let us solve the heat condition problem

$$\frac{\partial u}{\partial t} = a^2 \frac{\partial^2 u}{\partial x^2}, \qquad 0 < x < L, \quad 0 < t, \tag{9.4.1}$$

with the boundary conditions

$$u(0,t) = 0, \quad u(L,t) = f(t), \quad 0 < t, \tag{9.4.2}$$

and the initial condition

$$u(x,0) = 0, \quad 0 < x < L. \tag{9.4.3}$$

The solution of Equation 9.4.1 through Equation 9.4.3 is difficult because of the time-dependent boundary condition. Instead of solving this system directly, let us solve the easier problem

$$\frac{\partial A}{\partial t} = a^2 \frac{\partial^2 A}{\partial x^2}, \qquad 0 < x < L, \quad 0 < t, \tag{9.4.4}$$

with the boundary conditions

$$A(0,t) = 0, \quad A(L,t) = 1, \qquad 0 < t, \tag{9.4.5}$$

and the initial condition

$$A(x,0) = 0, \qquad 0 < x < L. \tag{9.4.6}$$

Separation of variables yields the solution

$$A(x,t) = \frac{x}{L} + \frac{2}{\pi} \sum_{n=1}^{\infty} \frac{(-1)^n}{n} \sin\left(\frac{n\pi x}{L}\right) \exp\left(-\frac{a^2 n^2 \pi^2 t}{L^2}\right). \tag{9.4.7}$$

Consider the following case. Suppose that we maintain the temperature at zero at the end $x = L$ until $t = \tau_1$ and then raise it to the value of unity. The resulting temperature distribution equals zero everywhere when $t < \tau_1$ and equals $A(x, t - \tau_1)$ for $t > \tau_1$. We have merely shifted our time axis so that the initial condition occurs at $t = \tau_1$.

Consider an analogous, but more complicated, situation of the temperature at the end position $x = L$ held at $f(0)$ from $t = 0$ to $t = \tau_1$, at which time we abruptly change it by the amount $f(\tau_1) - f(0)$ to the value $f(\tau_1)$. This temperature remains until $t = \tau_2$ when we again abruptly change it by an amount $f(\tau_2) - f(\tau_1)$. We can imagine this process continuing up to the instant $t = \tau_n$. Because of linear superposition, which we introduced in Section 8.3, the temperature distribution at any given time equals the sum of these temperature increments:

$$u(x,t) = f(0)A(x,t) + [f(\tau_1) - f(0)]A(x, t - \tau_1) + [f(\tau_2) - f(\tau_1)]A(x, t - \tau_2)$$
$$+ \cdots + [f(\tau_n) - f(\tau_{n-1})]A(x, t - \tau_n), \tag{9.4.8}$$

where τ_n is the time of the most recent temperature change. If we write

$$\Delta f_k = f(\tau_k) - f(\tau_{k-1}), \quad \text{and} \quad \Delta \tau_k = \tau_k - \tau_{k-1}, \tag{9.4.9}$$

Equation 9.4.8 becomes

$$u(x,t) = f(0)A(x,t) + \sum_{k=1}^{n} A(x, t - \tau_k) \frac{\Delta f_k}{\Delta \tau_k} \Delta \tau_k. \tag{9.4.10}$$

Consequently, in the limit of $\Delta \tau_k \to 0$, Equation 9.4.10 becomes

$$u(x,t) = f(0)A(x,t) + \int_0^t A(x, t - \tau) f'(\tau) \, d\tau, \tag{9.4.11}$$

assuming that $f(t)$ is differentiable. Equation 9.4.11 is the *superposition integral*. We can obtain alternative forms by integration by parts:

$$u(x,t) = f(t)A(x,0) - \int_0^t f(\tau) \frac{\partial A(x, t - \tau)}{\partial \tau} \, d\tau, \tag{9.4.12}$$

or

$$u(x,t) = f(t)A(x,0) + \int_0^t f(\tau) \frac{\partial A(x, t - \tau)}{\partial t} \, d\tau, \tag{9.4.13}$$

because

$$\frac{\partial A(x, t - \tau)}{\partial \tau} = -\frac{\partial A(x, t - \tau)}{\partial t}. \tag{9.4.14}$$

To illustrate[11] the superposition integral, suppose $f(t) = t$. Then, by Equation 9.4.11,

$$u(x,t) = \int_0^t \left\{ \frac{x}{L} + \frac{2}{\pi} \sum_{n=1}^{\infty} \frac{(-1)^n}{n} \sin\left(\frac{n\pi x}{L}\right) \exp\left[-\frac{a^2 n^2 \pi^2}{L^2}(t - \tau)\right] \right\} d\tau \tag{9.4.15}$$

$$= \frac{xt}{L} + \frac{2L^2}{a^2 \pi^3} \sum_{n=1}^{\infty} \frac{(-1)^n}{n^3} \sin\left(\frac{n\pi x}{L}\right) \left[1 - \exp\left(-\frac{a^2 n^2 \pi^2 t}{L^2}\right)\right]. \tag{9.4.16}$$

Consider now the heat conduction problem with time-dependent forcing and/or boundary conditions:

$$\frac{\partial u}{\partial t} = a^2 L(u) + F(P, t), \qquad 0 < t, \tag{9.4.17}$$

$$B(u) = g(Q, t), \qquad 0 < t, \tag{9.4.18}$$

---

[11] This occurs, for example, in McAfee, K. B., 1958: Stress-enhanced diffusion in glass. I. Glass under tension and compression. *J. Chem. Phys.*, **28**, 218–226. McAfee used an alternative method of guessing the solution.

and
$$u(P,0) = h(P), \tag{9.4.19}$$

where

$$L(u) = C_0 + C_1 \frac{\partial}{\partial x_1}\left(K_1 \frac{\partial u}{\partial x_1}\right) + C_2 \frac{\partial}{\partial x_2}\left(K_2 \frac{\partial u}{\partial x_2}\right) + C_3 \frac{\partial}{\partial x_3}\left(K_3 \frac{\partial u}{\partial x_3}\right), \tag{9.4.20}$$

$$B(u) = c_0 + c_1 \frac{\partial u}{\partial x_1} + c_2 \frac{\partial u}{\partial x_2} + c_3 \frac{\partial u}{\partial x_3}, \tag{9.4.21}$$

P denotes an arbitrary interior point at (x_1, x_2, x_3) of a region R, and Q is any point on the boundary of R. Here c_i, C_i, and K_i are functions of x_1, x_2, and x_3 only.

Bartels and Churchill[12] extended Duhamel's theorem to solve this heat conduction problem. They did this by first introducing the simpler initial-boundary-value problem:

$$\frac{\partial v}{\partial t} = a^2 L(v) + F(P, t_1), \qquad 0 < t, \tag{9.4.22}$$

$$B(v) = g(Q, t_1), \qquad 0 < t, \tag{9.4.23}$$

and

$$v(P,0) = h(P), \tag{9.4.24}$$

which has a constant forcing and boundary conditions in place of the time-dependent ones. Here t_1 denotes an arbitrary but *fixed* instant of time. Then Bartels and Churchill proved that the solution to the original problem is given by the convolution integral

$$u(P,t) = \frac{\partial}{\partial t}\left[\int_0^t v(P, t - \tau, \tau)\, d\tau\right]. \tag{9.4.25}$$

- **Example 7.4.1**

Let us resolve Equation 9.4.1 through Equation 9.4.3 using Equation 9.4.25.

We begin by solving the auxiliary problem Equation 9.4.22 through Equation 9.4.24:

$$\frac{\partial v}{\partial t} = a^2 \frac{\partial^2 v}{\partial x^2}, \qquad 0 < x < L, \quad 0 < t, t_1, \tag{9.4.26}$$

subject to the boundary conditions

$$v(0, t, t_1) = 0, \quad v(L, t, t_1) = f(t_1), \qquad 0 < t, t_1, \tag{9.4.27}$$

and the initial condition $v(x, 0, t_1) = 0$, $0 < x < L$.

The heat condition problem Equation 9.4.26 and Equation 9.4.27 can be solved using separation of variables where $v(x, t, t_1) = w(x, t_1) + \theta(x, t, t_1)$,

$$w''(x, t_1) = 0, \qquad w(0, t_1) = 0, \quad w(L, t_1) = f(t_1), \qquad 0 < x < L, \tag{9.4.28}$$

[12] Bartels, R. C. F., and R. V. Churchill, 1942: Resolution of boundary problems by the use of a generalized convolution. *Am. Math. Soc. Bull.*, **48**, 276–282.

and

$$\frac{\partial \theta}{\partial t} = a^2 \frac{\partial^2 \theta}{\partial x^2}, \qquad 0 < x < L, \quad 0 < t, t_1, \tag{9.4.29}$$

subject to the boundary conditions

$$\theta(0, t, t_1) = \theta(L, t, t_1) = 0, \qquad 0 < t, t_1, \tag{9.4.30}$$

and the initial condition $\theta(x, 0, t_1) = -w(x, t_1)$, $0 < x < L$. This yields the solution that

$$v(r, t, t_1) = \frac{f(t_1)x}{L} + \frac{2}{\pi} f(t_1) \sum_{n=1}^{\infty} \frac{(-1)^n}{n} \sin\left(\frac{n\pi x}{L}\right) \exp\left(-\frac{a^2 n^2 \pi^2 t}{L^2}\right). \tag{9.4.31}$$

Therefore,

$$u(x, t) = \frac{\partial}{\partial t}\left[\int_0^t v(x, t - \tau, \tau)\, d\tau\right]. \tag{9.4.32}$$

Using the Leibniz rule and the fact that $v(x, 0, t_1) = 0$,

$$u(x, t) = \int_0^t \frac{\partial v(x, t - \tau, \tau)}{\partial t}\, d\tau. \tag{9.4.33}$$

Substituting Equation 9.4.31 into Equation 9.4.33, we finally have that

$$u(x, t) = -\frac{2a^2 \pi}{L^2} \sum_{n=1}^{\infty} n \sin\left(\frac{n\pi x}{L}\right) \exp\left(-\frac{a^2 n^2 \pi^2 t}{L^2}\right) \int_0^t (-1)^n f(\tau) \exp\left(\frac{a^2 n^2 \pi^2 \tau}{L^2}\right)\, d\tau. \tag{9.4.34}$$

Problems

1. Solve the heat equation[13]

$$\frac{\partial u}{\partial t} = a^2 \frac{\partial^2 u}{\partial x^2}, \qquad 0 < x < L, \quad 0 < t,$$

subject to the boundary conditions $u(0, t) = u(L, t) = f(t)$, $0 < t$, and the initial condition $u(x, 0) = 0$, $0 < x < L$.

Step 1: First solve the heat conduction problem

$$\frac{\partial A}{\partial t} = a^2 \frac{\partial^2 A}{\partial x^2}, \qquad 0 < x < L, \quad 0 < t,$$

subject to the boundary conditions $A(0, t) = A(L, t) = 1$, $0 < t$, and the initial condition $A(x, 0) = 0$, $0 < x < L$. Show that

$$A(x, t) = 1 - \frac{4}{\pi} \sum_{n=1}^{\infty} \frac{\sin[(2n-1)\pi x/L]}{2n-1} e^{-a^2(2n-1)^2 \pi^2 t/L^2}.$$

---

[13] See Tao, L. N., 1960: Magnetohydrodynamic effects on the formation of Couette flow. *J. Aerosp. Sci.*, **27**, 334–338.

Step 2: Use Duhamel's theorem and show that

$$u(x,t) = \frac{4\pi a^2}{L^2} \sum_{n=1}^{\infty} (2n-1) \sin\left[\frac{(2n-1)\pi x}{L}\right] e^{-a^2(2n-1)^2\pi^2 t/L^2} \int_0^t f(\tau) e^{a^2(2n-1)^2\pi^2\tau/L^2}\, d\tau.$$

2. Solve the heat equation

$$\frac{\partial u}{\partial t} - \frac{\partial^2 u}{\partial x^2} = t\sin(x), \qquad 0 < x < \pi, \quad 0 < t,$$

subject to the boundary conditions $u(0,t) = u(\pi,t) = 0$, $0 < t$, and the initial condition $u(x,0) = 0$, $0 < x < \pi$.

Step 1: First solve the heat conduction problem

$$\frac{\partial v}{\partial t} - \frac{\partial^2 v}{\partial x^2} = t_1\sin(x), \qquad 0 < x < \pi, \quad 0 < t, t_1,$$

subject to the boundary conditions $v(0,t,t_1) = v(\pi,t,t_1) = 0$, $0 < t, t_1$, and the initial condition $v(x,0,t_1) = 0$, $0 < x < \pi$. Show that $v(x,t,t_1) = t_1\sin(x) - t_1\sin(x)e^{-t}$.

Step 2: Use Equation 9.4.25 to show that the solution to our original problem is $u(x,t) = (t - 1 + e^{-t})\sin(x)$.

9.5 NUMERICAL SOLUTION OF THE HEAT EQUATION

In addition to separation of variables, linear heat equations can also be solved using transform methods (see, for example, Section 6.8). However, when this analytic technique fails or we have a nonlinear heat equation, we must resort to numerical techniques. This section develops some of these techniques.

Starting with the heat equation

$$\frac{\partial u}{\partial t} = a^2\frac{\partial^2 u}{\partial x^2}, \tag{9.5.1}$$

we must first replace the exact derivatives with finite differences. Drawing upon our work in Section 8.5,

$$\frac{\partial u(x_m,t_n)}{\partial t} = \frac{u_m^{n+1} - u_m^n}{\Delta t} + O(\Delta t), \tag{9.5.2}$$

and

$$\frac{\partial^2 u(x_m,t_n)}{\partial x^2} = \frac{u_{m+1}^n - 2u_m^n + u_{m-1}^n}{(\Delta x)^2} + O[(\Delta x)^2], \tag{9.5.3}$$

where the notation u_m^n denotes $u(x_m,t_n)$. Figure 9.5.1 illustrates our numerical scheme when we hold both ends at the temperature of zero. Substituting Equation 9.5.2 and Equation 9.5.3 into Equation 9.5.1 and rearranging,

$$u_m^{n+1} = u_m^n + \frac{a^2\Delta t}{(\Delta x)^2}\left(u_{m+1}^n - 2u_m^n + u_{m-1}^n\right). \tag{9.5.4}$$

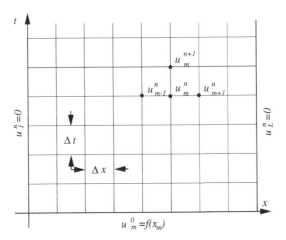

Figure 9.5.1: Schematic of the numerical solution of the heat equation when we hold both ends at a temperature of zero.

The numerical integration begins with $n = 0$ and the value of u_{m+1}^0, u_m^0, and u_{m-1}^0 are given by $f(m\Delta x)$.

Once again we must check the *convergence*, *stability*, and *consistency* of our scheme. We begin by writing u_{m+1}^n, u_{m-1}^n, and u_m^{n+1} in terms of the exact solution u and its derivatives evaluated at the point $x_m = m\Delta x$ and $t_n = n\Delta t$. By Taylor's expansion,

$$u_{m+1}^n = u_m^n + \Delta x \left.\frac{\partial u}{\partial x}\right|_n^m + \frac{1}{2}(\Delta x)^2 \left.\frac{\partial^2 u}{\partial x^2}\right|_n^m + \frac{1}{6}(\Delta x)^3 \left.\frac{\partial^3 u}{\partial x^3}\right|_n^m + \cdots, \qquad (9.5.5)$$

$$u_{m-1}^n = u_m^n - \Delta x \left.\frac{\partial u}{\partial x}\right|_n^m + \frac{1}{2}(\Delta x)^2 \left.\frac{\partial^2 u}{\partial x^2}\right|_n^m - \frac{1}{6}(\Delta x)^3 \left.\frac{\partial^3 u}{\partial x^3}\right|_n^m + \cdots, \qquad (9.5.6)$$

and

$$u_m^{n+1} = u_m^n + \Delta t \left.\frac{\partial u}{\partial t}\right|_n^m + \frac{1}{2}(\Delta t)^2 \left.\frac{\partial^2 u}{\partial t^2}\right|_n^m + \frac{1}{6}(\Delta t)^3 \left.\frac{\partial^3 u}{\partial t^3}\right|_n^m + \cdots. \qquad (9.5.7)$$

Substituting into Equation 9.5.4, we obtain

$$\frac{u_m^{n+1} - u_m^n}{\Delta t} - a^2 \frac{u_{m+1}^n - 2u_m^n + u_{m-1}^n}{(\Delta x)^2}$$
$$= \left.\left(\frac{\partial u}{\partial t} - a^2 \frac{\partial^2 u}{\partial x^2}\right)\right|_n^m + \frac{1}{2}\Delta t \left.\frac{\partial^2 u}{\partial t^2}\right|_n^m - \frac{1}{12}(a\Delta x)^2 \left.\frac{\partial^4 u}{\partial x^4}\right|_n^m + \cdots. \quad (9.5.8)$$

The first term on the right side of Equation 9.5.8 vanishes because $u(x,t)$ satisfies the heat equation. Thus, in the limit of $\Delta x \to 0$, $\Delta t \to 0$, the right side of Equation 9.5.8 vanishes and the scheme is *consistent*.

To determine the *stability* of the explicit scheme, we again use the Fourier method. Assuming a solution of the form:

$$u_n^m = e^{im\theta}e^{in\lambda}, \qquad (9.5.9)$$

we substitute Equation 9.5.9 into Equation 9.5.4 and find that

$$\frac{e^{i\lambda} - 1}{\Delta t} = a^2 \frac{e^{i\theta} - 2 + e^{-i\theta}}{(\Delta x)^2}, \qquad (9.5.10)$$

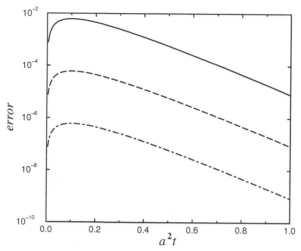

Figure 9.5.2: The growth of error $\|e_n\|$ as a function of a^2t for various resolutions. For the top line, $\Delta x = 0.1$; for the middle line, $\Delta x = 0.01$; and for the bottom line, $\Delta x = 0.001$.

or

$$e^{i\lambda} = 1 - 4\frac{a^2\Delta t}{(\Delta x)^2}\sin^2\left(\frac{\theta}{2}\right). \tag{9.5.11}$$

The quantity $e^{i\lambda}$ will grow exponentially unless

$$-1 \le 1 - 4\frac{a^2\Delta t}{(\Delta x)^2}\sin^2\left(\frac{\theta}{2}\right) < 1. \tag{9.5.12}$$

The right inequality is trivially satisfied if $a^2\Delta t/(\Delta x)^2 > 0$, while the left inequality yields

$$\frac{a^2\Delta t}{(\Delta x)^2} \le \frac{1}{2\sin^2(\theta/2)}, \tag{9.5.13}$$

leading to the stability condition $0 < a^2\Delta t/(\Delta x)^2 \le \frac{1}{2}$. This is a rather restrictive condition because doubling the resolution (halving Δx) requires that we reduce the time step by a quarter. Thus, for many calculations, the required time step may be unacceptably small. For this reason, many use an implicit form of the finite differencing (Crank-Nicholson implicit method[14]):

$$\frac{u_m^{n+1} - u_m^n}{\Delta t} = \frac{a^2}{2}\left[\frac{u_{m+1}^n - 2u_m^n + u_{m-1}^n}{(\Delta x)^2} + \frac{u_{m+1}^{n+1} - 2u_m^{n+1} + u_{m-1}^{n+1}}{(\Delta x)^2}\right], \tag{9.5.14}$$

although it requires the solution of a simultaneous set of linear equations. However, there are several efficient methods for their solution.

Finally we must check and see if our explicit scheme *converges* to the true solution. If we let e_m^n denote the difference between the exact and our finite differenced solution to the heat equation, we can use Equation 9.5.8 to derive the equation governing e_m^n and find that

$$e_m^{n+1} = e_m^n + \frac{a^2\Delta t}{(\Delta x)^2}\left(e_{m+1}^n - 2e_m^n + e_{m-1}^n\right) + O[(\Delta t)^2 + \Delta t(\Delta x)^2], \tag{9.5.15}$$

for $m = 1, 2, \ldots, M$. Assuming that $a^2\Delta t/(\Delta x)^2 \le \frac{1}{2}$, then

---

[14] Crank, J., and P. Nicholson, 1947: A practical method for numerical evaluation of solutions of partial differential equations of the heat-conduction type. *Proc. Cambridge. Philos. Soc.*, **43**, 50–67.

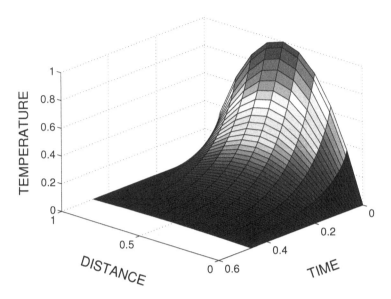

Figure 9.5.3: The numerical solution $u(x,t)$ of the heat equation with $a^2 \Delta t/(\Delta x)^2 = 0.47$ at various positions $x' = x/L$ and times $t' = a^2 t/L^2$ using Equation 9.5.4. The initial temperature $u(x,0)$ equals $4x'(1-x')$ and we hold both ends at a temperature of zero.

$$|e_m^{n+1}| \leq \frac{a^2 \Delta t}{(\Delta x)^2} |e_{m-1}^n| + \left[1 - 2\frac{a^2 \Delta t}{(\Delta x)^2}\right] |e_m^n| + \frac{a^2 \Delta t}{(\Delta x)^2} |e_{m+1}^n| + A[(\Delta t)^2 + \Delta t (\Delta x)^2]$$

$$\tag{9.5.16}$$

$$\leq ||e_n|| + A[(\Delta t)^2 + \Delta t (\Delta x)^2], \tag{9.5.17}$$

where $||e_n|| = \max_{m=0,1,\dots,M} |e_m^n|$. Consequently,

$$||e_{n+1}|| \leq ||e_n|| + A[(\Delta t)^2 + \Delta t (\Delta x)^2]. \tag{9.5.18}$$

Because $||e_0|| = 0$ and $n\Delta t \leq t_n$, we find that

$$||e_{n+1}|| \leq An[(\Delta t)^2 + \Delta t (\Delta x)^2] \leq At_n[\Delta t + (\Delta x)^2]. \tag{9.5.19}$$

As $\Delta x \to 0$, $\Delta t \to 0$, the errors tend to zero and we have convergence. We have illustrated Equation 9.5.19 in Figure 9.5.2 by using the finite difference equation, Equation 9.5.4, to compute $||e_n||$ during a numerical experiment that used $a^2 \Delta t/(\Delta x)^2 = 0.5$, and $f(x) = \sin(\pi x)$. Note how each increase of resolution by 10 results in a drop in the error by 100.

The following examples illustrate the use of numerical methods.

● **Example 9.5.1**

For our first example, we redo Example 9.3.1 with $a^2 \Delta t/(\Delta x)^2 = 0.47$ and 0.53. Our numerical solution was computed using the MATLAB script:

```
clear
coeff = 0.47; % coeff = a²Δt/(Δx)²
ncount = 1; dx = 0.1; dt = coeff * dx * dx;
```

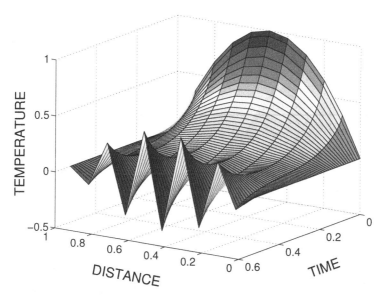

Figure 9.5.4: Same as Figure 9.5.3 except that $a^2 \Delta t/(\Delta x)^2 = 0.53$.

```
N = 99; x = 0:dx:1;
M = 1/dx + 1; % M = number of spatial grid points
tplot(1) = 0; u = zeros(M,N+1);
for m = 1:M; u(m,1)=4*x(m)*(1-x(m)); temp(m,1)=u(m,1); end
% integrate forward in time
for n = 1:N
  t = dt * n;
  for m = 2:M-1
    u(m,n+1) = u(m,n) + coeff*(u(m+1,n)-2*u(m,n)+u(m-1,n));
  end
  if mod(n+1,2) == 0
    ncount = ncount + 1; tplot(ncount) = t;
    for m = 1:M; temp(m,ncount) = u(m,n+1); end
end; end
% plot the numerical solution
X = x' * ones(1,length(tplot)); T = ones(M,1) * tplot;
surf(X,T,temp)
xlabel('DISTANCE','Fontsize',20); ylabel('TIME','Fontsize',20)
zlabel('TEMPERATURE','Fontsize',20)
```

As Figure 9.5.3 shows, the solution with $a^2 \Delta t/(\Delta x)^2 < 1/2$ performs well. On the other hand, Figure 9.5.4 shows small-scale, growing disturbances when $a^2 \Delta t/(\Delta x)^2 > 1/2$. It should be noted that for the reasonable $\Delta x = L/100$, it takes approximately *20,000* time steps before we reach $a^2 t/L^2 = 1$. $\qquad\square$

• **Example 9.5.2**

In this example, we redo the previous example with an insulated end at $x = L$. Using the centered differencing formula,

$$u_{M+1}^n - u_{M-1}^n = 0, \tag{9.5.20}$$

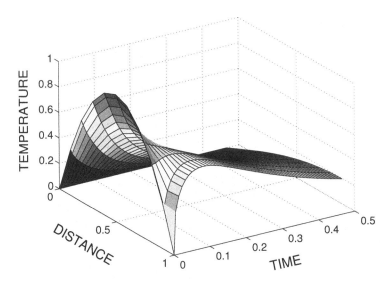

Figure 9.5.5: Same as Figure 9.5.3 except that we now have an insulated boundary condition $u_x(L, t) = 0$.

because $u_x(L, t) = 0$. Also, at $i = M$,

$$u_M^{n+1} = u_M^n + \frac{a^2 \Delta t}{(\Delta x)^2} \left(u_{M+1}^n - 2u_M^n + u_{M-1}^n \right). \qquad (9.5.21)$$

Eliminating u_{M+1}^n between the two equations,

$$u_M^{n+1} = u_M^n + \frac{a^2 \Delta t}{(\Delta x)^2} \left(2u_{M-1}^n - 2u_M^n \right). \qquad (9.5.22)$$

To implement this new boundary condition in our MATLAB script, we add the line
`u(M,n+1) = u(M,n) + 2 * coeff * (u(M-1,n) - u(M,n));`
after the lines
```
for m = 2:M-1
u(m,n+1) = u(m,n) + coeff * (u(m+1,n) - 2 * u(m,n) + u(m-1,n));
end
```
Figure 9.5.5 illustrates our numerical solution at various positions and times.

Project: Implicit Numerical Integration of the Heat Equation

The difficulty in using explicit time differencing to solve the heat equation is the very small time step that must be taken at moderate spatial resolutions to ensure stability. This small time step translates into an unacceptably long execution time. In this project you will investigate the Crank-Nicholson implicit scheme, which allows for a much more reasonable time step.

Step 1: Develop a MATLAB script that uses the Crank-Nicholson equation, Equation 9.5.14, to numerically integrate the heat equation. To do this, you will need a tridiagonal solver to find u_m^{n+1}. This is explained at the end of Section 3.1. However, many numerical methods books[15] actually have code already developed for your use. You might as well use this code.

---

[15] For example, Press, W. H., B. P. Flannery, S. A. Teukolsky, and W. T. Vetterling, 1986: *Numerical Recipes: The Art of Scientific Computing.* Cambridge University Press, Section 2.6.

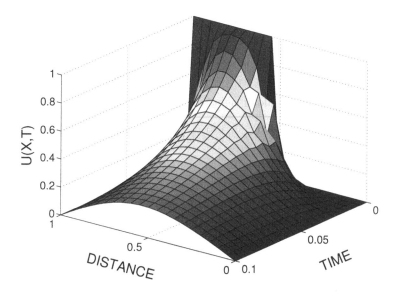

Figure 9.5.6: The numerical solution $u(x, t)$ of the heat equation $u_t = a^2 u_{xx}$ using the Crank-Nicholson method. The parameters used in the numerical solution are $a^2 \Delta t = 0.005$ and $\Delta x = 0.05$. Both ends are held at zero with an initial condition of $u(x, 0) = 0$ for $0 \leq x < \frac{1}{2}$, and $u(x, 0) = 1$ for $\frac{1}{2} < x \leq 1$.

Step 2: Test your code by solving the heat equation given the initial condition $u(x, 0) = \sin(\pi x)$, and the boundary conditions $u(0, t) = u(1, t) = 0$. Find the solution for various values of Δt with $\Delta x = 0.01$. Compare this numerical solution against the exact solution that you can find. How does the error (between the numerical and exact solutions) change with Δt? For small Δt, the errors should be small. If not, then you have a mistake in your code.

Step 3: Once you have confidence in your code, discuss the behavior of the scheme for various values of Δx and Δt for the initial condition $u(x, 0) = 0$ for $0 \leq x < \frac{1}{2}$, and $u(x, 0) = 1$ for $\frac{1}{2} < x \leq 1$ with the boundary conditions $u(0, t) = u(1, t) = 0$. See Figure 9.5.6. Although you can take quite a large Δt, what happens? Did a similar problem arise in Step 2? Explain your results.[16] Zvan et al.[17] have reported a similar problem in the numerical integration of the Black-Scholes equation (another parabolic partial differential equation) from mathematical finance.

Project: Saulyev's Explicit Methods for the Heat Equation

In 1957 V. K. Saulyev[18] suggested that Equation 9.5.3 be replaced by

$$\frac{\partial^2 u(x_m, t_n)}{\partial x^2} = \frac{u_{m+1}^n - u_m^n - u_m^{n+1} + u_{m-1}^{n+1}}{(\Delta x)^2},$$

[16] Luskin, M., and R. Rannacher, 1982: On the smoothing property of the Crank-Nicolson scheme. *Applicable Anal.*, **14**, 117–135.

[17] Zvan, R., K. Vetzal, and P. Forsyth, 1998: Swing low, swing high. *Risk*, **11(3)**, 71–75.

[18] Saulyev, V. K., 1964: *Integration of Equations of Parabolic Type by the Method of Nets*. Pergamon Press, 365 pp.

if we have a Dirichlet condition along left side, or

$$\frac{\partial^2 u(x_m, t_n)}{\partial x^2} = \frac{u_{m+1}^{n+1} - u_m^{n+1} - u_m^n + u_{m-1}^n}{(\Delta x)^2},$$

if we have a Dirichlet condition along right side. Therefore, Equation 9.5.4 becomes

$$(1+\theta)u_m^{n+1} = \theta u_{m-1}^{n+1} + (1-\theta)u_m^n + \theta u_{m+1}^n$$

in the first case, and

$$(1+\theta)u_m^{n+1} = \theta u_{m+1}^{n+1} + (1-\theta)u_m^n + \theta u_{m-1}^n$$

in the second case. Here $\theta = a^2 \Delta t/(\Delta x)^2$. These are actually *explicit* schemes since the value at u_{m-1}^{n+1} or u_{m+1}^{n+1} is either known from the boundary condition or has just been computed. It can be shown that[19] (1) these schemes are unconditionally stable and (2) consistency requires that Δt tends to 0 faster than Δx.

Step 1: Using Example 9.3.1, compare the exact solution Equation 9.3.25 with the numerical solution given by Saulyev's schemes when you scan from left to right as a function of θ. In particular, examine the relative error as a function of position and time.

Step 2: Redo Step 1 but now scan from right to left.

Step 3: Saulyev himself did not advise (on page 29 of his book) just employing one method or the other, but suggested using them alternatively, such as one for the odd (time) steps and the other in the even (time) steps. Take his advice and test out his suggestion by redoing Step 1 using his suggestion.

**Project: Numerical Solution of the Heat Equation
with a Non-Local Boundary Condition**

In the solution of the heat equation in this chapter, we have always had two boundary conditions, one at $x = 0$ and one at $x = L$. However, in the design of photoelectric cells, the electric signal generated in the cell is proportional to $\int_0^b u(x,t)\,dx$, where $u(x,t)$ denotes the concentration of the chemical present at the location x and time t. The concentration, in turn, depends upon the diffusion of the chemical as a light beam passes through the tube at right angles between $x = 0$ and $x = b$.

To model this concentration we must solve the diffusion equation

$$\frac{\partial u}{\partial t} = \frac{\partial^2 u}{\partial x^2}, \qquad 0 < x < L, \quad 0 < t,$$

subject to the constraints that

$$\int_0^b u(x,t)\,dx = M(t), \qquad \frac{\partial u(L,t)}{\partial x} = g(t), \qquad 0 < t,$$

---

[19] Østerby, O., 2017: On Saulyev's methods. DAIMI Report Series, 43(599), 11 pp. http: tidsskrift.dk /daimipb/article/view/26410/23231

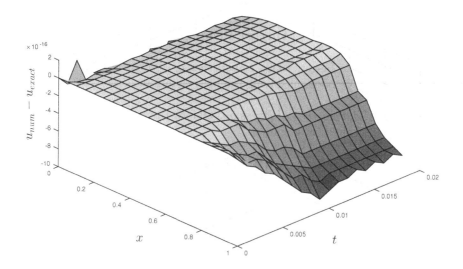

Figure 9.5.7: The error between a numerical and exact solution for a heat equation with a non-local boundary condition as a function of position x and time t. Here $L = 1$, $b = 1/2$, $\Delta x = 0.05$, and $\Delta t = 0.00025$ so that $\theta = 0.1$.

where $0 < b < L$. Of course, we have the initial condition that

$$u(x, 0) = f(x), \qquad 0 < x < L.$$

The integral constraint is referred to as a *non-local boundary condition*.

To keep this project simple, let us integrate the diffusion equation using an explicit time differencing scheme. Assuming the $x_i = i\Delta x$ and $t_n = n\Delta t$, where $i = 0, 1, 2, \ldots, M$ and $n = 0, 1, 2, \ldots$, we have that

$$u_i^{n+1} = u_i^n + \theta(u_{i+1}^n - 2u_i^n + u_{i-1}^n), \tag{9.5.23}$$

with $i = 1, 2, 3, \ldots, M - 1$, $n = 0, 1, 2, \ldots$ and $\Delta x = L/M$. Here $\theta = \Delta t/(\Delta x)^2$.

At $x = L$, we have that

$$\frac{u_{M+1}^n - u_{M-1}^n}{2\Delta x} = g(t_n).$$

This can be combined with the heat equation (Equation 9.5.23) to yield a predictive equation for the grid point $i = M$.

The interesting aspect of this project is how to predict the value of u_0^{n+1}. Employing Simpson's rule, the integral condition gives

$$u_0^{n+1} + 4u_1^{n+1} + 2u_2^{n+1} + 4u_3^{n+1} + \cdots + 4u_{J-1}^{n+1} + u_J^{n+1} = \frac{\Delta x}{3} M(t_{n+1}),$$

where $J = b/\Delta x$ and must be an even integer. In this scheme we first update the grid points from $i = 1$ to M and then use this equation to give u_0^{n+1}.

Step 1: Using this scheme, create a code to compute u_i^n. Compare this scheme against the exact solution $u(x, t) = t + x^2/2$, $g(t) = 1$ and $M(t) = bt + b^3/6$. Figure 9.5.7 shows the error between the numerical solution and the exact solutions.

Step 2: Simpson's rule is not the only choice for an integration scheme. Redo Step 1 but use the trapezoidal rule.

Further Readings

Carslaw, H. S., and J. C. Jaeger, 1959: *Conduction of Heat in Solids.* Oxford University Press, 510 pp. The source book on solving the heat equation.

Crank, J., 1970: *The Mathematics of Diffusion.* Oxford University Press, 347 pp. A source book on the solution of the heat equation.

Koshlyakov, N. S., M. M. Smirnov, and E. B. Gliner, 1964: *Differential Equations of Mathematical Physics.* North-Holland Publishing, 701 pp. See Part III. Nice presentation of mathematical techniques.

Morse, P. M., and H. Feshback, 1953: *Methods of Theoretical Physics.* McGraw-Hill Book Co., 997 pp. A portion of Chapter 12 is devoted to solving the heat equation.

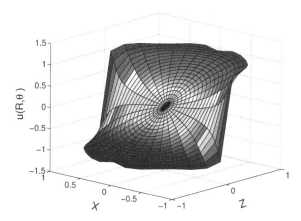

Chapter 10

Laplace's Equation

In the previous chapter we solved the one-dimensional heat equation. Quite often we found that the transient solution died away, leaving a steady state. The partial differential equation that describes the steady state for two-dimensional heat conduction is Laplace's equation

$$\frac{\partial^2 u}{\partial x^2} + \frac{\partial^2 u}{\partial y^2} = 0. \tag{10.0.1}$$

In general, this equation governs physical processes where *equilibrium* has been reached. It also serves as the prototype for a wider class of *elliptic equations*

$$a(x,t)\frac{\partial^2 u}{\partial x^2} + b(x,t)\frac{\partial^2 u}{\partial x \partial t} + c(x,t)\frac{\partial^2 u}{\partial t^2} = f\left(x,t,u,\frac{\partial u}{\partial x},\frac{\partial u}{\partial t}\right), \tag{10.0.2}$$

where $b^2 < 4ac$. Unlike the heat and wave equations, there are no initial conditions and the boundary conditions completely specify the solution. In this chapter we present some of the common techniques for solving this equation.

10.1 DERIVATION OF LAPLACE'S EQUATION

Imagine a thin, flat plate of heat-conducting material between two sheets of insulation. Sufficient time has passed so that the temperature depends only on the spatial coordinates x and y. Let us now apply the law of conservation of energy (in rate form) to a small rectangle with sides Δx and Δy.

If $q_x(x, y)$ and $q_y(x, y)$ denote the heat flow rates in the x- and y-direction, respectively, conservation of energy requires that the heat flow into the slab equals the heat flow out of the slab if there is no storage or generation of heat. Now

$$\text{rate in} = q_x(x, y + \Delta y/2)\Delta y + q_y(x + \Delta x/2, y)\Delta x, \qquad (10.1.1)$$

and

$$\text{rate out} = q_x(x + \Delta x, y + \Delta y/2)\Delta y + q_y(x + \Delta x/2, y + \Delta y)\Delta x. \qquad (10.1.2)$$

If the plate has unit thickness,

$$[q_x(x, y + \Delta y/2) - q_x(x + \Delta x, y + \Delta y/2)]\Delta y$$
$$+ [q_y(x + \Delta x/2, y) - q_y(x + \Delta x/2, y + \Delta y)]\Delta x = 0. \qquad (10.1.3)$$

Upon dividing through by $\Delta x \Delta y$, we obtain two differences quotients on the left side of Equation 10.1.3. In the limit as $\Delta x, \Delta y \to 0$, they become partial derivatives, giving

$$\frac{\partial q_x}{\partial x} + \frac{\partial q_y}{\partial y} = 0 \qquad (10.1.4)$$

for any point (x, y).

We now employ Fourier's law to eliminate the rates q_x and q_y, yielding

$$\frac{\partial}{\partial x}\left(a^2 \frac{\partial u}{\partial x}\right) + \frac{\partial}{\partial y}\left(a^2 \frac{\partial u}{\partial y}\right) = 0, \qquad (10.1.5)$$

if we have an isotropic (same in all directions) material. Finally, if a^2 is constant, Equation 10.1.5 reduces to

$$\frac{\partial^2 u}{\partial x^2} + \frac{\partial^2 u}{\partial y^2} = 0, \qquad (10.1.6)$$

which is the two-dimensional, steady-state heat equation (i.e., $u_t \approx 0$ as $t \to \infty$).

Solutions of Laplace's equation (called *harmonic functions*) differ fundamentally from those encountered with the heat and wave equations. These latter two equations describe the evolution of some phenomena. Laplace's equation, on the other hand, describes things at equilibrium. Consequently, any change in the boundary conditions affects to some degree the *entire* domain because a change to any one point causes its neighbors to change in order to reestablish the equilibrium. Those points will, in turn, affect others. Because all of these points are in equilibrium, this modification must occur instantaneously.

Further insight follows from the *maximum principle*. If Laplace's equation governs a region, then its solution cannot have a relative maximum or minimum *inside* the region unless the solution is constant.[1] If we think of the solution as a steady-state temperature distribution, this principle is clearly true because at any one point, the temperature cannot be greater than at all other nearby points. If that were so, heat would flow away from the hot point to cooler points nearby, thus eliminating the hot spot when equilibrium was once again restored.

It is often useful to consider the two-dimensional Laplace's equation in other coordinate systems. In polar coordinates, where $x = r\cos(\theta)$, $y = r\sin(\theta)$, and $z = z$, Laplace's equation becomes

$$\frac{\partial^2 u}{\partial r^2} + \frac{1}{r}\frac{\partial u}{\partial r} + \frac{\partial^2 u}{\partial z^2} = 0, \qquad (10.1.7)$$

---

[1] For the proof, see Courant, R., and D. Hilbert, 1962: *Methods of Mathematical Physics, Vol. 2: Partial Differential Equations.* Interscience, pp. 326–331.

Today we best remember Pierre-Simon Laplace (1749–1827) for his work in celestial mechanics and probability. In his five volumes *Traité de Mécanique céleste* (1799–1825), he accounted for the theoretical orbits of the planets and their satellites. Laplace's equation arose during this study of gravitational attraction. (Portrait courtesy of the Archives de l'Académie des sciences, Paris.)

if the problem possesses axisymmetry. On the other hand, if the solution is independent of z, Laplace's equation becomes

$$\frac{\partial^2 u}{\partial r^2} + \frac{1}{r}\frac{\partial u}{\partial r} + \frac{1}{r^2}\frac{\partial^2 u}{\partial \theta^2} = 0. \tag{10.1.8}$$

In spherical coordinates, $x = r\cos(\varphi)\sin(\theta)$, $y = r\sin(\varphi)\sin(\theta)$, and $z = r\cos(\theta)$, where $r^2 = x^2 + y^2 + z^2$, θ is the angle measured *down* to the point from the z-axis (colatitude) and φ is the angle made between the x-axis and the projection of the point on the xy plane. In the case of axisymmetry (no φ dependence), Laplace's equation becomes

$$\frac{\partial}{\partial r}\left(r^2\frac{\partial u}{\partial r}\right) + \frac{1}{\sin(\theta)}\frac{\partial}{\partial \theta}\left[\sin(\theta)\frac{\partial u}{\partial \theta}\right] = 0. \tag{10.1.9}$$

10.2 BOUNDARY CONDITIONS

Because Laplace's equation involves time-independent phenomena, we must only specify boundary conditions. As we discussed in Section 9.2, we can classify these boundary conditions as follows:

1. Dirichlet condition: u given

2. Neumann condition: $\dfrac{\partial u}{\partial n}$ given, where n is the unit normal direction

3. Robin condition: $u + \alpha\dfrac{\partial u}{\partial n}$ given

along any section of the boundary. In this chapter we will deal with Dirichlet and Neumann boundary conditions; problems with Robin conditions will be deferred to Chapter 11 because we must first study the Sturm-Liouville problem. In the case of Laplace's equation, if all of the boundaries have Neumann conditions, then the solution is not unique. This follows from the fact that if $u(x, y)$ is a solution, so is $u(x, y) + c$, where c is any constant.

Finally we note that we must specify the boundary conditions along each side of the boundary. These sides may be at infinity as in problems with semi-infinite domains. We must specify values along the entire boundary because we could not have an equilibrium solution if any portion of the domain was undetermined.

10.3 SEPARATION OF VARIABLES

As in the case of the heat and wave equations, separation of variables is the most popular technique for solving Laplace's equation. Although the same general procedure carries over from the previous two chapters, the following examples fill out the details.

• Example 10.3.1: Groundwater flow in a valley

Over a century ago, a French hydraulic engineer named Henri-Philibert-Gaspard Darcy (1803–1858) published the results of a laboratory experiment on the flow of water through sand. He showed that the *apparent* fluid velocity $\mathbf{q}$ relative to the sand grains is directly proportional to the gradient of the hydraulic potential $-k\nabla\varphi$, where the hydraulic potential φ equals the sum of the elevation of the point of measurement plus the pressure potential $(p/\rho g)$. In the case of steady flow, the combination of Darcy's law with conservation of mass $\nabla \cdot \mathbf{q} = 0$ yields Laplace's equation $\nabla^2\varphi = 0$ if the aquifer is isotropic (same in all directions) and homogeneous.

To illustrate how separation of variables can be used to solve Laplace's equation, we will determine the hydraulic potential within a small drainage basin that lies in a shallow valley. See Figure 10.3.1. Following Tóth,[2] the governing equation is the two-dimensional Laplace equation

$$\frac{\partial^2 u}{\partial x^2} + \frac{\partial^2 u}{\partial y^2} = 0, \quad 0 < x < L, \quad 0 < y < z_0, \tag{10.3.1}$$

along with the boundary conditions

$$u(x, z_0) = gz_0 + gcx, \tag{10.3.2}$$

$$u_x(0, y) = u_x(L, y) = 0, \quad \text{and} \quad u_y(x, 0) = 0, \tag{10.3.3}$$

where $u(x, y)$ is the hydraulic potential, g is the acceleration due to gravity, and c gives the slope of the topography. The conditions $u_x(L, y) = 0$, and $u_y(x, 0) = 0$ specify a no-flow condition through the bottom and sides of the aquifer. The condition $u_x(0, y) = 0$ ensures symmetry about the $x = 0$ line. Equation 10.3.1 gives the fluid potential at the water table,

    [2] Tóth, J., 1962: A theory of groundwater motion in small drainage basins in central Alberta, Canada. *J. Geophys. Res.*, **67**, 4375–4387.

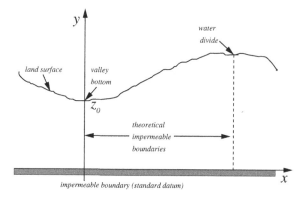

Figure 10.3.1: Cross section of a valley.

where z_0 is the elevation of the water table above the standard datum. The term gcx in Equation 10.3.2 expresses the increase of the potential from the valley bottom toward the water divide. On average it closely follows the topography.

Following the pattern set in the previous two chapters, we assume that $u(x, y) = X(x)Y(y)$. Then Equation 10.3.1 becomes

$$X''Y + XY'' = 0. \tag{10.3.4}$$

Separating the variables yields

$$\frac{X''}{X} = -\frac{Y''}{Y}. \tag{10.3.5}$$

Both sides of Equation 10.3.5 must be constant, but the sign of that constant is not obvious. From previous experience we anticipate that the ordinary differential equation in the x-direction leads to a Sturm-Liouville problem because it possesses homogeneous boundary conditions. Proceeding along this line of reasoning, we consider three separation constants.

Trying a positive constant (say, m^2), Equation 10.3.5 separates into the two ordinary differential equations

$$X'' - m^2 X = 0, \qquad \text{and} \qquad Y'' + m^2 Y = 0, \tag{10.3.6}$$

which have the solutions

$$X(x) = A \cosh(mx) + B \sinh(mx), \tag{10.3.7}$$

and

$$Y(y) = C \cos(my) + D \sin(my). \tag{10.3.8}$$

Because the boundary conditions, Equation 10.3.3, imply $X'(0) = X'(L) = 0$, both A and B must be zero, leading to the trivial solution $u(x, y) = 0$.

When the separation constant equals zero, we find a nontrivial solution given by $X_0(x) = 1$, and $Y_0(y) = \frac{1}{2}A_0 + B_0 y$. However, because $Y_0'(0) = 0$ from Equation 10.3.3, $B_0 = 0$. Thus, the particular solution for a zero separation constant is $u_0(x, y) = A_0/2$.

Finally, taking both sides of Equation 10.3.5 equal to $-k^2$,

$$X'' + k^2 X = 0, \qquad \text{and} \qquad Y'' - k^2 Y = 0. \tag{10.3.9}$$

The first of these equations, along with the boundary conditions $X'(0) = X'(L) = 0$, gives $X_n(x) = \cos(k_n x)$, with $k_n = n\pi/L$, $n = 1, 2, 3, \ldots$. The function $Y_n(y)$ for the same separation constant is

$$Y_n(y) = A_n \cosh(k_n y) + B_n \sinh(k_n y). \tag{10.3.10}$$

We must take $B_n = 0$ because $Y'_n(0) = 0$.

We now have the product solution $X_n(x)Y_n(y)$, which satisfies Laplace's equation and all of the boundary conditions except Equation 10.3.2. By the principle of superposition, which we introduced in Section 8.3, the general solution is

$$u(x, y) = \frac{A_0}{2} + \sum_{n=1}^{\infty} A_n \cos\left(\frac{n\pi x}{L}\right) \cosh\left(\frac{n\pi y}{L}\right). \tag{10.3.11}$$

Applying this equation, we find that

$$u(x, z_0) = gz_0 + gcx = \frac{A_0}{2} + \sum_{n=1}^{\infty} A_n \cos\left(\frac{n\pi x}{L}\right) \cosh\left(\frac{n\pi z_0}{L}\right), \tag{10.3.12}$$

which we recognize as a Fourier half-range cosine series such that

$$A_0 = \frac{2}{L} \int_0^L (gz_0 + gcx)\, dx, \tag{10.3.13}$$

and

$$\cosh\left(\frac{n\pi z_0}{L}\right) A_n = \frac{2}{L} \int_0^L (gz_0 + gcx) \cos\left(\frac{n\pi x}{L}\right) dx. \tag{10.3.14}$$

Performing the integrations,

$$A_0 = 2gz_0 + gcL, \tag{10.3.15}$$

and

$$A_n = -\frac{2gcL[1 - (-1)^n]}{n^2\pi^2 \cosh(n\pi z_0/L)}. \tag{10.3.16}$$

Finally, the complete solution is

$$u(x, y) = gz_0 + \frac{gcL}{2} - \frac{4gcL}{\pi^2} \sum_{m=1}^{\infty} \frac{\cos[(2m-1)\pi x/L] \cosh[(2m-1)\pi y/L]}{(2m-1)^2 \cosh[(2m-1)\pi z_0/L]}. \tag{10.3.17}$$

Figure 10.3.2 presents two graphs by Tóth for two different aquifers. We see that the solution satisfies the boundary condition at the bottom and side boundaries. Water flows from the elevated land (on the right) into the valley (on the left), from regions of high to low hydraulic potential. □

● **Example 10.3.2**

In the previous example, we had the advantage of homogeneous boundary conditions along $x = 0$ and $x = L$. In a different hydraulic problem, Kirkham[3] solved the more difficult problem of

$$\frac{\partial^2 u}{\partial x^2} + \frac{\partial^2 u}{\partial y^2} = 0, \quad 0 < x < L, \quad 0 < y < h, \tag{10.3.18}$$

[3] Kirkham, D., 1958: Seepage of steady rainfall through soil into drains. *Trans. Am. Geophys. Union*, **39**, 892–908.

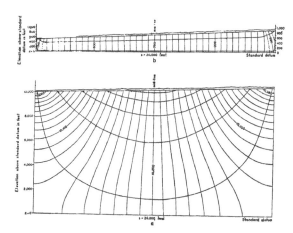

Figure 10.3.2: Two-dimensional potential distribution and flow patterns for different depths of the horizontally impermeable boundary.

subject to the Dirichlet boundary conditions and

$$u(0, y) = \begin{cases} 0, & 0 < y < a, \\ \frac{RL}{b-a}(y - a), & a < y < b, \\ RL, & b < y < h. \end{cases} \tag{10.3.20}$$

This problem arises in finding the steady flow within an aquifer resulting from the introduction of water at the top due to a steady rainfall and its removal along the sides by drains. The parameter L equals half of the distance between the drains, h is the depth of the aquifer, and R is the rate of rainfall.

The point of this example is: *We need homogeneous boundary conditions along either the x or y boundaries for separation of variables to work.* We achieve this by breaking the original problem into two parts, namely

$$u(x, y) = v(x, y) + w(x, y) + RL, \tag{10.3.21}$$

where

$$\frac{\partial^2 v}{\partial x^2} + \frac{\partial^2 v}{\partial y^2} = 0, \quad 0 < x < L, \quad 0 < y < h, \tag{10.3.22}$$

with

$$v(0, y) = v(L, y) = 0, \qquad v(x, h) = 0, \tag{10.3.23}$$

and

$$v(x, 0) = R(x - L); \tag{10.3.24}$$

$$\frac{\partial^2 w}{\partial x^2} + \frac{\partial^2 w}{\partial y^2} = 0, \quad 0 < x < L, \quad 0 < y < h, \tag{10.3.25}$$

with

$$w(x, 0) = w(x, h) = 0, \qquad w(L, y) = 0, \tag{10.3.26}$$

and

$$w(0, y) = \begin{cases} -RL, & 0 < y < a, \\ \frac{RL}{b-a}(y - a) - RL, & a < y < b, \\ 0, & b < y < h. \end{cases} \tag{10.3.27}$$

Employing the same technique as in Example 10.3.1, we find that

$$v(x,y) = \sum_{n=1}^{\infty} A_n \sin\left(\frac{n\pi x}{L}\right) \frac{\sinh[n\pi(h-y)/L]}{\sinh(n\pi h/L)}, \tag{10.3.28}$$

where

$$A_n = \frac{2}{L} \int_0^L R(x-L)\sin\left(\frac{n\pi x}{L}\right)\,dx = -\frac{2RL}{n\pi}. \tag{10.3.29}$$

Similarly, the solution to $w(x,y)$ is found to be

$$w(x,y) = \sum_{n=1}^{\infty} B_n \sin\left(\frac{n\pi y}{h}\right) \frac{\sinh[n\pi(L-x)/h]}{\sinh(n\pi L/h)}, \tag{10.3.30}$$

where

$$B_n = \frac{2}{h}\left[-RL\int_0^a \sin\left(\frac{n\pi y}{h}\right)\,dy + RL\int_a^b \left(\frac{y-a}{b-a}-1\right)\sin\left(\frac{n\pi y}{h}\right)\,dy\right] \tag{10.3.31}$$

$$= \frac{2RL}{\pi}\left\{\frac{h}{(b-a)n^2\pi}\left[\sin\left(\frac{n\pi b}{h}\right)-\sin\left(\frac{n\pi a}{h}\right)\right]-\frac{1}{n}\right\}. \tag{10.3.32}$$

The complete solution consists of substituting Equation 10.3.28 and Equation 10.3.30 into Equation 10.3.21. □

• Example 10.3.3: Poisson's integral formula

In this example we find the solution to Laplace's equation within a unit disc. The problem can be posed as

$$\frac{\partial^2 u}{\partial r^2} + \frac{1}{r}\frac{\partial u}{\partial r} + \frac{1}{r^2}\frac{\partial^2 u}{\partial \varphi^2} = 0, \quad 0 \le r < 1, \quad 0 \le \varphi \le 2\pi, \tag{10.3.33}$$

with the boundary condition $u(1,\varphi) = f(\varphi)$.

We begin by assuming the separable solution $u(r,\varphi) = R(r)\Phi(\varphi)$ so that

$$\frac{r^2 R'' + rR'}{R} = -\frac{\Phi''}{\Phi} = k^2. \tag{10.3.34}$$

The solution to $\Phi'' + k^2\Phi = 0$ is

$$\Phi(\varphi) = A\cos(k\varphi) + B\sin(k\varphi). \tag{10.3.35}$$

The solution to $R(r)$ is

$$R(r) = Cr^k + Dr^{-k}. \tag{10.3.36}$$

Because the solution must be bounded for all r and periodic in φ, we must take $D = 0$ and $k = n$, where $n = 0, 1, 2, 3, \ldots$. Then, the most general solution is

$$u(r,\varphi) = \tfrac{1}{2}a_0 + \sum_{n=1}^{\infty} [a_n\cos(n\varphi) + b_n\sin(n\varphi)]\, r^n, \tag{10.3.37}$$

where a_n and b_n are chosen to satisfy

$$u(1, \varphi) = f(\varphi) = \tfrac{1}{2}a_0 + \sum_{n=1}^{\infty} a_n \cos(n\varphi) + b_n \sin(n\varphi). \tag{10.3.38}$$

Because

$$a_n = \frac{1}{\pi} \int_{-\pi}^{\pi} f(\theta) \cos(n\theta)\, d\theta, \quad b_n = \frac{1}{\pi} \int_{-\pi}^{\pi} f(\theta) \sin(n\theta)\, d\theta, \tag{10.3.39}$$

we may write $u(r, \varphi)$ as

$$u(r, \varphi) = \frac{1}{\pi} \int_{-\pi}^{\pi} f(\theta) \left\{ \tfrac{1}{2} + \sum_{n=1}^{\infty} r^n \cos[n(\theta - \varphi)] \right\} d\theta. \tag{10.3.40}$$

If we let $\alpha = \theta - \varphi$, and $z = r[\cos(\alpha) + i\sin(\alpha)]$, then

$$\sum_{n=0}^{\infty} r^n \cos(n\alpha) = \Re\left(\sum_{n=0}^{\infty} z^n \right) = \Re\left(\frac{1}{1-z} \right) = \Re\left[\frac{1}{1 - r\cos(\alpha) - ir\sin(\alpha)} \right] \tag{10.3.41}$$

$$= \Re\left[\frac{1 - r\cos(\alpha) + ir\sin(\alpha)}{1 - 2r\cos(\alpha) + r^2} \right] \tag{10.3.42}$$

for all r such that $|r| < 1$. Consequently,

$$\sum_{n=0}^{\infty} r^n \cos(n\alpha) = \frac{1 - r\cos(\alpha)}{1 - 2r\cos(\alpha) + r^2} \tag{10.3.43}$$

$$\frac{1}{2} + \sum_{n=1}^{\infty} r^n \cos(n\alpha) = \frac{1 - r\cos(\alpha)}{1 - 2r\cos(\alpha) + r^2} - \frac{1}{2} \tag{10.3.44}$$

$$= \frac{1}{2} \frac{1 - r^2}{1 - 2r\cos(\alpha) + r^2}. \tag{10.3.45}$$

Substituting Equation 10.3.45 into Equation 10.3.40, we finally have that

$$u(r, \varphi) = \frac{1}{2\pi} \int_{-\pi}^{\pi} f(\theta) \frac{1 - r^2}{1 - 2r\cos(\theta - \varphi) + r^2}\, d\theta. \tag{10.3.46}$$

This solution to Laplace's equation within the unit circle is referred to as *Poisson's integral formula*.[4]

---

[4] Poisson, S. D., 1820: Mémoire sur la manière d'exprimer les fonctions par des séries de quantités périodiques, et sur l'usage de cette transformation dans la résolution de différens problèmes. *J. École Polytech.*, **18**, 417–489.

Problems

Solve Laplace's equation over the rectangular region $0 < x < a, 0 < y < b$ with the following boundary conditions. Illustrate your solution using MATLAB.

1. $u(x,0) = u(x,b) = u(a,y) = 0, u(0,y) = 1$

2. $u(x,0) = u(0,y) = u(a,y) = 0, u(x,b) = x$

3. $u(x,0) = u(0,y) = u(a,y) = 0, u(x,b) = x - a$

4. $u(x,0) = u(0,y) = u(a,y) = 0,$

$$u(x,b) = \begin{cases} 2x/a, & 0 < x < a/2, \\ 2(a-x)/a, & a/2 < x < a. \end{cases}$$

5. $u_y(x,0) = u_y(x,b) = 0, \ u(0,y) = u(a,y) = 1$

6. $u_y(x,0) = u(x,b) = 0, \ u(0,y) = u(a,y) = 1$ Hint: Set $u(x,y) = 1 + v(x,y)$.

7. $u_x(0,y) = 0, \ u(a,y) = u(x,0) = u(x,b) = 1$ Hint: Set $u(x,y) = 1 + v(x,y)$.

8. $u(a,y) = u(x,b) = 0, \ u(0,y) = u(x,0) = 1$ Hint: Set $u(x,y) = v(x,y) + w(x,y)$ with $v(0,y) = v(a,y) = v(x,b) = 0, v(x,0) = 1$ and $w(x,0) = w(x,b) = w(a,y) = 0, w(0,y) = 1$.

9. $u_x(0,y) = u_x(a,y) = 0, u(x,b) = u_1,$

$$u(x,0) = \begin{cases} f(x), & 0 < x < \alpha, \\ 0, & \alpha < x < a. \end{cases} \quad \text{Hint: Set } u(x,y) = u_1 + v(x,y).$$

10. Variations in the earth's surface temperature can arise as a result of topographic undulations and the altitude dependence of the atmospheric temperature. These variations, in turn, affect the temperature within the solid earth. To show this, solve Laplace's equation with the surface boundary condition that

$$u(x,0) = T_0 + \Delta T \cos(2\pi x/\lambda),$$

where λ is the wavelength of the spatial temperature variation. What must be the condition on $u(x,y)$ as we go towards the center of the earth (i.e., $y \to \infty$)?

11. Tóth[5] generalized his earlier analysis of groundwater in an aquifer when the water table follows the topography. Find the groundwater potential if it varies as $u(x,z_0) = g[z_0 + cx + a\sin(bx)]$ at the surface $y = z_0$, while $u_x(0,y) = u_x(L,y) = u_y(x,0) = 0$, where g is the acceleration due to gravity. Assume that $bL \neq n\pi$, where $n = 1, 2, 3, \ldots$.

12. During his study of fluid flow within a packed bed, Grossman[6] solved

$$\frac{\partial^2 u}{\partial x^2} + \frac{\partial^2 u}{\partial y^2} = 0, \quad 0 < x < 1, \quad 0 < y < L,$$

[5] Tóth, J. A., 1963: A theoretical analysis of groundwater flow in small drainage basins. *J. Geophys. Res.*, **68**, 4795–4812.

[6] Grossman, G., 1975: Stresses and friction forces in moving packed beds. *AICHE J.*, **21**, 720–730.

subject to the boundary conditions $u(x,0) = L$, $u(x,L) = 0$, $0 < x < 1$, and $u_x(0,y) = 0$, $u_x(1,y) = -\gamma$, $0 < y < L$. What should he have found? Hint: Introduce $u(x,y) = L - y + \gamma v(x,y)$.

Poisson's Integral Formula

13. Using the relationship

$$\int_0^{2\pi} \frac{d\varphi}{1 - b\,\cos(\varphi)} = \frac{2\pi}{\sqrt{1 - b^2}}, \qquad |b| < 1$$

and Poisson's integral formula, find the solution to Laplace's equation within a unit disc if $u(1,\varphi) = f(\varphi) = T_0$, a constant.

10.4 POISSON'S EQUATION ON A RECTANGLE

Poisson's equation[7] is Laplace's equation with a source term:

$$\frac{\partial^2 u}{\partial x^2} + \frac{\partial^2 u}{\partial y^2} = f(x,y). \tag{10.4.1}$$

It arises in such diverse areas as groundwater flow, electromagnetism, and potential theory. Let us solve it if $u(0,y) = u(a,y) = u(x,0) = u(x,b) = 0$.

We begin by solving a similar partial differential equation:

$$\frac{\partial^2 u}{\partial x^2} + \frac{\partial^2 u}{\partial y^2} = \lambda u, \quad 0 < x < a, \quad 0 < y < b, \tag{10.4.2}$$

by separation of variables. If $u(x,y) = X(x)Y(y)$, then

$$\frac{X''}{X} + \frac{Y''}{Y} = \lambda. \tag{10.4.3}$$

Because we must satisfy the boundary conditions that $X(0) = X(a) = Y(0) = Y(b) = 0$, we have the following solutions:

$$X_n(x) = \sin\left(\frac{n\pi x}{a}\right), \qquad Y_m(x) = \sin\left(\frac{m\pi y}{b}\right) \tag{10.4.4}$$

with $\lambda_{nm} = -n^2\pi^2/a^2 - m^2\pi^2/b^2$; otherwise, we would only have trivial solutions. The corresponding particular solutions are

$$u_{nm} = A_{nm} \sin\left(\frac{n\pi x}{a}\right) \sin\left(\frac{m\pi y}{b}\right), \tag{10.4.5}$$

---

[7] Poisson, S. D., 1813: Remarques sur une équation qui se présente dans la théorie des attractions des sphéroïdes. *Nouv. Bull. Soc. Philomath. Paris*, **3**, 388–392.

Siméon-Denis Poisson (1781–1840) was a product as well as a member of the French scientific establishment of his day. Educated at the École Polytechnique, he devoted his life to teaching, both in the classroom and with administrative duties, and to scientific research. Poisson's equation dates from 1813 when Poisson sought to extend Laplace's work on gravitational attraction. (Portrait courtesy of the Archives de l'Académie des sciences, Paris.)

where $n = 1, 2, 3, \ldots$, and $m = 1, 2, 3, \ldots$.

For a fixed y, we can expand $f(x, y)$ in the half-range Fourier sine series

$$f(x, y) = \sum_{n=1}^{\infty} A_n(y) \sin\left(\frac{n\pi x}{a}\right), \qquad (10.4.6)$$

where

$$A_n(y) = \frac{2}{a} \int_0^a f(x, y) \sin\left(\frac{n\pi x}{a}\right) \, dx. \qquad (10.4.7)$$

However, we can also expand $A_n(y)$ in a half-range Fourier sine series

$$A_n(y) = \sum_{m=1}^{\infty} a_{nm} \sin\left(\frac{m\pi y}{b}\right), \qquad (10.4.8)$$

where

$$a_{nm} = \frac{2}{b} \int_0^b A_n(y) \sin\left(\frac{m\pi y}{b}\right) \, dy = \frac{4}{ab} \int_0^b \int_0^a f(x, y) \sin\left(\frac{n\pi x}{a}\right) \sin\left(\frac{m\pi y}{b}\right) \, dx \, dy, \qquad (10.4.9)$$

and

$$f(x,y) = \sum_{n=1}^{\infty} \sum_{m=1}^{\infty} a_{nm} \sin\left(\frac{n\pi x}{a}\right) \sin\left(\frac{m\pi y}{b}\right). \tag{10.4.10}$$

In other words, we re-expressed $f(x,y)$ in terms of a *double Fourier series*.
Because Equation 10.4.2 must hold for each particular solution,

$$\frac{\partial^2 u_{nm}}{\partial x^2} + \frac{\partial^2 u_{nm}}{\partial y^2} = \lambda_{nm} u_{nm} = a_{nm} \sin\left(\frac{n\pi x}{a}\right) \sin\left(\frac{m\pi y}{b}\right), \tag{10.4.11}$$

if we now associate Equation 10.4.1 with Equation 10.4.2. Therefore, the solution to Poisson's equation on a rectangle where the boundaries are held at zero is the double Fourier series

$$u(x,y) = -\sum_{n=1}^{\infty} \sum_{m=1}^{\infty} \frac{a_{nm}}{n^2\pi^2/a^2 + m^2\pi^2/b^2} \sin\left(\frac{n\pi x}{a}\right) \sin\left(\frac{m\pi y}{b}\right). \tag{10.4.12}$$

Problems

1. The equation

$$\frac{\partial^2 u}{\partial x^2} + \frac{\partial^2 u}{\partial y^2} = -\frac{R}{T}, \quad 0 < x < a, \quad 0 < y < b,$$

describes the hydraulic potential (elevation of the water table) $u(x,y)$ within a rectangular island on which a recharging well is located at $(a/2, b/2)$. Here R is the rate of recharging and T is the product of the hydraulic conductivity and aquifer thickness. If the water table is at sea level around the island so that $u(0,y) = u(a,y) = u(x,0) = u(x,b) = 0$, find $u(x,y)$ everywhere in the island.

2. Solve

$$\frac{\partial^2 u}{\partial x^2} + \frac{\partial^2 u}{\partial y^2} = e^{2y} \sin(x), \quad 0 < x < \pi, \quad 0 < y < H,$$

subject to the boundary conditions that (1) $u(x,0) = 0$, $u(x,H) = f(x)$ for $0 < x < \pi$ and (2) $u(0,y) = u(\pi,y) = 0$ for $0 < y < H$.

Step 1: Setting $u(x,y) = v(x,y) + w(x,y)$, show that you can break the problem into two parts. The first part consists of solving the Laplace equation

$$\frac{\partial^2 v}{\partial x^2} + \frac{\partial^2 v}{\partial y^2} = 0, \quad 0 < x < \pi, \quad 0 < y < H,$$

subject to the boundary conditions that (1) $v(x,0) = 0$, $v(x,H) = f(x)$ for $0 < x < \pi$ and (2) $v(0,y) = v(\pi,y) = 0$ for $0 < y < H$. The second part involves solving

$$\frac{\partial^2 w}{\partial x^2} + \frac{\partial^2 w}{\partial y^2} = e^{2y} \sin(x), \quad 0 < x < \pi, \quad 0 < y < H,$$

subject to the boundary conditions that (1) $w(x,0) = w(x,H) = 0$ for $0 < x < \pi$ and (2) $w(0,y) = w(\pi,y) = 0$ for $0 < y < H$.

Step 2: Use separation of variables, and show that

$$v(x, y) = \sum_{n=1}^{\infty} A_n \sinh(ny) \sin(nx),$$

where

$$\sinh(nH) A_n = \frac{2}{\pi} \int_0^\pi f(x) \sin(nx)\, dx.$$

Step 3: Assuming that $w(x, y) = g(y) \sin(x)$, show that $g(y)$ is governed by the ordinary differential equation:

$$g''(y) - g(y) = e^{2y}, \qquad g(0) = g(H) = 0.$$

Step 4: Using the method of undetermined coefficients, show that the solution to Step 3 is

$$g(y) = \frac{e^{2y}}{3} + \frac{\sinh(y - H) - e^{2H} \sinh(y)}{3 \sinh(H)}.$$

Step 5: Show that the half-range Fourier sine expansion for e^{2y} is

$$e^{2y} = 2\pi \sum_{n=1}^{\infty} \frac{n}{n^2\pi^2 + 4H^2} \left[1 - (-1)^n e^{2H}\right] \sin\left(\frac{n\pi y}{H}\right), \qquad 0 < y < H.$$

Step 6: Using the results from Step 5 and assuming that

$$g(y) = \sum_{n=1}^{\infty} B_n \sin\left(\frac{n\pi y}{H}\right), \qquad 0 < y < H,$$

show that an alternative expression for $w(x.y)$ is

$$w(x, y) = 2\pi H^2 \sin(x) \sum_{n=1}^{\infty} \frac{n\left[(-1)^n e^{2H} - 1\right]}{(n^2\pi^2 + H^2)(n^2\pi^2 + 4H^2)} \sin\left(\frac{n\pi y}{H}\right).$$

The figure labeled Problem 2 illustrates $v(x, y)$ when $H = 2$.

3. Solve

$$\frac{\partial^2 u}{\partial x^2} + \frac{\partial^2 u}{\partial y^2} = -h, \qquad 0 < x < 1, \quad 0 < y < \infty,$$

subject to the boundary conditions that (1) $u(x, 0) = 0$, $\lim_{y \to \infty} |u(x, y)| < \infty$ for $0 < x < 1$, and (2) $u(0, y) = 0$, $u(1, y) = 1$ for $0 < y < \infty$.

Step 1: Setting $u(x, y) = x + hx(1 - x)/2 - v(x, y)$, show that the problem reduces to solving Laplace's equation:

$$\frac{\partial^2 v}{\partial x^2} + \frac{\partial^2 v}{\partial y^2} = 0, \qquad 0 < x < 1, \quad 0 < y < \infty,$$

Problem 2   Problem 3

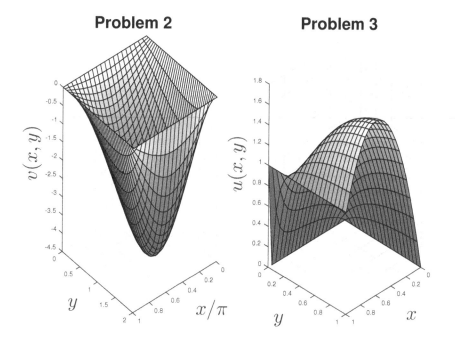

subject to the boundary conditions that (1) $v(0, y) = v(1, y) = 0$ for $0 < y < \infty$, and

$$v(x, 0) = x + hx(1 - x)/2, \quad \lim_{y \to \infty} |v(x, y)| < \infty, \qquad 0 < x < 1.$$

Step 2: Use separation of variables, and show that

$$v(x, y) = \frac{2}{\pi} \sum_{n=1}^{\infty} \frac{(-1)^{n+1}}{n} \exp(-n\pi y) \sin(n\pi x) + \frac{2h}{\pi^3} \sum_{n=1}^{\infty} \frac{[1 - (-1)^n]}{n^3} \exp(-n\pi y) \sin(n\pi x).$$

The figure labeled Problem 3 illustrates $u(x, y)$ when $h = 10$.

10.5 NUMERICAL SOLUTION OF LAPLACE'S EQUATION

In addition to the separation of variables, linear Laplace equations can be solved using conformal mapping or transform methods (see Section 6.7). However, when these analytics techniques fail us or we have a domain that is rather irregular in shape or we have a nonlinear equation, we must resort to numerical techniques. This section develops the popular numerical method of successive relaxation.

The numerical analysis of an elliptic partial differential equation begins by replacing the continuous partial derivatives by finite-difference formulas. Employing centered differencing,

$$\frac{\partial^2 u}{\partial x^2} = \frac{u_{m+1,n} - 2u_{m,n} + u_{m-1,n}}{(\Delta x)^2} + O[(\Delta x)^2], \qquad (\mathbf{10.5.1})$$

and

$$\frac{\partial^2 u}{\partial y^2} = \frac{u_{m,n+1} - 2u_{m,n} + u_{m,n-1}}{(\Delta y)^2} + O[(\Delta y)^2], \qquad (\mathbf{10.5.2})$$

where $u_{m,n}$ denotes the solution value at the grid point m, n. If $\Delta x = \Delta y$, Laplace's equation becomes the difference equation

$$u_{m+1,n} + u_{m-1,n} + u_{m,n+1} + u_{m,n-1} - 4u_{m,n} = 0. \tag{10.5.3}$$

Thus, we must now solve a set of simultaneous linear equations that yield the value of the solution at each grid point.

The solution of Equation 10.5.3 is best done using techniques developed by algebraists. A very popular method for directly solving systems of linear equations is Gaussian elimination. However, for many grids at a reasonable resolution, the number of equations is generally in the tens of thousands. Because most of the coefficients in the equations are zero, Gaussian elimination is unsuitable, both from the point of view of computational expense and accuracy. For this reason alternative methods have been developed that generally use successive corrections or iterations. The most common of these point iterative methods are the Jacobi method, unextrapolated Liebmann or Gauss-Seidel method, and extrapolated Liebmann or successive over-relaxation (SOR). None of these approaches is completely satisfactory because of questions involving convergence and efficiency. Because of its simplicity we will focus on the Gauss-Seidel method.

We may illustrate the Gauss-Seidel method by considering the system:

$$10x + y + z = 39, \tag{10.5.4}$$

$$2x + 10y + z = 51, \tag{10.5.5}$$

and

$$2x + 2y + 10z = 64. \tag{10.5.6}$$

An important aspect of this system is the dominance of the coefficient of x in the first equation of the set and that the coefficients of y and z are dominant in the second and third equations, respectively.

The Gauss-Seidel method may be outlined as follows:

• Assign an initial value for each unknown variable. If possible, make a good first guess. If not, any arbitrarily selected values may be chosen. The initial value will not affect the convergence but will affect the number of iterations until convergence.

• Starting with Equation 10.5.4, solve that equation for a new value of the unknown which has the largest coefficient in that equation, using the assumed values for the other unknowns.

• Go to Equation 10.5.5 and employ the same technique used in the previous step to compute the unknown that has the largest coefficient in that equation. Where possible, use the latest values.

• Proceed to the remaining equations, always solving for the unknown having the largest coefficient in the particular equation and always using the *most recently* calculated values for the other unknowns in the equation. When the last equation, Equation 10.5.6, has been solved, you have completed a single iteration.

• Iterate until the value of each unknown does not change within a predetermined value.

Usually a compromise must be struck between the accuracy of the solution and the desired rate of convergence. The more accurate the solution is, the longer it will take for the solution to converge.

To illustrate this method, let us solve our system, Equation 10.5.4 through Equation 10.5.6, with the initial guess $x = y = z = 0$. The first iteration yields $x = 3.9$, $y = 4.32$, and $z = 4.756$. The second iteration yields $x = 2.9924$, $y = 4.02592$, and $z = 4.996336$. As can be readily seen, the solution is converging to the correct solution of $x = 3$, $y = 4$, and $z = 5$.

Applying these techniques to Equation 10.5.3,

$$u_{m,n}^{k+1} = \tfrac{1}{4} \left(u_{m+1,n}^k + u_{m-1,n}^{k+1} + u_{m,n+1}^k + u_{m,n-1}^{k+1} \right), \qquad (10.5.7)$$

where we assume that the calculations occur in order of increasing m and n.

• **Example 10.5.1**

To illustrate the numerical solution of Laplace's equation, let us redo Example 10.3.1 with the boundary condition along $y = H$ simplified to $u(x, H) = 1 + x/L$.

We begin by finite-differencing the boundary conditions. The condition $u_x(0, y) = u_x(L, y) = 0$ leads to $u_{1,n} = u_{-1,n}$ and $u_{M+1,n} = u_{M-1,n}$ if we employ centered differences at $m = 0$ and $m = M$. Substituting these values in Equation 10.5.7, we have the following equations for the left and right boundaries:

$$u_{0,n}^{k+1} = \tfrac{1}{4} \left(2u_{1,n}^k + u_{0,n+1}^k + u_{0,n-1}^{k+1} \right) \qquad (10.5.8)$$

and

$$u_{M,n}^{k+1} = \tfrac{1}{4} \left(2u_{M-1,n}^{k+1} + u_{M,n+1}^k + u_{M,n-1}^{k+1} \right). \qquad (10.5.9)$$

On the other hand, $u_y(x, 0) = 0$ yields $u_{m,1} = u_{m,-1}$, and

$$u_{m,0}^{k+1} = \tfrac{1}{4} \left(u_{m+1,0}^k + u_{m-1,0}^{k+1} + 2u_{m,1}^k \right). \qquad (10.5.10)$$

At the bottom corners, Equation 10.5.8 through Equation 10.5.10 simplify to

$$u_{0,0}^{k+1} = \tfrac{1}{2} \left(u_{1,0}^k + u_{0,1}^k \right) \qquad (10.5.11)$$

and

$$u_{L,0}^{k+1} = \tfrac{1}{2} \left(u_{L-1,0}^{k+1} + u_{L,1}^k \right). \qquad (10.5.12)$$

These equations along with Equation 10.5.7 were solved with the Gauss-Seidel method using the MATLAB script:

```
clear
dx = 0.1; x = 0:dx:1; M = 1/dx+1; % M = number of x grid points
dy = 0.1; y = 0:dy:1; N = 1/dy+1; % N = number of y grid points
X = x' * ones(1,N); Y = ones(M,1) * y;
u = zeros(M,N); % create initial guess for the solution
% introduce boundary condition along y = H
for m = 1:M; u(m,N) = 1 + x(m); end
% start Gauss-Seidel method for Laplace's equation
for iter = 1:256
% do the interior first
  for n = 2:N-1; for m = 2:M-1;
    u(m,n) = (u(m+1,n)+u(m-1,n)+u(m,n+1)+u(m,n-1)) / 4;
```

```
   end; end
% now do the x = 0 and x = L sides
   for n = 2:N-1
      u(1,n) = (2*u( 2 ,n)+u(1,n+1)+u(1,n-1)) / 4;
      u(M,n) = (2*u(M-1,n)+u(M,n+1)+u(M,n-1)) / 4;
   end
% now do the y = 0 side
   for m = 2:M-1
      u(m,1) = (u(m+1,1)+u(m-1,1)+2*u(m,2)) / 4;
   end
% finally do the corners
   u(1,1) = (u(2,1)+u(1,2))/2; u(M,1) = (u(M-1,1)+u(M,2))/2;
% plot the solution
   if (iter == 4) subplot(2,2,1), [cs,h] = contourf(X,Y,u);
      clabel(cs,h,[0.2 0.6 1 1.4],'Fontsize',16)
      axis tight; title('after 4 iterations','Fontsize',20);
      ylabel('Y/H','Fontsize',20); end
   if (iter == 16) subplot(2,2,2), [cs,h] = contourf(X,Y,u);
      clabel(cs,h,'Fontsize',16)
      axis tight; title('after 16 iterations','Fontsize',20);
      ylabel('Y/H','Fontsize',20); end
   if (iter == 64) subplot(2,2,3), [cs,h] = contourf(X,Y,u);
      clabel(cs,h,'Fontsize',16)
      axis tight; title('after 64 iterations','Fontsize',20);
      xlabel('X/L','Fontsize',20); ylabel('Y/H','Fontsize',20);
   end
   if (iter == 256) subplot(2,2,4), [cs,h] = contourf(X,Y,u);
      clabel(cs,h,'Fontsize',16)
      axis tight; title('after 256 iterations','Fontsize',20);
      xlabel('X/L','Fontsize',20); ylabel('Y/H','Fontsize',20);
   end
end
```

The initial guess everywhere except along the top boundary was zero. In Figure 10.5.1 we illustrate the numerical solution after 4, 16, 64, and 256 iterations, where we have taken 11 grid points in the x and y directions.

Project: Successive Over-Relaxation

The fundamental difficulty with relaxation methods used in solving Laplace's equation is the rate of convergence. Assuming $\Delta x = \Delta y$, the most popular method for accelerating convergence of these techniques is *successive over-relaxation (SOR)*:

$$u_{m,n}^{k+1} = (1-\omega)u_{m,n}^k + \omega R_{m,n},$$

where

$$R_{m,n} = \tfrac{1}{4}\left(u_{m+1,n}^k + u_{m-1,n}^{k+1} + u_{m,n+1}^k + u_{m,n-1}^{k+1}\right).$$

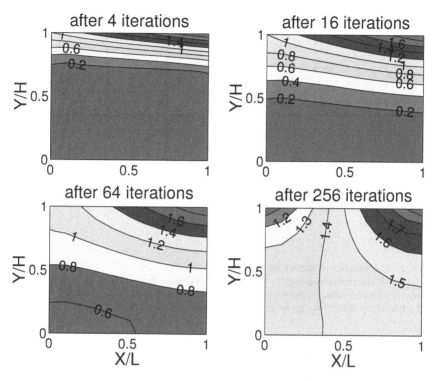

Figure 10.5.1: The solution to Laplace's equation by the Gauss-Seidel method. The boundary conditions are $u_x(0,y) = u_x(L,y) = u_y(x,0) = 0$, and $u(x,H) = 1 + x/L$.

Most numerical methods books dealing with partial differential equations discuss the theoretical reasons behind this technique;[8] the optimum value always lies between one and two. In the present case, a theoretical analysis[9] gives

$$\omega_{\text{opt}} = \frac{4}{2 + \sqrt{4 - c^2}},$$

where

$$c = \cos\left(\frac{\pi}{N}\right) + \cos\left(\frac{\pi}{M}\right),$$

and N and M are the number of mesh divisions on each side of the rectangular domain. Recently Yang and Gobbert[10] generalized the analysis and found the optimal relaxation parameter for the successive-overrelaxation method when it is applied to the Poisson equation in any space dimensions.

Step 1: Write a MATLAB script that uses the Gauss-Seidel method to numerically solve Laplace's equation for $0 \le x \le L$, $0 \le y \le z_0$ with the following boundary conditions: $u(x,0) = 0$, $u(x,z_0) = 1 + x/L$, $u(0,y) = y/z_0$, and $u(L,y) = 2y/z_0$. Because this solution will act as "truth" in this project, you should iterate until the solution does not change.

---

[8] For example, Young, D. M., 1971: *Iterative Solution of Large Linear Systems*. Academic Press, 570 pp.

[9] Yang, S., and M. K. Gobbert, 2009: The optimal relaxation parameter for the SOR method applied to the Poisson equation in any space dimensions. *Appl. Math. Letters*, **22**, 325–331.

[10] Ibid.

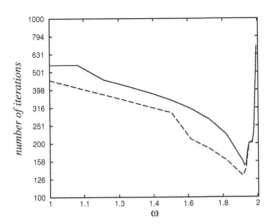

Figure 10.5.2: The number of iterations required so that $|R_{m,n}| \leq 10^{-3}$ as a function of ω during the iterative solution of the problem posed in the project. We used $\Delta x = \Delta y = 0.01$, and $L = z_0 = 1$. The iteration count for the boundary conditions stated in Step 1 is given by the solid line while the iteration count for the boundary conditions given in Step 2 is shown by the dotted line. The initial guess equaled zero.

Step 2: Now redo the calculation using successive over-relaxation. Count the number of iterations until $|R_{m,n}| \leq 10^{-3}$ for *all* m and n. Plot the number of iterations as a function of ω. How does the curve change with resolution Δx? How does your answer compare to the theoretical value? See Figure 10.5.2.

Step 3: Redo Steps 1 and 2 with the exception of $u(0, y) = u(L, y) = 0$. How has the convergence rate changed? Can you explain why? How sensitive are your results to the first guess?

Project: Finite Difference Solution of Poisson's Equation

The analytic solution of Laplace's and Poisson's equations are often not available and we must resort to numerical approximation. In this project you will use finite differences to solve the Poisson equation:

$$\frac{\partial^2 u}{\partial x^2} + \frac{\partial^2 u}{\partial y^2} = e^{2y} \sin(x), \qquad 0 < x < \pi, \quad 0 < y < \pi, \qquad (10.5.13)$$

subject to the boundary conditions that $u(x, 0) = u(x, \pi) = 0$ for $0 < x < \pi$ and $u(0, y) = w(\pi, y) = 0$ for $0 < y < \pi$. The exact solution is

$$u(x, y) = \sin(x) \left[\frac{e^{2y}}{3} + \frac{\sinh(y - \pi) - e^{2\pi} \sinh(y)}{3 \sinh(\pi)} \right]. \qquad (10.5.14)$$

This project requires a knowledge of matrix algebra (see Sections 3.1 and 3.4).

Step 1: If we introduce nodal points at $x_m = mh$ and $y_n = nh$, where $m = 0, 1, 2, \ldots, M+1$, $n = 0, 1, 2, \ldots, N+1$, and $h = \Delta x = \Delta y = \pi/(M+1)$. Using centered finite differencing, show that the partial differential equation can be approximated by

$$u_{m+1}^{n+1} + u_{m+1}^{n-1} + u_{m-1}^{n+1} + u_{m-1}^{n-1} - 4u_m^n = h^2 e^{2y_n} \sin(x_m), \quad n = 1, 2, \ldots, N, \ m = 1, 2, \ldots, M.$$

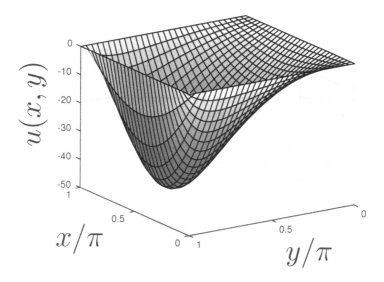

Figure 10.5.3: The numerical solution to Equation 10.5.13. Here $N = M = 29$ so that $h = \pi/30$. The largest relative error is 0.175.

Here $u_0^n = u_{M+1}^n = u_m^0 = u_m^{N+!} = 0$.

Step 2: We now want to cast Step 1 as the linear algebra problem $A\mathbf{u} = \mathbf{f}$. As a first step, let us choose $N = M = 3$ and $h = \pi/4$. This is a very crude approximation but we will soon increase the value of N and M. Show that this simple case yields 9 simultaneous equations which can be written in matrix form as

$$
\begin{pmatrix}
-4 & 1 & 0 & 1 & 0 & 0 & 0 & 0 & 0 \\
1 & -4 & 1 & 0 & 1 & 0 & 0 & 0 & 0 \\
0 & 1 & -4 & 1 & 0 & 1 & 0 & 0 & 0 \\
1 & 0 & 1 & -4 & 1 & 0 & 1 & 0 & 0 \\
0 & 1 & 0 & 1 & -4 & 1 & 0 & 1 & 0 \\
0 & 0 & 1 & 0 & 1 & -4 & 1 & 0 & 1 \\
0 & 0 & 0 & 1 & 0 & 1 & -4 & 1 & 0 \\
0 & 0 & 0 & 0 & 1 & 0 & 1 & -4 & 1 \\
0 & 0 & 0 & 0 & 0 & 1 & 0 & 1 & -4
\end{pmatrix}
\begin{pmatrix}
u_1^1 \\ u_1^2 \\ u_1^3 \\ u_2^1 \\ u_2^2 \\ u_2^3 \\ u_3^1 \\ u_3^2 \\ u_3^3
\end{pmatrix}
=
\begin{pmatrix}
h^2 f_1^1 - u_1^1 - u_1^0 \\
h^2 f_1^2 - u_0^2 \\
h^2 f_1^3 - u_0^3 - u_1^4 \\
h^2 f_2^1 \quad - u_2^0 \\
h^2 f_2^2 \\
h^2 f_2^3 \quad - u_2^4 \\
h^2 f_3^1 - u_4^1 - u_3^0 \\
h^2 f_3^2 - u_4^2 \\
h^2 f_3^3 - u_4^3 - u_3^4
\end{pmatrix}.
$$

In the present case, $u_0^n = u_4^n = u_m^0 = u_m^4 = 0$; in general, this will not be the case.

Step 3: Using MABLAB, solve for u_m^n for $m = 1, 2, 3$ and $n = 1, 2, 3$. Find that grid point which has the largest relative error $|u_{numerical}(x, y) - u_{exact}(x, y)|$ between the numerical and exact solutions.

Step 4: Now generalize your code to treat the general case and find the numerical solution to Poisson's equation. For a particular value of M, N, find the grid point which has the largest value of relative error.

Because the matrix A is a sparse matrix (most of its elements equal zero) this direct method for finding numerical solutions to Poisson's equation was rarely used because of its computational expense. However, with the advent of power of computational engines such as MATLAB this method is becoming competitive with other, more complicated methods.

Project: Mixed Boundary-Value Problems

In this chapter we focused on solving Laplace's equation when we have a Dirichlet or Neumann boundary condition. For example, $u(0, y) = u(\pi, y) = 0$. In Chapter 11 we will expand our knowledge by considering problems where we have one type of boundary condition along one boundary [for example, the Neumann condition $u_x(0, y) = 0$] and another type of boundary condition along another boundary [for example, the Dirichlet condition $u(\pi, y) = 0$]. Some call this situation a mixed boundary-value problem; we will *not*.

In this project we consider the case where we have one type of boundary condition along some portion of a boundary and another type along other parts of the *same* boundary. A simple example of a *mixed boundary-value problem* is

$$\frac{\partial^2 u}{\partial x^2} + \frac{\partial^2 u}{\partial y^2} = 0, \qquad 0 < x < \pi, \quad 0 < y < \infty, \tag{10.5.15}$$

subject to the boundary conditions that

$$\begin{cases} u(x, 0) = 1, & 0 \le x \le c, \\ u_y(x, 0) = 0, & c < x \le \pi, \end{cases} \quad \lim_{y \to \infty} u(x, y) \to 0, \qquad 0 < x < \pi, \tag{10.5.16}$$

and

$$u_x(0, y) = u(\pi, y) = 0, \qquad 0 < y < \infty, \tag{10.5.17}$$

where $0 < c < \pi$. Solving this challenging problem will require a knowledge of matrix algebra (see Sections 3.1 and 3.4).

Step 1: We begin as usual by using separation of variables. Show that

$$u(x, y) = \sum_{n=1}^{\infty} A_n \frac{\exp\left[-\left(n - \frac{1}{2}\right) y\right]}{n - \frac{1}{2}} \cos\left[\left(n - \frac{1}{2}\right) x\right] \tag{10.5.18}$$

solves Laplace's equation and the boundary conditions *except* along $y = 0$.

Step 2: Show that our solution must satisfy the dual conditions:

$$\sum_{n=1}^{\infty} \frac{A_n}{n - \frac{1}{2}} \cos\left[\left(n - \frac{1}{2}\right) x\right] = 1, \qquad 0 \le x \le c, \tag{10.5.19}$$

and

$$\sum_{n=1}^{\infty} A_n \cos\left[\left(n - \frac{1}{2}\right) x\right] = 0, \qquad c < x < \pi. \tag{10.5.20}$$

This is similar to other separation of variables problems where our final task was to determine the Fourier coefficients of a half-range expansion, except that we now have two series!!

Step 3: There is actually a closed form solution to this dual Fourier series Equation 10.5.19 and 10.5.20:

$$A_n = \frac{P_{n-1}[\cos(c)]}{K(\cos^2(c/2))},$$

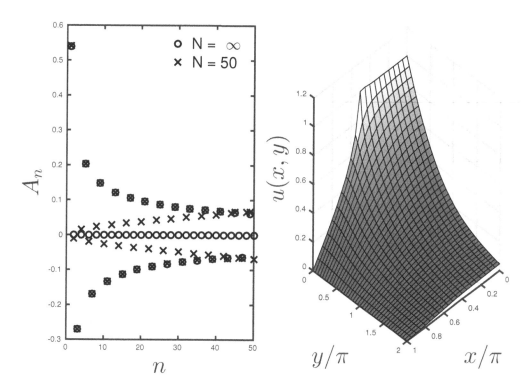

Figure 10.5.4: The numerical solution to the mixed boundary-value problem Equations 10.5.15 through 10.5.17. Here $c = \pi/2$.

where $P_n(\cdot)$ is the Legendre polynomial of order n (see Section 12.1) and $K(\cdot)$ is a complete elliptic integral of the first kind. The derivation of this result is beyond the scope of this book.[11]

What would we have done if this result did not exist? Let us replace the conditions Equations 10.5.19 and 10.5.20 with

$$\sum_{n=1}^{N} \frac{A_n}{n - \frac{1}{2}} \cos\left[\left(n - \tfrac{1}{2}\right) x_i\right] = 1, \qquad 0 \le x_i \le c,$$

and

$$\sum_{n=1}^{\infty} A_n \cos\left[\left(n - \tfrac{1}{2}\right) x_i\right] = 0, \qquad c < x_i < \pi,$$

where $x_i = (i - 1)\pi/N$ and $i = 1, 2, 3, \ldots, N$. This results in a system of $N \times N$ linear simultaneous equations.

Step 4: Using MATLAB, solve for the N values of A_n in Step 3. As a function of c, compare these values of A_n with the ones for $N = \infty$. Employing these new A_n's, compute $u(x, y)$. Does it satisfy the boundary conditions?

Step 5: Let's attack this problem using centered finite differencing. Using the relaxation method, write down a scheme to solve our problem. This problem poses several interesting

---

[11] See Duffy, D. G., 2008: *Mixed Boundary Value Problems*. Chapman & Hall/CRC, Example 3.1.1.

challenges. First, how do we deal with the fact that y extends from 0 to ∞? Second, how do we compute the solution for $c < x < \pi$ along $y = 0$ and for $0 < y < \infty$ for $x = 0$?

Step 6: Compute the solution using relaxation and compare it with the solution that you found in Step 4.

Further Readings

Koshlyakov, N. S., M. M. Smirnov, and E. B. Gliner, 1964: *Differential Equations of Mathematical Physics.* North-Holland Publishing, 701 pp. See Part II. Detailed presentation of mathematical techniques.

Morse, P. M., and H. Feshback, 1953: *Methods of Theoretical Physics.* McGraw-Hill Book Co., 997 pp. Chapter 10 is devoted to solving both Laplace's and Poisson's equations.

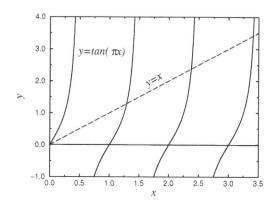

Chapter 11
The Sturm-Liouville Problem

In Chapters 1, 2, and 7 we solved ordinary differential equations where some quantity evolves in time from some *initial condition*. Although these *initial-value problems* are important in engineering and the physical sciences, they are not the only special class of problems involving differential equations.

In the three previous chapters we solved linear partial differential equations using the technique commonly called "separation of variables." Assuming that the solution could be written $u(x,t) = X(x)T(t)$ or $u(x,y) = X(x)Y(y)$, we solved the ordinary differential equation

$$X'' + \lambda X = 0, \qquad 0 < x < L, \quad 0 < \lambda, \qquad (11.0.1)$$

with the boundary conditions $X(0) = X(L) = 0$ or $X'(0) = X'(L) = 0$. This is an example of a *boundary-value problem* because the solution must satisfy boundary conditions at the end points rather than a single initial condition. Furthermore, unlike initial-value problems, the present boundary-value problem has an infinite number of possible solutions. For example, if $X(0) = X(L) = 0$, then we have

$$X_n(x) = \sin(n\pi x/L), \quad \text{with} \quad \lambda_n = n^2\pi^2/L^2 \quad \text{and} \quad n = 1, 2, 3, \dots. \qquad (11.0.2)$$

Prefiguring the material in this chapter, the λ_n's are called the eigenvalues and the $X_n(x)$'s are the corresponding eigenfunctions of this Sturm-Liouville boundary-value problem.

In this chapter we expand our ability to solve boundary-value problems. For example, we would like to solve the boundary-value problem

$$X'' + \lambda X = 0, \qquad 0 < x < L, \qquad (11.0.3)$$

with $X(0) = X'(L) = 0$ and $\lambda > 0$. The eigenfunctions for this problem will be shown to be $X_n(x) = \sin[(2n-1)\pi x/2]$ with a corresponding eigenvalue of $\lambda_n = (2n-1)^2\pi^2/4$.

This is not the first time that we have encountered the eigenvalue problem. In Section 3.5 we studied eigenvalues and eigenvectors in our solution of systems of ordinary differential equations. Indeed, had we used finite-difference techniques to solve the heat equation (for example), we would have obtained a system of ordinary differential equations and the eigenvectors would be the discrete approximation of the corresponding continuous eigenfunctions that would arise during the separation of variable solution.

Finally, just as we used half-range Fourier sine and cosine expansions to solve partial differential equations, we will construct linear sums of eigenfunctions from Sturm-Liouville problems to re-express a piece-wise continuous function $f(x)$. These eigenfunction expansions, in turn, will be used in solving linear partial differential equations.

11.1 EIGENVALUES AND EIGENFUNCTIONS

Repeatedly, in the next three chapters on partial differential equations, we will solve the following second-order linear differential equation:

$$\frac{d}{dx}\left[p(x)\frac{dy}{dx}\right] + [q(x) + \lambda r(x)]y = 0, \quad a \le x \le b, \tag{11.1.1}$$

together with the boundary conditions:

$$\alpha y(a) + \beta y'(a) = 0 \quad \text{and} \quad \gamma y(b) + \delta y'(b) = 0. \tag{11.1.2}$$

In Equation 11.1.1, $p(x)$, $q(x)$, and $r(x)$ are real functions of x; λ is a parameter; and $p(x)$ and $r(x)$ are functions that are continuous and positive on the interval $a \le x \le b$. Taken together, Equation 11.1.1 and Equation 11.1.2 constitute a regular *Sturm-Liouville problem*, named after the French mathematicians Sturm and Liouville[1] who first studied these equations in the 1830s. In the case when $p(x)$ or $r(x)$ vanishes at one of the endpoints of the interval $[a, b]$ or when the interval is of infinite length, the problem becomes a *singular Sturm-Liouville problem*.

Consider now the solutions to the regular Sturm-Liouville problem. Clearly there is the trivial solution $y = 0$ for all λ. However, nontrivial solutions exist only if λ takes on specific values; these values are called *characteristic values* or *eigenvalues*. The corresponding nontrivial solutions are called the *characteristic functions* or *eigenfunctions*; trivial solutions are not eigenfunctions by definition. In particular, we have the following theorems.

Theorem: *For a regular Sturm-Liouville problem with $p(x) > 0$, all of the eigenvalues are real if $p(x)$, $q(x)$, and $r(x)$ are real functions and the eigenfunctions are differentiable and continuous.*

---

[1] For the complete history as well as the relevant papers, see Lützen, J., 1984: Sturm and Liouville's work on ordinary linear differential equations. The emergence of Sturm-Liouville theory. *Arch. Hist. Exact Sci.*, **29**, 309–376.

By the time that Charles-François Sturm (1803–1855) met Joseph Liouville in the early 1830s, he had already gained fame for his work on the compression of fluids and his celebrated theorem on the number of real roots of a polynomial. An eminent teacher, Sturm spent most of his career teaching at various Parisian colleges. (Portrait courtesy of the Archives de l'Académie des sciences, Paris.)

Proof: Let $y(x) = u(x) + iv(x)$ be an eigenfunction corresponding to an eigenvalue $\lambda = \lambda_r + i\lambda_i$, where λ_r, λ_i are real numbers and $u(x), v(x)$ are real functions of x. Substituting into the Sturm-Liouville equation yields

$$\{p(x)[u'(x) + iv'(x)]\}' + [q(x) + (\lambda_r + i\lambda_i)r(x)][u(x) + iv(x)] = 0. \qquad (11.1.3)$$

Separating the real and imaginary parts gives

$$[p(x)u'(x)]' + [q(x) + \lambda_r]u(x) - \lambda_i r(x)v(x) = 0, \qquad (11.1.4)$$

and

$$[p(x)v'(x)]' + [q(x) + \lambda_r]v(x) + \lambda_i r(x)u(x) = 0. \qquad (11.1.5)$$

If we multiply Equation 11.1.4 by v and Equation 11.1.5 by u and subtract the results, we find that

$$u(x)[p(x)v'(x)]' - v(x)[p(x)u'(x)]' + \lambda_i r(x)[u^2(x) + v^2(x)] = 0. \qquad (11.1.6)$$

The derivative terms in Equation 11.1.6 can be rewritten so that it becomes

$$\frac{d}{dx}\{[p(x)v'(x)]u(x) - [p(x)u'(x)]v(x)\} + \lambda_i r(x)[u^2(x) + v^2(x)] = 0. \qquad (11.1.7)$$

Although educated as an engineer, Joseph Liouville (1809–1882) would devote his life to teaching pure and applied mathematics in the leading Parisian institutions of higher education. Today he is most famous for founding and editing for almost 40 years the *Journal de Liouville*. (Portrait courtesy of the Archives de l'Académie des sciences, Paris.)

Integrating from a to b, we find that

$$-\lambda_i \int_a^b r(x)[u^2(x) + v^2(x)]\, dx = \{p(x)[u(x)v'(x) - v(x)u'(x)]\}\big|_a^b. \qquad (\mathbf{11.1.8})$$

From the boundary conditions, Equation 11.1.2,

$$\alpha[u(a) + iv(a)] + \beta[u'(a) + iv'(a)] = 0, \qquad (\mathbf{11.1.9})$$

and

$$\gamma[u(b) + iv(b)] + \delta[u'(b) + iv'(b)] = 0. \qquad (\mathbf{11.1.10})$$

Separating the real and imaginary parts yields

$$\alpha u(a) + \beta u'(a) = 0, \quad \text{and} \quad \alpha v(a) + \beta v'(a) = 0, \qquad (\mathbf{11.1.11})$$

and

$$\gamma u(b) + \delta u'(b) = 0, \quad \text{and} \quad \gamma v(b) + \delta v'(b) = 0. \qquad (\mathbf{11.1.12})$$

Both α and β cannot be zero; otherwise, there would be no boundary condition at $x = a$. Similar considerations hold for γ and δ. Therefore,

$$u(a)v'(a) - u'(a)v(a) = 0, \quad \text{and} \quad u(b)v'(b) - u'(b)v(b) = 0, \qquad (\mathbf{11.1.13})$$

if we treat α, β, γ, and δ as unknowns in a system of homogeneous equations, Equation 11.1.11 and Equation 11.1.12, and require that the corresponding determinants equal zero. Applying Equation 11.1.13 to the right side of Equation 11.1.8, we obtain

$$\lambda_i \int_a^b r(x)[u^2(x) + v^2(x)]\, dx = 0. \tag{11.1.14}$$

Because $r(x) > 0$, the integral is positive and $\lambda_i = 0$. Since $\lambda_i = 0$, λ is purely real. This implies that the eigenvalues are real. □

If there is only one independent eigenfunction for each eigenvalue, that eigenvalue is *simple*. When more than one eigenfunction belongs to a single eigenvalue, the problem is *degenerate*.

Theorem: *The regular Sturm-Liouville problem has infinitely many real and simple eigenvalues λ_n, $n = 0, 1, 2, \ldots$, which can be arranged in a monotonically increasing sequence $\lambda_0 < \lambda_1 < \lambda_2 < \cdots$ such that $\lim_{n \to \infty} \lambda_n = \infty$. Every eigenfunction $y_n(x)$ associated with the corresponding eigenvalue λ_n has exactly n zeros in the interval (a, b). For each eigenvalue there exists only one eigenfunction (up to a multiplicative constant).*

The proof is beyond the scope of this book but may be found in more advanced treatises.[2]□

In the following examples we illustrate how to find these real eigenvalues and their corresponding eigenfunctions.

• **Example 11.1.1**

Let us find the eigenvalues and eigenfunctions of

$$y'' + \lambda y = 0, \tag{11.1.15}$$

subject to the boundary conditions

$$y(0) = 0, \quad \text{and} \quad y(\pi) - y'(\pi) = 0. \tag{11.1.16}$$

Our first task is to check to see whether the problem is indeed a regular Sturm-Liouville problem. A comparison between Equation 11.1.1 and Equation 11.1.15 shows that they are the same if $p(x) = 1$, $q(x) = 0$, and $r(x) = 1$. Similarly, the boundary conditions, Equation 11.1.16, are identical to Equation 11.1.2 if $\alpha = \gamma = 1$, $\delta = -1$, $\beta = 0$, $a = 0$, and $b = \pi$.

Because the form of the solution to Equation 11.1.15 depends on λ, we consider three cases: λ negative, positive, or equal to zero. The general solution[3] of the differential equation is

$$y(x) = A\cosh(mx) + B\sinh(mx), \quad \text{if} \quad \lambda < 0, \tag{11.1.17}$$

---

[2] See, for example, Birkhoff, G., and G.-C. Rota, 1989: *Ordinary Differential Equations.* John Wiley & Sons, Chapters 10 and 11; Sagan, H., 1961: *Boundary and Eigenvalue Problems in Mathematical Physics.* John Wiley & Sons, Chapter 5.

[3] In many differential equations courses, the solution to $y'' - m^2 y = 0$, $m > 0$ is written $y(x) = c_1 e^{mx} + c_2 e^{-mx}$. However, we can rewrite this solution as $y(x) = (c_1 + c_2)\frac{1}{2}(e^{mx} + e^{-mx}) + (c_1 - c_2)\frac{1}{2}(e^{mx} - e^{-mx}) = A\cosh(mx) + B\sinh(mx)$, where $\cosh(mx) = (e^{mx} + e^{-mx})/2$ and $\sinh(mx) = (e^{mx} - e^{-mx})/2$. The advantage of using these hyperbolic functions over exponentials is the simplification that occurs when we substitute the hyperbolic functions into the boundary conditions.

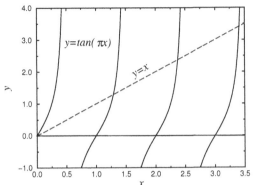

Figure 11.1.1: Graphical solution of $\tan(\pi x) = x$. The roots of $\tan(k\pi) = k$ occur where the two lines intersect.

$$y(x) = C + Dx, \quad \text{if} \quad \lambda = 0, \tag{11.1.18}$$

and

$$y(x) = E\cos(kx) + F\sin(kx), \quad \text{if} \quad \lambda > 0, \tag{11.1.19}$$

where for convenience $\lambda = -m^2 < 0$ in Equation 11.1.17 and $\lambda = k^2 > 0$ in Equation 11.1.19. Both k and m are real and positive by these definitions.

Turning to the condition that $y(0) = 0$, we find that $A = C = E = 0$. The other boundary condition $y(\pi) - y(\pi) = 0$ gives

$$B[\sinh(m\pi) - m\cosh(m\pi)] = 0, \tag{11.1.20}$$

$$D = 0, \tag{11.1.21}$$

and

$$F[\sin(k\pi) - k\cos(k\pi)] = 0. \tag{11.1.22}$$

Let us turn to Equation 11.1.20 first. For nontrivial solutions to exist, $B \neq 0$. This occurs only if $\sinh(m\pi) - m\cosh(m\pi) = 0$. Solving this equation numerically (see the next example), we find that $m_{-1} = 0.99618173$, yielding $\lambda_{-1} = -0.992378039$. I have used a negative index to remind ourselves that $\lambda < 0$.

Next, consider Equation 11.1.21. This clearly shows that there are only trivial solutions for $\lambda = 0$.

Finally, Equation 11.1.22 yields nontrivial solutions only if $\tan(k\pi) = k$. We can find these roots either numerically (see the next example) or graphically. Figure 11.1.1 illustrates the graphical method.

Let us now find the corresponding eigenfunctions. There are two classes: For $\lambda < 0$, we have $y_{-1}(x) = \sinh(m_{-1}x)$. On the other hand, for $\lambda > 0$ we have that $y_n(x) = \sin(k_nx)$, where k_n is the nth root of $\tan(k\pi) = k$. Figure 11.1.2 illustrates the first four eigenfunctions from this second class of eigenfunctions. Traditionally, eigenfunction solutions are written without their arbitrary amplitude constant. $\square$

● **Example 11.1.2: Numerical computation of the eigenvalues**

In the previous example we showed how you could use graphical methods to find k_n and the corresponding eigenvalue λ_n. With the advent of powerful computational engines such as MATLAB, no one uses graphical methods anymore; rather we now use numerical methods. For example, using the Newton-Raphson method, we can find m_{-1} via

$$m_{-1}^{(i+1)} = m_{-1}^{(i)} - f[m_{-1}^{(i)}]/f'[m_{-1}^{(i)}], \qquad i = 0, 1, 2, \ldots, \tag{11.1.23}$$

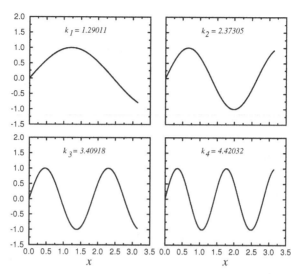

Figure 11.1.2: The first four eigenfunctions $\sin(k_n x)$ corresponding to the case of $\lambda_n = k_n^2 > 0$.

where $f(m) = \sinh(m\pi) - m \cosh(m\pi)$, and $f'(m) = (\pi - 1) \cosh(m\pi) - m \pi \sinh(m\pi)$. Similarly, to compute k_n, we can use the algorithm:

$$k_n^{(i+1)} = k_n^{(i)} - f[k_n^{(i)}]/f'[k_n^{(i)}], \qquad i = 0, 1, 2, \ldots, \tag{11.1.24}$$

where $f(k) = \sin(k\pi) - k \cos(k\pi)$, and $f'(k) = (\pi - 1) \cos(k\pi) + k \pi \sin(k\pi)$. Table 11.1.1 illustrates this calculation and shows that after two iterations the answer is accurate enough for most applications. The first guess equals $m_{-1}^{(0)}$ and $k_n^{(0)}$. ☐

● **Example 11.1.3**

For our second example, let us solve the Sturm-Liouville problem,[4]

$$y'' + \lambda y = 0, \tag{11.1.25}$$

with the boundary conditions

$$y(0) - y'(0) = 0, \quad \text{and} \quad y(\pi) - y'(\pi) = 0. \tag{11.1.26}$$

Once again the three possible solutions to Equation 11.1.25 are

$$y(x) = A \cosh(mx) + B \sinh(mx), \quad \text{if} \quad \lambda = -m^2 < 0, \tag{11.1.27}$$

$$y(x) = C + Dx, \quad \text{if} \quad \lambda = 0, \tag{11.1.28}$$

and

$$y(x) = E \cos(kx) + F \sin(kx), \quad \text{if} \quad \lambda = k^2 > 0. \tag{11.1.29}$$

---

[4] Sosov and Theodosiou [Sosov, Y., and C. E. Theodosiou, 2002: On the complete solution of the Sturm-Liouville problem $(d^2 X/dx^2) + \lambda^2 X = 0$ over a closed interval. *J. Math. Phys. (Woodbury, NY)*, **43**, 2831–2843] have analyzed this problem with the general boundary conditions, Equation 11.1.2.

Table 11.1.1: The Eigenvalues for the Problem Stated in Example 11.1.1 Found Using the Newton-Raphson Method. Here $\lambda_{-1} = -m_{-1}^2$ and $\lambda_n = k_n^2$ for $n = 1, 2, \ldots, 10$. The Boldface Digits Are Those that Differ from the Exact Answer.

	$m_{-1}^{(0)}$ or $k_n^{(0)}$	$m_{-1}^{(1)}$ or $k_n^{(1)}$	$m_{-1}^{(2)}$ or $k_n^{(2)}$	exact
m_{-1}	**1.00000000**	0.9962**2790**	0.9961817**4**	0.99618173
k_1	1.**50000000**	1.28779341	1.29011**264**	1.29010965
k_2	2.**50000000**	2.37267605	2.37305**303**	2.37305297
k_3	3.**50000000**	3.409**05432**	3.40917905	3.40917905
k_4	4.**50000000**	4.429**26447**	4.42932070	4.42932070
k_5	5.**50000000**	5.44212**548**	5.44215560	5.44215560
k_6	6.**50000000**	6.45102**925**	6.45104726	6.45104726
k_7	7.**50000000**	7.45755**868**	7.45757031	7.45757031
k_8	8.**50000000**	8.46255**178**	8.46255972	8.46255972
k_9	9.**50000000**	9.46649**370**	9.46649936	9.46649936
k_{10}	10.**50000000**	10.46968**477**	10.46968896	10.46968896

Let us first check and see if there are any nontrivial solutions for $\lambda < 0$. Two simultaneous equations result from the substitution of Equation 11.1.27 into Equation 11.1.26:

$$A - mB = 0, \tag{11.1.30}$$

and

$$[\cosh(m\pi) - m\sinh(m\pi)]A + [\sinh(m\pi) - m\cosh(m\pi)]B = 0. \tag{11.1.31}$$

The elimination of A between the two equations yields

$$\sinh(m\pi)(1 - m^2)B = 0. \tag{11.1.32}$$

If Equation 11.1.27 is a nontrivial solution, then $B \neq 0$, and

$$\sinh(m\pi) = 0, \quad \text{or} \quad m^2 = 1. \tag{11.1.33}$$

The condition $\sinh(m\pi) = 0$ cannot hold because it implies $m = \lambda = 0$, which contradicts the assumption used in deriving Equation 11.1.27 that $\lambda < 0$. On the other hand, $m^2 = 1$ is quite acceptable. It corresponds to the eigenvalue $\lambda = -1$ and the eigenfunction is

$$y_0 = \cosh(x) + \sinh(x) = e^x, \tag{11.1.34}$$

because it satisfies the differential equation

$$y_0'' - y_0 = 0, \tag{11.1.35}$$

and the boundary conditions

$$y_0(0) - y_0'(0) = 0, \quad \text{and} \quad y_0(\pi) - y_0'(\pi) = 0. \tag{11.1.36}$$

An alternative method of finding m, which is quite popular because of its use in more difficult problems, follows from viewing Equation 11.1.30 and Equation 11.1.31 as a system of homogeneous linear equations, where A and B are the unknowns. It is well known[5] that for Equation 11.1.30 and Equation 11.1.31 to have a nontrivial solution (i.e., $A \neq 0$ and/or $B \neq 0$) the determinant of the coefficients must vanish:

$$\begin{vmatrix} 1 & -m \\ \cosh(m\pi) - m\sinh(m\pi) & \sinh(m\pi) - m\cosh(m\pi) \end{vmatrix} = 0. \tag{11.1.37}$$

Expanding the determinant,

$$\sinh(m\pi)(1 - m^2) = 0, \tag{11.1.38}$$

which leads directly to Equation 11.1.33.

We consider next the case of $\lambda = 0$. Substituting Equation 11.1.28 into Equation 11.1.26, we find that

$$C - D = 0, \quad \text{and} \quad C + D\pi - D = 0. \tag{11.1.39}$$

This set of simultaneous equations yields $C = D = 0$ and we have only trivial solutions for $\lambda = 0$.

Finally, we examine the case when $\lambda > 0$. Substituting Equation 11.1.29 into Equation 11.1.26, we obtain

$$E - kF = 0, \tag{11.1.40}$$

and

$$[\cos(k\pi) + k\sin(k\pi)]E + [\sin(k\pi) - k\cos(k\pi)]F = 0. \tag{11.1.41}$$

The elimination of E from Equation 11.1.40 and Equation 11.1.41 gives

$$F(1 + k^2)\sin(k\pi) = 0. \tag{11.1.42}$$

If Equation 11.1.29 is nontrivial, $F \neq 0$, and

$$k^2 = -1, \quad \text{or} \quad \sin(k\pi) = 0. \tag{11.1.43}$$

The condition $k^2 = -1$ violates the assumption that k is real, which follows from the fact that $\lambda = k^2 > 0$. On the other hand, we can satisfy $\sin(k\pi) = 0$ if $k = 1, 2, 3, \ldots$; a negative k yields the same λ. Consequently, we have the additional eigenvalues $\lambda_n = n^2$.

Let us now find the corresponding eigenfunctions. Because $E = kF$, $y(x) = F\sin(kx) + Fk\cos(kx)$ from Equation 11.1.29. Thus, the eigenfunctions for $\lambda > 0$ are

$$y_n(x) = \sin(nx) + n\cos(nx). \tag{11.1.44}$$

Figure 11.1.3 illustrates some of the eigenfunctions given by Equation 11.1.34 and Equation 11.1.44. $\qquad\square$

• **Example 11.1.4**

Consider now the Sturm-Liouville problem

$$y'' + \lambda y = 0, \tag{11.1.45}$$

---

[5] See Chapter 3.

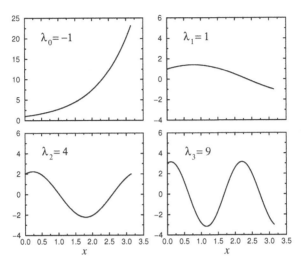

Figure 11.1.3: The first four eigenfunctions for the Sturm-Liouville problem, Equation 11.1.25 and Equation 11.1.26.

with

$$y(\pi) = y(-\pi), \quad \text{and} \quad y'(\pi) = y'(-\pi). \tag{11.1.46}$$

This is *not* a regular Sturm-Liouville problem because the boundary conditions are periodic and do not conform to the canonical boundary condition, Equation 11.1.2.

The general solution to Equation 11.1.45 is

$$y(x) = A\cosh(mx) + B\sinh(mx), \quad \text{if} \quad \lambda = -m^2 < 0, \tag{11.1.47}$$

$$y(x) = C + Dx, \quad \text{if} \quad \lambda = 0, \tag{11.1.48}$$

and

$$y(x) = E\cos(kx) + F\sin(kx), \quad \text{if} \quad \lambda = k^2 > 0. \tag{11.1.49}$$

Substituting these solutions into the boundary condition, Equation 11.1.46,

$$A\cosh(m\pi) + B\sinh(m\pi) = A\cosh(-m\pi) + B\sinh(-m\pi), \tag{11.1.50}$$

$$C + D\pi = C - D\pi, \tag{11.1.51}$$

and

$$E\cos(k\pi) + F\sin(k\pi) = E\cos(-k\pi) + F\sin(-k\pi), \tag{11.1.52}$$

or

$$B\sinh(m\pi) = 0, \quad D = 0, \quad \text{and} \quad F\sin(k\pi) = 0, \tag{11.1.53}$$

because $\cosh(-m\pi) = \cosh(m\pi)$, $\sinh(-m\pi) = -\sinh(m\pi)$, $\cos(-k\pi) = \cos(k\pi)$, and $\sin(-k\pi) = -\sin(k\pi)$. Because m must be positive, $\sinh(m\pi)$ cannot equal zero and $B = 0$. On the other hand, if $\sin(k\pi) = 0$ or $k = n$, $n = 1, 2, 3, \ldots$, we have a nontrivial solution for positive λ and $\lambda_n = n^2$. Note that we still have A, C, E, and F as free constants.

From the boundary condition, Equation 11.1.46,

$$A\sinh(m\pi) = A\sinh(-m\pi), \tag{11.1.54}$$

and

$$-E\sin(k\pi) + F\cos(k\pi) = -E\sin(-k\pi) + F\cos(-k\pi). \qquad (11.1.55)$$

The solution $y_0(x) = C$ identically satisfies the boundary condition, Equation 11.1.46, for all C. Because m and $\sinh(m\pi)$ must be positive, $A = 0$. From Equation 11.1.53, we once again have $\sin(k\pi) = 0$, and $k = n$. Consequently, the eigenfunction solutions to Equation 11.1.45 and Equation 11.1.46 are

$$\lambda_0 = 0, \qquad y_0(x) = 1, \qquad (11.1.56)$$

and

$$\lambda_n = n^2, \qquad y_n(x) = \begin{cases} \sin(nx), \\ \cos(nx), \end{cases} \qquad (11.1.57)$$

and we have a degenerate set of eigenfunctions to the Sturm-Liouville problem, Equation 11.1.45, with the periodic boundary condition, Equation 11.1.46.

Problems

Find the eigenvalues and eigenfunctions for each of the following:

1. $y'' + \lambda y = 0, \quad y'(0) = 0, \quad y(L) = 0$

2. $y'' + \lambda y = 0, \quad y'(0) = 0, \quad y'(\pi) = 0$

3. $y'' + \lambda y = 0, \quad y(0) + y'(0) = 0, \quad y(\pi) + y'(\pi) = 0$

4. $y'' + \lambda y = 0, \quad y'(0) = 0, \quad y(\pi) - y'(\pi) = 0$

5. $y^{(iv)} + \lambda y = 0, \quad y(0) = y''(0) = 0, \quad y(L) = y''(L) = 0$

Find an equation from which you could find λ and give the form of the eigenfunction for each of the following:

6. $y'' + \lambda y = 0, \quad y(0) + y'(0) = 0, \quad y(1) = 0$

7. $y'' + \lambda y = 0, \quad y(0) = 0, \quad y(\pi) + y'(\pi) = 0$

8. $y'' + \lambda y = 0, \quad y'(0) = 0, \quad y(1) - y'(1) = 0$

9. $y'' + \lambda y = 0, \quad y(0) + y'(0) = 0, \quad y'(\pi) = 0$

10. $y'' + \lambda y = 0, \quad y(0) + y'(0) = 0, \quad y(\pi) - y'(\pi) = 0$

11. Find the eigenvalues and eigenfunctions of the Sturm-Liouville problem:

$$\frac{d}{dx}\left(x\frac{dy}{dx}\right) + \frac{\lambda}{x}y = 0, \qquad 1 \le x \le e$$

for each of the following boundary conditions: (a) $u(1) = u(e) = 0$, (b) $u(1) = u'(e) = 0$, and (c) $u'(1) = u'(e) = 0$.

Step 1: Using the transformation $\eta = \ln(x)$, show that we can transform the differential equation into

$$\frac{d^2y}{d\eta^2} + \lambda y = 0, \qquad 0 \le \eta \le 1.$$

State the boundary conditions that will now occur at $\eta = 0$ and $\eta = 1$.

Step 2: Solve this new differential equation for the three separate cases of $\lambda < 0$, $\lambda = 0$ and $\lambda > 0$.

Step 3: Transform any nontrivial solutions back into the dependent variable x.

12. Find the eigenvalues and eigenfunctions of the following Sturm-Liouville problem:

$$x^2 y'' + 2xy' + \lambda y = 0, \qquad y(1) = y(e) = 0, \qquad 1 \le x \le e.$$

Step 1: Show that our Sturm-Liouville equation can be rewritten

$$x(xy'' + y') + xy' + \lambda y = 0, \qquad \text{or} \qquad x\frac{d}{dx}\left(x\frac{dy}{dx}\right) + x\frac{dy}{dx} + \lambda y = 0.$$

Step 2: Using the transformation $\eta = \ln(x)$, show that we can transform the second differential equation into

$$\frac{d^2y}{d\eta^2} + \frac{dy}{d\eta} + \lambda y = 0, \qquad 0 \le \eta \le 1.$$

State the boundary condition now at $\eta = 0$ and $\eta = 1$.

Step 3: Solve this new differential equation for the three separate cases of $\lambda < 0$, $\lambda = 0$ and $\lambda > 0$.

Step 4: Transform any nontrivial solutions back into the dependent variable x.

13. Find the eigenvalues and eigenfunctions of the following Sturm-Liouville problem:

$$\frac{d}{dx}(x^3 y') + \lambda xy = 0, \qquad y(1) = y(e^\pi) = 0, \qquad 1 \le x \le e^\pi.$$

Step 1: Show that our Sturm-Liouville equation can be rewritten

$$x(xy'' + y') + 2xy' + \lambda y = 0 \qquad \text{or} \qquad x\frac{d}{dx}\left(x\frac{dy}{dx}\right) + 2x\frac{dy}{dx} + \lambda y = 0.$$

Step 2: Using the transformation $\eta = \ln(x)$, show that we can transform the second differential equation in Step 1 into

$$\frac{d^2y}{d\eta^2} + 2\frac{dy}{d\eta} + \lambda y = 0, \qquad 0 \le \eta \le \pi.$$

State the boundary condition now at $\eta = 0$ and $\eta = 1$.

Step 3: Solve this new differential equation for the three separate cases of $\lambda < 0$, $\lambda = 0$ and $\lambda > 0$.

Step 4: Transform any nontrivial solutions back into the dependent variable x.

14. Find the eigenvalues and eigenfunctions of the following Sturm-Liouville problem:

$$\frac{d}{dx}\left(\frac{1}{x}y'\right) + \frac{\lambda}{x^3}y = 0, \qquad y(1) = y(e) = 0, \qquad 1 \leq x \leq e.$$

Step 1: Show that our Sturm-Liouville equation can be rewritten

$$x(xy'' + y') - 2xy' + \lambda y = 0 \qquad \text{or} \qquad x\frac{d}{dx}\left(x\frac{dy}{dx}\right) - 2x\frac{dy}{dx} + \lambda y = 0.$$

Step 2: Using the transformation $\eta = \ln(x)$, show that we can transform the second differential equation in Step 1 into

$$\frac{d^2y}{d\eta^2} - 2\frac{dy}{d\eta} + \lambda y = 0, \qquad 0 \leq \eta \leq 1.$$

State the boundary condition now at $\eta = 0$ and $\eta = 1$.

Step 3: Solve this new differential equation for the three separate cases of $\lambda < 0$, $\lambda = 0$ and $\lambda > 0$.

Step 4: Transform any nontrivial solutions back into the dependent variable x.

15. Find the eigenvalues and eigenfunctions of the following problem:

$$y'''' - \lambda^4 y = 0, \qquad y'''(0) = y''(0) = y''(1) = y'(1) = 0, \qquad 0 < x < 1.$$

Step 1: Show that if $\lambda^4 < 0$, we can write $\lambda^4 = k^4 e^{\pi i}$ with $k > 0$. In this case, the solution is $y(x) = Ae^{\lambda_1 x} + Be^{\lambda_2 x} + Ce^{\lambda_3 x} + De^{\lambda_4 x}$, where $\lambda_1 = ke^{\pi i/4}$, $\lambda_2 = ke^{3\pi i/4}$, $\lambda_3 = ke^{5\pi i/4}$, $\lambda_4 = ke^{7\pi i/4}$.

Step 2: By substituting the solution from Step 1 into the boundary conditions, show that you have a trivial solution for $\lambda^4 < 0$.

Step 3: Show that $\lambda^4 = 0$, then $y(x) = A + Bx + Cx^2 + Dx^3$.

Step 4: By substituting the solution from Step 3 into the boundary conditions, show that the eigenfunction is $y_0(x) = 1$.

Step 5: Show that for $\lambda^4 > 0$, then $y(x) = A\cosh(\lambda x) + B\sinh(\lambda x) + C\cos(\lambda x) + B\sin(\lambda x)$.
Step 6: By substituting the solution from Step 5 into the boundary conditions, show that the eigenfunction is $y_n(x) = \cosh(\lambda_n x) + \cos(\lambda_n x) - \tanh(\lambda_n)[\sinh(\lambda_n x) + \sin(\lambda_n x)]$, where λ_n satisfies the equation $\tanh(\lambda) = -\tan(\lambda)$ and $n = 1, 2, 3, \ldots$.

Project: Numerical Solution of the Sturm-Liouville Problem

You may have been struck by the similarity of the algebraic eigenvalue problem to the Sturm-Liouville problem. (See Section 3.5.) In both cases, nontrivial solutions exist only for characteristic values of λ. The purpose of this project is to further deepen your insight into these similarities. This project requires a knowledge of matrix algebra (see Sections 3.1 and 3.4).

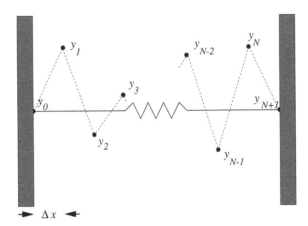

Figure 11.1.4: Schematic for finite-differencing a Sturm-Liouville problem into a set of difference equations.

Consider the Sturm-Liouville problem

$$y'' + \lambda y = 0, \qquad y(0) = y(\pi) = 0. \tag{11.1.58}$$

We know that it has the nontrivial solutions $\lambda_m = m^2$, $y_m(x) = \sin(mx)$, where $m = 1, 2, 3, \ldots$.

Step 1: Let us solve this problem numerically. Introducing centered finite differencing and the grid shown in Figure 11.1.4, show that

$$y'' \approx \frac{y_{n+1} - 2y_n + y_{n-1}}{(\Delta x)^2}, \qquad n = 1, 2, \ldots, N,$$

where $y_n = y(x_n)$, $x_n = n\Delta x$ and $\Delta x = \pi/(N+1)$. Show that the finite-differenced form of Equation 11.1.58 is

$$-h^2 y_{n+1} + 2h^2 y_n - h^2 y_{n-1} = \lambda y_n \tag{11.1.59}$$

with $y_0 = y_{N+1} = 0$, and $h = 1/(\Delta x)$.

Step 2: Solve Equation 11.1.59 as an algebraic eigenvalue problem using $N = 1, 2, \ldots$. Show that it can be written in the matrix form of

$$
\begin{pmatrix}
2h^2 & -h^2 & 0 & \cdots & 0 & 0 & 0 \\
-h^2 & 2h^2 & -h^2 & \cdots & 0 & 0 & 0 \\
0 & -h^2 & 2h^2 & \cdots & 0 & 0 & 0 \\
\vdots & \vdots & \vdots & \ddots & \vdots & \vdots & \vdots \\
0 & 0 & 0 & \cdots & -h^2 & 2h^2 & -h^2 \\
0 & 0 & 0 & \cdots & 0 & -h^2 & 2h^2
\end{pmatrix}
\begin{pmatrix}
y_1 \\ y_2 \\ y_3 \\ \vdots \\ y_{N-1} \\ y_N
\end{pmatrix}
= \lambda
\begin{pmatrix}
y_1 \\ y_2 \\ y_3 \\ \vdots \\ y_{N-1} \\ y_N
\end{pmatrix}. \tag{11.1.60}
$$

Note that the coefficient matrix is symmetric.

Step 3: You are now ready to compute the eigenvalues. For small N this could be done by hand. However, it is easier just to write a MATLAB program that will handle any $N \geq 2$. Table 11.1.2 has been provided so that you can check your program.

With your program, answer the following questions: How do your computed eigenvalues compare to the eigenvalues given by the Sturm-Liouville problem, Equation 11.1.58? What

Table 11.1.2: Eigenvalues λ_m, Ordered According to Increasing Magnitude, Computed from Equation 11.1.60 as a Numerical Approximation of the Sturm-Liouville Problem, Equation 11.1.58.

N	λ_1	λ_2	λ_3	λ_4	λ_5	λ_6	λ_7
1	0.81057						
2	0.91189	2.73567					
3	0.94964	3.24228	5.53491				
4	0.96753	3.50056	6.63156	9.16459			
5	0.97736	3.64756	7.29513	10.94269	13.61289		
6	0.98333	3.73855	7.71996	12.13899	16.12040	18.87563	
7	0.98721	3.79857	8.00605	12.96911	17.93217	22.13966	24.95100
8	0.98989	3.84016	8.20702	13.56377	19.26430	24.62105	28.98791
20	0.99813	3.97023	8.84993	15.52822	23.85591	33.64694	44.68265
50	0.99972	3.99498	8.97438	15.91922	24.80297	35.59203	48.24538

happens as you increase N? Which computed eigenvalues agree best with those given by the Sturm-Liouville problem, Equation 11.1.58? Which ones compare the worst?

Step 4: Let us examine the eigenvectors now. Starting with the smallest eigenvalue (number it $m = 1$), use MATLAB to plot the corresponding mth eigenvector $Cy_n^{(m)}$ as a function of x_n $n = 1, 2, \ldots, N$, $m = 1, 2, \ldots, N$, and C is chosen so that $C^2 \Delta x \sum_n [y_n^{(m)}]^2 = 1$. On the same plot, graph $y_m(x) = \sqrt{2/\pi} \sin(mx)$. Why did we choose C as we did? Which eigenvectors and eigenfunctions agree the best? Which eigenvectors and eigenfunctions agree the worst? Why? Why are there N eigenvectors and an infinite number of eigenfunctions?

Step 5: As we show in the next section, the most important property of eigenfunctions is orthogonality. But what do we mean by orthogonality in the case of eigenvectors? Recall from three-dimensional vectors we had the scalar dot product

$$\mathbf{a} \cdot \mathbf{b} = a_1 b_1 + a_2 b_2 + a_3 b_3.$$

For n-dimensional vectors, this dot product is generalized to the inner product

$$\mathbf{x} \cdot \mathbf{y} = \sum_{k=1}^{n} x_k y_k.$$

Orthogonality implies that $\mathbf{x} \cdot \mathbf{y} = 0$ if $\mathbf{x} \neq \mathbf{y}$. Are your eigenvectors orthogonal?

11.2 ORTHOGONALITY OF EIGENFUNCTIONS

In the previous section we saw how nontrivial solutions to the regular Sturm-Liouville problem consist of eigenvalues and eigenfunctions. The most important property of eigenfunctions is orthogonality.

Theorem: *Let the functions $p(x)$, $q(x)$, and $r(x)$ of the regular Sturm-Liouville problem, Equation 11.1.1 and Equation 11.1.2, be real and continuous on the interval $[a, b]$. If $y_n(x)$ and $y_m(x)$ are continuously differentiable eigenfunctions corresponding to the distinct eigenvalues λ_n and λ_m, respectively, then $y_n(x)$ and $y_m(x)$ satisfy the* orthogonality *condition:*

$$\int_a^b r(x) y_n(x) y_m(x)\, dx = 0, \qquad (11.2.1)$$

if $\lambda_n \neq \lambda_m$. When Equation 11.2.1 is satisfied, the eigenfunctions $y_n(x)$ and $y_m(x)$ are said to be *orthogonal* to each other with respect to the *weight function* $r(x)$. The term *orthogonality* appears to be borrowed from linear algebra where a similar relationship holds between two perpendicular or orthogonal vectors.

Proof: Let $y_n(x)$ and $y_m(x)$ denote the eigenfunctions associated with two different eigenvalues λ_n and λ_m. Then

$$\frac{d}{dx}\left[p(x)\frac{dy_n}{dx}\right] + [q(x) + \lambda_n r(x)] y_n(x) = 0, \qquad (11.2.2)$$

$$\frac{d}{dx}\left[p(x)\frac{dy_m}{dx}\right] + [q(x) + \lambda_m r(x)] y_m(x) = 0, \qquad (11.2.3)$$

and both solutions satisfy the boundary conditions. Let us multiply the first differential equation by y_m; the second by y_n. Next, we subtract these two equations and move the terms containing $y_n y_m$ to the right side. The resulting equation is

$$y_n \frac{d}{dx}\left[p(x)\frac{dy_m}{dx}\right] - y_m \frac{d}{dx}\left[p(x)\frac{dy_n}{dx}\right] = (\lambda_n - \lambda_m) r(x) y_n y_m. \qquad (11.2.4)$$

Integrating Equation 11.2.4 from a to b yields

$$\int_a^b \left\{ y_n \frac{d}{dx}\left[p(x)\frac{dy_m}{dx}\right] - y_m \frac{d}{dx}\left[p(x)\frac{dy_n}{dx}\right] \right\} dx = (\lambda_n - \lambda_m) \int_a^b r(x) y_n y_m\, dx. \qquad (11.2.5)$$

We can simplify the left side of Equation 11.2.5 by integrating by parts to give

$$\int_a^b \left\{ y_n \frac{d}{dx}\left[p(x)\frac{dy_m}{dx}\right] - y_m \frac{d}{dx}\left[p(x)\frac{dy_n}{dx}\right] \right\} dx$$

$$= [p(x) y_m' y_n - p(x) y_n' y_m]_a^b - \int_a^b p(x)[y_n' y_m' - y_n' y_m']\, dx. \qquad (11.2.6)$$

The second integral equals zero since the integrand vanishes identically. Because $y_n(x)$ and $y_m(x)$ satisfy the boundary condition at $x = a$,

$$\alpha y_n(a) + \beta y_n'(a) = 0, \qquad (11.2.7)$$

and

$$\alpha y_m(a) + \beta y_m'(a) = 0. \qquad (11.2.8)$$

These two equations are simultaneous equations in α and β. Hence, the determinant of the equations must be zero:

$$y_n'(a)y_m(a) - y_m'(a)y_n(a) = 0. \tag{11.2.9}$$

Similarly, at the other end,

$$y_n'(b)y_m(b) - y_m'(b)y_n(b) = 0. \tag{11.2.10}$$

Consequently, the right side of Equation 11.2.6 vanishes and Equation 11.2.5 reduces to Equation 11.2.1. □

• **Example 11.2.1**

Let us verify the orthogonality condition for the eigenfunctions that we found in Example 11.1.1.

Because $r(x) = 1$, $a = 0$, $b = \pi$, $y_{-1}(x) = \sinh(m_{-1}x)$ and $y_n(x) = \sin(k_n x)$, we find that

$$\int_a^b r(x)y_n y_m\,dx = \int_0^\pi \sinh(m_{-1}x)\sin(k_m x)\,dx \tag{11.2.11}$$

$$= \left. \frac{m_{-1}\cosh(m_{-1}x)\sin(k_m x) - k_m \sinh(m_{-1}x)\cos(k_m x)}{m_{-1}^2 + k_m^2}\right|_0^\pi \tag{11.2.12}$$

$$= \frac{m_{-1}\cosh(m_{-1}\pi)\sin(k_m \pi) - k_m \sinh(m_{-1}\pi)\cos(k_m \pi)}{m_{-1}^2 + k_m^2} \tag{11.2.13}$$

$$= 0, \tag{11.2.14}$$

since $\sinh(m_{-1}\pi) = m_{-1}\cosh(m_{-1}\pi)$ and $\sin(k_n\pi) = k_n\cos(k_n\pi)$. Similarly, if $n, m > 0$,

$$\int_a^b r(x)y_n y_m\,dx = \int_0^\pi \sin(k_n x)\sin(k_m x)\,dx \tag{11.2.15}$$

$$= \tfrac{1}{2}\int_0^\pi \left\{\cos[(k_n - k_m)x] - \cos[(k_n + k_m)x]\right\}dx \tag{11.2.16}$$

$$= \left.\frac{\sin[(k_n - k_m)x]}{2(k_n - k_m)}\right|_0^\pi - \left.\frac{\sin[(k_n + k_m)x]}{2(k_n + k_m)}\right|_0^\pi \tag{11.2.17}$$

$$= \frac{\sin[(k_n - k_m)\pi]}{2(k_n - k_m)} - \frac{\sin[(k_n + k_m)\pi]}{2(k_n + k_m)} \tag{11.2.18}$$

$$= \frac{\sin(k_n\pi)\cos(k_m\pi) - \cos(k_n\pi)\sin(k_m\pi)}{2(k_n - k_m)}$$

$$- \frac{\sin(k_n\pi)\cos(k_m\pi) + \cos(k_n\pi)\sin(k_m\pi)}{2(k_n + k_m)} \tag{11.2.19}$$

$$= \frac{k_n\cos(k_n\pi)\cos(k_m\pi) - k_m\cos(k_n\pi)\cos(k_m\pi)}{2(k_n - k_m)}$$

$$- \frac{k_n\cos(k_n\pi)\cos(k_m\pi) + k_m\cos(k_n\pi)\cos(k_m\pi)}{2(k_n + k_m)} \tag{11.2.20}$$

$$= \frac{(k_n - k_m)\cos(k_n\pi)\cos(k_m\pi)}{2(k_n - k_m)}$$

$$- \frac{(k_n + k_m)\cos(k_n\pi)\cos(k_m\pi)}{2(k_n + k_m)} = 0. \tag{11.2.21}$$

We used the relationships $k_n = \tan(k_n\pi)$, and $k_m = \tan(k_m\pi)$ to simplify Equation 11.2.19. Note, however, that

$$\int_0^\pi \sinh(m_{-1}x)\sinh(m_{-1}x)\,dx = \tfrac{1}{2}\int_0^\pi [\cosh(2m_{-1}x) - 1]\,dx \qquad (11.2.22)$$

$$= \frac{\sinh(2m_{-1}\pi)}{4m_{-1}} - \frac{\pi}{2} \qquad (11.2.23)$$

$$= \tfrac{1}{2}[\cosh^2(m_{-1}\pi) - \pi] > 0, \qquad (11.2.24)$$

since $\sinh(2A) = 2\sinh(A)\cosh(A)$ and $\sinh(m_{-1}\pi) = m_{-1}\cosh(m_{-1}\pi)$. Similarly,

$$\int_0^\pi \sin(k_n x)\sin(k_n x)\,dx = \tfrac{1}{2}\int_0^\pi [1 - \cos(2k_n x)]\,dx = \frac{\pi}{2} - \frac{\sin(2k_n\pi)}{4k_n} = \tfrac{1}{2}[\pi - \cos^2(k_n\pi)] > 0,$$
$$(11.2.25)$$

because $\sin(2A) = 2\sin(A)\cos(A)$, and $k_n = \tan(k_n\pi)$. That is, any eigenfunction *cannot* be orthogonal to itself.

In closing, we note that had we defined the eigenfunction in our example as

$$y_{-1}(x) = \frac{\sinh(m_{-1}x)}{\sqrt{[\cosh^2(m_{-1}\pi) - \pi]/2}} \qquad (11.2.26)$$

and

$$y_n(x) = \frac{\sin(k_n x)}{\sqrt{[\pi - \cos^2(k_n\pi)]/2}} \qquad (11.2.27)$$

rather than $y_{-1}(x) = \sinh(m_{-1}x)$ and $y_n(x) = \sin(k_n x)$, the orthogonality condition would read

$$\int_0^\pi y_n(x)y_m(x)\,dx = \begin{cases} 0, & m \neq n, \\ 1, & m = n, \end{cases} \qquad (11.2.28)$$

where $n, m = -1, 1, 2, 3, \ldots.$. This process of *normalizing* an eigenfunction so that the orthogonality condition becomes

$$\int_a^b r(x)y_n(x)y_m(x)\,dx = \begin{cases} 0, & m \neq n, \\ 1, & m = n, \end{cases} \qquad (11.2.29)$$

generates *orthonormal* eigenfunctions. We will see the convenience of doing this in the next section.

Problems

1. The Sturm-Liouville problem $y'' + \lambda y = 0$, $y(0) = y(L) = 0$ has the eigenfunction solution $y_n(x) = \sin(n\pi x/L)$. By direct integration, verify the orthogonality condition, Equation 11.2.1.

2. The Sturm-Liouville problem $y'' + \lambda y = 0$, $y'(0) = y'(L) = 0$ has the eigenfunction solutions $y_0(x) = 1$ and $y_n(x) = \cos(n\pi x/L)$. By direct integration, verify the orthogonality condition, Equation 11.2.1.

3. The Sturm-Liouville problem $y'' + \lambda y = 0$, $y(0) = y'(L) = 0$ has the eigenfunction solution $y_n(x) = \sin[(2n - 1)\pi x/(2L)]$. By direct integration, verify the orthogonality condition, Equation 11.2.1.

4. The Sturm-Liouville problem $y'' + \lambda y = 0$, $y'(0) = y(L) = 0$ has the eigenfunction solution $y_n(x) = \cos[(2n - 1)\pi x/(2L)]$. By direct integration, verify the orthogonality condition, Equation 11.2.1.

11.3 EXPANSION IN SERIES OF EIGENFUNCTIONS

In calculus we learned that under certain conditions we could represent a function $f(x)$ by a linear and infinite sum of polynomials $(x - x_0)^n$. In this section we show that an analogous procedure exists for representing a piece-wise continuous function by a linear sum of eigenfunctions. These *eigenfunction expansions* will be used to solve partial differential equations.

Let the function $f(x)$ be defined in the interval $a < x < b$. We wish to re-express $f(x)$ in terms of the eigenfunctions $y_n(x)$ given by a regular Sturm-Liouville problem. Assuming that the function $f(x)$ can be represented by a uniformly convergent series,[6] we write

$$f(x) = \sum_{n=1}^{\infty} c_n y_n(x).$$

(11.3.1)

The orthogonality relation, Equation 11.2.1, gives us the method for computing the coefficients c_n. First we multiply both sides of Equation 11.3.1 by $r(x)y_m(x)$, where m is a fixed integer, and then integrate from a to b. Because this series is uniformly convergent and $y_n(x)$ is continuous, we can integrate the series term by term, or

$$\int_a^b r(x)f(x)y_m(x)\,dx = \sum_{n=1}^{\infty} c_n \int_a^b r(x)y_n(x)y_m(x)\,dx.$$

(11.3.2)

The orthogonality relationship states that all of the terms on the right side of Equation 11.3.2 must disappear except the one for which $n = m$. Thus, we are left with

$$\int_a^b r(x)f(x)y_m(x)\,dx = c_m \int_a^b r(x)y_m(x)y_m(x)\,dx$$

(11.3.3)

or

$$c_n = \frac{\int_a^b r(x)f(x)y_n(x)\,dx}{\int_a^b r(x)y_n^2(x)\,dx},$$

(11.3.4)

---

[6] If $S_n(x) = \sum_{k=1}^n u_k(x)$, $S(x) = \lim_{n \to \infty} S_n(x)$, and $0 < |S_n(x) - S(x)| < \epsilon$ for all $n > M > 0$, the series $\sum_{k=1}^{\infty} u_k(x)$ is uniformly convergent if M is dependent on ϵ alone and not x.

if we replace m by n in Equation 11.3.3.

Usually, both integrals in Equation 11.3.4 are evaluated by direct integration. In the case when the evaluation of the denominator is very difficult, Lockshin[7] has shown that the denominator of Equation 11.3.4 always equals

$$\int_a^b r(x) y^2(x)\, dx = p(x) \left[\frac{\partial y}{\partial x} \frac{\partial y}{\partial \lambda} - y \frac{\partial^2 y}{\partial \lambda \partial x} \right] \Bigg|_a^b, \tag{11.3.5}$$

for a regular Sturm-Liouville problem with eigenfunction solution y, where $p(x)$, $q(x)$, and $r(x)$ are continuously differentiable on the interval $[a, b]$.

The series, Equation 11.3.1, with the coefficients found by Equation 11.3.4, is a *generalized Fourier series* of the function $f(x)$ with respect to the eigenfunction $y_n(x)$. It is called a generalized Fourier series because we generalized the procedure of re-expressing a function $f(x)$ by sines and cosines into one involving solutions to regular Sturm-Liouville problems. Note that if we had used an orthonormal set of eigenfunctions, then the denominator of Equation 11.3.4 would equal one and we reduce our work by half. The coefficients c_n are the *Fourier coefficients*.

One of the most remarkable facts about generalized Fourier series is their applicability even when the function has a finite number of bounded discontinuities in the range $[a, b]$. We may formally express this fact by the following theorem:

Theorem: *If both $f(x)$ and $f'(x)$ are piece-wise continuous in $a \le x \le b$, then $f(x)$ can be expanded in a uniformly convergent Fourier series, Equation 11.3.1, whose coefficients c_n are given by Equation 11.3.4. It converges to $[f(x^+) + f(x^-)]/2$ at any point x in the open interval $a < x < b$.*

The proof is beyond the scope of this book but can be found in more advanced treatises.[8] If we are willing to include stronger constraints, we can make even stronger statements about convergence. For example,[9] if we require that $f(x)$ be a continuous function with a piece-wise continuous first derivative, then the eigenfunction expansion, Equation 11.3.1, converges to $f(x)$ uniformly and absolutely in $[a, b]$ if $f(x)$ satisfies the same boundary conditions as does $y_n(x)$. □

In the case when $f(x)$ is discontinuous, we are not merely rewriting $f(x)$ in a new form. We are actually choosing the coefficients c_n so that the eigenfunction expansion fits $f(x)$ in the "least squares" sense that

$$\int_a^b r(x) \left| f(x) - \sum_{n=1}^{\infty} c_n y_n(x) \right|^2 dx = 0. \tag{11.3.6}$$

Consequently we should expect peculiar things, such as spurious oscillations, to occur in the neighborhood of the discontinuity. These are *Gibbs phenomena*,[10] the same phenomena discovered with Fourier series. See Section 5.2.

---

[7] Lockshin, J. L, 2001: Explicit closed-form expression for eigenfunction norms. *Appl. Math. Lett.*, **14**, 553–555.

[8] For example, Titchmarsh, E. C., 1962: *Eigenfunction Expansions Associated with Second-Order Differential Equations. Part 1.* Oxford University Press, pp. 12–16.

[9] Tolstov, G. P., 1962: *Fourier Series.* Dover Publishers, p. 255.

[10] Apparently first discussed by Weyl, H., 1910: Die Gibbs'sche Erscheinung in der Theorie der Sturm-Liouvilleschen Reihen. *Rend. Circ. Mat. Palermo*, **29**, 321–323.

• **Example 11.3.1**

To illustrate the concept of an eigenfunction expansion, let us find the expansion for $f(x) = x$ over the interval $0 < x < \pi$ using the solution to the regular Sturm-Liouville problem of

$$y'' + \lambda y = 0, \qquad y(0) = y(\pi) = 0. \tag{11.3.7}$$

This problem arises when we solve the wave or heat equation by separation of variables.

Because the eigenfunctions are $y_n(x) = \sin(nx)$, $n = 1, 2, 3, \ldots$, $r(x) = 1$, $a = 0$, and $b = \pi$, Equation 11.3.4 yields

$$c_n = \frac{\int_0^\pi x \sin(nx)\, dx}{\int_0^\pi \sin^2(nx)\, dx} = \frac{-x\cos(nx)/n + \sin(nx)/n^2 \big|_0^\pi}{x/2 - \sin(2nx)/(4n)\big|_0^\pi} = -\frac{2}{n}\cos(n\pi) = -\frac{2}{n}(-1)^n.$$
$$\tag{11.3.8}$$

Equation 11.3.1 then gives

$$f(x) = -2\sum_{n=1}^\infty \frac{(-1)^n}{n}\sin(nx). \tag{11.3.9}$$

This particular example is in fact an example of a half-range sine expansion.

Finally, we must state the values of x for which Equation 11.3.9 is valid. At $x = \pi$ the series converges to zero while $f(\pi) = \pi$. At $x = 0$ both the series and the function converge to zero. Hence this series expansion is valid for $0 \le x < \pi$. $\qquad\square$

• **Example 11.3.2**

For our second example let us find the expansion for $f(x) = x$ over the interval $0 \le x < \pi$ using the solution to the regular Sturm-Liouville problem of

$$y'' + \lambda y = 0, \qquad y(0) = y(\pi) - y'(\pi) = 0. \tag{11.3.10}$$

We will encounter this problem when we solve the heat equation with radiative boundary conditions by separation of variables.

Because $r(x) = 1$, $a = 0$, $b = \pi$, and the eigenfunctions are $y_{-1}(x) = \sinh(m_{-1}x)$ and $y_n(x) = \sin(k_n x)$, where $\sinh(m_{-1}) = m_{-1}\cosh(m_{-1})$ and $k_n = \tan(k_n\pi)$, Equation 11.3.4 yields

$$c_{-1} = \frac{\int_0^\pi x\sinh(m_{-1}x)\, dx}{\int_0^\pi \sinh^2(m_{-1}x)\, dx} = \frac{\int_0^\pi x\sinh(m_{-1}x)\, dx}{\frac{1}{2}\int_0^\pi [\cosh(2m_{-1}x) - 1]\, dx} \tag{11.3.11}$$

$$= 2\frac{x\cosh(m_{-1}x)/m_{-1} - \sinh(m_{-1}x)/m_{-1}^2 \big|_0^\pi}{\sinh(2m_{-1}x)/(2m_{-1}) - x\big|_0^\pi} \tag{11.3.12}$$

$$= \frac{2\pi\cosh(m_{-1}\pi)/m_{-1} - 2\sinh(m_{-1}\pi)/m_{-1}^2}{\sinh(2m_{-1}\pi)/(2m_{-1}) - \pi} \tag{11.3.13}$$

$$= \frac{2(\pi - 1)\cosh(m_{-1}\pi)/m_{-1}}{\cosh^2(m_{-1}\pi) - \pi}, \tag{11.3.14}$$

where we used the property that $\sinh(m_{-1}\pi) = m_{-1}\cosh(m_{-1}\pi)$; and

$$c_n = \frac{\int_0^\pi x\sin(k_n x)\, dx}{\int_0^\pi \sin^2(k_n x)\, dx} = \frac{\int_0^\pi x\sin(k_n x)\, dx}{\frac{1}{2}\int_0^\pi [1 - \cos(2k_n x)]\, dx} = \frac{2\sin(k_n x)/k_n^2 - 2x\cos(k_n x)/k_n \big|_0^\pi}{x - \sin(2k_n x)/(2k_n)\big|_0^\pi}$$
$$\tag{11.3.15}$$

$$= \frac{2\sin(k_n\pi)/k_n^2 - 2\pi\cos(k_n\pi)/k_n}{\pi - \sin(2k_n\pi)/(2k_n)} = \frac{2(1-\pi)\cos(k_n\pi)/k_n}{\pi - \cos^2(k_n\pi)}, \tag{11.3.16}$$

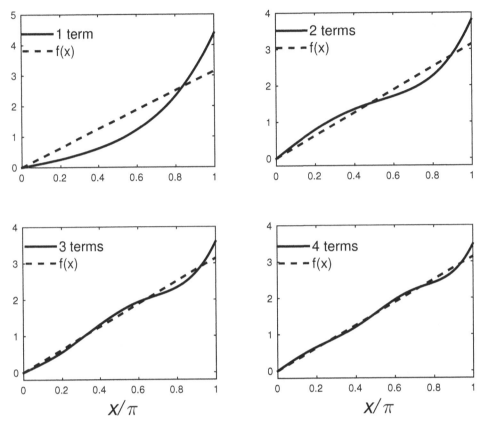

Figure 11.3.1: The eigenfunction expansion, Equation 11.3.17, for $f(x) = x$, $0 < x/\pi < 1$, when we truncate the series so that it includes only the first, first two, first three, and first four terms.

where we used the property that $\sin(k_n\pi) = k_n\cos(k_n\pi)$. Equation 11.3.1 then gives

$$
f(x) = 2(\pi - 1)\left\{\frac{\cosh(m_{-1}\pi)\sinh(m_{-1}x)}{m_{-1}[\cosh^2(m_{-1}\pi) - \pi]} - \sum_{n=1}^{\infty}\frac{\cos(k_n\pi)\sin(k_nx)}{k_n[\pi - \cos^2(k_n\pi)]}\right\}. \tag{11.3.17}
$$

Figure 11.3.1 illustrates our eigenfunction expansion of $f(x) = x$ in truncated form when we only include one, two, three, and four terms.

To illustrate the use of Equation 11.3.5, we note that

$$
y(x) = \sin\left(\sqrt{\lambda}\,x\right), \quad \frac{\partial y}{\partial x} = \sqrt{\lambda}\,\cos\left(\sqrt{\lambda}\,x\right), \quad \frac{\partial y}{\partial \lambda} = \frac{x}{2\sqrt{\lambda}}\cos\left(\sqrt{\lambda}\,x\right), \tag{11.3.18}
$$

and

$$
\frac{\partial^2 y}{\partial\lambda\partial x} = \frac{\partial^2 y}{\partial x\partial\lambda} = \frac{1}{2\sqrt{\lambda}}\cos\left(\sqrt{\lambda}\,x\right) - \frac{x}{2}\sin\left(\sqrt{\lambda}\,x\right). \tag{11.3.19}
$$

Therefore,

$$
\int_0^{\pi}\sin^2\left(\sqrt{\lambda}\,x\right)dx = \left\{\frac{x}{2}\cos^2\left(\sqrt{\lambda}\,x\right) - \sin\left(\sqrt{\lambda}\,x\right)\right.
$$
$$
\left.\times\left[\frac{1}{2\sqrt{\lambda}}\cos\left(\sqrt{\lambda}\,x\right) - \frac{x}{2}\sin\left(\sqrt{\lambda}\,x\right)\right]\right\}\Bigg|_0^{\pi} \tag{11.3.20}
$$

$$\int_0^\pi \sin^2\left(\sqrt{\lambda}\,x\right)\,dx = \left\{\frac{x}{2} - \frac{1}{2\sqrt{\lambda}}\sin\left(\sqrt{\lambda}\,x\right)\cos\left(\sqrt{\lambda}\,x\right)\right\}\Bigg|_0^\pi \qquad (11.3.21)$$

$$= \frac{\pi}{2} - \frac{1}{2\sqrt{\lambda}}\sin\left(\sqrt{\lambda}\,\pi\right)\cos\left(\sqrt{\lambda}\,\pi\right). \qquad (11.3.22)$$

If $\lambda_{-1} = -m_{-1}^2$, $\sqrt{\lambda_{-1}} = im_{-1}$ and we obtain Equation 11.2.24 if we use $\sinh(m_{-1}\pi) = m_{-1}\cosh(m_{-1}\pi)$. On the other hand, if $\lambda_n = k_n$, $\sqrt{\lambda_n} = k_n$ and we obtain Equation 11.2.25 if we use $\sin(k_n\pi) = k_n\cos(k_n\pi)$. $\qquad\qquad\qquad\qquad\qquad\qquad\qquad\Box$

- **Example 11.3.3: The separation of variables solution to the heat equation when the left wall is insulated**

The reason why we developed expansions in orthogonal functions is to solve linear partial differential equations with complicated boundary conditions using the technique of separation of variables. It is assumed that the reader already understands this technique.

Let us solve the heat equation

$$\frac{\partial u}{\partial t} = a^2 \frac{\partial^2 u}{\partial x^2}, \qquad 0 < x < L, \quad 0 < t, \qquad (11.3.23)$$

which satisfies the initial condition $u(x,0) = x$, $0 < x < L$, and the boundary conditions $u_x(0,t) = u(L,t) = 0$, $0 < t$. The condition $u_x(0,t) = 0$ expresses mathematically the constraint that no heat flows through the left boundary (insulated end condition).

Employing separation of variables, the positive and zero separation constants yield trivial solutions. For a negative separation constant, however,

$$X'' + k^2 X = 0, \qquad (11.3.24)$$

with $X'(0) = X(L) = 0$ because $u_x(0,t) = X'(0)T(t) = 0$, and $u(L,t) = X(L)T(t) = 0$. This regular Sturm-Liouville problem has the solution

$$X_n(x) = \cos\left[\frac{(2n-1)\pi x}{2L}\right], \qquad n = 1,2,3,\dots. \qquad (11.3.25)$$

The temporal solution then becomes

$$T_n(t) = B_n \exp\left[-\frac{a^2(2n-1)^2\pi^2 t}{4L^2}\right]. \qquad (11.3.26)$$

Consequently, a linear superposition of the particular solutions gives the total solution, which equals

$$u(x,t) = \sum_{n=1}^{\infty} B_n \cos\left[\frac{(2n-1)\pi x}{2L}\right]\exp\left[-\frac{a^2(2n-1)^2\pi^2}{4L^2}t\right]. \qquad (11.3.27)$$

Our final task remains to find the coefficients B_n. Evaluating Equation 11.3.27 at $t = 0$,

$$u(x,0) = x = \sum_{n=1}^{\infty} B_n \cos\left[\frac{(2n-1)\pi x}{2L}\right], \qquad 0 < x < L. \qquad (11.3.28)$$

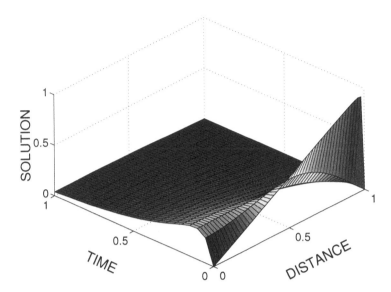

Figure 11.3.2: The temperature $u(x,t)/L$ within a thin bar as a function of position x/L and time a^2t/L^2 when we insulate the left end and hold the right end at the temperature of zero. The initial temperature equals x.

Equation 11.3.28 is not a half-range cosine expansion; it is an expansion in the orthogonal functions $\cos[(2n-1)\pi x/(2L)]$ corresponding to the regular Sturm-Liouville problem, Equation 11.3.24. Consequently, B_n is given by Equation 11.3.4 with $r(x) = 1$ as

$$B_n = \frac{\int_0^L x \cos[(2n-1)\pi x/(2L)]\,dx}{\int_0^L \cos^2[(2n-1)\pi x/(2L)]\,dx} \tag{11.3.29}$$

$$= \frac{\frac{4L^2}{(2n-1)^2\pi^2}\cos\left[\frac{(2n-1)\pi x}{2L}\right]\Big|_0^L + \frac{2Lx}{(2n-1)\pi}\sin\left[\frac{(2n-1)\pi x}{2L}\right]\Big|_0^L}{\frac{x}{2}\Big|_0^L + \frac{L}{2(2n-1)\pi}\sin\left[\frac{(2n-1)\pi x}{L}\right]\Big|_0^L} \tag{11.3.30}$$

$$= \frac{8L}{(2n-1)^2\pi^2}\left\{\cos\left[\frac{(2n-1)\pi}{2}\right] - 1\right\} + \frac{4L}{(2n-1)\pi}\sin\left[\frac{(2n-1)\pi}{2}\right] \tag{11.3.31}$$

$$= -\frac{8L}{(2n-1)^2\pi^2} - \frac{4L(-1)^n}{(2n-1)\pi}, \tag{11.3.32}$$

as $\cos[(2n-1)\pi/2] = 0$, and $\sin[(2n-1)\pi/2] = (-1)^{n+1}$. Consequently, the complete solution is

$$u(x,t) = -\frac{4L}{\pi}\sum_{n=1}^{\infty}\left[\frac{2}{(2n-1)^2\pi} + \frac{(-1)^n}{2n-1}\right]\cos\left[\frac{(2n-1)\pi x}{2L}\right]\exp\left[-\frac{(2n-1)^2\pi^2 a^2 t}{4L^2}\right].$$
$$\tag{11.3.33}$$

Figure 11.3.2 illustrates the evolution of the temperature field with time. It was generated using the MATLAB script:

```
clear
M = 200; dx = 0.02; dt = 0.05;
% compute Fourier coefficients
sign = -1;
```

```
for m = 1:M
  temp1 = 2*m-1;
  a(m) = 2/(pi*temp1*temp1) + sign/temp1;
  sign = - sign;
end
% compute grid and initialize solution
X = [0:dx:1]; T = [0:dt:1];
u = zeros(length(T),length(X));
XX = repmat(X,[length(T) 1]);
TT = repmat(T',[1 length(X)]);
% compute solution from Equation 11.3.33
for m = 1:M
  temp1 = (2*m-1)*pi/2;
  u = u + a(m) * cos(temp1*XX) .* exp(-temp1 * temp1 * TT);
end
u = - (4/pi) * u;
surf(XX,TT,u); axis([0 1 0 1 0 1]);
xlabel('DISTANCE','Fontsize',20); ylabel('TIME','Fontsize',20)
zlabel('SOLUTION','Fontsize',20)
```

Initially, heat near the center of the bar flows toward the cooler, insulated end, resulting in an increase of temperature there. On the right side, heat flows out of the bar because the temperature is maintained at zero at $x = L$. Eventually the heat that has accumulated at the left end flows rightward because of the continual heat loss on the right end. In the limit of $t \to \infty$, all of the heat has left the bar. □

- **Example 11.3.4: The separation of variables solution to the heat equation with a radiation boundary condition**

As our second example of how a Sturm-Liouville problem arises during the solution of a partial differential equation, we solve the heat equation by separation of variables when the temperature or flux of heat has been specified at the ends of the rod. In many physical applications, one or both of the ends may radiate to free space at temperature u_0. According to Stefan's law, the amount of heat radiated from a given area dA in a given time interval dt is $\sigma(u^4 - u_0^4) \, dA \, dt$, where σ is called the Stefan-Boltzmann constant. On the other hand, the amount of heat that reaches the surface from the interior of the body, assuming that we are at the right end of the bar, equals $-\kappa \, u_x \, dA \, dt$, where κ is the thermal conductivity. Because these quantities must be equal,

$$-\kappa \frac{\partial u}{\partial x} = \sigma(u^4 - u_0^4) = \sigma(u - u_0)(u^3 + u^2 u_0 + u u_0^2 + u_0^3). \tag{11.3.34}$$

If u and u_0 are nearly equal, we may approximate the second bracketed term on the right side of Equation 11.3.34 as $4u_0^3$. We write this approximate form of Equation 11.3.34 as

$$-\frac{\partial u}{\partial x} = h(u - u_0), \tag{11.3.35}$$

where h, the *surface conductance* or the *coefficient of surface heat transfer*, equals $4\sigma u_0^3/\kappa$. Equation 11.3.35 is a "radiation" boundary condition. Sometimes someone will refer to it as "Newton's law" because this equation is mathematically identical to Newton's law of cooling of a body by forced convection.

Table 11.3.1: The First Ten Roots of $\alpha + hL\tan(\alpha) = 0$ and C_n for $hL = 1$

n	α_n	Approximate α_n	C_n
1	2.0288	2.2074	118.9221
2	4.9132	4.9246	31.3414
3	7.9787	7.9813	27.7549
4	11.0855	11.0865	16.2891
5	14.2074	14.2079	14.9916
6	17.3364	17.3366	10.8362
7	20.4692	20.4693	10.2232
8	23.6043	23.6044	8.0999
9	26.7409	26.7410	7.7479
10	29.8786	29.8786	6.4626

Let us now solve the problem of a rod that we initially heat to the uniform temperature of 100. We then allow it to cool by maintaining the temperature at zero at $x = 0$ and radiatively cooling to the surrounding air at the temperature of zero[11] at $x = L$. We may restate the problem as

$$\frac{\partial u}{\partial t} = a^2 \frac{\partial^2 u}{\partial x^2}, \qquad 0 < x < L, \quad 0 < t, \tag{11.3.36}$$

with the initial condition $u(x,0) = 100$, $0 < x < L$, and the boundary conditions $u(0,t) = u_x(L,t) + hu(L,t) = 0$, $0 < t$.

Once again, we assume a product solution $u(x,t) = X(x)T(t)$ with a negative separation constant so that

$$\frac{X''}{X} = \frac{T'}{a^2 T} = -k^2. \tag{11.3.37}$$

We obtain for the x-dependence that

$$X'' + k^2 X = 0, \tag{11.3.38}$$

but the boundary conditions are now $X(0) = X'(L) + hX(L) = 0$. The most general solution of Equation 11.3.38 is $X(x) = A\cos(kx) + B\sin(kx)$. However, $A = 0$, because $X(0) = 0$. On the other hand,

$$k\cos(kL) + h\sin(kL) = kL\cos(kL) + hL\sin(kL) = 0, \tag{11.3.39}$$

if $B \neq 0$. The nondimensional number hL is the *Biot number* and depends completely upon the physical characteristics of the rod.

In Example 11.1.1 we saw how to find the roots of the transcendental equation $\alpha + hL\tan(\alpha) = 0$, where $\alpha = kL$. Consequently, if α_n is the nth root of this equation, then the eigenfunction is $X_n(x) = \sin(\alpha_n x/L)$. In Table 11.3.1, we list the first ten α_n's for $hL = 1$.

---

[11] Although this would appear to make $h = 0$, we have merely chosen a temperature scale so that the air temperature is zero and the absolute temperature used in Stefan's law is nonzero.

In general, we must find the roots either numerically or graphically. If α is large, however, we can find approximate values[12] by noting that

$$\cot(\alpha) = -hL/\alpha \approx 0, \quad \text{or} \quad \alpha_n = (2n-1)\pi/2, \tag{11.3.40}$$

where $n = 1, 2, 3, \ldots$. We can obtain a better approximation by setting $\alpha_n = (2n-1)\pi/2 - \epsilon_n$, where $\epsilon_n \ll 1$. Substituting into $\alpha + hL\tan(\alpha) = 0$, we find that

$$[(2n-1)\pi/2 - \epsilon_n]\cot[(2n-1)\pi/2 - \epsilon_n] + hL = 0. \tag{11.3.41}$$

We can simplify Equation 11.3.41 to

$$\epsilon_n^2 + (2n-1)\pi\epsilon_n/2 + hL = 0, \tag{11.3.42}$$

because $\cot[(2n-1)\pi/2 - \theta] = \tan(\theta)$, and $\tan(\theta) \approx \theta$ for $\theta \ll 1$. Solving for ϵ_n,

$$\epsilon_n \approx -\frac{2hL}{(2n-1)\pi}, \quad \text{and} \quad \alpha_n \approx \frac{(2n-1)\pi}{2} + \frac{2hL}{(2n-1)\pi}. \tag{11.3.43}$$

In Table 11.3.1 we compare the approximate roots given by Equation 11.3.43 with the actual roots.

Returning to the method of separation of variables, the temporal part equals

$$T_n(t) = C_n \exp\left(-k_n^2 a^2 t\right) = C_n \exp\left(-\frac{\alpha_n^2 a^2 t}{L^2}\right). \tag{11.3.44}$$

Consequently, the general solution is

$$u(x,t) = \sum_{n=1}^{\infty} C_n \sin\left(\frac{\alpha_n x}{L}\right) \exp\left(-\frac{\alpha_n^2 a^2 t}{L^2}\right), \tag{11.3.45}$$

where α_n is the nth root of $\alpha + hL\tan(\alpha) = 0$.

To determine C_n, we use the initial condition $u(x,0) = 100$ and find that

$$100 = \sum_{n=1}^{\infty} C_n \sin\left(\frac{\alpha_n x}{L}\right). \tag{11.3.46}$$

Equation 11.3.46 is an eigenfunction expansion of 100 employing the eigenfunctions from the Sturm-Liouville problem

$$X'' + k^2 X = 0, \tag{11.3.47}$$

---

[12] Using the same technique, Stevens and Luck [Stevens, J. W., and R. Luck, 1999: Explicit approximations for all eigenvalues of the 1-D transient heat conduction equations. *Heat Transfer Eng.*, **20(2)**, 35–41] have found approximate solutions to $\zeta_n \tan(\zeta_n) = Bi$. They showed that

$$\zeta_n \approx z_n + \frac{-B + \sqrt{B^2 - 4C}}{2},$$

where

$$B = z_n + (1 + Bi)\tan(z_n), \qquad C = Bi - z_n \tan(z_n),$$

$$z_n = c_n + \frac{\pi}{4}\left(\frac{Bi - c_n}{Bi + c_n}\right), \qquad c_n = \left(n - \frac{3}{4}\right)\pi.$$

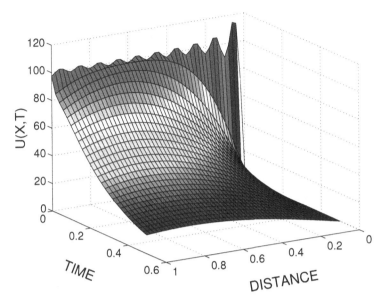

Figure 11.3.3: The temperature $u(x,t)$ within a thin bar as a function of position x/L and time a^2t/L^2 when we allow the bar to radiatively cool at $x = L$ while the temperature is zero at $x = 0$. Initially the temperature was 100.

with the boundary conditions $X(0) = X'(L) + hX(L) = 0$. Thus, the coefficient C_n is given by Equation 11.3.4 or

$$C_n = \frac{\int_0^L 100 \sin(\alpha_n x/L)\, dx}{\int_0^L \sin^2(\alpha_n x/L)\, dx}, \qquad (11.3.48)$$

as $r(x) = 1$. Performing the integrations,

$$C_n = \frac{100L[1 - \cos(\alpha_n)]/\alpha_n}{\frac{1}{2}[L - L\sin(2\alpha_n)/(2\alpha_n)]} = \frac{200[1 - \cos(\alpha_n)]}{\alpha_n[1 + \cos^2(\alpha_n)/(hL)]}, \qquad (11.3.49)$$

because $\sin(2\alpha_n) = 2\cos(\alpha_n)\sin(\alpha_n)$, and $\alpha_n = -hL\tan(\alpha_n)$. The complete solution is

$$u(x,t) = \sum_{n=1}^{\infty} \frac{200[1 - \cos(\alpha_n)]}{\alpha_n[1 + \cos^2(\alpha_n)/(hL)]} \sin\left(\frac{\alpha_n x}{L}\right) \exp\left(-\frac{\alpha_n^2 a^2 t}{L^2}\right). \qquad (11.3.50)$$

Figure 11.3.3 illustrates this solution for $hL = 1$ at various times and positions. It was generated using the MATLAB script:

```
clear
hL = 1; M = 200; dx = 0.02; dt = 0.02;
% create initial guess at alpha_n
zero = zeros(M,1);
for n = 1:M
  temp = (2*n-1)*pi; zero(n) = 0.5*temp + 2*hL/temp;
end;
% use Newton-Raphson method to improve values of alpha_n
for n = 1:M; for k = 1:10
  f = zero(n) + hL * tan(zero(n)); fp =1 + hL * sec(zero(n))^2;
```

```
   zero(n) = zero(n) - f / fp;
end; end;
% compute Fourier coefficients
for m = 1:M
   a(m) = 200*(1-cos(zero(m)))/(zero(m)*(1+cos(zero(m))^2/hL));
end
% compute grid and initialize solution
X = [0:dx:1]; T = [0:dt:0.5];
u = zeros(length(T),length(X));
XX = repmat(X,[length(T) 1]);
TT = repmat(T',[1 length(X)]);
% compute solution from Equation 11.3.50
for m = 1:M
   u = u + a(m) * sin(zero(m)*XX) .* exp(-zero(m)*zero(m)*TT);
end
surf(XX,TT,u)
xlabel('DISTANCE','Fontsize',20); ylabel('TIME','Fontsize',20)
zlabel('U(X,T)','Fontsize',20)
```

The heat lost to the environment occurs either because the temperature at an end is zero or because it radiates heat to space that has the temperature of zero.   □

• Example 11.3.5: Duhamel's theorem for the heat equation

In addition to finding solutions to heat conduction problems with time-dependent boundary conditions, we can also apply the superposition integral to the nonhomogeneous heat equation when the source depends on time. Jeglic[13] used this technique in obtaining the temperature distribution within a slab heated by alternating electric current. If we assume that the flat plate has a surface area A and depth L, then the heat equation for the plate when electrically heated by an alternating current of frequency ω is

$$\frac{\partial u}{\partial t} - a^2 \frac{\partial^2 u}{\partial x^2} = \frac{2q}{\rho C_p AL} \sin^2(\omega t), \qquad 0 < x < L, \quad 0 < t, \tag{11.3.51}$$

where q is the average heat rate caused by the current, ρ is the density, C_p is the specific heat at constant pressure, and a^2 is the diffusivity of the slab. We will assume that we insulated the inner wall so that $u_x(0,t) = 0$, $0 < t$, while we allow the outer wall to radiatively cool to free space at the temperature of zero or $\kappa u_x(L,t) + h u(L,t) = 0$, $0 < t$, where κ is the thermal conductivity and h is the heat transfer coefficient. The slab is initially at the temperature of zero $u(x,0) = 0$, $0 < x < L$.

To solve the heat equation, we first solve the simpler problem of

$$\frac{\partial A}{\partial t} - a^2 \frac{\partial^2 A}{\partial x^2} = 1, \qquad 0 < x < L, \quad 0 < t, \tag{11.3.52}$$

---

[13] Jeglic, F. A., 1962: An analytical determination of temperature oscillations in a wall heated by alternating current. *NASA Tech. Note No. D–1286*. In a similar vein, Al-Nimr and Abdallah (Al-Nimr, M. A., and M. R. Abdallah, 1999: Thermal behavior of insulated electric wires producing pulsating signals. *Heat Transfer Eng.*, **20(4)**, 62–74) found the heat transfer with an insulated wire that carries an alternating current.

Table 11.3.2: The First Six Roots of the Equation $k_n \tan(k_n) = h^*$

h^*	k_1	k_2	k_3	k_4	k_5	k_6
0.001	0.03162	3.14191	6.28334	9.42488	12.56645	15.70803
0.002	0.04471	3.14223	6.28350	9.42499	12.56653	15.70809
0.005	0.07065	3.14318	6.28398	9.42531	12.56677	15.70828
0.010	0.09830	3.14477	6.28478	9.42584	12.56717	15.70860
0.020	0.14095	3.14795	6.28637	9.42690	12.56796	15.70924
0.050	0.22176	3.15743	6.29113	9.43008	12.57035	15.71115
0.100	0.31105	3.17310	6.29906	9.43538	12.57432	15.71433
0.200	0.43284	3.20393	6.31485	9.44595	12.58226	15.72068
0.500	0.65327	3.29231	6.36162	9.47748	12.60601	15.73972
1.000	0.86033	3.42562	6.43730	9.52933	12.64529	15.77128
2.000	1.07687	3.64360	6.57833	9.62956	12.72230	15.83361
5.000	1.31384	4.03357	6.90960	9.89275	12.93522	16.01066
10.000	1.42887	4.30580	7.22811	10.20026	13.21418	16.25336
20.000	1.49613	4.49148	7.49541	10.51167	13.54198	16.58640
∞	1.57080	4.71239	7.85399	10.99557	14.13717	17.27876

with the boundary conditions $A_x(0,t) = \kappa A_x(L,t) + hA(L,t) = 0$, $0 < t$, and the initial condition $A(x,0) = 0$, $0 < x < L$. The solution $A(x,t)$ is the *indicial admittance* because it is the response of a system to forcing by the step function $H(t)$.

We solve Equation 11.3.52 by separation of variables. We begin by assuming that $A(x,t)$ consists of a steady-state solution $w(x)$ plus a transient solution $v(x,t)$, where

$$a^2 w''(x) = -1, \qquad 0 < x < L, \tag{11.3.53}$$

subject to the boundary conditions $w'(0) = \kappa w'(L) + hw(L) = 0$ and

$$\frac{\partial v}{\partial t} = a^2 \frac{\partial^2 v}{\partial x^2}, \qquad 0 < x < L, \tag{11.3.54}$$

with the boundary conditions $v_x(0,t) = \kappa v_x(L,t) + hv(L,t) = 0$, $0 < t$, and the initial condition that $v(x,0) = -w(x)$, $0 < x < L$.

Solving Equation 11.3.53,

$$w(x) = \frac{L^2 - x^2}{2a^2} + \frac{\kappa L}{ha^2}. \tag{11.3.55}$$

Turning to the transient solution $v(x,t)$, we use separation of variables and find that

$$v(x,t) = \sum_{n=1}^{\infty} C_n \cos\left(\frac{k_n x}{L}\right) \exp\left(-\frac{a^2 k_n^2 t}{L^2}\right), \tag{11.3.56}$$

where k_n is the nth root of the transcendental equation: $k_n \tan(k_n) = hL/\kappa = h^*$. Table 11.3.2 gives the first six roots for various values of hL/κ.

Our final task is to compute C_n. After substituting $t = 0$ into Equation 11.3.56, we are left with an orthogonal expansion of $-w(x)$ using the eigenfunctions $\cos(k_n x/L)$. From the theory of eigenfunction expansions, we have that

$$C_n = \frac{\int_0^L -w(x) \cos(k_n x/L)\, dx}{\int_0^L \cos^2(k_n x/L)\, dx} = \frac{-L^3 \sin(k_n)/(a^2 k_n^3)}{L[k_n + \sin(2k_n)/2]/(2k_n)} = -\frac{2L^2 \sin(k_n)}{a^2 k_n^2 [k_n + \sin(2k_n)/2]}. \tag{11.3.57}$$

Combining Equation 11.3.56 and Equation 11.3.57,

$$v(x,t) = -\frac{2L^2}{a^2} \sum_{n=1}^{\infty} \frac{\sin(k_n)\cos(k_n x/L)}{k_n^2[k_n + \sin(2k_n)/2]} \exp\left(-\frac{a^2 k_n^2 t}{L^2}\right). \tag{11.3.58}$$

Consequently, $A(x,t)$ equals

$$A(x,t) = \frac{L^2 - x^2}{2a^2} + \frac{\kappa L}{ha^2} - \frac{2L^2}{a^2} \sum_{n=1}^{\infty} \frac{\sin(k_n)\cos(k_n x/L)}{k_n^2[k_n + \sin(2k_n)/2]} \exp\left(-\frac{a^2 k_n^2 t}{L^2}\right). \tag{11.3.59}$$

We now wish to use the solution, Equation 11.3.59, to find the temperature distribution within the slab when it is heated by a time-dependent source $f(t)$. As in the case of time-dependent boundary conditions, we imagine that we can break the process into an infinite number of small changes to the heating, which occur at the times $t = \tau_1$, $t = \tau_2$, etc. Consequently, the temperature distribution at the time t following the change at $t = \tau_n$ and before the change at $t = \tau_{n+1}$ is

$$u(x,t) = f(0)A(x,t) + \sum_{k=1}^{n} A(x, t - \tau_k)\frac{\Delta f_k}{\Delta \tau_k} \Delta \tau_k, \tag{11.3.60}$$

where

$$\Delta f_k = f(\tau_k) - f(\tau_{k-1}), \quad \text{and} \quad \Delta \tau_k = \tau_k - \tau_{k-1}. \tag{11.3.61}$$

In the limit of $\Delta \tau_k \to 0$,

$$u(x,t) = f(0)A(x,t) + \int_0^t A(x, t - \tau)f'(\tau)\, d\tau = f(t)A(x,0) + \int_0^t f(\tau)\frac{\partial A(x, t - \tau)}{\partial \tau}\, d\tau. \tag{11.3.62}$$

In our present problem,

$$f(t) = \frac{2q}{\rho C_p A L} \sin^2(\omega t), \quad \text{and} \quad f'(t) = \frac{2q\omega}{\rho C_p A L} \sin(2\omega t). \tag{11.3.63}$$

Therefore,

$$u(x,t) = \frac{2q\omega}{\rho C_p A L} \int_0^t \sin(2\omega\tau)\left\{\frac{L^2 - x^2}{2a^2} + \frac{\kappa L}{ha^2}\right.$$

$$\left. -\frac{2L^2}{a^2} \sum_{n=1}^{\infty} \frac{\sin(k_n)}{k_n^2[k_n + \sin(2k_n)/2]} \cos\left(\frac{k_n x}{L}\right) \exp\left[-\frac{a^2 k_n^2(t - \tau)}{L^2}\right]\right\} d\tau \tag{11.3.64}$$

$$= -\frac{q}{\rho C_p A L} \left(\frac{L^2 - x^2}{2a^2} + \frac{\kappa L}{ha^2}\right) \cos(2\omega\tau)|_0^t \tag{11.3.65}$$

$$-\frac{4L^2 q\omega}{a^2 \rho C_p A L} \sum_{n=1}^{\infty} \frac{\sin(k_n)\exp(-a^2 k_n^2 t/L^2)}{k_n^2[k_n + \sin(2k_n)/2]} \cos\left(\frac{k_n x}{L}\right) \int_0^t \sin(2\omega\tau)\exp\left(\frac{a^2 k_n^2 \tau}{L^2}\right) d\tau$$

$$= \frac{qL}{a^2 A\rho C_p} \left\{\left[\frac{L^2 - x^2}{2L^2} + \frac{\kappa}{hL}\right][1 - \cos(2\omega t)]\right.$$

$$-\sum_{n=1}^{\infty} \frac{4\sin(k_n)\cos(k_n x/L)}{k_n^2[k_n + \sin(2k_n)/2][4 + a^4 k_n^4/(L^4\omega^2)]}$$

$$\left. \times \left[\frac{a^2 k_n^2}{\omega L^2} \sin(2\omega t) - 2\cos(2\omega t) + 2\exp\left(-\frac{a^2 k_n^2 t}{L^2}\right)\right]\right\}. \tag{11.3.66}$$

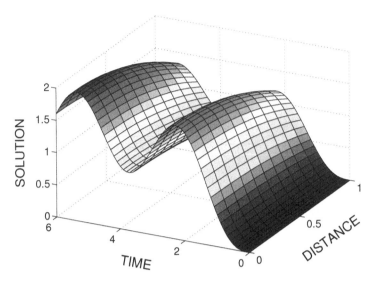

Figure 11.3.4: The nondimensional temperature $a^2 A\rho C_p u(x,t)/qL$ within a slab that we heat by alternating electric current as a function of position x/L and time $a^2 t/L^2$ when we insulate the $x = 0$ end and let the $x = L$ end radiate to free space at temperature zero. The initial temperature is zero, $hL/\kappa = 1$, and $a^2/(L^2\omega) = 1$.

Figure 11.3.4 illustrates Equation 11.3.66 for $hL/\kappa = 1$, and $a^2/(L^2\omega) = 1$. This figure was created using the MATLAB script:

```
clear
asq_over_omegaL2 = 1; h_star = 1; m = 0; M = 10;
dx = 0.1; dt = 0.1;
% create initial guess at k_n
zero = zeros(length(M));
for n = 1:10000
  k1 = 0.1*n; k2 = 0.1*(n+1);
  prod = k1 * tan(k1);
  y1 = h_star - prod; y2 = h_star - k2 * tan(k2);
  if (y1*y2 <= 0 & prod < 2 & m < M) m = m+1; zero(m) = k1; end;
end;
% use Newton-Raphson method to improve values of k_n
for n = 1:M; for k = 1:10
  f = h_star - zero(n) * tan(zero(n));
  fp = - tan(zero(n)) - zero(n) * sec(zero(n))^2;
  zero(n) = zero(n) - f / fp;
end; end;
% compute grid and initialize solution
X = [0:dx:1]; T = [0:dt:6];
temp1 = (0.5 + 1/h_star)*ones(1,length(X)) - 0.5*X.*X;
temp2 = ones(1,length(T)) - cos(2*T);
u = temp1' * temp2;
XX = X' * ones(1,length(T));
TT = ones(1,length(X))' * T;
% compute solution from Equation 11.3.66
for m = 1:M
```

```
   xtemp1 = zero(m) * zero(m);
   xtemp2 = 4 + asq_over_omegaL2*asq_over_omegaL2*xtemp1*xtemp1;
   xtemp3 = asq_over_omegaL2 * xtemp1;
   xtemp4 = zero(m) + sin(2*zero(m))/2;
   xtemp5 = asq_over_omegaL2 * xtemp1;
   aaaaa = 4 * sin(zero(m)) / (xtemp1 * xtemp2 * xtemp4);
   u = u - aaaaa * cos(zero(m)*X)' ...
      * (xtemp5 * sin(2*T) - 2 * cos(2*T) + 2 * exp(-xtemp5 * T));
end
surf(XX,TT,u)
xlabel('DISTANCE','Fontsize',20); ylabel('TIME','Fontsize',20)
zlabel('SOLUTION','Fontsize',20)
```

The oscillating solution, reflecting the periodic heating by the alternating current, rapidly reaches equilibrium. Because heat is radiated to space at $x = L$, the temperature is maximum at $x = 0$ at any given instant as heat flows from $x = 0$ to $x = L$. $\qquad\square$

• Example 11.3.6: Duhamel's theorem and heat conduction

Consider the following heat conduction problem with time-dependent forcing and/or boundary conditions:

$$\frac{\partial u}{\partial t} = a^2 L(u) + f(P,t), \qquad 0 < t, \tag{11.3.67}$$

$$B(u) = g(Q,t), \qquad 0 < t, \tag{11.3.68}$$

and

$$u(P,0) = h(P), \tag{11.3.69}$$

where

$$L(u) = C_0 + C_1 \frac{\partial}{\partial x_1}\left(K_1 \frac{\partial u}{\partial x_1}\right) + C_2 \frac{\partial}{\partial x_2}\left(K_2 \frac{\partial u}{\partial x_2}\right) + C_3 \frac{\partial}{\partial x_3}\left(K_3 \frac{\partial u}{\partial x_3}\right), \tag{11.3.70}$$

$$B(u) = c_0 + c_1 \frac{\partial u}{\partial x_1} + c_2 \frac{\partial u}{\partial x_2} + c_3 \frac{\partial u}{\partial x_3}, \tag{11.3.71}$$

P denotes an arbitrary interior point at (x_1, x_2, x_3) of a region R, and Q is any point on the boundary of R. Here c_i, C_i, and K_i are functions of x_1, x_2, and x_3 only.

Many years ago, Bartels and Churchill[14] extended Duhumel's theorem to solve this heat conduction problem. They did this by first introducing the simpler initial-boundary-value problem:

$$\frac{\partial v}{\partial t} = a^2 L(v) + f(P,t_1), \qquad 0 < t, \tag{11.3.72}$$

$$B(v) = g(Q,t_1), \qquad 0 < t, \tag{11.3.73}$$

and

$$v(P,0) = h(P), \tag{11.3.74}$$

---

[14] Bartels, R. C. F., and R. V. Churchill, 1942: Resolution of boundary problems by the use of a generalized convolution. *Am. Math. Soc. Bull.*, **48**, 276–282.

which has a constant forcing and boundary conditions in place of the time-dependent ones. Here t_1 denotes an arbitrary but *fixed* instant of time. Then Bartels and Churchill proved that the solution to the original problem is given by the convolution integral

$$u(P,t) = \frac{\partial}{\partial t} \left[\int_0^t v(P, t - \tau, \tau) \, d\tau \right]. \tag{11.3.75}$$

To illustrate[15] this technique, let us solve

$$\frac{\partial u}{\partial t} = a^2 \left(\frac{\partial^2 u}{\partial r^2} + \frac{2}{r} \frac{\partial u}{\partial r} \right) = \frac{a^2}{r} \frac{\partial^2 (ru)}{\partial r^2}, \qquad 0 < \alpha < r < \beta, \quad 0 < t, \tag{11.3.76}$$

subject to the boundary conditions $u(\alpha, t) = u_0 e^{-ct}$, $u_r(\beta, t) = 0$, $0 < t$, and the initial condition $u(r, 0) = u_0$, $\alpha < r < \beta$.

We begin by solving the alternative problem

$$\frac{\partial v}{\partial t} = a^2 \left(\frac{\partial^2 v}{\partial r^2} + \frac{2}{r} \frac{\partial v}{\partial r} \right) = \frac{a^2}{r} \frac{\partial^2 (rv)}{\partial r^2}, \qquad \alpha < r < \beta, \quad 0 < t, \tag{11.3.77}$$

subject to the boundary conditions $v(\alpha, t, t') = u_0 e^{-ct'}$, $v_r(\beta, t, t') = 0$, $0 < t$, and the initial condition $v(r, 0, t') = u_0$, $\alpha < r < \beta$, or equivalently

$$\frac{\partial w}{\partial t} = a^2 \left(\frac{\partial^2 w}{\partial r^2} + \frac{2}{r} \frac{\partial w}{\partial r} \right) = \frac{a^2}{r} \frac{\partial^2 (rw)}{\partial r^2}, \qquad \alpha < r < \beta, \quad 0 < t, \tag{11.3.78}$$

subject to the boundary conditions $w(\alpha, t, t') = 0$, $w_r(\beta, t, t') = 0$, and the initial condition $w(r, 0, t') = u_0(1 - e^{-ct'})$, $\alpha < r < \beta$, where $v(r, t, t') = u_0 e^{-ct'} + w(r, t, t')$.

The heat condition problem Equation 11.3.78 can be solved using separation of variables. This yields

$$r\, w(r, t, t') = \alpha u_0 (1 - e^{-ct'}) \sum_{n=1}^{\infty} \frac{\sin[k_n(r - \alpha)]}{k_n c_n} e^{-a^2 k_n^2 t}, \tag{11.3.79}$$

where k_n is the nth root of $\beta k = \tan[k(\beta - \alpha)]$, and $2c_n = \{\beta \sin^2[k_n(\beta - \alpha)] - \alpha\}$. Therefore,

$$r\, u(r, t) = \frac{\partial}{\partial t} \left\{ \int_0^t r\, u_0 e^{-c\tau} + \alpha u_0 (1 - e^{-c\tau}) \sum_{n=1}^{\infty} \frac{\sin[k_n(r - \alpha)]}{k_n c_n} e^{-a^2 k_n^2 (t - \tau)} \, d\tau \right\} \tag{11.3.80}$$

$$= \frac{r\, u_0}{c} \frac{\partial}{\partial t} \left[1 - e^{-ct} \right]$$

$$+ \alpha u_0 \sum_{n=1}^{\infty} \frac{\sin[k_n(r - \alpha)]}{k_n c_n} \frac{\partial}{\partial t} \left\{ \int_0^t \left[e^{-a^2 k_n^2 (t - \tau)} - e^{-a^2 k_n^2 (t - \tau) - c\tau} \right] \, d\tau \right\} \tag{11.3.81}$$

$$= r\, u_0 e^{-ct} + \alpha u_0 \sum_{n=1}^{\infty} \frac{\sin[k_n(r - \alpha)]}{k_n c_n} \frac{\partial}{\partial t} \left\{ \frac{1 - e^{-a^2 k_n^2 t}}{a^2 k_n^2} - \frac{e^{-ct} - e^{-a^2 k_n^2 t}}{a^2 k_n^2 - c} \right\} \tag{11.3.82}$$

$$= r\, u_0 e^{-ct} + \alpha c u_0 \sum_{n=1}^{\infty} \frac{\sin[k_n(r - \alpha)]}{k_n c_n} \frac{e^{-a^2 k_n^2 t} - e^{-ct}}{c - a^2 k_n^2}, \tag{11.3.83}$$

---

[15] See Reiss, H., and V. K. LaMer, 1950: Diffusional boundary value problems involving moving boundaries, connected with the growth of colloidal particles. *J. Chem. Phys.*, **18**, 1–12.

and the final answer is

$$u(r,t) = u_0 e^{-ct} + \frac{acu_0}{r} \sum_{n=1}^{\infty} \frac{e^{-a^2 k_n^2 t} - e^{-ct}}{(c - a^2 k_n^2) k_n c_n} \sin[k_n(r - a)].$$

(11.3.84)

Problems

1. The Sturm-Liouville problem $y'' + \lambda y = 0$, $y(0) = y(L) = 0$ has the eigenfunction solution $y_n(x) = \sin(n\pi x/L)$. Find the eigenfunction expansion for $f(x) = x$ using this eigenfunction.

2. The Sturm-Liouville problem $y'' + \lambda y = 0$, $y'(0) = y'(L) = 0$ has the eigenfunction solutions $y_0(x) = 1$, and $y_n(x) = \cos(n\pi x/L)$. Find the eigenfunction expansion for $f(x) = x$ using these eigenfunctions.

3. The Sturm-Liouville problem $y'' + \lambda y = 0$, $y(0) = y'(L) = 0$ has the eigenfunction solution $y_n(x) = \sin[(2n-1)\pi x/(2L)]$. Find the eigenfunction expansion for $f(x) = x$ using this eigenfunction.

4. The Sturm-Liouville problem $y'' + \lambda y = 0$, $y'(0) = y(L) = 0$ has the eigenfunction solution $y_n(x) = \cos[(2n-1)\pi x/(2L)]$. Find the eigenfunction expansion for $f(x) = x$ using this eigenfunction.

5. Consider the eigenvalue problem

$$y'' + (\lambda - a^2)y = 0, \qquad 0 < x < 1,$$

with the boundary conditions $y'(0) + ay(0) = 0$ and $y'(1) + ay(1) = 0$.

Step 1: Show that this is a regular Sturm-Liouville problem.

Step 2: Show that the eigenvalues and eigenfunctions are $\lambda_0 = 0$, $y_0(x) = e^{-ax}$ and $\lambda_n = a^2 + n^2\pi^2$, $y_n(x) = a\sin(n\pi x) - n\pi\cos(n\pi x)$, where $n = 1, 2, 3, \ldots$.

Step 3: Given a function $f(x)$, show that we can expand it as follows:

$$f(x) = C_0 e^{-ax} + \sum_{n=1}^{\infty} C_n \left[a\sin(n\pi x) - n\pi\cos(n\pi x) \right],$$

where

$$\left(1 - e^{-2a}\right) C_0 = 2a \int_0^1 f(x) e^{-ax}\, dx,$$

and

$$(a^2 + n^2\pi^2)C_n = 2 \int_0^1 f(x) \left[a\sin(n\pi x) - n\pi\cos(n\pi x) \right]\, dx.$$

6. Consider the eigenvalue problem

$$y'''' + \lambda y'' = 0, \qquad 0 < x < 1,$$

with the boundary conditions $y(0) = y'(0) = y(1) = y'(1) = 0$. Prove the following points:

Step 1: Show that the eigenfunctions are

$$y_n(x) = 1 - \cos(k_n x) + \frac{1 - \cos(k_n)}{k_n - \sin(k_n)}[\sin(k_n x) - k_n x],$$

where k_n denotes the nth root of $\sin(k/2)[\sin(k/2) - (k/2)\cos(k/2)] = 0$.

Step 2: Show that there are two classes of eigenfunctions: $\kappa_n = 2n\pi$ with $y_n(x) = 1 - \cos(2n\pi x)$, and $\tan(\kappa_n/2) = \kappa_n/2$ with $y_n(x) = 1 - \cos(\kappa_n x) + 2[\sin(\kappa_n x) - \kappa_n x]/\kappa_n$.

Step 3: Show that the orthogonality condition for this problem is

$$\int_0^1 y_n'(x) y_m'(x)\, dx = 0, \qquad n \neq m,$$

where $y_n(x)$ and $y_m(x)$ are two distinct eigenfunction solutions of this problem. Hint: Follow the proof in Section 11.2 and integrate repeatedly by parts to eliminate higher derivative terms.

Step 4: Show that we can construct an eigenfunction expansion for an arbitrary function $f(x)$ via

$$f(x) = \sum_{n=1}^{\infty} C_n y_n(x), \qquad 0 < x < 1, \quad \text{provided} \quad C_n = \frac{\int_0^1 f'(x) y_n'(x)\, dx}{\int_0^1 [y_n'(x)]^2\, dx}.$$

What are the condition(s) on $f(x)$?

Separation of Variables Solution to the Wave Equation

7. The differential equation for the longitudinal vibrations of a rod within a viscous fluid is

$$\frac{\partial^2 u}{\partial t^2} + 2h\frac{\partial u}{\partial t} = c^2\frac{\partial^2 u}{\partial x^2}, \qquad 0 < x < L, \qquad 0 < t,$$

where c is the velocity of sound in the rod and h is the damping coefficient. If the rod is fixed at $x = 0$ so that $u(0, t) = 0$, and allowed to freely oscillate at the other end $x = L$, so that $u_x(L, t) = 0$, find the vibrations for any location x and subsequent time t if the rod has the initial displacement of $u(x, 0) = x$ and the initial velocity $u_t(x, 0) = 0$ for $0 < x < L$. Assume that $h < c\pi/(2L)$. Why?

8. A closed pipe of length L contains air whose density is slightly greater than that of the outside air in the ratio of $1 + s_0$ to 1. Everything being at rest, we suddenly draw aside the disk closing one end of the pipe. We want to determine what happens *inside* the pipe after we remove the disk.

As the air rushes outside, it generates sound waves within the pipe. The wave equation

$$\frac{\partial^2 u}{\partial t^2} = c^2\frac{\partial^2 u}{\partial x^2}$$

governs these waves, where c is the speed of sound and $u(x, t)$ is the velocity potential. Without going into the fluid mechanics of the problem, the boundary conditions are

a. No flow through the closed end: $u_x(0,t) = 0$.

b. No infinite acceleration at the open end: $u_{xx}(L,t) = 0$.

c. Air is initially at rest: $u_x(x,0) = 0$.

d. Air initially has a density greater than the surrounding air by the amount s_0: $u_t(x,0) = -c^2 s_0$.

Find the velocity potential at all positions within the pipe and all subsequent times.

Separation of Variables Solution to the Heat Equation

For Problems 9–13, solve the heat equation $u_t = a^2 u_{xx}$, $0 < x < \pi$, $0 < t$, subject to the following boundary conditions and initial condition. Then plot your results using MATLAB.

9. $u_x(0,t) = u(\pi,t) = 0$, $0 < t$; $u(x,0) = x^2 - \pi^2$, $0 < x < \pi$

10. $u(0,t) = 0$, $u_x(\pi,t) = 0$, $0 < t$; $u(x,0) = 1$, $0 < x < \pi$

11. $u(0,t) = 0$, $u_x(\pi,t) = 0$, $0 < t$; $u(x,0) = x$, $0 < x < \pi$

12. $u(0,t) = 0$, $u_x(\pi,t) = 0$, $0 < t$; $u(x,0) = \pi - x$, $0 < x < \pi$

13. $u(0,t) = T_0$, $u_x(\pi,t) = 0$, $0 < t$; $u(x,0) = T_1 \neq T_0$, $0 < x < \pi$

14. It is well known that a room with masonry walls is often very difficult to heat. Consider a wall of thickness L, conductivity κ, and diffusivity a^2, which we heat by a surface heat flux at a constant rate H. The temperature of the outside (out-of-doors) face of the wall remains constant at T_0 and the entire wall initially has the uniform temperature T_0. Let us find the temperature of the inside face as a function of time.[16]
We begin by solving the heat conduction problem

$$\frac{\partial u}{\partial t} = a^2 \frac{\partial^2 u}{\partial x^2}, \qquad 0 < x < L, \quad 0 < t,$$

subject to the boundary conditions that $u_x(0,t) = -H/\kappa$, $u(L,t) = T_0$, $0 < t$, and the initial condition that $u(x,0) = T_0$, $0 < x < L$. Show that the temperature field equals

$$u(x,t) = T_0 + \frac{HL}{\kappa}\left\{1 - \frac{x}{L} - \frac{8}{\pi^2}\sum_{n=1}^{\infty}\frac{1}{(2n-1)^2}\cos\left[\frac{(2n-1)\pi x}{2L}\right]\exp\left[-\frac{(2n-1)^2\pi^2 a^2 t}{4L^2}\right]\right\}.$$

Therefore, the rise of temperature at the interior wall $x = 0$ is

$$\frac{HL}{\kappa}\left\{1 - \frac{8}{\pi^2}\sum_{n=1}^{\infty}\frac{1}{(2n-1)^2}\exp\left[-\frac{(2n-1)^2\pi^2 a^2 t}{4L^2}\right]\right\},$$

or

$$\frac{8HL}{\kappa\pi^2}\sum_{n=1}^{\infty}\frac{1}{(2n-1)^2}\left\{1 - \exp\left[-\frac{(2n-1)^2\pi^2 a^2 t}{4L^2}\right]\right\}.$$

---

[16] See Dufton, A. F., 1927: The warming of walls. *Philos. Mag., Ser. 7*, **4**, 888–889.

For $a^2t/L^2 \leq 1$, this last expression can be approximated[17] by $2Hat^{1/2}/\pi^{1/2}\kappa$. We thus see that the temperature will initially rise as the square root of time and diffusivity and inversely with conductivity. For an average rock, $\kappa = 0.0042$ g/cm-s, and $a^2 = 0.0118$ cm^2/s, while for wood (spruce) $\kappa = 0.0003$ g/cm-s, and $a^2 = 0.0024$ cm^2/s.

The same set of equations applies to heat transfer within a transistor operating at low frequencies.[18] At the junction ($x = 0$), heat is produced at the rate of H and flows to the transistor's supports ($x = \pm L$) where it is removed. The supports are maintained at the temperature T_0, which is also the initial temperature of the transistor.

15. We want to find the rise of the water table of an aquifer, which we sandwich between a canal and impervious rocks if we suddenly raise the water level in the canal h_0 units above its initial elevation and then maintain the canal at this level. The linearized Boussinesq equation

$$\frac{\partial u}{\partial t} = \frac{\partial^2 u}{\partial x^2}, \qquad 0 < x < L, \quad 0 < t,$$

governs the level of the water table with the boundary conditions $u(0,t) = h_0$, and $u_x(L,t) = 0$, and the initial condition $u(x,0) = 0$. Find the height of the water table at any point in the aquifer and any time $t > 0$.

16. Solve the nonhomogeneous heat equation

$$\frac{\partial u}{\partial t} - a^2 \frac{\partial^2 u}{\partial x^2} = e^{-x}, \qquad 0 < x < \pi, \quad 0 < t,$$

subject to the boundary conditions $u(0,t) = u_x(\pi,t) = 0$, $0 < t$, and the initial condition $u(x,0) = f(x)$, $0 < x < \pi$. Hint: First find the steady-state solution $w(x)$ and then write $u(x,t) = w(x) + v(x,t)$, where $v(x,t)$ is the transient solution so that $u(x,t)$ satisfies the initial condition.

---

[17] Let us define the function:

$$f(t) = \sum_{n=1}^{\infty} \frac{1 - \exp[-(2n-1)^2\pi^2 a^2 t/L^2]}{(2n-1)^2}.$$

Then

$$f'(t) = \frac{a^2\pi^2}{L^2} \sum_{n=1}^{\infty} \exp[-(2n-1)^2\pi^2 a^2 t/L^2].$$

Consider now the integral

$$\int_0^{\infty} \exp\left(-\frac{a^2\pi^2 t}{L^2}x^2\right) dx = \frac{L}{2a\sqrt{\pi t}}.$$

If we approximate this integral by using the trapezoidal rule with $\Delta x = 2$, then

$$\int_0^{\infty} \exp\left(-\frac{a^2\pi^2 t}{L^2}x^2\right) dx \approx 2\sum_{n=1}^{\infty} \exp[-(2n-1)^2\pi^2 a^2 t/L^2],$$

and $f'(t) \approx a\pi^{3/2}/(4Lt^{1/2})$. Integrating and using $f(0) = 0$, we finally have $f(t) \approx a\pi^{3/2}t^{1/2}/(2L)$. The smaller a^2t/L^2 is, the smaller the error will be. For example, if $t = L^2/a^2$, then the error is 2.4%.

[18] Mortenson, K. E., 1957: Transistor junction temperature as a function of time. *Proc. IRE*, **45**, 504–513. Equation 2a should read $T_x = -F/k$.

17. Solve the nonhomogeneous heat equation

$$\frac{\partial u}{\partial t} = a^2 \frac{\partial^2 u}{\partial x^2} + \frac{A_0}{c\rho}, \qquad 0 < x < L, \quad 0 < t,$$

where $a^2 = \kappa/c\rho$, with the boundary conditions that $u_x(0,t) = 0$, $\kappa u_x(L,t) + hu(L,t) = 0$, $0 < t$, and the initial condition that $u(x,0) = 0$, $0 < x < L$. Hint: First find the steady-state solution $w(x)$ and then write $u(x,t) = w(x) + v(x,t)$, where $v(x,t)$ is the transient solution so that $u(x,t)$ satisfies the initial condition.

18. Solve[19]

$$\frac{\partial u}{\partial t} + \kappa_1 u = \frac{\partial^2 u}{\partial x^2}, \qquad 0 < x < L, \quad 0 < t,$$

with the boundary conditions $u_x(0,t) = 0$, $a^2 u_x(L,t) + \kappa_2 u(L,t) = 0$, $0 < t$, and the initial condition $u(x,0) = u_0$, $0 < x < L$.

19. Solve

$$\frac{\partial u}{\partial t} + \frac{\partial u}{\partial x} = a^2 \frac{\partial^2 u}{\partial x^2}, \qquad 0 < x < 1, \quad 0 < t,$$

with the boundary conditions $a^2 u_x(0,t) = u(0,t)$, $u_x(1,t) = 0$, $0 < t$, and the initial condition $u(x,0) = 1$, $0 < x < 1$. Hint: Let $u(x,t) = v(x,t) \exp[(2x-t)/(4a^2)]$ so that the problem becomes

$$\frac{\partial v}{\partial t} = a^2 \frac{\partial^2 v}{\partial x^2}, \qquad 0 < x < 1, \quad 0 < t,$$

with the boundary conditions $2a^2 v_x(0,t) = v(0,t)$, $2a^2 v_x(1,t) = -v(1,t)$, $0 < t$, and the initial condition $v(x,0) = \exp[-x/(2a^2)]$, $0 < x < 1$.

20. Solve the heat equation in spherical coordinates

$$\frac{\partial u}{\partial t} = \frac{a^2}{r^2} \frac{\partial}{\partial r}\left(r^2 \frac{\partial u}{\partial r}\right) = \frac{a^2}{r} \frac{\partial^2 (ru)}{\partial r^2}, \qquad \alpha < r < \beta, \quad 0 < t,$$

subject to the boundary conditions $u(\alpha,t) = u_r(\beta,t) = 0$, $0 < t$, and the initial condition $u(r,0) = u_0$, $\alpha < r < \beta$.

21. Solve[20] the heat equation in spherical coordinates

$$\frac{\partial u}{\partial t} = a^2 \left(\frac{\partial^2 u}{\partial r^2} + \frac{2}{r} \frac{\partial u}{\partial r}\right) = \frac{a^2}{r} \frac{\partial^2 (ru)}{\partial r^2}, \qquad 0 \le r < b, \quad 0 < t,$$

subject to the boundary conditions $\lim_{r \to 0} u_r(r,t) \to 0$, $u_r(b,t) = -Au(b,t)/b$, $0 < t$, and the initial condition $u(r,0) = u_0$, $0 \le r < b$. Hint: Introduce the dependent variable $v(r,t) = r\,u(r,t)$.

[19] Motivated by problems solved in Gomer, R., 1951: Wall reactions and diffusion in static and flow systems. *J. Chem. Phys.*, **19**, 284–289.

[20] Zhou, H., S. Abanades, G. Flamant, D. Gauthier, and J. Lu, 2002: Simulation of heavy metal vaporization dynamics of a fluidized bed. *Chem. Eng. Sci.*, **57**, 2603–2614. See also Mantell, C., M. Rodriguez, and E. Martinez de la Ossa, 2002: Semi-batch extraction of anthocyanins from red grape pomace in packed beds: Experimental results and process modelling. *Chem. Eng. Sci.*, **57**, 3831–3838.

22. Use separation of variables to solve[21] the partial differential equation

$$\frac{\partial u}{\partial t} = \frac{\partial^2 u}{\partial x^2} + 2a\frac{\partial u}{\partial x}, \qquad 0 < x < 1, \quad 0 < t,$$

subject to the boundary conditions that $u_x(0,t) + 2au(0,t) = 0$, $u_x(1,t) + 2au(1,t) = 0$, $0 < t$, and the initial condition that $u(x,0) = 1$, $0 < x < 1$.

Step 1: Introducing $u(x,t) = e^{-ax}v(x,t)$, show that the problem becomes

$$\frac{\partial v}{\partial t} = \frac{\partial^2 v}{\partial x^2} - a^2 v, \qquad 0 < x < 1, \quad 0 < t,$$

subject to the boundary conditions that $v_x(0,t) + av(0,t) = 0$, $v_x(1,t) + av(1,t) = 0$, $0 < t$, and the initial condition that $u(x,0) = e^{ax}$, $0 < x < 1$.

Step 2: Assuming that $v(x,t) = X(x)T(t)$, show that the problem reduces to the ordinary differential equations

$$X'' + (\lambda - a^2)X = 0, \qquad X'(0) + aX(0) = 0, \quad X'(1) + aX(1) = 0,$$

and $T' + \lambda T = 0$, where λ is the separation constant.

Step 3: Solve the eigenvalue problem and show that $\lambda_0 = 0$, $X_0(x) = e^{-ax}$, $T_0(t) = A_0$, and $\lambda_n = a^2 + n^2\pi^2$, $X_n(x) = a\sin(n\pi x) - n\pi\cos(n\pi x)$, and $T_n(t) = A_n e^{-(a^2+n^2\pi^2)t}$, where $n = 1,2,3,\ldots$, so that

$$v(x,t) = A_0 e^{-ax} + \sum_{n=1}^{\infty} A_n \left[a\sin(n\pi x) - n\pi\cos(n\pi x)\right] e^{-(a^2+n^2\pi^2)t}.$$

Step 4: Evaluate A_0 and A_n and show that

$$u(x,t) = \frac{2ae^{-2ax}}{1 - e^{-2a}} + 4a\pi \sum_{n=1}^{\infty} \frac{n\left[1 - (-1)^n e^a\right]}{(a^2 + n^2\pi^2)^2} \left[a\sin(n\pi x) - n\pi\cos(n\pi x)\right] e^{-ax-(a^2+n^2\pi^2)t}.$$

23. Use separation of variables to solve[22] the partial differential equation

$$\frac{\partial^3 u}{\partial x^2 \partial t} = \frac{\partial^4 u}{\partial x^4}, \qquad 0 < x < 1, \quad 0 < t,$$

subject to the boundary conditions that $u(0,t) = u_x(0,t) = u(1,t) = u_x(1,t) = 0$, $0 < t$, and the initial condition that $u(x,0) = Ax/2 - (1-A)x^2\left(\frac{3}{2} - x\right)$, $0 \le x \le 1$.

Step 1: Assuming that $u(x,t) = X(x)T(t)$, show that the problem reduces to the ordinary differential equations $X'''' + k^2 X'' = 0$, $X(0) = X'(0) = X(1) = X'(1) = 0$, and $T' + k^2 T = 0$, where k^2 is the separation constant.

[21] See DeGroot, S. R., 1942: Théorie phénoménologique de l'effet Soret. *Physica*, **9**, 699–707.

[22] See Hamza, E. A., 1999: Impulsive squeezing with suction and injection. *J. Appl. Mech.*, **66**, 945–951.

Step 2: Solving the eigenvalue problem first, show that

$$X_n(x) = 1 - \cos(k_n x) + \frac{1 - \cos(k_n)}{k_n - \sin(k_n)}[\sin(k_n x) - k_n x],$$

where k_n denotes the nth root of $\sin(k/2)[\sin(k/2) - (k/2)\cos(k/2)] = 0$.

Step 3: Using the results from Step 2, show that there are two classes of eigenfunctions: $\kappa_n = 2n\pi$, $X_n(x) = 1 - \cos(2n\pi x)$, and $\tan(\kappa_n/2) = \kappa_n/2$, $X_n(x) = 1 - \cos(\kappa_n x) + 2[\sin(\kappa_n x) - \kappa_n x]/\kappa_n$.

Step 4: Consider the eigenvalue problem

$$X'''' + \lambda X'' = 0, \qquad 0 < x < 1,$$

with the boundary conditions $X(0) = X'(0) = X(1) = X'(1) = 0$. Show that the orthogonality condition for this problem is

$$\int_0^1 X_n'(x) X_m'(x) \, dx = 0, \qquad n \neq m,$$

where $X_n(x)$ and $X_m(x)$ are two distinct eigenfunctions of this problem. Then show that we can construct an eigenfunction expansion for an arbitrary function $f(x)$ via

$$f(x) = \sum_{n=1}^{\infty} C_n X_n(x), \qquad \text{provided} \qquad C_n = \frac{\int_0^1 f'(x) X_n'(x) \, dx}{\int_0^1 [X_n'(x)]^2 \, dx}$$

and $f'(x)$ exists over the interval $(0,1)$. Hint: Follow the proof in Section 11.2 and integrate repeatedly by parts to eliminate the higher derivative terms.

Step 5: Show that

$$\int_0^1 [X_n'(x)]^2 \, dx = 2n^2\pi^2,$$

if $X_n(x) = 1 - \cos(2n\pi x)$, and

$$\int_0^1 [X_n'(x)]^2 \, dx = \kappa_n^2/2,$$

if $X_n(x) = 1 - \cos(\kappa_n x) + 2[\sin(\kappa_n x) - \kappa_n x]/\kappa_n$. Hint: $\sin(\kappa_n) = \kappa_n[1 + \cos(\kappa_n)]/2$.

Step 6: Use the above results to show that

$$u(x,t) = \sum_{n=1}^{\infty} A_n[1 - \cos(2n\pi x)]e^{-4n^2\pi^2 t}$$

$$+ \sum_{n=1}^{\infty} B_n \left\{ 1 - \cos(\kappa_n x) - \frac{2}{\kappa_n}[\sin(\kappa_n x) - \kappa_n x] \right\} e^{-\kappa_n^2 t},$$

where A_n is the Fourier coefficient corresponding to the eigenfunction $1 - \cos(2n\pi x)$ while B_n is the Fourier coefficient corresponding to the eigenfunction $1 - \cos(\kappa_n x) - 2[\sin(\kappa_n x) - \kappa_n x]/\kappa_n$.

Step 7: Show that $A_n = 0$ and $B_n = 2(1 - A)/\kappa_n^2$, so that

$$u(x,t) = 2(1 - A) \sum_{n=1}^{\infty} \left\{ 1 - \cos(\kappa_n x) - \frac{2}{\kappa_n} [\sin(\kappa_n x) - \kappa_n x] \right\} \frac{e^{-\kappa_n^2 t}}{\kappa_n^2}.$$

Hint: $\sin(\kappa_n) = \kappa_n [1 + \cos(\kappa_n)]/2$, $\sin(\kappa_n) = 2[1 - \cos(\kappa_n)]/\kappa_n$, and $\cos(\kappa_n) = (4 - \kappa_n^2)/(4 + \kappa_n^2)$.

24. Solve the heat equation

$$\frac{\partial u}{\partial t} = \frac{\partial^2 u}{\partial x^2}, \qquad 0 < x < 1, \quad 0 < t,$$

subject to the boundary conditions $u(0,t) = f(t)$, $u_x(1,t) = -hu(1,t)$, $0 < t$, and the initial condition $u(x,0) = 0$, $0 < x < 1$.

Step 1: First solve the heat conduction problem

$$\frac{\partial A}{\partial t} = \frac{\partial^2 A}{\partial x^2}, \qquad 0 < x < 1, \quad 0 < t,$$

subject to the boundary conditions $A(0,t) = 1$, $A_x(1,t) = -hA(1,t)$, $0 < t$, and the initial condition $A(x,0) = 0$, $0 < x < 1$. Show that

$$A(x,t) = 1 - \frac{hx}{1+h} - 2\sum_{n=1}^{\infty} \frac{k_n^2 + h^2}{k_n \left(k_n^2 + h^2 + h\right)} \sin(k_n x) e^{-k_n^2 t},$$

where k_n is the nth root of $k \cot(k) = -h$.

Step 2: Use Duhamel's theorem and show that

$$u(x,t) = 2\sum_{n=1}^{\infty} \frac{k_n (k_n^2 + h^2)}{k_n^2 + h^2 + h} \sin(k_n x) e^{-k_n^2 t} \int_0^t f(\tau) e^{k_n^2 \tau} \, d\tau.$$

Separation of Variables Solution to Laplace's Equation

Solve Laplace's equation $u_{xx} + u_{yy} = 0$ over the rectangular region $0 < x < a$, $0 < y < b$ with the following boundary conditions. Illustrate your solution using MATLAB.

25. $u_x(0,y) = u(a,y) = u(x,0) = 0$, $u(x,b) = 1$

26. $u_y(x,0) = u(x,b) = u(a,y) = 0$, $u(0,y) = 1$

27. $u_x(a,y) = u_y(x,b) = 0$, $u(0,y) = u(x,0) = 1$ Hint: Let $u(x,y) = v(x,y) + w(x,y)$ with $v(0,y) = v_x(a,y) = v_y(x,b) = 0$, $v(x,0) = 1$ and $w(x,0) = w_y(x,b) = w_x(a,y) = 0$, $w(0,y) = 1$.

11.4 FINITE ELEMENT METHOD

In Section 1.7 we showed how to solve ordinary differential equations using finite differences. Here we introduce a popular alternative, the finite element method, and will use it to solve the Sturm-Liouville problem. One advantage of this approach is that we can focus on the details of the numerical scheme.

The finite element method breaks the global solution domain into a number of simply shaped subdomains, called *elements*. The global solution is then constructed by assembling the results from all of the elements. A particular strength of this method is that the elements do not have to be the same size; this allows us to have more resolution in regions where the solution is rapidly changing and fewer elements where the solution changes slowly. Overall, the solution of the ordinary differential equation is given by a succession of piecewise continuous functions.

Consider the Sturm-Liouville problem

$$-\frac{d}{dx}\left[p(x)\frac{dy}{dx}\right] + q(x)y(x) - \lambda r(x)y(x) = 0, \qquad a < x < b, \qquad (11.4.1)$$

with

$$y(a) = 0, \qquad \text{or} \qquad p(a)y'(a) = 0, \qquad (11.4.2)$$

and

$$y(b) = 0, \qquad \text{or} \qquad p(b)y'(b) = 0. \qquad (11.4.3)$$

Our formulation of the finite element approximation to the exact solution is called the *Galerkin weighted residual approach*. This is not the only possible way of formulating the finite element equations, but it is similar to the eigenfunction expansions that we highlighted in this chapter. Our approach consists of two steps: First we assume that $y(x)$ can be expressed over a particular element by

$$y(x) = \sum_{j=1}^{J} y_j \varphi_j(x), \qquad (11.4.4)$$

where $\varphi_j(x)$ is the jth *approximation* or *shape function*, and J is the total number of elements.

Let us define the residue

$$R(x) = -\frac{d}{dx}\left[p(x)\frac{dy}{dx}\right] + q(x)y(x) - \lambda r(x)y(x). \qquad (11.4.5)$$

We now require that for each element along the segment $\Omega_e = (x_n, x_{n+1})$,

$$\int_{\Omega_e} R(x)\varphi_i(x)\,dx = 0, \qquad i = 1, 2, 3, \ldots, J. \qquad (11.4.6)$$

The points x_n and x_{n+1} are known as *nodes*. Substituting Equation 11.4.5 into Equation 11.4.6,

$$\int_{\Omega_e}\left\{-\frac{d}{dx}\left[p(x)\frac{dy}{dx}\right] + q(x)y(x) - \lambda r(x)y(x)\right\}\varphi_i(x)\,dx = 0. \qquad (11.4.7)$$

Because

$$-\int_{x_n}^{x_{n+1}}\frac{d}{dx}\left[p(x)\frac{dy}{dx}\right]\varphi_i(x)\,dx = -p(x)\frac{dy}{dx}\varphi_i(x)\Big|_{x_n}^{x_{n+1}} + \int_{x_n}^{x_{n+1}} p(x)\frac{dy}{dx}\frac{d\varphi_i(x)}{dx}\,dx, \quad (11.4.8)$$

then

$$\int_{x_n}^{x_{n+1}} \left[p(x)\frac{dy}{dx}\frac{d\varphi_i(x)}{dx} + q(x)y(x)\varphi_i(x) - \lambda r(x)y(x)\phi_i(x) \right] dx = \left. p(x)\frac{dy}{dx}\varphi_i(x) \right|_{x_n}^{x_{n+1}}.$$
(11.4.9)

Upon using Equation 11.4.4 to eliminate $y(x)$ and reversing the order of summation and integration, our second step in the finite element method involves solving for y_j via

$$\sum_{j=1}^{J} \left\{ \int_{x_n}^{x_{n+1}} p(x)\frac{d\varphi_i(x)}{dx}\frac{d\varphi_j(x)}{dx} dx + \int_{x_n}^{x_{n+1}} q(x)\varphi_i(x)\varphi_j(x) dx \right.$$

$$\left. - \lambda \int_{x_n}^{x_{n+1}} r(x)\varphi_i(x)\varphi_j(x) dx \right\} y_j = \left. p(x)\frac{dy}{dx}\varphi_i(x) \right|_{x_n}^{x_{n+1}},$$
(11.4.10)

or using matrix notation

$$K\mathbf{y} - \lambda M\mathbf{y} = \mathbf{b},$$
(11.4.11)

where

$$K = \begin{pmatrix} K_{11} & K_{12} & \dots & K_{1J} \\ K_{21} & K_{22} & \dots & K_{2J} \\ \vdots & \vdots & \vdots & \vdots \\ K_{J1} & K_{J2} & \dots & K_{JJ} \end{pmatrix}, \quad M = \begin{pmatrix} M_{11} & M_{12} & \dots & M_{1J} \\ M_{21} & M_{22} & \dots & M_{2J} \\ \vdots & \vdots & \vdots & \vdots \\ M_{J1} & M_{J2} & \dots & M_{JJ} \end{pmatrix}$$
(11.4.12)

$$\mathbf{b} = \begin{pmatrix} b_1 \\ b_2 \\ \vdots \\ b_J \end{pmatrix}, \quad \mathbf{y} = \begin{pmatrix} y_1 \\ y_2 \\ \vdots \\ y_J \end{pmatrix},$$
(11.4.13)

$$b_i = p(x_{n+1})y'(x_{n+1})\varphi_i(x_{n+1}) - p(x_n)y'(x_n)\varphi_i(x_n),$$
(11.4.14)

$$K_{ij} = \int_{x_n}^{x_{n+1}} p(x)\frac{d\varphi_i(x)}{dx}\frac{d\varphi_j(x)}{dx} dx + \int_{x_n}^{x_{n+1}} q(x)\varphi_i(x)\varphi_j(x) dx,$$
(11.4.15)

and

$$M_{ij} = \int_{x_n}^{x_{n+1}} r(x)\varphi_i(x)\varphi_j(x) dx.$$
(11.4.16)

Why do we prefer to use Equation 11.4.10 rather than Equation 11.4.7? There are two reasons. First, it offers a convenient method for introducing the specified boundary conditions, Equation 11.4.2 and Equation 11.4.3. Second, it has lowered the highest-order derivatives from a second to a first derivative. This yields the significant benefit that $\varphi_i(x)$ must only be continuous but not necessarily a continuous slope at the nodes.

An important question is how we will evaluate the integrals in Equation 11.4.15 and Equation 11.4.16. Because $p(x)$, $q(x)$, and $r(x)$ are known, we could substitute these quantities along with $\varphi_i(x)$ and $\varphi_j(x)$ into Equation 11.4.15 and Equation 11.4.16 and perform the integration, presumably numerically. In a similar vein, we could develop curve fits for $p(x)$, $q(x)$, and $r(x)$ and again perform the integrations. However, we simply use their values at the midpoint between the nodes, $\overline{x}_n = (x_{n+1} + x_n)/2$, because $p(x)$, $q(x)$ and $r(x)$ usually vary slowly over the interval (x_n, x_{n+1}).

At this point we will specify J. The simplest case is $J = 2$ and we have the linear element:

$$\varphi_1(x) = \frac{x - x_1}{x_2 - x_1} \quad \text{and} \quad \varphi_2(x) = \frac{x_2 - x}{x_2 - x_1},$$
(11.4.17)

Table 11.4.1: The System Topology for 4 Finite-Element Segmentations When a Linear Interpolation Is Used

	Node Numbers	
Element	Local	Global
1	1	1
	2	2
2	1	2
	2	3
3	1	3
	2	4
4	1	4
	2	5

where x_1 and x_2 are *local* nodal points located at the end of the element. It directly follows that

$$\frac{dy}{dx} = \frac{d\varphi_1}{dx}y_1 + \frac{d\varphi_2}{dx}y_2 = \frac{y_2 - y_1}{x_2 - x_1}. \tag{11.4.18}$$

In other words, dy/dx equals the slope of the straight line connecting the nodes. Similarly,

$$\int_{x_1}^{x_2} y(x)\,dx = \tfrac{1}{2}\left(y_2 + y_1\right)\left(x_2 - x_1\right) \tag{11.4.19}$$

and we simply have the trapezoidal rule.

Substituting $\varphi_1(x)$ and $\varphi_2(x)$ into Equation 11.4.15 and Equation 11.4.16 and carrying out the integration, we obtain

$$K_{11} = \frac{p(x_c)}{L} + \frac{q(x_c)L}{3}, \quad K_{12} = -\frac{p(x_c)}{L} + \frac{q(x_c)L}{6}, \quad K_{21} = K_{12}, \quad \text{and} \quad K_{22} = K_{11}, \tag{11.4.20}$$

with $L = x_2 - x_1$ and $x_c = (x_1 + x_2)/2$. Similarly,

$$M_{11} = \frac{r(x_c)L}{3} = M_{22}, \quad \text{and} \quad M_{12} = \frac{r(x_c)L}{6} = M_{21}. \tag{11.4.21}$$

Finally, because $\varphi_1(x_1) = 0$, $\varphi_1(x_2) = 1$, $\varphi_2(x_1) = 1$, and $\varphi_2(x_2) = 0$, $b_1 = -p(x_1)y'(x_1)$ and $b_2 = p(x_2)y_2(x_2)$.

Having obtained the finite element representation for nodes 1 and 2, we would like to extend these results to an arbitrary number of additional nodes. This is done by setting up a look-up table that relates the global nodal points to the local ones. For example, suppose we would like 5 nodes between a and b with $x = x_1$, x_2, x_3, x_4, and x_5. Then Table 11.4.1 illustrates our look-up table.

Having developed the spatial layout, we are now ready to assemble the matrix for the entire interval (a, b). For clarity we will give the intermediate steps. Taking the first element into account,

$$K = \begin{pmatrix} K_{11}^{(1)} & K_{12}^{(1)} & 0 & 0 & 0 \\ K_{21}^{(1)} & K_{22}^{(1)} & 0 & 0 & 0 \\ 0 & 0 & 0 & 0 & 0 \\ 0 & 0 & 0 & 0 & 0 \\ 0 & 0 & 0 & 0 & 0 \end{pmatrix}, \quad M = \begin{pmatrix} M_{11}^{(1)} & M_{12}^{(1)} & 0 & 0 & 0 \\ M_{21}^{(1)} & M_{22}^{(1)} & 0 & 0 & 0 \\ 0 & 0 & 0 & 0 & 0 \\ 0 & 0 & 0 & 0 & 0 \\ 0 & 0 & 0 & 0 & 0 \end{pmatrix}, \quad (\mathbf{11.4.22})$$

$$\mathbf{b} = \begin{pmatrix} -p(x_1)y'(x_1) \\ p(x_2)y'(x_2) \\ 0 \\ 0 \\ 0 \end{pmatrix}, \quad \text{and} \quad \mathbf{y} = \begin{pmatrix} y_1 \\ y_2 \\ 0 \\ 0 \\ 0 \end{pmatrix}. \quad (\mathbf{11.4.23})$$

Here we have added a subscript (1) to K_{ij} and M_{ij} to denote that value for the first element should be used in computing $p(x_c)$, $q(x_c)$, $r(x_c)$, and L. Consequently, when we introduce the second element, K, M, and $\mathbf{y}$ become

$$K = \begin{pmatrix} K_{11}^{(1)} & K_{12}^{(1)} & 0 & 0 & 0 \\ K_{21}^{(1)} & K_{22}^{(1)} + K_{11}^{(2)} & K_{12}^{(2)} & 0 & 0 \\ 0 & K_{21}^{(2)} & K_{22}^{(2)} & 0 & 0 \\ 0 & 0 & 0 & 0 & 0 \\ 0 & 0 & 0 & 0 & 0 \end{pmatrix}, \quad (\mathbf{11.4.24})$$

$$M = \begin{pmatrix} M_{11}^{(1)} & M_{12}^{(1)} & 0 & 0 & 0 \\ M_{21}^{(1)} & M_{22}^{(1)} + M_{11}^{(2)} & M_{12}^{(2)} & 0 & 0 \\ 0 & M_{21}^{(2)} & M_{22}^{(2)} & 0 & 0 \\ 0 & 0 & 0 & 0 & 0 \\ 0 & 0 & 0 & 0 & 0 \end{pmatrix}, \quad (\mathbf{11.4.25})$$

$$\mathbf{b} = \begin{pmatrix} -p(x_1)y'(x_1) \\ 0 \\ p(x_3)y'(x_3) \\ 0 \\ 0 \end{pmatrix}, \quad \text{and} \quad \mathbf{y} = \begin{pmatrix} y_1 \\ y_2 \\ y_3 \\ 0 \\ 0 \end{pmatrix}. \quad (\mathbf{11.4.26})$$

Note that the numbering for y_i corresponds to ith *global* node number. Continuing with this process of adding additional elements to the system matrix, we finally have

$$K = \begin{pmatrix} K_{11}^{(1)} & K_{12}^{(1)} & 0 & 0 & 0 \\ K_{21}^{(1)} & K_{22}^{(1)} + K_{11}^{(2)} & K_{12}^{(2)} & 0 & 0 \\ 0 & K_{21}^{(2)} & K_{22}^{(2)} + K_{11}^{(3)} & K_{12}^{(3)} & 0 \\ 0 & 0 & K_{21}^{(3)} & K_{22}^{(3)} + K_{11}^{(4)} & K_{12}^{(4)} \\ 0 & 0 & 0 & K_{21}^{(4)} & K_{22}^{(4)} \end{pmatrix}, \quad (\mathbf{11.4.27})$$

$$M = \begin{pmatrix} M_{11}^{(1)} & M_{12}^{(1)} & 0 & 0 & 0 \\ M_{21}^{(1)} & M_{22}^{(1)} + M_{11}^{(2)} & M_{12}^{(2)} & 0 & 0 \\ 0 & M_{21}^{(2)} & M_{22}^{(2)} + M_{11}^{(3)} & M_{12}^{(3)} & 0 \\ 0 & 0 & M_{21}^{(3)} & M_{22}^{(3)} + M_{11}^{(4)} & M_{12}^{(4)} \\ 0 & 0 & 0 & M_{21}^{(4)} & M_{22}^{(4)} \end{pmatrix}, \quad (\mathbf{11.4.28})$$

Table 11.4.2: The Lowest Eigenvalue for Equation 11.4.30, Which Is Solved Using a Finite Element Method

L	λ
0.250	10.6745
0.100	10.2335
0.050	10.1717
0.020	10.1544
0.010	10.1520
0.002	10.1512

where

$$\mathbf{b} = \begin{pmatrix} -p(x_1)y'(x_1) \\ 0 \\ 0 \\ 0 \\ p(x_5)y'(x_5) \end{pmatrix}, \quad \text{and} \quad \mathbf{y} = \begin{pmatrix} y_1 \\ y_2 \\ y_3 \\ y_4 \\ y_5 \end{pmatrix}. \tag{11.4.29}$$

Let us examine the $\mathbf{b}$ vector more closely. In the final form of the finite element formulation, $\mathbf{b}$ has non-zero values only at the end points; the contributions from intermediate nodal points vanish because $p(x)y'(x)$ is continuous within the interval (a, b). Furthermore, if $y'(a) = y'(b) = 0$ from the boundary conditions, then the $\mathbf{b}$ vector becomes the zero vector. On the other hand, if $y(a) = y(b) = 0$, then $y_1 = y_5 = 0$ and the eigenvalue problem involves a 3×3 matrix with the unknowns y_2, y_3, and y_4. Similarly, if $y'(a) = y(b) = 0$, then we have a 4×4 matrix with the unknowns y_1, y_2, y_3, y_4, and $y_5 = 0$. Finally, if $y(a) = y'(b) = 0$, we again have a 4×4 matrix involving $y_1 = 0$ and the unknowns y_2, y_3, y_4, and y_5.

To illustrate this scheme, consider the Sturm-Liouville problem

$$y'' + (\lambda - x^2)y = 0, \qquad 0 < x < 1 \tag{11.4.30}$$

with $y(0) = y(1) = 0$. Here $p(x) = 1$, $q(x) = x^2$, and $r(x) = 1$.

The MATLAB code begins with the choice of the number of elements, N. Once that is done, L immediately follows because L = 1/(N-1). We will also need to have the value of x at the node points x(n) = L*(n-1) where n = 1:N.

With these preliminaries out of the way, we begin by setting up the matrices K and M given by Equation 11.4.27 and Equation 11.4.28. The corresponding MATLAB code is:

```
for i = 1:N-1
  x_c = 0.5*(x(i) + x(i+1));
  p = 1; q = x_c*x_c; r = 1;
  K_11 = p/L + q*L/3; K_22 = K_11;
  K_12 = -p/L + q*L/6; K_21 = K_12;
  M_11 = r*L/3; M_22 = M_11;
  M_12 = r*L/6; M_21 = M_12;
  KK( i , i ) = KK( i , i ) + K_11;
  KK( i ,i+1) = KK( i ,i+1) + K_12;
  KK(i+1, i ) = KK(i+1, i ) + K_21;
  KK(i+1,i+1) = KK(i+1,i+1) + K_22;
```

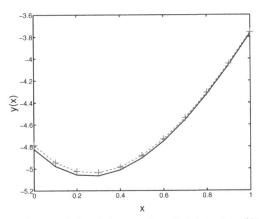

Figure 11.4.1: Numerical solution of $y'' + 2y' + y = 2x + 3\sin(x)$ with $y'(0) = -2$ and $y'(1) = 3$ using finite elements with $\Delta x = 0.1$. The crosses indicate the exact solution.

```
MM( i , i ) = MM( i , i ) + M_11;
MM( i ,i+1) = MM( i ,i+1) + M_12;
MM(i+1, i ) = MM(i+1, i ) + M_21;
MM(i+1,i+1) = MM(i+1,i+1) + M_22;
end
```
Note that the arrays KK and MM have already been defined as $N \times N$ arrays with all of their elements set to zero.

Finally, because y_1 and y_N are zero, we must extract that portion of K and M for which $y_m \neq 0$. This is done as follows:
```
for j = 1:N-2
for i = 1:N-2
  A(i,j) = KK(i+1,j+1);
  B(i,j) = MM(i+1,j+1);
end; end
```
Finally, the eigenvalues are found by `eig(A,B)`. If the corresponding eigenfunction is desired, then the corresponding eigenfunction gives y_j for $j = 2, 3, \ldots, N-1$ using Equation 11.4.4. Table 11.4.2 illustrates how the lowest eigenvalue for Equation 11.4.30 improves in accuracy as the number of nodes is increased.

Project: Finite Element Solution of Boundary-Value Problems

In addition to solving the Sturm-Liouville problem, finite element methods can be used to solve the standard boundary-value problem:

$$-\frac{d}{dx}\left[p(x) \frac{dy}{dx} \right] + q(x)y = f(x), \qquad 0 < x < 1,$$

where we specify $y(0)$ *or* $y'(0)$ at $x = 0$ and $y(1)$ *or* $y'(1)$ at $x = 1$. Although you could create your MATLAB code to solve this problem, MATLAB code has already been developed and is available online.[23] The purpose of this project is for you to become comfortable using this scheme.

---

[23] For example, http://people.sc.fsu.edu/~burkardt/m_src/fem1d/fem1d.html

Table 11.4.3: The Eigenvalues for the Sturm-Liouville Problem Stated in Example 11.1.1 Using the Finite Element Method for Various N's. The Boldface Digits Are Those that Differ from the Exact Answer.

$N = 5$	$N = 10$	$N = 25$	$N = 50$	exact
−0.9**6197352**	−0.98**441038**	−0.99**108552**	−0.992**05427**	−0.99237804
1.7**5605999**	1.6**8721832**	1.66**802975**	1.66**529435**	1.66438291
6.**69690282**	5.**89559315**	5.6**7320930**	5.6**4181953**	5.63138040
15.**81967372**	12.**76377241**	11.**80121128**	11.6**6700268**	11.62250178
26.**80475760**	22.**90921756**	20.**13008890**	19.**74581245**	19.61888190
	37.**11271111**	30.**78756638**	29.**90669715**	29.61705752
	55.**95775087**	43.**93943526**	42.**18874880**	41.61601074
	78.**82277983**	59.**78883650**	57.**64002877**	55.61535492
	102.**01056615**	78.**57612764**	73.**31730755**	71.61491703
	117.**83174892**	100.**57733062**	92.**28610692**	89.61461020
		126.**09956238**	113.**62082677**	109.61438688

Step 1: Using the method of undetermined coefficients, solve the boundary-value problem

$$y'' + 2y' + y = 2x + 3\sin(x), \qquad y'(0) = -2, \quad y'(1) = 3.$$

Show that

$$y(x) = 2x - 4 - \tfrac{3}{2}\cos(x) + \tfrac{1}{2}\left[3\sin(1) - 2\right]e^{1-x} + \tfrac{1}{2}\left[3\sin(1) - 2 - 8/e\right]xe^{1-x}.$$

Step 2: Show that the ordinary differential equation can be written as

$$\frac{d}{dx}\left(e^{2x}\frac{dy}{dx}\right) + e^{2x}y = 2xe^{2x} + 3e^{2x}\sin(x), \qquad y'(0) = -2, \quad y'(1) = 3.$$

Step 3: Find the numerical solution of the boundary-value problem using finite elements. Figure 11.4.1 illustrates the solution.

Project: Robin Boundary Condition[24]

In this section we showed how to find the eigenvalues and eigenfunctions for a Sturm-Liouville problem using finite element methods when we have Dirichlet and/or Neumann conditions. What do we do when we have a Robin boundary condition at one or both ends? Answering that question is the goal of this project.

---

[24] Suggested by Akano, T. T., and O. A. Fakinlede, 2015: Numerical computation of Sturm-Liouville problem with Robin boundary condition. *Int. J. Math. Comput. Sci.*, **9**, 690–694.

The difficulty here is that the vector **b** is not equal to zero. However, from Equation 11.1.2 we have $y'(a) = -\alpha y(a)/\beta$ and $y'(b) = -\gamma y(b)/\delta$ provided β and δ are nonzero. In this case we can rewrite **b** as

$$
\mathbf{b} = \begin{pmatrix} \alpha p(x_1) y(x_1)/\beta \\ 0 \\ 0 \\ 0 \\ -\gamma p(x_5) y(x_5)/\gamma \end{pmatrix} = G_{ij}\mathbf{y},
$$

where

$$
G_{ij} = \begin{cases} \alpha p(x_1)/\beta, & i = j = 1 \\ -\gamma p(x_5)/\delta, & i = j = 5 \\ 0, & \text{otherwise.} \end{cases}
$$

We can then combine G_{ij} with K_{ij} and then numerically solve the resulting classic eigenvalue problem.

Using the code that we developed to solve Equation 11.4.20, use the finite element technique to find the eigenvalues for Example 11.1.1. In this case we still have $y_1 = 0$ but $y'_N = y_N$. Table 11.4.2 shows some of the numerical results at different resolutions.

Further Readings

Akulenko, L. D., and S. V. Nesterov, 2004: *High-Precision Methods in Eigenvalue Problems.* CRC Press, 260 pp. A fairly new book on analytic, asymptotic and numerical methods in solving the Sturm-Liouville problem.

Clarlet, P. G., 1978: *The Finite Element Method for Elliptic Problems.* Elsevier North-Holland, 530 pp. The classic text on finite element methods.

Titchmarsh, E. C., 1946: *Eigenfunction Expansions Associated with Second Order Differential Equations.* Camp Press, 188 pp. A rigorous treatment of the Sturm-Liouville problem.

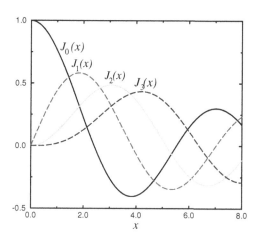

Chapter 12
Special Functions

In the previous chapter we studied how boundary-value problems of the form:

$$\frac{d}{dx}\left[p(x)\frac{dy}{dx}\right] + [q(x) + \lambda r(x)]y = 0, \quad a \le x \le b, \tag{12.0.1}$$

along with the boundary conditions:

$$\alpha y(a) + \beta y'(a) = 0 \quad \text{and} \quad \gamma y(b) + \delta y'(b) = 0, \tag{12.0.2}$$

the so-called Sturm-Liouville problem, could be used to re-express a continuous function $f(x)$ in terms of discrete solution to Equation 12.0.1. There we required that the real function $p(x)$ and $r(x)$ be continuous and positive on the interval $a \le x \le b$. In this chapter we consider the problem when $p(a)$ and/or $p(b)$ equal zero.

Let $y_n(x)$ and $y_m(x)$ denote the eigenfunctions associated with two different eigenvalues λ_n and λ_m. Then

$$\frac{d}{dx}\left[p(x)\frac{dy_n}{dx}\right] + [q(x) + \lambda_n r(x)]y_n(x) = 0, \tag{12.0.3}$$

$$\frac{d}{dx}\left[p(x)\frac{dy_m}{dx}\right] + [q(x) + \lambda_m r(x)]y_m(x) = 0. \tag{12.0.4}$$

Let us multiply the first differential equation by y_m; the second by y_n. Next, we subtract these two equations and move the terms containing $y_n y_m$ to the right side, resulting in

$$y_n\frac{d}{dx}\left[p(x)\frac{dy_m}{dx}\right] - y_m\frac{d}{dx}\left[p(x)\frac{dy_n}{dx}\right] = (\lambda_n - \lambda_m)r(x)y_n y_m. \tag{12.0.5}$$

Integrating Equation 12.0.5 from a to b yields

$$\int_a^b \left\{ y_n \frac{d}{dx}\left[p(x)\frac{dy_m}{dx} \right] - y_m \frac{d}{dx}\left[p(x)\frac{dy_n}{dx} \right] \right\} dx = (\lambda_n - \lambda_m) \int_a^b r(x)y_n y_m\, dx. \quad (\mathbf{12.0.6})$$

We can simplify the left side of Equation 12.0.6 by integrating by parts to give

$$\int_a^b \left\{ y_n \frac{d}{dx}\left[p(x)\frac{dy_m}{dx} \right] - y_m \frac{d}{dx}\left[p(x)\frac{dy_n}{dx} \right] \right\} dx$$

$$= [p(x)y'_m y_n - p(x)y'_n y_m]_a^b - \int_a^b p(x)[y'_n y'_m - y'_n y'_m]\, dx. \quad (\mathbf{12.0.7})$$

The second integral equals zero since the integrand vanishes identically. Combining Equation 12.0.6 and Equation 12.0.7 together, we find

$$(\lambda_n - \lambda_m) \int_a^b r(x)y_n y_m\, dx = [p(b)y'_m(b)y_n(b) - p(b)y'_n(b)y_m(b)$$

$$- p(a)y'_m(a)y_n(a) + p(a)y'_n(a)y_m(a)]. \quad (\mathbf{12.0.8})$$

From Equation 12.0.8 the right side vanishes and we preserve orthogonality (1) if $y_n(x)$ is finite and $p(x)y'_n(x)$ tends to zero at both endpoints (as in the case of Legendre polynomials in Section 12.1) or (2) if $y(a)$ is finite, $p(x)y'_n(x) \to 0$ as $x \to a$, and $y(x)$ satisfies Equation 12.0.2 at $x = b$ (as in the case of Bessel functions in Section 12.2). In this chapter we present two cases of these singular Sturm-Liouville problems and how to form expansions using the corresponding eigenfunctions.

• Example 12.0.1

Consider the zeroth-order Bessel equation:

$$xy'' + y' + \mu^2 xy = 0, \qquad 0 \le x < L. \quad (\mathbf{12.0.9})$$

(Equation 12.2.1 with $n = 0$.) Here $a = 0$, $b = L$, $p(x) = x$, $r(x) = x$ and $\lambda = \mu^2$. Because $p(0) = 0$, this is an example of a singular Sturm-Liouville problem.

In Section 12.2 we will show that Equation 12.0.9 has two linearly independent solutions: $y_1(x) = J_0(\mu x)$ and $y_2(x) = Y_0(\mu x)$. Can we use either of these in an eigenfunction expansion? To do so, it must satisfy an orthogonality condition. Therefore, a solution must satisfy the criteria stated above. From Equation 12.2.10 and Equation 12.2.12, we find that

$$y_1(x) = J_0(\mu x) = 1 - \frac{\mu^2 x^2}{4} + \cdots, \qquad y'_1(x) = -\frac{\mu^2 x}{2} + \cdots, \quad (\mathbf{12.0.10})$$

and

$$y_2(x) = \frac{2}{\pi} \ln(\mu x/2) + \cdots, \qquad y'_2(x) = \frac{2}{\pi x} + \cdots, \quad (\mathbf{12.0.11})$$

for small x. Hence,

$$\lim_{x \to 0} y_1(x) \to 1, \qquad \lim_{x \to 0} x\, y'_1(x) \to 0, \quad (\mathbf{12.0.12})$$

and

$$\lim_{x \to 0} y_2(x) \to -\infty, \qquad \lim_{x \to 0} x\, y'_2(x) \to \frac{2}{\pi}. \quad (\mathbf{12.0.13})$$

Born into an affluent family, Adrien-Marie Legendre's (1752–1833) modest family fortune was sufficient to allow him to devote his life to research in celestial mechanics, number theory, and the theory of elliptic functions. In July 1784 he read before the *Académie des sciences* his *Recherches sur la figure des planètes*. It is in this paper that Legendre polynomials first appeared. (Portrait courtesy of the Archives de l'Académie des sciences, Paris.)

Clearly only $y_1(x)$ satisfies the criteria and we must discard $y_2(x)$ from further discussion.

Turning to the other endpoint $x = L$, $y_1(x)$ must satisfy $\gamma y_1(L) + \delta y_1'(L) = 0$, where either γ or δ is nonzero. In Section 12.2 we will show that $y_1(x)$ also satisfies this condition if $y_1(x) = J_0(\mu_k x)$, where μ_k is the kth zero of $hJ_0(\mu L) + \mu J_0'(\mu L) = 0$ and $0 \le h < \infty$.

12.1 LEGENDRE'S POLYNOMIALS

Consider now Legendre's equation:

$$(1 - x^2)\frac{d^2y}{dx^2} - 2x\frac{dy}{dx} + n(n+1)y = 0, \qquad (\mathbf{12.1.1})$$

or

$$\frac{d}{dx}\left[(1 - x^2)\frac{dy}{dx}\right] + n(n+1)y = 0, \qquad (\mathbf{12.1.2})$$

where we set $a = -1$, $b = 1$, $\lambda = n(n+1)$, $p(x) = 1 - x^2$, $q(x) = 0$, and $r(x) = 1$. This equation arises in the solution of partial differential equations involving spherical geometry. Because $p(-1) = p(1) = 0$, we are faced with a singular Sturm-Liouville problem. However, as Equation 12.0.8 shows, orthogonality is preserved here.

Equation 12.1.1 does not have a simple general solution. [If $n = 0$, then $y(x) = 1$ is a solution.] Consequently we try to solve it with the power series:

$$y(x) = \sum_{k=0}^{\infty} A_k x^k, \quad y'(x) = \sum_{k=0}^{\infty} k A_k x^{k-1}, \quad \text{and} \quad y''(x) = \sum_{k=0}^{\infty} k(k-1) A_k x^{k-2}. \quad (\mathbf{12.1.3})$$

Substituting into Equation 12.1.1,

$$\sum_{k=0}^{\infty} k(k-1) A_k x^{k-2} + \sum_{k=0}^{\infty} \left[n(n+1) - 2k - k(k-1) \right] A_k x^k = 0, \quad (\mathbf{12.1.4})$$

which equals

$$\sum_{m=2}^{\infty} m(m-1) A_m x^{m-2} + \sum_{k=0}^{\infty} \left[n(n+1) - k(k+1) \right] A_k x^k = 0. \quad (\mathbf{12.1.5})$$

If we define $k = m - 2$ in the first summation, then

$$\sum_{k=0}^{\infty} (k+2)(k+1) A_{k+2} x^k + \sum_{k=0}^{\infty} \left[n(n+1) - k(k+1) \right] A_k x^k = 0. \quad (\mathbf{12.1.6})$$

Because Equation 12.1.6 must be true for any x, each power of x must vanish separately. It then follows that

$$(k+2)(k+1) A_{k+2} = [k(k+1) - n(n+1)] A_k, \quad (\mathbf{12.1.7})$$

or

$$A_{k+2} = \frac{[k(k+1) - n(n+1)]}{(k+1)(k+2)} A_k, \quad (\mathbf{12.1.8})$$

where $k = 0, 1, 2, \dots$. Note that we still have the two arbitrary constants A_0 and A_1 that are necessary for the general solution of Equation 12.1.1.

The first few terms of the solution associated with A_0 are

$$u_p(x) = 1 - \frac{n(n+1)}{2!} x^2 + \frac{n(n-2)(n+1)(n+3)}{4!} x^4$$
$$- \frac{n(n-2)(n-4)(n+1)(n+3)(n+5)}{6!} x^6 + \cdots, \quad (\mathbf{12.1.9})$$

while the first few terms associated with the A_1 coefficient are

$$v_p(x) = x - \frac{(n-1)(n+2)}{3!} x^3 + \frac{(n-1)(n-3)(n+2)(n+4)}{5!} x^5$$
$$- \frac{(n-1)(n-3)(n-5)(n+2)(n+4)(n+6)}{7!} x^7 + \cdots. \quad (\mathbf{12.1.10})$$

If n is an *even* positive integer (including $n = 0$), then the series, Equation 12.1.9, terminates with the term involving x^n: The solution is a polynomial of degree n. Similarly, if n is an *odd* integer, the series, Equation 12.1.10, terminates with the term involving x^n. Otherwise, for n noninteger the expressions are infinite series.

Table 12.1.1: The First Ten Legendre Polynomials

$$P_0(x) = 1$$

$$P_1(x) = x$$

$$P_2(x) = \tfrac{1}{2}(3x^2 - 1)$$

$$P_3(x) = \tfrac{1}{2}(5x^3 - 3x)$$

$$P_4(x) = \tfrac{1}{8}(35x^4 - 30x^2 + 3)$$

$$P_5(x) = \tfrac{1}{8}(63x^5 - 70x^3 + 15x)$$

$$P_6(x) = \tfrac{1}{16}(231x^6 - 315x^4 + 105x^2 - 5)$$

$$P_7(x) = \tfrac{1}{16}(429x^7 - 693x^5 + 315x^3 - 35x)$$

$$P_8(x) = \tfrac{1}{128}(6435x^8 - 12012x^6 + 6930x^4 - 1260x^2 + 35)$$

$$P_9(x) = \tfrac{1}{128}(12155x^9 - 25740x^7 + 18018x^5 - 4620x^3 + 315x)$$

$$P_{10}(x) = \tfrac{1}{256}(46189x^{10} - 109395x^8 + 90090x^6 - 30030x^4 + 3465x^2 - 63)$$

For reasons that will become apparent, we restrict ourselves to positive integers n. Actually, this includes all possible integers because the negative integer $-n - 1$ has the same Legendre's equation and solution as the positive integer n. These polynomials are *Legendre polynomials*[1] and we may compute them by the power series:

$$P_n(x) = \sum_{k=0}^{m} (-1)^k \frac{(2n - 2k)!}{2^n k!(n - k)!(n - 2k)!} x^{n-2k}, \qquad (\textbf{12.1.11})$$

where $m = n/2$, or $m = (n - 1)/2$, depending upon which is an integer. We chose to use Equation 12.1.11 over Equation 12.1.9 or Equation 12.1.10 because Equation 12.1.11 has the advantage that $P_n(1) = 1$. Table 12.1.1 gives the first ten Legendre polynomials.

The other solution, the infinite series, is the Legendre function of the second kind, $Q_n(x)$. Figure 12.1.1 illustrates the first four Legendre polynomials $P_n(x)$ while Figure 12.1.2 gives the first four Legendre functions of the second kind $Q_n(x)$. From this figure we see that $Q_n(x)$ becomes infinite at the points $x = \pm 1$. As shown earlier, this is important because we are only interested in solutions to Legendre's equation that are finite over the interval $[-1, 1]$. On the other hand, in problems where we exclude the points $x = \pm 1$, Legendre functions of the second kind will appear in the general solution.[2]

In the case that n is not an integer, we can construct a solution[3] that remains finite at $x = 1$ but not at $x = -1$. Furthermore, we can construct a solution that is finite at

---

[1] Legendre, A. M., 1785: Sur l'attraction des sphéroïdes homogénes. *Mém. math. phys. présentés à l'Acad. sci. pars divers savants*, **10**, 411–434. The best reference on Legendre polynomials is Hobson, E. W., 1965: *The Theory of Spherical and Ellipsoidal Harmonics*. Chelsea Publishing Co., 500 pp.

[2] See Smythe, W. R., 1950: *Static and Dynamic Electricity*. McGraw-Hill, Section 5.215, for an example.

[3] See Carrier, G. F., M. Krook, and C. E. Pearson, 1966: *Functions of the Complex Variable: Theory and Technique*. McGraw-Hill, pp. 212–213.

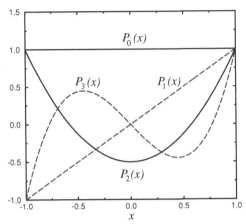

Figure 12.1.1: The first four Legendre functions of the first kind.

$x = -1$ but not at $x = 1$. Because our solutions must be finite at both endpoints so that we can use them in an eigenfunction expansion, we must reject these solutions from further consideration and are left only with Legendre polynomials. From now on, we will only consider the properties and uses of these polynomials.

Although we have the series, Equation 12.1.11, to compute $P_n(x)$, there are several alternative methods. We obtain the first method, known as *Rodrigues's formula*,[4] by writing Equation 12.1.11 in the form

$$P_n(x) = \frac{1}{2^n n!} \sum_{k=0}^{n} (-1)^k \frac{n!}{k!(n-k)!} \frac{(2n-2k)!}{(n-2k)!} x^{n-2k} \tag{12.1.12}$$

$$= \frac{1}{2^n n!} \frac{d^n}{dx^n} \left[\sum_{k=0}^{n} (-1)^k \frac{n!}{k!(n-k)!} x^{2n-2k} \right]. \tag{12.1.13}$$

The last summation is the binomial expansion of $(x^2 - 1)^n$ so that

$$P_n(x) = \frac{1}{2^n n!} \frac{d^n}{dx^n} (x^2 - 1)^n. \tag{12.1.14}$$

Another method for computing $P_n(x)$ involves the use of recurrence formulas. The first step in finding these formulas is to establish the fact that

$$(1 + h^2 - 2xh)^{-1/2} = P_0(x) + hP_1(x) + h^2 P_2(x) + \cdots. \tag{12.1.15}$$

The function $(1 + h^2 - 2xh)^{-1/2}$ is the *generating function* for $P_n(x)$. We obtain the expansion via the formal binomial expansion

$$(1 + h^2 - 2xh)^{-1/2} = 1 + \tfrac{1}{2}(2xh - h^2) + \tfrac{1}{2}\tfrac{3}{2}\tfrac{1}{2!}(2xh - h^2)^2 + \cdots. \tag{12.1.16}$$

---

[4] Rodrigues, O., 1816: Mémoire sur l'attraction des sphéroïdes. *Correspond. l'École Polytech.*, **3**, 361–385.

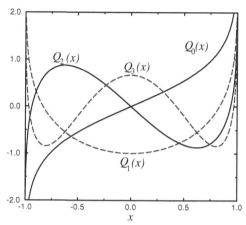

Figure 12.1.2: The first four Legendre functions of the second kind.

Upon expanding the terms contained in $2x - h^2$ and grouping like powers of h,

$$(1 + h^2 - 2xh)^{-1/2} = 1 + xh + (\tfrac{3}{2}x^2 - \tfrac{1}{2})h^2 + \cdots. \qquad (12.1.17)$$

A direct comparison between the coefficients of each power of h and the Legendre polynomial $P_n(x)$ completes the demonstration. Note that these results hold only if $|x|$ and $|h| < 1$.

Next we define $W(x, h) = (1 + h^2 - 2xh)^{-1/2}$. A quick check shows that $W(x, h)$ satisfies the first-order partial differential equation

$$(1 - 2xh + h^2)\frac{\partial W}{\partial h} + (h - x)W = 0. \qquad (12.1.18)$$

The substitution of Equation 12.1.15 into Equation 12.1.18 yields

$$(1 - 2xh + h^2)\sum_{n=0}^{\infty} nP_n(x)h^{n-1} + (h - x)\sum_{n=0}^{\infty} P_n(x)h^n = 0. \qquad (12.1.19)$$

Setting the coefficients of h^n equal to zero, we find that

$$(n + 1)P_{n+1}(x) - 2nxP_n(x) + (n - 1)P_{n-1}(x) + P_{n-1}(x) - xP_n(x) = 0, \qquad (12.1.20)$$

or

$$\boxed{(n + 1)P_{n+1}(x) - (2n + 1)xP_n(x) + nP_{n-1}(x) = 0} \qquad (12.1.21)$$

with $n = 1, 2, 3, \ldots$.

Similarly, the first-order partial differential equation

$$(1 - 2xh + h^2)\frac{\partial W}{\partial x} - hW = 0 \qquad (12.1.22)$$

leads to

$$(1 - 2xh + h^2)\sum_{n=0}^{\infty} P_n'(x)h^n - \sum_{n=0}^{\infty} P_n(x)h^{n+1} = 0, \qquad (12.1.23)$$

which implies

$$P'_{n+1}(x) - 2xP'_n(x) + P'_{n-1}(x) - P_n(x) = 0. \qquad (\mathbf{12.1.24})$$

Differentiating Equation 12.1.21, we first eliminate $P'_{n-1}(x)$ and then $P'_{n+1}(x)$ from the resulting equations and Equation 12.1.24. This gives two further recurrence relationships:

$$P'_{n+1}(x) - xP'_n(x) - (n+1)P_n(x) = 0, \quad n = 0, 1, 2, \dots, \qquad (\mathbf{12.1.25})$$

and

$$xP'_n(x) - P'_{n-1}(x) - nP_n(x) = 0, \quad n = 1, 2, 3, \dots. \qquad (\mathbf{12.1.26})$$

Adding Equation 12.1.25 and Equation 12.1.26, we obtain the more symmetric formula

$$P'_{n+1}(x) - P'_{n-1}(x) = (2n+1)P_n(x), \ n = 1, 2, 3, \dots. \qquad (\mathbf{12.1.27})$$

Given any two of the polynomials $P_{n+1}(x)$, $P_n(x)$, and $P_{n-1}(x)$, Equation 12.1.21 or Equation 12.1.27 yields the third.

- **Example 12.1.1**

Let us use Rodrigues' formula to compute $P_2(x)$. From Equation 12.1.14 with $n = 2$,

$$P_2(x) = \frac{1}{2^2 2!} \frac{d^2}{dx^2}[(x^2 - 1)^2] = \frac{1}{8}\frac{d^2}{dx^2}(x^4 - 2x^2 - 1) = \frac{1}{2}(3x^2 - 1). \quad (\mathbf{12.1.28})$$

$\square$

- **Example 12.1.2**

Let us compute $P_3(x)$ from a recurrence relation. From Equation 12.1.21 with $n = 2$,

$$3P_3(x) - 5xP_2(x) + 2P_1(x) = 0. \qquad (\mathbf{12.1.29})$$

But $P_2(x) = (3x^2 - 1)/2$, and $P_1(x) = x$, so that

$$3P_3(x) = 5xP_2(x) - 2P_1(x) = 5x[(3x^2 - 1)/2] - 2x = \tfrac{15}{2}x^3 - \tfrac{9}{2}x, \qquad (\mathbf{12.1.30})$$

or

$$P_3(x) = (5x^3 - 3x)/2. \qquad (\mathbf{12.1.31})$$

$\square$

- **Example 12.1.3**

We want to show that

$$\int_{-1}^{1} P_n(x)\,dx = 0, \qquad n > 0. \qquad (\mathbf{12.1.32})$$

From Equation 12.1.27,

$$(2n+1)\int_{-1}^{1} P_n(x)\,dx = \int_{-1}^{1} [P'_{n+1}(x) - P'_{n-1}(x)]\,dx \tag{12.1.33}$$

$$= P_{n+1}(x) - P_{n-1}(x)|_{-1}^{1} \tag{12.1.34}$$

$$= P_{n+1}(1) - P_{n-1}(1) - P_{n+1}(-1) + P_{n-1}(-1) = 0, \tag{12.1.35}$$

because $P_n(1) = 1$ and $P_n(-1) = (-1)^n$. $\qquad\qquad\square$

Having determined several methods for finding the Legendre polynomial $P_n(x)$, we now turn to the actual orthogonality condition.[5] Consider the integral

$$J = \int_{-1}^{1} \frac{dx}{\sqrt{1+h^2-2xh}\,\sqrt{1+t^2-2xt}}, \qquad |h|, |t| < 1 \tag{12.1.36}$$

$$= \int_{-1}^{1} [P_0(x) + hP_1(x) + \cdots + h^n P_n(x) + \cdots]$$
$$\times [P_0(x) + tP_1(x) + \cdots + t^n P_n(x) + \cdots]\,dx \tag{12.1.37}$$

$$= \sum_{n=0}^{\infty} \sum_{m=0}^{\infty} h^n t^m \int_{-1}^{1} P_n(x)P_m(x)\,dx. \tag{12.1.38}$$

On the other hand, if $a = (1+h^2)/2h$, and $b = (1+t^2)/2t$, the integral J is

$$J = \int_{-1}^{1} \frac{dx}{\sqrt{1+h^2-2xh}\,\sqrt{1+t^2-2xt}} \tag{12.1.39}$$

$$= \frac{1}{2\sqrt{ht}} \int_{-1}^{1} \frac{dx}{\sqrt{a-x}\,\sqrt{b-x}} = \frac{1}{\sqrt{ht}} \int_{-1}^{1} \frac{\frac{1}{2}\left(\frac{1}{\sqrt{a-x}} + \frac{1}{\sqrt{b-x}}\right)}{\sqrt{a-x}+\sqrt{b-x}}\,dx \tag{12.1.40}$$

$$= -\frac{1}{\sqrt{ht}}\ln\left(\sqrt{a-x}+\sqrt{b-x}\right)\Big|_{-1}^{1} = \frac{1}{\sqrt{ht}}\ln\left(\frac{\sqrt{a+1}+\sqrt{b+1}}{\sqrt{a-1}+\sqrt{b-1}}\right). \tag{12.1.41}$$

But $a+1 = (1+h^2+2h)/2h = (1+h)^2/2h$, and $a-1 = (1-h)^2/2h$. After a little algebra,

$$J = \frac{1}{\sqrt{ht}}\ln\left(\frac{1+\sqrt{ht}}{1-\sqrt{ht}}\right) = \frac{2}{\sqrt{ht}}\left[\sqrt{ht} + \frac{1}{3}\sqrt{(ht)^3} + \frac{1}{5}\sqrt{(ht)^5} + \cdots\right] \tag{12.1.42}$$

$$= 2\left(1 + \frac{ht}{3} + \frac{h^2 t^2}{5} + \cdots + \frac{h^n t^n}{2n+1} + \cdots\right). \tag{12.1.43}$$

As we noted earlier, the coefficient of $h^n t^m$ in this series is $\int_{-1}^{1} P_n(x)P_m(x)\,dx$. If we match the powers of $h^n t^m$, the orthogonality condition is

$$\boxed{\int_{-1}^{1} P_n(x)P_m(x)\,dx = \begin{cases} 0, & m \neq n, \\ \frac{2}{2n+1}, & m = n. \end{cases}} \tag{12.1.44}$$

[5] See Symons, B., 1982: Legendre polynomials and their orthogonality. *Math. Gaz.*, **66**, 152–154.

Some Useful Relationships Involving Legendre Polynomials

Rodrigues's formula

$$P_n(x) = \frac{1}{2^n n!} \frac{d^n}{dx^n} (x^2 - 1)^n$$

Recurrence formulas

$$(n+1)P_{n+1}(x) - (2n+1)xP_n(x) + nP_{n-1}(x) = 0, \qquad n = 1, 2, 3, \ldots$$

$$P'_{n+1}(x) - P'_{n-1}(x) = (2n+1)P_n(x), \qquad n = 1, 2, 3, \ldots$$

Orthogonality condition

$$\int_{-1}^{1} P_n(x)P_m(x)\, dx = \begin{cases} 0, & m \neq n, \\ \dfrac{2}{2n+1}, & m = n. \end{cases}$$

With the orthogonality condition, Equation 12.1.44, we are ready to show that we can represent a function $f(x)$, which is piece-wise differentiable in the interval $(-1, 1)$, by the series:

$$f(x) = \sum_{m=0}^{\infty} A_m P_m(x), \qquad -1 \leq x \leq 1. \tag{12.1.45}$$

To find A_m we multiply both sides of Equation 12.1.45 by $P_n(x)$ and integrate from -1 to 1:

$$\int_{-1}^{1} f(x)P_n(x)\, dx = \sum_{m=0}^{\infty} A_m \int_{-1}^{1} P_n(x)P_m(x)\, dx. \tag{12.1.46}$$

All of the terms on the right side vanish except for $n = m$ because of the orthogonality condition, Equation 12.1.44. For this reason, the coefficient A_n is

$$A_n \int_{-1}^{1} P_n^2(x)\, dx = \int_{-1}^{1} f(x)P_n(x)\, dx, \tag{12.1.47}$$

or

$$A_n = \frac{2n+1}{2} \int_{-1}^{1} f(x)P_n(x)\, dx. \tag{12.1.48}$$

In the special case when $f(x)$ and its first n derivatives are continuous throughout the interval $(-1, 1)$, we may use Rodrigues' formula to evaluate

$$\int_{-1}^{1} f(x)P_n(x)\, dx = \frac{1}{2^n n!} \int_{-1}^{1} f(x) \frac{d^n (x^2 - 1)^n}{dx^n}\, dx = \frac{(-1)^n}{2^n n!} \int_{-1}^{1} (x^2 - 1)^n f^{(n)}(x)\, dx$$

$$\tag{12.1.49}$$

by integrating by parts n times. Hence,

$$A_n = \frac{2n+1}{2^{n+1}n!} \int_{-1}^{1} (1-x^2)^n f^{(n)}(x)\, dx. \tag{12.1.50}$$

A particularly useful result follows from Equation 12.1.50 if $f(x)$ is a polynomial of degree k. Because all derivatives of $f(x)$ of order n vanish identically when $n > k$, $A_n = 0$ if $n > k$. It follows that any polynomial of degree k can be expressed as a linear combination of the first $k+1$ Legendre polynomials $[P_0(x), \ldots, P_k(x)]$. Another way of viewing this result is to recognize that any polynomial of degree k is an expansion in powers of x. When we expand in Legendre polynomials we are merely regrouping these powers of x into new groups that can be identified as $P_0(x), P_1(x), P_2(x), \ldots, P_k(x)$.

- **Example 12.1.4**

Let us express $f(x) = x^2$ in terms of Legendre polynomials. The results from Equation 12.1.50 mean that we need only worry about $P_0(x)$, $P_1(x)$, and $P_2(x)$:

$$x^2 = A_0 P_0(x) + A_1 P_1(x) + A_2 P_2(x). \tag{12.1.51}$$

Substituting for the Legendre polynomials,

$$x^2 = A_0 + A_1 x + \tfrac{1}{2} A_2 (3x^2 - 1), \tag{12.1.52}$$

and

$$A_0 = \tfrac{1}{3}, \quad A_1 = 0, \quad \text{and} \quad A_2 = \tfrac{2}{3}. \tag{12.1.53}$$

$\square$

- **Example 12.1.5**

Let us find the expansion in Legendre polynomials of the function:

$$f(x) = \begin{cases} 0, & -1 < x < 0, \\ 1, & 0 < x < 1. \end{cases} \tag{12.1.54}$$

We could have done this expansion as a Fourier series, but in the solution of partial differential equations on a sphere, we must make the expansion in Legendre polynomials.

In this problem, we find that

$$A_n = \frac{2n+1}{2} \int_0^1 P_n(x)\, dx. \tag{12.1.55}$$

Therefore,

$$A_0 = \tfrac{1}{2} \int_0^1 1\, dx = \tfrac{1}{2}, \qquad A_1 = \tfrac{3}{2} \int_0^1 x\, dx = \tfrac{3}{4}, \tag{12.1.56}$$

$$A_2 = \tfrac{5}{2} \int_0^1 \tfrac{1}{2}(3x^2 - 1)\, dx = 0, \quad \text{and} \quad A_3 = \tfrac{7}{2} \int_0^1 \tfrac{1}{2}(5x^3 - 3x)\, dx = -\tfrac{7}{16}, \tag{12.1.57}$$

so that

$$f(x) = \tfrac{1}{2}P_0(x) + \tfrac{3}{4}P_1(x) - \tfrac{7}{16}P_3(x) + \tfrac{11}{32}P_5(x) + \cdots. \tag{12.1.58}$$

Figure 12.1.3 illustrates the expansion, Equation 12.1.58, where we used only the first four terms. It was created using the MATLAB script:

```
clear;
x = [-1:0.01:1]; % create x points in plot
f = zeros(size(x)); % initialize function f(x)
for k = 1:length(x) % construct function f(x)
  if x(k) < 0; f(k) = 0; else f(k) = 1; end;
end
% initialize Fourier-Legendre series with zeros
flegendre = zeros(size(x));
% read in Fourier coefficients
a(1) = 1/2; a(2) = 3/4; a(3) = 0;
a(4) = -7/16; a(5) = 0; a(6) = 11/32;
clf % clear any figures
for n = 1:6
% compute Legendre polynomial
  N = n-1; P = legendre(N,x);
% compute Fourier-Legendre series
  flegendre = flegendre + a(n) * P(1,:);
% create plot of truncated Fourier-Legendre series
%     with n terms
  if n==1 subplot(2,2,1), plot(x,flegendre,x,f,'--');
    legend('one term','f(x)'); legend boxoff; end
  if n==2 subplot(2,2,2), plot(x,flegendre,x,f,'--');
    legend('two terms','f(x)'); legend boxoff; end
  if n==4 subplot(2,2,3), plot(x,flegendre,x,f,'--');
    legend('four terms','f(x)'); legend boxoff;
    xlabel('x','Fontsize',20); end
  if n==6 subplot(2,2,4), plot(x,flegendre,x,f,'--');
    legend('six terms','f(x)'); legend boxoff;
    xlabel('x','Fontsize',20); end
  axis([-1 1 -0.5 1.5])
end
```

As we add each additional term in the orthogonal expansion, the expansion fits $f(x)$ better in the "least squares" sense of Equation 11.3.6. The spurious oscillations arise from trying to represent a discontinuous function by four continuous, oscillatory functions. Even if we add additional terms, the spurious oscillations persist, although located nearer to the discontinuity. This is another example of *Gibbs phenomena*.[6] See Section 5.2.   □

So far we have explored the nature of Legendre polynomials and shown how we may expand a well-behaved function as an expansion in Legendre polynomials. We are now ready to show how they can be used to solve linear Fredholm integral and partial differential equations.

---

[6] Weyl, H., 1910: Die Gibbs'sche Erscheinung in der Theorie der Kugelfunktionen. *Rend. Circ. Mat. Palermo*, **29**, 308–321.

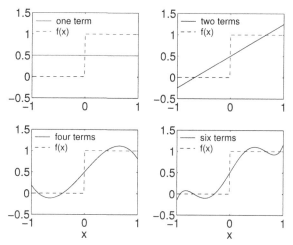

Figure 12.1.3: Representation of the function $f(x) = 1$ for $0 < x < 1$ and 0 for $-1 < x < 0$ by various partial summations of its Legendre polynomial expansion. The dashed lines denote the exact function.

• **Example 12.1.6: Numerical solution of linear Fredholm integral equations**

The integral equation

$$y(x) = e^{2x} - \frac{3x}{4e^2} - \frac{e^2 x}{4} + \int_{-1}^{1} xt\, y(t)\, dt \qquad (12.1.59)$$

is an example of a linear Fredholm integral equation of the second kind:

$$y(x) = f(x) + \lambda \int_{-1}^{1} K(x,t)\, y(t)\, dt, \qquad (12.1.60)$$

where $f(x) = e^{2x} - 3xe^{-2}/4 - e^2 x/4$, $\lambda = 1$, and $K(x,t) = xt$. A quick check shows that $y(x) = e^{2x}$ is the solution. In general, these equations must be solved numerically by first replacing the integral with some quadrature formula and then solving the resulting system of linear equations.

An alternative would be to assume that the solution can be written as an expansion in Legendre polynomials:

$$y(x) = \sum_{n=0}^{N} A_n P_n(x). \qquad (12.1.61)$$

This method is not new. S. Chandrasekhar[7] used it in the 1940s to solve the radiative transfer equation. However, we shall follow the work of Yalçınbas, Aynigül and Akkaya[8] who formulated the method using matrix methods. Because the method is elegant and serves as a refresher on linear algebra, we employ it here.

We begin by assuming both $f(x)$ and $y(x)$ can be written as Fourier-Legendre expansions:

$$y(x) = \sum_{n=0}^{N} a_n P_n(x) = \mathbf{P}_x \mathbf{a}^T, \qquad (12.1.62)$$

---

[7] See Chandrasekhar, S., 1944: On the radiative equilibrium of a stellar atmosphere. *Astrophys. J.*, **99**, 180–190.

[8] See Yalçınbas, S., M. Aynigül, and T. Akkaya, 2010: Legendre series solutions of Fredholm integral equations. *Math. Comput. Appl.*, **15**, 371–381.

and

$$f(x) = \sum_{n=0}^{N} f_n P_n(x) = \mathbf{P}_x \mathbf{f}^T, \qquad (12.1.63)$$

where we have the row vectors $\mathbf{a} = [a_0 \, a_1 \, \ldots \, a_N]$, $\mathbf{f} = [f_0 \, f_1 \, \ldots \, f_N]$, and $\mathbf{P}_x = [P_0(x) \, P_1(x) \, \ldots \, P_N(x)]$. In Problem 15 we show how to compute the Fourier-Legendre coefficients for e^x. Preforming the same exercise here for e^{2x}, we have that

$$f_0 = c_0/2, f_1 = 3c_1/2 - 0.75e^{-2} - 0.25e^2, \qquad \text{and} \qquad f_n = (2n+1)c_n/2, \qquad (12.1.64)$$

where $c_0 = \sinh(2)$, $c_1 = \cosh(2) - \sinh(2)/2$, and $c_{n+1} = c_{n-1} - (2n+1)c_n/2$ for $n \geq 1$.

Next, let us examine the kernel of the integration. We begin by noting that we can write the kernel as

$$K(x,t) = \sum_{r=0}^{N} \sum_{s=0}^{N} k_{r,s} P_r(x) P_s(t). = \mathbf{P}_x K \mathbf{P}_t^T. \qquad (12.1.65)$$

For example,

$$xt = P_1(x)P_1(t) = \begin{pmatrix} P_0(x) \\ P_1(x) \\ P_2(x) \\ P_3(x) \\ P_4(x) \\ P_5(x) \end{pmatrix}^T \begin{pmatrix} 0 & 0 & 0 & 0 & 0 & 0 \\ 0 & 1 & 0 & 0 & 0 & 0 \\ 0 & 0 & 0 & 0 & 0 & 0 \\ 0 & 0 & 0 & 0 & 0 & 0 \\ 0 & 0 & 0 & 0 & 0 & 0 \\ 0 & 0 & 0 & 0 & 0 & 0 \end{pmatrix} \begin{pmatrix} P_0(t) \\ P_1(t) \\ P_2(t) \\ P_3(t) \\ P_4(t) \\ P_5(t) \end{pmatrix} \qquad (12.1.66)$$

for the special case of $N = 5$.

Upon substituting for $\mathbf{y}$, $\mathbf{f}$ and $K(x,t)$ into Equation 12.1.60, we have that

$$\mathbf{P}_x \mathbf{a}^T = \mathbf{P}_x \mathbf{f}^T + \lambda \mathbf{P}_x K \left(\int_{-1}^{1} \mathbf{P}_t^T \mathbf{P}_t \, dt \right) \mathbf{a}^T, \qquad (12.1.67)$$

or

$$\mathbf{a}^T = \mathbf{f}^T + \lambda K \left(\int_{-1}^{1} \mathbf{P}_t^T \mathbf{P}_t \, dt \right) \mathbf{a}^T, \qquad (12.1.68)$$

or

$$\mathbf{a}^T = \mathbf{f}^T + \lambda K Q \mathbf{a}^T, \qquad (12.1.69)$$

where

$$Q = \int_{-1}^{1} \mathbf{P}_t^T \mathbf{P}_t \, dt = \begin{pmatrix} 2 & 0 & 0 & 0 & \cdots & 0 \\ 0 & 2/3 & 0 & 0 & \cdots & 0 \\ 0 & 0 & 2/5 & 0 & \cdots & 0 \\ 0 & 0 & 0 & 2/7 & \cdots & 0 \\ \vdots & \vdots & \vdots & \vdots & \ddots & \vdots \\ 0 & 0 & 0 & 0 & \cdots & 2/(2N+1) \end{pmatrix}. \qquad (12.1.70)$$

Finally, we can write 12.1.67 as

$$(1 - \lambda K Q) \mathbf{a}^T = \mathbf{f}^T. \qquad (12.1.71)$$

Then we can use any computational engine we wish to solve for $\mathbf{a}$. After that, we simply substitute into Equation 12.1.62.

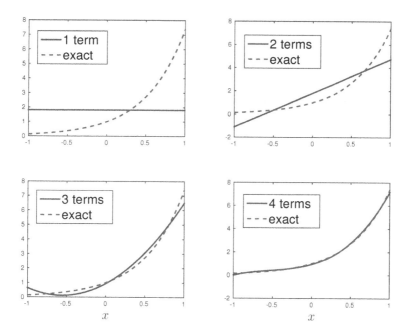

Figure 12.1.4: Comparison between the Legendre polynomial expansion and the exact solution to the linear Fredholm equation, Equation 12.1.59.

Figure 12.1.4 compares the numerical solution of a linear Fredholm equation, Equation 12.1.59, where we used a Legendre polynomial expansion against the exact solution when we include one, two, three and four terms in the expansion. In this particular problem the expansion works well. $\qquad\square$

• **Example 12.1.7: The potential within a conducting sphere**

Let us use the method of separation of variables to find the potential at any point P within a conducting sphere of radius a. The derivation of the Laplacian in polar spherical coordinates is given in Appendix B. At the surface, the potential is held at V_0 in the hemisphere $0 < \theta < \pi/2$, and $-V_0$ for $\pi/2 < \theta < \pi$.

Laplace's equation in spherical coordinates is

$$\frac{\partial}{\partial r}\left(r^2\frac{\partial u}{\partial r}\right) + \frac{1}{\sin(\theta)}\frac{\partial}{\partial \theta}\left[\sin(\theta)\frac{\partial u}{\partial \theta}\right] = 0, \quad 0 \le r < a, \quad 0 \le \theta \le \pi. \qquad (\mathbf{12.1.72})$$

To solve Equation 12.1.72 we set $u(r,\theta) = R(r)\Theta(\theta)$ by separation of variables. Substituting into this equation, we have that

$$\frac{1}{R}\frac{d}{dr}\left(r^2\frac{dR}{dr}\right) = -\frac{1}{\sin(\theta)\Theta}\frac{d}{d\theta}\left[\sin(\theta)\frac{d\Theta}{d\theta}\right] = k^2, \qquad (\mathbf{12.1.73})$$

or

$$r^2 R'' + 2rR' - k^2 R = 0, \qquad (\mathbf{12.1.74})$$

and

$$\frac{1}{\sin(\theta)}\frac{d}{d\theta}\left[\sin(\theta)\frac{d\Theta}{d\theta}\right] + k^2\Theta = 0. \qquad (\mathbf{12.1.75})$$

A common substitution replaces θ with $\mu = \cos(\theta)$. Then, as θ varies from 0 to π, μ varies from 1 to -1. With this substitution, Equation 12.1.75 becomes

$$\frac{d}{d\mu}\left[(1-\mu^2)\frac{d\Theta}{d\mu}\right] + k^2\Theta = 0. \tag{12.1.76}$$

This is Legendre's equation, which we examined earlier. Consequently, because the solution must remain finite at the poles, $k^2 = n(n+1)$, and

$$\Theta_n(\theta) = P_n(\mu) = P_n[\cos(\theta)], \tag{12.1.77}$$

where $n = 0, 1, 2, 3, \ldots$.

Turning to Equation 12.1.74, this equation is the equidimensional or Euler-Cauchy linear differential equation. One method of solving this equation consists of introducing a new independent variable s so that $r = e^s$, or $s = \ln(r)$. Because

$$\frac{d}{dr} = \frac{ds}{dr}\frac{d}{ds} = e^{-s}\frac{d}{ds}, \tag{12.1.78}$$

it follows that

$$\frac{d^2}{dr^2} = \frac{d}{dr}\left(e^{-s}\frac{d}{ds}\right) = e^{-s}\frac{d}{ds}\left(e^{-s}\frac{d}{ds}\right) = e^{-2s}\left(\frac{d^2}{ds^2} - \frac{d}{ds}\right). \tag{12.1.79}$$

Substituting into Equation 12.1.74,

$$\frac{d^2 R_n}{ds^2} + \frac{dR_n}{ds} - n(n+1)R_n = 0. \tag{12.1.80}$$

Equation 12.1.80 is a second-order, constant coefficient ordinary differential equation, which has the solution

$$R_n(r) = C_n e^{ns} + D_n e^{-(n+1)s} = C_n \exp[n\ln(r)] + D_n \exp[-(n+1)\ln(r)] \tag{12.1.81}$$
$$= C_n \exp[\ln(r^n)] + D_n \exp[\ln(r^{-1-n})] = C_n r^n + D_n r^{-1-n}. \tag{12.1.82}$$

A more convenient form of the solution is

$$R_n(r) = A_n \left(\frac{r}{a}\right)^n + B_n \left(\frac{r}{a}\right)^{-1-n}, \tag{12.1.83}$$

where $A_n = a^n C_n$ and $B_n = D_n/a^{n+1}$. We introduced the constant a, the radius of the sphere, to simplify future calculations.

Using the results from Equation 12.1.77 and Equation 12.1.83, the solution to Laplace's equation in axisymmetric problems is

$$u(r,\theta) = \sum_{n=0}^{\infty}\left[A_n \left(\frac{r}{a}\right)^n + B_n \left(\frac{r}{a}\right)^{-1-n}\right]P_n[\cos(\theta)]. \tag{12.1.84}$$

In our particular problem we must take $B_n = 0$ because the solution becomes infinite at $r = 0$ otherwise. If the problem had involved the domain $a < r < \infty$, then $A_n = 0$ because the potential must remain finite as $r \to \infty$.

Finally, we must evaluate A_n. Finding the potential at the surface,

$$u(a,\mu) = \sum_{n=0}^{\infty} A_n P_n(\mu) = \begin{cases} V_0, & 0 < \mu \le 1, \\ -V_0, & -1 \le \mu < 0. \end{cases} \qquad (12.1.85)$$

Upon examining Equation 12.1.85, it is merely an expansion in Legendre polynomials of the function

$$f(\mu) = \begin{cases} V_0, & 0 < \mu \le 1, \\ -V_0, & -1 \le \mu < 0. \end{cases} \qquad (12.1.86)$$

Hence, from Equation 12.1.86,

$$A_n = \frac{2n+1}{2} \int_{-1}^{1} f(\mu) P_n(\mu)\, d\mu. \qquad (12.1.87)$$

Because $f(\mu)$ is an odd function, $A_n = 0$ if n is even. When n is odd, however,

$$A_n = (2n+1) \int_{0}^{1} V_0 P_n(\mu)\, d\mu. \qquad (12.1.88)$$

We can further simplify Equation 12.1.88 by using the relationship that

$$\int_{x}^{1} P_n(t)\, dt = \frac{1}{2n+1} \left[P_{n-1}(x) - P_{n+1}(x) \right], \qquad (12.1.89)$$

where $n \ge 1$. In our problem, then,

$$A_n = \begin{cases} V_0[P_{n-1}(0) - P_{n+1}(0)], & n \text{ odd}, \\ 0, & n \text{ even}. \end{cases} \qquad (12.1.90)$$

The first few terms are $A_1 = 3V_0/2$, $A_3 = -7V_0/8$, and $A_5 = 11V_0/16$.

Figure 12.1.5 illustrates our solution. It was created using the MATLAB script:

```
clear
N = 51; dr = 0.05; dtheta = pi / 15;
% compute grid and set solution equal to zero
r = [0:dr:1]; theta = [0:dtheta:2*pi];
mu = cos(theta); Z = r' * mu;
for L = 1:2
  if L == 1 X = r' * sin(theta);
  else X = -r' * sin(theta); end
  u = zeros(size(X));
% compute solution from Equation 12.1.84
  rfactor = r;
  for n = 1:2:N
    A = legendre(n-1,0); B = legendre(n+1,0); coeff = A(1)-B(1);
    C = legendre(n,mu); Theta = C(1,:);
    u = u + coeff * rfactor' * Theta;
    rfactor = rfactor .* r .* r;
  end
  surf(Z,X,u); hold on; end
xlabel('Z','Fontsize',20); ylabel('X','Fontsize',20)
```

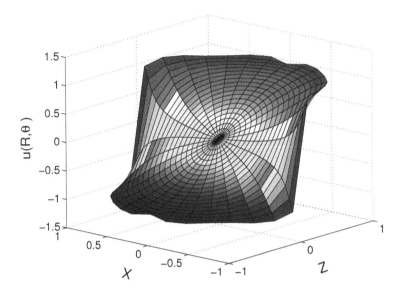

Figure 12.1.5: Electrostatic potential within a conducting sphere when the upper hemispheric surface has the potential 1 and the lower surface has the potential −1.

```
zlabel('u(R,\theta )','Fontsize',20);
```

Here we have the convergence of the equipotentials along the equator and at the surface. The slow rate at which the coefficients are approaching zero suggests that the solution suffers from Gibbs phenomena along the surface. □

• Example 12.1.8: Steady-state temperature within a metallic sphere

Using separation of variables, we now find the steady-state temperature field within a metallic sphere of radius a, which we place in direct sunlight and allow to radiatively cool. This classic problem, first solved by Rayleigh,[9] requires the use of spherical coordinates with its origin at the center of the sphere and its z-axis pointing toward the sun. The derivation of the Laplacian in spherical coordinates is given in Appendix B. With this choice for the coordinate system, the incident sunlight is

$$D(\theta) = \begin{cases} D(0)\cos(\theta), & 0 \le \theta \le \pi/2, \\ 0, & \pi/2 \le \theta \le \pi. \end{cases} \tag{12.1.91}$$

If heat dissipation takes place at the surface $r = a$ according to Newton's law of cooling and the temperature of the surrounding medium is zero, the solar heat absorbed by the surface dA must balance the Newtonian cooling at the surface plus the energy absorbed into the sphere's interior. This physical relationship is

$$(1 - \rho)D(\theta)\,dA = \epsilon u(a, \theta)\,dA + \kappa \frac{\partial u(a, \theta)}{\partial r}\,dA, \tag{12.1.92}$$

[9] Rayleigh, J. W., 1870: On the values of the integral $\int_0^1 Q_n Q_{n'}\,d\mu$, Q_n, $Q_{n'}$ being Laplace's coefficients of the orders n, n', with application to the theory of radiation. *Philos. Trans. R. Soc. London, Ser. A*, **160**, 579–590.

where ρ is the reflectance of the surface (the albedo), ϵ is the surface conductance or coefficient of surface heat transfer, and κ is the thermal conductivity. Simplifying Equation 12.1.92, we have that

$$\frac{\partial u(a,\theta)}{\partial r} = \frac{1-\rho}{\kappa}D(\theta) - \frac{\epsilon}{\kappa}u(a,\theta) \qquad (12.1.93)$$

for $r = a$.

If the sphere has reached thermal equilibrium, Laplace's equation describes the temperature field within the sphere. In the previous example, we showed that the solution to Laplace's equation in axisymmetric problems is

$$u(r,\theta) = \sum_{n=0}^{\infty}\left[A_n\left(\frac{r}{a}\right)^n + B_n\left(\frac{r}{a}\right)^{-1-n}\right]P_n[\cos(\theta)]. \qquad (12.1.94)$$

In this problem, $B_n = 0$ because the solution would become infinite at $r = 0$ otherwise. Therefore,

$$u(r,\theta) = \sum_{n=0}^{\infty}A_n\left(\frac{r}{a}\right)^n P_n[\cos(\theta)]. \qquad (12.1.95)$$

Differentiation gives

$$\frac{\partial u}{\partial r} = \sum_{n=0}^{\infty}A_n\frac{nr^{n-1}}{a^n}P_n[\cos(\theta)]. \qquad (12.1.96)$$

Substituting into the boundary condition leads to

$$\sum_{n=0}^{\infty}A_n\left(\frac{n}{a}+\frac{\epsilon}{\kappa}\right)P_n[\cos(\theta)] = \left(\frac{1-\rho}{\kappa}\right)D(\theta), \qquad (12.1.97)$$

or

$$D(\mu) = \sum_{n=0}^{\infty}\left[\frac{n\kappa+\epsilon a}{a(1-\rho)}\right]A_nP_n(\mu) = \sum_{n=0}^{\infty}C_nP_n(\mu), \qquad (12.1.98)$$

where

$$C_n = \left[\frac{n\kappa+\epsilon a}{a(1-\rho)}\right]A_n, \quad \text{and} \quad \mu = \cos(\theta). \qquad (12.1.99)$$

We determine the coefficients by

$$C_n = \frac{2n+1}{2}\int_{-1}^{1}D(\mu)P_n(\mu)\,d\mu = \frac{2n+1}{2}D(0)\int_{0}^{1}\mu P_n(\mu)\,d\mu. \qquad (12.1.100)$$

Evaluation of the first few coefficients gives

$$A_0 = \frac{(1-\rho)D(0)}{4\epsilon}, \quad A_1 = \frac{a(1-\rho)D(0)}{2(\kappa+\epsilon a)}, \quad A_2 = \frac{5a(1-\rho)D(0)}{16(2\kappa+\epsilon a)}, \quad A_3 = 0, \quad (12.1.101)$$

$$A_4 = -\frac{3a(1-\rho)D(0)}{32(4\kappa+\epsilon a)}, \quad A_5 = 0, \quad A_6 = \frac{13a(1-\rho)D(0)}{256(6\kappa+\epsilon a)}, \quad A_7 = 0, \quad (12.1.102)$$

$$A_8 = -\frac{17a(1-\rho)D(0)}{512(8\kappa+\epsilon a)}, \quad A_9 = 0, \quad \text{and} \quad A_{10} = \frac{49a(1-\rho)D(0)}{2048(10\kappa+\epsilon a)}. \qquad (12.1.103)$$

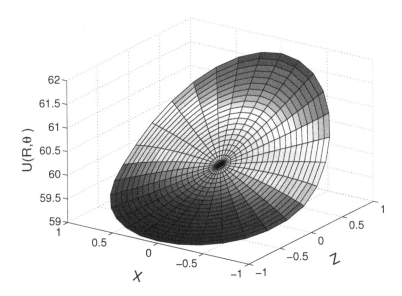

Figure 12.1.6: The difference (in °C) between the temperature field within a blackened iron surface of radius 0.1 m and the surrounding medium when we heat the surface by sunlight and allow it to radiatively cool.

Figure 12.1.6 illustrates the temperature field within the sphere with $D(0) = 1200$ W/m^2, $\kappa = 45$ W/m K, $\epsilon = 5$ W/m^2 K, $\rho = 0$, and $a = 0.1$ m. This corresponds to a cast iron sphere with blackened surface in sunlight. This figure was created by the MATLAB script:

```
clear
dr = 0.05; dtheta = pi / 15;
D_0 = 1200; kappa = 45; epsilon = 5; rho = 0; a = 0.1;
% compute grid and set solution equal to zero
r = [0:dr:1]; theta = [0:dtheta:pi];
mu = cos(theta); Z = r' * mu;
aaaa = (1-rho) * D_0 / ( 4 * epsilon);
aa(1) = a * (1-rho) * D_0 / ( 2 * ( kappa+epsilon*a));
aa(2) = 5 * a * (1-rho) * D_0 / ( 16 * (2*kappa+epsilon*a));
aa(3) = 0;
aa(4) = - 3 * a * (1-rho) * D_0 / ( 32 * (4*kappa+epsilon*a));
aa(5) = 0;
aa(6) = 13 * a * (1-rho) * D_0 / ( 256 * (6*kappa+epsilon*a));
aa(7) = 0;
aa(8) = -17 * a * (1-rho) * D_0 / ( 512 * (8*kappa+epsilon*a));
aa(9) = 0;
aa(10) = 49 * a * (1-rho) * D_0 / (2048 * (10*kappa+epsilon*a));
for L = 1:2
   if L == 1 X = r' * sin(theta);
   else X = -r' * sin(theta); end
   u = aaaa * ones(size(X));
% compute solution from Equation 12.1.95
   rfactor = r;
   for n = 1:10
```

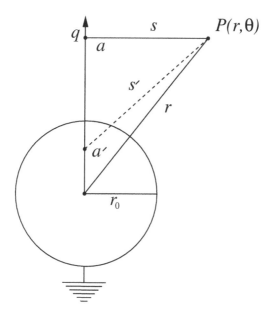

Figure 12.1.7: Point charge $+q$ in the presence of a grounded conducting sphere.

```
   A = legendre(n,mu); Theta = A(1,:);
   u = u + aa(n) * rfactor' * Theta;
   rfactor = rfactor .* r;
 end
surf(Z,X,u); hold on; end
xlabel('Z','Fontsize',20); ylabel('X','Fontsize',20);
zlabel('U(R,\theta )','Fontsize',20);
```

The temperature is quite warm with the highest temperature located at the position where the solar radiation is largest; the coolest temperatures are located in the shadow region. □

• **Example 12.1.9: Potential exterior to a conducting, grounded sphere**

In this example we find the potential at any point P exterior to a conducting, grounded sphere centered at $z = 0$ after we place a point charge $+q$ at $z = a$ on the z-axis. See Figure 12.1.7. The derivation of the Laplacian in spherical polar coordinates is given in Appendix B. From the principle of linear superposition, the total potential $u(r, \theta)$ equals the sum of the potential from the point charge and the potential $v(r, \theta)$ due to the induced charge on the sphere

$$u(r, \theta) = \frac{q}{s} + v(r, \theta). \qquad (12.1.104)$$

In common with the first term q/s, $v(r, \theta)$ must be a solution of Laplace's equation. In Example 12.1.7 we showed that the general solution to Laplace's equation in axisymmetric problems is

$$v(r, \theta) = \sum_{n=0}^{\infty} \left[A_n \left(\frac{r}{r_0} \right)^n + B_n \left(\frac{r}{r_0} \right)^{-1-n} \right] P_n[\cos(\theta)]. \qquad (12.1.105)$$

Because the solutions must be valid *anywhere* outside of the sphere, $A_n = 0$; otherwise, the solution would not remain finite as $r \to \infty$. Hence,

$$v(r,\theta) = \sum_{n=0}^{\infty} B_n \left(\frac{r}{r_0}\right)^{-1-n} P_n[\cos(\theta)]. \qquad (12.1.106)$$

We determine the coefficient B_n by the condition that $u(r_0,\theta) = 0$, or

$$\left.\frac{q}{s}\right|_{\text{on sphere}} + \sum_{n=0}^{\infty} B_n P_n[\cos(\theta)] = 0. \qquad (12.1.107)$$

We must expand the first term on the left side of Equation 12.1.107 in terms of Legendre polynomials. From the law of cosines,

$$s = \sqrt{r^2 + a^2 - 2ar\cos(\theta)}. \qquad (12.1.108)$$

Consequently, if $a > r$, then

$$\frac{1}{s} = \frac{1}{a}\left[1 - 2\cos(\theta)\frac{r}{a} + \left(\frac{r}{a}\right)^2\right]^{-1/2}. \qquad (12.1.109)$$

Earlier we showed that

$$(1 - 2xz + z^2)^{-1/2} = \sum_{n=0}^{\infty} P_n(x)z^n. \qquad (12.1.110)$$

Therefore,

$$\frac{1}{s} = \frac{1}{a}\sum_{n=0}^{\infty} P_n[\cos(\theta)]\left(\frac{r}{a}\right)^n. \qquad (12.1.111)$$

From Equation 12.1.107,

$$\sum_{n=0}^{\infty} \left[\frac{q}{a}\left(\frac{r_0}{a}\right)^n + B_n\right] P_n[\cos(\theta)] = 0. \qquad (12.1.112)$$

We can only satisfy Equation 12.1.112 if the square-bracketed term vanishes identically so that

$$B_n = -\frac{q}{a}\left(\frac{r_0}{a}\right)^n. \qquad (12.1.113)$$

On substituting Equation 12.1.113 back into Equation 12.1.106,

$$v(r,\theta) = -\frac{qr_0}{ra}\sum_{n=0}^{\infty} \left(\frac{r_0^2}{ar}\right)^n P_n[\cos(\theta)]. \qquad (12.1.114)$$

The physical interpretation of Equation 12.1.114 is as follows: Consider a point, such as a' (see Figure 12.1.7) on the z-axis. If $r > a'$, the Legendre expansion of $1/s'$ is

$$\frac{1}{s'} = \frac{1}{r}\sum_{n=0}^{\infty} P_n[\cos(\theta)]\left(\frac{a'}{r}\right)^n, \quad r > a'. \qquad (12.1.115)$$

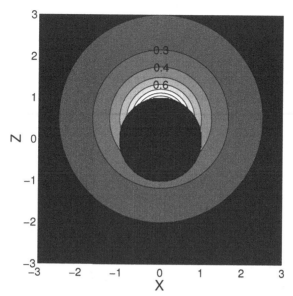

Figure 12.1.8: Electrostatic potential outside of a grounded conducting sphere in the presence of a point charge located at $a/r_0 = 2$. Contours are in units of $-q/r_0$.

Using Equation 12.1.115, we can rewrite it as

$$v(r,\theta) = -\frac{qr_0}{as'}, \tag{12.1.116}$$

if we set $a' = r_0^2/a$. Our final result is then

$$u(r,\theta) = \frac{q}{s} - \frac{q'}{s'}, \tag{12.1.117}$$

provided that q' equals $r_0 q/a$. In other words, when we place a grounded conducting sphere near a point charge $+q$, it changes the potential in the same manner as would a point charge of the opposite sign and magnitude $q' = r_0 q/a$, placed at the point $a' = r_0^2/a$. The charge q' is the *image* of q.

Figure 12.1.8 illustrates the solution, Equation 12.1.114, and was created using the MATLAB script:

```
clear
a_over_r0 = 2;
% set up x-z array
dx = 0.02; x = -3:dx:3; dz = 0.02; z = -3:dz:3;
u = 1000 * zeros(length(x),length(z));
X = x' * ones(1,length(z)); Z = ones(length(x),1) * z;
% compute r and theta
rr = sqrt(X .* X + Z .* Z);
theta = atan2(X,Z);
% find the potential
r_over_aprime = a_over_r0 * rr;
s = 1 + r_over_aprime .* r_over_aprime ...
    - 2 * r_over_aprime .* cos(theta);
for j = 1:length(z); for i = 1:length(x);
```

```
      if rr(i,j) >= 1; u(i,j) = 1 ./ sqrt(s(i,j)); end;
end; end
% plot the solution
[cs,h] = contourf(X,Z,u); colormap(hot); brighten(hot,0.5);
axis square; clabel(cs,h,'manual','Fontsize',16);
xlabel('X','Fontsize',20); ylabel('Z','Fontsize',20);
```

Because the charge is located directly above the sphere, the electrostatic potential for any fixed r is largest at the point $\theta = 0$ and weakest at $\theta = \pi$.

Problems

Find the first three nonvanishing coefficients in the Legendre polynomial expansion for the following functions:

1. $f(x) = \begin{cases} 0, & -1 < x < 0, \\ x, & 0 < x < 1. \end{cases}$

2. $f(x) = \begin{cases} 1/(2\epsilon), & |x| < \epsilon, \\ 0, & \epsilon < |x| < 1, \end{cases}$

3. $f(x) = |x|, \qquad |x| < 1.$

4. $f(x) = x^3, \qquad |x| < 1.$

5. $f(x) = \begin{cases} -1, & -1 < x < 0, \\ 1, & 0 < x < 1. \end{cases}$

6. $f(x) = \begin{cases} -1, & -1 < x < 0, \\ x, & 0 < x < 1. \end{cases}$

Then use MATLAB to illustrate various partial sums of the Fourier-Legendre series.

7. Use Rodrigues's formula to show that $P_4(x) = \frac{1}{8}(35x^4 - 30x^2 + 3)$.

8. Given $P_5(x) = \frac{63}{8}x^5 - \frac{70}{8}x^3 + \frac{15}{8}x$ and $P_4(x)$ from Problem 7, use the recurrence formula for $P_{n+1}(x)$ to find $P_6(x)$.

9. Show that (a) $P_n(1) = 1$, (b) $P_n(-1) = (-1)^n$, (c) $P_{2n+1}(0) = 0$, and (d) $P_{2n}(0) = (-1)^n(2n)!/(2^{2n}n!n!)$.

10. Prove that

$$\int_x^1 P_n(t)\,dt = \frac{1}{2n+1}[P_{n-1}(x) - P_{n+1}(x)], \qquad n > 0.$$

11. Given[10]

$$P_n[\cos(\theta)] = \frac{2}{\pi}\int_0^\theta \frac{\cos[(n+\frac{1}{2})x]}{\sqrt{2[\cos(x) - \cos(\theta)]}}\,dx = \frac{2}{\pi}\int_\theta^\pi \frac{\sin[(n+\frac{1}{2})x]}{\sqrt{2[\cos(\theta) - \cos(x)]}}\,dx,$$

show that the following generalized Fourier series holds:

$$\frac{H(\theta - t)}{\sqrt{2\cos(t) - 2\cos(\theta)}} = \sum_{n=0}^\infty P_n[\cos(\theta)] \cos\left[\left(n + \tfrac{1}{2}\right)t\right], \qquad 0 \le t < \theta \le \pi,$$

---

[10] Hobson, E. W., 1965: *The Theory of Spherical and Ellipsoidal Harmonics.* Chelsea Publishing Co., pp. 26–27.

if we use the eigenfunction $y_n(x) = \cos\left[\left(n + \frac{1}{2}\right)x\right]$, $0 < x < \pi$, $r(x) = 1$ and $H(\cdot)$ is Heaviside's step function, and

$$\frac{H(t - \theta)}{\sqrt{2\cos(\theta) - 2\cos(t)}} = \sum_{n=0}^{\infty} P_n[\cos(\theta)]\sin\left[\left(n + \frac{1}{2}\right)t\right], \quad 0 \le \theta < t \le \pi,$$

if we use the eigenfunction $y_n(x) = \sin\left[\left(n + \frac{1}{2}\right)x\right]$, $0 < x < \pi$, $r(x) = 1$ and $H(\cdot)$ is Heaviside's step function.

12. The series given in Problem 11 are also expansions in Legendre polynomials. In that light, show that

$$\int_0^t \frac{P_n[\cos(\theta)]\sin(\theta)}{\sqrt{2\cos(\theta) - 2\cos(t)}}\, d\theta = \frac{\sin\left[\left(n + \frac{1}{2}\right)t\right]}{n + \frac{1}{2}},$$

and

$$\int_t^\pi \frac{P_n[\cos(\theta)]\sin(\theta)}{\sqrt{2\cos(t) - 2\cos(\theta)}}\, d\theta = \frac{\cos\left[\left(n + \frac{1}{2}\right)t\right]}{n + \frac{1}{2}},$$

where $0 < t < \pi$.

13. (a) Use the generating function, Equation 12.1.15, to show that

$$\frac{1}{\sqrt{1 - 2tx + t^2}} = \sum_{n=0}^{\infty} t^{-n-1} P_n(x), \quad |x| < 1,\ 1 < |t|.$$

(b) Use the results from part (a) to show that

$$\frac{1}{\sqrt{\cosh(\mu) - x}} = \sqrt{2} \sum_{n=0}^{\infty} e^{-(n+\frac{1}{2})|\mu|} P_n(x), \quad |x| < 1.$$

Hint:

$$\frac{1}{\sqrt{\cosh(\mu) - x}} = \frac{\sqrt{2}}{\sqrt{e^{|\mu|} - 2x + e^{-|\mu|}}}.$$

14. The generating function, Equation 12.1.15, actually holds[11] for $|h| \le 1$ if $|x| < 1$. Using this relationship, show that

$$\sum_{n=0}^{\infty} P_n(x) = \frac{1}{\sqrt{2(1 - x)}}, \quad |x| < 1,$$

and

$$\sum_{n=0}^{\infty} \frac{P_n(x)}{n + 1} = \ln\left[\frac{1 + \sqrt{(1 - x)/2}}{\sqrt{(1 - x)/2}}\right], \quad |x| < 1.$$

Use these relationships to show that

$$\sum_{n=1}^{\infty} \frac{2n + 1}{n + 1} P_n(x) = 2\sum_{n=1}^{\infty} P_n(x) - \sum_{n=1}^{\infty} \frac{P_n(x)}{n + 1} = \frac{1}{\sqrt{(1 - x)/2}} - \ln\left[\frac{1 + \sqrt{(1 - x)/2}}{\sqrt{(1 - x)/2}}\right] - 1,$$

---

[11] Ibid., p. 28.

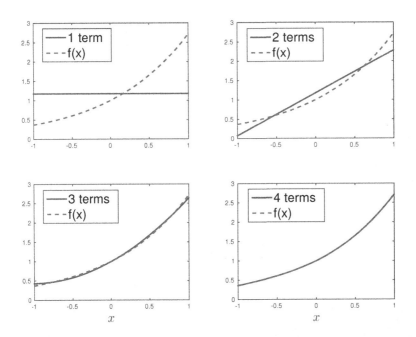

Figure 12.1.9: Representation of the function $f(x) = e^x$ for $-1 < x < 1$ by various partial summations of its Legendre polynomial expansion. The dashed lines denote the exact function.

if $|x| < 1$.

15. Find the Fourier-Legendre expansion for $f(x) = e^x$.

Step 1: Show that the Fourier-Legendre coefficients are given by

$$A_n = \frac{2n+1}{2}c_n, \qquad \text{where} \qquad c_n = \int_{-1}^{1} e^x P_n(x)\, dx.$$

Step 2: Using integration by parts, show that

$$c_n = e - (-1)^n e^{-1} - \int_{-1}^{1} e^x P_n'(x)\, dx.$$

Hint: Use Problem 9.

Step 3: Show that $c_{n+1} - c_{n-1} = -(2n+1)c_n$ for $n \geq 1$, where $c_0 = 2\sinh(1)$ and $c_1 = 2e^{-1}$. Hint: Use Equation 12.1.27. Figure 12.1.9 illustrates the expansion for several partial sums.

Step 4: Redo the problem for $f(x) = e^{-x}$.

Linear Fredholm Integral Equation

Utilizing Example 12.1.6, employ Fourier-Legendre expansions to solve the following linear Fredholm integral equations of the second kind:

16.

$$y(x) = (x+1)^2 + \int_{-1}^{1} (xt + x^2 t^2) y(t)\, dt$$

17.

$$y(x) = \frac{4x^2}{5} + \frac{1}{2}\int_{-1}^{1} x^2 t^2 y(t)\, dt$$

18.

$$y(x) = x^3 - \frac{2}{7}x^2 + 5 + \int_{-1}^{1} (x^2 t^3 + 1)y(t)\, dt$$

Separation of Variables Solution to Laplace's Equation

19. Find the steady-state temperature within a sphere of radius a if the temperature along its surface is maintained at the temperature $u(a,\theta) = 100[\cos(\theta) - \cos^5(\theta)]$.

20. Find the steady-state temperature within a sphere if the upper half of the exterior surface at radius a is maintained at the temperature 100 while the lower half is maintained at the temperature 0.

21. The surface of a sphere of radius a has a temperature of zero everywhere except in a spherical cap at the north pole (defined by the cone $\theta = \alpha$), where it equals T_0. Find the steady-state temperature within the sphere.

12.2 BESSEL FUNCTIONS

In the previous section we discussed the solutions to Legendre's equation, especially with regard to their use in orthogonal expansions. In this section we consider another classic equation, Bessel's equation[12]

$$x^2 y'' + xy' + (\mu^2 x^2 - n^2)y = 0, \tag{12.2.1}$$

or

$$\frac{d}{dx}\left(x\frac{dy}{dx}\right) + \left(\mu^2 x - \frac{n^2}{x}\right)y = 0. \tag{12.2.2}$$

Once again, our ultimate goal is the use of its solutions in orthogonal expansions. These orthogonal expansions, in turn, are used in the solution of partial differential equations in cylindrical coordinates.

A quick check of Bessel's equation shows that it conforms to the canonical form of the Sturm-Liouville problem: $p(x) = x$, $q(x) = -n^2/x$, $r(x) = x$, and $\lambda = \mu^2$. Restricting our attention to the interval $[0, L]$, the Sturm-Liouville problem involving Equation 12.2.2 is singular because $p(0) = 0$. From Equation 12.0.8 the eigenfunctions of a singular Sturm-Liouville problem will still be orthogonal over the interval $[0, L]$ if (1) $y(x)$ is finite and $xy'(x)$ is zero at $x = 0$, and (2) $y(x)$ satisfies the homogeneous boundary condition, Equation 12.0.2, at $x = L$. Thus, we only seek solutions that satisfy these conditions.

We cannot write down the solution to Bessel's equation in a simple closed form; as in the case with Legendre's equation, we must find the solution by power series. Because we intend to make the expansion about $x = 0$ and this point is a regular singular point, we must

[12] Bessel, F. W., 1824: Untersuchung des Teils der planetarischen Störungen, welcher aus der Bewegung der Sonne entsteht. *Abh. d. K. Akad. Wiss. Berlin*, 1–52. See Dutka, J., 1995: On the early history of Bessel functions. *Arch. Hist. Exact Sci.*, **49**, 105–134. The classic reference on Bessel functions is Watson, G. N., 1966: *A Treatise on the Theory of Bessel Functions*. Cambridge University Press, 804 pp.

It was Friedrich Wilhelm Bessel's (1784–1846) apprenticeship to the famous mercantile firm of Kulenkamp that ignited his interest in mathematics and astronomy. As the founder of the German school of practical astronomy, Bessel discovered his functions while studying the problem of planetary motion. Bessel functions arose as coefficients in one of the series that described the gravitational interaction between the sun and two other planets in elliptic orbit. (Portrait courtesy of Photo AKG, London, with permission.)

use the method of Frobenius, where n is an integer.[13] Moreover, because the quantity n^2 appears in Equation 12.2.2, we may take n to be nonnegative without any loss of generality.

To simplify matters, we first find the solution when $\mu = 1$; the solution for $\mu \neq 1$ follows by substituting μx for x. Consequently, we seek solutions of the form

$$y(x) = \sum_{k=0}^{\infty} B_k x^{2k+s}, \qquad y'(x) = \sum_{k=0}^{\infty} (2k+s)B_k x^{2k+s-1}, \qquad (\mathbf{12.2.3})$$

and

$$y''(x) = \sum_{k=0}^{\infty} (2k+s)(2k+s-1)B_k x^{2k+s-2}, \qquad (\mathbf{12.2.4})$$

where we formally assume that we can interchange the order of differentiation and summation. The substitution of Equation 12.2.3 and Equation 12.2.4 into Equation 12.2.1 with

---

[13] This case is much simpler than for arbitrary n. See Hildebrand, F. B., 1962: *Advanced Calculus for Applications*. Prentice-Hall, Section 4.8.

$\mu = 1$ yields

$$\sum_{k=0}^{\infty}(2k+s)(2k+s-1)B_kx^{2k+s} + \sum_{k=0}^{\infty}(2k+s)B_kx^{2k+s} + \sum_{k=0}^{\infty}B_kx^{2k+s+2} - n^2\sum_{k=0}^{\infty}B_kx^{2k+s} = 0,$$

$$(12.2.5)$$

or

$$\sum_{k=0}^{\infty}[(2k+s)^2 - n^2]B_kx^{2k} + \sum_{k=0}^{\infty}B_kx^{2k+2} = 0. \qquad (12.2.6)$$

If we explicitly separate the $k = 0$ term from the other terms in the first summation in Equation 12.2.6,

$$(s^2 - n^2)B_0 + \sum_{m=1}^{\infty}[(2m+s)^2 - n^2]B_mx^{2m} + \sum_{k=0}^{\infty}B_kx^{2k+2} = 0. \qquad (12.2.7)$$

We now change the dummy integer in the first summation of Equation 12.2.7 by letting $m = k + 1$ so that

$$(s^2 - n^2)B_0 + \sum_{k=0}^{\infty}\{[(2k+s+2)^2 - n^2]B_{k+1} + B_k\}x^{2k+2} = 0. \qquad (12.2.8)$$

Because Equation 12.2.8 must be true for all x, each power of x must vanish identically. This yields $s = \pm n$, and

$$[(2k+s+2)^2 - n^2]B_{k+1} + B_k = 0. \qquad (12.2.9)$$

Since the difference of the larger indicial root from the lower root equals the integer $2n$, we are only guaranteed a power series solution of the form given by Equation 12.2.3 for $s = n$. If we use this indicial root and the recurrence formula, Equation 12.2.9, this solution, known as the Bessel function of the first kind of order n and denoted by $J_n(x)$, is

$$J_n(x) = \sum_{k=0}^{\infty}\frac{(-1)^k(x/2)^{n+2k}}{k!(n+k)!}. \qquad (12.2.10)$$

To find the second general solution to Bessel's equation, the one corresponding to $s = -n$, the most economical method[14] is to express it in terms of partial derivatives of $J_n(x)$ with respect to its order n:

$$Y_n(x) = \left[\frac{\partial J_\nu(x)}{\partial \nu} - (-1)^n\frac{\partial J_{-\nu}(x)}{\partial \nu}\right]_{\nu=n}. \qquad (12.2.11)$$

Upon substituting the power series representation, Equation 12.2.10, into Equation 12.2.11,

$$Y_n(x) = \frac{2}{\pi}J_n(x)\ln(x/2) - \frac{1}{\pi}\sum_{k=0}^{n-1}\frac{(n-k-1)!}{k!}\left(\frac{x}{2}\right)^{2k-n}$$

$$-\frac{1}{\pi}\sum_{k=0}^{\infty}\frac{(-1)^k(x/2)^{n+2k}}{k!(n+k)!}[\psi(k+1) + \psi(k+n+1)], \qquad (12.2.12)$$

---

[14] See Watson, op. cit., Section 3.5, for the derivation.

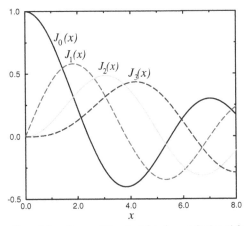

Figure 12.2.1: The first four Bessel functions of the first kind over $0 \le x \le 8$.

where

$$\psi(m+1) = -\gamma + 1 + \frac{1}{2} + \cdots + \frac{1}{m}, \qquad (\textbf{12.2.13})$$

$\psi(1) = -\gamma$, and γ is Euler's constant (0.5772157). In the case of $n = 0$, the first sum in Equation 12.2.12 disappears. This function $Y_n(x)$ is Neumann's Bessel function of the second kind of order n. Hence, the general solution to Equation 12.2.1 is

$$y(x) = AJ_n(\mu x) + BY_n(\mu x). \qquad (\textbf{12.2.14})$$

Although Equations 12.2.10 and 12.2.12 provide precise methods for computing $J_n(x)$ and $Y_n(x)$, they are not particularly insightful. Consider now the limit $x \to \infty$. It can be shown[15] that

$$J_n(x) \sim \sqrt{\frac{2}{\pi x}} \cos\left(x - \frac{n\pi}{2} - \frac{\pi}{4}\right), \qquad (\textbf{12.2.15})$$

and

$$Y_n(x) \sim \sqrt{\frac{2}{\pi x}} \sin\left(x - \frac{n\pi}{2} - \frac{\pi}{4}\right). \qquad (\textbf{12.2.16})$$

Clearly Bessel functions of the first kind are mathematical cousins to the trigonometric functions cosine and sine. I say "cousins" rather than "brothers and sisters" because they are distinctly different; their amplitudes decrease as $1/\sqrt{x}$ as x increases.

Turning to the limit of $x \to 0$ we find that

$$J_n(x) \approx \left(\tfrac{1}{2}x\right)^n / \Gamma(n+1), \qquad (\textbf{12.2.17})$$

and

$$Y_0(x) \approx -(2/\pi)\ln(x), \qquad Y_n(x) \approx -[\Gamma(n)/\pi]\left(\tfrac{1}{2}x\right)^{-n}, \qquad (\textbf{12.2.18})$$

where $\Gamma(\cdot)$ is the gamma function. We see that $J_n(x)$ is well behaved near $x = 0$ while $Y_n(x)$ tends to negative infinity near the origin. These results are confirmed in Figure 12.2.1 which

---

[15] Watson, op. cit., Chapter VII.

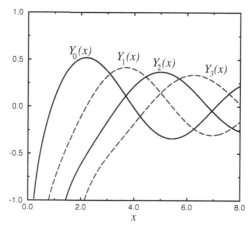

Figure 12.2.2: The first four Bessel functions of the second kind over $0 \le x \le 8$.

illustrates the functions $J_0(x)$, $J_1(x)$, $J_2(x)$, and $J_3(x)$ and Figure 12.2.2 which gives $Y_0(x)$, $Y_1(x)$, $Y_2(x)$, and $Y_3(x)$.

An equation that is very similar to Equation 12.2.1 is

$$x^2 \frac{d^2y}{dx^2} + x\frac{dy}{dx} - (n^2 + x^2)y = 0. \qquad (12.2.19)$$

It arises in the solution of partial differential equations in cylindrical coordinates. If we substitute $ix = t$ (where $i = \sqrt{-1}$) into Equation 12.2.19, it becomes Bessel's equation:

$$t^2 \frac{d^2y}{dt^2} + t\frac{dy}{dt} + (t^2 - n^2)y = 0. \qquad (12.2.20)$$

Consequently, we may immediately write the solution to Equation 12.2.20 as

$$y(x) = c_1 J_n(ix) + c_2 Y_n(ix), \qquad (12.2.21)$$

if n is an integer. Traditionally, the solution to Equation 12.2.20 has been written

$$y(x) = c_1 I_n(x) + c_2 K_n(x) \qquad (12.2.22)$$

rather than in terms of $J_n(ix)$ and $Y_n(ix)$, where

$$I_n(x) = \sum_{k=0}^{\infty} \frac{(x/2)^{2k+n}}{k!(k+n)!}, \qquad \text{and} \qquad K_n(x) = \frac{\pi}{2} i^{n+1} \left[J_n(ix) + iY_n(ix)\right]. \qquad (12.2.23)$$

The function $I_n(x)$ is the modified Bessel function of the first kind, of order n, while $K_n(x)$ is the modified Bessel function of the second kind, of order n.

Once again, it is useful to examine their behavior for large and small x. We find that

$$I_n(x) \sim \frac{e^x}{\sqrt{2\pi x}}, \qquad \text{and} \qquad K_n(x) \sim \sqrt{\frac{\pi}{2x}} e^{-x}, \qquad (12.2.24)$$

as $x \to \infty$, while

$$I_n(x) \approx \left(\tfrac{1}{2}x\right)^n / \Gamma(n+1), \qquad \text{and} \qquad K_0(x) \approx -\ln(x), \qquad K_n(x) \approx \tfrac{1}{2}\Gamma(n)\left(\tfrac{1}{2}x\right)^{-n}, \qquad (12.2.25)$$

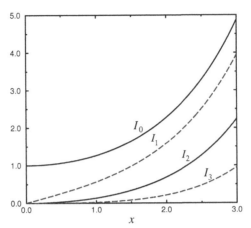

Figure 12.2.3: The first four modified Bessel functions of the first kind over $0 \le x \le 3$.

where $\Gamma(\cdot)$ is the gamma function. From these limits we see that $I_n(\cdot)$ and $K_n(\cdot)$ are mathematical cousins of e^x and e^{-x}, respectively. Furthermore, $I_n(\cdot)$ is well behaved near $x = 0$ while $K_n(\cdot)$ tends to infinity. These results are confirmed in Figure 12.2.3 which illustrates $I_0(x)$, $I_1(x)$, $I_2(x)$, and $I_3(x)$ and in Figure 12.2.4 which plots $K_0(x)$, $K_1(x)$, $K_2(x)$, and $K_3(x)$. Note that $K_n(x)$ has no real zeros while $I_n(x)$ equals zero only at $x = 0$ for $n \ge 1$.

As our derivation suggests, modified Bessel functions are related to ordinary Bessel functions via complex variables. In particular, $J_n(iz) = i^n I_n(z)$, and $I_n(iz) = i^n J_n(z)$ for z complex.

We must now find how $J_n(x)$ is related to $J_{n+1}(x)$ and $J_{n-1}(x)$. Assuming that n is a positive integer, we multiply the series, Equation 12.2.10, by x^n and then differentiate with respect to x. This gives

$$\frac{d}{dx}\left[x^n J_n(x)\right] = \sum_{k=0}^{\infty} \frac{(-1)^k (2n+2k) x^{2n+2k-1}}{2^{n+2k} k! (n+k)!} = x^n \sum_{k=0}^{\infty} \frac{(-1)^k (x/2)^{n-1+2k}}{k!(n-1+k)!} = x^n J_{n-1}(x)$$

$$(12.2.26)$$

or

$$\frac{d}{dx}\left[x^n J_n(x)\right] = x^n J_{n-1}(x) \qquad (12.2.27)$$

for $n = 1, 2, 3, \ldots$. Similarly, multiplying Equation 12.2.10 by x^{-n}, we find that

$$\frac{d}{dx}\left[x^{-n} J_n(x)\right] = -x^{-n} J_{n+1}(x) \qquad (12.2.28)$$

for $n = 0, 1, 2, 3, \ldots$. If we now carry out the differentiation on Equation 12.2.27 and Equation 12.2.28 and divide by the factors $x^{\pm n}$, we have that

$$J_n'(x) + \frac{n}{x} J_n(x) = J_{n-1}(x), \qquad (12.2.29)$$

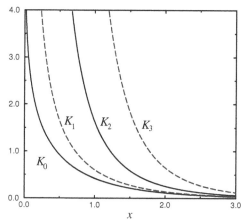

Figure 12.2.4: The first four modified Bessel functions of the second kind over $0 \leq x \leq 3$.

and

$$J_n'(x) - \frac{n}{x} J_n(x) = -J_{n+1}(x). \tag{12.2.30}$$

Equation 12.3.29 and Equation 12.3.30 immediately yield the *recurrence relationships*

$$J_{n-1}(x) + J_{n+1}(x) = \frac{2n}{x} J_n(x) \tag{12.2.31}$$

and

$$J_{n-1}(x) - J_{n+1}(x) = 2J_n'(x) \tag{12.2.32}$$

for $n = 1, 2, 3, \ldots$. For $n = 0$, we replace Equation 12.2.31 by $J_0'(x) = -J_1(x)$. Many of the most useful recurrence formulas are summarized in Table 12.2.1 for Bessel functions.

• **Example 12.2.1**

Starting with Bessel's equation, we show that the solution to

$$y'' + \frac{1-2a}{x} y' + \left(b^2 c^2 x^{2c-2} + \frac{a^2 - n^2 c^2}{x^2} \right) y = 0 \tag{12.2.33}$$

is

$$y(x) = A x^a J_n(bx^c) + B x^a Y_n(bx^c), \tag{12.2.34}$$

provided that $bx^c > 0$ so that $Y_n(bx^c)$ exists.

The general solution to

$$\xi^2 \frac{d^2\eta}{d\xi^2} + \xi \frac{d\eta}{d\xi} + (\xi^2 - n^2)\eta = 0 \tag{12.2.35}$$

Table 12.2.1: Some Useful Relationships Involving Bessel Functions of Integer Order

$$J_{n-1}(z) + J_{n+1}(z) = \frac{2n}{z} J_n(z), \qquad n = 1, 2, 3, \ldots$$

$$J_{n-1}(z) - J_{n+1}(z) = 2J_n'(z), \quad n = 1, 2, 3, \ldots; \quad J_0'(z) = -J_1(z)$$

$$\frac{d}{dz}\left[z^n J_n(z)\right] = z^n J_{n-1}(z), \qquad n = 1, 2, 3, \ldots$$

$$\frac{d}{dz}\left[z^{-n} J_n(z)\right] = -z^{-n} J_{n+1}(z), \qquad n = 0, 1, 2, 3, \ldots$$

$$I_{n-1}(z) - I_{n+1}(z) = \frac{2n}{z} I_n(z), \qquad n = 1, 2, 3, \ldots$$

$$I_{n-1}(z) + I_{n+1}(z) = 2I_n'(z), \quad n = 1, 2, 3, \ldots; \quad I_0'(z) = I_1(z)$$

$$K_{n-1}(z) - K_{n+1}(z) = -\frac{2n}{z} K_n(z), \qquad n = 1, 2, 3, \ldots$$

$$K_{n-1}(z) + K_{n+1}(z) = -2K_n'(z), \quad n = 1, 2, 3, \ldots; \quad K_0'(z) = -K_1(z)$$

$$J_n(ze^{m\pi i}) = e^{nm\pi i} J_n(z)$$

$$I_n(ze^{m\pi i}) = e^{nm\pi i} I_n(z)$$

$$K_n(ze^{m\pi i}) = e^{-mn\pi i} K_n(z) - m\pi i \frac{\cos(mn\pi)}{\cos(n\pi)} I_n(z)$$

$$I_n(z) = e^{-n\pi i/2} J_n(ze^{\pi i/2}), \qquad -\pi < \arg(z) \leq \pi/2$$

$$I_n(z) = e^{3n\pi i/2} J_n(ze^{-3\pi i/2}), \qquad \pi/2 < \arg(z) \leq \pi$$

is

$$\eta = AJ_n(\xi) + BY_n(\xi). \tag{12.2.36}$$

If we now let $\eta = y(x)/x^a$ and $\xi = bx^c$, then

$$\frac{d}{d\xi} = \frac{dx}{d\xi}\frac{d}{dx} = \frac{x^{1-c}}{bc}\frac{d}{dx}, \qquad \frac{d^2}{d\xi^2} = \frac{x^{2-2c}}{b^2 c^2}\frac{d^2}{dx^2} - \frac{(c-1)x^{1-2c}}{b^2 c^2}\frac{d}{dx}, \tag{12.2.37}$$

$$\frac{d}{dx}\left(\frac{y}{x^a}\right) = \frac{1}{x^a}\frac{dy}{dx} - \frac{a}{x^{a+1}}y, \qquad \text{and} \qquad \frac{d^2}{dx^2}\left(\frac{y}{x^a}\right) = \frac{1}{x^a}\frac{d^2y}{dx^2} - \frac{2a}{x^{a+1}}\frac{dy}{dx} + \frac{a(1+a)}{x^{a+2}}y. \tag{12.2.38}$$

Substituting Equation 12.2.37 and Equation 12.2.38 into Equation 12.2.35 and simplifying yields the desired result. $\square$

• Example 12.2.2

Let us find[16] the general solution to the nonhomogeneous differential equation

$$\frac{d^2y}{dr^2} + \frac{1}{r}\frac{dy}{dr} - k^2y = -S(r), \qquad (\mathbf{12.2.39})$$

where k is a real parameter.

The homogeneous solution is

$$y_H(r) = C_1 I_0(kr) + C_2 K_0(kr). \qquad (\mathbf{12.2.40})$$

Using variation of parameters, we assume that the particular solution can be written

$$y_p(r) = A(r)I_0(kr) + B(r)K_0(kr), \qquad (\mathbf{12.2.41})$$

where

$$A'(r) = \begin{vmatrix} 0 & K_0(kr) \\ -S(r) & kK_0'(kr) \end{vmatrix} \Bigg/ \begin{vmatrix} I_0(kr) & K_0(kr) \\ kI_0'(kr) & kK_0'(kr) \end{vmatrix}, \qquad (\mathbf{12.2.42})$$

and

$$B'(r) = \begin{vmatrix} I_0(kr) & 0 \\ kI_0'(kr) & -S(r) \end{vmatrix} \Bigg/ \begin{vmatrix} I_0(kr) & K_0(kr) \\ kI_0'(kr) & kK_0'(kr) \end{vmatrix}. \qquad (\mathbf{12.2.43})$$

Expanding the determinants, we find

$$A'(r) = S(r)K_0(kr)/\left\{k\left[I_0(kr)K_0'(kr) - I_0'(kr)K_0(kr)\right]\right\} \qquad (\mathbf{12.2.44})$$

and

$$B'(r) = -S(r)I_0(kr)/\left\{k\left[I_0(kr)K_0'(kr) - I_0'(kr)K_0(kr)\right]\right\}. \qquad (\mathbf{12.2.45})$$

Evaluating the Wronskian[17] for modified Bessel functions,

$$I_0(z)K_0'(z) - I_0'(z)K_0(z) = -1/z, \qquad (\mathbf{12.2.46})$$

$$A'(r) = -rS(r)K_0(kr) \qquad \text{and} \qquad B'(r) = rS(r)I_0(kr). \qquad (\mathbf{12.2.47})$$

Integrating the previous equations,

$$A(r) = -\int^r xS(x)K_0(kx)\,dx \qquad \text{and} \qquad B(r) = \int^r xS(x)I_0(kx)\,dx. \qquad (\mathbf{12.2.48})$$

Hence, the general solution is the sum of the particular and homogeneous solutions,

$$y(r) = C_1 I_0(kr) + C_2 K_0(kr) - I_0(kr)\int^r xS(x)K_0(kx)\,dx + K_0(kr)\int^r xS(x)I_0(kx)\,dx.$$

$$(\mathbf{12.2.49})$$

$\square$

---

[16] See Hassan, M. H. A., 1988: Ion distribution functions during ion cyclotron resonance heating at the fundamental frequency. *Phys. Fluids*, **31**, 596–599.

[17] Watson, op. cit., p. 80, Formula 19.

• **Example 12.2.3**

Let us show that

$$x^2 J_n''(x) = (n^2 - n - x^2) J_n(x) + x J_{n+1}(x). \tag{12.2.50}$$

From Equation 12.2.30
$$J_n'(x) = \frac{n}{x} J_n(x) - J_{n+1}(x), \tag{12.2.51}$$

$$J_n''(x) = -\frac{n}{x^2} J_n(x) + \frac{n}{x} J_n'(x) - J_{n+1}'(x), \tag{12.2.52}$$

and

$$J_n''(x) = -\frac{n}{x^2} J_n(x) + \frac{n}{x} \left[\frac{n}{x} J_n(x) - J_{n+1}(x) \right] - \left[J_n(x) - \frac{n+1}{x} J_{n+1}(x) \right] \tag{12.2.53}$$

after using Equation 12.2.29 and Equation 12.2.30. Simplifying,

$$J_n''(x) = \left(\frac{n^2 - n}{x^2} - 1 \right) J_n(x) + \frac{J_{n+1}(x)}{x}. \tag{12.2.54}$$

After multiplying Equation 12.2.54 by x^2, we obtain Equation 12.2.50.   □

• **Example 12.2.4**

Let us show that
$$\int_0^a x^5 J_2(x)\, dx = a^5 J_3(a) - 2a^4 J_4(a). \tag{12.2.55}$$

We begin by integrating Equation 12.2.55 by parts. If $u = x^2$, and $dv = x^3 J_2(x)\, dx$, then
$$\int_0^a x^5 J_2(x)\, dx = x^5 J_3(x) \Big|_0^a - 2 \int_0^a x^4 J_3(x)\, dx, \tag{12.2.56}$$

because $d[x^3 J_3(x)]/dx = x^2 J_2(x)$ by Equation 12.2.27. Finally,

$$\int_0^a x^5 J_2(x)\, dx = a^5 J_3(a) - 2x^4 J_4(x) \Big|_0^a = a^5 J_3(a) - 2a^4 J_4(a), \tag{12.2.57}$$

since $x^4 J_3(x) = d[x^4 J_4(x)]/dx$ by Equation 12.2.27.   □

• **Example 12.2.5: Finding the zeros of $J_0(\cdot)$**

Unlike sine and cosine, the zeros of the Bessel function $J_0(\cdot)$ cannot be written down analytically but must be found numerically. From Equation 12.2.15, a good first guess would be $x_m = n\pi/2 + (2m-1)\pi/2 + \pi/4$. We can then use the Newton-Raphson method to obtain the exact answer to any degree of accuracy. In the present case, the kth iteration equals
$$x_m^{(k+1)} = x_m^{(k)} - \frac{f(x_m^{(k)})}{f'(x_m^{(k)})} = x_m^{(k)} + \frac{J_0(x_m^{(k)})}{J_1(x_m^{(k)})}. \tag{12.2.58}$$

Table 12.2.2: The First Ten Zeros of $J_0(\cdot)$ Found Using the Newton-Raphson Method. The Boldface Digits Are Those that Differ from the Exact Answer.

m	first guess	$x_m^{(1)}$	$x_m^{(2)}$	exact
1	2.**35619449**	2.40**436809**	2.40482551	2.40482556
2	5.**49778714**	5.52003**672**	5.52007811	5.52007811
3	8.**63937980**	8.65371**699**	8.65372791	8.65372791
4	11.**78097245**	11.79153**010**	11.79153444	11.79153444
5	14.**92256510**	14.93091**557**	14.93091771	14.93091771
6	18.**06415776**	18.07106**276**	18.07106397	18.07106397
7	21.**20575041**	21.21163**588**	21.21163663	21.21163663
8	24.**34734307**	24.35247**104**	24.35247153	24.35247153
9	27.**48893572**	27.49347**879**	27.49347913	27.49347913
10	30.**63052837**	30.63460**622**	30.63460647	30.63460647

Table 12.2.2 presents results from a simple MATLAB code that compares the first guess and two successive iterations with the exact solution. Here $f(x) = J_0(x)$ and $f'(x) = J_0'(x) = -J_1(x)$. It shows that a mere two iterations give the correct zero. □

Let us now turn our attention to the method for re-expressing a well-behaved function $f(x)$ in terms of Bessel functions, a *Fourier-Bessel series*. Although we found solutions to Bessel's equation, Equation 12.2.1, as well as Equation 12.2.19, can we use any of them in an eigenfunction expansion? Referring back to the discussion following Equation 12.0.8 we see that the Bessel function must satisfy two conditions.

Consider the second condition now. The first part requires that x times the derivative of the Bessel function must tend to zero as $x \to 0$. From Figures 12.2.1–12.2.4 we see that $J_n(x)$ and $I_n(x)$ remain finite at $x = 0$ while $Y_n(x)$ and $K_n(x)$ do not. Furthermore, the products $xJ_n'(x)$ and $xI_n'(x)$ tend to zero at $x = 0$. Thus, both $J_n(x)$ and $I_n(x)$ satisfy the first part of the second requirement for a Fourier-Bessel expansion.

The second part requires that the Bessel function must satisfy the homogeneous boundary condition: $\gamma y_n(b) + \delta y_n'(b) = 0$ with $\gamma \neq 0$ and/or $\delta \neq 0$. From Figure 12.2.3 we see that $I_n(x)$ can never satisfy this condition, while from Figure 12.2.1, $J_n(x)$ can. For that reason, we discard $I_n(x)$ from further consideration and continue our analysis only with $J_n(x)$.

The exact form of the expansion depends upon the boundary condition at $x = L$. There are three possible cases. One of them is $y(L) = 0$ and results in the condition that $J_n(\mu_k L) = 0$. Another condition is $y'(L) = 0$ and gives $J_n'(\mu_k L) = 0$. Finally, if $hy(L) + y'(L) = 0$, then $hJ_n(\mu_k L) + \mu_k J_n'(\mu_k L) = 0$. In all of these cases, the eigenfunction expansion is the same, namely

$$f(x) = \sum_{k=1}^{\infty} A_k J_n(\mu_k x), \tag{12.2.59}$$

where μ_k is the kth positive solution of either $J_n(\mu_k L) = 0$, $J_n'(\mu_k L) = 0$, or $hJ_n(\mu_k L) + \mu_k J_n'(\mu_k L) = 0$.

We now need a mechanism for computing A_k. We begin by multiplying Equation 12.2.59 by $xJ_n(\mu_m x)\,dx$ and integrate from 0 to L. This yields

$$\sum_{k=1}^{\infty} A_k \int_0^L xJ_n(\mu_k x)J(\mu_m x)\,dx = \int_0^L xf(x)J_n(\mu_m x)\,dx. \qquad (\mathbf{12.2.60})$$

From the general orthogonality condition, Equation 12.0.8 with the right side equal to zero,

$$\int_0^L xJ_n(\mu_k x)J_n(\mu_m x)\,dx = 0, \qquad (\mathbf{12.2.61})$$

if $k \neq m$. Equation 12.2.60 then simplifies to

$$A_m \int_0^L xJ_n^2(\mu_m x)\,dx = \int_0^L xf(x)J_n(\mu_m x)\,dx, \qquad (\mathbf{12.2.62})$$

or

$$\boxed{A_k = \frac{1}{C_k} \int_0^L xf(x)J_n(\mu_k x)\,dx,} \qquad (\mathbf{12.2.63})$$

where

$$C_k = \int_0^L xJ_n^2(\mu_k x)\,dx, \qquad (\mathbf{12.2.64})$$

and k replaces m in Equation 12.2.62.

The factor C_k depends upon the nature of the boundary conditions at $x = L$. In all cases we start from Bessel's equation

$$[xJ_n'(\mu_k x)]' + \left(\mu_k^2 x - \frac{n^2}{x}\right) J_n(\mu_k x) = 0. \qquad (\mathbf{12.2.65})$$

If we multiply both sides of Equation 12.2.65 by $2xJ_n'(\mu_k x)$, the resulting equation is

$$(\mu_k^2 x^2 - n^2)\left[J_n^2(\mu_k x)\right]' = -\frac{d}{dx}[xJ_n'(\mu_k x)]^2. \qquad (\mathbf{12.2.66})$$

An integration of Equation 12.2.66 from 0 to L, followed by the subsequent use of integration by parts, results in

$$(\mu_k^2 x^2 - n^2)J_n^2(\mu_k x)\Big|_0^L - 2\mu_k^2 \int_0^L xJ_n^2(\mu_k x)\,dx = -[xJ_n'(\mu_k x)]^2\Big|_0^L. \qquad (\mathbf{12.2.67})$$

Because $J_n(0) = 0$ for $n > 0$, $J_0(0) = 1$ and $xJ_n'(x) = 0$ at $x = 0$, the contribution from the lower limits vanishes. Thus,

$$C_k = \int_0^L xJ_n^2(\mu_k x)\,dx = \frac{1}{2\mu_k^2}\left[(\mu_k^2 L^2 - n^2)J_n^2(\mu_k L) + L^2 J_n'^2(\mu_k L)\right]. \qquad (\mathbf{12.2.68})$$

Because

$$J_n'(\mu_k x) = \frac{n}{x} J_n(\mu_k x) - \mu_k J_{n+1}(\mu_k x) \qquad (12.2.69)$$

from Equation 12.2.30, C_k becomes

$$C_k = \tfrac{1}{2} L^2 J_{n+1}^2(\mu_k L), \qquad (12.2.70)$$

if $J_n(\mu_k L) = 0$. Otherwise, if $J_n'(\mu_k L) = 0$, then

$$C_k = \frac{\mu_k^2 L^2 - n^2}{2\mu_k^2} J_n^2(\mu_k L). \qquad (12.2.71)$$

Finally,

$$C_k = \frac{\mu_k^2 L^2 - n^2 + h^2 L^2}{2\mu_k^2} J_n^2(\mu_k L), \qquad (12.2.72)$$

if $\mu_k J_n'(\mu_k L) = -h J_n(\mu_k L)$.

All of the preceding results must be slightly modified when $n = 0$ and the boundary condition is $J_0'(\mu_k L) = 0$ or $\mu_k J_1(\mu_k L) = 0$. This modification results from the additional eigenvalue $\mu_0 = 0$ being present and we must add the extra term A_0 to the expansion. For this case the series reads

$$f(x) = A_0 + \sum_{k=1}^{\infty} A_k J_0(\mu_k x), \qquad (12.2.73)$$

where the equation for finding A_0 is

$$A_0 = \frac{2}{L^2} \int_0^L f(x)\, x\, dx, \qquad (12.2.74)$$

and Equation 12.2.63 and Equation 12.2.71 with $n = 0$ give the remaining coefficients.

• **Example 12.2.6**

Let us expand $f(x) = x$, $0 < x < 1$, in the series

$$f(x) = \sum_{k=1}^{\infty} A_k J_1(\mu_k x), \qquad (12.2.75)$$

where μ_k denotes the kth zero of $J_1(\mu)$. From Equation 12.2.63 and Equation 12.2.70,

$$A_k = \frac{2}{J_2^2(\mu_k)} \int_0^1 x^2 J_1(\mu_k x)\, dx. \qquad (12.2.76)$$

However, from Equation 12.2.27,

$$\frac{d}{dx}\left[x^2 J_2(x)\right] = x^2 J_1(x), \tag{12.2.77}$$

if $n = 2$. Therefore, Equation 12.2.76 becomes

$$A_k = \frac{2x^2 J_2(x)}{\mu_k^3 J_2^2(\mu_k)}\Bigg|_0^{\mu_k} = \frac{2}{\mu_k J_2(\mu_k)}, \tag{12.2.78}$$

and the resulting expansion is

$$x = 2\sum_{k=1}^{\infty} \frac{J_1(\mu_k x)}{\mu_k J_2(\mu_k)}, \qquad 0 \le x < 1. \tag{12.2.79}$$

Figure 12.2.5 shows the Fourier-Bessel expansion of $f(x) = x$ in truncated form when we only include one, two, three, and four terms. It was created using the MATLAB script:

```
clear;
x = [0:0.01:1]; % create x points in plot
f = x; % construct function f(x)
% initialize Fourier-Bessel series
fbessel = zeros(size(x));
% read in the first four zeros of J_1(mu) = 0
mu(1) =  3.83171; mu(2) =  7.01559;
mu(3) = 10.17347; mu(4) = 13.32369;
clf % clear any figures
for n = 1:4
% Fourier coefficient
  factor = 2 / (mu(n) * besselj(2,mu(n)));
% compute Fourier-Bessel series
  fbessel = fbessel + factor * besselj(1,mu(n)*x);
% create plot of truncated Fourier-Bessel series
%     with n terms
  subplot(2,2,n), plot(x,fbessel,x,f,'--')
  axis([0 1 -0.25 1.25])
  if n == 1 legend('1 term','f(x)'); legend boxoff;
    else legend([num2str(n) ' terms'],'f(x)'); legend boxoff;
  end
  if n > 2 xlabel('x','Fontsize',20); end
end
```
□

• **Example 12.2.7**

Let us expand the function $f(x) = x^2$, $0 < x < 1$, in the series

$$f(x) = \sum_{k=1}^{\infty} A_k J_0(\mu_k x), \tag{12.2.80}$$

where μ_k denotes the kth positive zero of $J_0(\mu)$. From Equation 12.2.63 and Equation 12.2.70,

$$A_k = \frac{2}{J_1^2(\mu_k)} \int_0^1 x^3 J_0(\mu_k x)\,dx. \tag{12.2.81}$$

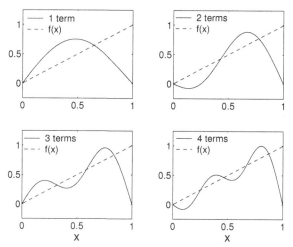

Figure 12.2.5: The Fourier-Bessel series representation, Equation 12.2.79, for $f(x) = x$, $0 < x < 1$, when we truncate the series so that it includes only the first, first two, first three, and first four terms.

If we let $t = \mu_k x$, the integration, Equation 12.2.81, becomes

$$A_k = \frac{2}{\mu_k^4 J_1^2(\mu_k)} \int_0^{\mu_k} t^3 J_0(t)\, dt. \tag{12.2.82}$$

We now let $u = t^2$ and $dv = t J_0(t)\, dt$ so that integration by parts results in

$$A_k = \frac{2}{\mu_k^4 J_1^2(\mu_k)} \left[t^3 J_1(t) \big|_0^{\mu_k} - 2 \int_0^{\mu_k} t^2 J_1(t)\, dt \right] = \frac{2}{\mu_k^4 J_1^2(\mu_k)} \left[\mu_k^3 J_1(\mu_k) - 2 \int_0^{\mu_k} t^2 J_1(t)\, dt \right], \tag{12.2.83}$$

because $v = t J_1(t)$ from Equation 12.2.27. If we integrate by parts once more, we find that

$$A_k = \frac{2}{\mu_k^4 J_1^2(\mu_k)} \left[\mu_k^3 J_1(\mu_k) - 2\mu_k^2 J_2(\mu_k) \right] = \frac{2}{J_1^2(\mu_k)} \left[\frac{J_1(\mu_k)}{\mu_k} - \frac{2 J_2(\mu_k)}{\mu_k^2} \right]. \tag{12.2.84}$$

However, from Equation 12.2.31 with $n = 1$, $J_1(\mu_k) = \frac{1}{2}\mu_k \left[J_2(\mu_k) + J_0(\mu_k) \right]$, or $J_2(\mu_k) = 2 J_1(\mu_k)/\mu_k$, because $J_0(\mu_k) = 0$. Therefore,

$$A_k = \frac{2(\mu_k^2 - 4) J_1(\mu_k)}{\mu_k^3 J_1^2(\mu_k)}, \qquad \text{and} \qquad x^2 = 2 \sum_{k=1}^{\infty} \frac{(\mu_k^2 - 4) J_0(\mu_k x)}{\mu_k^3 J_1(\mu_k)}, \qquad 0 < x < 1. \tag{12.2.85}$$

Figure 12.2.6 shows the representation of x^2 by the Fourier-Bessel series, Equation 12.2.85 when we truncate it so that it includes only one, two, three, or four terms. As we add each additional term in the orthogonal expansion, the expansion fits $f(x)$ better in the "least squares" sense of Equation 11.3.6. $\qquad \square$

• **Example 12.2.8**

Let us construct the Fourier-Bessel expansion for

$$\frac{I_0(Mr)}{I_0(M)} = \sum_{n=1}^{\infty} A_n J_0(k_n r), \qquad 0 \le r < 1, \tag{12.2.86}$$

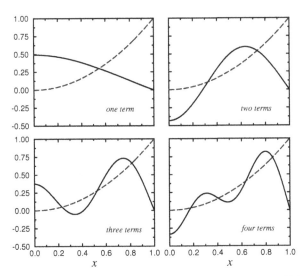

Figure 12.2.6: The Fourier-Bessel series representation, Equation 12.2.85, for $f(x) = x^2$, $0 < x < 1$, when we truncate the series so that it includes only the first, first two, first three, and first four terms.

where k_n denotes the nth positive zero of $J_0(k)$. From Equation 12.2.63 and Equation 12.2.70,

$$A_n = \frac{2}{J_1^2(k_n)} \int_0^1 \frac{I_0(Mr)}{I_0(M)} J_0(k_n r)\, r\, dr. \tag{12.2.87}$$

To evaluate the integral, we first write down the governing differential equations for $J_0(k_n r)$ and $I_0(Mr)$:

$$\frac{1}{r} \frac{d}{dr} \left[r \frac{dJ_0(k_n r)}{dr} \right] + k_n^2 J_0(k_n r) = 0, \tag{12.2.88}$$

and

$$\frac{1}{r} \frac{d}{dr} \left[r \frac{dI_0(Mr)}{dr} \right] - M^2 I_0(Mr) = 0. \tag{12.2.89}$$

Multiply Equation 12.2.89 by $J_0(k_n r)$ and subtracting the resulting equation from Equation 12.2.88 after we have multiplied it by $I_0(Mr)$, we find that

$$d[r I_0(Mr) J_0'(k_n r) - r J_0(k_n) I_0'(Mr)] + (M^2 + k_n^2) I_0(Mr) J_0(k_n r)\, r\, dr = 0. \tag{12.2.90}$$

Next, we integrate Equation 12.2.90 from 0 to 1. Using the facts that $J_0(k_n) = 0$ and $J_0'(k_n r) = -k_n J_1(k_n r)$, we obtain

$$\int_0^1 I_0(Mr) J_0(k_n r)\, r\, dr = \frac{k_n}{M^2 + k_n^2} I_0(M) J_1(k_n). \tag{12.2.91}$$

Finally, we substitute Equation 12.2.91 into Equation 12.2.87 and obtain

$$\frac{I_0(Mr)}{I_0(M)} = 2 \sum_{n=1}^{\infty} \frac{k_n J_0(k_n r)}{(M^2 + k_n^2) J_1(k_n)}. \tag{12.2.92}$$

Figure 12.2.7 illustrates various partial sums of this expansion.   □

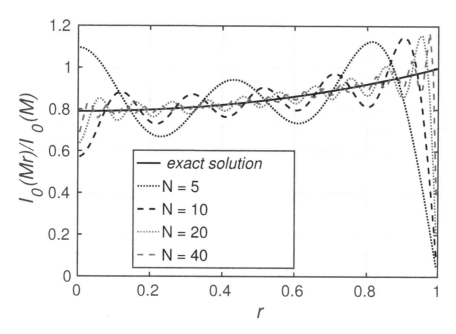

Figure 12.2.7: A plot of $I_0(Mr)/I_0(M)$ given by the Fourier-Bessel expansion Equation 12.2.92 for various partial summations when $M = 1$.

We now develop an understanding of Bessel functions and how we can expand a well-behaved function in terms of $J_n(\cdot)$. Let us now apply them to solving various linear partial differential equations.

- **Example 12.2.9: Axisymmetric vibrations of a circular membrane**

The wave equation

$$\frac{1}{c^2}\frac{\partial^2 u}{\partial t^2} = \frac{\partial^2 u}{\partial r^2} + \frac{1}{r}\frac{\partial u}{\partial r}, \qquad 0 \le r < a, \quad 0 < t \qquad (12.2.93)$$

governs axisymmetric vibrations of a circular membrane, where $u(r,t)$ is the vertical displacement of the membrane, r is the radial distance, t is time, c is the square root of the ratio of the tension of the membrane to its density, and a is the radius of the membrane. The conversion from rectangular to polar coordinates is given in Appendix A. We will solve Equation 12.2.93 when the membrane is initially at rest, $u(r,0) = 0$, and struck so that its initial velocity is

$$\frac{\partial u(r,0)}{\partial t} = \begin{cases} P/(\pi\epsilon^2\rho), & 0 \le r < \epsilon, \\ 0, & \epsilon < r < a. \end{cases} \qquad (12.2.94)$$

If this problem can be solved by separation of variables, then $u(r,t) = R(r)T(t)$. Following the substitution of this $u(r,t)$ into Equation 12.2.93, separation of variables leads to

$$\frac{1}{rR}\frac{d}{dr}\left(r\frac{dR}{dr}\right) = \frac{1}{c^2 T}\frac{d^2 T}{dt^2} = -k^2, \qquad (12.2.95)$$

or

$$\frac{1}{r}\frac{d}{dr}\left(r\frac{dR}{dr}\right) + k^2 R = 0, \qquad (12.2.96)$$

and

$$\frac{d^2T}{dt^2} + k^2 c^2 T = 0. \tag{12.2.97}$$

The separation constant $-k^2$ must be negative so that we obtain solutions that remain bounded in the region $0 \le r < a$ and can satisfy the boundary condition. This boundary condition is $u(a,t) = R(a)T(t) = 0$, or $R(a) = 0$.

The solutions of Equation 12.2.96 and Equation 12.2.97, subject to the boundary condition, are

$$R_n(r) = J_0\left(\frac{\lambda_n r}{a}\right), \tag{12.2.98}$$

and

$$T_n(t) = A_n \sin\left(\frac{\lambda_n ct}{a}\right) + B_n \cos\left(\frac{\lambda_n ct}{a}\right), \tag{12.2.99}$$

where λ_n satisfies the equation $J_0(\lambda) = 0$. Because $u(r,0) = 0$, and $T_n(0) = 0$, $B_n = 0$. Consequently, the product solution is

$$u(r,t) = \sum_{n=1}^{\infty} A_n J_0\left(\frac{\lambda_n r}{a}\right) \sin\left(\frac{\lambda_n ct}{a}\right). \tag{12.2.100}$$

To determine A_n, we use the condition

$$\frac{\partial u(r,0)}{\partial t} = \sum_{n=1}^{\infty} \frac{\lambda_n c}{a} A_n J_0\left(\frac{\lambda_n r}{a}\right) = \begin{cases} P/(\pi\epsilon^2 \rho), & 0 \le r < \epsilon, \\ 0, & \epsilon < r < a. \end{cases} \tag{12.2.101}$$

Equation 12.2.101 is a Fourier-Bessel expansion using the orthogonal function $J_0(\lambda_n r/a)$, where

$$\frac{\lambda_n c}{a} A_n = \frac{2}{a^2 J_1^2(\lambda_n)} \int_0^\epsilon \frac{P}{\pi\epsilon^2 \rho} J_0\left(\frac{\lambda_n r}{a}\right) r\, dr \tag{12.2.102}$$

from Equation 12.2.63 and Equation 12.2.70. Carrying out the integration,

$$A_n = \frac{2P J_1(\lambda_n \epsilon/a)}{c\pi\epsilon\rho\lambda_n^2 J_1^2(\lambda_n)}, \tag{12.2.103}$$

or

$$u(r,t) = \frac{2P}{c\pi\epsilon\rho} \sum_{n=1}^{\infty} \frac{J_1(\lambda_n \epsilon/a)}{\lambda_n^2 J_1^2(\lambda_n)} J_0\left(\frac{\lambda_n r}{a}\right) \sin\left(\frac{\lambda_n ct}{a}\right). \tag{12.2.104}$$

Figures 12.2.8, 12.2.9, and 12.2.10 illustrate the solution, Equation 12.2.104, for various times and positions when $\epsilon = a/4$, and $\epsilon = a/20$. They were generated using the MATLAB script:

```
% initialize parameters
clear; eps_over_a = 0.25; M = 20; dr = 0.02; dt = 0.02;
% load in zeros of J_0
zero( 1) =  2.40483; zero( 2) =  5.52008; zero( 3) =  8.65373;
zero( 4) = 11.79153; zero( 5) = 14.93092; zero( 6) = 18.07106;
zero( 7) = 21.21164; zero( 8) = 24.35247; zero( 9) = 27.49347;
zero(10) = 30.63461; zero(11) = 33.77582; zero(12) = 36.91710;
zero(13) = 40.05843; zero(14) = 43.19979; zero(15) = 46.34119;
```

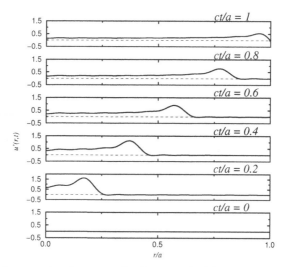

Figure 12.2.8: The axisymmetric vibrations $u'(r,t) = ca\rho u(r,t)/P$ of a circular membrane at various positions r/a at the times $ct/a = 0$, 0.2, 0.4, 0.6, 0.8, and 1 for $\epsilon = a/4$. Initially the membrane is struck by a hammer.

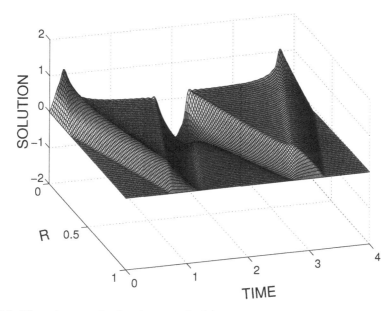

Figure 12.2.9: The axisymmetric vibrations $ca\rho u(r,t)/P$ of a circular membrane resulting from an initial hammer blow with $\epsilon = a/4$. The solution is plotted at various times ct/a and positions r/a.

```
zero(16) = 49.48261; zero(17) = 52.62405; zero(18) = 55.76551;
zero(19) = 58.90698; zero(20) = 62.04847;
% compute Fourier-Bessel coefficients
for m = 1:M
  a(m) = 2 * besselj(1,eps_over_a*zero(m)) ...
       / (eps_over_a*pi*zero(m)*zero(m)*besselj(1,zero(m))^2);
end
R = [0:dr:1]; T = [0:dt:4];
```

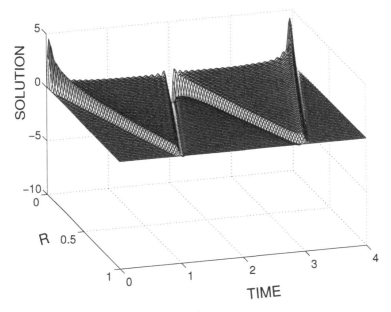

Figure 12.2.10: Same as Figure 12.2.9 except $\epsilon = a/20$.

```
u = zeros(length(T),length(R));
RR = repmat(R,[length(T) 1]);
TT = repmat(T',[1 length(R)]);
% compute solution from series solution
for m = 1:M
  u = u + a(m) .* besselj(0,zero(m)*RR) .* sin(zero(m)*TT);
end
% plot results
surf(RR,TT,u)
xlabel('R','Fontsize',20); ylabel('TIME','Fontsize',20)
zlabel('SOLUTION','Fontsize',20)
```

Figures 12.2.9 and 12.2.10 show that striking the membrane with a hammer generates a pulse that propagates out to the rim, reflects, inverts, and propagates back to the center. This process then repeats forever.   □

- **Example 12.2.10: Axisymmetric heat equation in an infinitely long cylinder**

In this example we illustrate how separation of variables can be employed in solving the axisymmetric heat equation in an infinitely long cylinder. In circular coordinates, the heat equation is

$$\frac{\partial u}{\partial t} = a^2 \left(\frac{\partial^2 u}{\partial r^2} + \frac{1}{r} \frac{\partial u}{\partial r} \right), \qquad 0 \le r < b, \quad 0 < t, \tag{12.2.105}$$

where r denotes the radial distance and a^2 denotes the thermal diffusivity. The conversion from rectangular to polar coordinates is given in Appendix A. Let us assume that we heated this cylinder of radius b to the uniform temperature T_0 and then allowed it to cool by having its surface held at the temperature of zero starting from the time $t = 0$.

We begin by assuming that the solution is of the form $u(r,t) = R(r)T(t)$ so that

$$\frac{1}{R}\left(\frac{d^2R}{dr^2} + \frac{1}{r}\frac{dR}{dr}\right) = \frac{1}{a^2T}\frac{dT}{dt} = -\frac{k^2}{b^2}. \qquad (\mathbf{12.2.106})$$

The only values of the separation constant that yield nontrivial solutions are negative. The nontrivial solutions are $R(r) = J_0(kr/b)$, where $J_0(\cdot)$ is the Bessel function of the first kind and zeroth order. A separation constant of zero gives $R(r) = \ln(r)$, which becomes infinite at the origin. Positive separation constants yield the modified Bessel function $I_0(kr/b)$. Although this function is finite at the origin, it cannot satisfy the boundary condition that $u(b,t) = R(b)T(t) = 0$, or $R(b) = 0$.

The boundary condition that $R(b) = 0$ requires that $J_0(k) = 0$. This transcendental equation yields an infinite number of constants k_n. For each k_n, the temporal part of the solution satisfies the differential equation

$$\frac{dT_n}{dt} + \frac{k_n^2 a^2}{b^2}T_n = 0, \qquad (\mathbf{12.2.107})$$

which has the solution

$$T_n(t) = A_n \exp\left(-\frac{k_n^2 a^2}{b^2}t\right). \qquad (\mathbf{12.2.108})$$

Hence, the product solutions are

$$u_n(r,t) = A_n J_0\left(k_n \frac{r}{b}\right)\exp\left(-\frac{k_n^2 a^2}{b^2}t\right). \qquad (\mathbf{12.2.109})$$

The total solution is a linear superposition of all of the particular solutions or

$$u(r,t) = \sum_{n=1}^{\infty} A_n J_0\left(k_n \frac{r}{b}\right)\exp\left(-\frac{k_n^2 a^2}{b^2}t\right). \qquad (\mathbf{12.2.110})$$

Our final task remains to determine A_n. From the initial condition that $u(r,0) = T_0$,

$$u(r,0) = T_0 = \sum_{n=1}^{\infty} A_n J_0\left(k_n \frac{r}{b}\right). \qquad (\mathbf{12.2.111})$$

From Equation 12.2.63 and Equation 12.2.70,

$$A_n = \frac{2T_0}{J_1^2(k_n)b^2}\int_0^b r J_0\left(k_n \frac{r}{b}\right)dr = \frac{2T_0}{k_n^2 J_1^2(k_n)}\left(\frac{k_n r}{b}\right)J_1\left(k_n \frac{r}{b}\right)\Big|_0^b = \frac{2T_0}{k_n J_1(k_n)} \qquad (\mathbf{12.2.112})$$

from Equation 12.2.27. Thus, the complete solution is

$$u(r,t) = 2T_0 \sum_{n=1}^{\infty} \frac{1}{k_n J_1(k_n)} J_0\left(k_n \frac{r}{b}\right)\exp\left(-\frac{k_n^2 a^2}{b^2}t\right). \qquad (\mathbf{12.2.113})$$

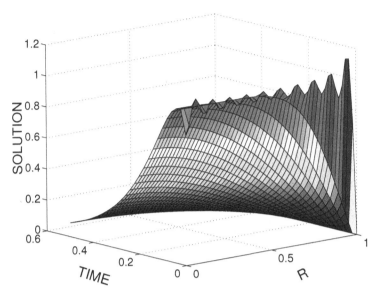

Figure 12.2.11: The temperature $u(r,t)/T_0$ within an infinitely long cylinder at various positions r/b and times a^2t/b^2 that we initially heated to the uniform temperature T_0 and then allowed to cool by forcing its surface to equal zero.

Figure 12.2.11 illustrates the solution, Equation 12.2.113, for various Fourier numbers a^2t/b^2. It was generated using the MATLAB script:

```
clear
M = 20; dr = 0.02; dt = 0.02;
% load in zeros of J_0
zero( 1) =  2.40482; zero( 2) =  5.52007; zero( 3) =  8.65372;
zero( 4) = 11.79153; zero( 5) = 14.93091; zero( 6) = 18.07106;
zero( 7) = 21.21164; zero( 8) = 24.35247; zero( 9) = 27.49347;
zero(10) = 30.63461; zero(11) = 33.77582; zero(12) = 36.91710;
zero(13) = 40.05843; zero(14) = 43.19979; zero(15) = 46.34119;
zero(16) = 49.48261; zero(17) = 52.62405; zero(18) = 55.76551;
zero(19) = 58.90698; zero(20) = 62.04847;
% compute Fourier coefficients
for m = 1:M
  a(m) = 2 / (zero(m)*besselj(1,zero(m)));
end
% compute grid and initialize solution
R = [0:dr:1]; T = [0:dt:0.5];
u = zeros(length(T),length(R));
RR = repmat(R,[length(T) 1]);
TT = repmat(T',[1 length(R)]);
% compute solution from Equation 12.2.113
for m = 1:M
  u = u + a(m)*besselj(0,zero(m)*RR).*exp(-zero(m)*zero(m)*TT);
end
surf(RR,TT,u)
xlabel('R','Fontsize',20); ylabel('TIME','Fontsize',20)
zlabel('SOLUTION','Fontsize',20)
```

Heat flows from the interior and is removed at the cylinder's surface where the temperature equals zero. The initial oscillations of the solution result from Gibbs phenomena because we have a jump in the temperature field at $r = b$. $\qquad\Box$

• Example 12.2.11: Heat conduction in an infinitely long, radiatively cooling cylinder

In this example[18] we find the evolution of the temperature field within a cylinder of radius b as it radiatively cools from an initial uniform temperature T_0. The conversion from rectangular to polar coordinates is given in Appendix A. The heat equation is then

$$\frac{\partial u}{\partial t} = a^2 \left(\frac{\partial^2 u}{\partial r^2} + \frac{1}{r} \frac{\partial u}{\partial r} \right), \qquad 0 \leq r < b, \quad 0 < t, \tag{12.2.114}$$

which we will solve by separation of variables $u(r,t) = R(r)T(t)$. Therefore,

$$\frac{1}{R} \left(\frac{d^2 R}{dr^2} + \frac{1}{r} \frac{dR}{dr} \right) = \frac{1}{a^2 T} \frac{dT}{dt} = -\frac{k^2}{b^2}, \tag{12.2.115}$$

because only a negative separation constant yields an $R(r)$ that is finite at the origin and satisfies the boundary condition. This solution is $R(r) = J_0(kr/b)$, where $J_0(\cdot)$ is the Bessel function of the first kind and zeroth order.

The radiative boundary condition can be expressed as

$$\frac{\partial u(b,t)}{\partial r} + hu(b,t) = T(t) \left[\frac{dR(b)}{dr} + hR(b) \right] = 0. \tag{12.2.116}$$

Because $T(t) \neq 0$,

$$kJ_0'(k) + hbJ_0(k) = -kJ_1(k) + hbJ_0(k) = 0, \tag{12.2.117}$$

where the product hb is the Biot number. The solution of the transcendental equation, Equation 12.2.117, yields an infinite number of distinct constants k_n. For each k_n, the temporal part equals the solution of

$$\frac{dT_n}{dt} + \frac{k_n^2 a^2}{b^2} T_n = 0, \tag{12.2.118}$$

or

$$T_n(t) = A_n \exp\left(-\frac{k_n^2 a^2}{b^2} t \right). \tag{12.2.119}$$

The product solution is, therefore,

$$u_n(r,t) = A_n J_0\left(k_n \frac{r}{b} \right) \exp\left(-\frac{k_n^2 a^2}{b^2} t \right) \tag{12.2.120}$$

and the most general solution is a sum of these product solutions

$$u(r,t) = \sum_{n=1}^{\infty} A_n J_0\left(k_n \frac{r}{b} \right) \exp\left(-\frac{k_n^2 a^2}{b^2} t \right). \tag{12.2.121}$$

---

[18] For another example of solving the heat equation with Robin boundary conditions, see Section 3.2 in Balakotaiah, V., N. Gupta, and D. H. West, 2000: A simplified model for analyzing catalytic reactions in short monoliths. *Chem. Eng. Sci.*, **55**, 5367–5383.

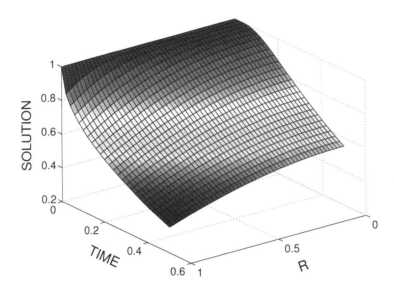

Figure 12.2.12: The temperature $u(r,t)/T_0$ within an infinitely long cylinder at various positions r/b and times $a^2 t/b^2$ that we initially heated to the temperature T_0 and then allowed to radiatively cool with $hb = 1$.

Finally, we must determine A_n. From the initial condition that $u(r,0) = T_0$,

$$u(r,0) = T_0 = \sum_{n=1}^{\infty} A_n J_0\left(k_n \frac{r}{b}\right), \tag{12.2.122}$$

where

$$A_n = \frac{2k_n^2 T_0}{b^2 [k_n^2 + b^2 h^2] J_0^2(k_n)} \int_0^b r J_0\left(k_n \frac{r}{b}\right) dr = \frac{2T_0}{[k_n^2 + b^2 h^2] J_0^2(k_n)} \left(\frac{k_n r}{b}\right) J_1\left(k_n \frac{r}{b}\right)\Big|_0^b \tag{12.2.123}$$

$$= \frac{2k_n T_0 J_1(k_n)}{[k_n^2 + b^2 h^2] J_0^2(k_n)} = \frac{2k_n T_0 J_1(k_n)}{k_n^2 J_0^2(k_n) + b^2 h^2 J_0^2(k_n)} = \frac{2k_n T_0 J_1(k_n)}{k_n^2 J_0^2(k_n) + k_n^2 J_1^2(k_n)} \tag{12.2.124}$$

$$= \frac{2T_0 J_1(k_n)}{k_n [J_0^2(k_n) + J_1^2(k_n)]}, \tag{12.2.125}$$

which follows from Equation 12.2.27, Equation 12.2.63, Equation 12.2.73, and Equation 12.2.117. Thus, the complete solution is

$$u(r,t) = 2T_0 \sum_{n=1}^{\infty} \frac{J_1(k_n)}{k_n [J_0^2(k_n) + J_1^2(k_n)]} J_0\left(k_n \frac{r}{b}\right) \exp\left(-\frac{k_n^2 a^2}{b^2} t\right). \tag{12.2.126}$$

Figure 12.2.12 illustrates the solution, Equation 12.2.126, for various Fourier numbers $a^2 t/b^2$ with $hb = 1$. It was created using the MATLAB script:

```
clear
hb = 1; m=0; M = 100; dr = 0.02; dt = 0.02;
% find k_n which satisfies hb J_0(k) = k J_1(k)
for n = 1:10000
```

```
   k1 = 0.05*n; k2 = 0.05*(n+1);
   y1 = hb * besselj(0,k1) - k1 * besselj(1,k1);
   y2 = hb * besselj(0,k2) - k2 * besselj(1,k2);
   if y1*y2 <= 0; m = m+1; zero(m) = k1; end;
end;
% use Newton-Raphson method to improve values of k_n
for n = 1:M; for k = 1:5
   term0 = besselj(0,zero(n));
   term1 = besselj(1,zero(n));
   term2 = besselj(2,zero(n));
   f = hb * term0 - zero(n) * term1;
   fp = 0.5*zero(n)*(term2-term0) - (1+hb)*term1;
   zero(n) = zero(n) - f / fp;
end; end;
% compute Fourier coefficients
for m = 1:M
   denom = zero(m)*(besselj(0,zero(m))^2+besselj(1,zero(m))^2);
   a(m) = 2 * besselj(1,zero(m)) / denom;
end
% compute grid and initialize solution
R = [0:dr:1]; T = [0:dt:0.5];
u = zeros(length(T),length(R));
RR = repmat(R,[length(T) 1]);
TT = repmat(T',[1 length(R)]);
% compute solution from Equation 12.2.126
for m = 1:M
   u = u + a(m)*besselj(0,zero(m)*RR).*exp(-zero(m)*zero(m)*TT);
end
surf(RR,TT,u)
xlabel('R','Fontsize',20); ylabel('TIME','Fontsize',20)
zlabel('SOLUTION','Fontsize',20)
```

Heat flows from the interior and is removed at the cylinder's surface where it radiates to space at the temperature zero. Note that we do *not* suffer from Gibbs phenomena here because there is no initial jump in the temperature distribution. □

• Example 12.2.12:   Temperature within an electrical cable

In the design of cable installations we need the temperature reached within an electrical cable as a function of current and other parameters. To this end,[19] let us solve the nonhomogeneous heat equation in cylindrical coordinates with a radiation boundary condition.

The derivation of the heat equation follows from the conservation of energy:

$$\text{heat generated} = \text{heat dissipated} + \text{heat stored},$$

---

[19] Iskenderian, H. P., and W. J. Horvath, 1946: Determination of the temperature rise and the maximum safe current through multiconductor electric cables. *J. Appl. Phys.*, **17**, 255–262.

or

$$I^2 RN\,dt = -\kappa \left[2\pi r \left. \frac{\partial u}{\partial r} \right|_r - 2\pi(r+\Delta r)\left. \frac{\partial u}{\partial r} \right|_{r+\Delta r} \right] dt + 2\pi r \Delta r c\rho\,du, \qquad (12.2.127)$$

where I is the current through each wire, R is the resistance of each conductor, N is the number of conductors in the shell between radii r and $r + \Delta r = 2\pi m r \Delta r/(\pi b^2)$, b is the radius of the cable, m is the total number of conductors in the cable, κ is the thermal conductivity, ρ is the density, c is the average specific heat, and u is the temperature. In the limit of $\Delta r \to 0$, Equation 12.2.127 becomes

$$\frac{\partial u}{\partial t} = A + \frac{a^2}{r}\frac{\partial}{\partial r}\left(r\frac{\partial u}{\partial r} \right), \qquad 0 \le r < b, \quad 0 < t, \qquad (12.2.128)$$

where $A = I^2 Rm/(\pi b^2 c\rho)$, and $a^2 = \kappa/(\rho c)$.

Equation 12.2.128 is the nonhomogeneous heat equation for an infinitely long, axisymmetric cylinder. Let us write the temperature as the sum of a steady-state and transient solution: $u(r,t) = w(r) + v(r,t)$. The steady-state solution $w(r)$ satisfies

$$\frac{1}{r}\frac{d}{dr}\left(r\frac{dw}{dr} \right) = -\frac{A}{a^2}, \qquad \text{or} \qquad w(r) = T_c - \frac{Ar^2}{4a^2}, \qquad (12.2.129)$$

where T_c is the (yet unknown) temperature in the center of the cable.

The transient solution $v(r,t)$ is governed by

$$\frac{\partial v}{\partial t} = a^2 \frac{1}{r}\frac{\partial}{\partial r}\left(r\frac{\partial v}{\partial r} \right), \qquad 0 \le r < b, \quad 0 < t, \qquad (12.2.130)$$

with the initial condition that $u(r,0) = T_c - Ar^2/(4a^2) + v(0,t) = 0$. At the surface $r = b$, heat radiates to free space so that the boundary condition is $u_r = -hu$, where h is the surface conductance. Because the temperature equals the steady-state solution when all transient effects die away, $w(r)$ must satisfy this radiation boundary condition regardless of the transient solution. This requires that

$$T_c = \frac{A}{a^2}\left(\frac{b^2}{4} + \frac{b}{2h} \right). \qquad (12.2.131)$$

Therefore, $v(r,t)$ must satisfy $v_r(b,t) = -hv(b,t)$ at $r = b$.

We find the transient solution $v(r,t)$ by separation of variables $v(r,t) = R(r)T(t)$. Substituting into Equation 12.2.130,

$$\frac{1}{rR}\frac{d}{dr}\left(r\frac{dR}{dr} \right) = \frac{1}{a^2 T}\frac{dT}{dt} = -k^2, \qquad (12.2.132)$$

or

$$\frac{d}{dr}\left(r\frac{dR}{dr} \right) + k^2 rR = 0, \qquad (12.2.133)$$

and

$$\frac{dT}{dt} + k^2 a^2 T = 0, \qquad (12.2.134)$$

with $R'(b) = -hR(b)$. The only solution of Equation 12.2.133 that remains finite at $r = 0$ and satisfies the boundary condition is $R(r) = J_0(kr)$, where $J_0(\cdot)$ is the zero-order

Bessel function of the first kind. Substituting $J_0(kr)$ into the boundary condition, the transcendental equation is

$$kbJ_1(kb) - hbJ_0(kb) = 0. \qquad (12.2.135)$$

For a given value of h and b, Equation 12.2.135 yields an infinite number of unique zeros k_n.

The corresponding temporal solution to the problem is

$$T_n(t) = A_n \exp(-a^2 k_n^2 t), \qquad (12.2.136)$$

so that the sum of the product solutions is

$$v(r,t) = \sum_{n=1}^{\infty} A_n J_0(k_n r) \exp(-a^2 k_n^2 t). \qquad (12.2.137)$$

Our final task remains to compute A_n. By evaluating Equation 12.2.137 at $t = 0$,

$$v(r,0) = \frac{Ar^2}{4a^2} - T_c = \sum_{n=1}^{\infty} A_n J_0(k_n r), \qquad (12.2.138)$$

which is a Fourier-Bessel series in $J_0(k_n r)$. Earlier we showed how to compute the coefficient of a Fourier-Bessel series that is expanded in $J_0(k_n r)$ and that has a boundary condition of the form given Equation 12.2.135. Applying those results here, we have that

$$A_n = \frac{2k_n^2}{(k_n^2 b^2 + h^2 b^2)J_0^2(k_n b)} \int_0^b r\left(\frac{Ar^2}{4a^2} - T_c\right) J_0(k_n r)\, dr \qquad (12.2.139)$$

from Equation 12.2.63 and Equation 12.2.72. Carrying out the indicated integrations,

$$A_n = \frac{2}{(k_n^2 + h^2)J_0^2(k_n b)} \left[\left(\frac{Ak_n b}{4a^2} - \frac{A}{k_n ba^2} - \frac{T_c k_n}{b}\right) J_1(k_n b) + \frac{A}{2a^2} J_0(k_n b) \right]. \qquad (12.2.140)$$

We obtained Equation 12.2.140 by using Equation 12.2.27 and integrating by parts as shown in Example 12.2.4.

To illustrate this solution, let us compute it for the typical parameters $b = 4$ cm, $hb = 1$, $a^2 = 1.14$ cm^2/s, $A = 2.2747$ °C/s, and $T_c = 23.94$°C. The value of A corresponds to 37 wires of #6 AWG copper wire within a cable carrying a current of 22 amps.

Figure 12.2.13 illustrates the solution as a function of the radius at various times. It was created using the MATLAB script:

```
clear
asq = 1.14; A = 2.2747; b = 4; dr = 0.02; dt = 0.02;
hb = 1; m=0; M = 10; T_c = 23.94;
const1 = A * b * b / (4 * asq); const2 = A * b * b / asq;
const3 = A * b * b / (2 * asq);
% find k_nb which satisfies hb J_0(kb) = kb J_1(kb)
for n = 1:10000
  k1 = 0.05*n; k2 = 0.05*(n+1);
  y1 = hb * besselj(0,k1) - k1 * besselj(1,k1);
  y2 = hb * besselj(0,k2) - k2 * besselj(1,k2);
  if y1*y2 <= 0; m = m+1; zero(m) = k1; end;
```

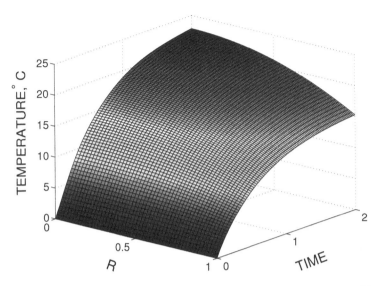

Figure 12.2.13: The temperature field (in degrees Celsius) within an electric copper cable containing 37 wires and a current of 22 amperes at various positions r/b and times a^2t/b^2. Initially the temperature was zero and then we allow the cable to cool radiatively as it is heated. The parameters are $hb = 1$ and the radius of the cable $b = 4$ cm.

```
end;
% use Newton-Raphson method to improve values of k_n
for n = 1:M; for k = 1:5
  term0 = besselj(0,zero(n));
  term1 = besselj(1,zero(n));
  term2 = besselj(2,zero(n));
  f = hb * term0 - zero(n) * term1;
  fp = 0.5*zero(n)*(term2-term0) - (1+hb)*term1;
  zero(n) = zero(n) - f / fp;
end; end;
for m = 1:M
  denom = (zero(m)*zero(m)+hb*hb)*besselj(0,zero(m))^2;
  a(m) = ((const1-T_c)*zero(m) ...
      - const2/zero(m))*besselj(1,zero(m)) ...
      + const3 * besselj(0,zero(m));
  a(m) = 2 * a(m) / denom;
end
% compute grid and initialize solution
R = [0:dr:1]; T = [0:dt:2];
u = T_c * ones(length(T),length(R));
RR = repmat(R,[length(T) 1]);
TT = repmat(T',[1 length(R)]);
% compute solution u(r,t) = w(r) + v(r,t)
u = u - const1 * RR .* RR;
for m = 1:M
  u = u + a(m)*besselj(0,zero(m)*RR).*exp(-zero(m)*zero(m)*TT);
end
surf(RR,TT,u); axis([0 1 0 2 0 25]);
```

```
xlabel('R','Fontsize',20); ylabel('TIME','Fontsize',20)
zlabel('TEMPERATURE,^\circ C','Fontsize',20)
```

From an initial temperature of zero, the temperature rises due to the constant electrical heating. After a short period of time, it reaches its steady-state distribution given by Equation 12.2.129. The cable is coolest at the surface where heat is radiating away. Heat flows from the interior to replace the heat lost by radiation. □

• Example 12.2.13: Electrostatic potential inside a closed cylinder of finite length

The *electrostatic potential* is defined as the amount of work that must be done against electric forces to bring a unit charge from a reference point to a given point. It is readily shown[20] that the electrostatic potential is described by Laplace's equation if there is no charge within the domain. Let us find the electrostatic potential $u(r, z)$ inside a closed cylinder of length L and radius a. The base and lateral surfaces have the potential 0 while the upper surface has the potential V.

Because the potential varies in only r and z, Laplace's equation in cylindrical coordinates reduces to

$$\frac{1}{r}\frac{\partial}{\partial r}\left(r\frac{\partial u}{\partial r}\right) + \frac{\partial^2 u}{\partial z^2} = 0, \qquad 0 \le r < a, \quad 0 < z < L, \tag{12.2.141}$$

subject to the boundary conditions $u(a, z) = u(r, 0) = 0$, and $u(r, L) = V$. The conversion from rectangular to polar coordinates is given in Appendix A. To solve this problem by separation of variables,[21] let $u(r, z) = R(r)Z(z)$ and

$$\frac{1}{rR}\frac{d}{dr}\left(r\frac{dR}{dr}\right) = -\frac{1}{Z}\frac{d^2 Z}{dz^2} = -\frac{k^2}{a^2}. \tag{12.2.142}$$

Only a negative separation constant yields nontrivial solutions in the radial direction. In that case, we have that

$$\frac{1}{r}\frac{d}{dr}\left(r\frac{dR}{dr}\right) + \frac{k^2}{a^2}R = 0. \tag{12.2.143}$$

The solutions of Equation 12.2.143 are the Bessel functions $J_0(kr/a)$ and $Y_0(kr/a)$. Because $Y_0(kr/a)$ becomes infinite at $r = 0$, the only permissible solution is $J_0(kr/a)$. The condition that $u(a, z) = R(a)Z(z) = 0$ forces us to choose values of k such that $J_0(k) = 0$. Therefore, the solution in the radial direction is $J_0(k_n r/a)$, where k_n is the nth root of $J_0(k) = 0$.

In the z direction,

$$\frac{d^2 Z_n}{dz^2} + \frac{k_n^2}{a^2}Z_n = 0. \tag{12.2.144}$$

The general solution to Equation 12.2.144 is

$$Z_n(z) = A_n \sinh\left(\frac{k_n z}{a}\right) + B_n \cosh\left(\frac{k_n z}{a}\right). \tag{12.2.145}$$

---

[20] For static fields, $\nabla \times \mathbf{E} = \mathbf{0}$, where $\mathbf{E}$ is the electric force. From Section 4.4, we can introduce a potential φ such that $\mathbf{E} = \nabla\varphi$. From Gauss' law, $\nabla \cdot \mathbf{E} = \nabla^2\varphi = 0$.

[21] Wang and Liu [Wang, M.-L., and B.-L. Liu, 1995: Solution of Laplace equation by the method of separation of variables. *J. Chinese Inst. Eng.*, **18**, 731–739] have written a review article on the solutions to Equation 12.2.143 based upon which order the boundary conditions are satisfied.

Because $u(r, 0) = R(r)Z(0) = 0$ and $\cosh(0) = 1$, B_n must equal zero. Therefore, the general product solution is

$$u(r, z) = \sum_{n=1}^{\infty} A_n J_0\left(\frac{k_n r}{a}\right) \sinh\left(\frac{k_n z}{a}\right). \qquad (12.2.146)$$

The condition that $u(r, L) = V$ determines the arbitrary constant A_n. Along $z = L$,

$$u(r, L) = V = \sum_{n=1}^{\infty} A_n J_0\left(\frac{k_n r}{a}\right) \sinh\left(\frac{k_n L}{a}\right), \qquad (12.2.147)$$

where

$$\sinh\left(\frac{k_n L}{a}\right) A_n = \frac{2V}{a^2 J_1^2(k_n)} \int_0^L r\, J_0\left(\frac{k_n r}{a}\right) dr \qquad (12.2.148)$$

from Equation 12.2.63 and Equation 12.2.70. Since

$$\sinh\left(\frac{k_n L}{a}\right) A_n = \frac{2V}{k_n^2 J_1^2(k_n)} \left(\frac{k_n r}{a}\right) J_1\left(\frac{k_n r}{a}\right)\Bigg|_0^a = \frac{2V}{k_n J_1(k_n)}, \qquad (12.2.149)$$

the solution is then

$$u(r, z) = 2V \sum_{n=1}^{\infty} \frac{J_0(k_n r/a)}{k_n J_1(k_n)} \frac{\sinh(k_n z/a)}{\sinh(k_n L/a)}. \qquad (12.2.150)$$

Figure 12.2.14 illustrates Equation 12.2.150 for the case when $L = a$ where we included the first 20 terms of the series. It was created using the MATLAB script:

```
clear
L_over_a = 1; M = 20; dr = 0.02; dz = 0.02;
% load in zeros of J_0
zero( 1) =  2.40482; zero( 2) =  5.52007; zero( 3) =  8.65372;
zero( 4) = 11.79153; zero( 5) = 14.93091; zero( 6) = 18.07106;
zero( 7) = 21.21164; zero( 8) = 24.35247; zero( 9) = 27.49347;
zero(10) = 30.63461; zero(11) = 33.77582; zero(12) = 36.91710;
zero(13) = 40.05843; zero(14) = 43.19979; zero(15) = 46.34119;
zero(16) = 49.48261; zero(17) = 52.62405; zero(18) = 55.76551;
zero(19) = 58.90698; zero(20) = 62.04847;
% compute Fourier coefficients
for m = 1:M
  a(m) = 2/(zero(m)*besselj(1,zero(m))*sinh(L_over_a*zero(m)));
end
% compute grid and initialize solution
R_over_a = [0:dr:1]; Z_over_a = [0:dz:1];
u = zeros(length(Z_over_a),length(R_over_a));
RR_over_a = repmat(R_over_a,[length(Z_over_a) 1]);
ZZ_over_a = repmat(Z_over_a',[1 length(R_over_a)]);
% compute solution from Equation 12.2.150
for m = 1:M
  u=u+a(m).*besselj(0,zero(m)*RR_over_a) .* sinh(zero(m)*ZZ_over_a);
end
surf(RR_over_a,ZZ_over_a,u)
```

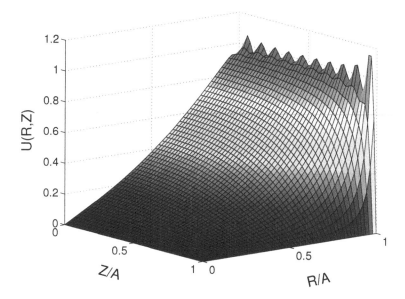

Figure 12.2.14: The steady-state potential (divided by V) within a cylinder of equal radius and height a when the top has the potential V while the lateral side and bottom are at potential 0.

```
xlabel('R/A','Fontsize',20); ylabel('Z/A','Fontsize',20)
zlabel('U(R,Z)','Fontsize',20)
```

Of particular interest are the ripples along the line $z = L$. Along that line, the solution must jump from V to 0 at $r = a$. For that reason our solution suffers from Gibbs phenomena along this boundary. As we move away from that region, the electrostatic potential varies smoothly. □

• **Example 12.2.14: Electrostatic potential inside a closed cylinder of finite length**

Let us now consider a similar, but slightly different, version of Example 12.2.13, where the ends are held at zero potential while the lateral side has the value V. Once again, the governing equation is Equation 12.2.141 with the boundary conditions $u(r,0) = u(r,L) = 0$ and $u(a,z) = V$.

Separation of variables yields

$$\frac{1}{rR}\frac{d}{dr}\left(r\frac{dR}{dr}\right) = -\frac{1}{Z}\frac{d^2Z}{dz^2} = \frac{k^2}{L^2} \qquad (12.2.151)$$

with $Z(0) = Z(L) = 0$. We chose a positive separation constant because a negative constant would give hyperbolic functions in z that cannot satisfy the boundary conditions. A separation constant of zero would give a straight line for $Z(z)$. Applying the boundary conditions gives a trivial solution. Consequently, the only solution in the z direction that satisfies the boundary conditions is $Z_n(z) = \sin(n\pi z/L)$.

In the radial direction, the differential equation is

$$\frac{1}{r}\frac{d}{dr}\left(r\frac{dR_n}{dr}\right) - \frac{n^2\pi^2}{L^2}R_n = 0. \qquad (12.2.152)$$

As we showed earlier, the general solution is

$$R_n(r) = A_n I_0\left(\frac{n\pi r}{L}\right) + B_n K_0\left(\frac{n\pi r}{L}\right), \tag{12.2.153}$$

where $I_0(\cdot)$ and $K_0(\cdot)$ are modified Bessel functions of the first and second kind, respectively, of order zero. Because $K_0(x)$ behaves as $-\ln(x)$ as $x \to 0$, we must discard it and our solution in the radial direction becomes $R_n(r) = A_n I_0(n\pi r/L)$. Hence, the product solution is

$$u_n(r,z) = A_n I_0\left(\frac{n\pi r}{L}\right) \sin\left(\frac{n\pi z}{L}\right), \tag{12.2.154}$$

and the general solution is a sum of these particular solutions, namely

$$u(r,z) = \sum_{n=1}^{\infty} A_n I_0\left(\frac{n\pi r}{L}\right) \sin\left(\frac{n\pi z}{L}\right). \tag{12.2.155}$$

Finally, we use the boundary conditions that $u(a,z) = V$ to compute A_n. This condition gives

$$u(a,z) = V = \sum_{n=1}^{\infty} A_n I_0\left(\frac{n\pi a}{L}\right) \sin\left(\frac{n\pi z}{L}\right), \tag{12.2.156}$$

so that

$$I_0\left(\frac{n\pi a}{L}\right) A_n = \frac{2}{L} \int_0^L V \sin\left(\frac{n\pi z}{L}\right) dz = \frac{2V[1 - (-1)^n]}{n\pi}. \tag{12.2.157}$$

Therefore, the final answer is

$$u(r,z) = \frac{4V}{\pi} \sum_{m=1}^{\infty} \frac{I_0[(2m-1)\pi r/L] \sin[(2m-1)\pi z/L]}{(2m-1) I_0[(2m-1)\pi a/L]}. \tag{12.2.158}$$

Figure 12.2.15 illustrates the solution, Equation 12.2.158, for the case when $L = a$. It was created using the MATLAB script:

```
clear
a_over_L = 1; M = 200; dr = 0.02; dz = 0.02;
% compute grid and initialize solution
R_over_L = [0:dr:1]; Z_over_L = [0:dz:1];
u = zeros(length(Z_over_L),length(R_over_L));
RR_over_L = repmat(R_over_L,[length(Z_over_L) 1]);
ZZ_over_L = repmat(Z_over_L',[1 length(R_over_L)]);

for m = 1:M
  temp = (2*m-1)*pi; prod1 = temp*a_over_L;
% compute modified Bessel functions in Equation 12.2.158
  for j = 1:length(Z_over_L); for i = 1:length(R_over_L);
    prod2 = temp*RR_over_L(i,j);
    if prod2 - prod1 > -10
      if prod2 < 20
        ratio(i,j) = besseli(0,prod2) / besseli(0,prod1);
      else
% for large values of prod, use asymptotic expansion
%          for modified bessel function
```

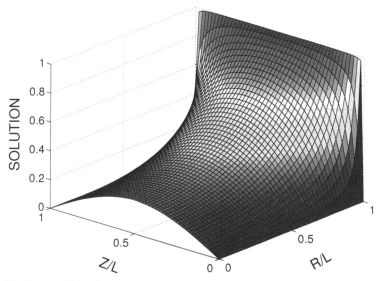

Figure 12.2.15: Potential (divided by V) within a conducting cylinder when the top and bottom have a potential 0 while the lateral side has a potential V.

```
      ratio(i,j) = sqrt(prod1/prod2) * exp(prod2-prod1); end;
   else
      ratio(i,j) = 0; end
   end; end;
% compute solution from Equation 12.2.158
   u = u + (4/temp) * ratio .* sin(temp*ZZ_over_L);
end
surf(RR_over_L,ZZ_over_L,u)
xlabel('R/L','Fontsize',20); ylabel('Z/L','Fontsize',20)
zlabel('SOLUTION','Fontsize',20)
```

Once again, there is a convergence of equipotentials at the corners along the right side. If we had plotted more contours, we would have observed Gibbs phenomena in the solution along the top and bottom of the cylinder. □

• **Example 12.2.15: Electrostatic potential inside a closed cylinder of semi-infinite length**

In the previous examples, the domain was always of finite extent. Assuming axial symmetry, let us now solve Laplace's equation

$$\frac{1}{r}\frac{\partial}{\partial r}\left(r\frac{\partial u}{\partial r}\right) + \frac{\partial^2 u}{\partial z^2} = 0, \quad 0 \le r < \infty, \quad 0 < z < \infty, \quad (\mathbf{12.2.159})$$

in the half-plane $z > 0$ subject to the boundary conditions

$$\lim_{z\to\infty}|u(r,z)| < \infty, \quad u(r,0) = \begin{cases} u_0, & r < a, \\ 0, & r > a, \end{cases} \quad (\mathbf{12.2.160})$$

$$\lim_{r\to 0}|u(r,z)| < \infty, \quad \text{and} \quad \lim_{r\to\infty}|u(r,z)| < \infty. \quad (\mathbf{12.2.161})$$

This problem gives the steady-state temperature distribution in the half-space $z > 0$ where the temperature on the bounding plane $z = 0$ equals u_0 within a circle of radius a and equals 0 outside of the circle.

As before, we begin by assuming the product solution $u(r, z) = R(r)Z(z)$ and separate the variables. Again, the separation constant may be positive, negative, or zero. Turning to the positive separation constant first, we have that

$$\frac{R''}{R} + \frac{1}{r}\frac{R'}{R} = -\frac{Z''}{Z} = m^2. \tag{12.2.162}$$

Focusing on the R equation,

$$\frac{R''}{R} + \frac{1}{r}\frac{R'}{R} - m^2 = 0, \quad \text{or} \quad r^2 R'' + rR - m^2 r^2 R = 0. \tag{12.2.163}$$

The solution to Equation 12.2.163 is

$$R(r) = A_1 I_0(mr) + A_2 K_0(mr), \tag{12.2.164}$$

where $I_0(\cdot)$ and $K_0(\cdot)$ denote modified Bessel functions of order zero and the first and second kind, respectively. Because $u(r, z)$, and hence $R(r)$, must be bounded as $r \to 0$, $A_2 = 0$. Similarly, since $u(r, z)$ must also be bounded as $r \to \infty$, $A_1 = 0$ because $\lim_{r \to \infty} I_0(mr) \to \infty$. Thus, there is only a trivial solution for a positive separation constant.

We next try the case when the separation constant equals 0. This yields

$$\frac{R''}{R} + \frac{1}{r}\frac{R'}{R} = 0, \quad \text{or} \quad r^2 R'' + rR = 0. \tag{12.2.165}$$

The solution here is

$$R(r) = A_1 + A_2 \ln(r). \tag{12.2.166}$$

Again, boundedness as $r \to 0$ requires that $A_2 = 0$. What about A_1? Clearly, for any arbitrary value of z, the amount of internal energy must be finite. This corresponds to

$$\int_0^\infty |u(r, z)| \, dr < \infty \quad \text{or} \quad \int_0^\infty |R(r)| \, dr < \infty \tag{12.2.167}$$

and $A_1 = 0$. The choice of the zero separation constant yields a trivial solution.

Finally, when the separation constant equals $-k^2$, the equations for $R(r)$ and $Z(z)$ are

$$r^2 R'' + rR + k^2 r^2 R = 0, \quad \text{and} \quad Z'' - k^2 Z = 0, \tag{12.2.168}$$

respectively. Solving for $R(r)$ first, we have that

$$R(r) = A_1 J_0(kr) + A_2 Y_0(kr), \tag{12.2.169}$$

where $J_0(\cdot)$ and $Y_0(\cdot)$ denote Bessel functions of order zero and the first and second kind, respectively. The requirement that $u(r, z)$, and hence $R(r)$, is bounded as $r \to 0$ forces us to take $A_2 = 0$, leaving $R(r) = A_1 J_0(kr)$. From the equation for $Z(z)$, we conclude that

$$Z(z) = B_1 e^{kz} + B_2 e^{-kz}. \tag{12.2.170}$$

Since $u(r, z)$, and hence $Z(z)$, must be bounded as $z \to \infty$, it follows that $B_1 = 0$, leaving $Z(z) = B_2 e^{-kz}$.

Presently our analysis follows closely those for a finite domain. However, we have satisfied all of the boundary conditions and yet there is still *no* restriction on k. Hence, we conclude that k is completely arbitrary and any product solution

$$u_k(r, z) = A_1 B_2 \, J_0(kr) \, e^{-kz} \qquad (\mathbf{12.2.171})$$

is a solution to our partial differential equation and satisfies the boundary conditions. From the principle of linear superposition, the most general solution equals the sum of *all* of the possible solutions, or

$$u(r, z) = \int_0^\infty A(k) \, k \, J_0(kr) \, e^{-kz} \, dk, \qquad (\mathbf{12.2.172})$$

where we have written the arbitrary constant $A_1 B_2$ as $A(k)k$. Our final task remains to compute $A(k)$.

Before we can find $A(k)$, we must derive an intermediate result. If we define our Fourier transform in an appropriate manner, we can write the two-dimensional Fourier transform pair as

$$f(x, y) = \frac{1}{2\pi} \int_{-\infty}^\infty \int_{-\infty}^\infty F(k, \ell) \, e^{i(kx+\ell y)} \, dk \, d\ell, \qquad (\mathbf{12.2.173})$$

where

$$F(k, \ell) = \frac{1}{2\pi} \int_{-\infty}^\infty \int_{-\infty}^\infty f(x, y) \, e^{-i(kx+\ell y)} \, dx \, dy. \qquad (\mathbf{12.2.174})$$

Consider now the special case where $f(x, y)$ is only a function of $r = \sqrt{x^2 + y^2}$, so that $f(x, y) = g(r)$. Then, changing to polar coordinates through the substitution $x = r \cos(\theta)$, $y = r \sin(\theta)$, $k = \rho \cos(\varphi)$, and $\ell = \rho \sin(\varphi)$, we have that

$$kx + \ell y = r\rho[\cos(\theta)\cos(\varphi) + \sin(\theta)\sin(\varphi)] = r\rho\cos(\theta - \varphi), \qquad (\mathbf{12.2.175})$$

and

$$dA = dx \, dy = r \, dr \, d\theta. \qquad (\mathbf{12.2.176})$$

Therefore, the integral in Equation 12.2.174 becomes

$$F(k, \ell) = \frac{1}{2\pi} \int_0^\infty \int_0^{2\pi} g(r) \, e^{-ir\rho\cos(\theta-\varphi)} r \, dr \, d\theta \qquad (\mathbf{12.2.177})$$

$$= \frac{1}{2\pi} \int_0^\infty r \, g(r) \left[\int_0^{2\pi} e^{-ir\rho\cos(\theta-\varphi)} \, d\theta \right] dr. \qquad (\mathbf{12.2.178})$$

If we introduce $\lambda = \theta - \varphi$, the integral

$$\int_0^{2\pi} e^{-ir\rho\cos(\theta-\varphi)} \, d\theta = \int_{-\varphi}^{2\pi-\varphi} e^{-ir\rho\cos(\lambda)} \, d\lambda = \int_0^{2\pi} e^{-ir\rho\cos(\lambda)} \, d\lambda = 2\pi J_0(\rho r).$$

$$(\mathbf{12.2.179})$$

The third integral in Equation 12.2.179 is equivalent to the second integral because the integral of a periodic function over one full period is the same regardless of where the integration begins. The final result follows from the integral definition of the Bessel function.[22] Therefore,

$$F(k, \ell) = \int_0^\infty r \, g(r) \, J_0(\rho r) \, dr. \qquad (\mathbf{12.2.180})$$

---

[22] Watson, G. N., 1966: *A Treatise on the Theory of Bessel Functions*. Cambridge University Press, Section 2.2, Equation 5.

Finally, because Equation 12.2.180 is clearly a function of $\rho = \sqrt{k^2 + \ell^2}$, $F(k, \ell) = G(\rho)$ and

$$G(\rho) = \int_0^\infty r \, g(r) \, J_0(\rho r) \, dr. \tag{12.2.181}$$

Conversely, if we begin with Equation 12.2.173, make the same substitution, and integrate over the $k\ell$-plane, we have that

$$f(x, y) = g(r) = \frac{1}{2\pi} \int_0^\infty \int_0^{2\pi} F(k, \ell) \, e^{ir\rho \cos(\theta - \varphi)} \rho \, d\rho \, d\varphi \tag{12.2.182}$$

$$= \frac{1}{2\pi} \int_0^\infty \rho \, G(\rho) \left[\int_0^{2\pi} e^{ir\rho \cos(\theta - \varphi)} \, d\varphi \right] d\rho \tag{12.2.183}$$

$$= \int_0^\infty \rho \, G(\rho) \, J_0(\rho r) \, d\rho. \tag{12.2.184}$$

Thus, we obtain the result that if $\int_0^\infty |F(r)| \, dr$ exists, then

$$g(r) = \int_0^\infty \rho \, G(\rho) \, J_0(\rho r) \, d\rho, \tag{12.2.185}$$

where

$$G(\rho) = \int_0^\infty r \, g(r) \, J_0(\rho r) \, dr. \tag{12.2.186}$$

Taken together, Equation 12.2.185 and Equation 12.2.186 constitute the *Hankel transform pair for Bessel function of order 0*. The function $G(\rho)$ is called the Hankel transform of $g(r)$.

Why did we introduce Hankel transforms? First, setting $z = 0$ in Equation 12.2.172, we find that

$$u(r, 0) = \int_0^\infty A(k) \, k \, J_0(kr) \, dk. \tag{12.2.187}$$

If we now compare Equation 12.2.187 with Equation 12.2.185, we recognize that $A(k)$ is the Hankel transform of $u(r, 0)$. Therefore,

$$A(k) = \int_0^\infty r \, u(r, 0) \, J_0(kr) \, dr = u_0 \int_0^a r \, J_0(kr) \, dr = \frac{u_0}{k} r \, J_1(kr) \Big|_0^a = \frac{au_0}{k} J_1(ka). \tag{12.2.188}$$

For this reason, the complete solution is

$$u(r, z) = au_0 \int_0^\infty J_1(ka) \, J_0(kr) \, e^{-kz} \, dk. \tag{12.2.189}$$

Figure 12.2.16 illustrates Equation 12.2.189.

Problems

1. Show from the series solution that

$$\frac{d}{dx} \left[J_0(kx) \right] = -k J_1(kx).$$

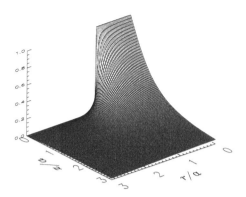

Figure 12.2.16: The axisymmetric potential $u(r,z)/u_0$ in the half-space $z > 0$ when $u(r,0) = u_0$ if $r < a$ and $u(r,0) = 0$ if $r > a$.

From the recurrence formulas, show the following relations:

2. $2J_0''(x) = J_2(x) - J_0(x)$ 3. $J_2(x) = J_0''(x) - J_0'(x)/x$

4.
$$J_0'''(x) = \frac{J_0(x)}{x} + \left(\frac{2}{x^2} - 1 \right) J_0'(x)$$

5.
$$\frac{J_2(x)}{J_1(x)} = \frac{1}{x} - \frac{J_0''(x)}{J_0'(x)} = \frac{2}{x} - \frac{J_0(x)}{J_1(x)} = \frac{2}{x} + \frac{J_0(x)}{J_0'(x)}$$

6.
$$J_4(x) = \left(\frac{48}{x^3} - \frac{8}{x} \right) J_1(x) - \left(\frac{24}{x^2} - 1 \right) J_0(x)$$

7.
$$J_{n+2}(x) = \left[2n + 1 - \frac{2n(n^2 - 1)}{x^2} \right] J_n(x) + 2(n+1)J_n''(x)$$

8.
$$J_3(x) = \left(\frac{8}{x^2} - 1 \right) J_1(x) - \frac{4}{x}J_0(x)$$

9.
$$4J_n''(x) = J_{n-2}(x) - 2J_n(x) + J_{n+2}(x)$$

10. Show that the maximum and minimum values of $J_n(x)$ occur when

$$x = \frac{nJ_n(x)}{J_{n+1}(x)}, \quad x = \frac{nJ_n(x)}{J_{n-1}(x)}, \quad \text{and} \quad J_{n-1}(x) = J_{n+1}(x).$$

Show that

11.
$$\frac{d}{dx}\left[x^2 J_3(2x)\right] = -x J_3(2x) + 2x^2 J_2(2x)$$

12.
$$\frac{d}{dx}\left[x J_0(x^2)\right] = J_0(x^2) - 2x^2 J_1(x^2)$$

13.
$$\int x^3 J_2(3x)\,dx = \tfrac{1}{3}x^3 J_3(3x) + C$$

14.
$$\int x^{-2} J_3(2x)\,dx = -\tfrac{1}{2}x^{-2} J_2(2x) + C$$

15.
$$\int x\ln(x) J_0(x)\,dx = J_0(x) + x\ln(x) J_1(x) + C$$

16.
$$\int_0^a x J_0(kx)\,dx = \frac{a^2 J_1(ka)}{ka}$$

17.
$$\int_0^1 x(1-x^2) J_0(kx)\,dx = \frac{4}{k^3} J_1(k) - \frac{2}{k^2} J_0(k)$$

18.
$$\int_0^1 x^3 J_0(kx)\,dx = \frac{k^2-4}{k^3} J_1(k) + \frac{2}{k^2} J_0(k)$$

19. Show that
$$1 = 2\sum_{k=1}^\infty \frac{J_0(\mu_k x)}{\mu_k J_1(\mu_k)}, \qquad 0 \le x < 1,$$

where μ_k is the kth positive root of $J_0(\mu) = 0$. Then use MATLAB to illustrate various partial sums of the Fourier-Bessel series.

20. Show that
$$\frac{1-x^2}{8} = \sum_{k=1}^\infty \frac{J_0(\mu_k x)}{\mu_k^3 J_1(\mu_k)}, \qquad 0 \le x \le 1,$$

where μ_k is the kth positive root of $J_0(\mu) = 0$. Then use MATLAB to illustrate various partial sums of the Fourier-Bessel series.

21. Show that

$$4x - x^3 = -16 \sum_{k=1}^{\infty} \frac{J_1(\mu_k x)}{\mu_k^3 J_0(2\mu_k)}, \qquad 0 \le x \le 2,$$

where μ_k is the kth positive root of $J_1(2\mu) = 0$. Then use MATLAB to illustrate various partial sums of the Fourier-Bessel series.

22. Show that

$$x^3 = 2 \sum_{k=1}^{\infty} \frac{(\mu_k^2 - 8) J_1(\mu_k x)}{\mu_k^3 J_2(\mu_k)}, \qquad 0 \le x \le 1,$$

where μ_k is the kth positive root of $J_1(\mu) = 0$. Then use MATLAB to illustrate various partial sums of the Fourier-Bessel series.

23. Show that

$$x = 2 \sum_{k=1}^{\infty} \frac{\mu_k J_2(\mu_k) J_1(\mu_k x)}{(\mu_k^2 - 1) J_1^2(\mu_k)}, \qquad 0 \le x \le 1,$$

where μ_k is the kth positive root of $J_1'(\mu) = 0$. Then use MATLAB to illustrate various partial sums of the Fourier-Bessel series.

24. Show that

$$1 - x^4 = 32 \sum_{k=1}^{\infty} \frac{(\mu_k^2 - 4) J_0(\mu_k x)}{\mu_k^5 J_1(\mu_k)}, \qquad 0 \le x \le 1,$$

where μ_k is the kth positive root of $J_0(\mu) = 0$. Then use MATLAB to illustrate various partial sums of the Fourier-Bessel series.

25. Show that

$$1 = 2\alpha L \sum_{k=1}^{\infty} \frac{J_0(\mu_k x/L)}{(\mu_k^2 + \alpha^2 L^2) J_0(\mu_k)}, \qquad 0 \le x \le L,$$

where μ_k is the kth positive root of $\mu J_1(\mu) = \alpha L J_0(\mu)$. Then use MATLAB to illustrate various partial sums of the Fourier-Bessel series.

26. Using the relationship[23]

$$\int_0^a J_\nu(\alpha r) J_\nu(\beta r)\, r\, dr = \frac{a\beta J_\nu(\alpha a) J_\nu'(\beta a) - a\alpha J_\nu(\beta a) J_\nu'(\alpha a)}{\alpha^2 - \beta^2},$$

show that

$$\frac{J_0(bx) - J_0(ba)}{J_0(ba)} = \frac{2b^2}{a} \sum_{k=1}^{\infty} \frac{J_0(\mu_k x)}{\mu_k (\mu_k^2 - b^2) J_1(\mu_k a)}, \qquad 0 \le x \le a,$$

where μ_k is the kth positive root of $J_0(\mu a) = 0$ and b is a constant.

---

[23] Watson, op. cit., Section 5.11, Equation 8.

27. Given the definite integral[24]

$$\int_0^1 \frac{x\, J_0(bx)}{\sqrt{1-x^2}}\, dx = \frac{\sin(b)}{b}, \qquad 0 < b,$$

show that

$$\frac{H(t-x)}{\sqrt{t^2-x^2}} = 2\sum_{k=1}^{\infty} \frac{\sin(\mu_k t)\, J_0(\mu_k x)}{\mu_k J_1^2(\mu_k)}, \qquad 0 < x < 1, \quad 0 < t \le 1,$$

where μ_k is the kth positive root of $J_0(\mu) = 0$ and $H(\cdot)$ is Heaviside's step function.

28. Using the same definite integral from the previous problem, show[25] that

$$\frac{H(a-x)}{\sqrt{a^2-x^2}} = \frac{2}{b}\sum_{n=1}^{\infty} \frac{\sin(\mu_n a/b)\, J_0(\mu_n x/b)}{\mu_n J_0^2(\mu_n)}, \qquad 0 \le x < b,$$

where $a < b$, μ_n is the nth positive root of $J_0'(\mu) = 0$, and $H(\cdot)$ is Heaviside's step function.

29. Given the definite integral[26]

$$\int_0^a \cos(cx)\, J_0\!\left(b\sqrt{a^2-x^2}\right) dx = \frac{\sin\!\left(a\sqrt{b^2+c^2}\,\right)}{\sqrt{b^2+c^2}}, \qquad 0 < b,$$

show that

$$\frac{\cosh\!\left(b\sqrt{t^2-x^2}\right)}{\sqrt{t^2-x^2}}\, H(t-x) = \frac{2}{a^2}\sum_{k=1}^{\infty} \frac{\sin\!\left(t\sqrt{\mu_k^2-b^2}\,\right) J_0(\mu_k x)}{\sqrt{\mu_k^2-b^2}\, J_1^2(\mu_k a)},$$

where $0 < x < a$, μ_k is the kth positive root of $J_0(\mu a) = 0$, $H(\cdot)$ is Heaviside's step function, and b is a constant.

30. Using the integral definition of the Bessel function[27] for $J_1(z)$:

$$J_1(z) = \frac{2}{\pi}\int_0^1 \frac{t\, \sin(zt)}{\sqrt{1-t^2}}\, dt, \qquad 0 < z,$$

show that

$$\frac{x}{t\sqrt{t^2-x^2}} H(t-x) = \frac{\pi}{L}\sum_{n=1}^{\infty} J_1\!\left(\frac{n\pi t}{L}\right) \sin\!\left(\frac{n\pi x}{L}\right), \qquad 0 \le x < L,$$

---

[24] Gradshteyn, I. S., and I. M. Ryzhik, 1965: *Table of Integrals, Series, and Products.* Academic Press, Section 6.567, Formula 1 with $\nu = 0$ and $\mu = -1/2$.

[25] See Wei, X. X., 2000: Finite solid circular cylinders subjected to arbitrary surface load. Part II– Application to double-punch test. *Int. J. Solids Struct.*, **37**, 5733–5744.

[26] Gradshteyn and Ryzhik, op. cit., Section 6.677, Formula 6.

[27] Gradshteyn and Ryzhik, Ibid., Section 3.753, Formula 5.

where $H(\cdot)$ is Heaviside's step function. Hint: Treat this as a Fourier half-range sine expansion.

31. Show that

$$\delta(x-b) = \frac{2b}{a^2} \sum_{k=1}^{\infty} \frac{J_0(\mu_k b/a) J_0(\mu_k x/a)}{J_1^2(\mu_k)}, \qquad 0 \le x, b < a,$$

where μ_k is the kth positive root of $J_0(\mu) = 0$ and $\delta(\cdot)$ is the Dirac delta function.

32. Show that

$$\frac{\delta(x)}{2\pi x} = \frac{1}{\pi a^2} \sum_{k=1}^{\infty} \frac{J_0(\mu_k x/a)}{J_1^2(\mu_k)}, \qquad 0 \le x < a,$$

where μ_k is the kth positive root of $J_0(\mu) = 0$ and $\delta(\cdot)$ is the Dirac delta function.

Separation of Variables Solution to the Wave Equation

33. Solve[28] the wave equation

$$\frac{\partial^2 u}{\partial t^2} = \frac{\partial}{\partial x}\left(x \frac{\partial u}{\partial x}\right), \qquad 0 \le x < 1, \qquad 0 < t,$$

subject to the boundary conditions $\lim_{x \to 0} |u(x,t)| < \infty$, $u(1,t) = 0$, $0 < t$, and the initial conditions

$$u(x,0) = 0, \quad 0 \le x \le 1, \qquad \frac{\partial u(x,0)}{\partial t} = \begin{cases} 1, & 0 \le x < a, \\ 0, & a < x \le 1, \end{cases}$$

where $a < 1$. Hint: Use the substitution $4x = r^2$.

Separation of Variables Solution to the Heat Equation

34. Solve[29] the heat equation in cylindrical coordinates

$$\frac{\partial u}{\partial t} = a^2 \left(\frac{\partial^2 u}{\partial r^2} + \frac{1}{r}\frac{\partial u}{\partial r}\right), \qquad 0 \le r < b, \quad 0 < t,$$

subject to the boundary conditions $\lim_{r \to 0} |u(r,t)| < \infty$, and $u(b,t) = u_0$, $0 < t$, and the initial condition $u(r,0) = 0$, $0 \le r < b$.

35. Solve the heat equation in cylindrical coordinates

$$\frac{\partial u}{\partial t} = \frac{a^2}{r} \frac{\partial}{\partial r}\left(r \frac{\partial u}{\partial r}\right), \qquad 0 \le r < b, \quad 0 < t,$$

subject to the boundary conditions $\lim_{r \to 0} |u(r,t)| < \infty$, and $u(b,t) = \theta$, $0 < t$, and the initial condition $u(r,0) = 1$, $0 \le r < b$.

---

[28] Solved in a slightly different manner by Bailey, H., 2000: Motions of a hanging chain after the free end is given an initial velocity. *Am. J. Phys.*, **68**, 764–767.

[29] See Destriau, G., 1946: Propagation des charges électriques sur les pellicules faiblement conductrices "problèm plan." *J. Phys. Radium*, **7**, 43–48.

36. Solve the heat equation in cylindrical coordinates

$$\frac{\partial u}{\partial t} = \frac{a^2}{r}\frac{\partial}{\partial r}\left(r\frac{\partial u}{\partial r}\right), \qquad 0 \le r < 1, \quad 0 < t,$$

subject to the boundary conditions $\lim_{r\to 0}|u(r,t)| < \infty$, and $u(1,t) = 0$, $0 < t$, and the initial condition

$$u(x,0) = \begin{cases} A, & 0 \le r < b, \\ B, & b < r < 1. \end{cases}$$

37. The equation[30]

$$\frac{\partial u}{\partial t} = \frac{G}{\rho} + \nu\left(\frac{\partial^2 u}{\partial r^2} + \frac{1}{r}\frac{\partial u}{\partial r}\right), \qquad 0 \le r < b, \quad 0 < t,$$

governs the velocity $u(r,t)$ of an incompressible fluid of density ρ and kinematic viscosity ν flowing in a long circular pipe of radius b with an imposed, constant pressure gradient $-G$. If the fluid is initially at rest, $u(r,0) = 0$, $0 \le r < b$, and there is no slip at the wall $u(b,t) = 0$, $0 < t$, find the velocity at any subsequent time and position. Hint: First find the steady-state solution $w(x)$ and then write $u(x,t) = w(x) + v(x,t)$, where $v(x,t)$ is the transient solution so that $u(x,t)$ satisfies the initial condition.

38. Solve the heat equation in cylindrical coordinates

$$\frac{\partial u}{\partial t} = \frac{a^2}{r}\frac{\partial}{\partial r}\left(r\frac{\partial u}{\partial r}\right), \qquad 0 \le r < b, \quad 0 < t,$$

subject to the boundary conditions $\lim_{r\to 0}|u(r,t)| < \infty$, and $u_r(b,t) = -h\,u(b,t)$, $0 < t$, and the initial condition $u(r,0) = b^2 - r^2$, $0 \le r < b$.

39. Solve[31] the heat equation in cylindrical coordinates

$$\frac{\partial u}{\partial t} = a^2\left(\frac{\partial^2 u}{\partial r^2} + \frac{1}{r}\frac{\partial u}{\partial r}\right) - \kappa u, \qquad 0 \le r < L, \quad 0 < t,$$

subject to the boundary conditions $\lim_{r\to 0}|u(r,t)| < \infty$, and $u_r(L,t) = -hu(L,t)$, $0 < t$, and the initial condition

$$u(r,0) = \begin{cases} 0, & 0 \le r < b, \\ T_0, & b < r \le L, \end{cases}$$

where $b < L$, and $0 < h, \kappa$. Hint: Introduce $u(r,t) = e^{-\kappa t}v(r,t)$.

40. Use separation of variables to solve[32] the partial differential equation

$$\frac{\partial u}{\partial t} = a^2\left(\frac{\partial^2 u}{\partial r^2} + \frac{1}{r}\frac{\partial u}{\partial r} - \frac{u}{r^2}\right), \qquad 0 \le r < 1, \quad 0 < t,$$

---

[30] See Szymanski, P., 1932: Quelques solutions exactes des équations de l'hydrodynamique du fluide visqueux dans le cas d'un tube cylindrique. *J. Math. Pures Appl., Ser. 9,* **11**, 67–107.

[31] Mack, W., M. Plöchl, and U. Gamer, 2000: Effects of a temperature cycle on an elastic-plastic shrink fit with solid inclusion. *Chinese J. Mech.,* **16**, 23–30.

[32] See Littlefield, D. L., 1991: Finite conductivity effects on the MHD instabilities in uniformly elongating plastic jets. *Phys. Fluids, Ser. A,* **3**, 1666–1673.

subject to the boundary conditions that $\lim_{r\to 0}|u(r,t)| < \infty$, $u(1,t) = K$, $0 < t$, and the initial condition that $u(r,0) = g(r)$, $0 < r < 1$.

Step 1: Setting $u(r,t) = Kr + v(r,t)$, show that the problem becomes

$$\frac{\partial v}{\partial t} = a^2\left(\frac{\partial^2 v}{\partial r^2} + \frac{1}{r}\frac{\partial v}{\partial r} - \frac{v}{r^2}\right), \qquad 0 \le r < 1, \quad 0 < t,$$

subject to the boundary conditions that $\lim_{r\to 0}|v(r,t)| < \infty$, $v(1,t) = 0$, $0 < t$, and the initial condition that $u(r,0) = g(r) - Kr$, $0 < r < 1$.

Step 2: Assuming that $v(r,t) = R(r)T(t)$, show that the problem reduces to the ordinary differential equations

$$R'' + \frac{1}{r}R' + \left(k^2 - \frac{1}{r^2}\right)R = 0,$$

and

$$T' + a^2 k^2 T = 0$$

with the boundary conditions $\lim_{r\to 0}|R(r)| < \infty$ and $R(1) = 0$, where k^2 is the separation constant.

Step 3: Solving the eigenvalue problem first, show that $R_n(r) = J_1(k_n r)$ where k_n denotes the nth root of $J_1(k) = 0$ and $n = 1, 2, 3, \ldots$.

Step 4: Show that $T_n(t) = A_n e^{-a^2 k_n^2 t}$ so that

$$v(r,t) = \sum_{n=1}^{\infty} A_n J_1(k_n r) e^{-a^2 k_n^2 t}.$$

Step 5: Using the initial condition, show that

$$v(r,0) = g(r) - Kr = \sum_{n=1}^{\infty} A_n J_1(k_n r).$$

Step 6: Evaluate A_n and show that it equals

$$A_n = \frac{2}{k_n J_0(k_n)}\left[K + \frac{k_n}{J_0(k_n)}\int_0^1 g(r)J_1(k_n r)\, r\, dr\right].$$

41. A thermometer measures temperature by the thermal expansion of a liquid (usually mercury or alcohol) stored in a bulb into a glass stem containing an empty cylindrical channel. Under normal conditions, temperature changes occur sufficiently slowly so that the temperature within the liquid is uniform. However, for rapid temperature changes (such as those that would occur during the rapid ascension of an airplane or meteorological balloon), significant errors could occur. In such situations the recorded temperature would lag behind the actual temperature because of the time needed for the heat to conduct in or out of the bulb. During his investigation of this question, McLeod[33] solved

$$\frac{\partial u}{\partial t} = a^2 \frac{1}{r}\frac{\partial}{\partial r}\left(r\frac{\partial u}{\partial r}\right), \qquad 0 \le r < b, \quad 0 < t,$$

---

[33] McLeod, A. R., 1919: On the lags of thermometers with spherical and cylindrical bulbs in a medium whose temperature is changing at a constant rate. *Philos. Mag., Ser. 6*, **37**, 134–144. See also Bromwich, T. J. I'A., 1919: Examples of operational methods in mathematical physics. *Philos. Mag., Ser. 6*, **37**, 407–419; McLeod, A. R., 1922: On the lags of thermometers. *Philos. Mag., Ser. 6*, **43**, 49–70.

subject to the boundary conditions $\lim_{r \to 0} |u(r,t)| < \infty$, and $u(b,t) = \varphi(t)$, $0 < t$, and the initial condition $u(r,0) = 0$, $0 \le r < b$. The analysis was as follows:

Step 1: First solve the heat conduction problem

$$\frac{\partial A}{\partial t} = \frac{a^2}{r} \frac{\partial}{\partial r} \left(r \frac{\partial A}{\partial r} \right), \qquad 0 \le r < b, \quad 0 < t,$$

subject to the boundary conditions $\lim_{r \to 0} |A(r,t)| < \infty$, and $A(b,t) = 1$, $0 < t$, and the initial condition $A(r,0) = 0$, $0 \le r < b$. Show that

$$A(r,t) = 1 - 2 \sum_{n=1}^{\infty} \frac{J_0(k_n r/b)}{k_n J_1(k_n)} e^{-a^2 k_n^2 t/b^2},$$

where $J_0(k_n) = 0$.

Step 2: Use Duhamel's theorem (see Section 9.4) and show that

$$u(r,t) = \frac{2a^2}{b^2} \sum_{n=1}^{\infty} \frac{k_n J_0(k_n r/b)}{J_1(k_n)} \int_0^t \varphi(\tau) e^{-a^2 k_n^2 (t-\tau)/b^2} \, d\tau.$$

Step 3: If $\varphi(t) = Gt$, show that

$$u(r,t) = 2G \sum_{n=1}^{\infty} \frac{J_0(k_n r/b)}{k_n J_1(k_n)} \left[t + \frac{b^2}{a^2 k_n^2} \left(e^{-a^2 k_n^2 t/b^2} - 1 \right) \right].$$

McLeod found that for a mercury thermometer of 10-cm length a lag of $0.01°$C would occur for a warming rate of $0.032°$C s^{-1} (a warming gradient of $1.9°$C per thousand feet and a descent of one thousand feet per minute). Although this is a very small number, when he included the surface conductance of the glass tube, the lag increased to $0.85°$C. Similar problems plague bimetal thermometers[34] and thermistors[35] used in radiosondes (meteorological sounding balloons).

Separation of Variables Solution to Laplace's Equation

42. Solve

$$\frac{\partial^2 u}{\partial r^2} + \frac{1}{r} \frac{\partial u}{\partial r} + \frac{\partial^2 u}{\partial z^2} = 0, \qquad 0 \le r < a, \quad -L < z < L,$$

with $\lim_{r \to 0} |u(r,z)| < \infty$, $u(a,z) = 0$, $-L < z < L$ and $u_z(r,-L) = u_z(r,L) = 1$, $0 < r < a$.

43. During their study of the role that diffusion plays in equalizing gas concentrations within that portion of the lung that is connected to terminal bronchioles, Chang et al.[36] solved Laplace's equation in cylindrical coordinates

$$\frac{\partial^2 u}{\partial r^2} + \frac{1}{r} \frac{\partial u}{\partial r} + \frac{\partial^2 u}{\partial z^2} = 0, \qquad 0 \le r < b, \quad -L < z < L,$$

---

[34] Mitra, H., and M. B. Datta, 1954: Lag coefficient of bimetal thermometer of chronometric radiosonde. *Indian J. Meteorol. Geophys.*, **5**, 257–261.

[35] Badgley, F. I., 1957: Response of radiosonde thermistors. *Rev. Sci. Instrum.*, **28**, 1079–1084.

[36] Chang, D. B., S. M. Lewis, and A. C. Young, 1976: A theoretical discussion of diffusion and convection in the lung. *Math. Biosci.*, **29**, 331–349.

subject to the boundary conditions that $\lim_{r \to 0} |u(r, z)| < \infty$, $u_r(b, z) = 0$, $-L < z < L$, and

$$\frac{\partial u(r, -L)}{\partial z} = \frac{\partial u(r, L)}{\partial z} = \begin{cases} A, & 0 \leq r < a, \\ 0, & a < r < b. \end{cases}$$

What should they have found?

44. Solve[37]

$$\frac{\partial^2 u}{\partial r^2} + \frac{1}{r}\frac{\partial u}{\partial r} + \frac{\partial^2 u}{\partial z^2} = 0, \qquad 0 \leq r < b, \quad 0 < z < L,$$

with the boundary conditions $\lim_{r \to 0} |u(r, z)| < \infty$, $u_r(b, z) = 0$, $0 \leq z \leq L$, and

$$\frac{\partial u(r, 0)}{\partial z} = \begin{cases} B, & 0 \leq r < a, \\ 0, & a < r < b, \end{cases} \qquad u(r, L) = A, \qquad 0 \leq r \leq b.$$

45. Solve

$$\frac{\partial^2 u}{\partial r^2} + \frac{1}{r}\frac{\partial u}{\partial r} + \frac{\partial^2 u}{\partial z^2} = 0, \qquad 0 \leq r < a, \quad 0 < z < h,$$

with the boundary conditions $\lim_{r \to 0} |u(r, z)| < \infty$, $u_r(a, z) = 0$, $0 < z < h$, and

$$u(r, h) = 0, \qquad 0 < r < a, \qquad \frac{\partial u(r, 0)}{\partial z} = \begin{cases} 1, & 0 \leq r < r_0, \\ 0, & r_0 < r < a. \end{cases}$$

46. Solve

$$\frac{\partial^2 u}{\partial r^2} + \frac{1}{r}\frac{\partial u}{\partial r} + \frac{\partial^2 u}{\partial z^2} = 0, \qquad 0 \leq r < 1, \quad 0 < z < d,$$

with the boundary conditions $\lim_{r \to 0} |u(r, z)| < \infty$, $u_r(1, z) = 0$, $0 < z < h$, and

$$\frac{\partial u(r, 0)}{\partial z} = 0, \quad 0 < r < a, \qquad u(r, d) = \begin{cases} -1, & 0 \leq r < a, \quad b < r < 1, \\ 1/(b^2 - a^2) - 1, & a < r < b. \end{cases}$$

47. Solve[38]

$$\frac{\partial^2 u}{\partial r^2} + \frac{1}{r}\frac{\partial u}{\partial r} - \frac{u}{r^2} + \frac{\partial^2 u}{\partial z^2} = 0, \qquad 0 \leq r < 1, \quad 0 < z < 1.$$

Take for the boundary conditions either (a) $\lim_{r \to 0} |u(r, z)| < \infty$, $u(1, z) = -1$, $0 < z < 1$, and $u_z(r, 0) = u(r, 1) = 0$, $0 < r < 1$; or (b) $\lim_{r \to 0} |u(r, z)| < \infty$, $u(1, z) = 0$, $0 < z < 1$, and $u_z(r, 0) = 0$, $u(r, 1) = r$, $0 < r < 1$.

48. Solve

$$\frac{\partial^2 u}{\partial r^2} + \frac{1}{r}\frac{\partial u}{\partial r} - \frac{u}{r^2} + \frac{\partial^2 u}{\partial z^2} = 0, \qquad 0 \leq r < a, \quad 0 < z < h,$$

with $\lim_{r \to 0} |u(r, z)| < \infty$, $u(a, z) = 0$, $0 < z < h$, and $u(r, 0) = 0$, $u_z(r, h) = Ar$, $0 \leq r < a$.

---

[37] See Keller, K. H., and T. R. Stein, 1967: A two-dimensional analysis of porous membrane transfer. *Math. Biosci.*, **1**, 421–437.

[38] See Muite, B. K., 2004: The flow in a cylindrical container with a rotating end wall at small but finite Reynolds number. *Phys. Fluids*, **16**, 3614–3626.

49. Solve

$$\frac{\partial^2 u}{\partial r^2} + \frac{1}{r}\frac{\partial u}{\partial r} - \frac{u}{r^2} + \frac{\partial^2 u}{\partial z^2} = 0, \qquad 0 \le r < a, \quad 0 < z < 1,$$

with $\lim_{r \to 0} |u(r,z)| < \infty$, $u(a,z) = z$, $0 < z < 1$, and $u(r,0) = u(r,1) = 0$, $0 \le r < a$.

50. Solve

$$\frac{\partial^2 u}{\partial r^2} + \frac{1}{r}\frac{\partial u}{\partial r} - \frac{u}{r^2} + \frac{\partial^2 u}{\partial z^2} = 0, \qquad 0 \le r < a, \quad 0 < z < h,$$

with $\lim_{r \to 0} |u(r,z)| < \infty$, $u_r(a,z) = 0$, $0 < z < h$, and $u(r,0) = 0$, $u_z(r,h) = r$, $0 \le r < a$.

51. Solve

$$\frac{\partial^2 u}{\partial r^2} + \frac{1}{r}\frac{\partial u}{\partial r} - \frac{u}{r^2} + \frac{\partial^2 u}{\partial z^2} = 0, \qquad 0 \le r < 1, \quad -a < z < a,$$

with the boundary conditions $\lim_{r \to 0} |u(r,z)| < \infty$, $u_r(1,z) = u(1,z)$, $-a < z < a$, and $-u_z(r,-a) = u_z(r,a) = r$, $0 \le r < 1$.

52. Solve[39]

$$\frac{\partial^2 u}{\partial r^2} + \frac{1}{r}\frac{\partial u}{\partial r} + \frac{\partial^2 u}{\partial z^2} - u = 0, \qquad 0 \le r < 1, \quad 0 < z < L,$$

subject to the boundary conditions that $\lim_{r \to 0} |u(r,z)| < \infty$, $u_r(1,z) = -h\,u(1,z)$, $0 < z < L$, and $u(r,0) = u_0$, $u(r,L) = 0$, $0 \le r < 1$, using the Fourier-Bessel series

$$u(r,z) = \sum_{n=1}^{\infty} A_n Z_n(z) J_0(k_n r),$$

where k_n is the nth root of $k\,J_0'(k) + h\,J_0(k) = h\,J_0(k) - k\,J_1(k) = 0$.

53. Solve[40] Laplace's equation in cylindrical coordinates

$$\frac{\partial^2 u}{\partial r^2} + \frac{1}{r}\frac{\partial u}{\partial r} + \frac{\partial^2 u}{\partial z^2} = 0, \qquad 0 \le r < a, \quad 0 < z < L,$$

subject to the boundary conditions that $\lim_{r \to 0} |u(r,z)| < \infty$, $-D u_r(a,z) = K u(a,z)$, $0 < z < L$, and $u(r,0) = u_0$, $u_z(r,L) = 0$, $0 \le r < a$.

54. Solve[41] the partial differential equation

$$\frac{\partial^2 u}{\partial r^2} + \frac{1}{r}\frac{\partial u}{\partial r} + \frac{\partial^2 u}{\partial z^2} = a^2 u, \qquad 0 \le r < 1, \quad 0 < z < 1,$$

[39] See Stripp, K. F., and A. R. Moore, 1955: The effects of junction shape and surface recombination on transistor current gain – Part II. *Proc. IRE*, **43**, 856–866.

[40] See Bischoff, K. B., 1966: Transverse diffusion in catalyst pores. *Indust. Engng. Chem. Fund.*, **5**, 135–136.

[41] See Gunn, D. J., 1967: Diffusion and chemical reaction in catalysis and absorption. *Chem. Engng. Sci.*, **22**, 1439–1455; Ho, T. C., and G. C. Hsiao, 1977: Estimation of the effectiveness factor for a cylindrical catalyst support: A singular perturbation approach. *Chem. Engng. Sci.*, **32**, 63–66.

subject to the boundary conditions that $\lim_{r \to 0} |u(r, z)| < \infty$, $u(1, z) = 1$, $0 < z < 1$, and $u(r, 0) = u(r, 1) = 1$, $0 \le r < 1$. Hint: Break the problem into three parts: $u(r, z) = u_1(r, z) + u_2(r, z) + u_3(r, z)$, where

$$\frac{\partial^2 u_1}{\partial r^2} + \frac{1}{r}\frac{\partial u_1}{\partial r} + \frac{\partial^2 u_1}{\partial z^2} = a^2 u_1, \qquad 0 \le r < 1, \quad 0 < z < 1,$$

subject to the boundary conditions that $\lim_{r \to 0} |u_1(r, z)| < \infty$, $u_1(1, z) = 1$, $0 < z < 1$, and $u_1(r, 0) = u_1(r, 1) = 0$, $0 \le r < 1$;

$$\frac{\partial^2 u_2}{\partial r^2} + \frac{1}{r}\frac{\partial u_2}{\partial r} + \frac{\partial^2 u_2}{\partial z^2} = a^2 u_2, \qquad 0 \le r < 1, \quad 0 < z < 1,$$

subject to the boundary conditions that $\lim_{r \to 0} |u_2(r, z)| < \infty$, $u_2(1, z) = 0$, $0 < z < 1$, and $u_2(r, 0) = 1$, $u_2(r, 1) = 0$, $0 \le r < 1$; and

$$\frac{\partial^2 u_3}{\partial r^2} + \frac{1}{r}\frac{\partial u_3}{\partial r} + \frac{\partial^2 u_3}{\partial z^2} = a^2 u_3, \qquad 0 \le r < 1, \quad 0 < z < 1,$$

subject to the boundary conditions that $\lim_{r \to 0} |u_3(r, z)| < \infty$, $u_3(1, z) = 0$, $0 < z < 1$, and $u_3(r, 0) = 0$, $u_3(r, 1) = 1$, $0 \le r < 1$.

55. Solve[42] Laplace's equation in cylindrical coordinates

$$\frac{\partial^2 u}{\partial r^2} + \frac{1}{r}\frac{\partial u}{\partial r} + \frac{\partial^2 u}{\partial z^2} = 0, \qquad 0 \le r < a, \quad -\infty < z < \infty,$$

subject to the boundary conditions that $\lim_{r \to 0} |u(r, z)| < \infty$,

$$u(a, z) = \begin{cases} -V, & |z| < d/2, \\ 0, & |z| > d/2, \end{cases}$$

and $\lim_{|z| \to \infty} u(r, z) \to 0$, $0 \le r < a$. Hint: Show that the solution can be written

$$u(r, z) = -V + \sum_{n=1}^{\infty} A_n \cosh\left(\frac{k_n z}{a}\right) J_0\left(\frac{k_n r}{a}\right), \qquad |z| < d/2,$$

and

$$u(r, z) = \sum_{n=1}^{\infty} B_n \exp\left(-\frac{k_n |z|}{a}\right) J_0\left(\frac{k_n r}{a}\right), \qquad |z| > d/2$$

with the additional conditions that $u(r, d^-/2) = u(r, d^+/2)$ and $u_z(r, d^-/2) = u_z(r, d^+/2)$, $0 \le r < a$, where k_n is the nth root of $J_0(k) = 0$. Then find A_n and B_n.

56. Solve Laplace's equation in cylindrical coordinates

$$\frac{\partial^2 u}{\partial r^2} + \frac{1}{r}\frac{\partial u}{\partial r} + \frac{\partial^2 u}{\partial z^2} = 0, \qquad 0 \le r < b, \quad 0 < z < \infty,$$

---

[42] See Striffler, C. D., C. A. Kapetanakos, and R. C. Davidson, 1975: Equilibrium properties of a rotating nonneutral E layer in a coupled magnetic field. *Phys. Fluids*, **18**, 1374–1382.

subject to the boundary conditions that $\lim_{r\to 0}|u(r,z)| < \infty$, $u_r(b,z) = 0$, $0 < z < \infty$, and $\lim_{z\to\infty}|u(r,z)| < \infty$,

$$u(r,0) = \begin{cases} A, & 0 \le r < a, \\ 0, & a < r < b. \end{cases}$$

57. Solve[43]

$$\frac{\partial^2 u}{\partial r^2} + \frac{1}{r}\frac{\partial u}{\partial r} + \frac{\partial^2 u}{\partial z^2} - \frac{\partial u}{\partial z} = 0, \quad 0 \le r < 1, \quad 0 < z < \infty,$$

with the boundary conditions $\lim_{r\to 0}|u(r,z)| < \infty$, $u_r(1,z) = -Bu(1,z)$, $0 < z$, and $u(r,0) = 1$, $\lim_{z\to\infty}|u(r,z)| < \infty$, $0 \le r < 1$, where B is a constant.

58. Solve[44]

$$\frac{\partial^2 u}{\partial r^2} + \frac{1}{r}\frac{\partial u}{\partial r} + \frac{\partial^2 u}{\partial z^2} - \frac{1}{H}\frac{\partial u}{\partial z} = 0, \quad 0 \le r < b, \quad 0 < z < \infty,$$

with the boundary conditions $\lim_{r\to 0}|u(r,z)| < \infty$, $u_r(b,z) = -hu(b,z)$, $0 < z$, along with $\lim_{z\to\infty}|u(r,z)| < \infty$, $0 \le r < b$, and

$$\frac{u(r,0)}{H} - u_z(r,0) = \begin{cases} Q, & 0 \le r < a, \\ 0, & a \le r < b, \end{cases}$$

where $b > a$.

Separation of Variables Solution for Poisson's Equation

59. Let us solve the axisymmetric Poisson equation inside a circular cylinder

$$\frac{1}{r}\frac{\partial}{\partial r}\left(r\frac{\partial u}{\partial r}\right) + \frac{\partial^2 u}{\partial z^2} = f(r,z), \quad 0 \le r < a, \quad |z| < b,$$

subject to the boundary conditions $\lim_{r\to 0}|u(r,z)| < \infty$, $u(a,z) = 0$, $|z| < b$, and $u(r,-b) = u(r,b) = 0$, $0 \le r < a$.

Step 1: Replace the original problem with

$$\frac{1}{r}\frac{\partial}{\partial r}\left(r\frac{\partial u}{\partial r}\right) + \frac{\partial^2 u}{\partial z^2} = \lambda u, \quad 0 \le r < a, \quad |z| < b,$$

subject to the same boundary conditions. Use separation of variables to show that the solution to this new problem is

$$u_{nm}(r,z) = A_{nm}J_0\left(k_n\frac{r}{a}\right)\cos\left[\frac{\left(m+\frac{1}{2}\right)\pi z}{b}\right],$$

where k_n is the nth zero of $J_0(k) = 0$, $n = 1,2,3,\ldots$, and $m = 0,1,2,\ldots$.

[43] See Kern, J., and J. O. Hansen, 1976: Transient heat conduction in cylindrical systems with an axially moving boundary. *Int. J. Heat Mass Transfer*, **19**, 707–714.

[44] See Smirnova, E. V., and I. A. Krinberg, 1970: Spatial distribution of the atoms of an impurity element in an arc discharge. I. *J. Appl. Spectroscopy*, **13**, 859–864.

Step 2: Show that $f(r, z)$ can be expressed as

$$f(r, z) = \sum_{n=1}^{\infty} \sum_{m=0}^{\infty} a_{nm} J_0\left(k_n \frac{r}{a}\right) \cos\left[\frac{\left(m + \frac{1}{2}\right) \pi z}{b}\right],$$

where

$$a_{nm} = \frac{2}{a^2 b J_1^2(k_n)} \int_{-b}^{b} \int_{0}^{a} f(r, z) J_0\left(k_n \frac{r}{a}\right) \cos\left[\frac{\left(m + \frac{1}{2}\right) \pi z}{b}\right] r \, dr \, dz.$$

Step 3: Show that the general solution is

$$u(r, z) = -\sum_{n=1}^{\infty} \sum_{m=0}^{\infty} a_{nm} \frac{J_0(k_n r/a) \cos\left[\left(m + \frac{1}{2}\right) \pi z/b\right]}{(k_n/a)^2 + \left[\left(m + \frac{1}{2}\right) \pi/b\right]^2}.$$

Appendix A: Derivation of the Laplacian in Polar Coordinates

In several of the problems and examples, it is convenient to express the Laplacian:

$$\nabla^2 u = \frac{\partial^2 u}{\partial x^2} + \frac{\partial^2 u}{\partial y^2} \tag{12.A.1}$$

in terms of polar coordinates $x = r \cos(\theta)$ and $y = r \sin(\theta)$.
    We begin by computing

$$\frac{\partial x}{\partial r} = \cos(\theta), \quad \frac{\partial x}{\partial \theta} = -r \sin(\theta), \quad \frac{\partial y}{\partial r} = \sin(\theta), \quad \text{and} \quad \frac{\partial y}{\partial \theta} = r \cos(\theta). \tag{12.A.2}$$

From the chain rule, we have that

$$\frac{\partial u}{\partial r} = \frac{\partial u}{\partial x} \frac{\partial x}{\partial r} + \frac{\partial u}{\partial y} \frac{\partial y}{\partial r} = \cos(\theta) \frac{\partial u}{\partial x} + \sin(\theta) \frac{\partial u}{\partial y}, \tag{12.A.3}$$

and

$$\frac{\partial^2 u}{\partial r^2} = \cos(\theta) \frac{\partial}{\partial r}\left(\frac{\partial u}{\partial x}\right) + \sin(\theta) \frac{\partial}{\partial r}\left(\frac{\partial u}{\partial y}\right) \tag{12.A.4}$$

$$= \cos(\theta) \left[\frac{\partial}{\partial x}\left(\frac{\partial u}{\partial x}\right) \frac{\partial x}{\partial r} + \frac{\partial}{\partial y}\left(\frac{\partial u}{\partial x}\right) \frac{\partial y}{\partial r}\right]$$

$$+ \sin(\theta) \left[\frac{\partial}{\partial x}\left(\frac{\partial u}{\partial y}\right) \frac{\partial x}{\partial r} + \frac{\partial}{\partial y}\left(\frac{\partial u}{\partial y}\right) \frac{\partial y}{\partial r}\right] \tag{12.A.5}$$

$$= \cos^2(\theta) \frac{\partial^2 u}{\partial x^2} + 2 \cos(\theta) \sin(\theta) \frac{\partial^2 u}{\partial x \partial y} + \sin^2(\theta) \frac{\partial^2 u}{\partial y^2}. \tag{12.A.6}$$

On the other hand,

$$\frac{\partial u}{\partial \theta} = \frac{\partial u}{\partial x} \frac{\partial x}{\partial \theta} + \frac{\partial u}{\partial y} \frac{\partial y}{\partial \theta} = -r \sin(\theta) \frac{\partial u}{\partial x} + r \cos(\theta) \frac{\partial u}{\partial y}, \tag{12.A.7}$$

and

$$\frac{\partial^2 u}{\partial \theta^2} = -r \cos(\theta)\frac{\partial u}{\partial x} - r \sin(\theta)\frac{\partial}{\partial \theta}\left(\frac{\partial u}{\partial x}\right) - r \sin(\theta)\frac{\partial u}{\partial y} + r \cos(\theta)\frac{\partial}{\partial \theta}\left(\frac{\partial u}{\partial y}\right) \quad \textbf{(12.A.8)}$$

$$= -r \cos(\theta)\frac{\partial u}{\partial x} - r \sin(\theta)\left[\frac{\partial}{\partial x}\left(\frac{\partial u}{\partial x}\right)\frac{\partial x}{\partial \theta} + \frac{\partial}{\partial y}\left(\frac{\partial u}{\partial x}\right)\frac{\partial y}{\partial \theta}\right]$$

$$- r \sin(\theta)\frac{\partial u}{\partial y} + r \cos(\theta)\left[\frac{\partial}{\partial x}\left(\frac{\partial u}{\partial y}\right)\frac{\partial x}{\partial \theta} + \frac{\partial}{\partial y}\left(\frac{\partial u}{\partial y}\right)\frac{\partial y}{\partial \theta}\right] \quad \textbf{(12.A.9)}$$

$$= -r \cos(\theta)\frac{\partial u}{\partial x} - r \sin(\theta)\left\{[-r \sin(\theta)]\frac{\partial^2 u}{\partial x^2} + r \cos(\theta)\frac{\partial^2 u}{\partial x \partial y}\right\}$$

$$- r \sin(\theta)\frac{\partial u}{\partial y} + r \cos(\theta)\left\{[-r \sin(\theta)]\frac{\partial^2 u}{\partial x \partial y} + r \cos(\theta)\frac{\partial^2 u}{\partial y^2}\right\} \quad \textbf{(12.A.10)}$$

$$= -r\left[\cos(\theta)\frac{\partial u}{\partial x} + \sin(\theta)\frac{\partial u}{\partial y}\right]$$

$$+ r^2\left[\sin^2(\theta)\frac{\partial^2 u}{\partial x^2} - 2\sin(\theta)\cos(\theta)\frac{\partial^2 u}{\partial x \partial y} + \cos^2(\theta)\frac{\partial^2 u}{\partial y^2}\right] \quad \textbf{(12.A.11)}$$

Dividing both sides of Equation 12.A.11 by r^2, we find that

$$\frac{1}{r^2}\frac{\partial^2 u}{\partial \theta^2} = -\frac{1}{r}\frac{\partial u}{\partial r} + \sin^2(\theta)\frac{\partial^2 u}{\partial x^2} - 2\sin(\theta)\cos(\theta)\frac{\partial^2 u}{\partial x \partial y} + \cos^2(\theta)\frac{\partial^2 u}{\partial y^2}. \quad \textbf{(12.A.12)}$$

Because $\sin^2(\theta) + \cos^2(\theta) = 1$, we obtain the final result that

$$\frac{\partial^2 u}{\partial x^2} + \frac{\partial^2 u}{\partial y^2} = \frac{\partial^2 u}{\partial r^2} + \frac{1}{r}\frac{\partial u}{\partial r} + \frac{\partial^2 u}{\partial \theta^2}. \quad \textbf{(12.A.13)}$$

Appendix B: Derivation of the Laplacian in Spherical Polar Coordinates

In problems involving spherical geometries, it is convenient to re-express the three-dimensional Laplacian:

$$\nabla^2 u = \frac{\partial^2 u}{\partial x^2} + \frac{\partial^2 u}{\partial y^2} + \frac{\partial^2 u}{\partial z^2} \quad \textbf{(12.B.1)}$$

in terms of spherical polar coordinates $x = r \cos(\varphi)\sin(\theta)$, $y = r \sin(\varphi)\sin(\theta)$, and $z = r \cos(\theta)$.

Because

$$\frac{\partial x}{\partial r} = \cos(\varphi)\sin(\theta), \quad \frac{\partial y}{\partial r} = \sin(\varphi)\sin(\theta), \quad \frac{\partial z}{\partial r} = \cos(\theta), \quad \textbf{(12.B.2)}$$

$$\frac{\partial x}{\partial \theta} = r \cos(\varphi)\cos(\theta), \quad \frac{\partial y}{\partial \theta} = r \sin(\varphi)\cos(\theta), \quad \frac{\partial z}{\partial \theta} = -r \sin(\theta), \quad \textbf{(12.B.3)}$$

and

$$\frac{\partial x}{\partial \varphi} = -r \sin(\varphi)\sin(\theta), \quad \frac{\partial y}{\partial \varphi} = r \cos(\varphi)\sin(\theta), \quad \frac{\partial z}{\partial \varphi} = 0, \quad \textbf{(12.B.4)}$$

we have that

$$\frac{\partial}{\partial r} = \frac{\partial x}{\partial r}\frac{\partial}{\partial x} + \frac{\partial y}{\partial r}\frac{\partial}{\partial y} + \frac{\partial z}{\partial r}\frac{\partial}{\partial z} = \cos(\varphi)\sin(\theta)\frac{\partial}{\partial x} + \sin(\varphi)\sin(\theta)\frac{\partial}{\partial y} + \cos(\theta)\frac{\partial}{\partial z}, \quad \textbf{(12.B.5)}$$

$$\frac{\partial}{\partial\theta} = \frac{\partial x}{\partial\theta}\frac{\partial}{\partial x} + \frac{\partial y}{\partial\theta}\frac{\partial}{\partial y} + \frac{\partial z}{\partial\theta}\frac{\partial}{\partial z} = r\,\cos(\varphi)\cos(\theta)\frac{\partial}{\partial x} + r\,\sin(\varphi)\cos(\theta)\frac{\partial}{\partial y} - r\,\sin(\theta)\frac{\partial}{\partial z},$$
(12.B.6)

and

$$\frac{\partial}{\partial\varphi} = \frac{\partial x}{\partial\varphi}\frac{\partial}{\partial x} + \frac{\partial y}{\partial\varphi}\frac{\partial}{\partial y} + \frac{\partial z}{\partial\varphi}\frac{\partial}{\partial z} = -r\,\sin(\varphi)\sin(\theta)\frac{\partial}{\partial x} + r\,\cos(\varphi)\sin(\theta)\frac{\partial}{\partial y}.$$
(12.B.7)

Solving for $\partial/\partial x$, $\partial/\partial y$, and $\partial/\partial z$, we have that

$$\frac{\partial}{\partial x} = \cos(\varphi)\sin(\theta)\frac{\partial}{\partial r} + \frac{\cos(\varphi)\cos(\theta)}{r}\frac{\partial}{\partial\theta} + \frac{\sin(\varphi)}{r\sin(\theta)}\frac{\partial}{\partial\varphi},$$
(12.B.8)

$$\frac{\partial}{\partial y} = \sin(\varphi)\sin(\theta)\frac{\partial}{\partial r} + \frac{\sin(\varphi)\cos(\theta)}{r}\frac{\partial}{\partial\theta} + \frac{\cos(\varphi)}{r\sin(\theta)}\frac{\partial}{\partial\varphi},$$
(12.B.9)

and

$$\frac{\partial}{\partial z} = \cos(\theta)\frac{\partial}{\partial r} - \frac{\sin(\theta)}{r}\frac{\partial}{\partial\theta}.$$
(12.B.10)

Next, let us introduce the unit vectors in r, θ and φ directions:

$$\widehat{\mathbf{r}} = \sin(\theta)\cos(\varphi)\widehat{\mathbf{x}} + \sin(\theta)\sin(\varphi)\widehat{\mathbf{y}} + \cos(\theta)\widehat{\mathbf{z}},$$
(12.B.11)

$$\widehat{\theta} = \cos(\theta)\cos(\varphi)\widehat{\mathbf{x}} + \cos(\theta)\sin(\varphi)\widehat{\mathbf{y}} - \sin(\theta)\widehat{\mathbf{z}},$$
(12.B.12)

$$\widehat{\varphi} = -\sin(\varphi)\widehat{\mathbf{x}} + \cos(\varphi)\widehat{\mathbf{y}}$$
(12.B.13)

or

$$\widehat{\mathbf{x}} = \sin(\theta)\cos(\varphi)\widehat{\mathbf{r}} + \cos(\theta)\cos(\varphi)\widehat{\theta} - \sin(\varphi)\widehat{\varphi},$$
(12.B.14)

$$\widehat{\mathbf{y}} = \sin(\theta)\sin(\varphi)\widehat{\mathbf{r}} + \cos(\theta)\sin(\varphi)\widehat{\theta} - \sin(\varphi)\widehat{\varphi},$$
(12.B.15)

$$\widehat{\mathbf{z}} = \cos\theta)\widehat{\mathbf{r}} - \sin(\theta)\widehat{\theta}.$$
(12.B.16)

Because

$$\nabla = \widehat{\mathbf{x}}\frac{\partial}{\partial x} + \widehat{\mathbf{y}}\frac{\partial}{\partial y} + \widehat{\mathbf{z}}\frac{\partial}{\partial z},$$
(12.B.17)

we find that

$$\nabla = \widehat{\mathbf{r}}\frac{\partial}{\partial r} + \widehat{\theta}\frac{1}{r}\frac{\partial}{\partial\theta} + \widehat{\varphi}\frac{1}{r\sin(\theta)}\frac{\partial}{\partial\varphi}.$$
(12.B.18)

Because the unit vectors $\widehat{\mathbf{r}}$, $\widehat{\theta}$ and $\widehat{\varphi}$ are not constant we will need the additional terms:

$$\frac{\partial\widehat{\mathbf{r}}}{\partial r} = 0,\quad \frac{\partial\widehat{\theta}}{\partial r} = 0,\quad \frac{\partial\widehat{\varphi}}{\partial r} = 0,\quad \frac{\partial\widehat{\mathbf{r}}}{\partial\theta} = \widehat{\theta},\quad \frac{\partial\widehat{\theta}}{\partial\theta} = -\widehat{\mathbf{r}},\quad \frac{\partial\widehat{\varphi}}{\partial\theta} = \mathbf{0},$$
(12.B.19)

$$\frac{\partial\widehat{\mathbf{r}}}{\partial\varphi} = -\sin(\theta)\widehat{\varphi},\quad \frac{\partial\widehat{\theta}}{\partial\varphi} = \cos(\theta)\widehat{\varphi},\quad \frac{\partial\widehat{\varphi}}{\partial\varphi} = -\sin(\theta)\widehat{\mathbf{r}} - \cos(\theta)\widehat{\theta}.$$
(12.B.20)

Using Equation 12.B.18 through Equation 12.B.20 we now compute $\nabla\cdot\nabla$. The final result is

$$\frac{\partial^2 u}{\partial x^2} + \frac{\partial^2 u}{\partial y^2} + \frac{\partial^2 u}{\partial z^2} = \frac{1}{r}\frac{\partial^2(ru)}{\partial r^2} + \frac{1}{r^2\sin(\theta)}\frac{\partial}{\partial\theta}\left[\sin(\theta)\frac{\partial u}{\partial\theta}\right] + \frac{1}{r^2\sin(\theta)}\frac{\partial^2 u}{\partial\varphi^2}.$$
(12.B.21)

Further Readings

Hobson, E. W., 1965: *The Theory of Spherical and Ellipsoidal Harmonics.* Chelsea Publishers, 500 pp. The classic treatise on Legendre polynomials.

Lebedev, N. N., 1972: *Special Functions and Their Applications.* Dover, 308 pp. A very practical guide to the special functions found in the natural sciences and engineering.

Watson, G. N., 1966: *A Treatise on the Theory of Bessel Functions.* Cambridge University Press, 804 pp. The standard reference on Bessel functions.

Answers
to the Odd-Numbered Problems

Section 1.1

1. first-order, linear   3. first-order, nonlinear   5. second-order, linear

7. third-order, nonlinear   9. second-order, nonlinear   11. first-order, nonlinear

13. first-order, nonlinear   15. second-order, nonlinear

Section 1.2

1. $y(x) = -\ln(C - x^2/2)$ 3. $y^2(x) - \ln^2(x) = 2C$ 5. $2 + y^2(x) = C(1 + x^2)$

7. $y(x) = -\ln(C - e^x)$ 9. $\sin^2(y) = 2\ln(x) + C$ 11. $y(x) = x + Ce^{-1/x}$

19. $[X] = \alpha \left[1 - e^{-(k_1+k_2)t}\right]$

Section 1.3

1. $\ln|y| - x/y = C$ 3. $|x|(x^2 + 3y^2) = C$ 5. $y = x\left(\ln|x| + C\right)^2$

7. $\sin(y/x) - \ln|x| = C$ 9. $\frac{2}{\sqrt{3}}\arctan\left(\frac{2y-x}{\sqrt{3}x}\right) = \ln|x| + C.$

Section 1.4

1. $xy^2 - \frac{1}{3}x^3 = C$ 3. $xy^2 - x + \cos(y) = C$ 5. $y/x + \ln(y) = C$

7. $\cos(xy) = C$ 9. $x^2y^3 + x^5y + y = C$ 11. $xy\ln(y) + e^x - e^{-y} = C$

13. $y - x + \frac{1}{2}\sin(2x + 2y) = C$

Section 1.5

1. $y(x) = \frac{1}{2}e^x + Ce^{-x}, \quad x \in (-\infty, \infty)$ 3. $y(x) = \ln(x)/x + Cx^{-1}, \quad x \neq 0$

5. $y(x) = 2x^3 \ln(x) + Cx^3, \quad x \in (-\infty, \infty)$

7. $e^{\sin(2x)}y(x) = C, \quad n\pi + \varphi < 2x < (n+1)\pi + \varphi$, where φ is any real and n is any integer.

9. $y(x) = \frac{4}{3} + \frac{11}{3}e^{-3x}, \quad x \in (-\infty, \infty)$ 11. $y(x) = (x + C)\csc(x)$

13. $y(x) = \dfrac{\cos^a(x)\, y(0)}{[\sec(x) + \tan(x)]^b} + \dfrac{c\, \cos^a(x)}{[\sec(x) + \tan(x)]^b} \displaystyle\int_0^x \dfrac{[\sec(\xi) + \tan(\xi)]^b}{\cos^{a+1}(\xi)}\, d\xi$

15. $y(x) = \dfrac{2ax - 1}{8a^2} + \dfrac{\omega^2 e^{-2ax}}{8a^2(a^2 + \omega^2)} - \dfrac{a\sin(2\omega x) - \omega\cos(2\omega x)}{8\omega(a^2 + \omega^2)}$

17. $y^2(x) = 2(x - x^{2/k})/(2 - k)$ if $k \neq 2$; $y^2(x) = x\ln(1/x)$ if $k = 2$

19. $[A](t) = [A]_0 e^{-k_1 t}, [B](t) = \dfrac{k_1[A]_0}{k_2 - k_1}\left[e^{-k_1 t} - e^{-k_2 t}\right],$

    $[C](t) = [A]_0\left(1 + \dfrac{k_1 e^{-k_2 t} - k_2 e^{-k_1 t}}{k_2 - k_1}\right)$

21. $y(x) = [Cx + x\ln(x)]^{-1}$ 23. $y(x) = \left[Cx^2 + \frac{1}{2}x^2\ln(x)\right]^2$

25. $y(x) = [Cx - x\ln(x)]^{1/2}$

Section 1.6

5. The equilibrium points are $x = 0, \frac{1}{2}$, and 1. The equilibrium at $x = \frac{1}{2}$ is unstable while the equilibriums at $x = 0$ and 1 are stable.

7. The equilibrium point for this differential equation is $x = 0$, which is stable.

Section 1.7

1. $x(t) = e^t + t + 1$ 3. $x(t) = [1 - \ln(t + 1)]^{-1}$

Section 2.0

1. $y_2(x) = A/x$ 3. $y_2(x) = Ax^{-4}$ 5. $y_2(x) = A(x^2 - x + 1)$

7. $y_2(x) = A\sin(x)/\sqrt{x}$ 9. $y(x) = C_2 e^{C_1 x}$ 11. $y(x) = \left(1 + C_2 e^{C_1 x}\right)/C_1$

13. $y(x) = -\ln|1 - x|$ 15. $y(x) = C_1 - 2\ln(x^2 + C_2)$

Section 2.1

1. $y(x) = C_1 e^{-x} + C_2 e^{-5x}$

3. $y(x) = C_1 e^x + C_2 x e^x$

5. $y(x) = C_1 e^{2x} \cos(2x) + C_2 e^{2x} \sin(2x)$

7. $y(x) = C_1 e^{-10x} + C_2 e^{4x}$

9. $y(x) = e^{-4x} [C_1 \cos(3x) + C_2 \sin(3x)]$

11. $y(x) = C_1 e^{-4x} + C_2 x e^{-4x}$

13. $y(x) = C_1 + C_2 x + C_3 \cos(2x) + C_4 \sin(2x)$

15. $y(x) = C_1 e^{2x} + C_2 e^{-x} \cos(\sqrt{3}\,x) + C_3 e^{-x} \sin(\sqrt{3}\,x)$

17. $y(x) = C_1 + (C_2 + C_3 x + C_4 x^2) e^x$

19. $y(t) = e^{-t/(2\tau)} \left\{ A \exp\left[t\sqrt{1 - 2A\tau}/(2\tau)\right] + B \exp\left[-t\sqrt{1 - 2A\tau}/(2\tau)\right] \right\}$

Section 2.2

1. $x(t) = 2\sqrt{26} \sin(5t + 1.7682)$

3. $x(t) = 2\cos(\pi t - \pi/3)$

5. $x(t) = s_0 \cos(\omega t) + \dfrac{v_0}{\omega} \sin(\omega t)$ and $v(t) = v_0 \cos(\omega t) - \omega s_0 \sin(\omega t)$, where $\omega^2 = Mg/(mL)$.

Section 2.3

1. $x(t) = 4e^{-2t} - 2e^{-4t}$

3. $x(t) = e^{-5t/2} \left[4\cos(6t) + \frac{13}{3}\sin(6t)\right]$

5. The roots are equal when $c = 4$.

Section 2.4

1. $y(x) = Ae^{-3x} + Be^{-x} + \frac{1}{3}x - \frac{1}{9}$

3. $y(x) = e^{-x}[A\cos(x) + B\sin(x)] + x^2 - x + 2$

5. $y(x) = A + Be^{-2x} + \frac{1}{2}x^2 + 2x + \frac{1}{2}e^{-2x}$

7. $y(x) = (A + Bx)e^{-2x} + \left(\frac{1}{9}x - \frac{2}{27}\right)e^x$

9. $y(x) = A\cos(3x) + B\sin(3x) + \frac{1}{12}x^2 \sin(3x) + \frac{1}{36}x \cos(3x)$

11. $y(x) = \dfrac{2ax - 1}{8a^2} + \dfrac{\omega^2 e^{-2ax}}{8a^2(a^2 + \omega^2)} - \dfrac{a\sin(2\omega x) - \omega\cos(2\omega x)}{8\omega(a^2 + \omega^2)}$

Section 2.5

1. $\gamma = 3$,

5. $I(t) = \dfrac{\omega_1 \omega^2 E_0}{L\omega^2 - 1/C} \sin(\omega_1 t) - \dfrac{\omega E_0 \sin(\omega t)}{L\omega^2 - 1/C}$

Section 2.6

1. $y(x) = Ae^x + Be^{3x} + \frac{1}{8}e^{-x}$

3. $y(x) = Ae^{2x} + Be^{-2x} - (3x + 2)e^x/9$

5. $y(x) = (A + Bx)e^{-2x} + x^3 e^{-2x}/6$

7. $y(x) = Ae^{2x} + Bxe^{2x} + \left(\frac{1}{2}x^2 + \frac{1}{6}x^3\right)e^{2x}$

9. $y(x) = Ae^x + Bxe^x + x\ln(x)e^x$

11. $y(x) = Ae^x + Be^{3x} + \frac{1}{5}\cos(x + 3) - \frac{2}{5}\sin(x + 3)$

Section 2.7

1. $y(x) = C_1 x + C_2 x^{-1}$

3. $y(x) = C_1 x^2 + C_2/x$

5. $y(x) = C_1/x + C_2 \ln(x)/x$

7. $y(x) = C_1 x \cos[2\ln(x)] + C_2 x \sin[\ln(x)]$

9. $y(x) = C_1 \cos[\ln(x)] + C_2 \sin[\ln(x)]$

11. $y(x) = C_1 x^2 + C_2 x^4 + C_3/x$

Section 2.8

1. The trajectories spiral outward from $(0,0)$.

3. The equilibrium points are $(x,0)$; they are unstable.

5. The equilibrium points are $v = 0$ and $|x| < 2$; they are unstable.

Section 3.1

1. $A + B = \begin{pmatrix} 4 & 5 \\ 3 & 4 \end{pmatrix} = B + A$ 3. $3A - 2B = \begin{pmatrix} 7 & 10 \\ -1 & 2 \end{pmatrix}$, $3(2A - B) = \begin{pmatrix} 15 & 21 \\ 0 & 6 \end{pmatrix}$

5. $(A + B)^T = \begin{pmatrix} 4 & 3 \\ 5 & 4 \end{pmatrix}$, $A^T + B^T = \begin{pmatrix} 4 & 3 \\ 5 & 4 \end{pmatrix}$

7. $AB = \begin{pmatrix} 11 & 11 \\ 5 & 5 \end{pmatrix}$, $A^T B = \begin{pmatrix} 5 & 5 \\ 8 & 8 \end{pmatrix}$, $BA = \begin{pmatrix} 4 & 6 \\ 8 & 12 \end{pmatrix}$, $B^T A = \begin{pmatrix} 5 & 8 \\ 5 & 8 \end{pmatrix}$

9. $BB^T = \begin{pmatrix} 2 & 4 \\ 4 & 8 \end{pmatrix}$, $B^T B = \begin{pmatrix} 5 & 5 \\ 5 & 5 \end{pmatrix}$ 11. $A^3 + 2A = \begin{pmatrix} 65 & 100 \\ 25 & 40 \end{pmatrix}$

13. yes $\begin{pmatrix} 27 & 11 \\ 2 & 5 \end{pmatrix}$ 15. yes $\begin{pmatrix} 11 & 8 \\ 8 & 4 \\ 5 & 3 \end{pmatrix}$ 17. no

19. $4A + 3A = \begin{pmatrix} 7 & 7 \\ 7 & 14 \\ 21 & 7 \end{pmatrix} = 7A$ 21. $(A^T)^T = \begin{pmatrix} 1 & 1 \\ 1 & 2 \\ 3 & 1 \end{pmatrix} = A$

23. $(AB)C = \begin{pmatrix} 2 & 2 \\ 1 & 1 \end{pmatrix} = A(BC)$ 25. $(A + B)C = \begin{pmatrix} 2 & 2 \\ 8 & 8 \end{pmatrix} = AC + BC$

27. $\begin{pmatrix} 0 & 1 & 0 \\ 1 & 0 & 0 \\ 0 & 0 & 1 \end{pmatrix} \begin{pmatrix} 0 & 1 & 0 \\ 1 & 0 & 0 \\ 0 & 0 & 1 \end{pmatrix} = \begin{pmatrix} 1 & 0 & 0 \\ 0 & 1 & 0 \\ 0 & 0 & 1 \end{pmatrix}$ 29. $\begin{pmatrix} 2 & 1 & 4 \\ 4 & 2 & 5 \\ 6 & -3 & 5 \end{pmatrix} \begin{pmatrix} x_1 \\ x_2 \\ x_3 \end{pmatrix} = \begin{pmatrix} 2 \\ 6 \\ 2 \end{pmatrix}$

Section 3.2

1. 7   3. 1   5. -24 7. 3

Section 3.3

1. $x_1 = \frac{9}{5}$, $x_2 = \frac{3}{5}$

3. $x_1 = -4$, $x_2 = -5$

5. $x_1 = 0$, $x_2 = 0$, $x_3 = -2$

7. $x_1 = -3$, $x_2 = -4/5$, $x_3 = 3/5$

Section 3.4

1. $x_1 = 1, x_2 = 2$

3. $x_1 = x_3 = \alpha, x_2 = -\alpha$

5. $x_1 = -1, x_2 = 2\alpha, x_3 = \alpha$

7. $x_1 = 1, x_2 = 2.6, x_3 = 2.2$

9. $A^{-1} = \begin{pmatrix} -1/13 & 5/13 \\ 2/13 & 3/13 \end{pmatrix}$

11. $A^{-1} = \begin{pmatrix} 1 & 2 & 5 \\ 0 & -1 & 2 \\ 2 & 4 & 11 \end{pmatrix}$

Section 3.5

1. $\lambda = 4$, $\quad \mathbf{x}_0 = \alpha \begin{pmatrix} 2 \\ 1 \end{pmatrix}$; $\qquad \lambda = -3$, $\quad \mathbf{x}_0 = \beta \begin{pmatrix} 1 \\ -3 \end{pmatrix}$

3. $\lambda = 1 + \sqrt{6}$, $\quad \mathbf{x}_0 = \alpha \begin{pmatrix} 3 \\ \sqrt{6} \end{pmatrix}$; $\qquad \lambda = 1 - \sqrt{6}$, $\quad \mathbf{x}_0 = \beta \begin{pmatrix} -3 \\ \sqrt{6} \end{pmatrix}$

5. $\lambda = 1$, $\quad \mathbf{x}_0 = \alpha \begin{pmatrix} -1 \\ 0 \\ 1 \end{pmatrix} + \beta \begin{pmatrix} 3 \\ 1 \\ 0 \end{pmatrix}$ $\quad \lambda = 0$, $\quad \mathbf{x}_0 = \gamma \begin{pmatrix} 1 \\ 1 \\ 1 \end{pmatrix}$

7. $\lambda = 1$, $\quad \mathbf{x}_0 = \alpha \begin{pmatrix} 1 \\ 0 \\ 0 \end{pmatrix} + \beta \begin{pmatrix} 0 \\ 1 \\ -1 \end{pmatrix}$ $\quad \lambda = 2$, $\quad \mathbf{x}_0 = \gamma \begin{pmatrix} 1 \\ 1 \\ 0 \end{pmatrix}$

9. $\lambda = 0$, $\quad \mathbf{x}_0 = \alpha \begin{pmatrix} 1 \\ 1 \\ 1 \end{pmatrix}$ $\quad \lambda = 1$, $\quad \mathbf{x}_0 = \beta \begin{pmatrix} 3 \\ 2 \\ 1 \end{pmatrix}$ $\quad \lambda = 2$, $\quad \mathbf{x}_0 = \gamma \begin{pmatrix} 7 \\ 3 \\ 1 \end{pmatrix}$

11 $\lambda = 0$, $\quad \mathbf{x}_0 = \alpha \begin{pmatrix} 0 \\ 1 \\ 0 \end{pmatrix}$ $\quad \lambda = 2$, $\quad \mathbf{x}_0 = \beta \begin{pmatrix} 2 \\ 1 \\ 0 \end{pmatrix}$ $\quad \lambda = 3$, $\quad \mathbf{x}_0 = \gamma \begin{pmatrix} 0 \\ 2 \\ 3 \end{pmatrix}$

13.

$$A^{23} = PD^{23}P^{-1} = \begin{pmatrix} 1 & 1 \\ -1 & 1 \end{pmatrix} \begin{pmatrix} -1 & 0 \\ 0 & 5^{23} \end{pmatrix} \begin{pmatrix} 0.5 & -0.5 \\ 0.5 & 0.5 \end{pmatrix}$$

$$= \begin{pmatrix} 0.5\ 5^{23} - 0.5 & 0.5\ 5^{23} + 0.5 \\ 0.5\ 5^{23} + 0.5 & 0.5\ 5^{23} - 0.5 \end{pmatrix}.$$

Section 3.6

1. $\mathbf{x} = c_1 \begin{pmatrix} 1 \\ -1 \end{pmatrix} e^{-t} + c_2 \begin{pmatrix} 1 \\ 1 \end{pmatrix} e^{3t}$

3. $\mathbf{x} = c_1 \begin{pmatrix} 1 \\ 2 \end{pmatrix} e^{3t} + c_2 \begin{pmatrix} 1 \\ -2 \end{pmatrix} e^{-t}$

5. $\mathbf{x} = c_1 \begin{pmatrix} 1 \\ -1 \end{pmatrix} e^{t/2} + c_2 \begin{pmatrix} t \\ -1/2 - t \end{pmatrix} e^{t/2}$

7. $\mathbf{x} = c_1 \begin{pmatrix} 1 \\ -1 \end{pmatrix} e^{2t} + c_2 \begin{pmatrix} -1 + t \\ -t \end{pmatrix} e^{2t}$

9. $\mathbf{x} = c_3 \begin{pmatrix} -3\cos(2t) - 2\sin(2t) \\ \cos(2t) \end{pmatrix} e^t + c_4 \begin{pmatrix} 2\cos(2t) - 3\sin(2t) \\ \sin(2t) \end{pmatrix} e^t$

11. $\mathbf{x} = c_3 \begin{pmatrix} 2\cos(t) \\ 7\cos(t) + \sin(t) \end{pmatrix} e^{-3t} + c_4 \begin{pmatrix} 2\sin(t) \\ 7\sin(t) - \cos(t) \end{pmatrix} e^{-3t}$

13. $\mathbf{x} = c_3 \begin{pmatrix} -\cos(2t) + \sin(2t) \\ \cos(2t) \end{pmatrix} e^t + c_4 \begin{pmatrix} -\cos(2t) - \sin(2t) \\ \sin(2t) \end{pmatrix} e^t$

15. $\mathbf{x} = c_1 \begin{pmatrix} -1 \\ 2 \end{pmatrix} e^{3t} + c_2 \begin{pmatrix} -3 \\ 2 \end{pmatrix} e^{-t}$ 17. $\mathbf{x} = c_1 \begin{pmatrix} 0 \\ 1 \\ 0 \end{pmatrix} + c_2 \begin{pmatrix} 0 \\ 2 \\ 1 \end{pmatrix} e^t + c_3 \begin{pmatrix} 2 \\ 1 \\ 0 \end{pmatrix} e^{2t}$

19. $\mathbf{x} = c_1 \begin{pmatrix} 3 \\ -2 \\ 12 \end{pmatrix} e^{-t} + c_2 \begin{pmatrix} 1 \\ 0 \\ 2 \end{pmatrix} e^t + c_3 \begin{pmatrix} 0 \\ 1 \\ 0 \end{pmatrix} e^{2t}$

Section 3.7

1. $\frac{1}{2} \begin{pmatrix} 3e^{4t} - e^{2t} & -3e^{4t} + 3e^{2t} \\ e^{4t} - e^{2t} & -e^{4t} + 3e^{2t} \end{pmatrix}$ 3. $\begin{pmatrix} e^{3t} & 5te^{3t} \\ 0 & e^{3t} \end{pmatrix}$ 5. $\begin{pmatrix} e^{2t} & 3te^{2t} & 4te^{2t} + \frac{9}{2}t^2 e^{2t} \\ 0 & e^{2t} & 3te^{2t} \\ 0 & 0 & e^{2t} \end{pmatrix}$

7.
$$x_1(t) = \left[e^t \cos(2t) + e^t \sin(2t) \right] x_1(0) - e^t \sin(2t) x_2(0) + \frac{e^t}{2} \sin(2t)$$

$$x_2(t) = 2e^t \sin(2t) x_1(0) + \left[e^t \cos(2t) - e^t \sin(2t) \right] x_2(0) + \frac{e^t}{2} \left[1 + \sin(2t) - \cos(2t) \right]$$

9.
$$x_1(t) = e^{2t} \cos(2t) x_1(0) + \tfrac{1}{2} e^{2t} \sin(2t) x_2(0)$$

$$x_2(t) = -2e^{2t} \sin(2t) x_1(0) + e^{2t} \cos(2t) x_2(0) - te^{2t}$$

11.
$$x_1(t) = [\cos(t) + 2\sin(t)] x_1(0) - \sin(t) x_2(0) + t\cos(t) + t\sin(t)$$

$$x_{2H}(t) = 5\sin(t) x_1(0) + [\cos(t) - 2\sin(t)] x_2(0) + t\cos(t) + 3t\sin(t) - \sin(t)$$

13.
$$x_1(t) = e^t x_1(0) + \left[e^{2t} - e^t \right] x_2(0) + \left[e^{3t} - e^{2t} \right] x_3(0) + \tfrac{1}{3} e^{3t} + \tfrac{1}{2} e^{2t} - e^t - t + \tfrac{1}{6}$$

$$x_2(t) = e^{2t} x_2(0) + \left[e^{3t} - e^{2t} \right] x_3(0) + \tfrac{1}{3} e^{3t} + \tfrac{1}{2} e^{2t} - \tfrac{5}{6},$$

$$x_3(t) = e^{3t} x_3(0) + \tfrac{1}{3} e^{3t} - t - \tfrac{1}{3}$$

15.
$$x_1(t) = \left(\tfrac{2}{3} e^{2t} + \tfrac{1}{3} e^{-t} \right) x_1(0) + \left(\tfrac{2}{3} e^{2t} - \tfrac{2}{3} e^{-t} \right) x_3(0) + \tfrac{4}{3} e^{2t} + \tfrac{1}{6} e^{-t} - \tfrac{3}{2} e^t$$

$$x_2(t) = e^t x_2(0) + te^t$$

$$x_3(t) = \left(\tfrac{1}{3} e^{2t} - \tfrac{1}{3} e^{-t} \right) x_1(0) + \left(\tfrac{1}{3} e^{2t} + \tfrac{2}{3} e^{-t} \right) x_3(0) + \tfrac{2}{3} e^{2t} - \tfrac{1}{6} e^{-t} - \tfrac{1}{2} e^t$$

Section 4.1

1. $\mathbf{a} \times \mathbf{b} = -3\mathbf{i} + 19\mathbf{j} + 10\mathbf{k}$ 3. $\mathbf{a} \times \mathbf{b} = \mathbf{i} - 8\mathbf{j} + 7\mathbf{k}$ 5. $\mathbf{a} \times \mathbf{b} = -3\mathbf{i} - 2\mathbf{j} - 5\mathbf{k}$

9. $\nabla f = y^2/z^3\,\mathbf{i} + 2xy/z^3\,\mathbf{j} - 3xy^2/z^4\,\mathbf{k}$

11. $\nabla f = 2x/(x^2 + y^2 + z^2)\,\mathbf{i} + 2y/(x^2 + y^2 + z^2\,\mathbf{j} + 2z/(x^2 + y^2 + z^2)\,\mathbf{k}$

13. $\nabla f = 2\mathbf{i} - 2y\mathbf{j} + 2z\mathbf{k}$

15. Plane parallel to the xy-plane at height of $z = 3$. $\mathbf{n} = \mathbf{k}$

17. Paraboloid, $\mathbf{n} = -2x\,\mathbf{i}/\sqrt{1 + 4z} - 2y\,\mathbf{j}/\sqrt{1 + 4z} + \mathbf{k}/\sqrt{1 + 4z}$

19. A plane, $\mathbf{n} = \mathbf{j}/\sqrt{2} - \mathbf{k}/\sqrt{2}$

21. A parabola of infinite extent along the y-axis, $\mathbf{n} = -2x\,\mathbf{i}/\sqrt{1 + 4x^2} + \mathbf{k}/\sqrt{1 + 4x^2}$.

23. $x = y - 1$ and $y = z$ 25. $x = 2y/(7y - 6)$ and $z = 3y/(4y - 3)$

27. $y = -\ln\left(3 - \frac{1}{2}x^2\right)$ and $z = 6 - \frac{1}{2}x^2$

Section 4.2

1. $\nabla \cdot \mathbf{F} = 2xz + z^2$, $\nabla \times \mathbf{F} = (2xy - 2yz)\mathbf{i} + (x^2 - y^2)\mathbf{j}$, $\nabla(\nabla \cdot \mathbf{F}) = 2z\mathbf{i} + (2x + 2z)\mathbf{k}$

3. $\nabla \cdot \mathbf{F} = 2(x - y) - xe^{-xy} + xe^{2y}$, $\nabla \times \mathbf{F} = 2xze^{2y}\mathbf{i} - ze^{2y}\mathbf{j} + [2(x - y) - ye^{-xy}]\,\mathbf{k}$
$\nabla(\nabla \cdot \mathbf{F}) = \left(2 - e^{-xy} + xye^{-xy} + e^{2y}\right)\mathbf{i} + \left(x^2 e^{-xy} + 2xe^{2y} - 2\right)\mathbf{j}$

5. $\nabla \cdot \mathbf{F} = 0$, $\nabla \times \mathbf{F} = -x^2\mathbf{i} + (5y - 9x^2)\mathbf{j} + (2xz - 5z)\mathbf{k}$, $\nabla(\nabla \cdot \mathbf{F}) = \mathbf{0}$

7. $\nabla \cdot \mathbf{F} = e^{-y} + z^2 - 3e^{-z}$, $\nabla \times \mathbf{F} = -2yz\mathbf{i} + xe^{-y}\mathbf{k}$, $\nabla(\nabla \cdot \mathbf{F}) = -e^{-y}\mathbf{j} + (2z + 3e^{-z})\mathbf{k}$

9. $\nabla \cdot \mathbf{F} = yz + x^3 z e^z + xye^z$,
$\nabla \times \mathbf{F} = (xe^z - x^3 ye^z - x^3 yze^z)\mathbf{i} + (xy - ye^z)\mathbf{j} + (3x^2 yze^z - xz)\mathbf{k}$,
$\nabla(\nabla \cdot \mathbf{F}) = \left(3x^2 ze^z + ye^z\right)\mathbf{i} + \left(z + xe^z\right)\mathbf{j} + \left(y + x^3 e^z + x^3 ze^z + xye^z\right)\mathbf{k}$

11. $\nabla \cdot \mathbf{F} = y^2 + xz^2 - xy\sin(z)$,
$\nabla \times \mathbf{F} = [x\cos(z) - 2xyz]\mathbf{i} - y\cos(z)\mathbf{j} + (yz^2 - 2xy)\mathbf{k}$,
$\nabla(\nabla \cdot \mathbf{F}) = [z^2 - y\sin(z)]\mathbf{i} + [2y - x\sin(z)]\mathbf{j} + [2xz - xy\cos(z)]\mathbf{k}$

13. $\nabla \cdot \mathbf{F} = y^2 + xz - xy\sin(z)$,
$\nabla \times \mathbf{F} = [x\cos(z) - xy]\mathbf{i} - y\cos(z)\mathbf{j} + (yz - 2xy)\mathbf{k}$,
$\nabla(\nabla \cdot \mathbf{F}) = [z - y\sin(z)]\mathbf{i} + [2y - x\sin(z)]\mathbf{j} + [x - xy\cos(z)]\mathbf{k}$

Section 4.3

1. $16/7 + 2/(3\pi)$ 3. $e^2 + 2e^8/3 + e^{64}/2 - 13/6$ 5. -4π 7. 0 9. 2π

Section 4.4

1. $\varphi(x, y, z) = x^2 y + y^2 z + 4z + \text{constant}$ 3. $\varphi(x, y, z) = xyz + \text{constant}$

5. $\varphi(x, y, z) = x^2 + 5x + y^3 + \ln(z) + \text{constant}$ 7. $\varphi(x, y, z) = xe^{2z} + y^3 + \text{constant}$

9. $\varphi(x, y, z) = xy + xz + \text{constant}$

Section 4.5

1. $1/2$ 3. 0 5. $27/2$ 7. 5

9. 0 11. $40/3$ 13. $86/3$ 15. 96π

Section 4.6

1. -5 3. 1 5. 0 7. 0 9. -16π 11. -2

Section 4.7

1. -10 3. 2 5. π 7. $45/2$

Section 4.8

1. 3 3. -16 5. 4π 7. $5/12$

Section 5.1

1. $f(t) = \dfrac{1}{2} - \dfrac{2}{\pi} \displaystyle\sum_{m=1}^{\infty} \dfrac{\sin[(2m-1)t]}{2m-1}$

3. $f(t) = -\dfrac{\pi}{4} + \displaystyle\sum_{n=1}^{\infty} \dfrac{(-1)^n - 1}{n^2\pi} \cos(nt) + \dfrac{1 - 2(-1)^n}{n} \sin(nt),$

5. $f(t) = \dfrac{\pi}{8} + \dfrac{2}{\pi} \displaystyle\sum_{n=1}^{\infty} \dfrac{2\cos(n\pi/2)\sin^2(n\pi/4)}{n^2} \cos(nt) + \dfrac{\sin(n\pi/2)}{n^2} \sin(nt)$

7. $f(t) = \dfrac{\sinh(aL)}{aL} + 2aL\sinh(aL) \displaystyle\sum_{n=1}^{\infty} \dfrac{(-1)^n}{a^2L^2 + n^2\pi^2} \cos\left(\dfrac{n\pi t}{L}\right)$

$\qquad\qquad - 2\pi\sinh(aL) \displaystyle\sum_{n=1}^{\infty} \dfrac{n(-1)^n}{a^2L^2 + n^2\pi^2} \sin\left(\dfrac{n\pi t}{L}\right)$

9. $f(t) = \dfrac{1}{\pi} + \dfrac{1}{2}\sin(t) - \dfrac{2}{\pi} \displaystyle\sum_{m=1}^{\infty} \dfrac{\cos(2mt)}{4m^2 - 1}$

11. $f(t) = \dfrac{a}{2} - \dfrac{4a}{\pi^2} \displaystyle\sum_{m=1}^{\infty} \dfrac{1}{(2m-1)^2} \cos\left[\dfrac{(2m-1)\pi t}{a}\right] - \dfrac{2a}{\pi} \displaystyle\sum_{n=1}^{\infty} \dfrac{(-1)^n}{n} \sin\left(\dfrac{n\pi t}{a}\right)$

13. $f(t) = \dfrac{\pi-1}{2} + \dfrac{1}{\pi} \displaystyle\sum_{n=1}^{\infty} \dfrac{\sin(n\pi t)}{n}$ 15. $f(t) = \dfrac{4a\cosh(a\pi/2)}{\pi} \displaystyle\sum_{m=1}^{\infty} \dfrac{\cos[(2m-1)t]}{a^2 + (2m-1)^2}$

Section 5.3

1. $f(x) = \dfrac{\pi}{2} - \dfrac{4}{\pi} \displaystyle\sum_{m=1}^{\infty} \dfrac{\cos[(2m-1)x]}{(2m-1)^2}, \quad f(x) = \dfrac{2}{\pi} \displaystyle\sum_{n=1}^{\infty} \dfrac{(-1)^{n+1}}{n} \sin(nx)$

3. $f(x) = \dfrac{a^3}{6} - \dfrac{a^2}{\pi^2} \displaystyle\sum_{m=1}^{\infty} \dfrac{1}{m^2} \cos\left(\dfrac{2m\pi x}{a}\right), \quad f(x) = \dfrac{8a^2}{\pi^3} \displaystyle\sum_{m=1}^{\infty} \dfrac{1}{(2m-1)^3} \sin\left[\dfrac{(2m-1)\pi x}{a}\right]$

5. $f(x) = \dfrac{1}{4} - \dfrac{2}{\pi^2} \displaystyle\sum_{m=1}^{\infty} \dfrac{\cos[2(2m-1)\pi x]}{(2m-1)^2}$, $\quad f(x) = \dfrac{4}{\pi^2} \displaystyle\sum_{m=1}^{\infty} \dfrac{(-1)^{m+1}}{(2m-1)^2} \sin[(2m-1)\pi x]$

7. $f(x) = \dfrac{2\pi^2}{3} - 4 \displaystyle\sum_{n=1}^{\infty} \dfrac{(-1)^n}{n^2} \cos(nx)$, $\quad f(x) = 2\pi \displaystyle\sum_{n=1}^{\infty} \dfrac{\sin(nx)}{n} + \dfrac{8}{\pi} \displaystyle\sum_{m=1}^{\infty} \dfrac{\sin[(2m-1)x]}{(2m-1)^3}$

9. $f(x) = \dfrac{a}{6} + \dfrac{4a}{\pi^2} \displaystyle\sum_{m=1}^{\infty} \dfrac{(-1)^m \sin[(2m-1)\pi/6]}{(2m-1)^2} \cos\left[\dfrac{(2m-1)\pi x}{a}\right]$

$f(x) = \dfrac{a}{\pi^2} \displaystyle\sum_{m=1}^{\infty} \dfrac{(-1)^m \sin(m\pi/3)}{m^2} \sin\left(\dfrac{2m\pi x}{a}\right) - \dfrac{2a}{3\pi} \displaystyle\sum_{n=1}^{\infty} \dfrac{(-1)^n}{n} \sin\left(\dfrac{n\pi x}{a}\right)$

11. $f(x) = \dfrac{3}{4} + \dfrac{1}{\pi} \displaystyle\sum_{m=1}^{\infty} \dfrac{(-1)^m}{2m-1} \cos\left[\dfrac{(2m-1)\pi x}{a}\right]$

$f(x) = \dfrac{1}{\pi} \displaystyle\sum_{n=1}^{\infty} \dfrac{1 + \cos(n\pi/2) - 2(-1)^n}{n} \sin\left(\dfrac{n\pi x}{a}\right)$

13. $f(x) = \dfrac{3a}{8} + \dfrac{2a}{\pi^2} \displaystyle\sum_{n=1}^{\infty} \dfrac{\cos(n\pi/2) - 1}{n^2} \cos\left(\dfrac{n\pi x}{a}\right)$,

$f(x) = \dfrac{a}{\pi} \displaystyle\sum_{n=1}^{\infty} \left[\dfrac{2}{n^2\pi} \sin\left(\dfrac{n\pi}{2}\right) - \dfrac{(-1)^n}{n}\right] \sin\left(\dfrac{n\pi x}{a}\right)$

Section 5.4

1. $f(t) = \dfrac{8}{\pi} \displaystyle\sum_{n=1}^{\infty} \dfrac{n}{4n^2 - 1} \sin(2n\pi t)$, $\quad f(t) = \dfrac{8}{\pi} \displaystyle\sum_{n=1}^{\infty} \dfrac{n}{4n^2 - 1} \cos(2n\pi t - \pi/2)$

3. $f(t) = \dfrac{3}{2} + \dfrac{2}{\pi} \displaystyle\sum_{n=1}^{\infty} \dfrac{1}{2n-1} \cos\left\{\dfrac{(2n-1)\pi t}{2} + [1 - (-1)^n]\dfrac{\pi}{2}\right\}$,

$f(t) = \dfrac{3}{2} + \dfrac{2}{\pi} \displaystyle\sum_{n=1}^{\infty} \dfrac{1}{2n-1} \sin\left[\dfrac{(2n-1)\pi t}{2} + (-1)^n \dfrac{\pi}{2}\right]$

5. $f(t) = \dfrac{\pi}{2} + \dfrac{4}{\pi} \displaystyle\sum_{n=1}^{\infty} \dfrac{\cos[(2n-1)t + \pi]}{(2n-1)^2}$, $\quad f(t) = \dfrac{\pi}{2} + \dfrac{4}{\pi} \displaystyle\sum_{n=1}^{\infty} \dfrac{\sin[(2n-1)t - \pi/2]}{(2n-1)^2}$

7. Defining

$$A_n = B_n = \dfrac{2}{n\pi} \sqrt{16 \sin^2\left(\dfrac{n\pi}{2}\right) + \left[1 - \cos\left(\dfrac{n\pi}{2}\right)\right]^2},$$

$$f(t) = 16 + \displaystyle\sum_{n=1}^{\infty} A_n \cos\left(\dfrac{n\pi t}{2} + \varphi_n\right), \quad \varphi_n = \tan^{-1}\left\{\dfrac{1}{4}\left[1 - \cos\left(\dfrac{n\pi}{2}\right)\right] \Big/ \sin\left(\dfrac{n\pi}{2}\right)\right\},$$

and φ_n lies in the first and second quadrants; and

$$f(t) = 16 + \displaystyle\sum_{n=1}^{\infty} B_n \sin\left(\dfrac{n\pi t}{2} + \varphi_n\right), \quad \varphi_n = \tan^{-1}\left\{-4 \sin\left(\dfrac{n\pi}{2}\right) \Big/ \left[1 - \cos\left(\dfrac{n\pi}{2}\right)\right]\right\},$$

and φ_n lies in the second quadrant near π.

Section 5.5

1. $f(t) = \dfrac{\pi}{2} - \dfrac{2}{\pi} \displaystyle\sum_{m=-\infty}^{\infty} \dfrac{e^{i(2m-1)t}}{(2m-1)^2}$

3. $f(t) = 1 + \dfrac{i}{\pi} \displaystyle\sum_{\substack{n=-\infty \\ n \neq 0}}^{\infty} \dfrac{e^{n\pi i t}}{n}$

5. $f(t) = \dfrac{1}{2} - \dfrac{i}{\pi} \displaystyle\sum_{m=-\infty}^{\infty} \dfrac{e^{2(2m-1)it}}{2m-1}$

7. $f(t) = -2 - \dfrac{3i}{\pi} \displaystyle\sum_{\substack{n=-\infty \\ n \neq 0}}^{\infty} \dfrac{\exp(n\pi t i/3)}{n}$

Section 5.6

1. $y(t) = A\cosh(t) + B\sinh(t) - \dfrac{1}{2} - \dfrac{2}{\pi} \displaystyle\sum_{n=1}^{\infty} \dfrac{\sin[(2n-1)t]}{(2n-1) + (2n-1)^3}$

3. $y(t) = Ae^{2t} + Be^{t} + \dfrac{1}{4} + \dfrac{6}{\pi} \displaystyle\sum_{n=1}^{\infty} \dfrac{\cos[(2n-1)t]}{[2-(2n-1)^2]^2 + 9(2n-1)^2}$

$\qquad + \dfrac{2}{\pi} \displaystyle\sum_{n=1}^{\infty} \dfrac{[2-(2n-1)^2]\sin[(2n-1)t]}{(2n-1)\{[2-(2n-1)^2]^2 + 9(2n-1)^2\}}$

5. $y(x) = \displaystyle\sum_{n=-\infty}^{\infty} \dfrac{c_n}{in+k} e^{inx}$

7. $y_p(t) = \dfrac{\pi}{8} - \dfrac{2}{\pi} \displaystyle\sum_{n=-\infty}^{\infty} \dfrac{e^{i(2n-1)t}}{(2n-1)^2[4-(2n-1)^2]}$

9. $q(t) = \displaystyle\sum_{n=-\infty}^{\infty} \dfrac{\omega^2 \varphi_n}{(in\omega_0)^2 + 2i\alpha n\omega_0 + \omega^2} e^{in\omega_0 t}$

Section 5.7

1. $f(t) = \dfrac{3}{2} - \cos(\pi x/2) - \sin(\pi x/2) - \dfrac{1}{2}\cos(\pi x)$

Section 6.3

1. $\pi e^{-|\omega/a|}/|a|$

3. $F(\omega) = 4e^{-6i\omega}\sin(\omega)/\omega$

Section 6.4

1. $f(t) = -t/(1+t^2)^2$ 3. $f(t) = e^{-|t|}/2$ 5. $f(t) = e^{-t}H(t) - e^{-t/2}H(t) + \dfrac{1}{2}te^{-t/2}H(t)$

Section 6.6

1. $y(t) = [(t-1)e^{-t} + e^{-2t}]H(t)$

3. $y(t) = \dfrac{1}{9}e^{-t}H(t) + \left[\dfrac{1}{9}e^{2t} - \dfrac{1}{3}te^{2t}\right]H(-t)$

Section 6.7

1. $u(x,y) = \dfrac{1}{\pi}\left[\tan^{-1}\left(\dfrac{1-x}{y}\right) + \tan^{-1}\left(\dfrac{x}{y}\right)\right]$

3. $u(x,y) = \dfrac{T_0}{\pi}\left[\dfrac{\pi}{2} - \tan^{-1}\left(\dfrac{x}{y}\right)\right]$

5. $u(x,y) = \dfrac{T_0}{\pi}\left[\tan^{-1}\left(\dfrac{1-x}{y}\right) + \tan^{-1}\left(\dfrac{1+x}{y}\right)\right] + \dfrac{T_1 - T_0}{2\pi}\, y \ln\left[\dfrac{(x-1)^2 + y^2}{x^2 + y^2}\right]$
$\qquad + \dfrac{T_1 - T_0}{\pi}\, x\left[\tan^{-1}\left(\dfrac{1-x}{y}\right) + \tan^{-1}\left(\dfrac{x}{y}\right)\right]$

Section 6.8

1. $u(x,t) = \frac{1}{2}\,\text{erf}\left(\dfrac{b-x}{\sqrt{4a^2 t}}\right) + \frac{1}{2}\,\text{erf}\left(\dfrac{b+x}{\sqrt{4a^2 t}}\right)$

3. $u(x,t) = \frac{1}{2}T_0\,\text{erf}\left(\dfrac{b-x}{\sqrt{4a^2 t}}\right) + \frac{1}{2}T_0\,\text{erf}\left(\dfrac{x}{\sqrt{4a^2 t}}\right)$

Section 7.1

1. $F(s) = s/(s^2 - a^2)$ 3. $F(s) = 1/s + 2/s^2 + 2/s^3$ 5. $F(s) = \left[1 - e^{-2(s-1)}\right]/(s-1)$

7. $F(s) = 2/(s^2 + 1) - s/(s^2 + 4) + \cos(3)/s - 1/s^2$

9. $f(t) = e^{-3t}$

11. $f(t) = \frac{1}{3}\sin(3t)$

13. $f(t) = 2\sin(t) - \frac{15}{2}t^2 + 2e^{-t} - 6\cos(2t)$ 17. $F(s) = 1/(2s) - sT^2/[2(s^2 T^2 + \pi^2)]$

Section 7.2

1. $f(t) = (t-2)H(t-2) - (t-2)H(t-3)$ 3. $y'' + 3y' + 2y = H(t-1)$

5. $y'' + 4y' + 4y = tH(t-2)$ 7. $y'' - 3y' + 2y = e^{-t}H(t-2)$

9. $y'' + y = \sin(t)[1 - H(t-\pi)]$

Section 7.3

1. $F(s) = 2/(s^2 + 2s + 5)$ 3. $F(s) = 2e^{-s}/s^3 + 2e^{-s}/s^2 + e^{-s}/s$

5. $F(s) = 1/(s-1)^2 + 3/(s^2 - 2s + 10) + (s-2)/(s^2 - 4s + 29)$

7. $F(s) = 2/(s+1)^3 + 2/(s^2 - 2s + 5) + (s+3)/(s^2 + 6s + 18)$

9. $F(s) = 2e^{-s}/s^3 + 2e^{-s}/s^2 + 3e^{-s}/s + e^{-2s}/s$

11. $F(s) = 4(s+3)/(s^2 + 6s + 13)^2$ 13. $F(s) = 1/(s^2 + 1) + e^{-s\pi}/(s^2 + 1)$

15. $f(t) = \frac{1}{6}t^3 e^{-2t}$ 17. $f(t) = e^{-t}\cos(t) - e^{-t}\sin(t)$

19. $f(t) = te^{-t} - \frac{1}{2}t^2 e^{-t} + \cos(t)e^{-t}$

21. $f(t) = te^{-2t} - t^2 e^{-2t} + \cos(t)e^{-2t} + 2\sin(t)e^{-2t}$

23. $f(t) = (t-2)e^{-(t-2)}H(t-2)$ 25. $f(t) = e^{-2(t-4)}\sin(t-4)H(t-4)$

27. $f(t) = \frac{1}{2}\sin[2(t-1)]H(t-1) + \left[(t-3)^2/2 - (t-3)^3/6\right]H(t-3)$

29. $F(s) = e^{-(s-1)}/(s-1)^2 + e^{-(s-1)}/(s-1) - e^{-2(s-1)}/(s-1)^2 - 2e^{-2(s-1)}/(s-1)$

31. $f(t) = \frac{1}{2}t - \frac{1}{2}tH(t-2) = \frac{1}{2}t - (t-2)H(t-2)/2 - H(t-2)$
$\quad F(s) = 1/(2s^2) - e^{-2s}/(2s^2) - e^{-2s}/2$

33. $f(t) = t - 2(t-2)H(t-2) + (t-4)H(t-4)$
$\quad F(s) = 1/s^2 - 2e^{-2s}/s^2 + e^{-4s}/s^2$

35. $Y(s) = e^{-s}/[s(s+1)(s+2)]$

37. $Y(s) = 2/(s+2)^2 + e^{-2s}/[s^2(s+2)^2] + 2e^{-2s}/[s(s+2)^2]$

39. $Y(s) = 2(s-3)/[(s-1)(s-2)] + e^{-2s-2}/[(s+1)(s-1)(s-2)]$

41. $Y(s) = 1/(s^2+1)^2 + e^{-s\pi}/(s^2+1)^2$

43. $F(s) = 1/s^2$, $\lim_{s\to\infty} sF(s) = \lim_{s\to\infty} 1/s = 0 = f(0)$

45. $F(s) = 1/(s+1)^2$, $\lim_{s\to\infty} sF(s) = \lim_{s\to\infty} s/(s+1)^2 = 0 = f(0)$

47. No

48. Yes, $\lim_{s\to 0} sF(s) = 1 = f(\infty)$

49. Yes, $\lim_{s\to 0} sF(s) = 0 = f(\infty)$

51. Yes, $\lim_{s\to 0} sF(s) = 1 = f(\infty)$

Section 7.4

1. $F(s) = \coth(s\pi/2)\Big/(s^2+1)$ 3. $F(s) = \dfrac{1-(1+as)e^{-as}}{s^2(1-e^{-2as})}$

Section 7.5

1. $f(t) = e^{-t} - e^{-2t}$ 3. $f(t) = \frac{5}{4}e^{-t} - \frac{6}{5}e^{-2t} - \frac{1}{20}e^{3t}$

5. $f(t) = e^{-2t}\cos\left(t + \frac{3\pi}{2}\right)$ 7. $f(t) = 2.3584\cos(4t + 0.5586)$

9. $f(t) = \frac{1}{2} + \frac{\sqrt{2}}{2}\cos\left(2t + \frac{5\pi}{4}\right)$

Section 7.6

11. $f(t) = e^t - t - 1$

Section 7.7

1. $y(t) = \frac{5}{4}e^{2t} - \frac{1}{4} + \frac{1}{2}t$ 3. $y(t) = e^{3t} - e^{2t}$ 5. $y(t) = -\frac{3}{4}e^{-3t} + \frac{7}{4}e^{-t} + \frac{1}{2}te^{-t}$

7. $y(t) = \frac{3}{4}e^{-t} + \frac{1}{8}e^t - \frac{7}{8}e^{-3t}$ 9. $y(t) = (t-1)H(t-1)$

11. $y(t) = e^{2t} - e^t + \left[\frac{1}{2} + \frac{1}{2}e^{2(t-1)} - e^{t-1}\right]H(t-1)$

13. $y(t) = \left[1 - e^{-2(t-2)} - 2(t-2)e^{-2(t-2)}\right]H(t-2)$

15. $y(t) = \left[\frac{1}{3}e^{2(t-2)} - \frac{1}{2}e^{t-2} + \frac{1}{6}e^{-(t-2)}\right]H(t-2)$

17. $y(t) = 1 - \cos(t) - [1 - \cos(t-T)]H(t-T)$

19. $y(t) = e^{-t} - \frac{1}{4}e^{-2t} - \frac{3}{4} + \frac{1}{2}t - \left[e^{-(t-a)} - \frac{1}{4}e^{-2(t-a)} - \frac{3}{4} + \frac{1}{2}(t-a)\right]H(t-a)$
    $+ a\left[\frac{1}{2}e^{-2(t-a)} + (t-a)e^{-(t-a)} - \frac{1}{2}\right]H(t-a)$

21. $y(t) = te^t + 3(t-2)e^{t-2}H(t-2)$

23. $y(t) = 3\left[e^{-2(t-2)} - e^{-3(t-2)}\right]H(t-2) + 4\left[e^{-3(t-5)} - e^{-2(t-5)}\right]H(t-5)$

25. $x(t) = \cos(\sqrt{2}t)\,e^{3t} - \sin(\sqrt{2}t)\,e^{3t}/\sqrt{2}$, $y(t) = \sin(\sqrt{2}t))e^{3t}/\sqrt{2}$

27. $x(t) = t - 1 + e^{-t}\cos(t), \quad y(t) = t^2 - t + e^{-t}\sin(t)$

29. $x(t) = 3F_1 - 2F_2 - F_1\cosh(t) + F_2 e^t - 2F_1\cos(t) + F_2\cos(t) - F_2\sin(t)$
$y(t) = F_2 - 2F_1 + F_1 e^{-t} - F_2\cos(t) + F_1\cos(t) + F_1\sin(t)$

Section 8.3

1. $u(x,t) = \dfrac{4L}{c\pi^2}\displaystyle\sum_{m=1}^{\infty}\dfrac{1}{(2m-1)^2}\sin\left[\dfrac{(2m-1)\pi x}{L}\right]\sin\left[\dfrac{(2m-1)\pi ct}{L}\right]$

3. $u(x,t) = \dfrac{9h}{\pi^2}\displaystyle\sum_{n=1}^{\infty}\dfrac{1}{n^2}\sin\left(\dfrac{2n\pi}{3}\right)\sin\left(\dfrac{n\pi x}{L}\right)\cos\left(\dfrac{n\pi ct}{L}\right)$

5. $u(x,t) = \sin\left(\dfrac{\pi x}{L}\right)\cos\left(\dfrac{\pi ct}{L}\right)$
$\quad + \dfrac{4aL}{\pi^2 c}\displaystyle\sum_{n=1}^{\infty}\dfrac{(-1)^{n+1}}{(2n-1)^2}\sin\left[\dfrac{(2n-1)\pi}{4}\right]\sin\left[\dfrac{(2n-1)\pi x}{L}\right]\sin\left[\dfrac{(2n-1)\pi ct}{L}\right]$

7. $u(x,t) = \dfrac{4L}{\pi^2}\displaystyle\sum_{n=1}^{\infty}\dfrac{(-1)^{n+1}}{(2n-1)^2}\sin\left[\dfrac{(2n-1)\pi x}{L}\right]\cos\left[\dfrac{(2n-1)\pi ct}{L}\right]$

Section 8.4

1. $u(x,t) = \sin(2x)\cos(2ct) + \cos(x)\sin(ct)/c$

3. $u(x,t) = \dfrac{1 + x^2 + c^2 t^2}{(1 + x^2 + c^2 t^2)^2 + 4x^2 c^2 t^2} + \dfrac{e^x\sinh(ct)}{c}$

5. $u(x,t) = \cos\left(\dfrac{\pi x}{2}\right)\cos\left(\dfrac{\pi ct}{2}\right) + \dfrac{\sinh(ax)\sinh(act)}{ac}$

Section 9.3

1. $u(x,t) = \dfrac{4A}{\pi}\displaystyle\sum_{m=1}^{\infty}\dfrac{\sin[(2m-1)x]}{2m-1}e^{-a^2(2m-1)^2 t}$

3. $u(x,t) = -2\displaystyle\sum_{n=1}^{\infty}\dfrac{(-1)^n}{n}\sin(nx)e^{-a^2 n^2 t}$

5. $u(x,t) = \dfrac{4}{\pi}\displaystyle\sum_{m=1}^{\infty}\dfrac{(-1)^{m+1}}{(2m-1)^2}\sin[(2m-1)x]e^{-a^2(2m-1)^2 t}$

7. $u(x,t) = \dfrac{\pi}{2} - \dfrac{4}{\pi}\displaystyle\sum_{m=1}^{\infty}\dfrac{\cos[(2m-1)x]}{(2m-1)^2}e^{-a^2(2m-1)^2 t}$

9. $u(x,t) = \dfrac{\pi}{2} - \dfrac{4}{\pi}\displaystyle\sum_{n=1}^{\infty}\dfrac{\cos[(2n-1)x]}{(2n-1)^2}e^{-a^2(2n-1)^2 t}$

11. $u(x,t) = T_0 + \dfrac{4(T_1 - T_0)}{\pi}\displaystyle\sum_{m=1}^{\infty}\dfrac{\sin[(2m-1)x]}{2m-1}e^{-a^2(2m-1)^2 t}$

13. $u(x,t) = h_1 + \dfrac{(h_2 - h_1)x}{L} + \dfrac{2(h_2 - h_1)}{\pi} \displaystyle\sum_{n=1}^{\infty} \dfrac{(-1)^n}{n} \sin\left(\dfrac{n\pi x}{L}\right) \exp\left(-\dfrac{a^2 n^2 \pi^2}{L^2} t\right)$

15. $u(x,t) = \dfrac{1}{3} - t - \dfrac{2}{\pi^2} \displaystyle\sum_{n=1}^{\infty} \dfrac{(-1)^n}{n^2} \cos(n\pi x) e^{-a^2 n^2 \pi^2 t}$

17. $u(x,t) = \dfrac{4}{\pi} \displaystyle\sum_{n=1}^{\infty} \dfrac{(-1)^{n+1}}{(2n-1)^4} \sin[(2n-1)x] \left[1 - e^{-(2n-1)^2 t}\right]$

19. $u(x,t) = 2\pi \displaystyle\sum_{n=1}^{\infty} \dfrac{n}{L^2 + n^2 \pi^2} \sin\left(\dfrac{n\pi x}{L}\right) \left\{\exp\left[-\left(\dfrac{n^2 \pi^2}{L^2} + 1\right) t\right] - 1\right\}$

Section 10.3

1. $u(x,y) = \dfrac{4}{\pi} \displaystyle\sum_{m=1}^{\infty} \dfrac{\sinh[(2m-1)\pi(a-x)/b] \sin[(2m-1)\pi y/b]}{(2m-1) \sinh[(2m-1)\pi a/b]}$

3. $u(x,y) = -\dfrac{2a}{\pi} \displaystyle\sum_{n=1}^{\infty} \dfrac{\sinh(n\pi y/a) \sin(n\pi x/a)}{n \, \sinh(n\pi b/a)}$

5. $u(x,y) = 1$ 7. $u(x,y) = 1$

9. $u(x,y) = u_1 + B_0(y - b)/2 + \displaystyle\sum_{n=1}^{\infty} B_n \sinh[n\pi(y - b)/a] \cos(n\pi x/a)$, where

$B_0 = 2u_1/b - 2\displaystyle\int_0^\alpha f(x)\,dx/(ab)$ and $B_n = -2\displaystyle\int_0^\alpha f(x) \cos(n\pi x/a)\,dx/[a \, \sinh(n\pi b/a)]$

11. $u(x,y) = \tfrac{1}{2}A_0 + \displaystyle\sum_{n=1}^{\infty} A_n \cosh(n\pi y/L) \cos(n\pi x/L)$,

where $A_0 = 2g\left[Lz_0 + \tfrac{1}{2}cL^2 + a/b - a\cos(bL)/b\right]/L$

and $\cosh(n\pi z_0/L)\, A_n = 2gcL[(-1)^n - 1]/(n^2 \pi^2) - 2abgL[(-1)^n \cos(bL) - 1]/(b^2 L^2 - n^2 \pi^2)$

13. $u(r,\varphi) = T_0$

Section 10.4

1. $u(x,y) = \dfrac{64R}{\pi^4 T} \displaystyle\sum_{n=1}^{\infty} \sum_{m=1}^{\infty} \dfrac{(-1)^{n+1}(-1)^{m+1}}{(2n-1)(2m-1)}$
$\times \dfrac{\cos[(2n-1)\pi x/2a] \cos[(2m-1)\pi y/b]}{(2n-1)(2m-1)[(2n-1)^2/a^2 + (2m-1)^2/b^2]}$

Section 11.1

1. $\lambda_n = (2n-1)^2 \pi^2/(4L^2)$, $y_n(x) = \cos[(2n-1)\pi x/(2L)]$

3. $\lambda_0 = -1$, $y_0(x) = e^{-x}$ and $\lambda_n = n^2$, $y_n(x) = \sin(nx) - n\cos(nx)$

5. $\lambda_n = -n^4 \pi^4/L^4$, $y_n(x) = \sin(n\pi x/L)$

7. $\lambda_n = k_n^2$, $y_n(x) = \sin(k_n x)$ with $k_n = -\tan(k_n)$

9. $\lambda_0 = -m_0^2$, $y_0(x) = \sinh(m_0 x) - m_0 \cosh(m_0 x)$ with $\coth(m_0 \pi) = m_0$; $\lambda_n = k_n^2$, $y_n(x) = \sin(k_n x) - k_n \cos(k_n x)$ with $k_n = -\cot(k_n \pi)$

11.
(a) $\qquad\qquad\qquad\qquad \lambda_n = n^2 \pi^2 \qquad\qquad\qquad\qquad y_n(x) = \sin[n\pi \ln(x)]$
(b) $\qquad\qquad\qquad\qquad \lambda_n = (2n-1)^2 \pi^2/4 \qquad\qquad y_n(x) = \sin[(2n-1)\pi \ln(x)/2]$
(c) $\quad \lambda_0 = 0 \quad y_0(x) = 1 \quad \lambda_n = n^2 \pi^2 \qquad\qquad\qquad y_n(x) = \cos[n\pi \ln(x)]$

13. $\lambda_n = n^2 + 1$, $y_n(x) = \sin[n \ln(x)]/x$

15. $\lambda = 0$, $y_0(x) = 1$; $y_n(x) = \cosh(\lambda_n x) + \cos(\lambda_n x) - \tanh(\lambda_n)[\sinh(\lambda_n x) + \sin(\lambda_n x)]$, where $n = 1, 2, 3, \ldots$, and λ_n is the nth root of $\tanh(\lambda) = -\tan(\lambda)$.

Section 11.3

1. $f(x) = \dfrac{2}{\pi} \displaystyle\sum_{n=1}^{\infty} \dfrac{(-1)^{n+1}}{n} \sin\left(\dfrac{n\pi x}{L}\right)$ $\qquad$ 3. $f(x) = \dfrac{8L}{\pi^2} \displaystyle\sum_{n=1}^{\infty} \dfrac{(-1)^{n+1}}{(2n-1)^2} \sin\left[\dfrac{(2n-1)\pi x}{2L}\right]$

7. $u(x,t) = \dfrac{8L}{\pi^2} e^{-ht} \displaystyle\sum_{n=1}^{\infty} \dfrac{(-1)^{n+1}}{(2n-1)^2} \sin\left[\dfrac{(2n-1)\pi x}{2L}\right] \left\{ \cos\left[t\sqrt{\lambda_n c^2 - h^2}\right] \right.$

$\left. + h \sin\left[t\sqrt{\lambda_n c^2 - h^2}\right] \middle/ \sqrt{\lambda_n c^2 - h^2} \right\}$, where $\lambda_n = (2n-1)^2 \pi^2/(4L^2)$

9. $u(x,t) = \dfrac{32}{\pi} \displaystyle\sum_{n=1}^{\infty} \dfrac{(-1)^n}{(2n-1)^3} \cos\left[\dfrac{(2n-1)x}{2}\right] e^{-a^2(2n-1)^2 t/4}$

11. $u(x,t) = \dfrac{8}{\pi} \displaystyle\sum_{n=1}^{\infty} \dfrac{(-1)^{n+1}}{(2n-1)^2} \sin\left[\dfrac{(2n-1)x}{2}\right] e^{-a^2(2n-1)^2 t/4}$

13. $u(x,t) = T_0 + \dfrac{4(T_1 - T_0)}{\pi} \displaystyle\sum_{n=1}^{\infty} \dfrac{\sin[(2n-1)x/2]}{2n-1} e^{-a^2(2n-1)^2 t/4}$

15. $u(x,t) = h_0 - \dfrac{4h_0}{\pi} \displaystyle\sum_{n=1}^{\infty} \dfrac{1}{2n-1} \sin\left[\dfrac{(2n-1)\pi x}{2L}\right] \exp\left[-\dfrac{a^2(2n-1)^2 \pi^2 t}{4L^2}\right]$

17. $u(x,t) = \dfrac{A_0(L^2 - x^2)}{2\kappa} + \dfrac{A_0 L}{h}$

$\qquad - \dfrac{2L^2 A_0}{\kappa} \displaystyle\sum_{n=1}^{\infty} \dfrac{\sin(\beta_n)}{\beta_n^3 [1 + \kappa \sin^2(\beta_n)/hL]} \cos\left(\dfrac{\beta_n x}{L}\right) \exp\left(-\dfrac{a^2 \beta_n^2 t}{L^2}\right)$,
where $\beta_n \tan(\beta_n) = hL/\kappa$.

19. $v(x,t) = 16a^4 \displaystyle\sum_{n=1}^{\infty} \dfrac{k_n [\sin(k_n x) + 2a^2 k_n \cos(k_n x)]}{(1 + 4a^4 k_n^2)(1 + 4a^2 + 4a^4 k_n^2)} e^{-a^2 k_n^2 t}$, and $u(x,t)$ equals $v(x,t)$ multiplied by $\exp[(2x - t)/(4a^2)]$ and $\tan(k_n) = 4a^2 k_n/(4a^4 k_n^2 - 1)$.

21. $u(r,t) = \dfrac{4bu_0}{r} \displaystyle\sum_{n=1}^{\infty} \dfrac{\sin(k_n) - k_n\cos(k_n)}{k_n[2k_n - \sin(2k_n)]} \sin\left(\dfrac{k_n r}{b}\right) e^{-a^2 k_n^2 t/b^2}$,

where $k_n \cot(k_n) = 1 - A$, $n\pi < k_n < (n+1)\pi$.

25. $u(x,y) = \dfrac{4}{\pi} \displaystyle\sum_{n=1}^{\infty} (-1)^{n+1} \dfrac{\sinh[(2n-1)\pi y/2a]\cos[(2n-1)\pi x/2a]}{(2n-1)\sinh[(2n-1)\pi b/2a]}$

27. $u(x,y) = \dfrac{4}{\pi} \displaystyle\sum_{n=1}^{\infty} \dfrac{\cosh[(2n-1)\pi(y-b)/2a]\sin[(2n-1)\pi x/2a]}{(2n-1)\cosh[(2n-1)\pi b/2a]}$

$\quad\quad + \dfrac{4}{\pi} \displaystyle\sum_{n=1}^{\infty} \dfrac{\cosh[(2n-1)\pi(x-a)/2b]\sin[(2n-1)\pi y/2b]}{(2n-1)\cosh[(2n-1)\pi a/2b]}$

Section 12.1

1. $f(x) = \frac{1}{4}P_0(x) + \frac{1}{2}P_1(x) + \frac{5}{16}P_2(x) + \cdots$ 3. $f(x) = \frac{1}{2}P_0(x) + \frac{5}{8}P_2(x) - \frac{3}{16}P_4(x) + \cdots$

5. $f(x) = \frac{3}{2}P_1(x) - \frac{7}{8}P_3(x) + \frac{11}{16}P_5(x) + \cdots$

17. $y(x) = x^3 - 5$

19. $u(r,\theta) = 400\left\{ \dfrac{r}{7a}P_1[\cos(\theta)] - \dfrac{1}{9}\left(\dfrac{r}{a}\right)^3 P_3[\cos(\theta)] - \dfrac{2}{63}\left(\dfrac{r}{a}\right)^5 P_5[\cos(\theta)] \right\}$

21. $u(r,\theta) = \frac{1}{2}T_0 \displaystyle\sum_{n=0}^{\infty} \{P_{n-1}[\cos(\alpha)] - P_{n+1}[\cos(\alpha)]\} \left(\dfrac{r}{a}\right)^n P_n[\cos(\theta)]$

Section 12.2

33. $u(r,t) = \sqrt{a} \displaystyle\sum_{n=1}^{\infty} \dfrac{J_1(2k_n\sqrt{a})J_0(k_n r)}{k_n^2 J_1^2(2k_n)} \sin(k_n t)$

or $u(x,t) = \sqrt{a} \displaystyle\sum_{n=1}^{\infty} \dfrac{J_1(2k_n\sqrt{a})J_0(2k_n\sqrt{x})}{k_n^2 J_1^2(2k_n)} \sin(k_n t)$,

where k_n is the nth solution of $J_0(2k) = 0$.

35. $u(r,t) = \theta + 2(1-\theta) \displaystyle\sum_{n=1}^{\infty} \dfrac{J_0(k_n r/b)}{k_n J_1(k_n)} \exp\left(-\dfrac{a^2 k_n^2}{b^2}t\right)$, where $J_0(k_n) = 0$

37. $u(r,t) = \dfrac{G}{4\rho\nu}(b^2 - r^2) - \dfrac{2Gb^2}{\rho\nu} \displaystyle\sum_{n=1}^{\infty} \dfrac{J_0(k_n r/b)}{k_n^3 J_1(k_n)} \exp\left(-\dfrac{\nu k_n^2}{b^2}t\right)$, with $J_0(k_n) = 0$

39. $u(r,t) = \dfrac{2T_0}{L^2} e^{-\kappa t} \displaystyle\sum_{n=1}^{\infty} \dfrac{[LJ_1(k_n L) - bJ_1(k_n b)]J_0(k_n r)}{k_n[J_0^2(k_n L) + J_1^2(k_n L)]} e^{-a^2 k_n^2 t}$

with $k_n J_1(k_n L) = h J_0(k_n L)$

43. $u(r,z) = \dfrac{Aa^2 z}{b^2} + \dfrac{2Aa}{b^2} \displaystyle\sum_{n=1}^{\infty} \dfrac{\sinh(k_n z)\, J_1(k_n a)\, J_0(k_n r)}{k_n^2 \cosh(k_n L) J_0^2(k_n b)}$ with $J_1(k_n b) = 0$

45. $u(r,z) = \dfrac{(z-h)r_0^2}{a^2} + 2r_0 \displaystyle\sum_{n=1}^{\infty} \dfrac{\sinh[k_n(z-h)/a] J_1(k_n r_0/a) J_0(k_n r/a)}{k_n^2 \cosh(k_n h/a) J_0^2(k_n)}$ with $J_1(k_n) = 0$

47. For case (a), $u(r,z) = \dfrac{4}{\pi} \displaystyle\sum_{n=1}^{\infty} \dfrac{I_1[(2n-1)\pi r/2]}{I_1[(2n-1)\pi/2]} \dfrac{\sin[(2n-1)\pi(z-1)/2]}{2n-1}$

For case (b), $u(r,z) = -2 \displaystyle\sum_{n=1}^{\infty} \dfrac{\cosh(k_n z)}{\cosh(k_n)} \dfrac{J_1(k_n r)}{k_n J_0(k_n)}$ with $J_1(k_n) = 0$

49. $u(r,z) = -\dfrac{2}{\pi} \displaystyle\sum_{n=1}^{\infty} \dfrac{(-1)^n I_1(n\pi r) \sin(n\pi z)}{n\, I_1(n\pi a)}$

51. $u(r,z) = 2 \displaystyle\sum_{n=1}^{\infty} \dfrac{\sinh(k_n z) J_1(k_n r)}{k_n^3 \cosh(k_n a) J_1(k_n)}$ with $k_n J_0(k_n) = J_1(k_n)$

53. $u(r,z) = 2u_0 \displaystyle\sum_{n=1}^{\infty} \dfrac{J_1(\mu_n) J_0(\mu_n r/a)}{\mu_n[J_0^2(\mu_n) + J_1^2(\mu_n)]} \dfrac{\cosh[\mu_n(L-z)/a]}{\cosh(\mu_n L/a)}$ with $\mu_n J_1(\mu_n) = \beta J_0(\mu_n)$

55. $u(r,z) = -V\left\{ 1 - 2 \displaystyle\sum_{n=1}^{\infty} \dfrac{\cosh(k_n z/a)}{k_n J_1(k_n)} \exp\left(-\dfrac{k_n d}{2a}\right) J_0\left(\dfrac{k_n r}{a}\right) \right\}$ if $|z| < d/2$,

and

$u(r,z) = -2V \displaystyle\sum_{n=1}^{\infty} \dfrac{\sinh[k_n d/(2a)]}{k_n J_1(k_n)} \exp\left(-\dfrac{k_n |z|}{a}\right) J_0\left(\dfrac{k_n r}{a}\right)$ if $|z| > d/2$

57. $u(r,z) = 2B \displaystyle\sum_{n=1}^{\infty} \dfrac{\exp[z(1 - \sqrt{1 + 4k_n^2})/2] J_0(k_n r)}{(k_n^2 + B^2) J_0(k_n)}$ with $k_n J_1(k_n) = B J_0(k_n)$

Index

Printed in the United States
by Baker & Taylor Publisher Services